Low-Cost Physics Experiments Using New Technologies

Low-Cost Physics Experiments Using New Technologies

Salvador Gil
Universidad Nacional de San Martín, Argentina

World Scientific

NEW JERSEY • LONDON • SINGAPORE • BEIJING • SHANGHAI • HONG KONG • TAIPEI • CHENNAI • TOKYO

Published by

World Scientific Publishing Co. Pte. Ltd.
5 Toh Tuck Link, Singapore 596224
USA office: 27 Warren Street, Suite 401-402, Hackensack, NJ 07601
UK office: 57 Shelton Street, Covent Garden, London WC2H 9HE

British Library Cataloguing-in-Publication Data
A catalogue record for this book is available from the British Library.

LOW-COST PHYSICS EXPERIMENTS USING NEW TECHNOLOGIES

ISBN 978-981-12-7775-7 (hardcover)
ISBN 978-981-12-7776-4 (ebook for institutions)
ISBN 978-981-12-7777-1 (ebook for individuals)

For any available supplementary material, please visit
https://www.worldscientific.com/worldscibooks/10.1142/13452#t=suppl

Desk Editor: Rhaimie Wahap

Typeset by Stallion Press
Email: enquiries@stallionpress.com

To

Rodrigo, Eugenio, Mandy.
To my dear siblings Susy, Stella, Diego, and Inés, and in the memory of my parents.

Preface

Homines dum docent discunt.
"Men learn while they teach."

Seneca (4 BC — 65 AD)

Several years ago, I found this quote by Seneca in a textbook belonging to one of my children. It summarized very well my own experience as a teacher. The same idea seems to have inspired great teachers, as Richard Feynman: "*If you want to master something, teach it. A great way to learn is to teach.*" Over the years, the more I tried to explain a topic to my students, the deeper the understanding I often gained. In this sense, this book is the diary of a student, albeit somewhat advanced in years, in search of meaning and harmony in the world around us. The search for order and harmony in the apparent chaos in which we often find ourselves immersed is an adventure — with its achievements and failures — that has given meaning and satisfaction to a large part of my life. This book is thus an attempt to share this adventure and joy with my readers.

This work is the result of collective learning, achieved over many years with my students and colleagues from different universities, with whom I have enjoyed long hours of work. In this text, I have tried to record some of those experiences, which I hope will be useful and stimulating to new students, physics instructors, and science enthusiasts in general. About 5 years ago, I published "*Experimentos de Física de bajo costo, usando nuevas tecnologías*," which was very well received in several Spanish-speaking countries. The present text in English is an updated version of the original in Spanish that was made possible thanks to the encouragement and hard work of *Giacomo Torzo* (Physics Dept. of Padova University), who has contributed to improving and enhancing the goals of this book. Also, I am deeply grateful to *Catherine Connon* for expert idiomatic revision.

Purpose of the book

In recent years, the quality of personal computers (PCs) and smartphones has increased significantly, making it possible to transform almost any PC or smartphone into a sophisticated mini-laboratory. This book takes advantage of these developments, and several experiments require no other equipment than the features that are regularly present in standard personal computers or smartphones, such as digital cameras, video recording, and sound recording. This enables many schools and universities, even with very limited resources, to perform challenging experiments that provide *meaningful learning*, while fostering the joy of research and science.

The aim of this book is to provide a series of physics experiments that use Information and Communication Technologies (ICT) as tools, highlighting the methodological aspects of physics and science in general. The experiments are intended mainly for college students of science and engineering, although some of them may be used in a high school setting. The proposed projects are intended for students to answer the following questions: How do we know? Why do we believe it? These questions illustrate the nature of scientific thinking. This book is complemented by an Internet portal Fisicarecrativa.com (www.fisicarecreativa.com) with several links to sites of interest to students and teachers of physics, as well as reports on projects similar to those proposed in this book, carried out by students from various universities.

How to use this book

The experimental projects proposed herein are organized around topics related to methodology and metrology, mechanics, electromagnetism, thermodynamics, optics, modern physics, and astrophysics. In particular, the proposed experiments intend to illustrate the phenomena that support the basic paradigms of physics, such as the laws of mechanics, the principles of conservation of energy, Maxwell's equations, the concept of waves, and quantum mechanics. We have tried to make the projects self-contained, i.e., each of them can be done without necessarily having done the ones that precede them in the text.

In a way, we have attempted to offer a set of scientific activities as in a "fork buffet" or "finger buffet", where each teacher or student can choose the experiments that interest them the most and that best fit their goals. This makes the book useful for different courses and for students with different levels of training. The activities identified with the symbol ♣ require a level of knowledge comparable to that of students in first-year university physics. Activities identified with ♣♣ denote more complex experiments, and those with ♣♣♣ include slightly more advanced topics, suitable for students seeking greater challenges.

Each chapter includes a short introduction reviewing the relevant conceptual framework for the experiments. Due to space limitations, the chapters are generally concise, but they all provide a list of references for a more comprehensive treatment of each topic. They cite teaching-oriented journals which are generally accessible to undergraduate students, such as the *American Journal of Physics*, *European Journal of Physics*, *Physics Teacher*, and *Latin-American Journal of Physics Education*, among others. We strongly recommend that the reader consult these sources frequently and use them to complement their experimental work and analysis. One of the great advantages of using ICT is access to specialized magazines and resources. Students' familiarity with this type of literature enables them to tap into sources of knowledge and relate directly to the process of creation and development of science. Many students can be encouraged to take part in this process and try to publish their own ideas. These are valuable activities that contribute significantly to training professionals, engineers, and scientists.

Another goal is that it should be possible to carry out the experiments with low-cost equipment. This broadens the possibilities of them being performed, since in places such as Latin America, Africa, and Asia, schools more often than not have laboratories with few resources. This book shows how a wide variety of experiments can be performed with very modest resources, nonetheless, posing interesting challenges to students and providing a meaningful and fun learning opportunity. Nevertheless, given the increasing availability of some elaborate modern equipment in many work environments, several experiments involving the use of more sophisticated devices such as gamma radiation detectors, multichannel, and "lock-in" amplifiers have also been included.

In the introductory experiments, we took a *constructivist* approach. Many of these experiments are designed to enable students to *discover or rediscover phenomena.* Similarly, students are induced through questions to "construct" the conceptual structure that explains the observations. In some activities, "riddles" are proposed so that by using the fundamental paradigms of physics discussed in conventional courses, students may explain the results they have discovered in the laboratory. This enables them to experience first-hand the various paths that science follows in its development and evolution. It is thus intended that students should learn physics — including its dynamics and development — by *immersion.*

This book intends to bridge the gap between traditional teaching approaches and new forms of research or active learning. The text is structured in modules that follow the blocks into which physics is traditionally divided in schools of science and engineering: mechanics, thermodynamics, waves, electromagnetism, etc. Each module proposes projects that include elements of *active learning* or inquiry but which can be included in a conventional curriculum in the desired proportion. Adequate articulation of both approaches makes them complementary and mutually enhancing.

We have also tried to promote the development of skills and analytical abilities such as the following:

✓ **Experimental and analytical skills:** Handling laboratory equipment, ability to measure a physical quantity carefully, analysis of measurement errors, and choice of the most suitable instrumentation for each purpose.

✓ **Critical analysis of results**, their implications, and generalizations, comparing results with theoretical or *a priori* expectations, and formulating hypotheses and new experiments.

✓ **Using computers for data collection and analysis of results**: It is currently possible to measure several variables and experiment in real time, recording this information for further analysis. Computers are a very useful tool in several steps of an experiment, analysis of results, and report writing.

✓ **Becoming familiar with current literature**, especially enjoyable, accessible journals, such as the *American Journal of Physics*, *Physics Teacher*, and *Latin-American Journal of Physics Education*.

✓ **Developing the ability to communicate results in writing** and drafting reports that follow internationally adopted models for scientific and technical publications.

Philosophical framework

Perhaps one of the most conspicuous features of the times we live in is constant technological, economic, political, and social change. The experience of the last few decades makes it clear how terribly limited is our ability to predict the direction of these changes. Despite the great breakthroughs and development of science and technology, the world is confronted with challenges that seem unexpected and beyond the forecast of the most advanced minds and institutions. Consider, for instance, the COVID-19 pandemic, which has caught the world completely off-guard. Faced with these realities and limitations, the question naturally arises: How can we prepare our students in science and technology, when we are almost certain that during their professional careers, they will face problems and use techniques and equipment that are unknown to us today? Furthermore, will the problems, techniques, and equipment used in their education become obsolete before they graduate from our colleges and universities? Obviously, the answers to these questions are highly complex, and clearly, we do not expect to have an answer. However, the exercise of trying to outline answers to these questions is an unavoidable challenge for all of us involved in education.

One possible answer to this dilemma of education is to emphasize the development of skills and attitudes that are as basic and broad as possible so that students acquire the ability to adapt to new and changing situations. In this sense, the teaching of basic sciences, such as physics in this case, can make a valid contribution to professional training, provided that such training and methodological aspects are emphasized together with specific information contents. So, for example, when we discuss and study the pendulum in the laboratory, what is essential is not necessarily its laws. It is unlikely that anyone will end up working with a pendulum during their professional career and, of course, there is plenty of information on the subject in the literature, which can be consulted at any time. However, the methodology that we use to study the behavior of a pendulum, test our hypotheses, test our explanations, and how we critically analyze the results and seek information to gain better understanding of the problem, are common to many areas of engineering and scientists of today and the future. Therefore, in addition to presenting **some basic information contents**, this project intends to develop in students the ability to face new problems with openness and rigor. In other words, students should realize **how we learn new things** (learn to learn) and face new problems with confidence and good judgment. If these goals are met, the educational goal of this book will have been achieved.

Pedagogical approach adopted in this book

Science immersion learning

A laboratory class is not necessarily a place where every concept discussed in a text or theoretical lesson is illustrated and demonstrated. Time, equipment, and personnel constraints would certainly make this goal impossible to achieve. Good textbooks, classroom demonstrations, videos, and discussions with teachers may meet this need more efficiently and economically. However, the laboratory fulfills an essential, irreplaceable mission that is much more viable and profitable: students should learn how scientific knowledge itself is generated.

Therefore, one of the important aims of this proposal is to introduce students to **learning** and **understanding** science in general and more specifically, physics. The aspect of *understanding* science is emphasized here over the aspect of *scientific information*, i.e., the procedural aspects of physics are prioritized. This is based on the conviction that scientists are characterized not by what they believe in but by the reasons that lead them to believe. Every scientific theory is based on empirical facts. As time goes by, new facts are discovered, while others are changed or even proved wrong. Consequently, our scientific conceptions must be revised and modified. Scientific knowledge is thus, by its very nature, provisional knowledge that can be refuted or falsified.

We have endeavored to encourage students to develop a **critical attitude** toward knowledge in general and scientific knowledge in particular. Science is a very powerful tool for understanding and changing our world, but it is also limited. Recognizing its limitations is therefore also an essential aspect of the learning process.

To achieve these goals, we have opted to focus on *some fundamental topics* where the basic assumptions and empirical facts that support the relevant theories are carefully discussed, prioritizing the depth of the treatment of the arguments over the extension and methodology over mere information.

A laboratory is an excellent pedagogical tool, and in many respects an essential environment, for teaching introductory-level science. The laboratory offers students the opportunity *to learn from their experiences.* It can and should be used to *stimulate curiosity* and *pleasure in research and discovery.* It offers students the opportunity to explore, manipulate, suggest hypotheses, make *mistakes, recognize them, and then learn from them.*

We have attempted to encourage the formulation of reasonable hypotheses to explain the observations made (i.e., the development of models that can explain the observations). We believe that finding unexpected results stimulates the learning process and keeps students interested. This is more constructive than using laboratory sessions simply to verify the results already discussed in textbooks or in the classroom. Solutions to experimental problems are not provided at the end of the book, so it is a challenge for students to trust their own judgment and gain confidence in their knowledge.

Many of the experiments proposed here call for the use of personal computers, data collection, and analysis systems. These technologies have become easily accessible and very widespread in recent years and provide the possibility of conducting more quantitative and accurate experiments. By improving the accuracy of the measurements, it is easy to identify the need to improve established theories. Likewise, the limitations of the proposed models become more evident. This type of experience can hardly be internalized in an environment other than the laboratory.

Stimulating creativity is another fundamental goal that can and must be achieved in the laboratory. By accepting and encouraging variations to the given problems, it is very gratifying to see how many students find new ways to achieve a certain goal or may even find a new goal, perhaps more valuable than the one originally conceived by the instructor. Data analysis and laboratory report preparation are also very important in the learning process, helping students to organize and summarize their observations and experiences. In the report, students should describe their results and compare them with theoretical expectations. It is also important that students should appreciate the degree of agreement or disagreement, draw conclusions, etc. Several of these skills are useful not only to STEM[a] students but also to social scientists.

There are also important by-products of these last steps, such as developing the ability to write reports, display results graphically, and design presentations. This expertise can be put to use in a multiplicity of working environments in the modern world. Additionally, students will hopefully learn how to use computers to acquire and/or analyze data and gain experience in basic statistical concepts from discussions of experimental errors and the level of significance of their

[a]STEM is a common term often used to mean Science, Technology, Engineering and Mathematics.

observations. These tools will enable them to expand their observation and measurement skills, which can be used in many walks of life.

Most experimental projects, by their nature, must be carried out in groups, which promotes cooperation and teamwork. Not all experimental projects have a "happy ending", where all the data recorded fully agree with the theoretical expectations. There are various reasons for this: systematic errors, the approximate nature of the theories presented in the textbooks, or complexities not well understood. This limitation may be a useful learning experience, to understand the problematic nature of science and why scientific theories need to be constantly revised and tested with experimental studies.

In short, the laboratory naturally provides an excellent opportunity to simulate situations in which not only the science but also a great number of modern professional and business activities develop and perhaps life itself.

Acknowledgments

This book is the result of the cooperative effort of many people. The writer of these lines is in a certain sense a chronicler of this experience. Numerous students from various Argentine universities inspired and performed most of the experiments presented in this book. I offer them my sincere thanks. Many teachers have been a source of inspiration throughout my career. In particular, L. C. de Cudmani, R. Vandenbosch, and Alejandro García. I also thank Prof. E. Rodríguez, with whom I wrote an earlier book. Different experiments were used in various experimental physics courses of the Physics Department of the Faculty of Exact and Natural Sciences of the University of Buenos Aires (UBA), the National University of General San Martín (UNSAM), the Favaloro University (UF), the National University of General Sarmiento (UNGS), the National University of the South (UNS), and numerous physics seminars held in Argentina, Uruguay, and Colombia. In all these courses, I have shared experiences with numerous students and colleagues. I thank them all for their collaboration and support in undertaking this educational project and for providing valuable contributions. In particular, I am grateful to the National Atomic Energy Commission of Argentina (CNEA) for giving me the opportunity to work, grow, and get started in experimental physics. The years I spent at the CNEA's TANDAR Laboratory had profound influence on my life, which is also reflected in this book. I also appreciate the years of work and training at the University of Washington Nuclear Physics Laboratory, Seattle.

I thank the many collaborators with whom over the years I developed some of the experiments included in this book, especially Dina Tobia, Martín Saleta, Hernán Reisin, Carlos Sendra, Guillermo Solovey, José Flores, Mariano Mayochi, José di Laccio, Silvia Calderón, Pablo Núñez, and Leila Iannelli. Finally, I sincerely thank *Giacomo Torzo* of the University of Padova, who not only encouraged me to write the present version in English but also helped develop it through multiple corrections and inspiring discussions. I am greatly indebted to him and deeply grateful for his help and friendship. I also thank Catherine Connon for her assistance and patience.

And last but not least, I thank my family for their understanding, encouragement, and affection over all these years, to whom I dedicate this book.

Contents

Experimental Projects

MODULE I
EXPERIMENTS IN SCIENCES

Chapter 1

ROLE OF LABORATORY IN SCIENCE LEARNING

"One thing I have learned in a long life: that all our science, measured against reality, is primitive and childlike — and yet it is the most precious thing we have..."

A. Einstein

Why are observations and experiments so important in science? To answer this question, we must examine and reflect on the nature of scientific knowledge. In this chapter, we briefly discuss some basic aspects of the scientific method, which is used in factual or empirical sciences, such as Physics, Chemistry, and Biology.

An experiment does not end until a report has been written analyzing and discussing the results. This chapter highlights the importance and characteristics of such reports.

When conducting an experiment, it is important to keep in mind safety rules and precautions. This chapter briefly reviews some of them.

Topics

- ✓ What is science?
- ✓ Why do we conduct experiments?
- ✓ Formal and factual sciences
- ✓ Falsificationism
- ✓ Paradigms
- ✓ Normal science
- ✓ Scientific revolution
- ✓ Scientific reports
- ✓ Safety in the laboratory

1.1 Why do we conduct experiments?

In principle, we could say that we conduct experiments to discover new phenomena and test our theories of the world in order to develop a conceptual framework that enables us to understand nature and use this knowledge to improve our quality of life [1, 2].

In this regard, it is important to distinguish two categories of sciences: the formal and the factual or empirical. Formal sciences, such as logic or mathematics, are not concerned with facts. Their objects of study are self-referential. For example, mathematics deals with the study of ideal entities, such as points, lines, planes, and numbers, regardless of their similarity to reality. The elements used by mathematics are ideal objects. Formal sciences are based on axioms accepted *a priori* from which consequences or theorems are deduced by applying the rules of logic.

The factual or empirical sciences, on the other hand, are concerned with the study of real objects that are external to science itself. Examples of factual sciences are Physics, Chemistry, Biology, Economics, Geology, and Medicine, among others. These sciences must derive general principles from observation and experimentation. In these disciplines, experiments play an essential role because it is through observations and experiments that we accept or discard hypotheses and theories. It is also through observations and experiments that we endeavor to infer the laws of nature. In a way, science operates like an observer who does not know the rules of chess and tries to discover them by watching experienced players.

Science is constructed through a process of trial and error. The method we use in science is not without criticism and objections. Epistemology and the philosophy of science address this interesting and arduous problem.

Toward the end of the nineteenth and early twentieth centuries, a broadly accepted and widespread current of thought, known as *positivism*, developed in Europe (mainly Vienna). For the followers of this school of thought, the only reliable and genuine knowledge is scientific knowledge, and only those statements that are liable to empirical verification lead to meaningful and useful knowledge. Put simply, we could say that within this conception:

✓ Science begins with observation.
✓ Observation provides a secure basis for deriving or to reach knowledge.

However, many thinkers criticize the status and role of observations themselves and contest these assumptions. It is known that two people who observe the same object from the same place and under the same circumstances do not necessarily have identical interpretations, even if the images they observe are almost identical (Figure 1.1). What an observer "sees" depends in part on culture, experience, expectations, knowledge, and circumstances. Added to this is the fact that sometimes theories precede observations, i.e., observations and experiments are conducted within the framework of a certain theory. It is therefore not entirely true that science begins with observation. Very often, observations are made in the light of certain theories. For example, many of the experiments currently underway at CERN's Large Hadron Collider (LHC) are designed to test existing physical theories, which means that in these cases,

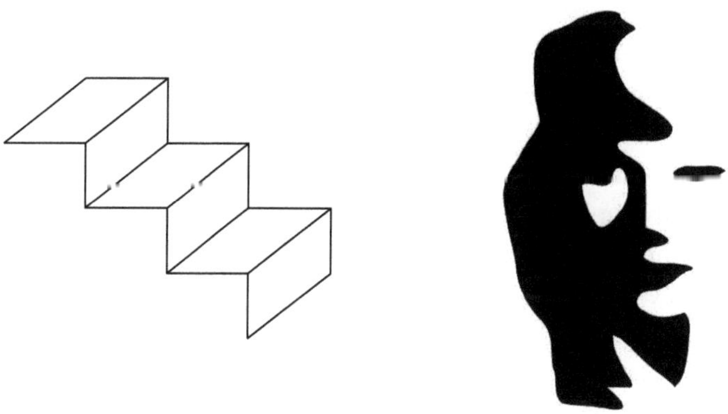

(a) Up or down? (b) What do you see in this picture?

Figure 1.1 Observations and insights are not always entirely objective. (a) Do these steps go up or down? Is it a staircase seen from above or from below? (b) What do you see in this figure? A horn player or a woman's face?

the theory precedes and guides the experiment. Among the original objectives of this great project were to discover the "Higgs boson" and to explore the physics prevalent in the early moments of the universe. Both goals depend on two important theories: the standard model of elementary particles and the Big Bang theory.

Modern positivists make a difference between the way we discover a theory and the way we justify it. They admit that theories can be conceived in different ways, for example, from a preexisting theory, following a moment of inspiration, accidentally, or after periods of observation. But its validation and test are always empirical.

In the first half of the twentieth century, Karl Popper proposed a criterion for testing scientific theories which he called *falsifiability*. For Popper, scientific theories are hypotheses from which it is possible to deduce verifiable claims. If the experimental observations do not agree with these statements, the hypothesis is rejected. If a hypothesis overcomes the effort to prove its falsity, it can be accepted, provisionally. According to this conception, no scientific theory can be established conclusively. In other words, it is not possible to "prove" a scientific theory. We never know if the observations we made were sufficient, as the following observation could contradict all the previous ones. Thus, for example, the statement; "All elephants are gray" cannot be proved. If we set out to test this claim experimentally by simply detecting an elephant that is not gray, we would discard or *falsify* the hypothesis. On the other hand, if after observing n elephants, it turns out that they are all gray, this does not prove that "all" elephants are gray. Nothing prevents an albino elephant from being born tomorrow or a year from now. All we can conclude from the observation "the n elephants observed are gray" is that our observations agree with the starting hypothesis, and therefore we can hold it temporarily, until we find an anomaly. Thus, when scientists are confronted with the hypothesis "*all swans are white*" they would not go out to find *white swans*, but rather, search for *non-white swans*, if they found one, the theory would be disproved or *falsified*. So, for Popper, science is more a question of falsifications than of confirmations.

In this regard, it is important to pay attention to this criterion when we conduct an experiment to test a hypothesis, avoiding

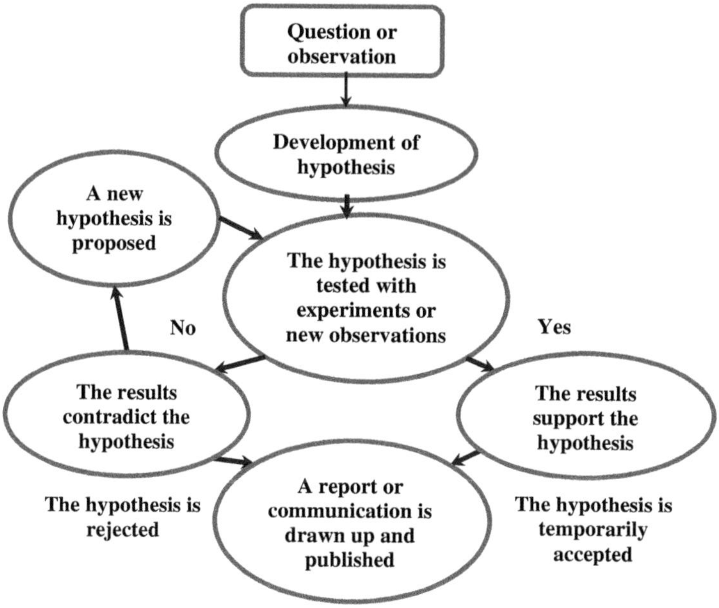

Figure 1.2 Schematic diagram of the method of factual sciences.

statements such as "this result *proves* the starting hypothesis." It would be more prudent and adequate to state "our observations agree with the proposed hypothesis." In factual sciences, we must be very careful to remember the *temporary or transitory* character of our theories. Figure 1.2 schematically illustrates this procedure.

For Popper, a scientific theory must always be formulated in such a way that it can be falsified.[a] This is known as the demarcation criterion. For example, a non-falsifiable statement would be "it might rain tomorrow," as no result can refute it. On the contrary, the statement "when light passes close to the Sun, it curves" is falsifiable, as it can be tested by simply making an observation, such as the apparent movement of stars during an eclipse. In this sense, for Popper, the more "risky" or refutable a theory is — i.e., the more ways we have to test its conclusions — the more "robust" it is. For Popper, the irrefutability of a scientific theory is not a

[a]https://en.wikipedia.org/wiki/Falsifiability.

virtue but a vice, which identifies it as pseudoscientific. According to falsificationism, it would seem that the most powerful tool of science is its ability to discard hypotheses. However, even this is not quite as forceful and clear as it might seem at first glance. Consider, for example, the different hypotheses proposed on the nature of light: corpuscular or wave. The interference and diffraction experiments would seem to falsify the corpuscular model, while the photoelectric effect would in principle falsify the wave model and support the corpuscular model. It has been possible to overcome this apparent contradiction only with the development of quantum mechanics, which eliminates this dichotomy in the nature of light. Popper also proposes a rational and objective model of the evolution of science, in a manner disconnected from the social context.

Another significant contribution to our view of science was made by T. Kuhn [3], who draws attention to the historical, sociological, and cultural aspects of science. In a given historical context, the scientific community shares a set of theories, techniques, and values that are called *paradigms*. This set of paradigms constitutes what Kuhn calls "a *normal science*." Scientists study facts and phenomena, some of which are "enigmatic," and try to explain them using current paradigms. Thus, for example, classical mechanics or classical electromagnetism or thermodynamics are examples of the "paradigms" that prevailed toward the end of the 19th century. When they tried to explain black body radiation (enigma), physicists found an "anomaly" in that black body radiation that could not be explained in the light of classical theories. When the anomalies of a theory accumulate, a "period of crisis" ensues.

The "crisis" is overcome when a new paradigm is developed, which explains both the events included in the previous paradigms and the anomalies that led to the crisis. In other words, according to Kuhn, a *scientific revolution* takes place, which resolves the anomalies of the previous theory. In physics, this happened with quantum mechanics and the theory of relativity. Following their advent, these new theories have become part of the current paradigms, they have become the new "normal" science, which is taught to students and used by scientists to solve their puzzles.

However, it is clear that this pattern of scientific evolution continues, and indeed, one of the greatest ambitions of many scientists and students of physics is to discover an anomaly and ultimately generate a "scientific revolution." In this book, we have tried to formulate several "riddles" or "enigmas" that we hope will challenge the students. We hope that they may serve as intellectual gymnastics that will prepare and encourage students to find and solve the puzzles they may encounter in the future. Perhaps some of them can identify anomalies in our current theories or participate in future revolutions. So, welcome to this adventure of thought.

Stay hungry. Stay foolish, Steve Jobs, Stanford, 2005.

1.2 Preparing laboratory reports

An experiment is not complete until a report has been written presenting and discussing the results. It is also useful to attempt to interpret these results in light of some relevant model or theory. The standardized format for scientific reports is described in more detail in Appendix A of this book. These reports follow the same general rules as most scientific publications. Indeed, writing and publishing scientific results is an intrinsic part of the scientific method. Through these publications, other researchers can replicate our results or perform tests under new conditions to validate or refute them. This procedure ultimately makes our results and theories more solid and reliable. Furthermore, brainstorming with colleagues is an exciting and enriching exercise.

1.3 Safety in the laboratory

A fundamental rule of great importance in laboratory work is to take care of your own safety, that of other people working with you, and the equipment. It is therefore important to remember some general basic rules:

✓ Don't turn on or plug in any instrument or device until your instructor has explained how it works and given you permission to do so. The same applies to handling substances. Before plugging in any device, read the instructions manual carefully and make sure you understand its operation, precautions for use, and safety recommendations.

✓ If you use hazardous substances, make sure that a qualified person instructs you on their use and consider all necessary precautions.

✓ Pay close attention to electrical connections, lamps, lasers, radioactive sources, etc. Always read the safety recommendations and don't use them without authorization from your lab supervisor or instructor.

✓ Remember that the lab is a place to learn science as well as safety regulations that minimize the risk of accidents. The laboratory should be a place where you can enjoy learning new and surprising things, but under no circumstance should you place at risk any person or the physical integrity of the facility.

Appendix B describes safety rules for several common situations, including safety measures to consider when working with certain materials. In addition, for any new equipment or substance you use, you should always read the user manual, be adequately informed of the specific safety rules, and follow the guidelines of instructors and others at the institution where you work or study.

References

[1] A. F. Chalmers, *What Is This Thing Called Science?*, Indianapolis: Hackett Pub., 1999, p. 266.

[2] Wikipedia, "Philosophy of science," 2021. Available: http://en.wikipedia.org/wiki/Philosophy_of_science.

[3] T. Kuhn, *The Structure of Scientific Revolutions* 3rd Ed., Chicago: Chicago University Press, 2012.

Chapter 2

INTRODUCTION TO GRAPHICAL ANALYSIS

In many circumstances, we need to find the dependence between two variables: x and y. To do so, we observe the variation of one of them, say y, as a function of the other, x. The variable x could be, for example, the stretching of a spring and y the force applied to stretch it. This chapter presents some qualitative methods of graphical analysis for interpreting experimental data and determining the relationships or regularities implicit in them. There are several cases in which these variables are connected by a linear, potential, or exponential relationship. Here, we discuss some examples of these types. There are some useful criteria for selecting the most appropriate scales for representing data in a graph. The usefulness of linear, semi-logarithmic, or double logarithmic scales is discussed, as well as the procedure of "linearization" of the graphical representations of different variables.

Goals

✓ Graphical representation of results
✓ Discovery of an empirical law
✓ Linear, exponential, and potential relationships
✓ Linear and logarithmic scales
✓ Pseudovariables

2.1 Graphical representation of results

One of the resources most often used by researchers to assess the degree of agreement between the results of an experiment and a theoretical model is to represent both sets of data in the same graph. In Figure 2.1, two interpretative models of the results are represented by lines (solid and dashed) and compared with the measured values, represented by circles.

At first glance, we can see that Model B provides a better description of data than Model A. There is consensus in the scientific literature on the use of *symbols* to represent experimental results or observations and *lines* to describe theoretical expectations.

The graphic representation of experimental data is often a source of inspiration for interpreting them or developing explanatory models and theories. Indeed, the graphical representation of the results of measurements is a widely used tool not only in experimental sciences but also in many academic disciplines and in various areas of daily life. Figure 2.2 shows the variation in Argentina's total energy consumption and gross domestic product (GDP) as a function of time between 1970 and 2006. The connection or correlation between these variables is evident in this graph and suggests that we explore the dependence between them.

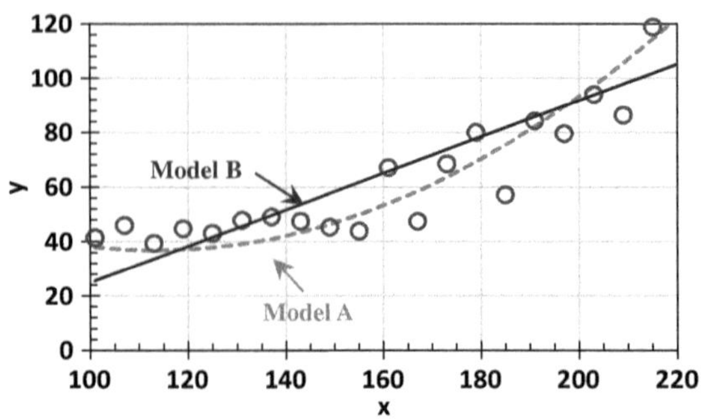

Figure 2.1 Data representation of variable y as a function of another variable x. The red circles represent the observed results. The same plot includes two interpretive models: Model A (dashed green line) and Model B (solid blue line).

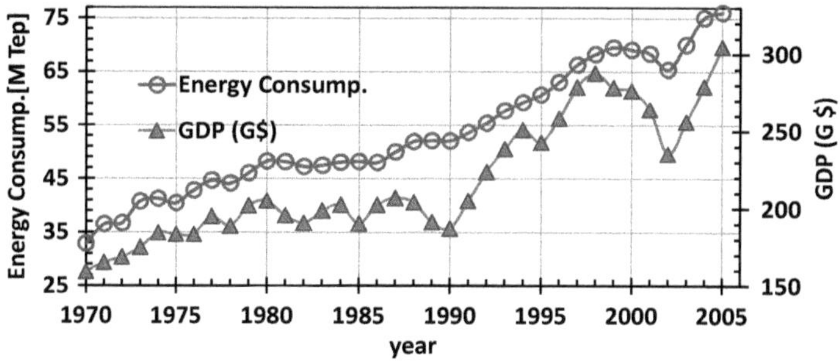

Figure 2.2 Representation of Argentina's total energy consumption as a function of time with circles referring to the left vertical axis. The unit of energy is millions of tons of oil equivalent, MTep. This figure also represents the change in gross domestic product (GDP) in G $ (Giga Pesos at constant values, i.e., corrected for inflation) as a function of time with triangles referring to the right vertical axis. The figure reveals a link between energy consumption and GDP.

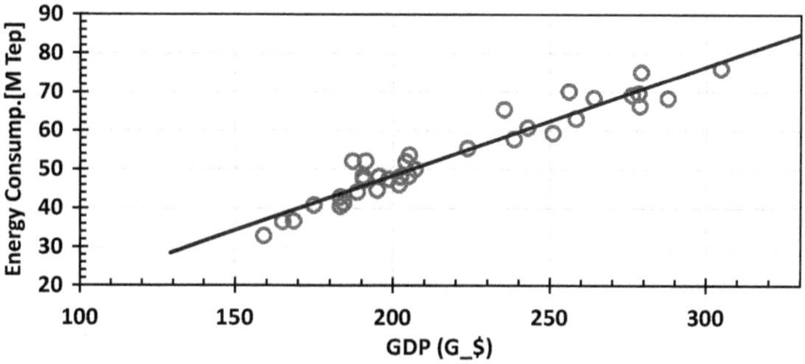

Figure 2.3 Representation of Argentina's total energy consumption as a function of GDP between the years 1970 and 2006. This graphic representation reveals the approximately linear dependence between these variables during that time.

Figure 2.3 shows that when we plot energy consumption as a function of GDP, the roughly linear dependence between these variables appears easily, suggesting a first interpretation of the problem at hand. This observation enables us to develop the first model of energy consumption for the country. Note how the graphical representation often leads us to "discover" the implicit relationship in the observed data. This visualization of the relationship between two variables is rarely so transparent and striking in a data table. Furthermore, a critical attitude should always be maintained toward these preliminary conclusions. There may be situations that make this dependence vary in subsequent years. For example, if a country engages in an energy efficiency program, GDP might continue to rise while energy consumption remains constant or decreases.

Data analysis is an activity that has developed greatly in recent times, as it is applicable to many academic, scientific, and economic disciplines and social activities. In fact, these techniques are the core of a new discipline commonly referred to as data science or data mining, a highly demanded professional career.

2.2 Choice of variables

In many situations, we want to investigate the variation or dependence of a given attribute of a system (which we will designate as variable Y) as a function of another variable of the system (which we will call X). Our goal is to find out whether there is a relationship that links variable X with variable Y. It is customary to call the variables that we control or that determine the state of the system *independent*. The variables determined by others are the *dependent* variables [1, 2]. In the example shown in Figure 2.2, the independent variable would be time (years), while both GDP and energy consumption would be dependent variables. This division is not always clear or possible to make. In real-world problems, there are generally several independent and several dependent variables.

For the sake of simplicity, we will assume in what follows that the system contains only two variables: the independent variable X and the dependent variable Y. The importance of this case is that in many practical situations, it is possible to design a study or experiment such that only one parameter at a time varies, while the rest remain constant. When this is possible, the analysis is greatly simplified.

2.2.1 *Linear law*

We say the relation of Y on X is *linear* if the observed data can be adequately described with a relationship of the following type:

$$Y = a \cdot X + b. \tag{2.1}$$

Figure 2.3 shows an example of this type. Parameter a is the slope of the line and b is the ordinate at the origin or the intersection of the line with the vertical y axis. A linear relationship between two variables is easy to identify with the naked eye. However, it is not so easy to distinguish whether the variables have a potential, exponential, or other kind of relationship [3, 4].

2.2.2 *Power law*

The variables X and Y are related by a *power law* if

$$Y = aX^b, \tag{2.2}$$

where a and b are constants other than zero. This *power law* is very common in natural sciences, economics, and many other applications.

To facilitate the task of finding both the scale exponent b and the constant a, it is convenient to represent $\log(Y)$ as a function of $\log(X)$. If we take the logarithm of each side of Eq.(2.2), we have

$$\log(Y) = \log(aX^b) = \log(a) + b\log(X). \tag{2.3}$$

If we make a change of variable, $\tilde{Y} \equiv \log(Y)$ and $\tilde{X} = \log(X)$, then

$$\tilde{Y} = \log(a) + b\tilde{X}. \tag{2.4}$$

So we see that the graphical representation of $\tilde{Y} versus \tilde{X}$ is a straight line with slope b and Y-intersection equal to $\log(a)$.

With most spreadsheets and math packages such as Microsoft Excel®, Originlab Origin®, Wolfram Mathematica®, and MathWorks Matlab®, one does not need to take the logarithm of the data to visualize the relationship indicated by Eq.(2.4). By simply selecting log scale for the original variables Y and X, the relation between the variables is clearly visible, as shown in Figures 2.4 and 2.5. When the relationship between two variables is not linear, it is useful to change one or both axes to log scale to see if the plot becomes "linear". If the plot becomes linear when we change both axes to log scale, as shown in Figure 2.5, it indicates that the relationship between them is a power law.

Double log graphs such as those in Figure 2.5 are also called *log-log plots*. Note that on a logarithmic scale, the decades are equidistant,

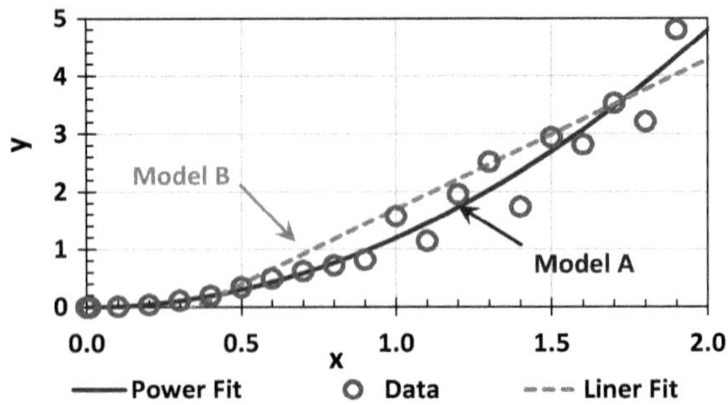

Figure 2.4 Representation of a data series with power law on a linear plot. The dashed line is a linear interpolation, and the solid curve is a power interpolation ($Y = ax^b$).

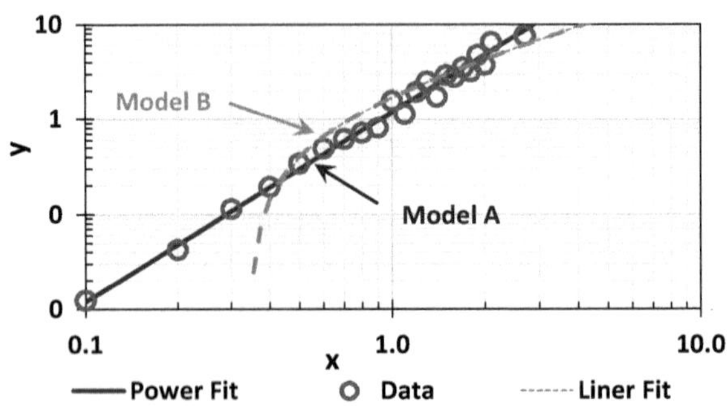

Figure 2.5 The same data as in Figure 2.4 shown using logarithmic scales or log-log scale. Note that the linear fit, represented by the dashed line, is no longer a straight line on these scales, while the power fit (solid line) appears as a straight line. It is interesting to carefully compare the lines in this figure with the same ones in Figure 2.4.

i.e., the distance between 0.1 and 1 is the same as that between 1 and 10, and so on. This is very different from what happens on linear or normal scales, where the distance between 0 and 1 is the same as that between 1 and 2 and so on, as seen in Figures 2.4 and 2.5.

On the other hand, if we suspect that the observed values (X, Y) have a power dependency, by plotting them on a log-log graph, the data align. In other words, if the data are "linearized" on a log-log scale, then we can infer that the relationship that connects them is a power law. The graph thus enables us to discover the underlying law that binds the variables X and Y.

Example: The period T of a simple pendulum is measured for different lengths L. In the case of small oscillation amplitudes, using Newton's laws, it follows that the variables are related by

$$T = 2\pi\sqrt{\frac{L}{g}}, \tag{2.5}$$

where g is the acceleration due to gravity. In this case, L represents the distance from the suspension point to the bob center of mass, see Figure 2.6. The expression (2.5) can be written as follows:

$$T = aL^b, \tag{2.6}$$

which is a power law with $a = 2\pi/\sqrt{g}$ and $b = 1/2$.

Thus, a graph of the measured values of T as a function of L, on a log-log scale, enables us to know whether Eq. (2.6) provides a good description of our data. If so, the graphical representation enables us to obtain the constants a and b. The value of a enables us to determine the value of g. Similarly, we could have plotted T^2 as a function of L, in which case the relationship between these two variables should be linear. From the slope of this linear plot ($a^2 = 4\pi^2/g$), we can find g.

Note that if we had inadvertently taken L_{measured} as the length of the pendulum, see Figure 2.6, the graph of T^2 as a function of L_{measured} would give a straight line of the type $T^2 = cL_{\text{measured}} + d$. From the interpolation of the data, we could find both $c = 4\pi^2/g$ and $d = 4\pi^2\Delta L/g$. From these values, we could deduce both the value

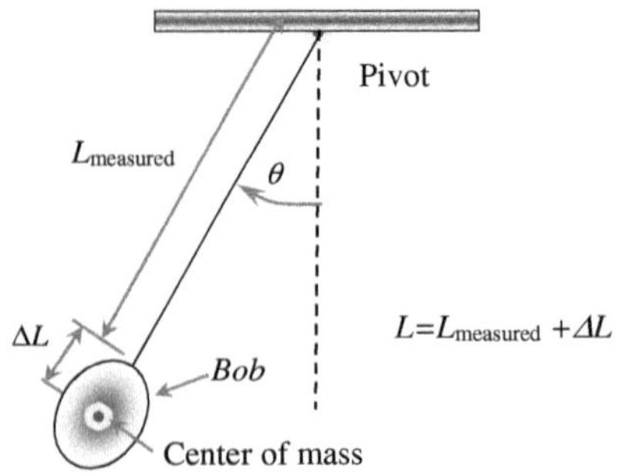

Figure 2.6 Simple pendulum. L_{measured} is the distance from the suspension (fulcrum-pivot) and P is a point at distance ΔL from the center of gravity.

of g and the missing distance ΔL of the point P from the center of gravity.

2.3 Exponential relationship

Another case of great interest is the exponential relationship between two variables. Suppose we have two variables, Y_1 and Y_2, as a function of t. If the relationships between these variables were

$$Y_1(t) = Ae^{-\lambda_1 t} \tag{2.7}$$

and

$$Y_2(t) = A(1 - e^{-\lambda_2 t}), \tag{2.8}$$

their graphical representations would be as shown in Figure 2.7.

It is easy to see that the first of these relations (Y_1) is "linearized" on a semi-logarithmic scale, while the second (Y_2) is not.

In this case, it is convenient to take the derivative of both expressions, i.e.,

$$\frac{dY_1(t)}{dt} = -A\lambda_1 e^{-\lambda_1 t} = -\lambda_1 Y_1(t) \tag{2.9}$$

and

$$\frac{dY_2(t)}{dt} = A\lambda_2 e^{-\lambda_2 t} = \lambda_2(A - Y_2(t)). \tag{2.10}$$

Thus, plotting the *derivative (dY/dt) as a function of the dependent variable* (Y) provides a straight line in both cases. From the values

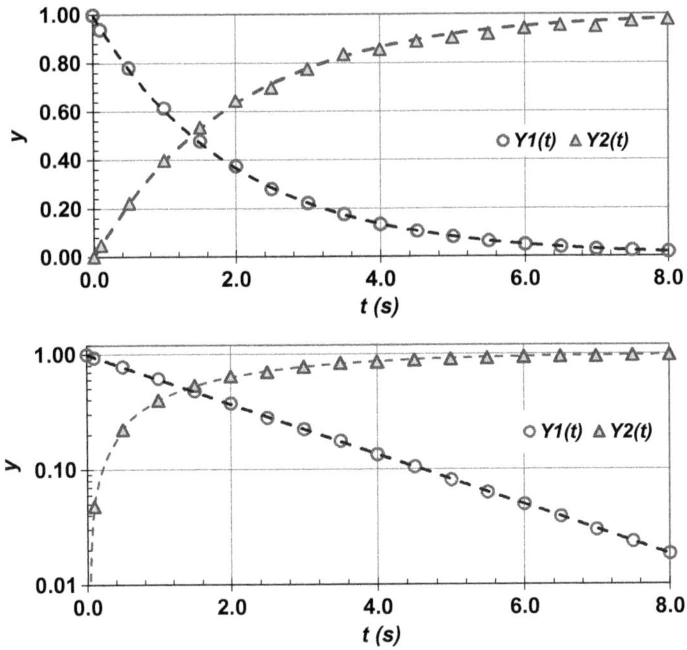

Figure 2.7 Plots of the functions (2.7) and (2.8) on a linear scale (upper panel) and on a semi-logarithmic scale (lower panel). Note that only the representation of $Y_1(t)$ is linearized on a semi-logarithmic scale.

of the slopes and their ordinates at the origin, we can estimate the parameters A and λ, as shown by Eqs. (2.9) and (2.10). Figure 2.8 shows the same functions as Figure 2.7. As can be seen, both relations became linear this time. It is clear that these alternative plots are very useful for this type of situation.

A difficulty with this representation is that it requires the derivative of the function in question to be known, for which we must use a numerical procedure. If the values of Y_1 and Y_2 are the results of measurements for different times t, we can estimate the derivative by computing the finite differences using consecutive data pairs:

$$\frac{dY(t_i)}{dt} \approx \frac{Y_{i+1} - Y_i}{t_{i+1} - t_i} \quad \text{or better} \quad \frac{dY(t_i)}{dt} \approx \frac{Y_{i+1} - Y_{i-1}}{t_{i+1} - t_{i-1}}. \tag{2.11}$$

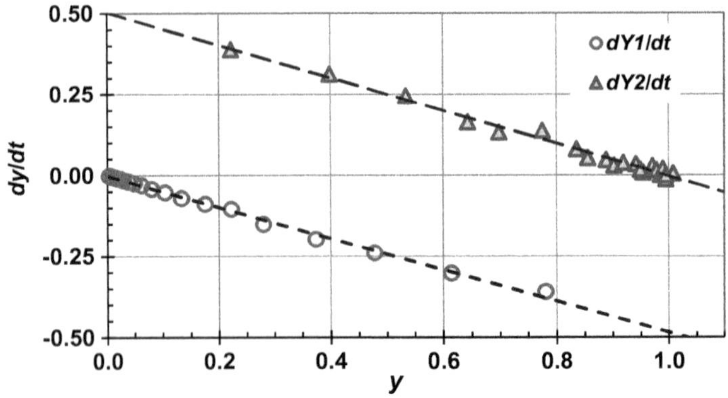

Figure 2.8 Linear scale representation of derivatives dY_1/dt and dY_2/dt as a function of dependent variables Y_1 and Y_2, respectively. In this example, $\lambda_1 = \lambda_2$, thus the data are aligned (linearized) on two parallel lines.

However, since the data obtained from measurements often contain errors, their differences $(Y_{i+1} - Y_i)$ or $(Y_{i+1} - Y_i)$ sometimes magnify their errors. As a result of this effect, figures of dY/dt as a function of Y can present large dispersion. One way to improve the estimation of the derivative is to use a set of data that are in a range where not much variation in the derivative is expected *a priori*. By using a group of values, we approximate a line that crosses them all, whose slope m we take as an estimate of the slope of the curve in the environment of that data, i.e., we make a local estimate of the derivative dY/dt using a group of values instead of using consecutive pairs. This operation can easily be done in a spreadsheet and often reduces the dispersion of the derivatives. It is convenient to carry out this operation with an odd number of data, which includes an equal number of major and minor abscissas at each point. For example, if five data are taken to obtain the slope, two correspond to data with a greater abscissa and two correspond to data with a lower abscissa with respect to the point in question. This avoids possible bias in derivatives.

2.4 Variable transformation — Pseudovariables

There are many functional forms that are not linearized using logarithmic scales, for example:

$$Y = a + b/X. \tag{2.12}$$

It is not possible to linearize this function by taking logarithmic scales. However, if we do the transformation:

$$Z = 1/X \quad \text{that gives} \quad Y = a + b \cdot Z, \tag{2.13}$$

we obtain a linear relationship between Y and the new *pseudovariable* Z. Moreover, if we represent Y as a function of Z, we obtain the values of the parameters a and b.

2.5 Tips for generating charts

Most spreadsheets and graphing software available on computers include among their options graphs with different types of scales described above. To obtain clear, suggestive plots, it is advisable to observe the following suggestions:

✓ Identify the axes with well-placed *labels*, indicating which *variables* are represented and what *units* are used.

✓ When representing *data* that are *results of measurements*, it is convenient to use *symbols* (squares, circles, rhombuses, etc.) as far as possible with their corresponding *uncertainties* or *absolute errors* (in the form of bars indicating the size of the errors or uncertainties). When different data sets are included in the same graph, it is recommended to use different symbols for each data set.

✓ When representing *models* or *theoretical predictions*, it is customary to use lines (solid, dotted, dashed, etc.).

✓ Include a caption that briefly describes what is represented and provides some additional information.

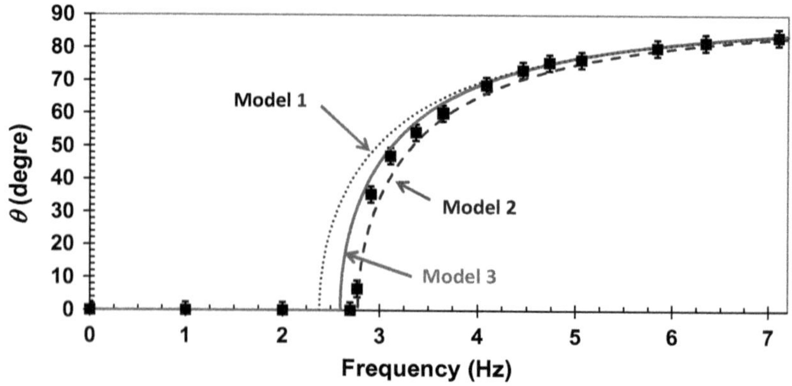

Figure 2.9 Example of a plot of experimental data and theoretical prediction. The solid squares correspond to the measured results, and the lines are predictions from different models that aim to describe the behavior of the data.

✓ Include a text box inside the graph with related complementary information to provide an understanding of the context in which the data are displayed or the particular experimental conditions in which they were recorded.

✓ Select easily distinguishable symbols to indicate different data series. The same goes for the shapes of the lines that correspond to the different models proposed to describe the behavior of the data.

This procedure is illustrated in Figure 2.9.

2.6 Summary of important concepts

The following questions are proposed for discussion among students. Try to provide a concise explanation of the following concepts and, when possible, provide an appropriate example.

✓ Why is it useful to graph data instead of presenting them as tables?

✓ Explain why the power and exponential functions are linearizable in log scales.

✓ Why is it useful to make scale changes (go to log scale) in some graphs?

✓ What can be said about the relation between two variables, if their graphical representation is linearized using semi-logarithmic scales?
✓ What can be concluded if a graphical representation is linearized using double logarithmic scales?
✓ Why is it useful to use pseudovariables?

2.7 Exercises and problems

The purpose of these exercises is to develop skills in the use of spreadsheets. One option is to use Microsoft Excel®, although programs like Google spreadsheet, MathWorks Matlab®, or Wolfram Mathematica® may be suitable for solving these exercises. You can download Excel spreadsheets with the solutions of a series of exercises presented here from the website *www.fisicarecreativa.com* (Physics Experiments Book).

In these exercises, we propose to

✓ generate mathematical functions of different types in a spreadsheet and represent them graphically,
✓ use graphs to find solutions to algebraic equations,
✓ graph functions on different types of scales (linear, logarithmic, or others).

(1) Using any spreadsheet, generate pairs of (x, y) values corresponding to the following functions. Use a separate sheet for each case:

(i) $y = 2x + 3$,
(ii) $y = 2x^2 + x$,
(iii) $y = x + 3$ and $y = 2x + 3$,
(iv) $y = 3x^3 - x$ and $y = 2\exp(x) - 1$,
(v) $y = 2\exp(2x)$ $y = 2x + 2$ and $y = 2x^{0,75}$,
(vi) $y = 2\exp(2x)$, $y = 2\exp(2x) + 3$ $y = 2x + 2$, and $y = 2x^{0,75} + 2$.

(a) Group the (x, y) pairs into columns and create a graphical representation of each function. Select the domain of the independent variable so that the shape of each function can be clearly observed; do the same with the interval of the dependent variable.

(b) For functions (iv), (v), and (vi), change the scale of the axes from linear to logarithmic (x-axis only, y-axis only, and both simultaneously). Discuss and justify in which cases the graphical representation of functions is linearized by changing the scale of the axes.

(c) For the systems of equations represented by (iii) and (iv), find the roots or solutions of the system of equations they represent. Remember that the roots of a system of equations are the pairs (x, y) that simultaneously satisfy all the equations that make up the system. Graphically, they are the coordinates of the intersection points of the curves. Could you find the solutions analytically in both cases (iii) and (iv)? If possible, compare the results with graphical methods. Discuss the advantages and disadvantages of each of these methods.

(2) Propose the type of graph (linear, log-log, or semi-log) and the corresponding variables or pseudovariables that enable the graphical representation of each of the following functions to be linearized. In each case, indicate the procedure to find, from the graph, the values of the constants a and b.

(i) $y = ax^2 + b$,
(ii) $y = ax^b$,
(iii) $y = \frac{a}{x} + b$,
(iv) $y^2 = ax^3 + b$,
(v) $y = a\exp(-bx)$,
(vi) $y = a\exp\left(-\frac{b}{x}\right)$,
(vii) $y = a\log(bx)$.

Consider the expression ${}^1/_x + {}^1/_y = {}^1/_a$. Show that there are two ways to linearize the graphical representation:

(a) by presenting ${}^1/_y$ as a function of ${}^1/_x$,
(b) by representing the product xy as a function of the sum $x+y$.

Explain how the value of the constant a can be obtained from each graph.

(3) In 1965, Gordon F. Moore, one of the founders of Intel (a company that makes computer microprocessors) made an interesting observation, which became known as *Moore's law*. He noticed

that the number of transistors that could be integrated into a chip was growing rapidly. This led him to argue that the number of transistors per inch in chips was doubling every year and that the trend would continue for decades to come. Subsequently, in 1975, he amended his own law stating that the rate would slow down and that integration capacity would double roughly every 24 months.

(a) Based on the data reproduced in Table 2.1, analyze the validity of Moore's law.
(b) What can you say about the validity of the first and second laws enunciated? Discuss and argue the reasons behind your conclusions.
(c) What do you predict will happen after 2005? Has it really happened?

(4) A group of researchers analyzed the evolution of the world population over the last fifty years to evaluate action plans for different organizations, both economic (IMF, World Bank, and WTO, among others) and social and human development organizations

Table 2.1 Characteristics of the computer chips that Intel has produced in recent decades.

Chip	**Year of production**	**No. of transistors per square inch**	**No. of years from 1970**
4004	1971	2.28×10^3	1
8088	1979	3.00×10^4	9
80286	1982	1.49×10^5	12
Intel 386	1985	2.94×10^5	15
Intel 486	1989	1.33×10^6	19
Pentium	1993	3.68×10^6	23
Pentium Pro	1995	5.98×10^6	25
Pentium II	1997	8.40×10^6	27
Pentium III	1999	3.44×10^7	29
Pentium 4	2001	5.07×10^7	31
Itanium	2001	2.83×10^7	31
Pentium III	2002	2.77×10^8	32
Itanium 215	2004	1.34×10^8	34
Itanium 2.9	2004	7.69×10^8	34

(UNESCO, UNICEF, ILO, and WHO, among others). From the data provided by an official body, the following table was made:

Year	**World population** $\times$ **10^{-9}**	**Year**	**World population** $\times$ **10^{-9}**
1950	2,557	1980	4,453
1951	2,594	1981	4,529
1952	2,636	1982	4,608
1953	2,682	1983	4,690
1954	2,730	1984	4,770
1955	2,781	1985	4,852
1956	2,834	1986	4,935
1957	2,890	1987	5,021
1958	2,947	1988	5,108
1959	2,999	1989	5,195
1960	3,041	1990	5,283
1961	3,082	1991	5,367
1962	3,138	1992	5,451
1963	3,207	1993	5,533
1964	3,278	1994	5,613
1965	3,347	1995	5,694
1966	3,418	1996	5,773
1967	3,487	1997	5,852
1968	3,559	1998	5,930
1969	3,634	1999	6,006
1970	3,709	2000	6,082
1971	3,786	2001	6,156
1972	3,863	2002	6,230
1973	3,939	2003	6,303
1974	4,013	2004	6,377

Source: https://www.census.gov/popclock/.

(a) Graph the population P as a function of time and the derivative of the population, dP/dt, as a function of time.

(b) Plot the percentage of population growth C_P as a function of time, i.e., $C_P = (dP/dt)/P(t)$. How does growth vary over time? Based on your results, how do you expect world population growth to continue over the next 50 years?
(c) Find the best function that represents both dP/dt and $P(t)$ as a function of time. How is the world population expected to evolve over the next 100 years?
(d) Can you predict how many inhabitants there will be in 2050 and 2100? Compare your predictions to other sources. How would you explain your results? See https://ourworldindata.org/future-population-growth and https://www.un.org/en/observances/world-population-day/resources.

(5) Imagine you have to report to the headquarters of the company you work for the company's earnings since its establishment in the country. The following table has been prepared from the data provided by the accounting management:

Activity month	Earnings (in thousands of USD)	Activity month	Earnings (in thousands of USD)
1	159	31	475
4	252	34	495
7	287	37	524
10	329	40	512
13	352	43	529
16	381	46	534
19	395	49	538
22	434	52	536
25	434	55	588
28	456	58	599

(a) Find the best function that represents the company's income as a function of time.
(b) Try to fit the best functional form to these data.

(c) What does each of the parameters of the selected function represent?
(d) Can you predict what the company income might be between months 80 and 1000? How would you make that prediction?

References

[1] D. C. Baird, *Experimentation: An Introduction to Measurement Theory and Experiment Design*, 3rd Ed., NY: Pearson, 1995.
[2] C. Deacon, "The importance of graphs in undergraduate physics," *Phys. Teach.*, vol. 37, p. 270, 1999.
[3] S. Gil and E. *Rodriguez, Física re-Creativa*, 1st Ed., Buenos Aires: Prentice Hall, 2001, p. 469.
[4] C. E. Swartz, *Used Math for the First Two Years of College Science*, 2nd Ed., NY: AAP. Teachers, 1993.

Chapter 3

INTRODUCTION TO GRAPHIC ANALYSIS

The world is characterized by great variability and diversity, in both its natural and cultural aspects. Scientists try to find regularity and order in this apparent "chaos" in which we live. In the following projects, we use graphic analysis techniques to help find regularities and deduce "empirical laws" that enable us to describe and systematize the observations. Of course, this is just one of the many tools available to help scientists discover these laws. In particular, we intend to find empirical scaling laws by analyzing data from different sources, including the fields of Physics, Astronomy, Biology, Linguistics, and Mathematics [1–4].

Goals

✓ Discovery of empirical laws
✓ Graphic analysis
✓ Study of scaling laws
✓ Allometric laws
✓ Importance of dimensions in biology
✓ Conservation laws
✓ Allometric laws in fractal systems
✓ Kleiber's law
✓ Benford's law
✓ Zipf's law

3.1 Scaling laws

Scaling laws are important in many branches of science. In particular, in biology, the so-called *allometric* [5] laws are frequent. These laws describe the relationships between anatomical, physiological, or behavioral characteristics and dimensions or shapes. These relations, in general, are described mathematically by potential expressions:

$$y = A_0 x^b, \tag{3.1}$$

where x is an independent variable and y the dependent variable. A_0 and b are two characteristic parameters of the system under consideration, an example of the relationship between respiratory rate and size in vertebrates. In this case, the variable y and represents the breathing rhythm or rate and x can be the mass or length of the animal [6, 7]. Other examples of power laws as (3.1) in Physics are the relationship between the period of a simple pendulum, T, and its length L:

$$T = (2\pi/\sqrt{g}) \cdot L^{1/2} = A_0 \cdot L^{0.5}, \tag{3.2}$$

or the relationship between the period T of the planets and their average distance from the Sun, d, known as Kepler's third law:

$$T = k \cdot d^{3/2}. \tag{3.3}$$

These scaling laws are also present in many other simple systems. For example, in a cube, its area A is proportional to the square of its side L, that is,

$$A = 6L^2 \Rightarrow A \propto L^2, \tag{3.4}$$

and its volume V:

$$V = L^3. \tag{3.5}$$

In these cases, the dependent variables A and V vary with different powers of L.

As we saw in Chapter 2, Eq. (3.1) becomes *linearized* when $\log y$ is plotted as a function of $\log x$:

$$\log y = \log A_0 + b \cdot \log x, \tag{3.6}$$

or even when we choose a *log-log scale* to represent y as a function of x.

The reverse is also true, if a set of experimental data (x_i, y_i) are aligned when they are plotted on a log-log scale, we can infer that the relationship that binds x is a power law.

Incidentally, although biological systems are some of the most complex systems in nature, many of their fundamental properties can be expressed by extremely simple allometric laws as a function of size or mass. An example of this type of relationship is the correspondence

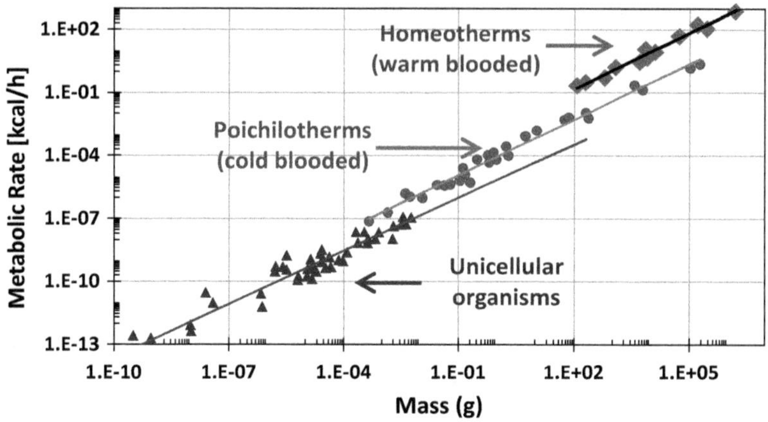

Figure 3.1 Kleiber's law, metabolic rate as a function of mass for various organisms on a log-log scale. Note that this relationship covers 22 orders of magnitude by mass. Figure taken from Ref. [3].

between an animal's basal metabolic rate (BMR) and its mass M. The quantity BMR is the minimum energy consumption per unit of time or minimum power for an animal to stay alive. This relationship is known as Kleiber's law [8, 9] (see Figure 3.1) and is expressed as follows:

$$\text{BMR} = A_0 \cdot M^{3/4}. \tag{3.7}$$

This law is satisfied for a wide variety of species, covering a range of mass variations of over 22 orders of magnitude. This pattern or system holds true for species ranging from bacteria to blue whales. Other examples of this type of relationship are the correlation between respiratory rate and mass or between an animal's longevity and mass. Scaling laws challenge and at the same time guide researchers in the search for patterns that attempt to describe them. Indeed, these surprising relationships and their implications have received considerable attention and have become one of the most active frontiers of research in recent years [8, 10].

Another simple observation relates the height of trees with the diameter of their trunks. These biomechanical principles have been studied in many systems and are very useful for understanding the architecture of plants. Estimation of tree mass has many important practical applications. Wood is often commercialized by weight or volume. Hence, estimating the weight and volume of a forest is critical for the economical, sustainable management of a forest. Moreover, since tree biomass is about 50% carbon, estimation of tree's mass and volume is crucial to calculating the planet's carbon budget [11, 12].

Potential relationships such as that of Eq. (3.1) have also been found in various areas of technology, mathematics, and language. An example is the relationship between cruising speed and the size of the wings of almost all animals and machines that fly [6].

Similar relationships can be studied in fractal systems (Figure 3.2). These widespread forms in nature are characterized by the similarity between a small part of the system and the whole, i.e., they show self-similarity for all scales. This type of structure can be observed in a tree, a fern, the circulatory or renal system, fruits, and plants, such as broccoli and cauliflower. The study of biological systems that exhibit this type of structure, as we shall see, reveals very particular allometric relationships that are characteristic of them [1, 10].

In many sets of statistical data, such as the number of people living in small towns and cities around the world, the presence of the first digit in these data does not occur randomly but follows a well-defined relationship. More specifically, if we select the first digit of this data set from a population, we observe that the digit 1 is more

Figure 3.2 Examples of fractal figures. On the left is a Sierpinski triangle. On the right, a fern leaf. Note how each part of these figures is a replica of the whole.

likely to appear than 2, etc., following a well-defined relationship, described by Benford's law [7]. Similarly, in Linguistics, it is found that, in a given text, in almost all languages, there are words that are repeated. If the words which are repeated several times are ordered and their position in this repetition ranking is designated by n, it is found that the number of times that a word appears, i.e., its frequency of occurrence f, is inversely proportional to n, i.e., $f \propto n^{-1}$. This relationship is known as Zipf's law [13] and is very simple to observe and analyze.

The following graphic analysis activities will enable us to find simple relationships in different natural and cultural systems [1] using spreadsheets and the techniques described in Chapter 2.

3.2 Allometry of plants and animals

3.2.1 *Project 1. Mass-length ratio of the leaves of a plant*

This activity explores the relationship between the dimensions of the leaves of the same plant, represented by their length and mass. For this exercise, Table 3.1 provides the observed values of a set of leaves from the plant *Nerium oleander* or Bay Laurel [1]. The goal is to find out the law behind this data set and if such a law exists. Of course, this exercise could be carried out with the leaves of any plant.

Suggested task:

✓ Graph the width of the leaf as a function of length using the data in Table 3.1. Is it possible to describe *width* as a function of *length* through a linear relationship? If we assume that the leaves of the same plant are similar to each other, we would expect proportionality between width and length by similarity, i.e., $a = \kappa L$. Does the data in Table 3.1 validate this expectation?

✓ Plot the mass of the leaves as a function of their length. By making appropriate changes to the scale, try to linearize the graphical representation between these variables. That is, by changing the scale of the axes from linear to logarithmic, etc., try to make the data appear aligned in some graphical representation. Is it possible to describe leaf mass as a function of length through a power law?

Table 3.1 Relation between the mass m, the width a, and the length l of the different leaves of a bay laurel or oleander.

Variety of pink laurel		
l (cm)	**a (mm)**	**m (g)**
8.5	11.0	0.278
8.5	11.0	0.283
9.4	13.0	0.395
10.0	14.0	0.435
10.4	15.0	0.454
11.6	16.0	0.602
12.1	17.0	0.656
12.3	17.0	0.671
14.0	18.5	0.718
14.4	23.0	1.033
16.0	26.0	1.478
16.5	24.0	1.263
17.3	26.0	1.308
20.5	31.0	2.045
20.3	30.0	1.946
22.2	36.0	2.424
25.0	41.0	3.246

✓ Most spreadsheets, such as Microsoft Excel®, have curve fitting tools, which means they can estimate the parameters of functions that best fit a data set. They generally use the least squares technique, which we will study later. Using these tools, determine the curve that best describes this dependence between leaf mass and length (Figure 3.3).

✓ Try to theoretically justify the results found. To do this, assume that the leaves are of medium thickness δ and density ρ, which we assume constant. If A represents the area of the leaf, its mass m will be proportional to $\rho A\delta$. If it is true that the width a is proportional to the length l, i.e., $a = \kappa l$ and $A = \kappa l^2$, we can write $m = kl^2\delta$, with $k(= \kappa\rho)$ being a constant of proportionality. Therefore, if we graph the pseudovariable (m/l^2) as a function of l, we can find out whether or not δ depends on l. In particular, consider whether the thickness δ dependence on l can be approximated by a power function of the type $\delta = A_0 l^\beta$.

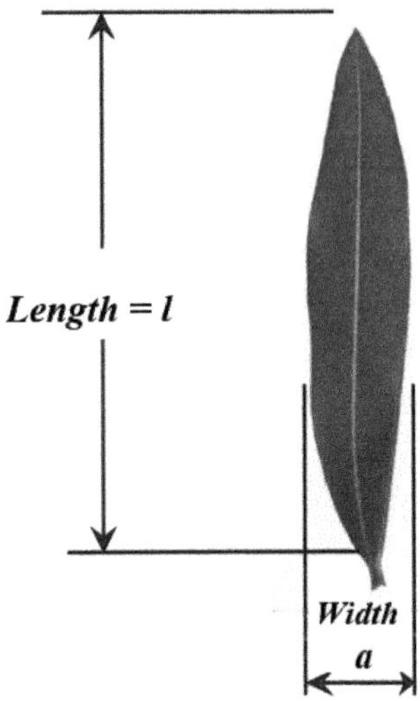

Figure 3.3 *Nerium* or oleander leaf.

Here A_0 is another constant of proportionality. Using the data in Table 3.1, test these assumptions and, if possible, determine the parameter β. From the analysis performed, try to answer the following questions:

- Does thickness δ vary with size or is it fairly constant?
- Does leaf thickness increase or decrease with its size, represented by l? How do you arrive at this conclusion? (Hint: If leaf thickness were strictly constant, the mass of the leaves would depend on the dimensions l as $m = kl^2$. If the volume as a whole increased proportionally to l, the expected relationship would be $m = kl^3$. If the thickness decreases with the size, $m = kl^\varepsilon$, with $\varepsilon < 2$).
- If the found dependence were $m = kl^{2+\beta}$, with $0 < \beta < 1$, it would imply that the leaf tends to maximize its surface compatibly with its mechanical strength. Discuss in light of the results found.

✓ Galileo Galilei made an interesting observation. Suppose we have a wooden plank, which is supported on a table as a cantilever (most of the plank protruding from the table like a balcony). If we build longer and longer planks with the same wood and the same width and thickness, there will come a time when the plank will no longer support its weight and break. This observation applied to trees suggests that as they grow in height, trees must also grow in diameter in order to support their weight. The same idea applied to leaves suggests that as they increase in size, they must increase in thickness to maintain their stiffness. Do your data agree with or contradict this statement or hypothesis?

Photosynthesis is the process by which plants capture and use the sun's energy (light) to transform inorganic matter from their external environment (nutrients) into organic matter which they use for growth and development. From this point of view, it could be hypothesized that plants would benefit from having the largest leaves possible, i.e., with the largest possible surface area, to optimize or maximize the photosynthesis process. One way to achieve this with the lowest material cost would be for leaf mass to vary with length at a power of less than 2. But an exponent of less than 2 would imply that leaf thickness would decrease with size. An exponent equal to 2 would imply that leaf thickness was constant. However, if the large leaves were the same thickness as the small ones, their mechanical strength to support their weight, and that of the water droplets that always adhere to them, would decrease with size. In order for mechanical strength to increase with the size of the leaf, its thickness must increase as it grows. Therefore, mass would be expected to vary according to size with an exponent greater than 2 but less than 3. Do your data agree with this argument?

3.2.2 *Project 2. Experiments with real plants*

(a) Relation between the size of a leaf and its mass

Recommended equipment: A scale with a range of approximately 100 g (or greater) and a sensitivity of 0.1 g or better. A ruler graduated in millimeters longer than the largest of the leaves used.

Select a plant whose leaves you want to examine. Collect leaves of different sizes, from the smallest to the largest you can find. Make sure that all the leaves come from the same plant and have been cut at the same time. Don't use leaves that have been cut on different days because old leaves lose mass due to the evaporation of the water they contain. For this experiment, it is best to have a scale with a sensitivity of 0.1 g or better. Obviously, this depends on the plant variety chosen. For the leaves of common plants, the mass of the leaves is of the order of fractions of a gram, so to observe significant variations, the appreciation of the scale must be sufficiently smaller than the variations in mass between leaves of different sizes [1].

Suggested task:

✓ Graph the change in mass as a function of length. By making appropriate changes to the scale, try to linearize the graphical representation of these variables.

✓ Fit the curve that best describes this dependence between variables.

✓ From the dependence of the mass found as a function of length, analyze the variation in leaf thickness with length. Does thickness vary with size or is it fairly constant?

✓ Are your data consistent with the statement: "The leaves tend to maximize their surface compatible with their mechanical strength"?

(b) Relation between the size of a fruit and its mass

Recommended equipment: A scale with a range of 1 kg (or more) and a sensitivity of 1 g. A ruler graduated in millimeters longer than the largest fruit.

Select a fruit for which you can get specimens of different sizes, preferably of the same variety. One possibility would be to get a bunch of bananas with small and large specimens. Carrots, pumpkins, etc., are also found in a wide range of sizes. Any fruit or nut that can be found in different sizes can be used for this activity.

Figure 3.4 Geometrically similar broccoli plants, i.e., the a/b ratio is approximately the same for all samples.

Another possibility would be to use a broccoli or cauliflower plant, which can be divided to produce specimens of different sizes, but which maintain the geometric similarity among each other (see Figure 3.4).

Take one of its dimensions as a reference, preferably the largest, to make the measurement easier and have greater variation and sensitivity in the characterization of the size. Measure this quantity, which we will call L, and the mass m of each sample [1].

Suggested task:

✓ Graph the variation of m as a function of L. Using appropriate changes of scale, try to linearize the graphical representation of these variables.

✓ Fit the curve that best describes this dependence between variables.

✓ If it is possible to adapt the data by means of an allometric law of the type $m = A_0 L^b$, determine the best values of parameters A_0 and b. Estimate the corresponding errors. Is b closer to 2 or 3 this time?

(c) Relationship between the size of a mammal species and its length

> **Recommended equipment:** A scale with a range of 1 kg (or more) and a sensitivity of 1 g. A ruler graduated in millimeters longer than the largest animal.

Select a mammal species for which you can obtain specimens of different sizes. One possibility would be to take a sample of humans. If you work in a group, you could each record the height h and mass m of different members of your families and friends. Build as long a table as possible of masses, heights, ages, sexes, and builds (divide this category into three types of physical build: slim, medium, and plump). Make sure to include small children and adults in the group. Combine your data with those recorded by the rest of the group. Analyze each individual build separately. Another possibility, if you have access to a rat nursery, would be to use rats of the same species, with a variety of sizes, from the smallest to the largest. In this case, h would represent the length of the animal taken from the end of the tail to the tip of the nose. For simplicity, there is no need to distinguish between gender and build in this case.

Suggested tasks:

- ✓ If you use human data, graph the change in m as a function of h, for a given gender and body type. In a separate graph, include all individuals, regardless of gender and build. If you use animals, you needn't distinguish between gender and build. Evaluate the possibility of linearizing the graphical representations through appropriate resizing. Fit the curve that best describes the dependency between the variables in each case.
- ✓ If it is possible to adapt the data by means of an allometric law of the type $m = A_0L^b$, determine the best values of parameters A_0 and b and estimate the corresponding errors.

✓ If the dependence found were $m = kl^{3+\beta}$, with $\beta \approx 0$, it would imply that the species under examination has three-dimensional growth, i.e., both width and thickness are proportional to length. Discuss in the light of the results found.

3.3 Conservation laws in nature

3.3.1 *Project 3. Search for conservation laws in nature*

Recommended equipment: A two-meter tape measure graduated in millimeters.

When we observe a branching tree or artery (or vein), the similarity of the shapes is remarkable, as can be seen in Figure 3.5. Perhaps the most notable common feature of these systems is their fractal structure, i.e., the fact that some parts of them have the same characteristics as the whole. In particular, when a branch of a tree or an artery forks, the branches that emerge are always thinner than the trunk from which they originate.

It is tempting to conjecture that perhaps the diameter d (or perimeter p) of the original trunk is related to the diameters of the

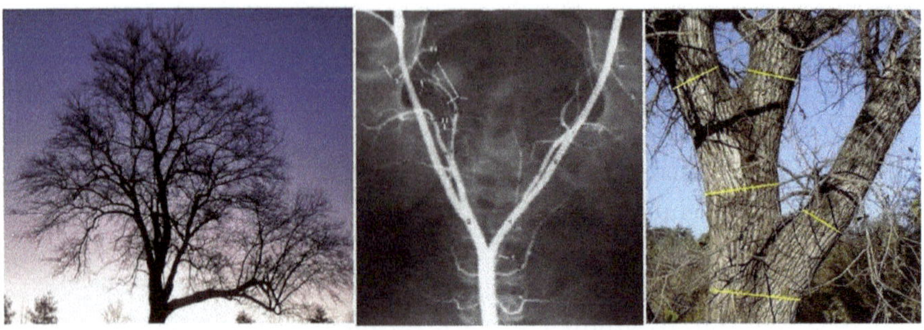

Figure 3.5 Photographs of trees (right and left). An angiography of an artery (center). Here we can see the fractal character of a tree and an artery in their branches. In the picture on the right, we can see the variation in the diameters of the branches, before and after a fork.

emerging branches, following some relation of the shape:

$$d_{\text{Trunk}} = \sum_i d_{\text{Branch}}(i) \quad \text{and also} \quad d^n_{\text{Trunk}} = \sum_i d^n_{\text{Branch}}(i), \tag{3.8}$$

where n is an exponent to be determined. For a branch or trunk with a quasi-circular section, the perimeter is approximately proportional to the average diameter of the branch or trunk. Therefore, expressions similar to (3.8) would also be valid for the perimeter p. Depending on which parameter is easier to measure, we can use one or the other. In the case of a tree, due to irregularities in shape, it is suggested to measure the perimeter, which provides a better estimate of the average diameter.

The objective of this activity is to investigate whether a "conservation law" of the type (3.8) holds for a given tree species. First, a relationship with an exponent $n = 2$ would seem reasonable. If tree nutrients (sap) flow through the trunk, the cross-sectional area can be expected to be preserved, such that all parts of the plant would be equally nourished. If the areas were to increase or decrease, the branches could receive too few or too many nutrients. In trees with a very long, unbranched trunk (e.g., a palm tree), trunk diameter decreases upward continuously, perhaps as a result of nutrient uptake by the trunk itself, causing fewer nutrients to reach levels further from the ground. Gravity would also make it difficult for nutrients to rise to high altitudes. However, at close range, it can be assumed that the areas of the trunk and branches are the same, i.e., the area is preserved as long as the sections immediately before and after a branch are considered. Even though there are trees with very different geometries, reflecting the complexity of biological systems, a relationship like (3.8) can be found in many varieties.

The objective of this activity is to test these hypotheses (conservation of the area) for a given species. The idea is therefore to measure in a tree or group of trees of the same species, the perimeters (or diameters) of a main branch and the perimeter of all the emerging branches, immediately after the branching. It is best to measure large and small branches for each tree used and avoid pruned branches, which can distort the effect we are trying to identify.

Suggested task:

✓ Combine all the available data to build a graph of $\sum_i d^n_{\text{branch}}(i)$ as a function of d^n_{trunk}, leaving the exponent n as a variable parameter. Then, by varying n, find the value for which the data are best aligned, with the smallest dispersion (i.e., so that R^2 is closest to 1).
✓ What can be concluded from these graphs and observations? What about the general validity of relations such as Eq. (3.8)?
✓ What can be said about the "conservation law" in tree branching?
✓ How would we investigate a similar conservation law for an artery from an angiogram like the one in Figure 3.5?
✓ How would we justify the existence of a conservation law in arteries if the sections of the arteries were preserved? Tip: For an incompressible liquid, the conservation of mass means that the flow rate of the fluid (i.e., the volume that passes per unit of time) must be the same along a pipe and its branches, in particular, before and after a branch [9, 10].

3.3.2 *Project 4. Importance of dimensions in biology*

Galileo made several interesting observations on the relationship between the height of trees and the diameter of the trunk, as well as the relationship between the size of animals and some of their properties. In ensuing centuries, and especially throughout the twentieth century, remarkable relationships have been found between the different properties of many animals and their size, which relate to the behaviors of such dissimilar species as bacteria and blue whales within the same system [2, 4, 8]. The sizes, masses, heart rates, and half-lives of various mammals are listed in Table 3.2.

Suggested task:

✓ From the data in Table 3.2, graph the heart rate dependence and average lifespan of these species as a function of their masses.
✓ Vary the scales of the graphs and try to linearize the graph representations. What dependence do we find for these data?
✓ Plot the product of heart rate and average lifespan as a function of mass for these species. Check whether this product remains

Table 3.2 Size, mass, heart rate, and average lifespan of various mammals [8, 9, 14].

Species	Mass (kg)	Heart Rate (1/min)	Lifespan (years)
Hamster	0.06	450	3
Rabbit	1.0	205	9
Chicken	1.5	275	15
Cat	2.0	150	15
Small dog	2.0	100	10
Medium dog	5	90	15
Monkey	5	190	15
Large dogs	8	75	17
Human	90	60	70
Pig	150	70	25
Cow	800	65	22
Giraffe	900	65	20
Horse	1,200	44	40
Elephant	5,000	30	70
Large whale	120,000	20	80

approximately constant for all species. If so, we can establish a conservation law for this product. Discuss the physical or biological significance of the heart rate product for the average lifespan. How would you estimate the total number of heartbeats of an animal throughout its life?

✓ In physics, when a quantity does not change during a transformation or a process, we say that this quantity is conserved. For example, the momentum in a collision, the electric charge, and so on. What can you say about the possible conservation laws in biology?

3.4 Zipf's law and Benford's law

3.4.1 *Project 5. Word frequency in languages: Zipf's law*

Texts in most modern languages include repeated words. It is interesting to perform a statistic of repeated words in a given text, ordering them according to the frequency of occurrence and calculating their probability of occurrence. This statistic for word repetition is called a unigram. The purpose of this activity is to study the histogram of the occurrence of words or, more precisely, the distribution of probabilities followed by unigrams [13, 15]. Take a text with at least 3,000 words and identify how many times the words are repeated. This can be done manually, though it's much easier to use a program. There are many programs available on the Internet that can do this analysis. One of them is the hermetic word frequency counter [15].

Suggested task:

✓ Using the procedure you consider most convenient, choose a text of at least 3000 words and build a unigram of the most frequently repeated words.

✓ On a spreadsheet, enter the words that are repeated multiple times in one column and, in the cell immediately to the right, the *frequency of occurrence* (f) of each. Sort the rows in descending order of *frequency of occurrence.* In a contiguous column, enter a number that indicates the *order of repetition (r)*, from highest to lowest.

✓ Plot *frequency of occurrence* according to the *order of repetition.* By changing the scales, try to linearize this graph. Which mathematical relationship best describes your unigram?

✓ Plot the product of *frequency of occurrence* by *order of appearance:* $f \times r$. Indicate whether this product is approximately constant for the chosen text. If so, we could state this result by establishing a conservation law: *the product of the frequency of occurrence of a word times its repetition order is a constant.* Discuss whether this statement is true.

✓ Zipf's law, formulated by George Kingsley Zipf (1902–1950), [13] professor of linguistics at Harvard University, states that the frequency f of a word's occurrence as a function of the order of repetition r can be expressed as $f = k/r$, where k is a constant. A few years later, Mandelbrot (1953) [16] proposed the following relationship:

$$f = k/(\alpha + r)^{\beta}, \tag{3.9}$$

where α and β are two additional constants. Which of the two expressions best describes your data?

3.4.2 *Project 6. Why is the first page of a manual or table in a library generally the most worn? Newcomb–Benford's law*

There is an interesting non-intuitive property that occurs with many series of numbers. The first digit of these series has a higher probability of being a 1, followed, in order of probability by $2, 3, \ldots, 9$. More specifically, this probability distribution can be expressed as follows [17–19]:

$$P(i) = \log_{10}(1 + 1/i), \tag{3.10}$$

where i represents the value of the first digit $(1, 2, 3, \ldots, 9)$ and $P(i)$ is its probability of occurrence. This is known as Benford's law and was discovered in 1938. However, the origin of this discovery dates back to 1881, when Simon Newcomb noted that in many libraries, the first page of logarithm tables was much more worn by use than the others, thus, for this reason, it is also referred to as the Newcomb–Benford law. This is somehow unexpected and surprising because, in a table that is used as a guide for looking up a number (such as a telephone directory or a logarithm table), it is counterintuitive that one page should be more worn than the others. Naively, one would expect the probability of the first digit of the number being looked up to be completely random [2, 7, 20].

This property is quite different from another category of numbers, the so-called *normal*. Examples of this type of numbers are π and e. In these numbers, all the digits occur with the same probability, i.e., their distribution is uniform. Note that while Benford's law refers only to the *first digit* of a series of numbers, normality is associated with *all digits* that form an irrational number. Benford's law has been successfully applied to many natural, cultural, and social phenomena: bill amounts, share prices, number of inhabitants of world populations, length of rivers, etc.

Suggested tasks:

(a) Benford's law in the power of a number

✓ Using an Excel spreadsheet or any other spreadsheet, place the natural numbers from 1 to 110 in the first column (A); these numbers are designated as n.

✓ Define a new natural number kk from 1 to 10 and place it in one of the top cells of the spreadsheet. This procedure is illustrated in Figure 3.6.

✓ In the second column (B), raise the number kk to the power n ($= A_n$ and $B_n = kk^\wedge n$).

✓ In the third column (C), transform the numbers in the second column into text: $txt_n = C_n$ =TEXT(B_n,0).

✓ In the fourth column (D), extract the first character of the text in the third column: =EXTRACT(txt_n,1,1) =EXTRACT(C_n,1,1). The function EXTRACT(C_n,i, m) takes from the text contained in the cell C_n, m digits from the i-th starting from the left. In other words, EXTRACT (C_n,1,1) takes the first digit of the text contained in C_n.

✓ Now, with the first digits of the fourth column (D), make a histogram of the first digits from 1 to 9. For them, in the sixth

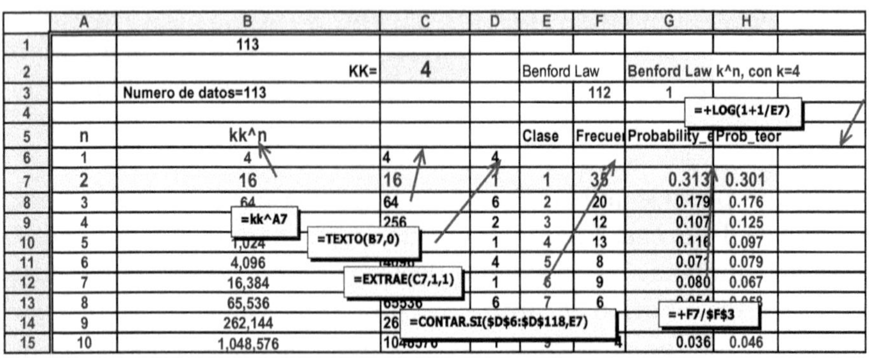

Figure 3.6 A section of the worksheet for studying Benford's law. An example of this type of spreadsheet can be downloaded from www.fisicarecreativa.com. In this example, the data to be analyzed is in column B from rows 6 to 118.

column, define the classes (histogram interval) from 1 to 9, as shown in Figure 3.6 (see also Chapter 5 for more information on histograms).

✓ In the fifth column (E), count the times each digit of the class occurs, i.e., the digits 1–9.

✓ In the sixth column (F), count the times each digit from 1 to 9 occurs in the fourth column D. COUNTIF ($ D $ 7: $ D $ 118, E7), the COUNTIF (range, F7) counts how many times in the *range* there is a match with the data in cell E7, here range = $ E $ 6: $ E $ 118.

✓ In the next column (G), normalize these frequencies of occurrences, dividing the result of the seventh column (F) by the number of all digits in column D. In this way, the sum of the data in column G should add up to 1.

✓ In column H, the probabilities predicted by Benford's law are calculated, Eq. (3.10).

✓ On the same graph, represent the values of the empirical (column G) and theoretical (column H) probabilities of the occurrence of each digit as a function of the digit (column A).

✓ What can you conclude from this comparison?

✓ Following the idea proposed in this activity, analyze the probability of occurrence of the second and third digits of the number kk^n. What can you conclude from this case?

(b) Benford's law and Fibonacci sequence

The Fibonacci sequence [21] or series is obtained in recurring form from two integers that we take as *seeds*, and the subsequent terms are obtained by adding the two previous ones. Thus, if a_0 and a_1 are the *seeds*, the following terms are obtained from the recursive relation:

$$a_{n+1} = a_n + a_{n-1}. \qquad (3.11)$$

For example, if we take the numbers 1 and 2 as seeds, the following terms are $3, 5, 8, 13, 21, 34, 55, \ldots$. If we change the *seeds*, the series changes. However, there are several interesting properties that are preserved in this sequence. One is that the quotient of two successive terms converges rapidly in the golden ratio [21]:

$$\lim_{n\to\infty}\left(\frac{a_{n+1}}{a_n}\right) = \varphi \equiv \frac{1+\sqrt{5}}{2} \approx 1.6180339\ldots. \tag{3.12}$$

The Fibonacci sequence appears frequently in nature: the number of rabbits that a pair generates over time, the arrangement of plant leaves, the pattern in a sunflower, the scales of a pineapple, etc. Another interesting property of this sequence is that if many terms ($n > 500$) are taken, the first digits of any Fibonacci sequence follow Benford's law, and it is precisely this property of the Fibonacci sequence that we are interested in analyzing.

Suggested task:

- ✓ Using an Excel spreadsheet or any other spreadsheet, enter the natural numbers from 1 to 1,000 in the first column (A); these numbers are designated as n.
- ✓ In the next column (B), enter two arbitrary integers in the first two rows, the seeds. The terms in the following lines are obtained using the recursive relation of Eq. (3.11).
- ✓ In the third column (C), define the quotient between successive terms of the series, i.e., $C_n = B_n/B_{n-1}$. Verify that by varying the seeds, the quotient always converges to the number φ.
- ✓ In the fourth column (D), transform the terms of the sequence into text, D_n =TEXT(B_n,0), as in the previous example.
- ✓ In the fifth column (E), extract the first digit of each member of the Fibonacci sequence, following the same technique as in the previous example.
- ✓ Analyze the probability of occurrence of the first digits. Compare your "experimental" results with the predictions of Benford's law, Eq. (3.10).
- ✓ What can you conclude from this comparison?

(c) Explore other number series

Many other data sets follow Benford's law. For example, you can explore what happens regarding the population of towns and cities (including large and small) in a country with population in the eight-figure range. Another example you can use for this study is the population of all the countries of the world.

✓ A population database of countries can be found on the Internet (http://www.indexmundi.com).
✓ Analyze the probability of occurrence of the first digits. Compare your "experimental" results with the predictions of Benford's law Eq. (3.10).
✓ What can you conclude from this comparison?

References

[1] P. Nuñez, S. Calderón and S. Gil, "Search for order and harmony in nature, discovering laws of scale in the classroom," *Lat. Am. J. Phys. Educ.*, vol. 4, no. 1, pp. 118–125, Jan 2010.
[2] K. Wiesenfeld, "Resource letter: ScL-1: Scaling laws," *Am. J. Phys.*, vol. 69, pp. 938–942, 2001.
[3] G. West and J. Brown, "Life's universal scaling laws," *Phys. Today*, pp. 36–42, 2004.
[4] G. West, "The surprising math of cities and corporations," 2011. Available: http://www.ted.com/talks/lang/eng/geoffrey_west_the_surprising_math_of_cities_and_corporations.html.
[5] Wikipedia, "Allometry," 2021. Available: https://en.wikipedia.org/wiki/Allometry.
[6] *The Simple Science of Flight, From Insect to Jumbo Jets*, MA: MIT Press, 1997.
[7] J. R. Bradley and D. L. Farnsworth, "What is Benford's law?," *Teach. Stat.*, vol. 31, no. 1, pp. 2–5, 2009.
[8] J. T. Bonner, *From Bacteria to Blue Wales, Why Size Matters*, Princeton NJ: Princeton Univ. Press, 2006.
[9] E. Yorke, "Energy cost and animal size," *Am. J. Phys.*, vol. 41, no. 11, p. 1286, 1973.
[10] B. J. Erquist, G. B. West and E. L. Charnov, "Allometric scaling of production and life-history variation in vascular plants," *Nature*, vol. 401, p. 907, 1999.
[11] J. E. Smith and L. S. Heath, *Forest Volume-to-Biomass Models and Estimates of Mass for Live and Standing Dead Trees of U.S. Forests*, Washignton, DC, 2002.
[12] H. E. Burkhart and M. Tomé, "Tree weight and biomass estimation," in *Modeling Forest Trees and Stands*, Dordrecht: Springer, 2012.

[13] Wikipedia, "Zipf's law," 2021. Available: http://en.wikipedia.org/wiki/Zipf%27s_law.
[14] San José State University, "Animal longevity and scale," 2021. Available: https://www.sjsu.edu/faculty/watkins/longevity.htm.
[15] Peter Meyer — Hermetic Systems, "Hermetic word frequency counter — Counts frequencies of different words in a file," 2021. Available: https://www.hermetic.ch/wfc/wfc.php.
[16] Wikipedia, "Benoit Mandelbrot," 2021. Available: https://en.wikipedia.org/wiki/Benoit_Mandelbrot.
[17] Wikipedia, "Benford's law," 2021. Available: http://en.wikipedia.org/wiki/Benford's_law.
[18] Wolfram MathWorld, " Benford's law, interactive demonstrations," 2021. Available: http://mathworld.wolfram.com/BenfordsLaw.html.
[19] A. A. Burgos and A. Santos, "The Newcomb–Benford law: Scale invariance and a simple Markov process based on it," *Am. J. Phys.*, vol. 89, pp. 851–861, 2021.
[20] T. Hill, "The first digit phenomenon," *American Scientist*, vol. 86, no. 4, pp. 358–364, July–August 1998.
[21] Wolfram MathWorld, "Fibonacci number," 2021. Available: http://mathworld.wolfram.com/FibonacciNumber.html.

Chapter 4

ERROR THEORY: UNCERTAINTY OF MEASUREMENT RESULTS

This chapter presents the basic concepts associated with measurement processes: physical magnitude, measurement errors or uncertainties, instrument precision, and accuracy. The concepts of measurement errors are introduced according to their source or origin: error of appreciation, error of accuracy, error of interaction, and error of definition. Statistical, systematic, and spurious errors are presented. The concept of significant figures is also discussed.

Goals

✓ Measurements, magnitudes, and measuring
✓ Measuring instruments and units
✓ Measurement errors or uncertainties
✓ Interaction, definition, and calibration
✓ Precision and accuracy
✓ Statistical, systematic, and spurious errors
✓ Absolute and relative errors
✓ Significant figures

4.1 Uncertainty of Measurement Results

A *physical quantity* is an attribute of a body, a phenomenon, or a substance which can be measured directly or indirectly. Examples of quantities are length, mass, power, speed, etc. A *measurand* is a specific magnitude of an object that we want to measure. For example, if we are interested in measuring the length of a bar, that specific length will be the *measurand* [1, 2].

The goal of a measurement is to compare and determine the value of the measurand. This process requires the choice of *measuring instruments* and a system of *measurement units*. For example, if we want to measure the length of a rod, the measuring instrument could be a ruler and if we choose the International System of Units (SI), the unit will be the meter. The ruler to be used must be calibrated in meters or in some submultiple thereof. The measurement method will consist of determining how many times the unit and fractions of it are contained in the value of the measurand [3, 4].

In general, the result of a measurement is only an approximation or estimate of the value of the measurand because of the limitations of the nature of measurement process due mainly to the following:

✓ the sensitivity and accuracy of the instruments used,
✓ the interaction of the measurement method with the measurand,
✓ the intrinsic definition of the object to be measured,
✓ the influence of the observer or observers performing the measurement.

These imperfections lead to an *error* or *uncertainty* in the measurement result. The term *error* is often used colloquially to mean *mistake*, but in science and engineering, the error of a measurement is more closely associated with the concept of uncertainty in the determination of a result. More precisely, what we seek in every measurement is to know the limits or probabilistic limits of these uncertainties. We seek to establish an interval:

$$\bar{x} - \Delta x \leq x \leq \bar{x} + \Delta x, \tag{4.1}$$

like the one illustrated in Figure 4.1, where we can say, with some probability, that the *best value*, or *most representative value*, of the magnitude x can be found. In other words, the objective of the measurement is to establish a *confidence interval* $(\bar{x} - \Delta x, \bar{x} + \Delta x)$ where with a certain probability we can ensure that the most representative value of the measurement can be found. The half-width of the interval is called *absolute uncertainty* (or *absolute error*) of the measurement.

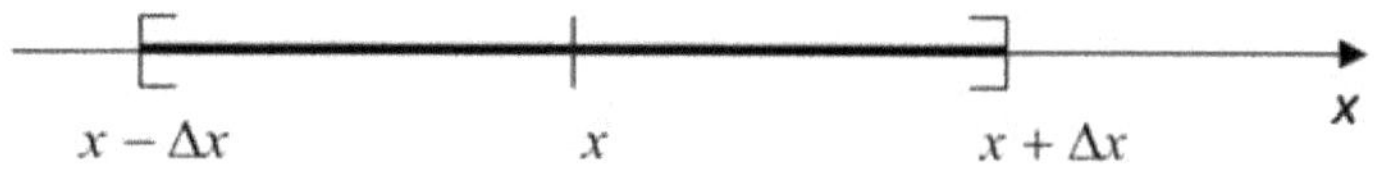

Figure 4.1 Interval associated with the result of a measurement. We call *best value* of the measurand the representative value of the interval (x); this value could be the center of the interval. The half-width Δx is called *absolute uncertainty* or *absolute error* of measurement.

The sensitivity of an instrument is associated with the minimum variation in the magnitude that it can detect with it. For example, with a ruler graduated in millimeters, we cannot detect variations smaller than about a millimeter so that its sensitivity is one millimeter. Measurement instruments have a finite sensitivity: the minimum variation that can be detected is called the *nominal sensitivity* of the instrument and generally coincides with the smallest division of its scale (see Figure 4.2).

The *interaction* of the measurement method with the measurand generates an uncertainty in the measurement. When we use a thermometer to measure a temperature, some heat flows from the object to the thermometer (or vice versa), so as a consequence of the measurement processes, the object's temperature is modified with respect to the original value we wished to determine, due to the inevitable interaction required. It is clear that this interaction may or may not be significant. If we are measuring the temperature of a cubic meter of water, the amount of heat transferred to the thermometer may not be significant, but it would be if the volume in question were that of an ant. In general, whenever we perform a measurement, we interact with the measurand object.

In turn, the quantities to be measured are not defined with infinite precision either. Imagine that we want to measure the length of a strip of wood. It is possible that by using increasingly precise instruments, we might begin to note the typical irregularities of cutting edges or, going even further, we might finally detect the atomic or molecular nature of the material that constitutes it. At this point, the length will no longer be well defined. In practice, it is possible that long before these extreme cases, the lack of parallelism

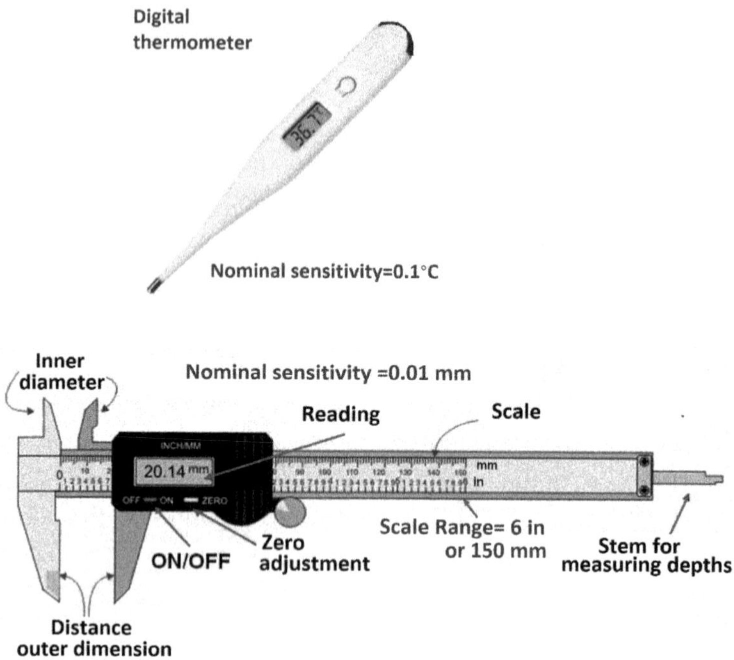

Figure 4.2 Top: Digital thermometer with a nominal sensitivity of 0.1°C. Bottom: Digital caliper with nominal sensitivity of 0.01 mm (with 150 mm range).

between its edges will cause the concept of the "length of the object" to become less and less clearly defined. This intrinsic limitation contributes an *intrinsic uncertainty* due to the lack of definition of the quantity in question.

Another example of lack of definition of a physical quantity is the case in which the quantity of alpha particles emitted by a radioactive source is counted in a given time interval, e.g., 5 seconds. Successive measurements of the same magnitude, for the same source and with the same instruments, will yield different results (similar, but generally different). In this case, again, we are facing a manifestation of an intrinsic uncertainty associated with the magnitude *number of particles emitted in five seconds*, rather than the uncertainties that originate from the imperfections of the instruments or the observer. In fact, this uncertainty is intrinsic to the statistical character of nature itself [4, 5].

All these limitations derive from the fact that we cannot obtain *the true* value of a measurand with certainty, but rather, we can only establish a possible range of values, where it can be reasonably contained, which we do by evaluating and reporting the measurement uncertainty. In this sense, the measurement process in physics is similar to the *interval estimation* that is performed in statistics [6].

One way to express the result of a measurement is as follows:

$$x = (\bar{x} \pm \Delta x) \text{ unit}, \tag{4.2}$$

where $\bar{x}$ is the *best value* of our measurement and Δx the uncertainty or *absolute error*. Here unit indicates the *unit of measurement adopted.*

The following concepts are also very useful:

- ✓ The relative uncertainty or relative error $\varepsilon_x = \Delta x/\bar{x}$, which expresses how significant the uncertainty is compared to the measured value or best value,
- ✓ The percentual relative uncertainty $\varepsilon\% = \varepsilon_x\, 100\%$.

$l \approx 20$ cm and $d \approx 1$ mm; since the nominal appreciation of the ruler is 1 mm, both magnitudes will have the same absolute error ($\Delta d \approx \Delta l \approx 1$ mm). However, it is clear that the quality of length measurement is better than the quality of diameter measurement, as is clearly described by $\Delta d/d \approx 100\%$ while $\Delta l/l \approx 5\%$.

Another common way of expressing a result and its uncertainty is the following concise notation: *measured value* (*uncertainty*), for example:

$$L = 21.1(1)\,\text{cm} \quad \text{which means } L = 21.1 \pm 0.1\,\text{cm}$$

or also

$$B = 5.076(5) \times 10^{-11}\ \text{m} \quad \text{which means}$$
$$B = (5.076 \pm 0.005) \times 10^{-11}\,\text{m}.$$

In both cases, the value in parentheses (*uncertainty*) refers to the last digit of the reported value (*measured value*).

Example 1. The radius of the Earth R_T, the distance of the Earth to the Sun d_{ST}, and the Sun-Mars distance d_{SM} were measured. The results were

(a) $R_T = (6.38 \pm 0.02) \times 10^6$ m,
(b) $d_{ST} = (1.50 \pm 0.02) \times 10^{11}$ m,
(c) $d_{SM} = (2.28 \pm 0.02) \times 10^{11}$ m.

Compare the absolute and relative errors of these measurements. Which of these measurements shows "better quality"? What is the parameter that has been most accurately measured?

The relative and absolute errors for each case are

$$\frac{\Delta R_T}{R_T} = \frac{0.02}{6.38} \approx 0.003, \quad \text{that is,} \quad \varepsilon_{R_T} \approx 0.3\% \quad \text{and}$$

$$\Delta R_T = 2 \times 10^4 \text{m},$$

$$\frac{\Delta R_{ST}}{R_{ST}} = \frac{0.02}{1.5} \approx 0.01, \quad \text{that is,} \quad \varepsilon_{R_{ST}} \approx 1\% \quad \text{and}$$

$$\Delta R_{ST} = 2 \times 10^9 \text{m},$$

$$\frac{\Delta R_{SM}}{R_{SM}} = \frac{0.02}{2.28} \approx 0.009, \quad \text{that is,} \quad \varepsilon_{SM} \approx 0.9\% \quad \text{and}$$

$$\Delta R_{SM} = 2 \times 10^9 \text{m}.$$

The radius of the Earth (R_T) is the parameter that has the "best quality" among them since its relative error is the smallest of the three. Moreover, R_T, is the measurand that was measured with the greatest precision since it has the lowest absolute error.

4.2 Sensitivity, precision, and accuracy

As we have seen, the *sensitivity of an instrument or a measurement method* is associated with the smallest variation in the magnitude that can be detected with a given instrument or method. Thus, we say that a micrometric screw (with a nominal sensitivity of 10 μm) is more sensitive than a ruler graduated in millimeters and that a stopwatch with a sensitivity of 10 ms is more sensitive than an ordinary clock.

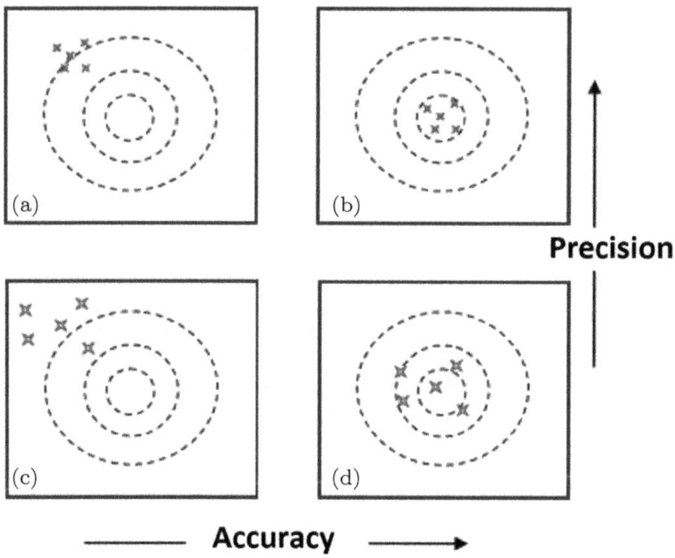

Figure 4.3 Schematic illustration of the concepts of precision and accuracy of a set of measurements. The centers of the circles indicate the position of the "best value" of the measurand and the crosses indicate the values of various measurements. The *dispersion* of the points provides an idea of the *precision*, while their effective center (*centroid*) is associated with the *accuracy*. (a) It is a precise but inaccurate determination, while (d) is more accurate but less precise; (b) shows a more accurate and precise determination; (c) is less precise than (a).

The *precision of a set of measurements* refers to the dispersion among the values found for the same measurand. The precision of a series of measurements is associated with their *repeatability*, i.e., whether or not repeated measurements of the same measurand yield similar results. Figure 4.3 illustrates this aspect of precision and its relationship to accuracy [3].

The *accuracy* of an instrument or measurement method is associated with the quality of its calibration with respect to the standard units taken as reference (standard kilogram, standard meter, etc.). When we talk about the *accuracy of a set of measurements*, it refers to how much the mean value of these measurements approaches or deviates from the best value of it. This is related to the greater or lesser bias of the measurements made with a given measurement method or instrument.

For example, if we make a set of length measurements with a thermally enlarged ruler, regardless of its precision, the set of measurements will be skewed with respect to its best value.

Imagine a stopwatch capable of determining one-hundredth of a second, but that gains two minutes per hour, while an ordinary wristwatch does not. In this case, the stopwatch is more sensitive than the ordinary watch but less accurate.

An instrument's accuracy is a measure of the quality of its calibration against internationally accepted *measurement standards.* In general, instruments are calibrated, but within certain limits. The calibration of an instrument should be as good as its sensitivity.

4.3 Source of errors

The sources of errors have diverse origins and can be classified as follows:

4.3.1 *Errors introduced by the instrument*

✓ **Reading error**, $\boldsymbol{\sigma_{\text{read}}}$**:** If the instrument is correctly calibrated, the uncertainty when making a measurement will be associated with the minimum division of its scale that we can detect with some measurement method. Note that the reading error is established as the minimum discernible division and not as the *minimum division of the instrument.* The *reading error* may be greater or less than the *nominal sensitivity*, depending on the skill (or lack thereof) of the observer. Thus, a trained observer may be able to appreciate fractions of a millimeter with an ordinary ruler, while another observer with poor eyesight using the same ruler may only be able to appreciate 2 mm. Nominal sensitivity is a characteristic of the instrument, but the reading error depends on both the instrument and the observer. The reading error is closely related to the *sensitivity* of the instrument or measurement method.

✓ **Accuracy error, $\sigma_{\mathbf{acc}}$:** This represents the absolute error with which the instrument has been calibrated against reliable standards.

✓ **Interaction error, $\sigma_{\mathbf{int}}$:** This comes from the interaction of the measurement method with the object to be measured. Its determination depends on the measurement that is made, and its value can be estimated from a careful analysis of the method used to carry out the measurement. For example, when the temperature of a body is measured, i.e., the volume of liquid of about 2 ml with a given thermometer, the interaction error must be estimated. For example, we can use two samples of the same liquid of very different volumes obtained from the same sample, one with a volume of about 2 ml and another with a much larger volume, let's say 400 ml. By measuring the temperature of both the samples with the same thermometer, that we know came from the same original source with a given temperature, we can estimate the interaction error.

✓ **Blur or lack of definition in the object subject to measurement, $\sigma_{\mathbf{def}}$:** This originates from the fact that the quantities to be measured are not defined with infinite precision. σ_{def} designates the uncertainty associated with the *lack of definition* of the object to be measured and represents its intrinsic uncertainty.

In general, all these sources of error will be present in a measurement, so it is useful to define the *uncertainty* or *nominal error* of a measurement, σ_{nom}, as the combination of all identified uncertainties:

$$\sigma^2_{\text{nom}} = \sigma^2_{\text{read}} + \sigma^2_{\text{def}} + \sigma^2_{\text{int}} + \sigma^2_{\text{acc}} + \dots \tag{4.3}$$

This procedure of adding the squares is a result of statistics and comes from assuming that the different sources of error are all independent of each other [7]. The ellipsis indicates the contributions of other possible sources of error.

Example 1. A time measurement with a manual stopwatch is affected by the operator's *reaction time*. In this case, a term must be included in Eq. (4.3) to take into account this new contribution.

Example 2. We want to determine the diameter of a tree trunk, d, and its cross-sectional area, A. How would we proceed and what are the main sources of uncertainty in this determination?

One method could be to measure the perimeter, P, with a tape measure and then determine the diameter from the relation $P = \pi d$, using this value to calculate the area. In this case, the greatest contribution to uncertainty comes from the lack of definition of the diameter. One way to estimate the uncertainty would be to determine the maximum and minimum values of the diameter using a series of measurements and take as $\sigma_{\text{diámeter}}$ the half-difference of these values:

$$\sigma_{\text{def}} = \sigma_{\text{diameter}} \cong 1/2(D_{\max} - D_{\min}). \tag{4.4}$$

4.4 Classification of errors

Depending on their nature, errors can be classified as *systematic*, *statistical*, and *illegitimate or spurious*.

(I) **Systematic errors**: They are caused by imperfections in measurement methods. For example, let's think about a clock that is fast or slow, an expanded ruler, the parallax error, etc. The errors introduced by these instruments or imperfect methods will always affect our results *in the same direction or sense* [2, 4].

(II) Accuracy errors are a source of systematic error, although they are neither the only ones nor all the same. Imagine the case of a well-calibrated scale for weighing people at shopping centers or other businesses. When people weigh themselves in public, they are normally fully dressed, so the resulting values will be affected by a systematic error due to the mass of their clothes. The only way to detect and correct systematic errors is by comparing our measurements with other alternative methods and performing a critical analysis of the instruments and procedures used. It is therefore advisable to insert reliable standards in the measurement process that enable the instrument to be calibrated during the measurement.

(III) **Statistical errors**: They are the ones that occur randomly, generally due to multiple and fortuitous causes. They occur when, for example, we make a mistake in counting the number of divisions of a ruler or are in the wrong position in front of an analog scale with a pointer. These errors can be made with equal probability by shortfall or by excess, so they may be reduced considerably by measuring several times and averaging the results. It is to this type of error that the statistical theory of measurement errors commonly refers to, which we will briefly formulate in the following. We will designate these errors as σ_{St}[2, 4].

(IV) **Illegitimate or spurious errors**: They are those we commit by mistake or carelessness. Suppose that to calculate the volume (V) of a spherical object we determine its diameter (d) but inadvertently use the wrong expression $V = 4\pi d^3/3$ instead of the correct one $V = \pi d^3/6$. If when entering the value of the diameter in the formula we make a mistake in the number entered or use incorrect units or the wrong expression for volume, we will clearly have "erred". This time, the error is the result of a mistake. We call these types of errors illegitimate or spurious. There is no possible theoretical treatment for them, and the only way to avoid them is to pay close attention to the execution and analysis of the procedures involved in the measurements.

An error of this type can lead to serious and even dramatic situations. For example, NASA's Mars Climate Orbiter space mission failed in December 1999 due to a mistake made in changing from English units to metric units in the formulas used to drive its navigation system. This error caused the probe to be destroyed by friction with the planet's atmosphere.

The final expression of the uncertainty Δx of a measurement takes into account all the different contributions of different origins and types. The usual prescription is to combine them as follows:

$$\Delta x = \sigma_{\text{ef}} = \sqrt{\sigma_{\text{est}}^2 + \sigma_{\text{nom}}^2} = \sqrt{\sigma_{\text{est}}^2 + \sigma_{\text{read}}^2 + \sigma_{\text{def}}^2 + \sigma_{\text{int}}^2 + \sigma_{\text{acc}}^2 + \dots} \tag{4.5}$$

This Δx is called the *combined uncertainty* or *effective measurement error.*

In 1993, the International Organization for Standardization (ISO) published the world's first official guide for the expression of measurement uncertainty [2]. In this guide, statistical uncertainties are called *type A uncertainties*, while those that are not corrected by repeated measurements are usually associated with *type B uncertainties*, [1, 2] which include systematic errors and any other uncertainty factors that the experimenter considers important and are not corrected by repeated measurements of the same measurand. According to this guide, the measured value should be reported along with an estimate of the combined total of type A and B uncertainties of the value. The total uncertainty or effective error is found by combining the components of the uncertainty, as described in Eq. (4.5).

In many practical applications and scientific publications, the uncertainties of each type are expressed separately, in order to indicate their respective effects on the result.

However, if you want to compare the measurements of the same parameter or measurand from two or more different methods or experiments, it is advisable to define an *effective uncertainty* that encompasses both sources, in order to be able to verify whether or not there is any discrepancy between measurements. In this case, to obtain the effective uncertainty, the uncertainties of each type are summed in quadrature, as shown in Eq. (4.5).

4.5 Significant figures

The result of a measurement, expressed in the form $\bar{x} \pm \Delta x$, must be consistent in the number of figures to be reported for best value x and uncertainty Δx. This refers to the number of significant figures that we include in each of them.

Consider a measurement made with a ruler graduated in millimeters. If we are careful, we will be able to ensure our result down to the millimeter figure or, in the best of cases, a fraction of a millimeter but not more. Thus, our result could be

$$L = (95.2 \pm 0.5)\ \text{mm} \tag{4.6}$$

or also

$$L = (95 \pm 1)\ \text{mm}. \tag{4.7}$$

In the first case, we say that the measurement has three *significant figures*, and in the second case, it has only two. The number of significant figures is equal to the number of digits contained in the measurement that are located to the left of the first digit affected by the error, including this digit. The first digit, i.e., the one furthest to the left, is the *most significant* (9 in our case), while the last is the *least significant.* Note that it makes no sense to include more figures in our L result than those where we are uncertain. Thus, it is not correct to express the result as, for example,

$$\cancel{L = (95.321 \pm 1)\ \text{mm}} \tag{4.8}$$

because if we have an uncertainty of the order of 1 mm, we cannot ensure values of hundredths and thousandths of a millimeter in the result. Operationally, once we have calculated the uncertainty of the measurement, we round the value of the measurand (which can come from an average and have many figures) and adapt the number of significant figures to make it compatible with the value of the uncertainty.

Uncertainties or errors are usually expressed with a single *most significant figure*, using more than one only in exceptional cases and when there is a clear basis to do so. In certain cases, it is justified to use more than one least significant figure. For instance, if we measure length with a ruler graduated in inches ($1'' = 1$ inch $= 25.4$ mm) with $\frac{1}{16}''$ divisions, and the result is $2\frac{3}{16}'' \pm \frac{1}{16}''$, when we report this measurement in the metric system, we can say that the result is 55.6 ± 1.6 mm, since the uncertainty of $\frac{1}{16}''$ equals 1.6 mm.

If the uncertainty of a result is not explicitly indicated, it is usual to consider it to be of the order of the least significant figure. For example, if we only have the information that the mass of a body is

$m = 52.4$ g, we can assume that the uncertainty is of the order of tenths of a gram, i.e., $m = 52.4 \pm 0.1$ g.

A possible ambiguity arises when a change of units is made. For example, if we want to express the length $L = (95 \pm 1)$ mm in mm, how many significant figures should the result have, after unit conversion? If we write $L = 95{,}000\mu$m, the conversion will have increased the number of significant figures from two to five, suggesting that we have measured with an instrument that appreciates microns, which is not true. Note that in science and engineering, 95 mm $\neq 95{,}000\mu$m. By the way, it would be interesting to compare the costs of the instruments to make these two kinds of determinations. Scientific notation is used to avoid this ambiguity. The conversion of values thereby involves only the transformation of the unit, preserving the number of significant figures in the original values. When we apply this to $L = (95 \pm 1)$ mm, we have

$$L = (95 \pm 1) \text{ mm} = (95 \pm 1) \times 10^3 \mu\text{m} = (9.5 \pm 0.1) \times 10^4 \mu\text{m}. \tag{4.9}$$

Indeed, the values 95 mm and $9.5 \times 10^4 \mu$m have the same number of significant figures. The uncertainty of 1 mm has been written as $1 \times 10^3 \mu$m, with one significant figure in both cases.

4.6 Determination of measurement errors

Single direct measurement: The discussion presented so far is useful to characterize the error or uncertainty of a quantity that is measured directly only once, e.g., to determine the time it takes for the Moon to cross the Earth's shadow: the duration of an eclipse. A single measurement of a measurand is perhaps the simplest situation that can arise in practice.

If x is the measured quantity that is being measured only once, its result can be expressed as follows:

$$x \pm \Delta x \quad \text{with} \quad \varepsilon_x\% = 100 \cdot \frac{\Delta x}{x}, \tag{4.10}$$

where x is the measured value and Δx its *effective measurement error* given by Eq. (4.5).

Direct repeated measurements: It is often possible and desirable to perform multiple measurements of a given magnitude. This technique makes it possible, among other things, to minimize statistical or random errors, which are always present in a measurement. This important aspect of measurement and how to optimize this measurement process are discussed in more detail in the following chapter. If N measurements of the same measurand are made, the result is expressed as follows:

$$x = \bar{x} \pm \Delta x \quad \text{with} \quad \varepsilon_x\% = 100 \cdot \frac{\Delta x}{\bar{x}}, \tag{4.11}$$

where $\bar{x}$ is the average of the measurements and Δx the combination of effective error and statistical error, which will be discussed in greater detail in the following chapter.

Indirect measurements: There are numerous cases in which the magnitude of interest is not measured directly but calculated from others that are measured directly. Imagine that we want to know the volume of a solid sphere. One way to do so is to measure its diameter and calculate the volume from it. The diameter error is characterized following the guidelines discussed above but determining the error in the volume requires the use of error propagation techniques, which are discussed in Chapter 6. For example, if x and y are the quantities that are measured directly and Z is calculated from them, we have the following:

$Z = x \pm y$	$\Delta Z^2 = \Delta x^2 + \Delta y^2$
$Z = xy$ or $Z = x/y$	$\left(\frac{\Delta Z}{Z}\right)^2 = \left(\frac{\Delta x}{x}\right)^2 + \left(\frac{\Delta y}{y}\right)^2$
$Z = f(x, y)$	$\Delta Z^2 = \left(\frac{df}{dx}\right)^2 \Delta x^2 + \left(\frac{df}{dy}\right)^2 \Delta y^2$

Measurement of fitting parameters: There are cases in which the variable of interest results from the fit of a line or other function to a set of directly measured data. For example, the constant k of a spring that follows Hooke's law $F = kx$, where F is the force applied to the spring and x is its stretch. In this case, we measure the variables F_i and x_i for different applied forces and their corresponding stretching. From the graph of F as a function of x, we determine the slope, k, of the line that best fits these data. The question now is how to calculate the error of this slope. This important example is discussed in Chapter 7.

4.7 Nonius, vernier, or caliper

Petrus Nonius [8] and Pierre Vernier [9] developed a very useful and versatile instrument for measuring angles with a precision of fractions of degrees. The same procedure was soon used to measure the length with fractions of mm. The device consists of two similar opposing rules, as shown in Figures 4.2 and 4.4.

The small sliding scale, *nonius* or *vernier*, has n divisions, which coincide with K divisions of the major scale (calibrated

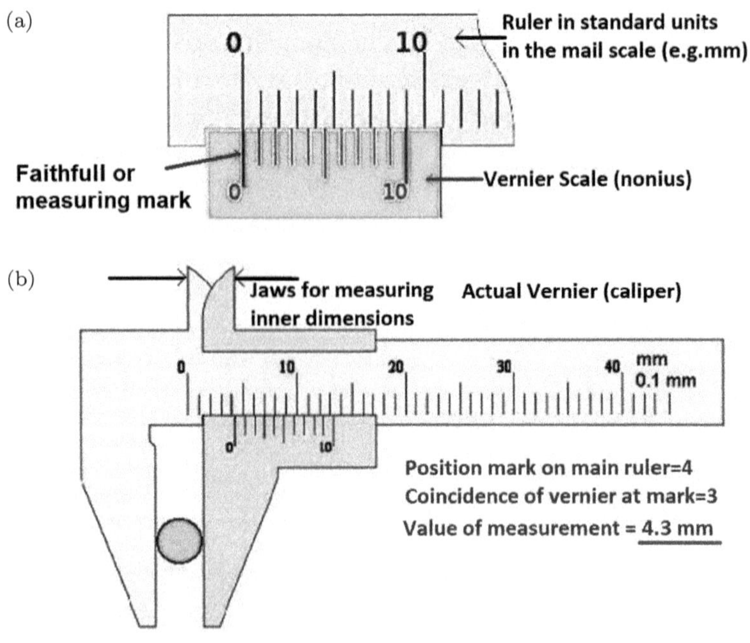

Figure 4.4 Illustration of a *nonius* or vernier.

standard rule). Typically, n is a decimal multiple (10, 20, 50) and $K = n - 1$. For example, if $n = 20$, these 20 divisions of the vernier occupy 19 mm. Thus, the distance between two consecutive divisions of the vernier is $(n - 1)/n$ units. If the division j of the vernier coincides with a division of the major rule, then to the value indicated by the main or pointer line, we must add a fraction j/n of the minimum division of the rule and the nominal appreciation of the vernier is $1/n$ of the least division.

In the case of the vernier in Figure 4.4(b): $K = 9$, $n = 10$, the minimum division of the rule is 1 mm, and the assessment of this vernier or nonius is 0.1 mm. In the example in the figure, the position of the pointer line is between 4 mm and 5 mm and $j = 3$, so the value measured by the vernier in the figure corresponds to 4.3 mm.

A more complete description of this device and simulation programs to practice its reading and using it can be found on the Internet: https://en.wikipedia.org/wiki/Vernier_scale.

4.8 Summary of important concepts

It is suggested that the reader give a concise explanation of the following concepts and, whenever possible, provide an example that illustrates each one:

- ✓ physical magnitude and measurand,
- ✓ errors of appreciation and accuracy of the instruments, nominal appreciation of the measuring instruments,
- ✓ errors of appreciation, accuracy, interaction, and definition of a measurand,
- ✓ nominal error of a magnitude that is measured only once,
- ✓ statistical, systematic, and spurious errors,
- ✓ significant figures in a measurement.

4.9 Exercises and problems

(1) Briefly describe what systematic, statistical, and spurious errors are. Provide an example of each of them. The same for the case of errors of definition, interaction, accuracy, and appreciation.

(2) How would you classify the following errors?

(a) A clock that gains 1 min/week.
(b) A student who reports the measurements of a ruler graduated in centimeters as inches.

(3) What are the most common sources of error when measuring?

(a) The thickness of a soft cotton rope with a caliper.
(b) The radius of a tree.
(c) The width of your table with a metal ruler graduated in mm.
(d) The diameter of a steel bearing ball about 2 cm in diameter with a caliper.

(4) Briefly state the procedure you would use to measure the mean diameter of a tree trunk and estimate the uncertainty of this determination without cutting the tree.
(5) Identify the number of significant figures of the following values:

Value	Number of significant figures
72.00	
0.72	
0.0072	
3.80×10^{-3}	
3.141592	
−300,000	
300,000.00	
0.300000	
5,670.00	
−0.09900	

(6) Why do we say that 75 m $\neq$ 75,000 mm? Why is 75 m $\neq$ 75.00 m?
(7) Express the result of the determination of a volume, from the values obtained by calculation (apply truncation and rounding to express the best value and the uncertainty with a number of compatible significant figures):

(a) Best value $V = 534.5376\,\text{cm}^3$ and absolute uncertainty $\Delta V = 0.03491\,\text{cm}^3$.
(b) $V = (\pm)\,\text{cm}^3$. Express the same result in the *best value* form (*uncertainty*).

(8) Correctly express the results of the following measurements:

Measurement	25.231	41.352	0.8923	253.33	655.3	120.2
Absolute Error	0.0258	0.258	0.0128	36.25	258.3	11.25

In each case, indicate the relative percent errors and indicate which of all these determinations has the best quality.

(9) The length of an object was measured only once with a micrometer screw. The measured length was $L = 15.12$ mm.
(a) Give an estimate of the absolute and relative error of this measurement.
(b) Express the result of this measurement in mm, m, and km, respecting the number of significant figures. What are the significant figures in this case? Justify your answer.

(10) Two clocks are available. Clock A has a second hand (which makes one complete turn in one minute), its face is divided into 60 units, and it loses 10 min per day. Clock B has a second hand, but its dial has only 24 divisions, and it neither gains nor loses more than 5 min in 10 days.
(a) Estimate the errors of appreciation and accuracy of both clocks.
(b) If you had to measure times on the order of 50 min with an error of less than 0.1%, which one would you use and why?

(11) The length of an object was measured once with a caliper with nominal sensitivity of 1/20 mm. The measured length was $L = 15.17$ mm. Give an estimate of the absolute and relative errors of this measurement. Write the best value of the length and its error.

(12) You have made a series of measurements which you must report in the forms $\langle A\rangle \pm \Delta$ A and *best value (uncertainty)*. Say how

would you do it, taking into account the number of significant figures of the best value and the uncertainty:

(a) $\langle V \rangle = 22.32323 \qquad \Delta V = 0.002352$.
(b) $\langle W \rangle = 2.233259 \times 10^{-2} \qquad \Delta W = 1.235 \times 10^{-3}$.
(c) $\langle X \rangle = 2.269 \qquad \Delta X = 0.022$.
(d) $\langle Y \rangle = 10002,909 \qquad \Delta Y = 23.230$.
(e) $\langle Z \rangle = 100.00234 \qquad \Delta Z = 0.0921$.

(13) The measurements indicated in the graphs were taken with a caliper. Say what values have been measured and what their nominal errors are.

(a) (b) (c)

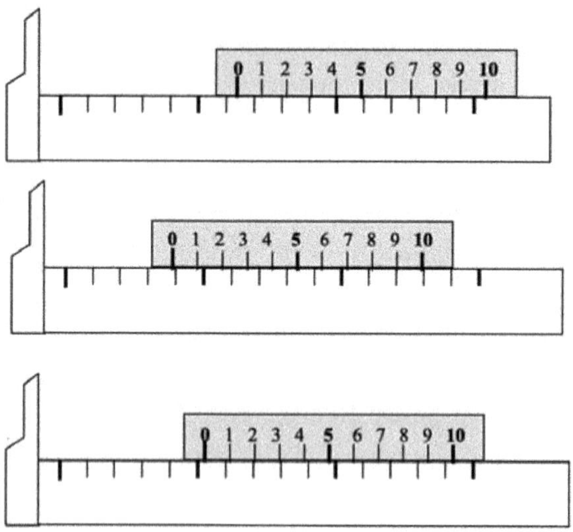

[Answers to problems 13: (a) 6.4 mm, (b) 3. mm, (c) 5.2 or 5.3 mm, nominal error 1/20 mm]

References

[1] NIST, *NIST Constants, Units & Uncertainty — Essential of Expressing Measurement Uncertainty*, Washington, DC, 2017.
[2] ISO, "ISO/IEC GUIDE 98-3:2008, Uncertainty of measurement — Part 3: Guide to the expression of uncertainty in measurement," 2017. Available: https://www.iso.org/standard/50461.html

[3] S. Allie and et al., "Teaching measurement in the introductory physics laboratory," *Phys. Teach.*, vol. 41, pp. 394–401, 2003.
[4] P. Bevington and D. K. Robinson, *Data Reduction and Error Analysis for the Physical Sciences*, 3rd Ed., 2003, p. 230.
[5] J. P. Paz, *Einstein contra la mecánica cuántica... el azar, la ignorancia y nuestra ignorancia sobre el azar*, Buenos Aires, 2006.
[6] C. T. Angell, *et al.*, "Concepts in nuclear science illustrated by experiments with radon," *Am. J. Phys.*, vol. 80, no. 1, pp. 61–65, January 2012.
[7] Wikipedia, "Estimation theory," 2012. Available: http://en.wikipedia.org/wiki/Estimation_theory.
[8] D. C. Baird, *Experimentation*, 3rd Ed., NJ: Prentice-Hall, Inc., 1995.
[9] Wikipedia, "Petro Nonius," 2021. Available: https://en.wikipedia.org/wiki/Pedro_Nunes.
[10] Wikipedia, "Pierre Vernier," 2021. Available: https://en.wikipedia.org/wiki/Pierre_Vernier.

Chapter 5

STATISTICAL DATA PROCESSING

The goals of this chapter are to discuss the statistical treatment of experimental data, review the notion of statistical distribution of a random variable, and present techniques for constructing a histogram. This chapter describes the statistical parameters that characterize a population and a sample. We discuss the error of a quantity measured N times and the concept of best value and statistical uncertainty. It also deals with the problem of the optimal number of measurements to be made in a given measurement process. Finally, it shows a way to combine independent measurements of the same measurand and discusses the question of discrepancy among two or more measurements.

Goals

✓ Histograms and statistical distribution. Graphic analysis
✓ Localization parameters of a distribution: mean, median, and mode
✓ Dispersion parameters: standard deviation
✓ Normal distribution
✓ Quantity measured N times
✓ Optimal number of measurements
✓ Combination of independent measurements
✓ Discrepancy between measurements

5.1 Samples and populations

Statistics is a science based on mathematics, in which the goal is to collect and interpret the data obtained in an experiment or analyze *a sample or population* [1, 2]. One of its goals is to learn or infer something about an entire group (*population*) by looking at data from a subset of its members, called *a sample* [2, 3]. Statistics is a very useful tool for analyzing the data obtained in a measurement process, especially when repeated measurements are taken of the same measurand.

5.2 Histograms and statistical distribution

Suppose we wanted to analyze the distribution of heights of all the people living in a given city. We could either measure the height of every person in the city or take a representative sample of people upon which to base our determinations.

This type of study is a typical statistical problem. If we take a sample of size N and measure each individual's height, the experiment will produce N results: $x_1, x_2, x_3, \ldots, x_N$. All these data will be included in a range of heights $(x_{\min}, x_{\max})$ between the lowest and highest measured heights.

A useful way to view the characteristics of this data set consists of dividing the range $(x_{\min}, x_{\max})$ into $m(<N)$ sub-intervals delimited by points $(y_1, y_2, y_3, \ldots, y_m)$. We will call these sub-ranges a *class interval or bin range*. Next, count the number n_1 of individuals in the sample whose heights are in the first range $[y_1, y_2)$, the number n_j of individuals in the sample that are in the jth interval $[y_{j-1}, y_{\mathrm{j}})$, etc., up to the mth sub-interval. Here we have used the usual notation of square brackets $[\ldots]$ to denote a closed interval (include the extreme) and round brackets $(\ldots)$ to denote an open interval (exclude the extreme).

With these values, we define the distribution function f_j, which is defined for each subrange as follows:

$$f_j = \frac{n_j}{\sum n_j} = \frac{n_j}{N}. \tag{5.1}$$

This distribution function is normalized, i.e.,

$$\sum f_j = 1. \tag{5.2}$$

The *bar graph* of f_j as a function of $z_j[z_j = (y_{j-1} + y_j)/2]$ provides a clear idea of how the heights of the individuals in the study sample are distributed. This type of chart is called a histogram [4] and most commercial software spreadsheets (Microsoft Excel®, OriginLab Origin®, etc.) have tools to perform the operations described and the resulting chart. Figure 5.1 illustrates two typical histograms.

In general, variable x, which describes the possible results of an experiment, can be regarded as a *random variable*. If the possible results of an experiment are finite (or countable), such as the results

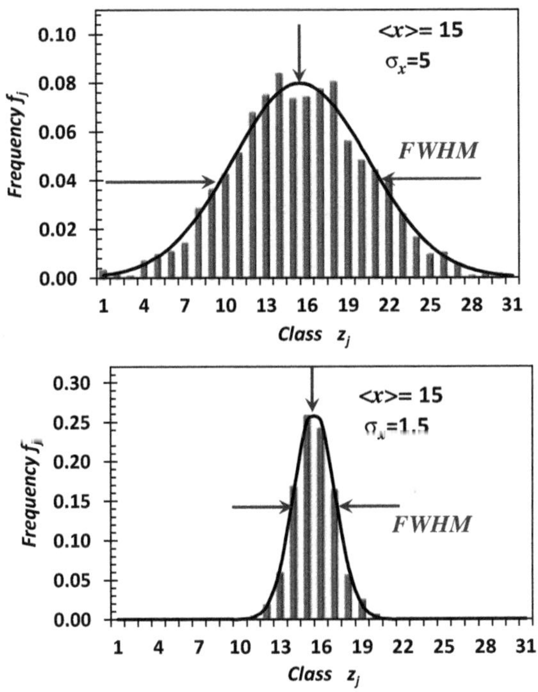

Figure 5.1 *Bar charts* or *histograms* of two samples with the same mean value but different degrees of dispersion. In this figure, solid line curves are two Gaussian or normal distributions, which attempt to describe the implicit distributions in the histograms, of which they are therefore not an intrinsic part. FWHM represents full width at half maximum.

of rolled dice, we denote them with x_i; the random variable *is discrete* and characterized by a *distribution* f_i. If the possible results x of the experiment are uncountable or form a continuum, such as the results of measuring a length or a voltage, the *distribution function* is described by a function $f(x)$.

Three important parameters of a population are as follows: [1, 5]

➢ The *average value* : $m = \sum x_j \cdot f_j = \frac{1}{N_{\text{population}}} \cdot \sum x_i$

$$= \int_x x \cdot f(x) dx. \tag{5.3}$$

This parameter m provides an idea of the location of the *center of mass* of the distribution. The average value corresponds to the "center of gravity" of the distribution.

➢ The *variance* : $\text{Var}(x) = \sigma_{x^2} = \frac{\sum (x_j - m)^2}{N_{\text{population}}} = \int_x (x-m)^2 f(x) dx.$ (5.4)

Here, $N_{\text{population}}$ and the sums refer to all individuals in the population. An integral is used when the random variable is continuous, and a sum is performed when the random variable is discrete; $f(x)$ or f_j are the corresponding distributions associated with the population.

➢ The *standard deviation* : $\sigma_x = \sqrt{\text{Var}(x)}$. (5.5)

The standard deviation σ_x and the variance are parameters that characterize the dispersion of the data around the mean or center of mass. The more concentrated the distribution of values around m, the lower will be σ_x, and vice versa. Figure 5.1 shows two distributions with the same mean value but different standard deviations σ_x.

Since data for a population are generally unknown, we need to deduce them from information from a sample. The *average value* of a sample (also called *mean value*) is usually indicated by the following symbols: $\langle x \rangle$ or $\bar{x}$. The average of a sample is defined as follows:

$$\langle x \geq \bar{x} = \frac{\sum_{i=1}^{N} x_i}{N}, \tag{5.6}$$

where N is the size of the sample. The mean value given by Eq. (5.3) is often called *the population mean* and the mean given by Eq. (5.6) the *sample mean.* In general, the sample mean is a very good estimate of the population mean.

If a sample is used to estimate variance or standard deviation, a good estimate is [2, 5]

$$S_x^2 = \frac{\sum (x_i - \bar{x})^2}{(N-1)}. \tag{5.7}$$

Note the difference between the denominators of expressions Eq. (5.4) and Eq. (5.7). If the sample is large, this difference is clearly not significant, but in general it is important to distinguish whether we are calculating the variance (or standard deviation) of a sample or a population. In general, in this chapter about measurements of a given measurand, we assume that the population (total number of possible measurements) is infinite. Therefore, we obtain estimators of these parameters from a sample of measurements. N represents the repeated measurements of a given measurand that we preform out of the infinite possible measurements that could hypothetically be taken.

5.3 Position parameters of a distribution

The usual parameters for characterizing the position or location of a distribution are as follows [3, 5]:

(a) *mean,*
(b) *median*
(c) *mode.*

The *median* is the value of the (random) variable that separates the data into two equal halves: those that are smaller and those that are larger than this value. In other words, half of the population or sample data is to the right of the median and the other half to the left.

The *mode* (or *modal value*) is the value of the (random) variable where the distribution function has a maximum. In a histogram or distribution, the mode is the value of the variable in which a peak or

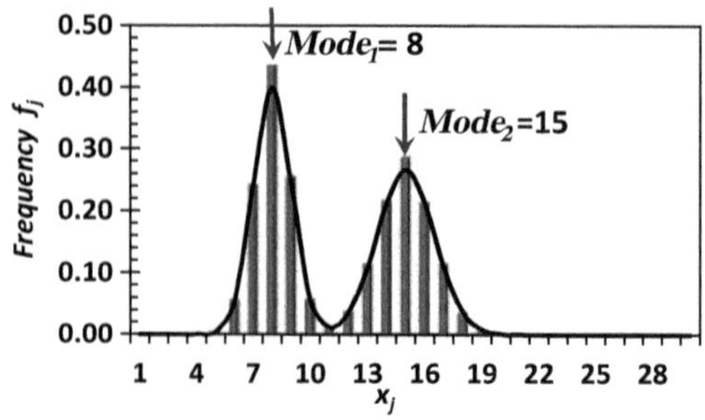

Figure 5.2 Histogram showing the mass distribution of an animal population. This is an example of a bimodal distribution in which two peaks or maxima (modes) are clearly seen.

a maximum is present. If a distribution has two maxima, it is called a bimodal distribution, if it has three, trimodal distribution, and so on. See Figure 5.2.

To estimate the median of a sample, we need to look at the list of data sorted from the lowest to the highest and find the central value in the list. If the number of data is odd, the median is exactly the middle value. If the number of data is even, the median is estimated as $1/2\ (X_{N/2} + X_{N/2+1})$. In a given distribution, a vertical line drawn from the median divides the distribution into two equivalent areas.

Average, mode, and median do not necessarily coincide. Nonetheless, they do coincide in unimodal and symmetrical distributions. This is the case for a Gaussian distribution like the one shown in Figure 5.1. In the case of an asymmetric distribution, the differences between mode, mean, and median can be substantial, as illustrated in Figure 5.3.

It is important to be able to recognize the most appropriate parameter to use in a given situation. Consider the distribution of household income in a given country. The presence of millionaires, even if there are relatively few, has an effect on the average that counteracts the many members of the population at the lower end of the wage scale. Thus, mode and media differ substantially. In this case, the population's income may be more faithfully represented by the mode than by the average. Indeed, the mode or the median of the population's income would vary little if a millionaire moved in or out, whereas the mean would change substantially. Statistical

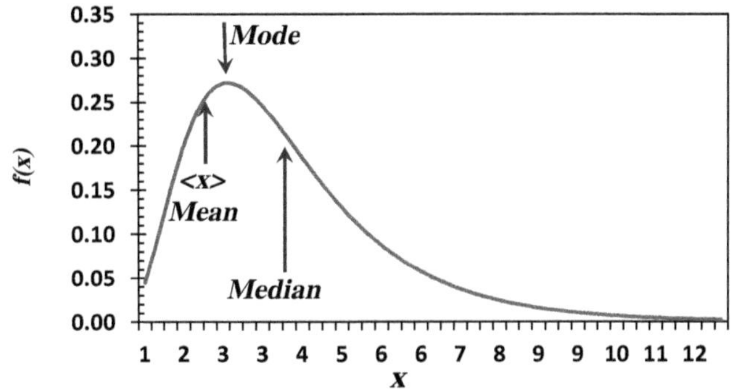

Figure 5.3 Example of unimodal asymmetric distribution. Note that here, the mode, the median, and the mean do not coincide, unlike what happens in a symmetric distribution such as the one shown in Figure 5.1.

data can often be interpreted in various ways. The following example illustrates the different interpretations that can be drawn from a statistical data set.

Example 1. A small business analyzes the need to discuss wages. The monthly payroll is as follows:

Director	\$15,000
Manager	\$10,000
Section head	\$3,000
7 workers	\$1,000 each

The employers argue that the salary should be discussed based on its average value. The union delegate argues that the salary should be discussed based on the median.

In this example, the average salary is \$3,500. The mode (most likely the salary) is \$1,000. The median is also \$1,000. If the Director and Manager doubled their own salaries, leaving the workers' salaries unchanged, the average would be \$6,000, while the mode and the median would not change. On the other hand, if all salaries were increased by a fixed sum of \$500, the average would be \$4,000 and the mode and median would be \$1,500. Workers' salaries would increase by 50%. It is thus clear why each party holds a different point of view.

5.4 Statistical parameters of dispersion — Standard deviation

When analyzing statistical data, it is important to distinguish whether we are referring to a population or a sample.

The standard deviation σ indicates how the individuals of a population are distributed with respect to the mean value m. Statistical theory shows that the sample mean Eq. (5.6) is a good estimate of the population mean Eq. (5.3). Likewise, the sample standard deviation S_x Eq. (5.7) is a good estimate of the population standard deviation σ given by Eq. (5.4).

Performing N measurements of a measurand is equivalent to extracting a sample of size N from a population of infinite size, i.e., of all the measurements that we could in principle make. With these N measurements, we try to estimate the mean values and the population standard deviation of all possible measurements, which we will not perform.

5.5 Normal or Gaussian distribution

A common (continuous) probability distribution in various fields is the *Gaussian or normal distribution* [2, 6], which is bell-shaped, as shown with solid lines in Figure 5.1. The mathematical expression of this distribution (whose mean value is m and standard deviation is σ) is as follows [2]:

$$f(x) = N(x; m, \sigma) = \frac{1}{\sqrt{2\pi\sigma^2}} \cdot e^{-\frac{(x-m)^2}{2\sigma^2}}. \qquad (5.8)$$

This function represents a *Gaussian bell* that is centered on m and has a width determined by the standard deviation σ. This distribution is normalized, which means that the area of this curve between $-\infty$ and $+\infty$ is equal to 1. The inflection points of the curve are located at $x - \sigma$ and $x + \sigma$. The area of this curve between these two points constitutes 68.3% of the total area. The area between $x - 2\sigma$ and $x + 2\sigma$ is 96% of the total. It is useful for this function to characterize the width at half its maximum height (FWHM, for "Full Width Half Maximum"), which is related to σ through FWHM $= 2.35\sigma$.

To compare a non-normalized histogram with a normal curve, we need to count the total number of data N_t, their mean value, and the standard deviation of the data (sample) S_x. To compare the histogram with the normal curve, we need to multiply the distribution given by Eq. (5.8) by the factor $N_{t.}\Delta x$, where Δx is the width of the range of classes or beans that we assume to be identical for each range.

Although the Gaussian distribution occurs naturally in many processes, it is certainly not unique, and there are many types of distributions that occur regularly in nature [6, 7].

5.6 Quantity measured *N* times

When we can only measure one quantity once ($N = 1$), as we have seen in the previous chapter, the best value of the measurement is simply this value, and its uncertainty is given by the nominal uncertainty, σ_{nom}, which takes into account the errors of the instrument, method, object definition, etc. as was discussed in Eq. (4.5). Therefore,

$$\Delta x = \sigma_{\text{nom}} = \sqrt{\sigma_{\text{est}}^2 + \sigma_{\text{read}}^2 + \sigma_{\text{def}}^2 + \sigma_{\text{int}}^2 + \cdots}. \tag{5.9}$$

Whenever possible, a quantity should be measured repeatedly, since averaging several measurements reduces statistical or random errors. This section discusses several strategies for optimizing this procedure.

In many practical cases, it is possible to perform N measurements of the quantity of interest. Given the random nature of statistical errors, the mean will be less affected than individual values by statistical variations.

However, the repeated measurement procedure is not applicable to reduce *systematic* errors, much less *spurious* ones. If a watch gains time, all the values it measures will be overestimated, no matter how many measurements we take.

Returning to the case of statistical errors, suppose we have made *N measurements of the same magnitude* that produce the results $x_1, x_2, \ldots, x_j, \ldots, x_N$. These N determinations can be considered a *sample* of all possible measurements that could be made (*population*). In very general conditions, it can be shown that the *best estimate* of quantity x is given by the average of the values [6, 7]:

$$\bar{x} = \frac{\sum x_j}{N}. \tag{5.10}$$

This result is also called the *best value* or the *most probable value* of the measurand. We will call this as follows:

$$\Delta x_j = x_j - \bar{x} \qquad j = 1, 2, \ldots, N, \tag{5.11}$$

the deviation of each measurement from $\bar{x}$. We also define the *standard deviation* of this sample S_x or the root mean square deviation of the individual measurements as follows:

$$S_x = \sqrt{\frac{\sum_{j=1}^{N}(x_j - \bar{x})^2}{N-1}}. \tag{5.12}$$

This value is a sample estimate and provides an indication of the magnitude of the overall deviation of each result, x_j, around the mean $\bar{x}$.

It is important to note that S_x does not depend on N but on the measurement process. If the observer increases the number of measurements N, in the expression (5.12), both the numerator and the denominator increase, but the result does not change significantly. If measurements are taken carefully, S_x should decrease, but if not, we can expect a wide distribution of measurements and large S_x. In the case of careful or more precise measurements, we expect the distribution of measurements to be optimized around the mean $\bar{x}$ (smaller S_x). Note that S_x has the same physical dimensions as $\bar{x}$, which enables direct comparison with it via the quotient $S_x/\bar{x}$. The lower the value of $S_x/\bar{x}$ (which is generally a constant of the measurement process and does not depend on N), the higher will be the quality of the measurement process.

If we perform several series of measurements of x, and calculate the mean value $\bar{x}$ for each series, we also expect these mean values ($\bar{x}$) to have their own distribution (since they will differ from each other) but with less dispersion than individual measurements. It can be shown that as the number N of measurements increases, the distribution of x will, in general, be a *normal distribution* with a standard deviation given by [7]

$$\sigma_{\rm st} = \sigma_{\rm average} = \sigma_m = \sqrt{\frac{\sum_{j=1}^{N}(x_j - \bar{x})^2}{N(N-1)}} = \frac{S_x}{\sqrt{N}}. \tag{5.13}$$

σ_{est} is called the *standard deviation of the mean* and in an experiment, it is a measure of the *statistical uncertainty associated with the best value of* $\bar{x}$ in the process of measuring the same quantity N times. As $\sigma_{\rm st} = S_x/\sqrt{N}$, $\sigma_{\rm St}$ will decrease progressively as N increases, since S_x does not depend on N.

📖 Expressing the result of a measurement as $(\bar{x} \pm \sigma_{\rm est})$ is equivalent to saying that the value of x is contained in the interval $(\bar{x} - \Delta x, \bar{x} + \Delta x)$ with a level of confidence or probability $p_0 = 0.68(68\%)$. It is equivalent to write

$$P(\bar{x} - \Delta x < x < \bar{x} + \Delta x) = p_0, \tag{5.14}$$

which is interpreted as "the probability that the best estimate of x is between $\bar{x} - \Delta x$ and $\bar{x} + \Delta x$ is equal to p_0". The p_0 value is known as the confidence coefficient or confidence level (CL) and interval $(\bar{x} - \Delta x, \bar{x} + \Delta x)$ determines a confidence interval for x.

📖 It follows from the above discussion that when we perform a series of N measurements of the same measurand, we realize that it is only a sample of the infinite measurements that could be made. The objective of the measurement is to estimate the mean value and its respective uncertainty with a certain level of confidence (CL), say 68%. If the distribution of the measurements has a Gaussian distribution, the estimators of these parameters will be as follows:

- ✓ estimate of the average value $\rightarrow \bar{x} = \sum x_j/N$,
- ✓ estimate of the uncertainty of the mean value (CL = 68%) $\rightarrow \sigma_{\rm st} = S_x/\sqrt{N}$.

5.7 Optimal number of measurements

Remember that S_x measures the dispersion of the distribution of the single measurements and does not depend on N but on the quality of the measurements, while $\sigma_{\rm st}$ decreases as N increases. As we have seen in Eq. (4.3), the *actual error of a measurement* process is given by the combination of nominal and statistical errors, that is,

$$\Delta x^2 = \sigma_{\text{nom}}^2 + \sigma_{\text{st}}^2. \tag{5.15}$$

It may at first be tempting to think that if we measure a quantity many times, we can overlook the contribution of statistical uncertainty in Eq. (5.15). Certainly, σ_{est} decreases as N increases, but the cost in time and money increases constantly with N. According to Eq. (5.15), it makes sense to decrease σ_{St} only until it is about equal to σ_{nom}, which is determined by the instrumentation and the measurement method. Further decreases in σ_{est} would not lead to an appreciable decrease in actual error Δx, Eq. (5.15), and therefore would not be profitable.

Equation (5.15) shows that it is reasonable to decrease σ_{st} until $\sigma_{\text{st}} \approx \sigma_{\text{nom}}$. This provides a criterion for deciding what the optimal number of measurements is. Since S_x is independent of N, the idea is to make a small number N_{prel} of preliminary measurements — say between 5 and 10 — and then calculate S_x. From the characteristics of the tool and the procedures used, we can find σ_{nom}. From the condition $\sigma_{\text{st}}(N_{\text{op}}) = S_x/\sqrt{N}_{\text{op}} \approx \sigma_{\text{nom}}$, we can estimate the optimal number of measurements to be performed:

$$N_{\text{op}} \approx 1 + \left(\frac{S_x}{\sigma_{\text{nom}}}\right)^2. \tag{5.16}$$

The one, 1, in the second member of Eq. (5.16) assures us that it is always necessary to perform at least one measurement. If $N_{\text{op}} > N_{\text{prel}}$, the measurements are completed to obtain N_{op} values and σ_x is recalculated. If $N_{\text{op}} < N_{\text{prel}}$, only preliminary measurements are performed, and they are all used. Finally, in all cases, the combined absolute uncertainty Δx will be calculated using Eq. (5.15).

5.8 Ten practical steps — Decalogue

In summary, the steps for measuring a physical quantity x are as follows:

(1) Analyze the possible sources of systematic errors and attempt to minimize them.

(2) Estimate the nominal uncertainty σ_{nom}.
(3) Take 5–10 preliminary measurements and determine the standard deviation of this sample, S_x, Eq. (5.12).
(4) Determine the optimal number of measurements N_{op}, Eq. (5.16).
(5) Complete N_{op} measurements of x.
(6) Calculate the mean and the statistical uncertainty σ_{est}.
(7) Evaluate the absolute *actual error* of the measurement by combining the nominal and statistical uncertainties, using Eq. (5.15).
(8) Express the result in the form $x = \bar{x} \pm \Delta x$ with the *corresponding unit*, making sure that the number of significant digits is correct, see Chapter 4.
(9) It is useful to calculate and indicate the relative percentage uncertainty of the measurement $x = 100.\ \Delta x/\bar{x}$, which can be used for comparisons with the results of other researchers or with other methods.
(10) If we want to study the statistical distribution of the results (e.g., whether or not it is normal), the histogram of the experimental data distribution can be compared with the corresponding normal curve, i.e., with a normal distribution of mean $\langle x \rangle$ and standard deviation $S_{x.}$

Sometimes it is useful to discriminate statistical errors from systematic errors and make both values explicit. In these cases, the result is expressed as $x = x \pm \sigma_{\text{st}} \pm \Delta x_{\text{sist}}$.

Example 1. The following five measurements of the diameter of a cylinder were made using a caliper with nominal appreciation $\sigma_{\text{nom}} = 0.1$ mm. Are these measurements sufficient? What should be the optimal number of measurements?

d(mm)=	10.2	10.8	11.0	10.0	10.1

In this case, the average value is 10.42 mm and $\sigma_{\text{st}} = 0.2$ mm, which is double the nominal appreciation. We should therefore perform more measurements so that $\sigma_{\mathrm{st}} \approx \sigma_{\mathrm{nom}}$. Using Eq. (5.16), we get $N_{\mathrm{op}} = 21$ measurements.

5.9 ♣ Combination of independent measurements

A frequent situation in science is the determination of the best value of a given quantity using different values from independent measurements (obtained by different authors, with different techniques or tools, etc.). Each of these independent measurements can have several associated uncertainties. Thus, we have a set of M measurements, each characterized by a pair (x_k, σ_k), with $k = 1, 2, \ldots, M$. Our goal is to find the best value for the size in question. It is clear that when we combine the different results to find the best value, $\langle x \rangle$, we need to take into account the respective uncertainties so that the more precise measurements contribute more ("weigh more") in the final result. In these cases, the best value $\langle x \rangle$ is given by [6, 7]

$$\langle x \rangle = \frac{\sum \frac{x_k}{\sigma_{k^2}}}{\sum \frac{1}{\sigma_{k^2}}}, \tag{5.17}$$

with absolute uncertainty $\Delta x \equiv \sigma_{<x>}$ defined by [6, 7]

$$\frac{1}{\sigma^2_{\langle x \rangle}} = \sum \frac{1}{\sigma_{k^2}}. \tag{5.18}$$

🕮 A special case of interest is when we have N determinations of the measurand, all with the same uncertainty σ. As can be easily deduced from Eq. (5.17), the mean will be

$$\langle x \rangle = \frac{\sum x_k}{N}, \tag{5.19}$$

which, as might be expected, corresponds to the expression (5.10). The uncertainty associated with this value will be, according to Eq. (5.18),

$$\sigma_{\langle x \rangle} = \frac{\sigma}{\sqrt{N}}, \tag{5.20}$$

which coincides with Eq. (5.13). Furthermore, the meaning of σ is illustrated as a measure of the dispersion associated with each individual measurement and $\sigma_{\langle x \rangle}$ as the dispersion associated with the best value.

5.10 Discrepancy

If a physical quantity is measured by two or more methods, or by different observers, it is possible — and very likely — that the results will not coincide. We say there is a *discrepancy* in the results.

The term *repeatability* is used to describe the agreement or otherwise between multiple measurements made by the same observer with the same method. *Reproducibility* is associated with the agreement or otherwise of measurements made by different observers or different methods.

The important thing is to know whether or not the discrepancy is significant. The following criterion is often applied. If the results of the two observations that are compared are independent (normal case) and have as results

$$\text{Measurement 1:} \qquad X_1 = \bar{X}_1 \pm \Delta X_1,$$

$$\text{Measurement 2:} \qquad X_2 = \bar{X}_2 \pm \Delta X_2,$$

we define

$$\Delta X^2 = \Delta X_1^2 + \Delta X_2^2. \tag{5.21}$$

If the data have a normal distribution, we say that with a confidence limit of 68% the measurements are different if

$$|\bar{X}_1 - \bar{X}_2| \geq \Delta X \tag{5.22}$$

and that with a confidence limit of 96% the measurements are different if

$$|\bar{X}_1 - \bar{X}_2| \geq 2 \cdot \Delta X. \tag{5.23}$$

These criteria can be generalized to broader confidence intervals in a similar way. They also apply when comparing laboratory-obtained values to tabulated or published values. Note the difference between discrepancy and uncertainty. Discrepancy is associated with the lack of overlap of two intervals (uncertainties) of two different results.

5.11 Summary of important concepts

It is suggested that the reader provide a concise explanation for the following items and, when possible, give an example:

✓ Describe what a histogram is and what is meant by statistical distribution of an attribute of a population.
✓ Explain how to find the error of a quantity that is measured N times.
✓ What is the best value of a measurement?
✓ Discuss the difference between the standard deviation of the individual measurements and the standard deviation of the mean.
✓ How is the statistical error obtained from a series of measurements of the same measurand?
✓ How do you determine the *optimal number* of measurements of a given quantity?
✓ How do you combine independent measurements of the same measurand?
✓ When do we say there is *discrepancy* among two or more measurements?

5.12 Exercises and problems

(1) Why do we say that when we make a measurement, we take a sample? What is the standard deviation of a sample?
(2) What is a histogram? What information does it provide?
(3) The diameter d of a sphere is measured several times with a caliper and the results are as follows:

						Average	Standard deviation
d (mm)	51.1	52.1	53.2	52.4	53.2	52.4	0.87

(a) Using your own judgment, what is the error of appreciation of the instrument used to measure this diameter?

(b) What is the nominal error of each of these measurements?
(c) What is the best value of each of these quantities?
(d) Analyze whether the number of measurements of d is adequate. What is the optimal number of measurements of d compatible with the instrument used?
(e) What are the absolute and relative errors of d?

(4) Two students took six measurements of the same length with a 0.5 mm graduated ruler. Their results, expressed in cm, were as follows:

Student A	11.2	11.5	11.6	10.5	11.9	11.0
Student B	11.5	11.6	11.4	11.5	11.5	11.6

(a) Say how each student should express their final results. Which of the measurements is of better quality and why?
(b) One of the students argues that both measurements have the same quality, as they both used the same rule. What would you say about this argument?
(c) Did both students find the same results? Is there a discrepancy between them? Justify.

(5) Several people independently measure the gravitational acceleration g value in a given location and find 9.80 ± 0.05 m/s^2, 9.81 ± 0.09 m/s^2, and 9.8 ± 0.1 m/s^2. How can you express the best value of g by combining these independent quantities?

(6) Indicate whether or not there is significant discrepancy between the following pairs of measurement results for the same physical quantity (use the criteria provided by (5.16) and (5.17)):

(a) $m_1 = 54.3 \pm 0.3\,\text{g}$ $\quad m_2 = 54.8 \pm 0.1\,\text{g}$,
(b) $v_1 = 100 \pm 3\,\text{m/s}$ $\quad v_2 = 105 \pm 3\,\text{m/s}$,
(c) $g_1 = 9.82 \pm 0.05\,\text{m/s}^2$ $\quad g_2 = 10.00 \pm 0.05\,\text{g}$,
(d) $Q_1 = 77.0 \pm 0.3\,\text{m}^3/\text{s}$ $\quad Q_2 = 78.0 \pm 0.5\,\text{g}$.

5.13 Histograms

5.13.1 *Goal*

The goal of these experiments is to analyze a series of measurements of physical quantity by using basic statistical concepts and constructing a histogram.

5.13.2 *Introduction*

When N measurements of the same quantity x are made under conditions of repeatability (i.e., when independent measurements are made under the same conditions, using the same method and by the same observer), statistical analysis of the data is required. For this activity, perform a statistical analysis of the data and express the measurement result in terms of the *average value* of the statistical estimators $\langle x \rangle$, *standard deviation* S_x, and *standard deviation of the mean* $\sigma_{\mathbf{est}}$. The data obtained can be represented in a histogram, in which the distribution of the values can be seen. The same type of analysis can be used in a quality control process when studying a batch of a product and analyzing the degree of dispersion of some of its properties around an average value.

5.13.3 *Project 7. Construction of histograms and study of empiric distributions*

Using the data from the Excel "*Histo1.xls*" spreadsheets that you can download from the Internet (https://www.fisicarecreativa.com/ajp/soft_sg.htm), for each sheet, build a histogram and calculate the parameters: mean, median, mode, and standard deviation.

(a) In each case, discuss what kind of distribution your data show.
(b) Indicate whether or not the distributions are symmetric.
(c) Are they unimodal?
(d) If the distribution is symmetric and unimodal, superimpose on the histogram the normal (or Gaussian) curve that best fits the observed distribution.

(e) If the first 50 data of each sheet were the result of measurements of a given quantity, indicate in each case their best value and its corresponding error.

(f) If the significant figures on each sheet indicate what the nominal error was for those measurements, indicate whether the first 50 measurements are an appropriate number or if more or perhaps fewer are needed. Justify all your answers.

5.13.4 *Project 8. Histogram obtained by hand*

With a ruler that does not exceed 20 cm, take about 100 measurements of the length of the table you're using or the height of a door. Take your measurements as quickly as possible. Divide the work among your team members.

✓ With the data collected by each observer, create a histogram showing the frequency of occurrence of each measurement.

✓ For each series of measurements, determine the best value of the length $\langle x \rangle$, the sample standard deviation (or the standard deviation of each measurement) S_x, and the standard deviation of the mean σ_x. If you are using Microsoft Excel®, the STDEV[a] function [8] directly calculates the sample standard deviation, which is S_x.

✓ Also collect all measurements in a single histogram and determine the mean value of all obtained values, the standard deviation, and the standard deviation of the mean.

✓ Using the mean and standard deviation values for each data set, plot the Gaussian curves corresponding to these parameters on each of the histograms. Note: When you want to compare a non-normalized histogram (i.e., a histogram whose area is not the unit)

[a] *STDEV function* estimates standard deviation based on a sample. The standard deviation is a measure of how widely values are dispersed from the average value (the mean). For more information about the functions of Excel, see [8]. If the data in the arrays of references contain numbers and text, only the numbers are considered and counted. Empty cells, logical values, text, or error values in the array or reference are ignored.

with a normal curve, you need to calculate the total number of data N_t as a whole, their average value, $\bar{x}$, and the standard deviation of said data, σ_x. If we assume that the range of classes is equally spaced with a separation $\Delta x (= x_i - x_{i-1})$, in order to compare the histogram with the normal curve, we have to multiply the distribution (5.10) by the factor $N_t \cdot \Delta x$.

✓ What can you say about the character of the distribution of the results obtained in your measurements? Are the values normally distributed?

References

[1] Wikipedia, "Statistical population," 2021. Available: https://en.wikipedia.org/wiki/Statistical_population.
[2] G. Blom, *Probabitily and Statistics: Theory and Applications*, NY: Springer, 2011.
[3] Wikipedia, "Statistics," 2021. Available: https://en.wikipedia.org/wiki/Statistics.
[4] Wikipedia, "Histogram," 2021. Available: https://en.wikipedia.org/wiki/Histogram.
[5] M. R. Spiegel and L. J. Stephens, "Statistics," *de Schaum's Outline of Theory and Problems*, NY: Mc Graw Hill, 1999.
[6] D. C. Baird, *Experimentation: An Introduction to Measurement Theory and Experiment Design*, 3rd Ed., NY: Pearson, 1995.
[7] P. Bevington and D. K. Robinson, *Data Reduction and Error Analysis for the Physical Sciences*, 3rd Ed., 2003, p. 230.
[8] Microsoft, "Statistical functions (reference)," 2021. Available: https://support.microsoft.com/en-us/office/statistical-functions-reference-624dac86-a375-4435-bc25-76d659719ffd.

Chapter 6

♣ INDIRECT MEASUREMENTS

This chapter introduces the concept of indirect measurement and presents the problem of error propagation. It discusses different techniques for presenting values of measurements and results, and ways of *truncating* and *rounding* figures. Finally, it presents criteria for choosing the most appropriate instruments for performing a measurement with a given required uncertainty.

Goals

✓ Indirect measurement
✓ Error propagation
✓ Truncation and rounding of values
✓ Choice of instruments and tools

6.1 Error propagation

Some quantities are not determined directly but derive from others that are measured directly. For example, to find the area of a rectangle, we can measure the lengths of its sides, or to determine the speed of a vehicle, we could measure the distances and time intervals independently. The question we want to answer here is how errors (uncertainties) in directly measured quantities will propagate or contribute to the error (uncertainty) of the derived quantity that is calculated using some mathematical expression. We will give only the results. For further details, see the bibliography cited at the end of this chapter [1, 2].

Figure 6.1 illustrates the concept of error propagation: the quantity y is computed from x by the mathematical function $y = f(x)$. From direct measurements, we know the best value x_0

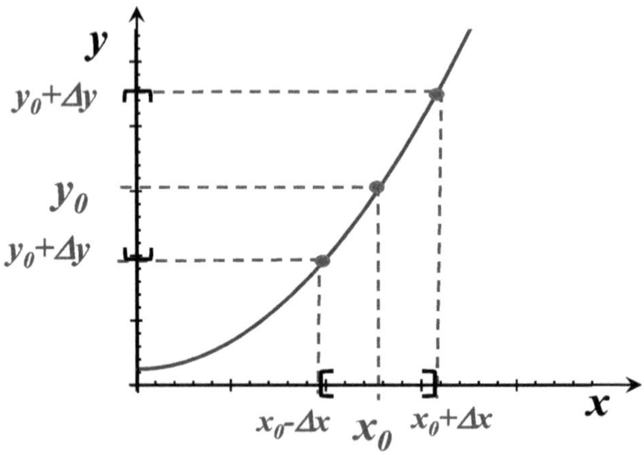

Figure 6.1 Influence of the error of a quantity x on the determination of the error of a derived quantity y.

and its uncertainty Δx; we want to know the best value of y and its uncertainty Δy.

Based on differential calculus, [3] we can write the following:

$$\Delta y \approx |f'(x_0)| \cdot \Delta x = \left|\frac{dy}{dx}\right|_{x=x_0} \cdot \Delta x. \tag{6.1}$$

The absolute value is used because we are estimating uncertainties, which are always positive quantities.

If we consider a more general case in which a quantity V is a function of several quantities $x, y, z, \ldots,$

$$V = V(x, y, z, \ldots), \tag{6.2}$$

where $x, y, z, \ldots$ are *all independent values* that have been measured directly, and we know their best values $x_0, y_0, z_0, \ldots$ and their corresponding uncertainties $\Delta x, \Delta y, \Delta z, \ldots$ It can be shown [1, 2, 4, 5] that the uncertainty of $V, \Delta V$ is given by

$$\Delta V = \sqrt{\left(\frac{\partial V}{\partial x}\right)^2 \Delta x^2 + \left(\frac{\partial V}{\partial y}\right)^2 \Delta y^2 + \left(\frac{\partial V}{\partial z}\right)^2 \Delta z^2 + \cdots}. \tag{6.2}$$

This equation is the *error propagation formula.*[a] The notation $\frac{\partial V}{\partial x}, \frac{\partial V}{\partial y}, \frac{\partial V}{\partial z}, \ldots$ indicates the partial derivation of the function V with respect to independent variables $x, y, z, \ldots$ and the formula is evaluated for the values $\langle x \rangle, \langle y \rangle, \langle z \rangle, \ldots$

Using the inequality

$$(|a| + |b|)^2 = a^2 + b^2 + 2|a||b| \geq a^2 + b^2, \tag{6.3}$$

from expression (6.2), we get

$$\Delta V^2 = \left(\frac{\partial V}{\partial x}\right)^2 \Delta x^2 + \left(\frac{\partial V}{\partial y}\right)^2 \Delta y^2 + \left(\frac{\partial V}{\partial z}\right)^2 \Delta z^2 + \cdots \leq \left(\left|\frac{\partial V}{\partial x}\right| \Delta x + \left|\frac{\partial V}{\partial y}\right| \Delta y + \left|\frac{\partial V}{\partial z}\right| \Delta z + \cdots\right)^2, \tag{6.4}$$

therefore, an upper bound for the uncertainty of V can be written as follows:

$$\Delta V \leq \left|\frac{\partial V}{\partial x}\right| \Delta x + \left|\frac{\partial V}{\partial y}\right| \Delta y + \left|\frac{\partial V}{\partial z}\right| \Delta z + \cdots \tag{6.5}$$

In the particular case where the function $V(x, y, z, \ldots)$ is *factorizable*[b] in powers of $x, y, z, \ldots$, the above expression can be put very simply. Suppose we have the function

$$V(x, y, z) = a\frac{x^n y^m}{z^l}, \tag{6.6}$$

where a, n, m, and l are different constants or real numbers. Applying the propagation Eq. (6.4), we obtain

$$\left(\frac{\Delta V}{V}\right)^2 = n^2 \cdot \left(\frac{\Delta x}{x}\right)^2 + m^2 \cdot \left(\frac{\Delta y}{y}\right)^2 + l^2 \cdot \left(\frac{\Delta z}{z}\right)^2. \tag{6.7}$$

For preliminary calculations, this expression can be approximated by

$$\frac{\Delta V}{V} \leq n \cdot \left|\frac{\Delta x}{x}\right| + m \cdot \left|\frac{\Delta y}{y}\right| + l \cdot \left|\frac{\Delta z}{z}\right|. \tag{6.8}$$

[a]In these first sections, we implicitly assume that the errors of the variables $x, y, z, \ldots$ are statistically independent of each other, i.e., there are no correlations between them. If there were, Eq.(3.1) would have to include terms comprising these correlations. See Refs. [1, 2, 4–8] and the last section in this chapter.

[b]Factorizable means that the expression of $V(x, y, z, \ldots)$ contains the independent variables in terms that are multiplied, such as the expression (6.6).

Another special case of interest is

$$Z = x \pm y. \tag{6.9}$$

Using Eq.(6.2), we get

$$(\Delta Z)^2 = (\Delta x)^2 + (\Delta y)^2. \tag{6.10}$$

📖 **Note:** Sometimes there is no analytic function that expresses the dependent variable y as a function of the independent one, which is measured directly. To be more precise, suppose we want to determine the mass, m, of an acrylic sphere, whose density, ρ, depends on the temperature, T, in an unknown way, but we have a table of ρ as a function of T. The diameter, d, of the sphere is measured directly, so we assume d and known Δd or

$$m(\rho, d) = \frac{\pi}{6} d^3 \cdot \rho(T). \tag{6.11}$$

From Eq.(6.7), we have

$$\left(\frac{\Delta m}{m}\right)^2 = \left(\frac{\Delta \pi}{\pi}\right)^2 + 3^2 \left(\frac{\Delta d}{d}\right)^2 + \left(\frac{\Delta \rho}{\rho}\right)^2. \tag{6.12}$$

The problem is that we do not know the function $\rho(T)$, so we cannot estimate $\Delta\rho$ starting from ΔT. What we do know are T_0 and its uncertainty ΔT. From the table of ρ as a function of T, we can find $\rho(T_0)$ and also $\rho(T_1)$ and $\rho(T_2)$, where $T_1 = T_0 - \Delta T$ and $T_2 = T_0 + \Delta T$, therefore $\Delta\rho = |\rho(T_1) - \rho(T_2)|$. Thus, by using Eq.(6.12), we can determine the uncertainty in the mass of the sphere under consideration.

6.2 Truncation of numbers

Consider the following example: to determine the density of a body, we measure its volume, $V = (3.5 \pm 0.2)\text{cm}^3$ (with $\varepsilon_V\% = 6\%$), and its mass, $m = (22.7 \pm 0.1)$g (with $\varepsilon_m\% = 0.4\%$). Using the definition of density, we have $\rho = m/V$. If we create this quotient with a calculator or a computer, the result could have 10 or more digits, for example,

$$\rho = 22.7/3.5 = 6.485714286\,\text{g/cm}^3. \tag{6.13}$$

The question is to know how many figures to report, since most of them are probably not significant. To answer this question, we propagate the errors in the variables m and V into the values of ρ, using the density definition: $\rho = m/V$. From Eq.(6.7), we have

$$\Delta\rho/\rho \approx \sqrt{0.004^2 + 0.06^2} = 0.06 \quad \text{and} \quad \Delta\rho \approx 0.4\,\text{g/cm}^3. \qquad (6.14)$$

Therefore, in expression (6.13), the only significant digit in the object's density is the first digit to the right of the decimal point, i.e., the result of Eq.(6.12) should be reported as follows:

$$\rho = (6.5 \pm 0.4)\text{g/cm}^3 \quad \text{and} \quad \varepsilon_\rho\% = 6\%. \qquad (6.15)$$

It is important to take this truncation criterion into account whenever we perform an operation using a calculator or computer and in reporting any of the results.

6.3 Choosing the best equipment or tools

Before taking a measurement, it is important to select the most appropriate tools or instruments with the required tolerance or uncertainty. Ignoring this step can lead to significant loss of work, time, and money. Imagine that we are producing cylinders for an internal combustion engine. Exceeding the required error would imply a waste of effort and resources because too many cylinders would be outside the preset range of tolerance and would have to be rejected. On the other hand, if the measurement is made with much less uncertainty than what is required, it might be acceptable, but the cost of producing the cylinders would be so much higher that they could be uncompetitive and impossible to sell. So, we need to determine in advance the most appropriate tools for this measurement, to meet the required tolerance or uncertainty with the minimum cost and effort.

To see how to approach this selection of instruments and tools, let's use the following example. Suppose our problem is to determine with an accuracy of 1% the volume of a wire with diameter $d \approx 3$ mm and length $L \approx 50$ cm. What tools should we use to achieve our goal with the lowest cost?

The required objective is $\Delta V/V \approx 0.01$. Since $V = \pi d^2 L/4$, we get

$$\begin{aligned} \frac{\Delta V}{V} &\approx \frac{\Delta \pi}{\pi} + 2\frac{\Delta d}{d} + \frac{\Delta L}{L} \\ 0.01 &\approx 0.001 + 0.006 + 0.002. \end{aligned} \tag{6.16}$$

The first expression is an application of Eq.(6.8), and this approximation is useful and sufficient for this preliminary analysis. The assignment of the values in the second line is somewhat arbitrary, but we observed that the relative uncertainty in the volume V does not exceed the required 1%. To the number π we assign a small relative uncertainty, and with this, we will determine how many digits to use with π, without the truncation error of π significantly affecting the determination of ρ. Note that the quality of the diameter measurement was given a greater incidence than that of the length L because the volume is proportional to d^2 and only proportional to L to the first power, therefore the relative uncertainty of d is multiplied by 2. This is why we have assigned a larger tolerance (more error) to the measurement of d than to the measurement of L. With this preliminary assignment, we decide which are the most appropriate measuring instruments (in general, the most appropriate are those that make measuring easier, quicker, and cheaper while meeting the established requirements).

Since

$$\begin{aligned} \frac{\Delta d}{d} &\approx 0.003 \Rightarrow \Delta d \approx 0.003 \cdot d = 0.003 \cdot 3\,\text{mm} \\ &\approx 0.009\,\text{mm} \approx 0.01\,\text{mm}, \end{aligned} \tag{6.17}$$

we must use, at least, a micrometric screw to measure d.

Similarly, we have for the measurement of L,

$$\frac{\Delta L}{L} \approx 0.002 \Rightarrow \Delta L \approx 0.002 \cdot L = 0.002 \cdot 50\,\text{cm} \approx 1\,\text{mm}, \tag{6.18}$$

therefore, an ordinary ruler graduated in millimeters will be enough to measure L.

For π, we have

$$\frac{\Delta\pi}{\pi} \approx 0.001 \Rightarrow \Delta\pi \approx 0.001 \cdot \pi = 0.001 \cdot 3 \approx 0.003, \tag{6.19}$$

indicating that we should use π with 3 or more decimal places so that the truncation error has a negligible effect. Therefore, the choice $\pi = 3.142$ would be appropriate in this case.

Note that so far, everything is preliminary, and we have only chosen the measuring instruments. Once the tools have been selected, we can proceed to use them to measure d and L. Note also that to choose the instruments, we must know the approximate values to be measured. Although this may seem paradoxical, for this preliminary analysis, we only need to have an idea of the orders of magnitude, not an exact value. This order of magnitude can be assessed by visual inspection or quick preliminary measurement with whatever instrument we have on hand. Finally, once d and L have been measured, we must use the correct propagation formula from Eq. (6.2) to calculate the combined uncertainty ΔV.

6.4 ♣♣ Propagation of errors with correlated variables

So far, we have assumed that the set of variables is measured directly and determine the value of another, z, of which we want to estimate the uncertainty. We also assume that the directly measured variables are independent of each other and that there is no correlation among them. However, there are many cases in which these assumptions are not satisfied and the correlation between the variables cannot be disregarded [1, 2]. To see this, suppose we want to know the best value for the uncertainty of a variable z:

$$z = z(u, v), \tag{6.20}$$

which we do not measure directly but calculate from the measurements of u and v. Consider the case where the variables u and v are not independent. If we made a long series of simultaneous measurements of u and v, we would get a series of N data $\{u_i, v_i\}$. We define the following quantities:

$$\bar{u} \equiv \langle u \rangle = \frac{\sum_i u_i}{N}, \quad \bar{v} \equiv \langle v \rangle = \frac{\sum_i v_i}{N}, \tag{6.21}$$

$$\sigma_u^2 = \frac{\sum_i (u_i - \bar{u})^2}{N} = \langle u^2 \rangle - \langle u \rangle^2,$$

$$\sigma_v^2 = \frac{\sum_i (v_i - \bar{v})^2}{N} = \langle v^2 \rangle - \langle v \rangle^2, \tag{6.22}$$

and

$$\sigma_{uv} = \frac{\sum_i (u_i - \bar{u})(v_i - \bar{v})}{N} = \langle uv \rangle - \langle u \rangle \langle v \rangle. \tag{6.23}$$

The definitions of mean values and variances have the usual meaning, as discussed in the previous chapter.[c] The covariance of u and v, σ_{uv}, is a measure of the correlation between variables.

When there is a correlation between the variables u and v, the expression (6.2) can be generalized as follows: [1, 6]

$$\Delta z^2 \equiv \sigma_z^2 = \left(\frac{\partial z}{\partial u}\right)^2 \cdot \sigma_u^2 + \left(\frac{\partial z}{\partial v}\right)^2 \cdot \sigma_v^2 + 2 \cdot \left(\frac{\partial z}{\partial u} \cdot \frac{\partial z}{\partial v}\right) \cdot \sigma_{uv}. \tag{6.24}$$

If the variables u and v were really independent of each other, it is clear that the signs of $(u_i - \bar{u})$ and $(v_i - \bar{v})$ would occur randomly and the value of the covariance would be $\sigma_{uv} = 0$; in this case, expression (6.24) is reduced to (6.2).

On the other hand, if there were a relationship between these variables, for example, if it were

$$v = c \cdot u + K, \rightarrow \langle v \rangle = c \cdot \langle u \rangle + K. \tag{6.25}$$

[c] In this section, we will consider only the statistical component of the uncertainties in the measured variables. Furthermore, we assume large samples that allow us to disregard the difference between N and $N - 1$ in the expressions of the variance and covariance of the samples.

Using Eq. (6.22), we get

$$\begin{aligned}\sigma_{uv} &= \langle uv\rangle - \langle u\rangle\langle v\rangle \\ &= c\langle uu\rangle + K\langle u\rangle - c\langle u\rangle^2 - K\langle u\rangle \quad \text{or} \\ \sigma_{uv} &= c\sigma_u^2. \end{aligned} \tag{6.26}$$

The problem in many practical applications is that, in general, the expression of the correlation coefficient or covariance between the variables used is not known. However, even in these cases, it is possible to obtain an upper bound for the uncertainty of z. According to Schwartz's inequality, [3] we have

$$\langle uv\rangle^2 \leq \langle u^2\rangle\langle v^2\rangle. \tag{6.27}$$

Therefore,

$$0 \leq |\sigma_{uv}| \leq \sigma_u \cdot \sigma_v. \tag{6.28}$$

We see that, in general, it is possible to obtain an upper limit for the amplitude of the uncertainty of z given by [6]

$$|\Delta z| \equiv |\sigma_z| \leq \left|\frac{\partial z}{\partial u}\right| \cdot \sigma_u + \left|\frac{\partial z}{\partial v}\right| \cdot \sigma_v, \tag{6.29}$$

which coincides with the approximate expression (6.5) and is valid whether or not there is a correlation between the variables measured directly. When the correlation can be estimated, expression (6.24) should be used to propagate the uncertainties.

6.5 Summary of important concepts

It is suggested that the reader give a concise explanation of the following concepts and, where possible, provide an example:

- ✓ What is an indirect measurement? Discuss the concept of error propagation.
- ✓ Explain how number truncation is performed.
- ✓ Discuss the criteria used for choosing instruments when you want to measure a quantity with a given error.

6.6 Exercises and problems

(1) The diameter of a sphere is $d = (99.1 \pm 0.8)$ cm. Calculate the surface area and volume of the sphere and find their respective absolute and relative errors.

(a) Which determination has the "best quality", area or volume?
(b) Explain why the quality of all these determinations is not the same, even though they are all based on a single datum: the diameter.
(c) Express the value of the volume and area of the sphere, indicating the best values and their corresponding absolute and relative errors.

(2) The sides a and b of a rectangle are measured: $a = (23.45 \pm 0.02)$ m and $b = (11.40 \pm 0.03)$ m. Calculate the perimeter and area of the rectangle and express the results with their respective uncertainties.
(3) Suppose we want to determine the volume of the lead extracted from a pencil. Determine the tools needed to measure the volume of the lead with a relative error of 2%. How many decimal places does it take for π?
(4) We want to know the density of a rubber sphere about 5 cm in diameter with an error of less than 5%. Indicate the accuracy of the instruments to be used (including the scales). Rubber has a density of approximately 1.5 g/cm^3.
(5) To find the surface of a sphere, the diameter d is measured several times with a caliper, providing the following values:

						Standard deviation
d(mm)	51.1	52.1	53.2	52.4	53.2	0.87

(a) Indicate with your own judgment the error of appreciation of the instrument used to measure this diameter, assuming that all decimals shown in the table are the results of measurements.

(b) What is the nominal error of each of these measurements?
(c) What is the best value of each of these quantities?
(d) Analyze whether the number of measurements of d is adequate. What is the optimal number of measurements of d? (See Chapter 2.)
(e) What are the absolute and relative errors of d?
(f) How many decimal places should be used for p to calculate the area?
(g) Determine the best value for the area of the sphere, its absolute error, and its relative error.

(6) The edges of a rectangular prism were measured, and the following results were obtained for the sides of the rectangle (a and b) and its height c:

	a (cm)	b (cm)	c (cm)
	4.8	11.1	21.7
	4.4	8.2	20.6
	5.1	12.7	22.3
	5.6	15.8	23.4
	5.6	15.6	23.3
	5.9	17.2	23.9
			19.6
			21.7
			23.2
			20.8
			21.2
			20.1
			20.9
Average (cm)	5.233	13.433	21.746
Standard deviation (cm)	0.57	3.40	1.38
Nominal error (cm)	0.2	0.4	0.1

(a) Indicate whether the number of available measurements for a, b, and c is sufficient. If not, how many more measurements of each quantity are necessary?
(b) Calculate the best value of the volume and its absolute error.

(7) To determine the viscosity η of a liquid, the following experiment is performed: the flow rate Q of the liquid outlet is measured through a horizontal pipe of length L and internal radius R, with a pressure drop ΔP at the ends of the pipe. Assume laminar flow which obeys the Hagen–Poiseuille law:

$$Q = \frac{\Delta P \pi R^4}{8\eta L}.$$

(a) Find the viscosity of the liquid by measuring $Q = 500 \pm 2\ \text{cm}^3/\text{s}$ when the pressure drop is $\Delta P = 123 \pm 1\ \text{N/m}^2$, in a tube length $L = 80.5 \pm 0.1$ cm and $R = 1000 \pm 1$ mm.
(b) Use the propagation formula to calculate the uncertainty of Q.
(c) Compare the relative errors of the different quantities that must be measured to obtain η. Which of these quantities is most critical in determining η and why?

(8) The deflection d of the free end of a horizontal cantilevered beam with the other end clamped to a table (rectangular section: sides a and b and length L) depends on the cantilever weight P and on the modulus of elasticity Y:

$$d = \frac{4PL^3}{Yab^3}.$$

The modulus of elasticity can be found with this simple laboratory experiment. What is the value of the Y modulus if the measured deflection is $d = (2.00 \pm 0.01)$ mm for a beam of dimensions $a = (12.40 \pm 0.02)$ mm, $b = (24.20 \pm 0.01)$ mm, $L = (50.00 \pm 0.01)$ cm, and weight $P = (12.67 \pm 0.05)$ N? Use the propagation formula to determine the uncertainty of Y.

References

[1] P. Bevington and D. K. Robinson, *Data Reduction and Error Analysis for the Physical Sciences*, 3rd Ed., 2003, p. 230.

[2] D. C. Baird, *Experimentation: An Introduction to Measurement Theory and Experiment Design*, 3rd Ed., NY: Pearson, 1995.

[3] R. Wrede and M. Spiegel, *Advanced Calculus-Third Edition (Schaum's Outlines)* 3rd Ed., NY: McGraw Hill, 2010.

[4] ISO, "Uncertainty of measurement — Part 3: Guide to the expression of uncertainty in measurement (GUM:1995) ISO/IEC," 1993.

[5] NIST, *NIST Constants, Units & Uncertainty — Essential of Expressing Measurement Uncertainty*, Washington DC, 2017.

[6] J. Taylor, "Simple examples of correlations in error propagation," *Am. J. Phys.*, vol. 53, pp. 663–667, 1985.

[7] G. Blom, *Probabitily and Statistics: Theory and Applications*, NY: Springer, 2011.

[8] ISO, "ISO/IEC GUIDE 98-3:2008, Uncertainty of measurement — Part 3: Guide to the expression of uncertainty in measurement," 2017. Available: https://www.iso.org/standard/50461.html.

Chapter 7

♣ LEAST SQUARES METHOD AND LINEAR REGRESSION

This chapter presents the method of least squares for finding the optimal parameters of a curve or theoretical model that fits a set of data (x_i, y_i). It discusses ways of quantifying the quality or goodness of a fit and considers the meaning of the correlation coefficient and the statistical parameter *chi square* (χ^2). When evaluating the quality of a model that describes a data set, it is important to consider the parsimony or *Occam's razor principle*, which states that complexity should not be introduced without necessity. Finally, it discusses some useful precautions to take into account in the analysis of experimental data.

Goals

✓ Least squares method
✓ χ^2 analysis
✓ Goodness of a fit
✓ Principle of parsimony
✓ Significance of parameters of a fit

7.1 Least squares method — Linear regression

Chapter 2 discussed the importance of graphical representations to discover or visualize mathematical expressions that describe or explain a set of empirical data (x_i, y_i). Quite often, the data suggest a linear relationship between two variables, namely $y = a \cdot x + b$. The question we want to answer is as follows: What are the parameters a and b of the line that best fit or describe the data?

The *method of least squares* is a procedure that enables us to answer this question. When the relationship between the variables x and y is linear, the least squares method also called the *linear regression method* is used [1, 2]. This chapter discusses the linear regression method, leaving for Appendices C and D the discussion of more general cases of least squares when the model that describes the relation between x and y is nonlinear and the data are also affected by errors [3].

In the case illustrated in Figure 7.1, the data have an approximately linear trend. Our goal is to find the *best line* that fits them, i.e., the values a and b of the line

$$y(x) = ax + b \tag{7.1}$$

that best describe the observed data. For this, it is useful to define the function χ^2 (chi square) [1, 2].

$$\chi^2(a, b) = \sum_i [y_i - (ax_i + b)]^2. \tag{7.2}$$

This function, χ^2, is a measure of the total squared deviation, χ^2, $[y_i - y(x_i)]^2$, of the observed values y_i with respect to those predicted by the linear model $a.x_i + b$. In other words, χ^2 is a measure of the total (vertical) distance of all the data (x_i, y_i) to the line. For a given

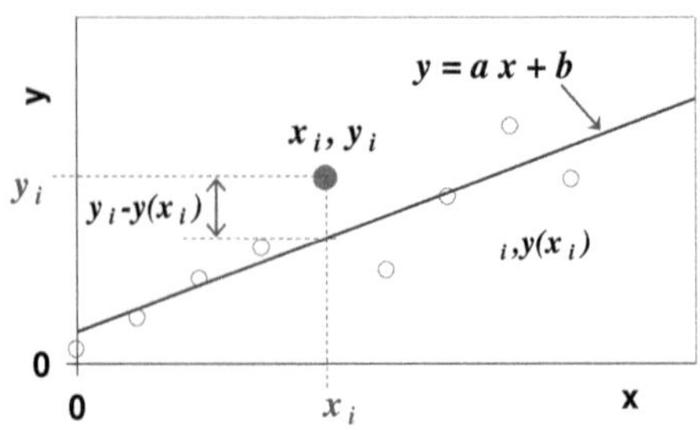

Figure 7.1 Graphical representation of (x_i, y_i) with a linear trend. The circles represent observed values. The line is the representation of the model $y(x) = ax + b$. The quantity $y_i - y(x_i)$ is the deviation of each observation of y_i from the value predicted by the model $y(x_i)$.

data set (x_i, y_i), the value of χ^2 depends on the parameters of the line, a and b. The least squares method assumes that the values of the slope a and the y-intercept b which best fit the data are those that minimize this total deviation, i.e., those that minimize the function $\chi^2(a, b)$. The minimization problem is reduced to solving the pair of equations:

$$\frac{d\chi^2}{da} = 0 \quad \text{and} \quad \frac{d\chi^2}{db} = 0\,. \tag{7.3}$$

Solving these equations enables us to find the optimal values of a and b. The results are straightforward [1, 2]:

$$a_0 = \frac{N\sum x_i y_i - \sum x_i \sum y_i}{N\sum x_i^2 - \left(\sum x_i\right)^2} \tag{7.4}$$

and

$$b_0 = \frac{\sum x_i^2 \sum y_i - \sum x_i \sum x_i y_i}{N\sum x_i^2 - \left(\sum x_i\right)^2}. \tag{7.5}$$

Here, the symbol $\sum$ refers to the sum $\sum_{i=1}^{i=N}$ on the N observed data.

The line obtained with these coefficients is called the *regression line*. Since the sums of Eqs. (7.4) and (7.5) occur very frequently, it is helpful to use the following notation:

$$\langle x^k \rangle = \frac{\sum x_i^k}{N}, \quad \langle y^k \rangle = \frac{\sum y_i^k}{N} \quad \text{with } k = 0, 1, 2, \ldots, N \tag{7.6}$$

and

$$\langle xy \rangle = \frac{\sum x_i \cdot y_i}{N}. \tag{7.7}$$

With this notation, expressions (7.4) and (7.5) can be written as follows:

$$a_0 = \frac{\langle xy \rangle - \langle x \rangle \langle y \rangle}{\langle x^2 \rangle - \langle x \rangle^2} \tag{7.8}$$

and

$$b_0 = \langle y \rangle - a_0 \langle x \rangle. \tag{7.9}$$

These results for a_0 and b_0 are built into most data analysis programs and spreadsheets. Programs such as Microsoft Excel®, Originlab Origin®, MathWorks Matlab®, etc. can readily provide this calculation, which is often performed using the "linear regression" or "linear fit" tools. The results (7.4) and (7.5) apply when all the data of the dependent variable have the same absolute uncertainty and the uncertainty of the independent variable is considered negligible. Appendix D discusses the case in which both variables have errors [4].

A measure of the quality or goodness of the fit performed is given by the *Pearson correlation coefficient* R^2 between the variables x and y, which takes on values between 0 and 1 and characterizes the dispersion of the data around the least squares line [1–3].

Consider the deviations of the observed points (x_i, y_i): (a) with respect to the line obtained from least squares and (b) with respect to the horizontal line $y = \bar{y}$, where $\bar{y}$ is the average of the y_i values. If the least squares line is a good description of the data, the values (x_i, y_i) are clustered along the regression line. The sum of the squares of the deviations from this line, represented by χ^2, should be less than the sum of the squares of the deviations from the horizontal line $y = \bar{y}$. The squared correlation coefficient, R^2, is defined as follows:

$$R^2 = \frac{\sum(y_i - \bar{y})^2 - \sum[y_i - (ax_i + b)]^2}{\sum(y_i - \bar{y})^2} = \frac{\sum(y_i - \bar{y})^2 - \chi^2}{\sum(y_i - \bar{y})^2}. \quad (7.10)$$

The first term in the numerator is the sum of the squares of the deviations of the points from the horizontal line that passes through $\bar{y}$. The second term is the sum of the squares of the deviations of the points from the regression line $y = a_0 x + b_0$, i.e., χ^2, defined by Eq. (7.2). Note that R^2 is dimensionless. If the data fall exactly on the regression line, there is perfect correlation, the second term is approximately zero ($\chi^2 \approx 0$), and $R^2 \approx 1$. On the other hand, the worse the fit, the higher the value of χ^2. The maximum value that χ^2 can reach is in the order of $\sum(y_i - \bar{y})^2$, in which case there is no correlation between the variables x and y, and the numerator of Eq. (7.19) is zero, i.e., $R^2 \approx 0$.

When the results of Eqs. (7.4) and (7.5) are introduced in Eq. (7.10), we get for R^2

$$R^2 = \left[\frac{\sum(x_i - \bar{x})(y_i - \bar{y})}{N\sigma_x\sigma_y}\right]^2 = \frac{\sigma_{xy}^2}{\sigma_x^2 \cdot \sigma_y^2}, \quad (7.11)$$

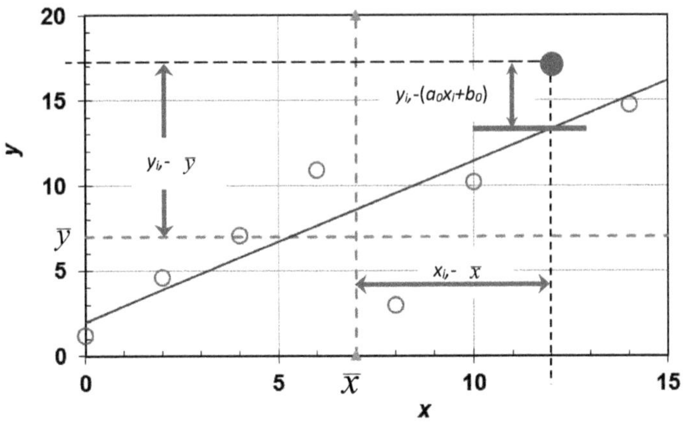

Figure 7.2 Empirical data (circles) that are grouped along a line and $= a_0 x + b_0$. The deviations of the points from the least squares line and the deviations of the points from the horizontal line serve to define the coefficient R^2 (Eq. (7.10)).

where

$$\sigma_x^2 = \langle x^2 \rangle - \langle x \rangle^2, \quad \sigma_x^2 = \langle y^2 \rangle - \langle y \rangle^2, \tag{7.12}$$

$$\sigma_{xy} = \langle xy \rangle - \langle x \rangle \langle y \rangle = \frac{\sum (x_i - \bar{x})(y_i - \bar{y})}{N}. \tag{7.13}$$

It is also possible to write

$$R^2 = 1 - \frac{\chi^2}{N \cdot \mathrm{Var}(y)}. \tag{7.14}$$

If $R^2 \approx 1$, we say that the linear model is adequate to describe the experimental data and there is ***good*** (linear) correlation between the x and y data. When $R^2 \approx 0$, we say that the linear expression is not an adequate description of the data. In this case, the graph should be analyzed carefully to find out if there is any nonlinear relationship that better approximates the dependence of y on x.

If $R^2 \approx 0$, it may indicate that there is no correlation between the variables. However, a value of $R \approx 0$ does not necessarily imply that there is no correlation between the variables, it only means that the *linear relationship* between them is not adequate. For example, if the pairs of points (x, y) describe a circle, we will have $R \approx 0$ (see Figure 7.3(d)). Of course, if the pairs (x, y) are not correlated, we will also have $R \approx 0$ (as illustrated in Figure 7.3(c)).

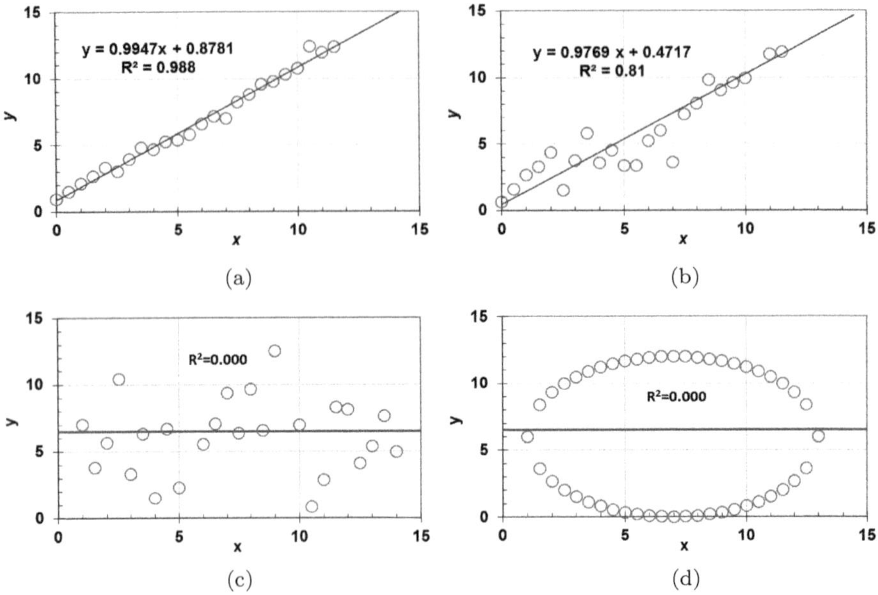

Figure 7.3 Fitting of experimental data by a linear model. (a) Case with a good linear correlation, (b) case with acceptable correlation, (c) case with practically no correlation, (d) case with good correlation but the linear model is inadequate.

7.2 Correlation and causality

When we study the relationship between the length L of a bar and its temperature T, a causal connection is often observed between them. In other words, the temperature T determines the value of L. An important observation to bear in mind is that a correlation between two sets of data x and y does not always imply a causal relationship between them. In other words, if $R^2 \approx 1$, it does not necessarily mean that and depends *causally* on x or vice versa. The correlation between the variables is a *necessary but not sufficient condition* for the existence of causal dependence between them. This logical fallacy of attributing causality to two events occurring at the same time is known as "*cum hoc ergo propter hoc*" ("If they appear together, they are cause and effect"). The following example illustrates this fallacy. The bigger the fire, the greater the number of firefighters fighting it. If the size of the fire were plotted as a function of the number of firefighters, a good correlation would surely be found. An erroneous statement which could be drawn would be to conclude that the number of firefighters determines or is the "*cause*" of the size of the fire.

A statistical correlation is an indication of a *possible* causal relationship. Establishing a causal connection between the variables requires a much more careful analysis. Failure to observe these criteria has led to notable errors in the past. One example is the observation that *Children who sleep with the light on are more likely to develop myopia in adulthood.* This study was conducted at a University of Pennsylvania medical center and was published in the prestigious journal *Nature* in May 1999 [5]. Later studies found a fallacy in this conclusion. More careful observations show that inheritance is important in myopia. Since myopic parents require good lighting to see, they tend to leave more lights on in their children's rooms [6, 7].

7.3 Uncertainty in the adjustment parameters

Often the goal of an experimental study is to find the parameters of a fit. One example is determining the elastic constant k of a spring from measurements of the applied forces F_i and their respective stretches x_i, assuming that there is a linear relationship between these variables of the type $F = kx$. In this case, the value of k will be precisely the slope of the line that best fits the data of F_i as a function of x_i. Another example is finding the electrical resistance R of a conductor from voltage measurements V_i and the currents I_i that flow through it. From the slope of V_i as a function of I_i, we obtain R. The question we want to answer now is as follows: What are the errors or uncertainties in these parameters obtained by least squares?

It is useful to have a method to estimate the *uncertainties* associated with the calculated values of the parameters a_0 and b_0

of Eqs. (7.4) and (7.5) [4], which we will denote with the symbols σ_a and σ_b. This section only presents the results; the interested reader will find a more exhaustive treatment and its justification in the references [1–4]. The uncertainties of the fit parameters are given by the following expressions:

$$\sigma_a = \sqrt{\frac{\chi_N^2}{N \cdot Var(x)}}, \tag{7.15}$$

$$\sigma_b = \sqrt{\frac{\chi_N^2 \cdot \sum_{i=1}^{N} x_i^2}{N \cdot Var(x)}} = \sigma_a \cdot \sqrt{\langle x^2 \rangle}, \tag{7.16}$$

where χ_N^2, known as the *chi-square value* per degree of freedom, is given by

$$\chi_N^2 = \frac{1}{N-2} \cdot \chi^2 \tag{7.17}$$

and

$$Var(x) = \frac{\sum_{i=1}^{N} x_i^2}{N} - \left(\frac{\sum_{i=1}^{N} x_i}{N}\right)^2 = \langle x^2 \rangle - \langle x \rangle^2. \tag{7.18}$$

The uncertainties of the parameters a_0 and b_0 can also be written in terms of the *correlation coefficient* R^2 as follows [4]:

$$\sigma_a = a_0 \cdot \sqrt{\frac{1}{(N-2)} \cdot \left(\frac{1}{R^2} - 1\right)}, \tag{7.19}$$

$$\sigma_b = \sigma_a \sqrt{\langle x^2 \rangle}, \tag{7.20}$$

with:

$$\langle x^2 \rangle = \frac{\sum x_i^2}{N}. \tag{7.21}$$

These expressions are of interest since they enable the estimation of the values of σ_a and σ_b from a_0 and R^2, which can be found using different spreadsheets and data fitting programs.

7.4 Occam's razor or criterion of parsimony

Occam's razor[a] establishes that when a theory or explanation of a phenomenon is being developed, no assumptions should be made beyond the minimum necessary. Descriptions should be kept as simple as possible until they are shown to be inappropriate. This philosophical principle also known as the *criterion of parsimony* is an idea that underlies all scientific and philosophical thought. It is also relevant to consider it when developing explanatory models.

If the behavior of a phenomenon can be explained with few explanatory variables, and if the relevant explanatory theory is not strong enough to suggest that other variables should be included, why introduce more variables? For example, if a phenomenon can be explained by a linear relationship, why use a polynomial of degree 5? If a set of data is well fitted by a line of the type $y = ax + b$, with a value of b close to zero, we should ask whether our data can be effectively explained by a relationship of the type $y = ax$. Note that this latter expression has only one free parameter (a), while the former has two (a and b), therefore ($y = ax$) is 50% simpler than $y = ax + b$. Of course, if $y = ax$ provides a poor fit, and $y = ax + b$ improves it, we opt for the expression with the two parameters. The important issue of the significance of the parameters of a fit is discussed in more detail in Appendix D.

Karl Popper [8] proposes an interesting justification for this principle. He claims that a theory or hypothesis is better when there are more ways to falsify or test it, i.e., when the hypothesis is "more falsifiable" (see Chapter 1). Imagine that we are studying a phenomenon that relates two observable variables (which can be

[a]*Pluralitas non est ponenda sine neccesitate* or *Complexity should not be introduced unnecessarily.* This statement is said by the English Franciscan philosopher and monk William of Ockham (or Occam) (ca. 1285–1349). As a good Franciscan, Occam was a minimalist, idealizing life in simplicity and poverty in the style of St. Francis of Assisi. Occam was excommunicated by Pope John XXII.

measured) X and Y. Imagine that we have two sets of observed data for these variables:

Data	X	Y
Measurement 1	1	1
Measurement 2	3	5

To explain this phenomenon, the following three alternative hypotheses are proposed, which are illustrated in Figure 7.4:

Hypothesis	
H1	$Y = aX + b$
H2	$Y = \alpha X^3 + \beta X^2 + \gamma X + \delta$
H3	$Y = cX^4 + dX^3 + eX^2 + fX + g$

This example clearly shows that a single additional observational data (one additional measurement) could in principle falsify hypothesis H1, i.e., if the new data point does not fall on the straight line. While to falsify hypothesis H2, at least 2 more measured data are required, and for hypothesis H3 at least 3 more data would be necessary. Therefore, since the simplest hypothesis (H1) is the most easily falsifiable, in the absence of other information about the phenomenon, this is the one we should choose.

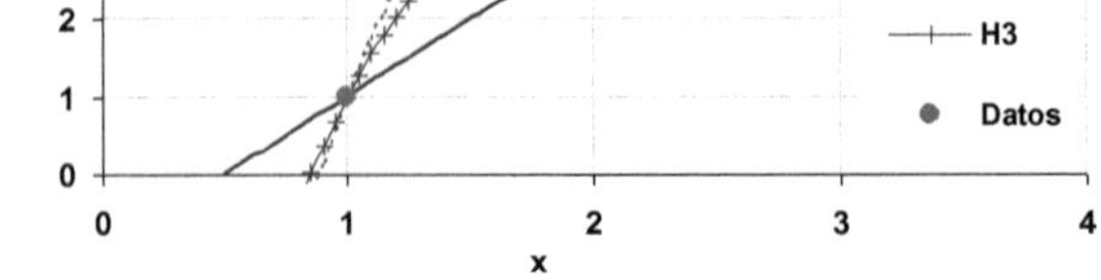

Figure 7.4 The curves represent three alternative hypotheses to explain the two pairs of observed data, represented by two filled circles.

🕮 **Example:** Upon studying the relationship between the stretch of a spring as a function of the applied force (see Figure 7.5), it is found that the line that best fits the data is

$$F(N) = 90.252\Delta x(m) + 0.0417 \quad \text{with } R^2 = 0.9923. \tag{7.22}$$

A question that we must always ask ourselves, according to the principle of parsimony, is whether the coefficients we have obtained are significant, i.e., whether it might be possible to find an even simpler functional relationship. Something simpler would be to fit the data with a line that necessarily crosses the origin. If we do this with the data in Figure 7.5, we get

$$F(N) = 89.61\Delta x(m) \quad \text{with } R^2 = 0.9977. \tag{7.23}$$

We see that in the latter case, the correlation coefficient is as good or better than that in the previous hypothesis, but the mathematical expression has half the amount of free parameters as before (only the slope a). Therefore, based on the criterion of parsimony, we are left with the latter fit, which is as good as the former but simpler.

Another way to analyze this problem consists of estimating the errors of the parameters of the first fit using expressions (7.19) and

ΔX(m)	ΔY(m)
0	0
0.02	1.65
0.033	3.2
0.045	3.6
0.06	5.5
0.075	7
0.091	8

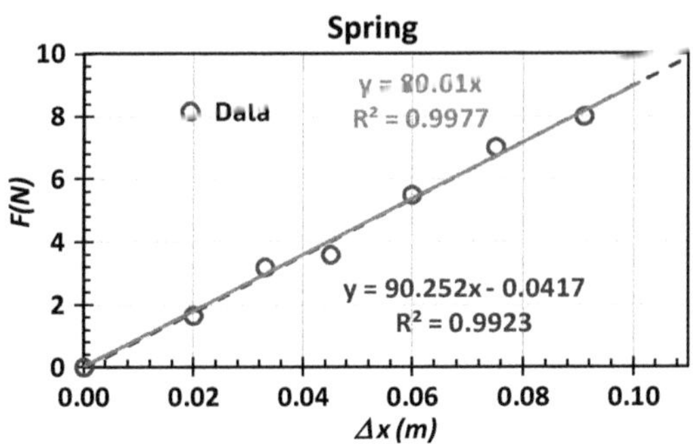

Figure 7.5 Example of data plot resulting from studying the relationship between the spring stretch ΔX(m) as a function of the applied force F, measured in Newtons.

(7.20). This provides the result:

$$F = a_0 \Delta x + b_0 \quad \text{with } R^2 = 0.992 \tag{7.24}$$

with

$$\begin{aligned} a_0 &= 90.25\,\text{N/m}, \quad \Delta a = \sigma_a = 3.5\,\text{N/m}, \\ b_0 &= -0.04\,\text{N}, \quad \text{and} \quad \Delta b = \sigma_a = 0.19\,\text{N}. \end{aligned} \tag{7.25}$$

In other words,

$$a_0 = (90 \pm 4)\,\text{N/m} \quad \text{and} \quad b_0 = (-0.04 \pm 0.2)\,\text{N}. \tag{7.26}$$

Therefore, we see that the coefficient b_0 is *compatible with zero* since its error (0.2 N) is greater than its absolute value (0.04 N). When the absolute error of a parameter is of the same order of magnitude or greater than its absolute value, we say that the parameter *is not significant* and is compatible with zero. In other words, the data are compatible with a null value of the parameter b_0. Therefore, according to the parsimony criterion, we can neglect this parameter and keep the simplest expression: $F = a_0 \Delta x$. Note how $\Delta a \ll a_0$, and the parameter a_0 is significant. The significance of the parameters of a fit is discussed in more detail in Appendix D.

7.5 Summary of important concepts

Provide a concise explanation of the following concepts and whenever possible, give an appropriate example:

✓ Review the fundamentals of the least square method.
✓ What interpretation could you suggest for the linear correlation coefficient R^2?
✓ Summarize the ideas underlying the parsimony principle.
✓ What parameters can be used to assess the quality or goodness of a fit? In particular, discuss similarities and differences between R^2 and χ^2.

7.6 Exercises and problems

(1) Pediatricians use charts of children's average growth as a function of age to assess whether their patients are progressing as expected. The pediatric staff who deal with these tables distinguish between the growth of boys and girls. Following is a table for the average height of boys from age one month to 6 years:

Age (months)	Average height for male (cm)
1	54
2	57.09
3	60.4
4	62.25
5	65
6	66.74
7	68.01
8	69.6
9	71.11
10	72.3
11	73.65
12	75.01
15	78.2
18	81.3
21	84
24	86.7
30	91.1
36	95.2
42	95.2
48	102.5
54	105.7
60	108.7
66	111.8
72	114.1

(a) Find the best relation that represents children's height as a function of time.
(b) Test whether a polynomial or potential function provides a good description of the results.
(c) Try to extend the prediction of these two models for 20 months into the future. Which description is more consistent with the expectation of a child's growth?
(d) Can you predict how tall an average child will be at the age of 10 years? How would you make that prediction? Analyze the degree of validity of your prediction.

(2) The following table shows weights and heights of Mexican boys and girls:

(a) Plot the heights and weights of girls and boys and try to linearize the graphs.
(b) Is it reasonable to extrapolate heights to 30 and 40 years? Justify your answer and compare it with what is known about human growth.

	Weight and height in girls		**Weight and height in boys**	
Age (months)	**weight (Kg)**	**length/ height (cm)**	**weight (Kg)**	**length/ height (cm)**
1	3.98	53.5	4.3	54.6
2	4.72	56.4	5.11	57.8
3	5.4	59.5	6	61.1
4	6.21	62	6.65	63.5
5	6.81	64.1	7.23	66
6	7.21	65.9	7.85	67.8
7	7.8	67.6	8.2	69.4
8	8.19	69.4	8.6	70.8
9	8.56	70.4	9.18	72.3

(Continued)

Age (months)	Weight and height in girls: weight (Kg)	Weight and height in girls: length/height (cm)	Weight and height in boys: weight (Kg)	Weight and height in boys: length/height (cm)
10	8.95	72	9.39	73.5
11	9.25	73.2	9.7	74.7
12	9.53	74.3	10.15	76.1
15	10.4	77.5	10.7	79
18	10.8	80.9	11.4	82.4
21	11.7	83.3	12	84.8
24	11.9	86.5	12.5	87.6
30	12.9	91.3	13.67	92.3
36	13.9	95.6	14.69	96.5
42	15	97.9	15.68	99.1
48	15.9	101.6	16.6	102.9
54	16.8	105	17.6	106.6
60	17.6	108.4	20.1	109.9
96		120		125
132		130		140
192		152		160
240		162		170
288		162		170
300		162		170

(3) Following is a table of the names of some dinosaurs with their estimated length and mass. From this table, what can you say about the dependence on the mass of dinosaurs as a function of their lengths? Reference: http://en.wikipedia.org/wiki/Dinosaur_size.

Name	Size (m)	Weight (kg)
Compsognathus	0.6	0.26
Compsognathus (1)	0.6	0.5
Compsognathus (2)	0.6	3.2
Juravenator	0.75	0.34
Juravenator (2)	0.75	0.34
Compsognathus (1)	1.4	3.50
Carcharodontosaurus (1)	12	6,500
Tyrannosaurus (1)	12.5	6,000
Tyrannosaurus (2)	13	13,800
Carcharodontosaurus(2)	13	13,800
Spinosaurus	14.3	7,000
Spinosaurus	18	20,900
Apatosaurus (deceptive lizard)	22.9	29,937
Argentinosaurus	30	80,000
Ultrasauros (ultra lizard)	30.2	63,350
Brachiosaurus (arm lizard)	30.5	45,360
Supersaurus (super lizard)	30.5	54,432
Diplodocus	33.5	38,000
Seismosaurus (tremor lizard)	45.7	90720

(1) and (2) correspond to data obtained from two different specimens of the same species.

(4) Kepler's laws: This table shows the mean distance to the Sun of the different planets in the Solar System and their respective periods of rotation around the Sun (length of the year).

(a) Using the data in the table, graph the duration of the planetary year or period, T_p, according to its distance to the Sun, d_S. Using linear and logarithmic scales on both axes, try to describe the dependence of T_p on d_S.

Planets	**Distance to the sun (km)**	**Period (days)**
Mercury	5.79×10^7	87.97
Venus	1.08×10^8	224.70
Earth	1.50×10^8	365.26
Mars	2.28×10^8	686.98
Jupiter	7.78×10^8	4,328.90
Saturn	1.43×10^9	10,752.90
Uranus	2.87×10^9	30,663.65
Neptune	4.50×10^9	60,152.00
Pluto	5.91×10^9	90,717.10

(b) Kepler's third law[b] postulates that T_p^2 is proportional to d_S^3. Do your results support or refute this law?

(c) In many areas of science, there are conservation laws, such as the laws of conservation of energy or momentum. These laws establish that certain quantities do not vary before or after a given process or are the same over time or when passing from one object to another in a system. In this regard, Kepler's third law establishes that the relationship $T_p^2\ /d_S^3$ is a constant for all the planets of the Solar System, i.e., this quantity is "conserved". From the data in the table, establish whether this statement is valid or not. Does your data support or refute this law?

(5) A hypothesis consistent with Newton's laws is to assume that if the planets move in circular or quasi-circular orbits, there is a force exerted by the Sun on the planets that attracts them toward the center (centripetal force). Suppose that this gravitational force has a dependence on the distance r from the Sun of the

[b]Johannes Kepler (1571–1630) was a German astronomer and mathematician known for his laws on the movement of the planets around the Sun. He was a collaborator of Tycho Brahe, whom he replaced as imperial mathematician of Rudolf II.

form $F_{\text{grav}} = k/r^n$, where n is a real number, for now unknown. As we know, for an object to move in a circular orbit of radius r and of period T, the centripetal force necessary to keep it in that orbit is

$$F_{\text{cent}} = F_{\text{grav}} \Rightarrow m\omega^2 r = m4\pi^2 r/T^2 = k/r^n, \tag{7.27}$$

where m represents the mass of the planet under study and r its distance from the Sun. From this relationship, it follows that

$$r^{n+1}/T^2 = 4\pi^2 k/m. \tag{7.28}$$

(a) Comparing this result with that of your graphical analysis in the previous problem, deduce the value of the exponent n of the gravitational force and how the constant k must depend on m so that Kepler's third law is fulfilled for all the planets of the Solar System.
(b) Compare your results with what is known about the gravitational force. What can you deduce from the conclusion drawn from Kepler's laws?
(c) What can you conclude for the n exponent in the gravitational law $F_{\text{grav}} = k/r^n$?

(6) We want to know the constant k of a spring. The measurements of hanging weights P (N) versus stretching X (m) afforded the following values:

P (N)	0	20	30	50	100	150	170
X (m)	0	0.0051	0.0071	0.0125	0.0256	0.031	0.0402

✓ Determine the best values of the slope and y-intercept of the line that best fits your data. What are the errors in these parameters?
✓ Is the value of the y-intercept significantly different from zero or consistent with this value?
✓ Estimate the value of k and its relative and absolute uncertainty.

References

[1] D. C. Baird, *Experimentation: An Introduction to Measurement Theory and Experiment Design*, 3rd Ed., NY: Pearson, 1995.

[2] P. Bevington and D. K. Robinson, *Data Reduction and Error Analysis for the Physical Sciences*, 3rd Ed., 2003, p. 230.

[3] S. L. Meyer, *Data Analysis for Scientists and Engineers*, NY: John Willey & Sons, Inc., 1975.

[4] J. Higbie, "Uncertainty in the linear regression slope," *Am. J. Phys.*, vol. 59, p. 184, 1991.

[5] G. E. Quinn, G. E. Shin, *et al.*, "Myopia and ambient lighting at night," *Nature*, vol. 399, pp. 113–114, 1999.

[6] Ohio State University, "Night lights don't lead to nearsightedness, study suggests," 2000. Availble: https://www.sciencedaily.com/releases/2000/03/000309074442.htm.

[7] K. Zadnik, *et al.*, "Myopia and ambient night-time lighting," *Nature*, vol. 404, pp. 143–144, 2000.

[8] Wikipedia, "Karl Popper," 2021. Availble: https://en.wikipedia.org/wiki/Karl_Popper.

Chapter 8

DENSITY MEASUREMENT

This chapter addresses Archimedes' principle and its connection to experiments to measure the density of different bodies. The projects illustrate the procedure of falsifying a hypothesis. Finally, an inference is made of the composition of the Earth based on the analysis of the density of elements present in its crust.

Goals

✓ Archimedean principle
✓ Falsifiability or refutability of a hypothesis
✓ Density measurement
✓ Composition of the Earth

8.1 Archimedes' principle: Determination of densities

The *density*, ρ, of a body is defined as the mass per unit volume. Similarly, specific gravity is defined as *weight per unit volume*. For a homogeneous body, i.e., when its properties are the same throughout, density is a characteristic of the substance of which it consists of. Density is a typical *intensive* quantity [1], i.e., a quantity that *does not depend on the amount of matter* that makes up the body but only on its composition. Other examples of intensive quantities are temperature, pressure, etc. Unlike intensive quantities, *extensive* quantities are those that vary proportionally to the amount of matter that makes up the body. They include mass, weight, volume, and number of molecules.

Every pure substance has a characteristic density ρ. For example, all pure gold objects have the same density ($\rho_{\mathrm{Au}} = 19.3\,\mathrm{g/cm^3}$),

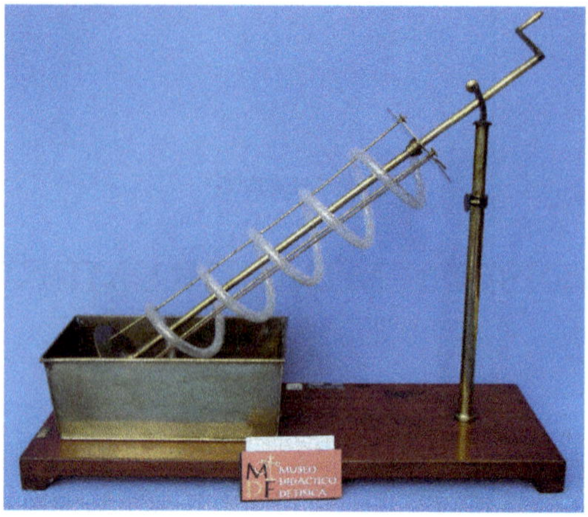

Figure 8.1 Archimedes' screw, a system for lifting water and cereals. Archimedes of Syracuse (287 BC–212 BC) was a Greek mathematician, physicist, engineer, inventor, and astronomer.

and the same goes for aluminum ($\rho_{\mathrm{Al}} = 2.7\,\mathrm{g/cm}^3$), iron ($\rho_{\mathrm{Fe}} = 7.8\,\mathrm{g/cm}^3$), and water at a given temperature ($\rho_{\mathrm{H_2O}} = 1.00 \pm 0.01\,\mathrm{g/cm}^3$, from 1°C to 35°C). This means that density is a very useful property to find out, easily and quickly, what an object is made of. According to tradition, in the third century BC, Archimedes used this property to find out whether a crown of King Hiero of Syracuse was made of solid gold. Archimedes' principle states the following: [2, 3]

> *Any body submerged in a fluid suffers an upward force (thrust) whose value is equal to the weight of the fluid dislodged by the body.*

This principle explains why corks, boats, and balloons float, and why it is easier to lift a person inside a pool full of water than outside of it. We will use this principle to measure densities.

Modern electronic scales are different from traditional two-arm balance scales: [4–6] among other things, they only have one plate. Based on Newton's third law, it is possible to use these balances to determine densities using Archimedes' principle.

If we have a glass of water on a single-pan scale, as illustrated in Figure 8.2, and immerse a body in it, the water will exert a push

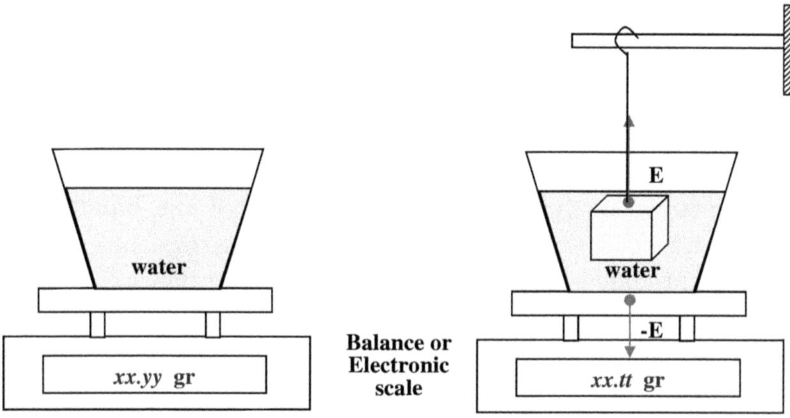

Figure 8.2 Measurement of the density of a body by the Archimedean method using a single-pan balance (standard electronic balance).

E on the body. According to Archimedes' principle, the modulus of this thrust will be

$$E = V_{\text{body}} \cdot \rho_{\text{water}} \cdot g = \frac{m_{\text{body}}}{\rho_{\text{body}}} \cdot \rho_{\text{water}} \cdot g. \tag{8.1}$$

It is useful to define the equivalent mass $m_E = V_{\text{body}} \cdot \rho_{\text{water}}$ as the mass whose weight would be equal to the thrust $E = m_E\, g$. According to Newton's third law (principle of action and reaction), if a thrust E acts on the body, it will react on the water (and the glass) with an equal and opposite force. In other words, when a body hanging from a string or support is submerged in water, as illustrated on the right side of Figure 8.2, the balance will detect an "increase in mass" (or weight) whose value is equal to m_E. To do this, place a glass of water on the scale and zero (*tare*), then slowly lower the body, hanging by a thread, until it is completely submerged in the water, without touching the bottom or sides of the glass. The mass increase recorded on the balance is m_E. Most balances are calibrated in units of mass and detect the increase in force on the pan as an increase in mass. The equivalent mass is

$$m_E = \frac{E}{g} = V_{\text{body}} \cdot \rho_{\text{water}} = \frac{m_{\text{body}}}{\rho_{\text{body}}} \cdot \rho_{\text{water}}. \tag{8.2}$$

Thus,

$$\rho_{\text{body}} = \frac{m_{\text{body}}}{m_E} \cdot \rho_{\text{water}}, \tag{8.3}$$

which enables us to determine the density of a body by weighing it twice: once in air (m_{body}), i.e., on the pan of the balance (not hanging), and again submerged in water (m_E), this time hanging as in the right panel of Figure 8.2. Knowing the density of water, ρ_{water}, Eq. (8.3) enables us to calculate ρ_{body}. The advantage of this method, in comparison to the use of the density definition $\rho_{\text{body}} = m_{\text{body}}/V_{\text{body}}$, is that it enables density to be measured without needing to know the volume of the body.

This is important for at least two reasons: (a) mass measurements are generally much more accurate than volume measurements and (b) if the shape of the body is irregular, it is difficult to determine its volume accurately. Equation (8.3) overcomes these two drawbacks and enables the density of a body to be determined easily and very precisely.

8.1.1 *Project 9. Archimedes' principle I — Falsifying a hypothesis*

Recommended equipment: An electronic balance with capacity of about 300 g (or greater) and sensitivity of 0.1 g or better. A 200 ml glass of water and a metal or plastic cylinder of about 2–3 cm in diameter and about 10 cm in height.

The goal of this activity is to experimentally test or falsify Archimedes' principle, Eq. (8.1). We could use a cylinder of constant cross-section and base area A_{base}, to which a lateral scale is attached to show the height $h_{\text{submerged}}$ of the submerged fraction. It could be a solid cylinder or a hollow cylinder containing ballast (sand). We propose to measure the thrust E or m_E as a function of the submerged volume, $V_{\text{submerged}} (= A_{\text{base}} \cdot h_{\text{submerged}})$. Submerge the cylinder progressively, hanging from a wire, as shown in Figure 8.3. Plot the measured values of m_E as a function of submerged volume

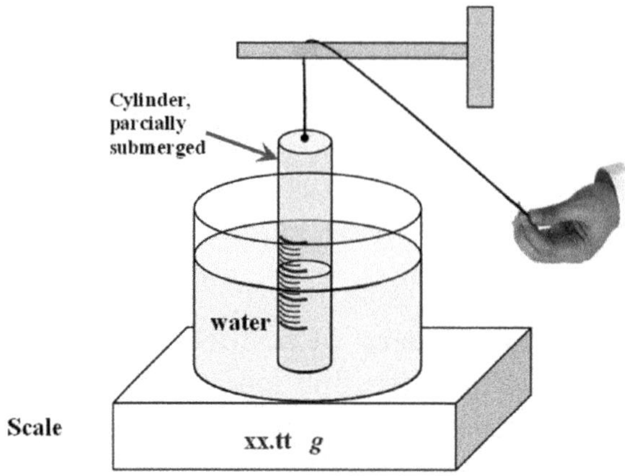

Figure 8.3 Setup to test Archimedes' principle using a single-pan scale (standard electronic scale). A cylinder of uniform section, marked with a scale, is immersed in the water up to a height $h_{\text{submerged}}$ and the thrust E — or equivalently m_E — is measured with the scale.

and compare graphically to the predictions of Archimedes' principle, i.e., Eq. (8.1). The value of m_E is given by

$$m_E = \frac{E}{g} = \rho_{\text{water}} \cdot V_{\text{submerged}} = \rho_{\text{water}} \cdot A_{\text{base}} \cdot h_{\text{submerged}}. \quad (8.4)$$

According to Eq. (8.1), m_E is expected to vary linearly with $V_{\text{submerged}}$ or with $h_{\text{submerged}}$. Therefore, according to Archimedes' principle, the data from m_E as a function of $V_{\text{submerged}}$ would be expected to have linear dependence, with slope equal to ρ_{water}.

Suggested tasks:

✓ Experimentally verify that by placing a cylinder or your finger into the water, without touching the container, the scale measurement effectively increases. If instead of a body you insert a finger, check that the force of the push is felt on the finger.

✓ From the graph of m_E as a function of $V_{\text{submerged}}$, check the linearity of your data. If the data show linear behavior, find the slope and its error (see Chapter 7). Compare the value of the slope to the density of the water. To make this comparison, it is important to use a consistent unit system, for example, the

MKS system or International System of Units (SI), commonly known as the metric system. It is also useful to know (measure) the temperature of the water since water density depends on temperature.

8.1.2 *Project 10. Archimedes' method to determine densities*

Measurement of the density of a body denser than water: Using a body denser than water, first measure its mass m_{body}. Then place a glass of water on the scale pan, where you can submerge the whole body without spilling water (try this first with the glass of water off the balance). Tare the balance with the glass of water (*tare* means to zero the balance with a given load). Put the body entirely inside the water but hanging from a support using a wire and determine the value of m_E (if your scale does not have tare, obtain the value of m_E by difference). From Eq. (8.2), the density of the body is obtained as

$$\rho_{\text{body}} = \frac{m_{\text{body}}}{m_E} \cdot \rho_{\text{water}}. \tag{8.5}$$

Suggested tasks:

✓ Determine the density of bodies whose composition you know, and which are usually fairly pure substances, e.g., aluminum, copper, and lead. Compare the results obtained experimentally with the corresponding values of these pure elements, which you can find in a periodic table or the table of physicochemical properties provided in Ref. [7]. Construct a bar graph to compare the measured density values to the corresponding values from tables. Include all the elements used in the same graph. Discuss the goodness of the proposed method for measuring densities.

✓ Construct the same bar graph but include the error bars associated with your measurements of density. Use the techniques discussed in Chapter 6.

✓ Determine the density of a ring (or a medal) that is supposed to be gold or silver. From your measurements, discuss whether the ring or medal is indeed pure gold or silver. Compare your results to values for the densities of the corresponding pure elements.

✓ The procedure followed constitutes one of the first practical problems that Archimedes supposedly solved: finding out whether or not a royal crown was made of pure gold. Archimedes did not use a balance to perform this test, he only measured volumes. Consult the literature to find out what method Archimedes used to solve this problem and compare it to the one proposed in this project.

Density of a body less dense than water: If the body is less dense than water, such as one made of wood or cork, the above method, Eq. (8.5), can also be used to determine its density. In this case, the total immersion of the body can only be done by forcing it with the help of a needle or a rigid rod. The important thing is to ensure that the fraction of the volume of this needle or rod immersed in water is negligible compared to the volume of the body in question.

Suggested tasks:

✓ Using this technique, determine the density of a piece of cork or wood or another object less dense than water. Take care not to let the object touch the sides or bottom of the glass.

8.2 Journey to the interior of the Earth

The volume of the Earth has been known since the third century BC from the measurement of its radius by Eratosthenes (see Chapter 12). From Cavendish's experiment in about 1799, we also know the mass of the Earth [2]. Using the current values of mass and volume, we can find the average density of the Earth [8–11].

8.2.1 *Project 11. Study of the density and internal composition of the Earth*

Use some of the techniques described in the previous projects to measure the density of some varieties of rocks collected from your

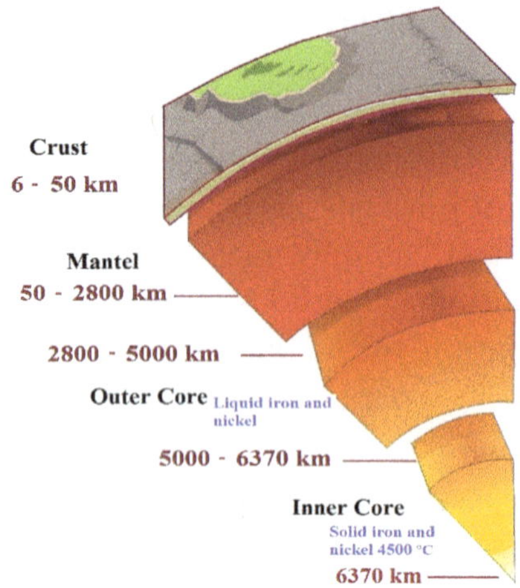

Mass	5.9736×10^{24} kg
Radius	6.3781×10^{6} m
Volume	1.087×10^{20} m^3
Density	5.5×10^{3} kg/m^3 ≈ 5.5 g/cm^3

Structure
Crust: Approximately 6 to 50 km, composed of rocks
Mantle: 2,800 km, composed mostly of solid siliceous rocks with magnesium, oxygen, iron, etc.
Outer core: 2,200 km thick, composed of molten iron and nickel
Inner core: 1,370 km, believed to be solid iron and nickel

Figure 8.4 Schematic diagram of the Earth's interior.
Source: http://pubs.usgs.gov/gip/interior/.

environment. We will take these measurements as the starting point to investigate the composition of the Earth.

Suggested tasks:

✓ Determine the density of at least two types of rocks and the sand you can find. Estimate the uncertainties in these determinations.

✓ The Earth is known to have mass $M_t = 5.98 \times 10^{24}$ kg and average radius $R_t = 6.37 \times 10^3$ km (errors in these amounts are at their last significant figure). Estimate the average density of the Earth using these data.

✓ Compare the values of the densities of the rocks and sand you measured with the average value of the Earth's density.

✓ As you may have noted, the densities of surface samples (rock and sand) are much lower than the average for Earth. Therefore, it is clear that the composition of the Earth deep inside cannot be the same as on the surface. In the interior of the Earth, there must be substances with much higher density than its surface. Using a table of densities, suggest possible elements that could make up most of the Earth's core. Hint: It may be helpful to remember that a frequent variety of meteorites that reach Earth are metallic meteorites, whose composition is iron and nickel, as you may verify by visiting some natural science museum [12].

8.3 Summary of important concepts

✓ Explain what the density of a body is and indicate at least two methods for measuring it.

✓ Why is it often not appropriate to measure mass and volume and then calculate the quotient to determine density? Could this method be used for a ring or a stone? Analyze the possible errors associated with these determinations.

✓ Explain Archimedes' principle. How do you use this principle to measure density?

✓ Some unreliable popular books, claim that the Earth is hollow. Based on your studies, how would you refute this hypothesis?

✓ What is the composition of the intermediate-size meteorites that periodically fall to Earth?

✓ Briefly discuss how the Earth's interior is composed and describe the facts that support these hypotheses.

References

[1] Wikipedia, "Intensive quantity," 2021. Available: https://en.wikipedia.org/wiki/Intensive_and_extensive_properties.

[2] R. Halliday, D. Resnick and M. Krane, *Fundamentals of Physics*, NY: John Wiley & Sons, Inc., 2002.

[3] H. D. Young, R. A. Freedman, *et al.*, *College Physics*, Vol. II, Addison-Wesley, 2009.

[4] Wikipedia, "Weighing scale," 2021. Available: https://en.wikipedia.org/wiki/Weighing_scale.

[5] D. C. Baird, *Experimentation: An Introduction to Measurement Theory and Experiment Design*, 3rd Ed., NY: Pearson, 1995.

[6] P. Bevington and D. K. Robinson, *Data Reduction and Error Analysis for the Physical Sciences*, 3rd Ed., 2003, p. 230.

[7] D. R. Lide, *CRC Handbook of Chemistry and Physics, Reference Book of Chemical and Physical Data*, 80th Ed., Springer, 1999.

[8] Wikipedia, "History of Earth," 2021. Available: https://en.wikipedia.org/wiki/History_of_Earth.

[9] "US Geological Service," 2021. Available: https://www.usgs.gov/.

[10] C. Allégre and S. Schneider, "The evolution of the Earth," *Scie. Am.*, October 1994.
[11] D. York, "The earliest history of the Earth," *Sci. Am.*, p. 60, January 1993.
[12] Natural History Museum, "Types of meteorites," 2021. Available: https://www.nhm.ac.uk/discover/types-of-meteorites.html.

MODULE II

EXPERIMENTS WITH MECHANICAL SYSTEMS

Chapter 9

GALILEO'S EXPERIMENTS USING NEW TECHNOLOGIES

This chapter discusses different alternatives to explore the basic laws of the pendulum experimentally and ways to determine the acceleration of gravity, g, by using this device. It also analyzes the characteristics of a crucial experiment: the free fall experiment. We attempt to recreate some of Galileo's experiments using Information Communication Technology (ICT). We explore the use of photogates connected to a personal computer (PC); these devices are used to measure times with great precision, which enables reconstruction of the kinematics of the movement of a body.

Goals

- ✓ Laws of the pendulum
- ✓ Discovery of empirical laws
- ✓ Determination of the acceleration of gravity
- ✓ Using photoswitches to measure times

9.1 Introductory experiments in mechanics: Photogates

This chapter begins by studying the laws of a simple pendulum. Next, it analyzes the free fall of a body, an experiment that recreates a milestone in the development of physics: the paradigm shift from Aristotelian physics [1] to the physics of Galileo and Newton. It also analyzes how the falling motion of a body may depend on its mass. Finally, it uses the experimental devices employed for studying free fall to analyze the conservation of energy. In all cases, it applies the graphical analysis techniques discussed in Chapter 2.

Photogates: A photogate (or photoswitch) is a device with a light emitter and a light detector (infrared) that generates an electrical signal (usually 5 V) which is activated or deactivated when the light beam is interrupted [2, 3] (see Figure 9.1). Photogates are connected to a port on the PC (USB or an interface) or to a smartphone via Bluetooth and through a specific program to measure the times that the interruptions last or the intervals between several successive interruptions. They can be implemented in smartphones as well. [4] These data are recorded in a file and can be used to analyze the results of an experiment. Photogates provide the following advantages:

- ✓ They eliminate observer reaction time (electronic response times are in the order of μs, much smaller than a typical reaction time of a person, of the order of 0.1 s).
- ✓ The interruption does not generate interference with the experiment (only the light is interrupted).
- ✓ The precision of time measurement is fractions of an *ms*.
- ✓ The data are accumulated and recorded on the computer for subsequent analysis.

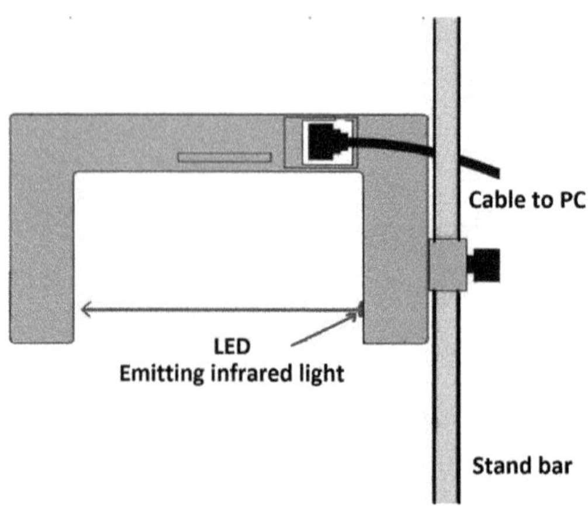

Figure 9.1 Diagram of a photogate.

✓ There are several firms that provide them and they can also be homemade [2, 3, 5].
✓ They are very low cost.

Photogates can be used to determine position, speed, acceleration, period, etc. of various mechanical systems.

9.1.1 *Project 12. Discovering the laws of the pendulum — Dependence of the period as a function of the length of the pendulum*

Recommended equipment: A bifilar pendulum of variable length, see Figure 9.1, a photoswitch connected to a computer or a stopwatch.

This activity is intended to study the properties of a simple pendulum by measuring the periods of oscillation as a function of its length and mass. We will use the techniques for graphical analysis of experimental data studied in the previous chapters.

Build a simple pendulum and investigate the dependence of the oscillation period T on its length L. One way to make the pendulum swing on a fixed plane is to build a bifilar pendulum, as illustrated in Figure 9.2.

Here, the pendulum hangs from a horizontal bar by two wires forming a V. At the lower vertex, the body that will swing (bob)

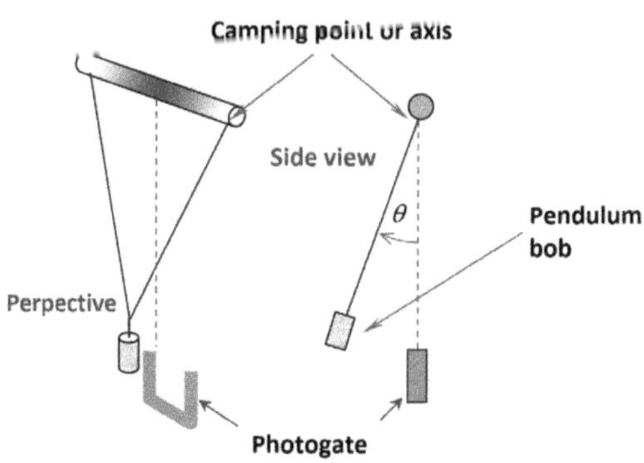

Figure 9.2 Diagram of a bifilar pendulum. θ is the amplitude of the oscillation. The photogate measures the periods.

is hung. In this way, the pendulum can only swing on a plane perpendicular to the axis of the horizontal bar. To measure the periods, it is convenient to use a photogate, as shown in Figure 9.2. If a photogate is not available, a stopwatch can be used and 10 or 15 oscillations in a row can be measured to reduce errors. The period will be the quotient between the measured time t and the number N of complete oscillations, i.e., $T = t/N$. Also, when using a photogate, it is advisable to measure several oscillations (10 or 15) and take the average as the best value and the standard deviation as the statistical error (see Chapter 5). If you use a photogate, make sure the program actually measures the period. To do this, slowly move the bob by hand and verify that you are effectively measuring the period of the pendulum. These experiments can also be carried out using a smartphone [6, 7].

Make sure that the oscillation amplitudes are "small", which in practice means that the maximum deviation angle with respect to the vertical is less than 10°. Remember that the period T is defined as the time it takes for the pendulum to travel between two successive occurrences of the same state in an oscillation (departure to one side until it returns to the same position and is moving in the same direction). More generally, we would say that T is the time it takes to reach two points with the same phase, i.e., the same angle and moving in the same direction.

Remember also that what we call the *length of the pendulum* generally does not match the length of the string. More precisely, L is the distance between the point of attachment of the pendulum (pivot) and the *center of mass* of the bob hanging from the string.

Suggested tasks:

✓ Build a bifilar pendulum of adjustable length and, if possible, use photogates to measure the periods. If the periods are measured manually with a stopwatch, measure several oscillations to reduce the errors.

✓ In each measurement, carefully measure the length of the pendulum, which is the distance from the axis of oscillation (clamping axis) to the center of mass of the bob. Sometimes the center of mass is not known or difficult to calculate, so you may temporarily assume as "measured length" L_m the distance between the top of the bob (reference point P) and the supporting bar.

✓ Graph T as a function of L_m, using linear and logarithmic scales.

✓ Also, represent T^2 as a function of L_m. Check whether it is possible to describe the trend observed in the data by the relationship $T^2 = aL_m + b$. If so, check whether the parameters a and b have statistical significance, i.e., make sure that $a/\Delta a \gg 1$ and $b/\Delta b \gg 1$ (see Chapter 7 and Appendix D). Note that if, for example, $b/\Delta b \approx 1$ or >1, then the value of b is comparable to its error and therefore we cannot state that $b \neq 0$. If b is significantly different from 0, then it is possible to define a new length $L_{\text{ef}} = L_m + \Delta L$, so that $T^2 = c\, L_{\text{ef}}$, and $L_{\text{ef}} = L + \Delta L$, with $\Delta L = b/a$. From your data, determine the values of ΔL and c. Consider that L_{ef} is the effective length of the pendulum and represents the equivalent length of an *ideal pendulum*(point mass) that would have the same period as the pendulum used, corresponding to the distance between the axis of rotation and the bob's center of mass.

✓ Using the analytical expression that best fits the values of T^2 as a function of L_m, with the least squares method, determine the value of the constants involved in the analytical expression and their respective uncertainties.

✓ Discuss the physical meaning of the effective length L_{ef} and verify that this distance best approximates the distance from the pendulum attachment point to the bob's center of mass. To do this, identify the center of mass of the bob and verify that the value $\Delta L = b/a$ approximates the distance between the adopted reference point P and the center of mass.

✓ Using the laws of mechanics, show that for a simple pendulum that oscillates with small amplitudes, the period is given by (see Annex A): [8, 9]

$$T = 2\pi \cdot \sqrt{\frac{L}{g}} \quad \text{or} \quad T^2 = \frac{4\pi^2}{g} \cdot L = c \cdot L, \tag{9.1}$$

where g is the acceleration of gravity and L the length of the pendulum ($L = L_{\text{ef}}$). Using your results from T^2 as a function of

L_{ef} and the analytical expressions that best fit the data, determine the value of g and estimate the value of its uncertainty, Δg.

Variation of the pendulum period with its mass

✓ Investigate the dependence of the period on the mass m of the pendulum. Make sure that the length of the wires remains constant and that the change in masses does not significantly affect the shape of the pendulum. Measure T for different masses m and plot $T(m)$ on a graph.

✓ Does the period of a pendulum depend on the mass of the bob?

9.2 Free fall experiment: Determination of g

In this section, we will study the free fall movement of a body using graphic analysis techniques. Times will be measured with a photogate connected to a computer. A smartphone can also be used to carry out this experiment. [10] We will analyze the characteristics of this crucial experiment.

The fall of bodies is one of the key experiments in the evolution of the scientific and philosophical thought. To understand its significance, you need to place yourself in the paradigm of Aristotelian physics, developed around 350 BC. This was the prevalent conception before the scientific revolution that occurred toward the end of the Renaissance, in which the main protagonists were Copernicus, Galileo, Kepler, and Newton (who of course worked independently at different times and in different places).

In Aristotle's conception [1], the four constituent elements of all material bodies were *fire, air, water*, and *earth*, each of which had motion properties intrinsic to its own nature. Thus, when released, a piece of earth had "natural" downward vertical motion toward the center of the Earth (which coincided with the very center of the Universe), while *fire* had "natural" ascending vertical motion. Thus, *earth* was naturally a heavy object and *fire* was naturally light. *Air* and *water* occupied an intermediate position between these extremes.

For a heavy object (Earth) to begin to move, it was necessary to apply a force to it. Even the words *animate* (with a soul) and *inanimate* (without a soul) reflect this conception. In this way, *what moves, moves through another.* The notion that characterizes the speed of movement is the time it takes to travel a given distance, which we could assimilate into our current concept of speed. At this point, it is important to note that modern concepts are not totally assimilable to those of Aristotle's time, so with this caveat, we will try to use modern mathematical language to explain Aristotle's ideas more simply. By the way, the use of mathematics to express physical laws began with Galileo.

Using an anachronism, we can say that in Aristotle's conception, the speed v acquired by a body is proportional to the applied force F and inversely proportional to the resistance R of the *medium*. That is, $v = F/R$. If a log is dragged by one horse, it will take a given time T to travel a certain distance d. If the same log is dragged by two horses (doubling the force), the same distance could be covered in about half the time, $T/2$ (the speed doubles). This law also explains why it is easier to run in air than in water. Water has more resistance than air (higher R). As can be seen, Aristotle's ideas are consistent with the physics of common sense and by no means absurd. They explain in a simple manner many phenomena that we observe in daily life. Consider, for instance, what should be done to increase the speed of a boat: clearly, we would use more rowers or the same number of rowers but stronger, i.e., we would seek to increase F. Also, to increase v, we would try to reduce the resistance R by making the boat narrower. These conceptions are still prevalent among the general public as they are very "intuitive".

Within this scheme, it was clear that the Earth should be immobile. If all its parts (any piece of Earth, for example, a stone) move naturally toward the center, it is clear that as a whole, the Earth itself must be spherical and centered on that point, the very center of the Universe. Thus, we see that within Aristotelian physics, it is not easy to transform the Earth into just another planet. Motion *is not relative* within this paradigm.

Likewise, if what moves is moved by another, and so on, successively, this cannot continue indefinitely, as Aristotle argued well. There must be an ultimate cause of the motion. This was the ideal point at which to place a divinity, and Saint Thomas Aquinas did not hesitate to use this argument to prove the existence of God.

Within the Aristotelian conception, heavier bodies are attracted to the Earth with greater force than light ones. Therefore, following Aristotle's argument ($v = F/R.$), heavier falling bodies should reach the ground faster than light ones. Galileo, in the 16th century, is said to have dropped two spheres of different masses from the Leaning Tower of Pisa and observed that they took the same time to reach the ground. He also conducted more controlled experiments using an inclined plane to study falling bodies. All his results indicated that light and heavy bodies fell in the same way. These results contributed to show a serious anomaly in the Aristotelian conception, which ended with one of the most important scientific revolutions in history.

In Galileo and Newton's conception, as long as the friction with air is negligible, all bodies fall with the same acceleration, regardless of their mass.

Experiment

This is an experiment to study the motion of a falling body by measuring the traveled distances as a function of time. To determine the times, we could use a photogate connected to a computer or a smartphone. By performing the experiment with bodies of different masses, we can compare the ideas of Aristotle and Galileo and recreate a crucial experiment in the history of physics: an experiment that contributed significantly to changing the paradigm of physics.

9.2.1 *Project 13. Study of free fall motion*

Recommended equipment: A transparent plastic or acrylic plate with evenly spaced dark stripes, called a "zebra" (see Figure 9.3). A photoswitch connected to a computer. Alternatively, a smartphone could also be used. A plastic jar hung from the bottom of the zebra and some ballast to vary its mass.

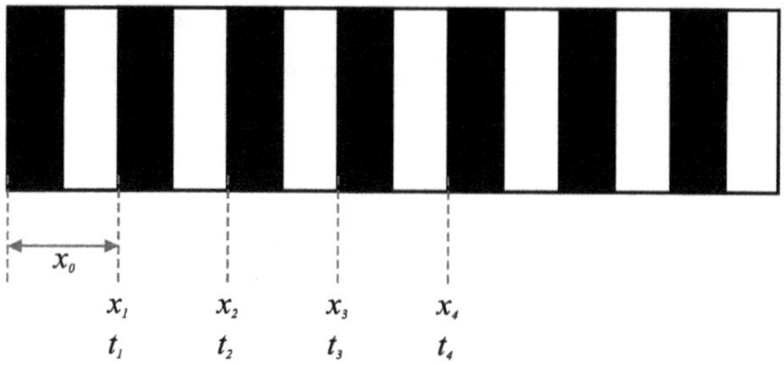

Figure 9.3 "Zebra" constructed of transparent plastic or acrylic with evenly spaced stripes. The distance (pitch) between two consecutive dark stripes is x_0.

This experiment uses a transparent plastic or acrylic plate to which are attached regularly spaced opaque stripes, as shown in Figure 9.3. When the plate is released from rest and passes between the arms of a photogate, the passage of the opaque stripes obstructs a beam of light and triggers the timing. A suitable program records the times that are associated with the distances traveled $x_1 = x_0$, $x_2 = 2x_0$, $x_3 = 3x_0$, etc.

Suggested tasks:

✓ To the best of your knowledge, describe the type of movement you expect the zebra to make as it falls.

✓ Study the characteristics of the operation of the photogate when the plate passes through it. To do this, move the plate slowly between the arms of the photogate and describe in detail what the times you are measuring are associated with.

✓ Determine the spatial period x_0 of the device and the distances x_1, x_2, all with their respective uncertainties, to be able to determine the times $t_1, t_2, \ldots$, etc., or some equivalent interval, such as $(t_2 - t_1), (t_3 - t_2), \ldots$

✓ Drop the zebra while the program measures the times t_i. Represent graphically the variation of the distance traveled by the zebra as a function of time, i.e., x_i as a function of t_i. From these data, calculate the speed of the zebra as a function of time. What can you say about the type of movement (uniform, uniformly accelerated, etc.) that it describes in its fall? Does this observation agree with your theoretical expectations? Of course, if a smartphone is used for this experiment, there is no need for the zebra.

✓ Hang weights of different values from the zebra and analyze the movement of the assembly in free fall. A simple way to do this is to hang a plastic container from the zebra, for example, an empty drinkable yogurt container. Place stones inside the container to increase its weight without modifying the shape of the falling system. Graph the *acceleration* as a function of the *mass* of the set. Do you see any correlation between these quantities? Can you conclude whether or not the acceleration depends on the mass of the falling body?

✓ From the analysis of acceleration as a function of mass, discuss with which of the paradigms your results best agree: that of Aristotle or that of Galileo and Newton.

9.2.2 *Project 14. Determination of g*

Recommended equipment: A transparent plastic or acrylic plate with evenly spaced dark stripes (zebra). A photogate connected to a computer.

This project proposes to determine the acceleration of gravity, g, and its uncertainty.

Suggested tasks:

➢ Using the same plastic zebra as in the previous activity, construct graphs of the following:

  ✓ speed as a function of time $v(t)$,
  ✓ space versus time $x(t)$.

From these graphs, $x(t)$ and $v(t)$, determine the value of the acceleration of gravity, g, and the respective uncertainties. Compare the value of g obtained from the graph $x(t)$ with the value of g obtained from the graph $v(t)$. Are these two values of g consistent? Discuss your results.

9.3 Precautions during analysis

An important aspect to take into account in this experiment is that the speed determined for each space period of the zebra is the *average speed* for this interval. The question is the choice of the time that is assigned to this speed. At the end of the nth spatial period, the plastic zebra will have fallen a distance x_n. The time it took to travel this distance, from the start of the first dark band, will be

$$t_n = t_{n-1} + \Delta t_n, \tag{9.2}$$

where $\Delta t = t_n - t_{n-1}$ is the time interval measured with the photogate that corresponds to the passage time of the nth spatial interval. This is schematically illustrated in Figure 9.4. Therefore, it is reasonable to represent x_n graphically as a function of t_n, and, from a fit of the data, by means of a second-order polynomial ($x = x_0 + v_0 t + gt^2/2$), find the acceleration g from the fit. However, *it is not correct* to do the same from the graph $v_n(t_n)$. This is because v_n is the average speed in the nth interval and must therefore be associated with an intermediate time value t_n^c defined as

$$t_n^c = \frac{t_n + t_{n-1}}{2} \tag{9.3}$$

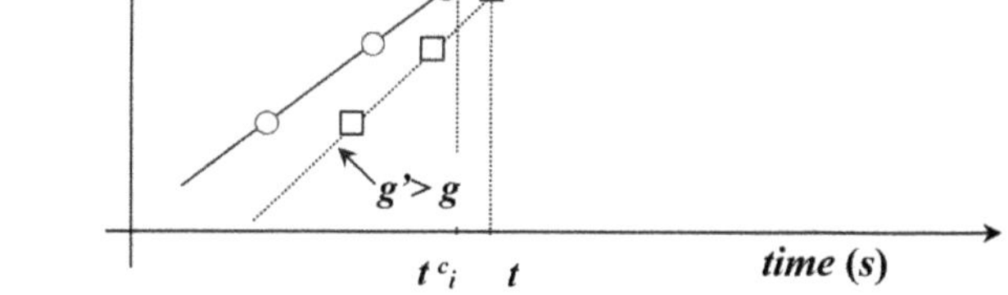

Figure 9.4 Schematic plot that illustrates the variation in the slope of the function $v(t)$ when representing v_i as a function of t_i^c (squares) and v_i as a function of t_i (circles). It is clear that for the latter case, a slope g' is greater than the best value obtained from a more adequate analysis, i.e., the slope obtained from the circles.

and not with the time t_n, which is associated with the interval in which the nth fall ends.

To try to clarify this argument, imagine for a moment that the motion of the zebra occurs at constant acceleration a. The equation that describes the speed of the zebra as a function of time is $v(t_i) = v_0 + a.t_i$. Therefore, the average velocity in the interval t_{n-1} to t_n will be

$$\begin{aligned}\bar{v}_n &= \frac{1}{2}(v_n + v_{n-1}) = \frac{1}{2}(2v_0 + a(t_n + t_{n-1})) \\ &= v_0 + a\frac{(t_n + t_{n-1})}{2} = v_0 + a \cdot t_n^c.\end{aligned}$$

In short, the graphs of $x_n(t_n)$ and $v_n(t_n^c)$ are equivalent. A more detailed discussion of this topic can be found in Refs. [11, 12].

If the time intervals were all equal, the value of the slope in the graphs of v_n as a function of t_n^c and of v_n as a function of t_n would be the same but, since the movement is not uniform, this hypothesis is not fulfilled. Using your data, compare the results of g by performing both types of analysis.

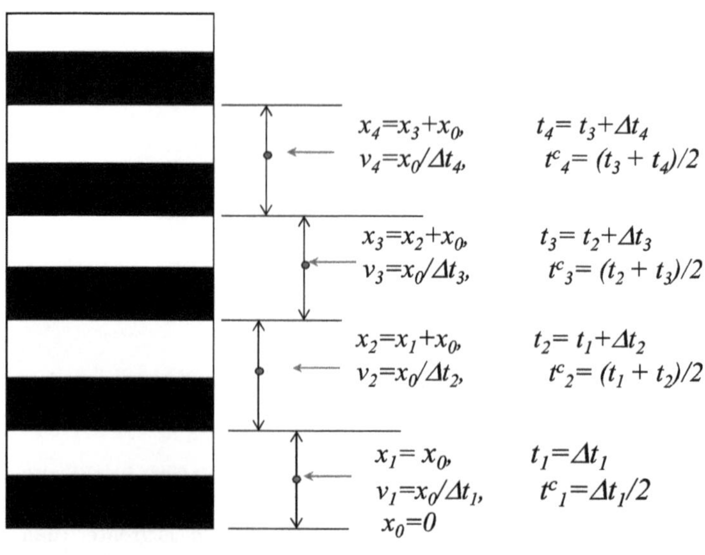

Figure 9.5 Correction scheme for the times assigned to each interval.

From your measurements, construct an acceleration graph of falling bodies of different weights using at least 4 or 5 different weights. From these plots,

✓ What can you say about the variation of the acceleration of fall, with respect to the weight of the bodies?
✓ What can you say about the fall time of bodies of different weights from a given height?
✓ Evaluate and discuss the implications of this experiment in relation to the hypotheses underlying the paradigms of Galileo and Aristotle.
✓ State your conclusions.

9.4 Evaluation questions

✓ Does the falling acceleration of a body depend on the mass of the falling object? Why does this result contradict Aristotelian physics? And the physics of Galileo and Newton?
✓ How do the values of g found with the pendulum experiment and the free fall experiment compare? Which has the least error?

9.4.1 *Project 15. Energy conservation*

Recommended equipment: A clear plastic or acrylic plate with evenly spaced dark stripes (zebra). A photoswitch connected to a computer.

This is an experiment to study the conservation of mechanical energy in the free fall of a body.

One of the most general principles of physics is the principle of conservation of energy [8, 9]. In its most general form, this conservation law is known as the first principle of thermodynamics. The objective of this experiment is to test its validity in a simple mechanical process: the free fall of a body.

The change in gravitational potential energy E_p undergone by a body of mass m when it is raised to a height Δy is given by

$$\Delta E_P = mg\Delta y, \tag{9.4}$$

where g is the value of the acceleration due to gravity. On the other hand, the kinetic energy of a body of mass m moving at speed v is given by

$$E_c = \frac{1}{2}mv^2. \tag{9.5}$$

So if a body falls from a certain height, it loses potential energy, and as the free fall movement is accelerated, the kinetic energy increases. The question we want to investigate is what happens to the combined value of both forms of energy. We designate the sum of the kinetic energy plus potential energy as mechanical energy, $E_{\text{mec}} = E_c + E_p$.

In order to perform this experiment quantitatively, we must simultaneously measure the height of the body (or its variation in height Δy) and its speed (or speed variation Δv).

9.4.2 *Exercise*

One way to simultaneously measure the velocity v and position (height y) of a falling body is to use a plastic zebra like the one shown in Figure 9.6.

The mass M of the system is made up of the zebra and the weight added to it. This value can be known by direct weighing. The distance between two consecutive dark stripes, x_0, can be known by direct measurement. Using a photogate, it is possible to measure the average speed in each section of the zebra. Also, using the measured value of x_0, we can determine the values of $y_1, y_2, \ldots, y_n$ of the zebra. As discussed in the previous section, the average speed, $\bar{v}_n$, must be associated with the coordinate $z_n = (y_n + y_{n-1})/2$.

If we assume that mechanical energy is conserved, which is the hypothesis to be tested (falsified) in this experiment, we have

$$\Delta E_{\text{mec}} = \Delta E_p + \Delta E_c = 0, \tag{9.6}$$

that is,

$$\frac{1}{2}M\bar{v}_1^2 + Mz_1 g = \frac{1}{2}M\bar{v}_n^2 + Mz_n g, \tag{9.7}$$

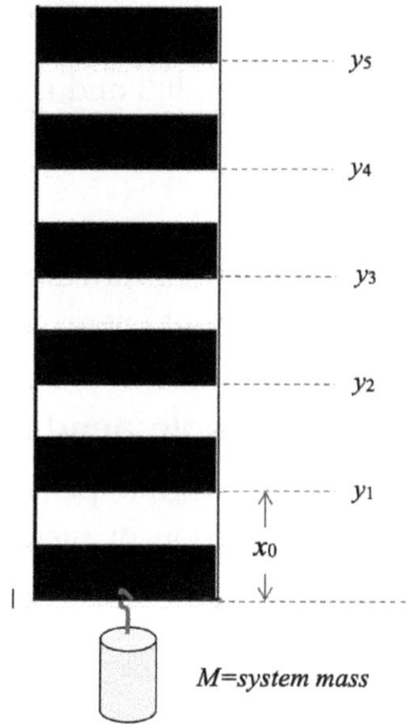

Figure 9.6 Plastic "zebra" for the free fall experiment.

where the subscript 1 refers to the first interval and n to any other. Hence (9.7) implies that

$$\frac{1}{2}\bar{v}_1^2 + z_1 g = \frac{1}{2}\bar{v}_n^2 + z_n g = A = \text{constant}. \tag{9.8}$$

Thus, the hypothesis that mechanical energy is conserved leads us to

$$\bar{v}_n^2 = A - 2gz_n. \tag{9.9}$$

If we graph the measured values $\bar{v}_n^2$ as a function of z_n, we should obtain a line with a negative slope equal to $2g$, independently of the mass of the system. This provides a criterion to falsify the hypothesis and analyze whether or not mechanical energy is conserved in the free fall process. That is, we have reduced the hypothesis of conservation of energy to testing whether or not $\bar{v}_n^2$ as a function of z_n is linear with slope $2g$.

Suggested tasks:

✓ Using the device shown in Figure 9.6 and a photogate to measure times, study the movement of the zebra from which different weights are hung.

✓ For each mass used, graph the experimental values: measured values of $\bar{v}_n^2$ as a function of z_n. From this graph, what can you conclude about the conservation of energy?

9.5 Annex A: Period of a simple pendulum

If we assume that the friction of a simple pendulum is negligible, by applying the principle of conservation of mechanical energy to the system illustrated in Figure 9.7, we have

$$E_{\text{Total}} = \frac{1}{2}mv^2 + mgh = mgh_0, \tag{9.10}$$

where it is assumed that h_0 represents the maximum height reached by the pendulum in its movement and that it is associated with the maximum angle θ_0 for $h_0 = l(1 - \cos\theta_0)$. Furthermore, we have the following relationship between the velocity of the bob v and the angular velocity:

$$\omega = d\theta/dt \quad \text{and} \quad v = ld\theta/dt.$$

From Eq. (9.10), we have

$$v^2 = l^2\left(\frac{d\theta}{dt}\right)^2 = g(h_0 - h) = gl(\cos\theta_0 - \cos\theta). \tag{9.11}$$

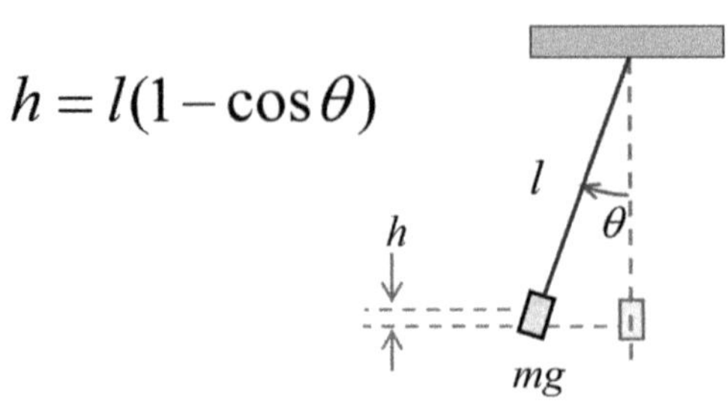

Figure 9.7 Diagram of a simple pendulum, where h is the height of the center of mass of the hanging body measured from its lowest position (position of stable equilibrium).

If we derive this equation with respect to time,

$$2l^2\left(\frac{d\theta}{\mathrm{dt}}\right)\frac{d^2\theta}{\mathrm{dt}^2} = gl(-\mathrm{sin}\theta)\left(\frac{d\theta}{\mathrm{dt}}\right), \tag{9.12}$$

that is,

$$\frac{d^2\theta}{\mathrm{dt}^2} = -\frac{g}{l}\mathrm{sin}\theta = -\omega_0^2\mathrm{sin}\theta. \tag{9.13}$$

Here we have defined the parameter $\omega_0^2 = g/l$. This is the general equation for a simple pendulum. The presence of $\sin\theta$ makes this differential equation *nonlinear*. However, if the pendulum amplitudes are *small*, i.e., if the approximation $\sin\theta \approx \theta$ is valid, then Eq. (9.13) becomes

$$\frac{d^2\theta}{dt^2} = -\omega_0^2\theta, \tag{9.14}$$

which is *linear* and says that the function $\theta(t)$ that we are looking for is a function such that differentiating it twice gives us the same function with a changed sign multiplied by the factor ω_0^2. The functions that fulfill this requirement are the sinusoidal functions. Therefore, the solution of Eq. (9.14) is

$$\theta(t) = A\cdot\sin(\omega_0 t + \varphi), \tag{9.15}$$

where A and φ are two constants that depend on the initial conditions, i.e., on the values of θ and $d\theta/dt$ at the instant $t = 0$. On the other hand, we see that after a certain time T, called the period, the elongation repeats its value, which is

$$\theta(t+T) = \theta(t), \tag{9.16}$$

$$\begin{aligned} A\sin[\omega_0(t+T)+\varphi] &= A\sin[\omega_0 t + \omega_0 T + \varphi] \\ &= A\sin[\omega_0 t + \varphi], \end{aligned} \tag{9.17}$$

it follows that $\omega_0 T = 2\pi$ and the period is given by

$$T = \frac{2\pi}{\omega_0} = 2\pi\sqrt{l/g}, \tag{9.18}$$

where we have substituted ω_0 with its value $\omega_0 = \sqrt{g/l}$. Note that, according to the laws of mechanics, for small amplitudes, the period of the simple pendulum should depend on $l^{1/2}$ and should not depend on its mass (mass of the bob), as long as the thread is inextensible. These two predictions can be tested with the proposed experiment.

References

[1] Wikipedia, "Aristotelian physics," 2021. Available: https://en.wikipedia.org/wiki/Aristotelian_physics.

[2] Vernier Software & Technology, "Photogates," 2021. Available: https://www.vernier.com/product/photogate/.

[3] C. A. Pasco, "Photogates, fences, and timers," 2021. [Online]. Available: https://www.pasco.com.

[4] K. Forinash and R. F. Wisman, "Photogate timing with a smartphone," *Phys. Teach.*, vol. 53, p. 234, 2015.

[5] IO Rodeo, "Making a photogate," 2021. Available: http://public.iorodeo.com/docs/photogate/photogate_assembly.html.

[6] P. Vogt and J. Kuhn, "Analyzing simple pendulum phenomena with a smartphone acceleration sensor," *Phys. Teach.*, vol. 50, p. 439, 2012.

[7] D. Li, L. Liu and S. Zhou, "Exploration of large pendulum oscillations and damping using a smartphone," *Phys. Teach.*, vol. 58, p. 634, 2020.

[8] D. Giancoli, *Physics Principles with Applications*, 7th Ed., NY: Pearson, 2014.

[9] J. D. Wilson, *College Physics*, 7th Ed., NY: Pearson, 2009.

[10] P. Vogt and J. Kuhn, "Analyzing free fall with asmartphone acceleration sensor," *Phys.Teach.*, vol. 50, no. 1, p. 182, 2012.

[11] W. J. Leonard, "Danger of automated data analysis," *Phys. Teach.*, vol. 35, p. 220, 1997.

[12] J. Wolbeck, "Instantaneous velocity using photogate timers," *Phys. Teach.*, vol. 48, no. 4, p. 262, 2010.

Chapter 10

THE DIGITAL CAMERA AND THE SMARTPHONE AS A MEASURING INSTRUMENT

Digital cameras and video recording equipment can be used as measurement instruments in the laboratory. This chapter proposes several simple experiments using a digital camera, whose results can be compared to relevant theoretical models. Specifically, we will study the geometric shapes of the shadows cast by a lamp on a wall and the shape of a water jet issuing from a hose. These phenomena can be studied using images taken with a digital camera, such as a smartphone camera. This chapter also analyzes the kinematics of a body in one and two dimensions, more specifically, the cases of free fall and projectile motion, using a digital camera in video mode. Subsequent chapters will use the techniques learned here to study other physical phenomena.

Goals

- ✓ Digital cameras as measuring instruments
- ✓ Study of shadow shapes
- ✓ Shapes of the paths of water jets
- ✓ Air friction force
- ✓ Study of ballistic body without friction
- ✓ Ballistic body with friction

10.1 The digital camera

A digital image is made up of a matrix of "dots" or elementary cells. Each cell is called a *pixel* (from the contraction of the words picture and cell) and is the smallest unit of a digital image. Each pixel (px) stores a number that represents its color. Pixels that store only one bit (i.e., 0 or 1) can be used to form black and white images, as illustrated in Figure 10.1.

The color of each pixel can also be encoded with an 8-bit (byte) number that supports $2^8 = 254$ color variations. Similarly, there are images in which each pixel can have 3 bytes of colors (24 bits), i.e., $2^{24} = 16777216$ color options, etc. Increasing the number of colors per pixel will increase the nuances or shades in an image and also image size, i.e., the space it occupies in bytes.

The resolution of a digital image depends on the number of pixels in the image. A 480×640 *pixel* image equals 307200 *pixels* ($\approx$307 kpx). A digital image of $2048 \times 1536 = 3145728$ equates to a resolution of 3.1 Mpx. This notation is very common in digital camera specifications. As the resolution of the image improves and the number of colors per pixel increases, the quality of the photogram is higher and the space it occupies grows. Actually, this brief description applies to BMP-type digital files. There are also many other compressed formats which allow a large amount of information to be stored in smaller files, such as *jpg* and *tif.* They follow protocols other than those mentioned above but equivalent to them [1–3].

1	1	1	1	1	1	1	1	1	1
1	1	1	1	1	1	1	1	1	1
1	1	1	0	0	0	0	1	1	1
1	1	1	0	1	1	1	1	1	1
1	1	1	0	1	1	1	1	1	1
1	1	1	0	0	0	0	1	1	1
1	1	1	0	1	1	1	1	1	1
1	1	1	0	1	1	1	1	1	1
1	1	1	0	0	0	0	1	1	1
1	1	1	1	1	1	1	1	1	1
1	1	1	1	1	1	1	1	1	1

Figure 10.1 10×11 pixel digital image, monochrome, each pixel stores 1 bit, $1 =$ white, $0 =$ black.

The pixels in a digital image can be easily observed by enlarging the image sufficiently using the *zoom* option. Pixels appear as small squares of color, or in black and white, or in shades of gray. The images are formed as a rectangular array of pixels. The origin of the *pixel* coordinates generally coincides with the upper left corner of the frame, as illustrated in Figure 10.2. The *pixel* coordinates of a frame can be read using almost any image viewing program, such as Microsoft PhotEd®, Microsoft Paint®, or Corel Draw®, among others. The coordinates of the mouse position can be read in these programs as a pair of numbers at the bottom of the image. See Figure 10.1. Thus, by placing the mouse in the lower right corner of the image, we can find out the resolution in pixels of the image.

The experiments proposed here can be performed with an ordinary commercial digital camera (with a resolution of 480×640 pixels or better), a webcam, or the camera of almost any smartphone. A video-capable digital camera or camcorder with a resolution of 480×640 pixels that can record at least 25 *frames per second* (25 fps)

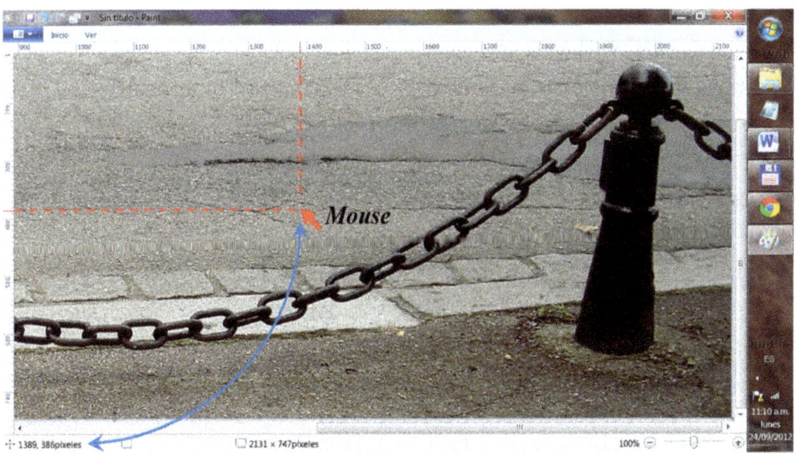

Figure 10.2 Digital photo of a chain. The actual height in meters of the post allows the pixel coordinates of the photo to be transformed into metric dimensions. Each point in a digital photo (such as at the point indicated by the arrow) has well-defined pixel coordinates, which are shown in the lower-left corner of the image (1389, 386). At the bottom, you can see the image size in pixels (2131×747).

may be suitable for studying the kinematics of objects in two dimensions. A video recorded with a smartphone camera can also work very well.

If a digital image is viewed on a computer, the coordinates in pixels of the position of the mouse pointer can be obtained [3–5], as illustrated in Figure 10.2.

If an object of known dimension is inserted in the image, it is possible to transform the coordinates in pixels into conventional coordinates, i.e., in *mm*, *cm*, and *m*. Of course, this procedure can be done in a traditional manner by noting the coordinates in *pixels* of the object to be measured and then using the coordinates in *pixels* of the object inserted in the frame as a calibration reference. Fortunately, there are multiple programs to transform the coordinates in *pixels* to standard units of length, very simply. Once the "calibration" process of the digital photo has been completed, the coordinates of any object in the frame can be obtained in real coordinates. Some programs that can do this are xyExtract Graph Digitizer [6], Tracker [7], LoggerPro from the Vernier company [8], etc.

Another alternative to quantitatively study a phenomenon recorded in a digital photo consists of cropping the frame in well-known, real dimensions. This can be achieved by placing a grid in the background of the photo as illustrated in Figure 10.3. These crosslines allow the image to be cropped to a size of known dimensions. When the image is cropped in this way, it is entered as the background in the chart plot area of a spreadsheet, the coordinate dimensions of the chart are chosen to match the actual dimensions of the photo, as shown in Figure 10.3, and the image is automatically calibrated in standard units. In this way, theoretical curves can be drawn that can be compared and superimposed on the phenomenon recorded in the photo. This technique is used in many cases [4, 5, 9], and some examples are available on the Internet [10].

Digital images are particularly useful for studying phenomena in which the most significant features of the problem are contained in a plane, for example, the shape of a shadow, the shape of the path of a water jet, or the shape of a hanging chain, as shown in Figure 10.3.

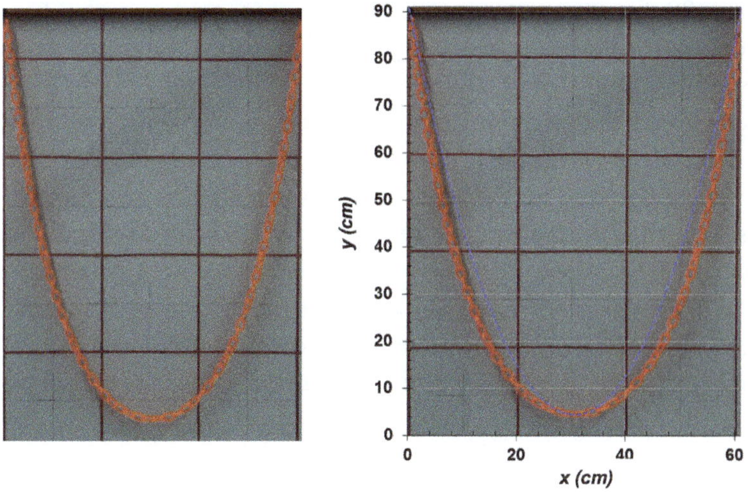

Figure 10.3 On the left is the image of a hanging chain. A 0.2 m × 0.2 m grid (solid red lines) was placed in the background to "calibrate" the dimensions of the photo. This grid also allows the frame to be cropped so that it has known dimensions −0.6 m × 0.9 m in this case. The image on the right shows how the frame was embedded in the background of a standard graph of a spreadsheet. The grid in dotted white lines and the curve in dashed blue line belong to the spreadsheet graph and can be easily compared to the image of the chain.

To avoid possible distortions produced by the perspective of a photo, it is desirable and advisable that the phenomenon or object under study should be contained in a plane. Also, the object or scale that will be used as a calibration must be in the same plane or very close to it. The camera should be located on a line perpendicular to the main plane of the image and that passes near the center of the image. Of course, the main plane under study and the lens plane should be as parallel as possible to avoid perspective distortion effects. It is also suggested that the distance to the object to be photographed should be greater than its dimensions.

In the case of a moving projectile, for example, a ball or a balloon, the digital camera should be used in video mode, and preferentially mounted on a fixed tripod. In a digital video, each video frame is a digital image that can be treated as discussed above. As the frames of a video are taken at fixed time intervals, for example, at a rate of 25 or 30 frames per second, each frame contains the time, in addition to the coordinates (x, y) of the frame. In other words, it is possible to reconstruct the trajectory of a moving body from the video. Again, there are multiple programs that greatly facilitate this task [7, 8]. In general, each digital image or video analysis software contains a tutorial. There are also cameras in which the time between frames, i.e., the fps (frames per second), can be selected. Some camera models can be connected to an "*intervalometer*" or a *time lapser* or self-timer, which is an external device that regulates the time and duration of each frame, and there are even smartphone apps that can do this. Thus, it is possible to program, for example, 10 shots per hour of a certain duration each.

10.2 Geometric shapes formed by the shadow of a lamp

The shadows cast by a lamp on a wall or screen can present regular and interesting patterns to study, as illustrated in Figure 10.4. The light emerging from a lamp forms a cone, with the vertex at the position of the light bulb (or light source) and an angular opening defined by the edge of the lampshade. Depending on the orientation

Figure 10.4 Digital photo of the shadow cast by a lamp on a wall. The real size of the diameter of the upper opening of the lampshade enables the coordinates in pixels of the digital photo to be transformed into real coordinates.

of the axis of the light cone with respect to the wall, we expect to observe a shadow with conical shapes: hyperbolas, parabolas, or others. It is simple to test this expectation experimentally using a digital photograph [11].

Once the coordinates of the shadow are known, it is easy to compare the results of the observed shadow to the corresponding theoretical predictions. Figure 10.4 shows the case in which the axis of the cone is parallel to the wall. In this case, we expect the shadow to form a hyperbola of the following equation:

$$\frac{(y-y_0)^2}{b^2} - \frac{(x-x_0)^2}{a^2} = 1, \tag{10.1}$$

where (x_0, y_0) are the coordinates of the center or origin of the hyperbola (coinciding with the filament of the light bulb) and a and b are the semi-lengths of the major and minor axes of the hyperbola, respectively. The adjustable parameters of the model are x_0, y_0, a, and b. The equations of the asymptotes are

$$\frac{(y-y_0)}{b} \pm \frac{(x-x_0)}{a} = 0. \tag{10.2}$$

We can also analyze what happens if the *generatrix* of the light cone is parallel to the wall. We can create this condition by tilting the lamp so that the edge of the shadow on the ceiling is exactly above the lamp. In this case, the shape of the shadow on the wall would be parabolic:

$$y - y_0 = c \cdot (x - x_o)^2 + k, \tag{10.3}$$

where (x_0, y_0) are the coordinates of the vertex of the parabola and c and k are two characteristic constants and are the adjustable parameters of the model.

10.2.1 *Project 16. Study of the shadow of a lamp*

Recommended basic equipment: A 480 × 640 *pixel* resolution digital camera or better. A lamp next to a wall or screen.

Place the lamp next to the wall or screen, making sure it casts a sharp shadow. Place an object of known size on the wall or screen (graduated ruler) and make sure that the axis of the cone of light thrown by the lamp is as vertical as possible, parallel to the wall. If possible, use a tripod to take the photos or make sure to hold the camera steady and that the vertical lines remain vertical in the photo.

Suggested tasks:

- ✓ With these precautions, evaluate whether the figures cast by the shadows can be explained by quadratic or conical shapes.
- ✓ Draw the observed data of the shadow with the theoretical expectations on the same graph and evaluate whether the theoretical expectations are met. What can you conclude from your analysis?

10.2.2 *Project 17. Path of a water jet*

> **Recommended equipment**: A hose connected to a water faucet, with a nozzle that allows us to choose a well-defined angle of departure of the jet, and a background grid with a sharp, clear scale near the plane of the water jet. A 480 × 640 pixel resolution digital camera or better.

One way to describe the shape of a water jet like the one in Figure 10.5 is to assume that the water particles follow the same path as a solid projectile, which is simple and usually discussed in most basic physics texts [12, 13]. The time evolution for the x (horizontal) and y(vertical) coordinates can be written as follows:

$$x(t) = x_0 + v_0 \cos\theta_0 \cdot t,$$

$$y(t) = y_0 + v_0 \sin\theta_0 \cdot t - \frac{1}{2} g \cdot t^2, \tag{10.4}$$

where (x_0, y_0) are the coordinates at $t = 0$ and v_0 is the modulus of the initial velocity of the fired projectile (water jet) that forms an angle θ_0 with the horizontal (axis x). Eliminating time t from

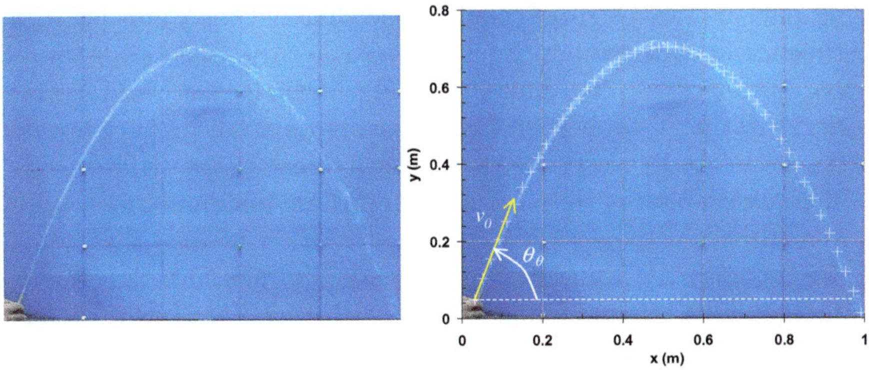

Figure 10.5 On the left is the image of a water jet issuing from the lower left corner of the frame. A 0.2 m × 0.2 m grid was placed in the background. This grid is used to "calibrate" the dimensions of the photo and allows the frame to be cut so that it has known dimensions, 0.8 m × 1 m in this case. The image on the right shows how the frame was embedded in the background of a standard spreadsheet graph. The grid in dotted lines and crosses belongs to the spreadsheet graph and can be easily compared to the water jet image. The crosses are predictions of the theoretical model.

Eq. (10.4), we obtain the expression of the trajectory:

$$y(x) = y_0 + \tan\theta_0(x - x_0) - \frac{g}{2 \cdot v_0^2 \cos^2\theta_0}(x - x_0)^2. \qquad (10.5)$$

This theoretical expression can be directly compared with the real water jet trajectory recorded in the digital photo [4, 9].

Choose lighting conditions (artificial or natural) that make the water jet (as well as the reference scale) clearly visible in the digital image. Use the faucet to regulate the water flow so that the entire jet path fits into the frame of the photograph for different exit angles. Once the faucet setting has been chosen, keep it the same for all the shots in the experiment, which is equivalent to maintaining v_0 constant.

Suggested tasks:

✓ Photograph the water jet for different angles of departure, θ_0, keeping the flow (speed v_0) of the jet outlet constant.

✓ Using a graphic analysis program (for example, xyExtract [6], or the method of introducing an image of known dimensions into the graphic area of a figure) compare the shape of the water jet with theoretical expectations, Eq. (10.5). To do this, use as adjustment parameters θ_0, v_0, x_0, and y_0, trying to achieve the best possible fit. Perform this operation for all the angles studied. The Physics Experiments Resources on www.fisicarecreativa.com [10] provide some examples of how to perform these manipulations with spreadsheets.

✓ To estimate the value of v_0 independently, measure the time it takes to fill a bucket of approximately 5 l. Weigh the bucket before and after filling and determine the volume of water loaded. Using the measured filling time, determine the outgoing water flow, Q_0. Estimate the cross-section A_0 of the hose nozzle. With these data, find an approximate value of $v_0 = Q_0/A_0$ and compare it to the value found from the water jet settings.

✓ One of the hypotheses on which this analysis was based is that the drops of water and the jet of water follow the same laws as solid material objects (particles). Based on your observations, what can you conclude regarding this hypothesis?

10.3 Viscous friction force of air

We propose to study the characteristics of the friction force that air exerts on different objects. The *friction force* or *air resistance* on falling objects is important because it is part of our daily experience. Falling leaves or raindrops do not appear to have acceleration nor do balloons inflated with air [14, 15], etc. This experiment proposes to use air-inflated party balloons with attached weights (e.g., paper clips). Another option is to use stacked (basket-shaped) paper coffee filters. By varying the number of stacked filters, their weight changes but the cross-section remains constant [14]. There are several programs that can be very useful to make this type of analysis very straightforward; in fact, most programs can directly provide the *fps* of the video [7, 8, 16].

Theoretical considerations: When an object falls, the forces exerted on it are the air thrust (Archimedes' buoyancy force E),

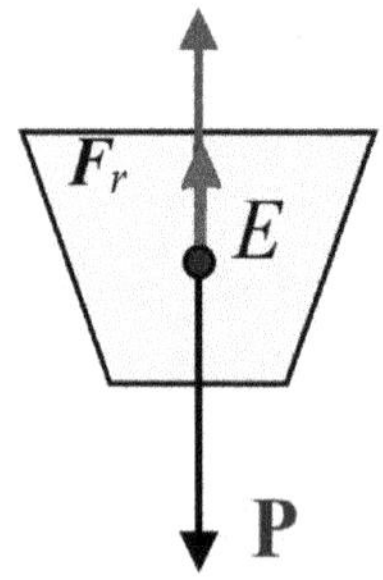

Figure 10.6 Free-body diagram of a coffee filter when it falls through the air.

the friction force with the air (F_{frict}), and its weight (P). If we use a positive downward-oriented frame of reference from Newton's second law, we have

$$P - E - F_{\text{frit}}(v) = m\frac{dv}{dt}. \tag{10.6}$$

The effective weight of the object is defined as $P_{\text{ef}} = P - E = mg$. (Here, if we use an inflated balloon, m is the effective mass of the balloon, weighed directly on a scale when the balloon is inflated.)

Our hypothesis consists of assuming that the friction force depends on the speed v raised to an unknown power n, i.e., $F_{\text{frict}} \propto v^n$. Expression (10.6) shows that, as the body accelerates, the friction force increases until it equals the effective weight. From that moment on, the object moves with constant speed, i.e., it reaches a limit speed v_{fin}. According to Eq. (10.6), we have

$$F_{\text{fict}}(v) = kv_{\text{fin}}^n = P_{\text{ef}} = mg, \tag{10.7}$$

where k is a constant of proportionality. By stacking coffee filters inside each other, or hanging different numbers of paper clips from a balloon, it is possible to vary the effective weight, m, without significantly changing the shape. By measuring v_{fin} for different groups of coffee filters or paper clips (several P_{ef} or m) and plotting P_{ef} as a function of v_{fin}, it is possible to verify whether Eq. (10.7) is fulfilled and, from these graphs, to find n.

10.3.1 *Project 18. ♣♣ Using video to study the kinematics of a body — Viscous friction force in air*

Recommended equipment: Several basket-shaped coffee filters or a party balloon with several paper clips that will be used to vary its weight. A digital camera that records video at 480 × 640 pixel resolution or better or a camcorder with about 25 fps.

The proposed experimental method uses a digital camera or smartphone in video mode or camcorder. To obtain a calibration of the pixels on a real scale, place an object of known dimensions close to the fall line of the bodies (coffee filters or balloons). Weigh the filters or inflated balloons (measure P_{ef}) before each video recording.

The video can be analyzed using any video analysis software, such as LoggerPro® Vernier Software [8], and Tracker [7]. Any of these programs can be used to find the vertical position y and horizontal displacement x at each instant of time t, i.e., the triplet (t, x, y). Always avoid air drafts so that $x \approx$ *constant*.

The values of (t, x, y) are used to construct graphs of y and v_y as a function of time t. In general, one would expect graphs of y as a function of t as shown in Figure 10.7, in which, after a certain time,

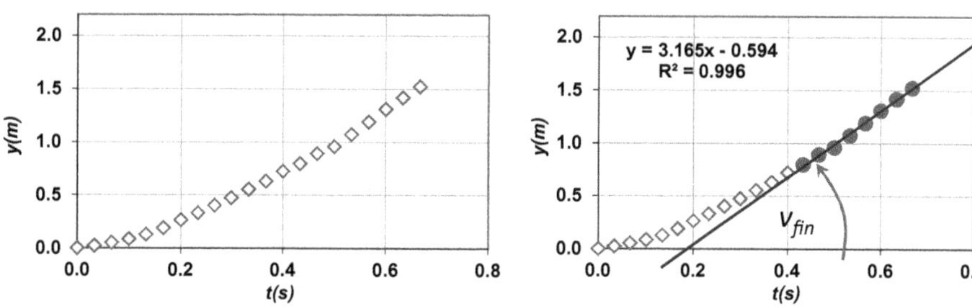

Figure 10.7 Graph of y as a function of time t. After a certain time, the data show linear behavior, indicating that a speed limit has been reached. The two graphs contain the same information, but in the one on the right, aligned data have been marked with filled circles. From the slope in this range, we can find v_{fin}.

a linear trend indicatives that a limit speed is reached. From this slope, we can find the value of v_{fin} for the different weights.

Once the limiting velocities v_{fin} for the different weights P_{ef} have been found, plot P_{ef} as a function of v_{fin}. If this graph, on a log-log scale, is linearized, it means that the relationship between these magnitudes is a power law, corroborating the hypothesis expressed in Eq. (10.7). From the fit of these data with a potential function, we can find the parameters k and n, following the procedure discussed in Chapter 2. The uncertainty of the exponent n can be calculated by graphing $\log(P_{\text{ef}})$ as a function of $\log(v_{\text{fin}})$. The value of n indicates the type of motion of the falling body.

The Reynolds number is defined as [17]

$$\text{Re} = \frac{\rho_m \cdot v \cdot d}{\eta_m}, \tag{10.8}$$

where ρ_m is the density of the medium (air), η_m its viscosity, v is the speed of the object, and d represents its characteristic dimension, in our case, the diameter of the balloons or coffee filters. If $\text{Re} < 2000$, the movement of the fluid around the object follows a *laminar regime* and the friction force is expected to be *proportional* to the velocity ($n = 1$). If $Re > 4{,}000$, the movement of the fluid around the object follows a *turbulent regime* and the friction force is expected to be *proportional to the square* of the velocity ($n = 2$). Annex A presents a heuristic argument that justifies the latter result. In Annex B, the equation of motion of a body that falls into a medium with turbulent friction is solved.

Suggested tasks:

- ✓ Record (film) the fall of the selected object with the different weights.
- ✓ For each weight, P_{ef}, find the speed limit v_{fin}.
- ✓ From the graph of the limiting velocity v_{fin} as a function of P_{ef}, find the parameters of the expression (10.7), i.e., k and n. In particular, if possible, estimate them with their corresponding error.
- ✓ Discuss the dependence of friction force on limiting speed. What expression does this force follow as a function of velocity?

✓ Calculate the value of Re when the body reaches the velocity limit and compare the results obtained with what is expected according to the movement of the fluid around the object. With what type of regime (laminar or turbulent) does your object move in the air?

✓ **Optional**: Using the theoretical expression of y as a function of t, discussed in Annex B, compare your experimental results to those obtained theoretically. Does the proposed model, Eq. (10.27), explain the results of $y(t)$?

10.3.2 *Project 19.* ♣♣ *Study of projectile motion*

Recommended equipment: A wooden sphere about 2 cm in diameter. A ping-pong ball. A projectile launcher. A digital camera that records 480 × 640 pixel resolution video or a camcorder, with around 25 fps.

This activity proposes to use a digital camera in video mode to study the kinematics of a two-dimensional moving body to study projectile motion in two different situations: first, when the friction effects with the air are negligible and second, when they are relevant. The first case is relatively simple, and the theoretical model is the same as described in the previous activity. The second case — when friction effects are important — is more challenging because it involves numerically integrating the equations of motion.

A projectile launcher such as any of the ones illustrated in Figure 10.8 or some similar device can be constructed to fire the projectiles. The device in Figure 10.8(c) consists of two tubes, one inside the other. The tube with a larger diameter has a small cross-shaped cut. An elastic band passes through it to hold the ball before the shot. Another rubber band makes the second tube, which functions as a plunger, hit the ball when released [18, 19].

Film the launch, and by observing the video frame by frame, measure the position (x, y) of the object as a function of time. The objective is to find, for each launch, the experimental motion equations $x(t)$ and $y(t)$, as well as the experimental determination of the trajectory, $y(x)$.

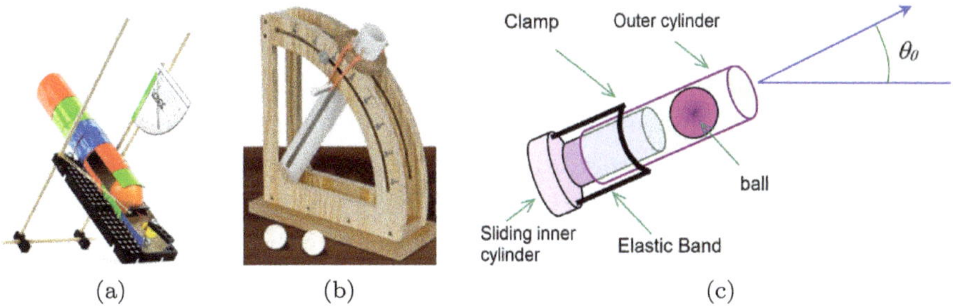

Figure 10.8 Three types of projectile launchers. There are serval possible models [14, 18, 19].

In the first case (negligible friction), you can use a wooden ball 1–2 cm in diameter or the ball that comes in some "roll-on" deodorants. In the second case, the projectile may be a ping-pong ball. The fundamental difference between these two projectiles is their mass and, consequently, their initial kinetic energies. The magnitude of these initial energies compared to the energy losses with air determines the type of trajectory of each body. For heavy balls, the energy losses due to friction with the air are generally negligible compared to the initial kinetic energy. For light balls, the opposite is true. It can be observed experimentally that in the launch of the heavy ball, with an approximate range of 1.50 m, its movement can be studied assuming the effects of friction with the air to be negligible since its trajectory is described very well by a parabola (Eq. 10.5). On the other hand, when launching a light ball with a range of approximately 4 m, it is necessary to include the effects of friction to describe its trajectory adequately, since this time the trajectory is not symmetrical and cannot be described by a parabola.

When the friction force with air is important, the movement is described theoretically through the numerical resolution of integrals that do not have a simple analytical solution. The solution consists of numerically integrating the equations of motion for small time intervals, as explained in the following:

Experiment 1: Case in which friction is negligible: In this case, film the launch of projectiles for different angles θ_0.

Suggested tasks:

✓ Determine the experimental path $y(x)$.

✓ On the same graph, draw the experimental data of the trajectory and the theoretical expectations (Eq. 10.5). By varying the values of the parameters of the theoretical expression, θ_0, v_0, x_0, and y_0, try to achieve the best possible fit. Perform this operation for all the angles studied.

✓ Find the experimental motion equations $x(t)$ and $y(t)$. Graph your results, including in the same figure the theoretical expectations, Eq. (10.5), with the parameters found previously.

✓ What can you conclude about the proposed theoretical model to explain your data?

♣♣ **Experiment 2: Case in which friction is appreciable**: When the friction force of air is important and cannot be neglected, it is generally not possible to find an analytical expression for the trajectory of the body.

Numerical integration of the motion equations: When the body moves within a fluid, the direction of the friction or drag force is opposite to the speed. The characteristic of this drag force depends on the shape of the body and the speed at which it moves through the medium. At low speeds, the fluid motion regime is laminar and the drag force is dominated by viscous effects that are proportional to the body velocity. At higher speeds [12, 17], the fluid motion regime is turbulent and the drag force is dominated by inertial effects. In this case, the dependence of this force on the velocity is quadratic. If Re is less than 2000, the flow around the body is essentially laminar and the friction force is proportional to the speed, i.e.,

$$F_r = k_L v. \tag{10.9}$$

The magnitude k_L depends on the size and shape of the object and the viscosity of the fluid. In particular, if the body is spherical and $R_e \leq 1$, friction force is given by Stokes' formula:

$$F_r = 6\pi r \eta v. \tag{10.10}$$

At high Reynolds numbers, $Re > 4000$, the fluid regime is turbulent. In these cases, a reasonable approximation of the friction force is given by [14, 15, 17]

$$F_r = \frac{1}{2} C_d \rho_{\text{air}} A v^2 = k \cdot v^2, \tag{10.11}$$

where C_d is a numerical coefficient whose value depends on Re and the shape of the object, A is the cross-sectional area of the body (πr^2 for a sphere), ρ_{air} is the density of the fluid, and v is the velocity of the projectile. These considerations should be taken into account when discussing pendulum, ping-pong ball, or soccer ball problems, for which air friction rarely has a laminar regime. In other words, the friction force with the air, in general, is not proportional to the speed, $F_r = kv$.

The equations of motion of a body that includes a friction force proportional to v^2 do not generally have a simple analytical resolution. But it is possible to find a numerical solution to the equation of motion. Choosing the vertical y-axis direction (Figure 10.9), we have

$$F_x = m\frac{dv_x}{dt} = -kv_x|v| = -kv_x\sqrt{v_x^2 + v_y^2}, \tag{10.12}$$

$$F_y = m\frac{dv_y}{dt} = -mg - kv_y|v| = -mg - kv_y\sqrt{v_x^2 + v_y^2}. \tag{10.13}$$

Here mg is the weight of the body and we have assumed that the friction force is given by $F_r = kv^2$.

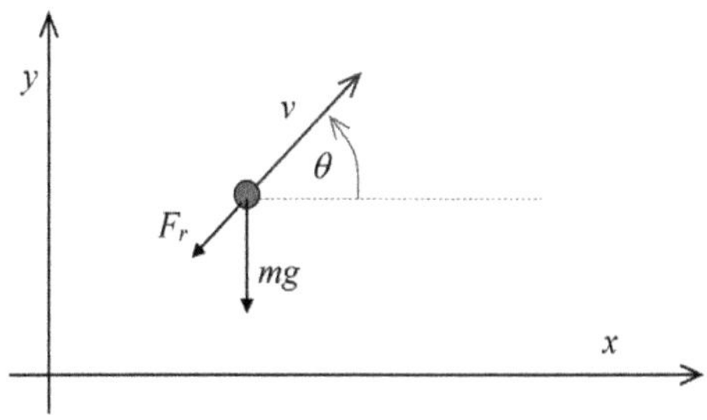

Figure 10.9 Object moving with velocity v. The weight is mg and the friction force F_r has the opposite direction with respect to the speed v.

If θ is the angle that the velocity v makes with the horizontal direction, we have

$$F_x = -F_r \cos\theta = -F_r \frac{v_x}{v} \tag{10.14}$$

and

$$F_y = -F_r \sin\theta - mg = -F_r \frac{v_y}{v} mg. \tag{10.15}$$

Thus,

$$a_x = -\frac{k}{m}\sqrt{v_x^2 + v_y^2} \cdot v_x \tag{10.16}$$

and

$$a_y = -\frac{k}{m}\sqrt{v_x^2 + v_y^2} \cdot v_y - g. \tag{10.17}$$

The approximation used is to assume that in a small Δt, the body moves as follows:

$$\Delta x = v_x \Delta t \quad \text{and} \quad \Delta y = v_y \Delta t. \tag{10.18}$$

Using Euler's approximation [20], an acceptable approximation of the data can be obtained with $\Delta t \approx 0.01$ s, although a better approximation would be obtained using the Runge–Kutta method [21]. From expressions (10.16) and (10.17), the values of v_y and v_x for each Δt are

$$v_x(t + \Delta t) = v_x(t) + a_x(t)\Delta t, \tag{10.19}$$

$$v_y(t + \Delta t) = v_y(t) + a_y(t)\Delta t. \tag{10.20}$$

By iterating this procedure, it is possible to find $v_x(t), v_y(t)$, $x(t)$, and $y(t)$. The algorithms described by Eqs. (10.18)–(10.20) can be easily implemented in any spreadsheet [14]. The Physics Experiments Resources on www.fisicarecreativa.com [10] provide examples of how to perform these algorithms with spreadsheets.

Experiment The experimental method is identical to that used in the case of the heavy ball, discussed previously, except that here, a ping-pong ball is used. Film the movement with a digital camera in video mode, following the same recommendations as in the previous activity.

Suggested tasks:

✓ Film the balls launched at different angles θ_0.
✓ Determine the experimental trajectory $y(x)$.
✓ On the same graph, draw the experimental data of the trajectory and the theoretical expectations, Eqs. (10.16)–(10.18). By varying the values of the parameters of the theoretical expression, θ_0, v_0, x_0, and y_0, try to achieve the best possible fit. Perform this operation for all the angles studied.
✓ Find the experimental motion equations $x(t)$ and $y(t)$. Graph your results, including the theoretical expectations Eqs. (10.16)–(10.18) with the parameters found previously.
✓ What can you conclude about the proposed theoretical model to explain your data?

10.4 Annex A: Laminar and turbulent regimes

An important characteristic of real fluids is that they have viscosity [12, 17]. A notable consequence of this property of fluids is that a force is required to move an object in a fluid.

In fluid mechanics, two regimes of fluid motion are identified: laminar flow and turbulent flow.

In *laminar flow*, the fluid particles move in an orderly manner, following stable trajectories and approximately parallel to each other. This is the regime observed in a slow-flowing channel. See Figure 10.10. In the *turbulent regime*, the fluid particles move in a disorderly way, changing their path irregularly or in vortices, so that the velocity of the fluid at a given point in space seems to vary constantly over time. These two regimes can be easily observed in incense or cigarette smoke. At first, the flow is laminar, but as the smoke rises, the flow gradually becomes turbulent. It is observed that the transition between these two regimes is largely associated with the value of a dimensionless quantity, known as the *Reynolds number* (Re).

Figure 10.10 Transition from laminar to turbulent flow. Upstream, the flow is approximately laminar, in the constriction produced by the rocks, the flow of the river becomes turbulent.

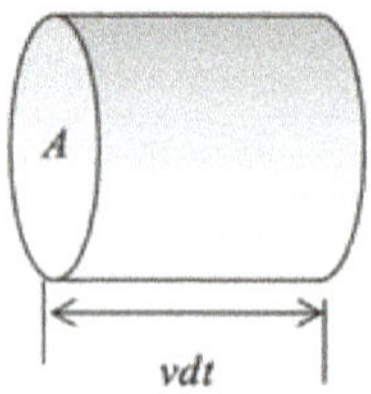

Figure 10.11 Volume swept by a body of cross-sectional area A moving in a fluid medium with velocity v.

Imagine a sphere of diameter d, moving in a fluid with speed v. The cross-sectional area of the sphere is $\pi d^2/4$.

As the fluid impacts the sphere, there is a momentum change in the fluid particles, which exert a force on the sphere. By Newton's third law, a reaction force equal to and opposite to that exerted on the fluid particles will be exerted on the sphere. The result of the fluid reaction is to generate a drag force on the sphere that tends to slow it down. We will try to estimate this drag force. Assuming that after hitting the sphere, the fluid particles depart in a direction perpendicular to their original direction, their change in momentum in dt will be

$$dp_x = \rho(A\, v\, dt)v. \tag{10.21}$$

Therefore, the reaction force against the moving object, i.e., the friction force, will be

$$F_r = \frac{dp_x}{dt} \approx \rho\, A\, v^2. \tag{10.22}$$

This is the friction force which the medium exerts on the moving body within a fluid in a turbulent regime.

10.5 Annex B: Falling motion in a fluid medium with friction proportional to v^2

Consider an object of mass m falling into a fluid medium. Assume that the friction force is proportional to the square of the velocity. Taking the axis direction and pointing downward, the equation of motion is

$$\begin{aligned} m\frac{dv}{dt} &= mg - m_a \cdot g - \beta \cdot A \cdot \rho \cdot v^2 \\ &= m\left(1 - \frac{\rho}{\rho_c}\right) \cdot g - \beta \cdot A \cdot \rho \cdot v^2, \end{aligned} \tag{10.23}$$

where g is the acceleration of gravity, m_a is the mass of air displaced by the body $(=m \cdot \rho/\rho_c)$, ρ is the air density, ρ_c is the density of the body, A is the area that the body passes through when moving, and β is a constant of proportionality. Equation (10.23) can be written as follows:

$$\frac{dv}{v_f^2 - v^2} = \frac{g}{v_f^2} \cdot (1 - \rho/\rho_c) \cdot dt, \tag{10.24}$$

where $v_f^2 = m \cdot g \cdot (1 - \rho/\rho_c)/(\beta \cdot A \cdot \rho)$. Integrating this expression, we obtain

$$v(t) = v_f \cdot \tan h\left(\frac{g \cdot (1 - \rho/\rho_c)}{v_f^2} \cdot t + c\right). \tag{10.25}$$

If for $t = 0$, $v(t) = 0$, the constant of integration is $c = 0$.

If we define $k_a = g \cdot (1 - \rho/\rho_c)/v_f^2 = \beta A\rho/m$,

$$v(t) = v_f \cdot \tan h(k_a \cdot t). \tag{10.26}$$

Integrating this last expression once more, we obtain

$$\begin{aligned} y(t) &= y_0 + \frac{v_f}{k_a} \cdot \ln[\cos h(k_a \cdot t)] \\ &= y_0 + \frac{v_f^3}{g \cdot (1 - \rho/\rho_c)} \cdot \ln[\cos h(k_a t)]. \end{aligned} \tag{10.27}$$

Here, y_0 is the position of the body for $t = 0$. In the balloon or coffee-filter-dropping experiments, what is measured is precisely $y(t)$. Thus, Eq. (10.27) can be compared directly to the experimental data. As the only unknown parameters in this expression are the limit or final speed v_f and k_a, their values can be found by fitting the theoretical curve to the experimental data.

References

[1] A. J. Almaliki, *et al.*, "Review on techniques and file formats of image compression," *Bull. Electr. Eng. Inf.*, vol. 9, no. 2, pp. 602–610, 2020.

[2] wikipedia, "Image file formats," 2021. Available: https://en.wikipedia.org/wiki/Image_file_formats.

[3] Cornell University, "Digital imaging tutorial," 2021. Available: http://preservationtutorial.library.cornell.edu/contents.html.

[4] S. Gil, H. D. Reisin and E. Rodríguez, "Using a digital camera as a measuring device," *Am. J. Phys.*, vol. 74, p. 768, 2006.

[5] M. E. Saleta, D. Tobia and S. Gil, "Experimental study of Bernoulli's equation with losses," *Am. J. Phys.*, vol. 73, p. 598, 2005.

[6] "xyExtract graph digitizer," 2021 Shareup LLC, 2021. Available: http://www.shareup.com/xyExtract-download-19171.html.

[7] Tracker Projects — Compadre, "Tracker," 2021. Available: https://physlets.org/tracker/.

[8] Vernier Software, "Logger pro and vernier video analysis," 2021. Available: www.vernier.com.

[9] S. Calderon, P. A. Nuñez and S. Gil, "Cámara digital como instrumento de laboratorio — estudio del tiro oblicuo," *Lat. Am. J. Phys. Educ.*, vol. 3, no. 1, pp. 84–92, 2009.

[10] S. Gil, "Fisica re-Creativa (Cap10)," 2021. Available: https://www. https://www.fisicarecreativa.com/exp_fisica/exp_fisica_recur.htm.

[11] K. E. Horst, "The shape of lamp shade shadows," *Phys. Teach.*, vol. 39, pp. 139–140, 2001.
[12] D. Giancoli, *Physics Principles with Applications*, 7th Ed., NY: Pearson, 2014.
[13] J. D. Wilson, *College Physics*, 7th Ed., NY: Pearson, 2009.
[14] S. Calderon, *et al.*, "Determinación de la fuerza de roce con el aire usando nuevas tecnología," *Revista de Enseñanza de la Física de la Asociación de Prof. de Física de la Rep. Arg.*, vol. 20, no. 1, pp. 55–64, 2007.
[15] K. Takahashi and D. Thompson, "Measuring air resistance in a computerized laboratory," *Am. J. Phys.*, vol. 67, no. 8, pp. 709–711, 1999.
[16] D. Brown, "Tracker," 2021. Available: https://www.youtube.com/watch?v=La3H7JywgX0.
[17] B. Munson, D. Young and T. Okiishi, *Fundamentals of Fluid Mechanics*, New York: Wiley, 1994.
[18] Scrollers, "Ping pong ball launchers pattern," 2021. Available: http://www.scrolleronline.com/product/Ping-Pong-Ball-Launchers-Pattern/354.
[19] "Make a cotton ball launcher," 2021. Available: https://www.sciencebuddies.org/stem-activities/cotton-ball-launcher.
[20] Wikipedia, "Euler's method," 2021. Available: https://en.wikipedia.org/wiki/Euler_method.
[21] Wikipedia, "Runge-Kutta method," 2021. Available: https://en.wikipedia.org/wiki/Runge%E2%80%93Kutta_methods.

Chapter 11

SOUND RECORDING AS A MEASURING INSTRUMENT

The aim of this chapter is to introduce sound recording using a smartphone or a personal computer (PC) as a measurement instrument in the laboratory, proposing some simple experiments to illustrate it. First, it discusses an experiment to measure the fall time of a body in order to estimate the value of the acceleration of gravity. Then it describes an activity for visualizing the basic characteristics of sound waves. In particular, it proposes an experiment to measure the speed of sound using a sound pulse that travels through a tube and is reflected at the tube ends.

Goals

✓ The sound card as a measuring instrument

✓ Determination of the speed of sound

✓ Measurement of fall times of a body

✓ Reflection of sound waves

11.1 Digital Sound recording

Most smartphones, laptops, and PCs are equipped with a *sound processing hardware* that enables them to record input sound signals and emit audio output, using various programs. Although sound recording technology changes constantly, it preserves the same basic principles of traditional *sound cards*. Sound recording and sound generation are used in multimedia applications and games, and for listening to and recording sounds or music, editing audio files, etc.

An *Analog-to-Digital Converter* (ADC) converts sounds into digital data. These digital data can, in turn, be transformed into analog audio signals through a *Digital-to-Analog Converter* (DAC). The analog signals can be amplified and heard through conventional headphones or speakers. Simpler cards record and play sounds alternately, while *full-duplex* sound cards are required to record and play *simultaneously*. Some devices, mainly laptops and PCs, have the following audio inputs and outputs (I/O) with a standardized color code:

- ✓ Pink connector: Analog microphone input. 3.5 mm connector (jack).
- ✓ Blue connector: "Line-in" analog input, audio signals from other audio devices (recorders, radios, etc.). 3.5 mm jack.
- ✓ Green connector: Main signal analog output (stereo) that can be connected to speakers. 3.5 mm jack.
- ✓ Game port connector or "Joystick input port": DB15 connector (15-pin female), not always available in new equipment.

Figure 11.1 shows a typical personal computer sound card, including its connectors. Various devices can be connected to the game port, e.g., joystick and photoswitches, among others. In modern computers, a game port is not usually available and is replaced by a USB port. Nonetheless, the line-in, microphone input, and audio output have remained in modern computers, usually with the same color codes.

Most PCs, laptops, and smartphones have programs installed that can record (e.g., Microsoft SoundRecorder®) and play sounds (e.g., Microsoft Windows Multimedia® or Phyphox app [1], among many others, for smartphones). There is also a wide variety of programs, several of which are freely available, such as Audacity[a] and Sound Forge.[b] They provide various options for processing audio signals and work as audio signal acquisition systems. Most of these programs allow us to record and hear audio signals, obtain frequency spectra,

---

[a] www.audacity.com, and http://en.wikipedia.org/wiki/Audacity
[b] http://en.wikipedia.org/wiki/Sound_Forge

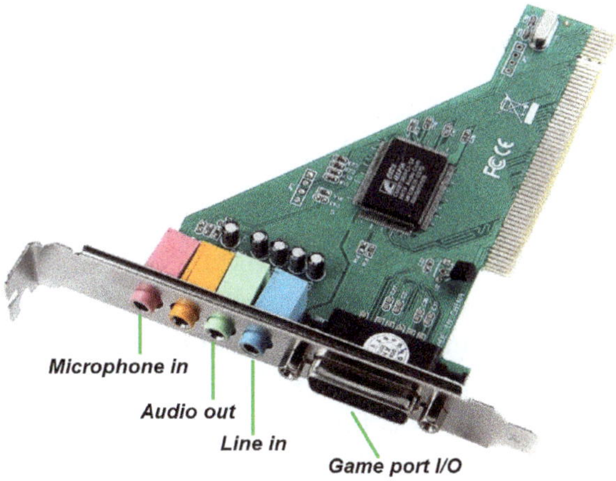

Figure 11.1 Standard PCI sound card used in traditional PCs.

edit, filter, trim, etc., and support a wide variety of audio formats (WAV, MPEG, MP3, and MIDI). There is also a great variety of programs that allow us to transform our PC into an oscilloscope or audio data acquisition system, e.g., MacScopeII,[c] Virtins Sound Card Oscilloscope,[d] and WinScope,[e] which are available on the Web. These programs allow us to acquire audio signals with a PC through the input lines ("line-in" or headphone) or a smartphone [2].

The sound recording facilities of almost all PCs, laptops, or smartphones can be used to convert any of these devices into a data acquisition system, a function generator, or a very low-cost oscilloscope [3–5]. There are several free access programs that enable these functions [6]. Similarly, programs such as MATLAB, in its Data Acquisition Toolbox, offer function generators and oscilloscopes called Daqscope and SoftScope.

In general, the maximum allowed voltage at the input of a sound card is about 3 V, so it is advisable to use a protection circuit, which is very easy to build, using four diodes. See Ref. [3].

[c]http://www.physics2000.com/Pages/MacScope.html
[d]http://www.virtins.com/
[e]http://www.comap.cz/questionnaire/winscope/

11.1.1 *Project 20. Determining the acceleration of gravity using audio signals*

Recommended equipment: A metal ball 1–3 cm in diameter or a heavy metal coin. A microphone connected to a PC or smartphone.

A heavy object falling from a height of the order of 1 meter takes about half a second to reach the ground. This fall time is too short to measure with a manually operated stopwatch since a person's reaction time is approximately 0.2 s. For falls from greater heights, the friction effects may not be negligible. To measure the time of the order of the second with a precision of the order of 1%, optical methods such as photogates (see Chapter 9) are required for starting and stopping interval measurement. An alternative way of measuring short time intervals is by recording audio signals and then analyzing the acoustic signal [7–10]. A simple way to achieve a clear acoustic signal that indicates the beginning of the fall of an object such as a large coin is to place it on a taut horizontal paper strip, turn on the computer's sound recorder, and a moment later, tear the strip of paper by means of a sudden jerk, producing a clearly audible sound as the strip breaks. Another alternative is to snip the paper with scissors, producing a clear sound. When the strip breaks, the coin drops, generating a new sound pulse when it reaches the ground. After the fall, stop the recording and analyze the recorded sound.

An alternative way to achieve the same effect is to build a rigid pendulum that produces a clear sound pulse when it strikes a metal ball laterally, causing the ball to fall without giving it an initial velocity in the vertical direction. Figure 11.2 shows a possible experimental setup [7]. Upon reaching the ground, the ball produces another sound pulse.

Record these sounds with the computer and then analyze the audio signal. By viewing the acoustic signal as a function of time, the fall time Δt can be determined with good precision. Figure 11.3 shows schematically the shape of the audio signal recorded during a

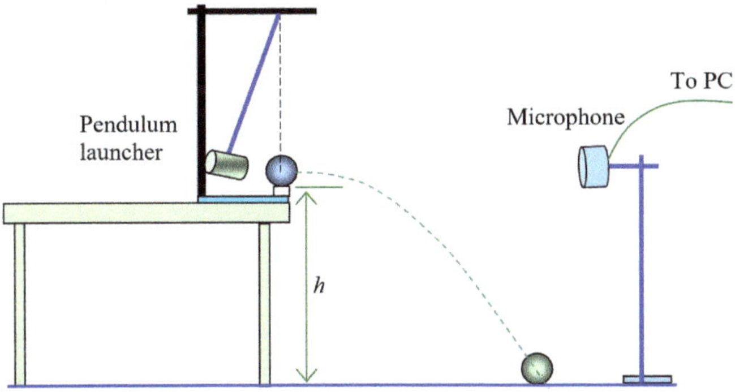

Figure 11.2 Experimental setup to measure the time of fall of a metal ball.

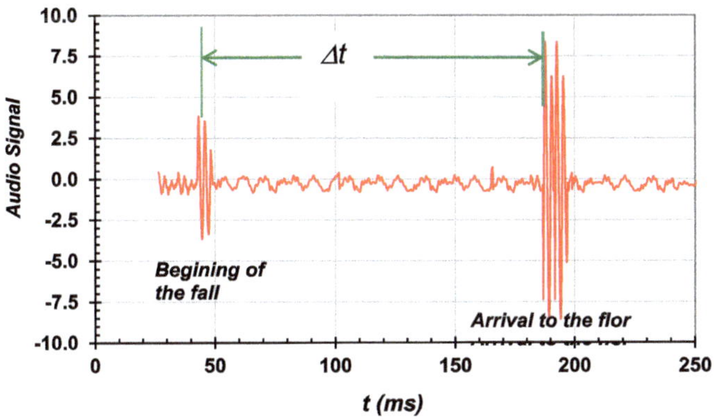

Figure 11.3 Characteristics of the audio signal recorded when a ball is dropped. The uncertainty in the determination of the beginning of the pulses allows the measurement errors of Δt to be estimated.

drop. Knowing the initial height h of the falling body, find the value of g, assuming a uniformly accelerated motion, as

$$g = \frac{2h}{\Delta t^2}. \tag{11.1}$$

Suggested tasks:

✓ Choose the setup that best suits your working conditions. Perform some preliminary tests to investigate the most suitable conditions to clearly register the sound pulses that indicate the beginning of the fall and the arrival to the ground.

✓ Determine the value of the drop time Δt for at least 5 falls from the same height h. Estimate the mean value of the interval and its uncertainty.
✓ Determine h with a relative error comparable to that of Δt.
✓ Find the best value of g and its uncertainty.

11.2 Sound waves

In classical physics, there are essentially two categories of objects in the universe: matter and waves. The transmission of energy from one point to another can be done by the transport of matter or by wave propagation [11]. An example of the former is the kinetic energy associated with a moving particle or body. In contrast, when a sound reaches our ears, the energy is associated with vibrations that propagate in the air, i.e., through acoustic waves. Sound waves are a particular example of mechanical waves that propagate in an elastic medium (solid, liquid, or gas). They are vibrations that cause variations in pressure and displacement of the particles of the medium in which they propagate. In sound waves, the displacement of the particles occurs in the same direction as the propagation of the wave, therefore they are longitudinal waves.

The propagation speed of a mechanical wave depends on the stiffness or compressibility of the medium and its density. In the case of sound waves in a gas, the pressure variations are so fast that the compressions and decompressions can be well described by an adiabatic process [11, 12]. Within this approximation, the speed of sound in a gas can be written as follows [12–14]:

$$c = \sqrt{\frac{\gamma RT}{M}}, \tag{11.2}$$

where γ is the adiabatic coefficient of the gas (equal to the quotient of the molar heat capacities of the gas at constant pressure and volume, $\gamma = C_P/C_V$), R is the universal gas constant, and T and M are the absolute temperature and molecular mass of the gas, respectively.

When a wave that is traveling through a medium encounters a boundary where the properties of the medium change (compressibility or density), or the geometric restrictions imposed on the wave vary, the waves are partially transmitted and partially reflected, according to the characteristics of the interface [14]. The boundary can be the interface between two media, such as air and a solid or liquid. A change in geometric constraint can be achieved in a tube with an open or closed end, or a change in tube diameter [12, 15].

A consequence of the finite diameter of a tube is that its *effective length* is different from its *geometric length* [12, 15]. Thus, the effective length L_{ef} of a tube with an open end is greater than its nominal length L_0. Empirically, it is found that the effective length of a tube with an open end is

$$L_{\text{ef}} = L_0 + f\,d, \tag{11.3}$$

where d is the diameter of the tube and the coefficient $f \approx 0.3$ [12, 15]. If both ends of the tube are open, $L_{\text{eff}} = L_0 + 2\,f\,d$.

If we consider a cylindrical tube with both ends A and B open, a sound pulse emitted from one of its ends (A) will be reflected at the other (B) and captured by the microphone. In reference to the experimental arrangement shown in Figure 11.4, the first pulse detected by the microphone is the one that comes directly from the emitter, before entering the tube. The next pulse that the microphone detects has a reflection. In turn, the third pulse that reaches the

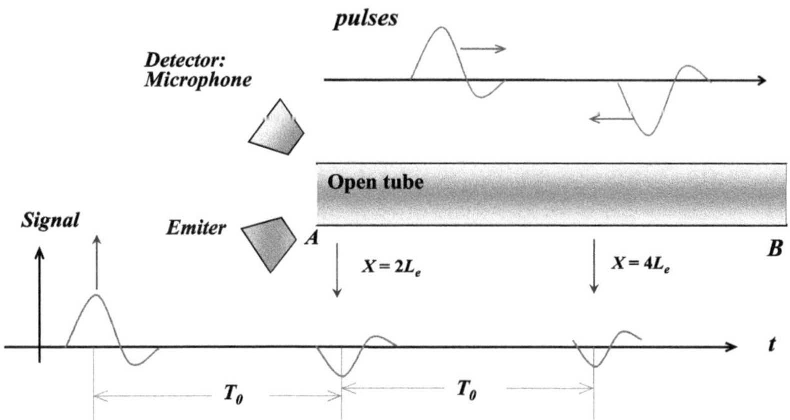

Figure 11.4 Diagram of a sound pulse propagating through an open tube. The upper panel shows schematically how the pulses move in the tube. Their inversion is a consequence of the reflection at the open ends. The lower panel shows how the signal would look at the microphone as a function of time. The delay of the pulses is given by the travel time of the pulse, $T_0 = 2L_{\text{ef}}/c$.

microphone experiences two additional reflections with respect to the second. All the following pulses have the same behavior as the third, i.e., they travel a distance $2L_{\mathrm{ef}}$ more than the previous one. Since not all of the pulse energy is reflected, the pulses decrease in amplitude as they undergo more reflections.

11.2.1 *Project 21. Determining the speed of sound*

Recommended equipment: A microphone connected to a PC or smartphone. Drainage pipes (PVC) from 1 to 7 cm in diameter and lengths of 1–5 m.

The proposed experiment makes use of a speaker or earphone (transmitter) and a microphone (detector). A PVC tube of diameter 1–7 cm and lengths of 1–5 m may be suitable. Construction discharge tubes are inexpensive and suitable for this experiment. In addition, we can buy couplings to connect them easily and make a set of tubes with different lengths and the same diameter. We will also need one or, better, two computers or two smartphones. A useful program for data acquisition could be Audacity® [16], which is freely accessible, but any other similar sound acquisition programs could also be used as well. With smartphones, simply record the sound.

Connect the microphone (receiver) to one of the PCs and the speaker or headphone (emitter) to the other. In the computer connected to the speaker, generate the sound signal using Audacity or a similar program. Give this wave the desired shape, frequency, and duration. One possibility would be to use sinusoidal pulses of 1500 Hz of $\Delta t = 3$ ms in duration, as illustrated in Figure 11.5. The idea is that the duration of the pulse should be much shorter than the travel time through the tube, so the different reflected signals can be well separated. In addition, the choice of the shape of the signal

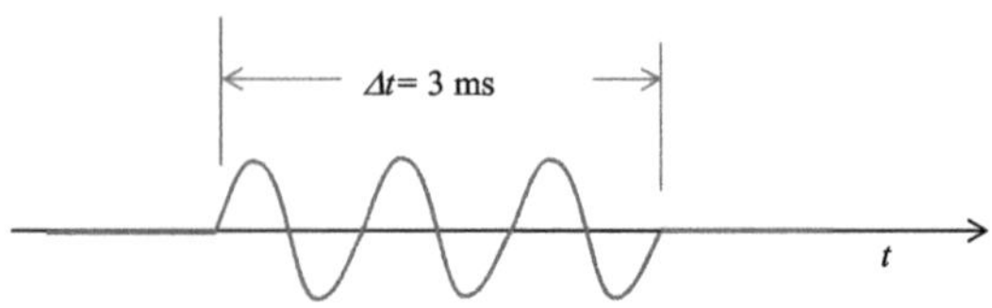

Figure 11.5 Signal that is emitted at the end of the tube.

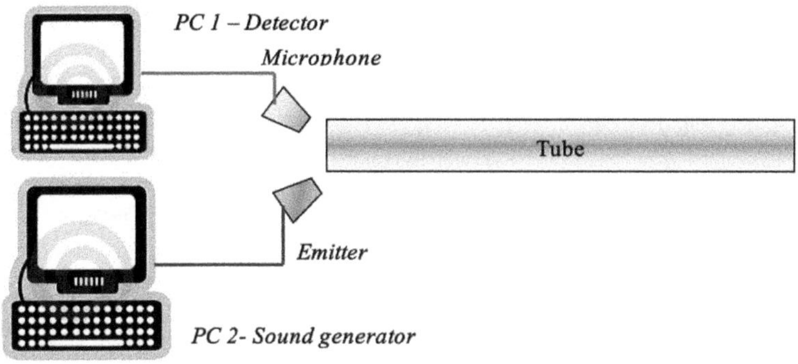

Figure 11.6 Experimental setup. The microphone is connected to one PC and the headphone or speaker to the other PC. The tube can be closed at the end opposite to where the signal is emitted and recorded.

allows it to be clearly distinguished notwithstanding the electronic noise captured by the microphone connected to the other PC.

With the PC connected to the microphone, record (digitally) the emitted sound and its reflections using Audacity or an equivalent program. Figure 11.6 illustrates an experimental setup that could be used. This experimental design using two computers can be simplified by using only one PC connected to the microphone, in which case the sound pulse can be generated by striking two pieces of wood or snapping your fingers. With a little practice, it is possible to generate short, clear pulses. However, using two PCs provides a clear signal that is easily distinguishable from any possible noise that the microphone might pick up.

Activity

Figures 11.7 and 11.8 show the characteristics of the incident and reflected pulses for an open tube. The arrival of the reflected pulses can be clearly observed. By analyzing this type of figure, it is possible to find the transit times of the pulses reflected 1, 2, and 3 times. These transit times are associated with travel distances of the pulses corresponding to distances $X = 2L_e$, $X = 4L_e$, and $X = 6L_e$, respectively. To determine the speed of sound, it is advisable to use tubes of different lengths. For each length, determine the effective length (L_e) of the tube Eq. (11.3) and measure the arrival times of the different reflected pulses.

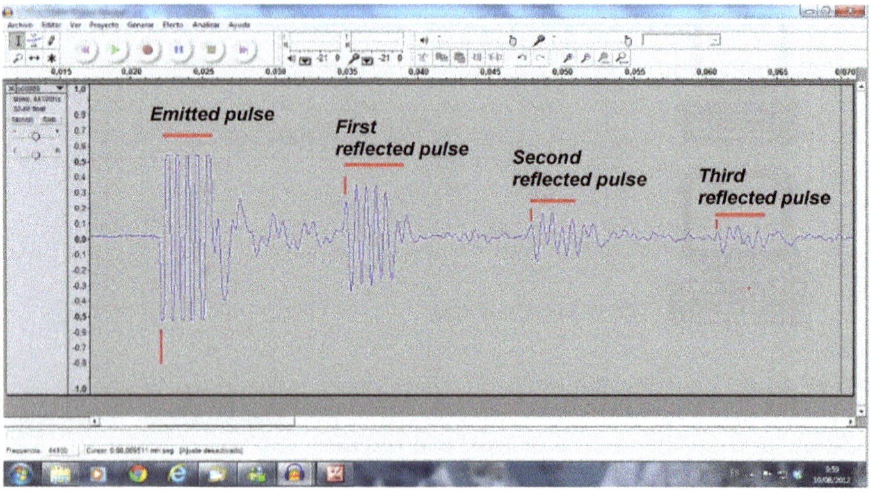

Figure 11.7 Example of the sound signal of an emitted pulse and its successive reflections in the open tube. This signal was recorded using the Audacity program.

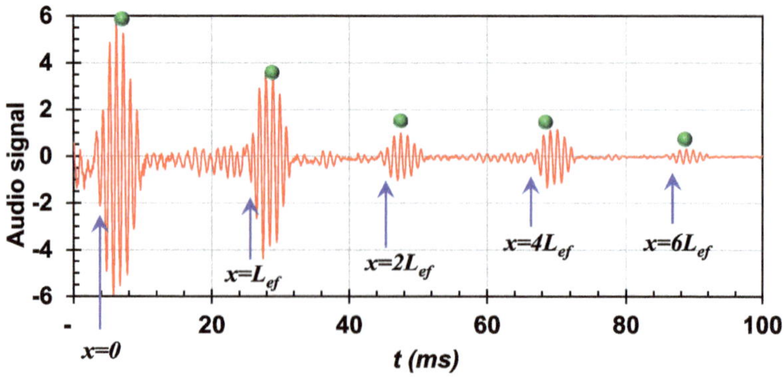

Figure 11.8 Original signal emitted by the earphone and the successive reflections. The dots mark the centroid of the reflected pulses.

Suggested tasks:

✓ For each length of the tubes chosen, determine the distances traveled by the different pulses and their travel times.

✓ Construct a graph of the distances traveled by the different pulses and their corresponding bearings or reflection pulses, as shown

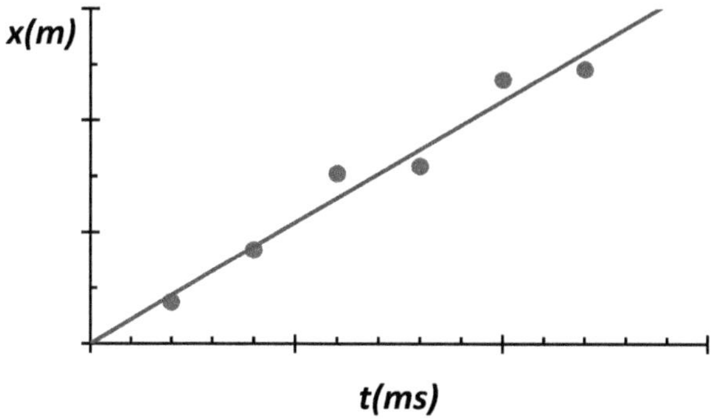

Figure 11.9 Distance traveled by the different pulses and their corresponding delays. The slope on the $x(t)$ graph allows us to determine the speed of sound.

schematically in Figure 11.9. The slope of this graph is the speed of sound.

✓ Using the data from all the tubes tested, find the best value for the speed of sound and estimate its uncertainty.

✓ Measure the temperature of the laboratory and use tables to find the expected value of sound at the working temperature. How do the values of your measurements compare to the expected values from the tables?

References

[1] Google Play-LuxDeLux, "Google play: Frequency sound generator," 2020. Available: https://play.google.com/store/apps/details?id=com.luxdelux.frequencygenerator&hl=es_UY&gl=US.

[2] C. J. Pereyra, *et al.*, "Fourier analysis of a vibrating string through a low-cost experimental setup and a smartphone," *Phys. Educ.*, vol. 53, p. 045019, 2018.

[3] U. Hassan, S. Pervaiz and M. Sabieh Anwar, "Inexpensive data acquisition with a sound card," *Phys. Teach.*, vol. 49, p. 537, 2011.

[4] W. C. Magno, *et al.*, "Probing a resonant circuit with a PC sound card," *Am. J. Phys.*, vol. 75, p. 161, 2007.

[5] J. A. Sans, *et al.*, "Smartphone: A new device for teaching physics," in *1st International Conference on Higher Education Advances, HEAd'15 Universitat Politècnica de València*, Valencia, 2015.

[6] Daqarta, "Data acquisition and real-time analysis scope — Spectrum — Spectrogram — Signal generator," 2021. Available: http://www.daqarta.com/.

[7] J. A. White, *et al.*, "A measurement of g listening to falling balls," *Phys. Teach.*, vol. 45, no. 3, p. 175, 2007.

[8] A. D. Bernstein, "Listening to the coefficient of restitution," *Am. J. Phys.*, vol. 45, pp. 41–44, 1977.

[9] P. A. Smith, C. D. Spencer and D. E. Jones, "Microcomputer listens to the coefficient of restitution," *Am. J. Phys.*, vol. 49, p. 136, 1981.

[10] P. Vogt and J. Kuhn, "Analyzing free fall with a smartphone acceleration sensor," *The Phys. Teach.*, vol. 50, no. 1, p. 182, 2012.

[11] R. Feynman, *et al.*, *The Feynman Lectures on Physics, Boxed Set: The New Millennium Edition*, Vol. 1, Basic Books, 2011.

[12] F. S. Crawford Jr., *Waves (Berkeley Physics Course, Vol. 3)*, Cambridge University Press, 2013.

[13] D. Giancoli, *Physics Principles with Applications*, 7th Ed., NY: Pearson, 2014.

[14] H. D. Young, R. A. Freedman, *et al.*, *College Physics*, Vol. II, Addison-Wesley, 2009.

[15] S. E. Calderón, P. Núñez and S. Gil, "Experimentos en el aula utilizando la tarjeta de sonido de una PC," *Lat. Am. J. Phys. Educ.*, vol. 4, no. 1, p. 188, 2010.

[16] Audacity, "A free multi-track audio editor and recorder," 2021. Available: http://Audacity.Sourceforge.Net.

[17] D. Potter, "Phase changes in reflected sound waves," *Phys. Teach.*, vol. 41, p. 12, 2003.

Chapter 12

MEASURING THE SOLAR SYSTEM FROM THE CLASSROOM

This chapter proposes to use classic geometric methods, combined with the tools provided by new information and communication technologies (ICT), to estimate astronomical distances from simple observations. The goals are to determine the radius of the Earth and the distance from the Earth to the Moon, the Sun, and some inner planets. It also proposes to estimate the diameters of the Sun and the Moon. It describes two activities to obtain Kepler's third law from solar system data and to determine the Hubble constant. The techniques used for making these estimates are based on digitized images of eclipses available on the Internet or taken during a lunar eclipse. The goal is to try to answer the following questions: *How do we know this? Why do we believe that?* The approach illustrates a way to combine the methods of classical antiquity with modern technologies.

Goals

- ✓ Estimation of astronomical distances
- ✓ Estimation of the radius of the Earth
- ✓ Size of the Moon
- ✓ Earth-Moon distance
- ✓ Distances in the Solar System
- ✓ Kepler's third law
- ✓ Hubble constant and estimation of the age of the universe

12.1 The size of the Moon and the Earth

During the observation of a lunar eclipse, the shadow cast by the Earth on the Moon reveals the spherical shape of the Earth and provides an idea of the size of our planet compared to the Moon [1] (see Figure 12.1).

This was one of the observations that Aristotle (384 BC–322 BC) relied on to argue for the roundness of the Earth. Aristarchus of Samos (310 BC–250 BC) was another of the great Greek philosophers who contributed significantly to the understanding of the Earth's position in the cosmos. He was a forerunner of the heliocentric model of the Solar System and served as director of the famous Library of Alexandria, one of the most important centers of study in the classical world. Aristarchus devised an ingenious method to measure the size of the Moon and its distance from the Earth by comparing the size of the Earth's shadow to the size of the Moon during a lunar eclipse. By simple visual observation, Aristarchus estimated that the diameter of the Earth's shadow was approximately twice the diameter of the moon.

Given that eclipses can be photographed and the availability of eclipse photos on the Internet [1, 2], it is very simple and instructive to use modern technology to make (and improve) Aristarchus's estimates. Figure 12.1 shows a photo of a lunar eclipse taken on February 20, 2008, in Buenos Aires City, using an ordinary

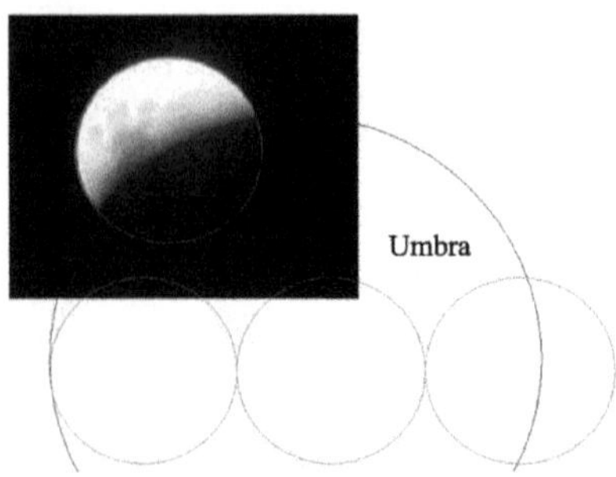

Figure 12.1 Photograph of the lunar eclipse of February 20, 2008. The Moon is in the umbra of the Earth. A circumference has been drawn to complete the shadow of the Earth.

digital camera. Using common graphing programs installed on most personal computers, it is possible to take measurements from frames like the one in Figure 12.1.

12.1.1 *Project 22. Determination of the size of the Moon and its distance from Earth — Aristarchus*

Recommended equipment: A digital photograph of a partial lunar eclipse. You can photograph an eclipse or download a photo of one from the Internet.

Using the technique described above (see Figure 12.1), determine the relationship between the size (diameter) of the Earth's shadow (umbra[a]) and the Moon.

Note that the radius of the Earth R_E is not the same as the radius R'_E of the shadow of the Earth. Due to the large size of the Sun compared to Earth, the darkest area of the shadow (*umbra*) of the Earth has a conical shape, as illustrated in Figure 12.2.

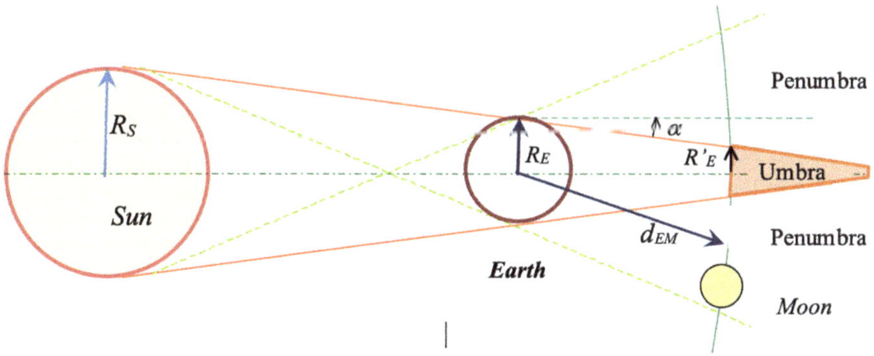

Figure 12.2 Sun, Earth, and Moon system. R_E is the radius of the Earth and R'_E is the radius of its shadow at a distance equal to that from the Earth to the Moon (d_{EM}). α is the opening angle of the Earth's shadow cone (*umbra*).

[a]https://en.wikipedia.org/wiki/Umbra,_penumbra_and_antumbra#Penumbra

Around 240 BC, Eratosthenes measured the Earth's radius, R_E. The currently accepted value for the terrestrial radius is $R_E = 6.371\,\text{km}$ [3].

During a total solar eclipse [4, 5], the Moon manages to eclipse the Sun very tightly, so it is clear that the apparent diameters (or subtended angles) of the Moon and the Sun are approximately the same, as shown in Figure 12.3. Thus,

$$\alpha = R_M/d_{\text{EM}} = R_S/d_{\text{ES}}. \tag{12.1}$$

Here, R_S and d_{ES} are the radius of the Sun and the Earth-Sun distance, respectively, and 2α is the angle subtended by the Moon. One way to estimate α is to take an opaque screen (metal or plastic) and drill a circular hole of known diameter d_h. A simple, practical way to manufacture the screen is to pierce a plastic card (such as a credit card) with a drill bit of known size (1 or 2 mm).

By holding the screen and looking through the hole at the Moon, the subtended angle is achieved when the whole Moon matches the size of the hole. The distance L_h is measured from the screen to the eye and the angle α is calculated as follows:

$$\alpha = \frac{d_h}{2L_h}\text{rad}. \tag{12.2}$$

Thus, we have

$$2\alpha = 2R_L/d_{\text{TL}}. \tag{12.3}$$

Knowing the size of the Moon, R_M, its distance to Earth, d_{EM}, can also be estimated.

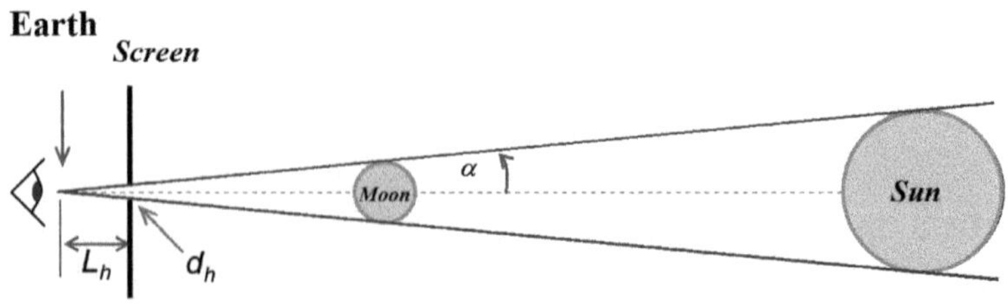

Figure 12.3 Apparent size or angle subtended by the Moon and by the Sun from the Earth.

Activity

- ✓ Determine the best value of α, the angle subtended by the Moon.
- ✓ From the analysis of one or more photos of a lunar eclipse, and the values of α and R_E provided, estimate the Moon's radius, R_M, and its distance to Earth, d_{EM}. In this case, you can assume that the size of the Earth's shadow is equal to that of the Earth, i.e., $R_E \approx R'_E$. (See Figure 12.2.)
- ✓ From the analysis of the photo of an eclipse similar to Figure 12.1, determine the best value of the quotient $k = R'_E/R_M$ and estimate its uncertainty. Using the geometric layout illustrated in Figure 12.2, show that

$$R_E = R'_E + \alpha d_{EM}. \qquad (12.4)$$

  Combining this result with (12.3), find again the values of R_M and d_{EM}.
- ✓ Compare the values you find to the accepted values for these quantities [3]. What can you conclude from this analysis?

12.2 Estimation of the terrestrial radius

Eratosthenes was also the director of the library of Alexandria during the third century BC. He was a philosopher, astronomer, historian, geographer, poet, theater critic, mathematician, and sportsman. It is said that on one of his trips from Greece to Alexandria, he was struck by seafarers' ability to navigate by night using the stars and by the fact that the star usually taken as a reference, in particular the north pole star, looked higher in Greece than in Alexandria, while the opposite was true of the Sun at noon.

In the library of Alexandria, Eratosthenes read about the city of Syene (today Aswan [24° 05′N, 32°54′E], located approximately 800 km south-east of Alexandria [31°11′N, 29°55′E]), where at noon on June 21 (summer solstice), the Sun fully illuminated (without shadow) a deep well. This did not happen in Alexandria. Based on this observation (absence of shadow at midday of the summer solstice in Syene), plus the angle of the Sun at the same time in Alexandria and the distance between the two cities, he was able to make the first determination of the radius of the Earth. His experiment is now a classic activity, often repeated in schools [6, 7].

This section proposes a very simple method, which can be carried out from a tall building (7 floors or more) or from a hill near the sea or a large lake [8]. The layout of the experiment is illustrated in Figure 12.4.

In Figure 12.4, h is the height of the building or hill with respect to the sea level. This value can be determined using an altimeter, which is calibrated by placing it on the beach at sea level. Another alternative is to use the Google Earth program. When the mouse is placed on the location of the building or hill, it provides the latitude, longitude, and elevation of the place with respect to sea level. The height of a building can be obtained from blueprints of a building or by direct measurement. Use whatever method is most convenient to estimate the height h (Figure 12.5), e.g., using a weight attached to a string or measuring the length of the building's overhang compared to the shadow of a vertical pole of known height.

The other important parameter to measure is the angle between the horizontal and the horizon, θ_f, see Figure 12.5. To do this, an experimental arrangement consisting of a crosshair telescopic

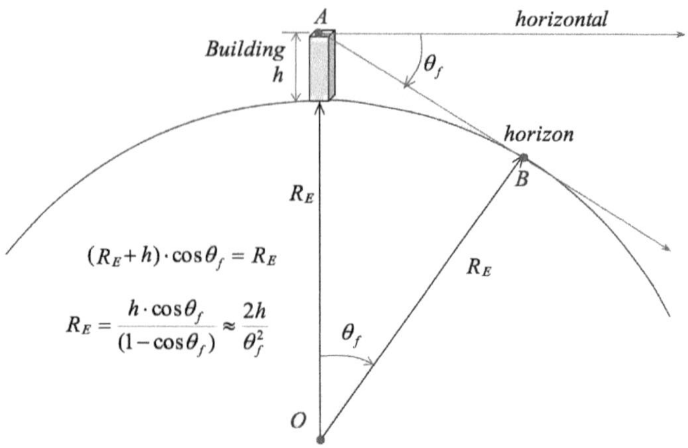

Figure 12.4 Experimental layout to measure the terrestrial radius R_E.

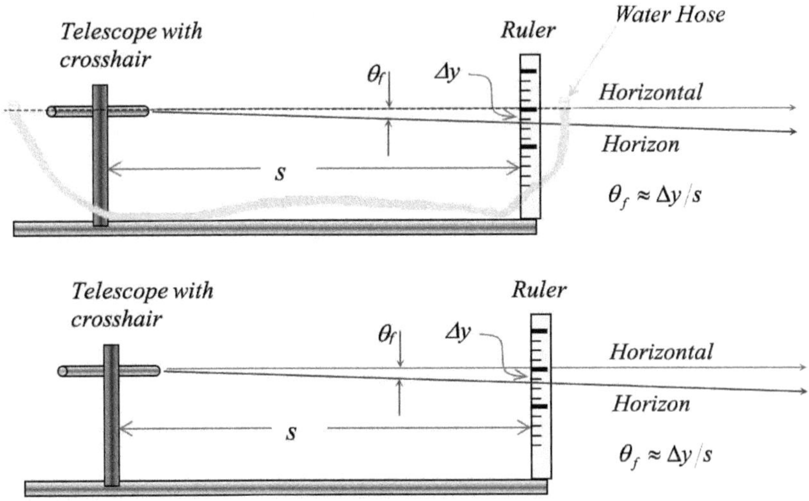

Figure 12.5 Experimental setup to measure the terrestrial radius.

sight[b] (such as that of a rifle, or a telescope or theodolite, mounted on a tripod) is suggested. Place a ruler graduated in millimeters at a known distance s a few meters away from the telescopic sight. With the aid of a flexible transparent plastic hose filled with water, make a mark on the ruler that defines the *horizontal line*. Next, position the cross of the reticle of the telescopic sight on the *horizon* (sea-air dividing line). Vary the focus of the sight, without modifying its angle, until the ruler can be seen clearly, thus enabling a new mark to be made on the ruler, which defines the distance Δy on it. This value can be used to estimate θ_f through the relation $\tan\theta_f = \Delta y/s$ and its uncertainty $\Delta\theta_f$.

From the geometry of the problem, see Figure 12.4, we have

$$(R_E + h)\cos\theta_f = R_E, \quad \text{thus,} \quad R_E = \frac{h\cos\theta_f}{1-\cos\theta_f} \approx \frac{2h}{\theta_f^2}. \qquad (12.5)$$

In this way, knowing h and θ_f, we can calculate the Earth's radius, R_E.

[b]https://en.wikipedia.org/wiki/Telescopic_sight

There is, however, an important point to keep in mind when looking from the top of a building or mountain to the horizon: the rays of the lines of sight that reach the eyepiece or the eye do not necessarily follow a straight line. This is because the refractive index of air varies with pressure. It is well known that atmospheric pressure varies with height. In fact, the most common altimeters use this property to determine heights. For altitudes less than about 4 km above sea level, pressure can be calculated using an isothermal atmosphere model [9, 10]:

$$P(z) = P_0 \cdot \exp(-z/h_{00}), \tag{12.6}$$

where P_0 is the pressure at sea level (≈ 1 atm), z is the vertical height in meters, and $h_{00} = \mathrm{RT/Mg} \approx 8420$ m at an average temperature of about 15°C, where R is the universal gas constant, T is the absolute temperature of the atmosphere, M is the (average) molecular mass of air, and g is the acceleration of gravity. If $z \ll h_{00}$, i.e., we are close to the surface of the Earth, $P(z) = P_0(1 - z/h_{00})$, i.e., the barometric pressure decreases at a rate of 12 Pa/m.

On the other hand, the refractive index, n, of a gas can be written as follows [9, 10]:

$$n(z) = \sqrt{K} = \sqrt{1 + 2c \cdot P} \approx 1 + cP(z). \tag{12.7}$$

Here, K is the dielectric constant of air, which depends on the number of molecules per unit volume, which in turn depends on the barometric pressure (absolute), and the constant c depends on the system of units (if P is measured in atmospheres $c_0 = 1.00029$). This means that as the refractive index varies with height, the ray will follow a slightly curved path, as shown in Figure 12.6 (see Annex A). The angle θ_ι measured with the telescope, as described above, will generally be somewhat less than the ideal value θ_f.

There are theoretical expressions that can be used to estimate the value of the ideal angle θ_f (the angle formed by a ray traveling in a straight line to the horizon) from the measured value θ_i. One result

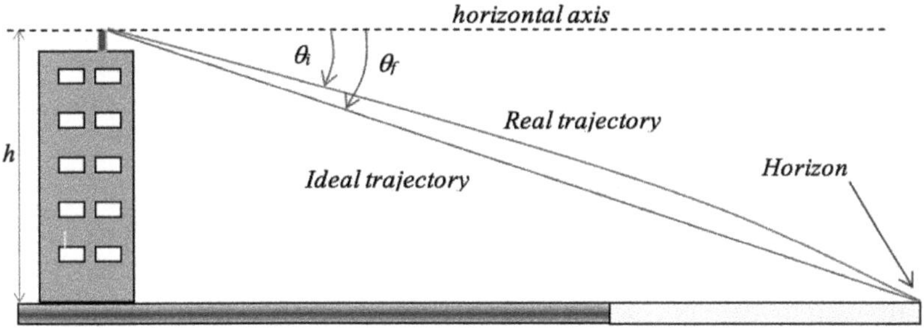

Figure 12.6 Schematic path of light rays.

that can be used is [8]

$$\delta = \theta_f - \theta_i = \text{Sin}^{-1}\left\{\text{Sin}(\alpha_i)\left[\frac{(1+\text{D}e^{-z_i/h_{00}})}{(1+\text{D}e^{-\beta\cdot z_f/h_{00}})}e^{(z_i-z_f)/R_T}\right]\right\}$$
$$- \text{Sin}^{-1}\left\{\text{Sin}(\alpha_i)\left[e^{(z_i-z_f)/R_T}\right]\right\}, \quad (12.8)$$

where $D = P_0\ c_0, z_i$ and z_f are the initial heights, that is, $z_i = h_0$ and $z_f = 0$, and R_T is a preliminary estimate of the terrestrial radius ($R_T \approx 6371\,\text{km}$). This correction considerably improves the estimation of the radius of the Earth. Without correction (12.8), expression (12.4) overestimates RT by almost 15%, whereas using the correction can improve the result considerably. A heuristic justification for this correction is provided in Annex A of this chapter.

12.2.1 *Project 23. Earth's Radius*

Recommended equipment: A telescopic sight with a crosshair reticle mounted on a tripod. A millimeter ruler about 40 cm long. A level or a transparent plastic hose with water. Access to a hill or building at least 25 or 30 m high overlooking the sea.

✓ Using the technique discussed above, determine the value of the Earth's radius R_E and estimate the uncertainty of its determination.

12.3 Hipparchus method — Size of the Moon

Some years after Aristarchus made the first estimates of the size and distance to the Moon, Hipparchus of Nicaea (190 BC–120 BC) devised a more precise method for the calculation of the lunar diameter. Hipparchus was born shortly before the death of Eratosthenes and later succeeded him as director of the Library of Alexandria. The idea of Hipparchus consists of inferring the relation of the size of the Moon with respect to the shadow of the Earth, based on the observation of the times involved in a lunar eclipse instead of on the relative sizes, as considered by Aristarchus [11].

Hipparchus's method is based on the fact that during a lunar eclipse, the modulus of the speed v_M of the Moon is constant. To obtain consistent results with the Hipparchus method, it is important that the trajectory of the Moon should pass through the center of the Earth's shadow, a condition that is not always met.

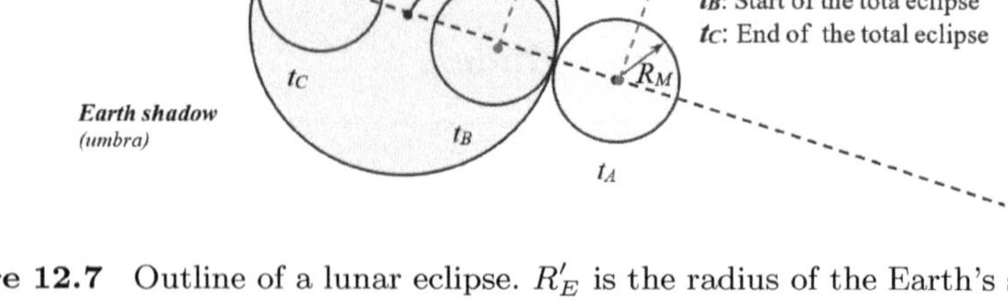

Figure 12.7 Outline of a lunar eclipse. R'_E is the radius of the Earth's shadow and R_M is the radius of the Moon. To apply the Hipparchus method, it is important that the trajectory of the Moon should pass through the center of the Earth's shadow.

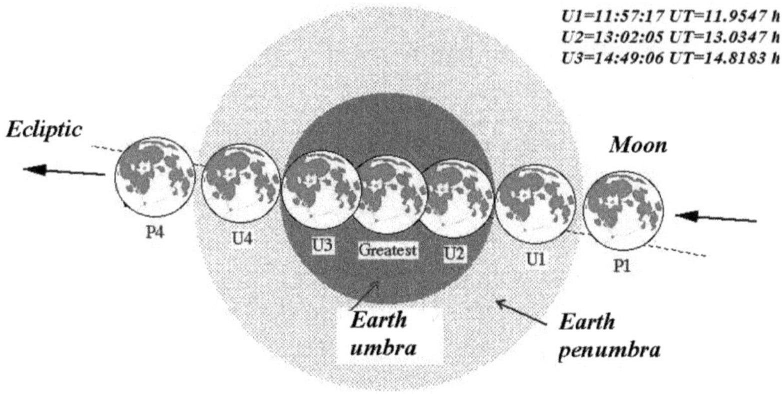

Figure 12.8 Information on the eclipse of July 16, 2000, collected by NASA [1]. The time from U1 to U2 ($\Delta t_1 = 1.080\,\text{h}$) is the dark time of the Moon. The time between U2 and U3 is the duration of the total eclipse ($\Delta t_2 = 1.7836\,\text{h}$). In this eclipse, the trajectory of the Moon passes very close to the center of the Earth's shadow.

It is easy to find an eclipse that meets this condition, like the one represented in Figure 12.8, in NASA's catalog of lunar eclipses [1, 5, 12]. Analyzing Figure 12.8 and considering the constancy of the Moon's velocity modulus, we have

$$\nu_M = \frac{2R'_E - 2R_L}{(t_C - t_B)} = \frac{2R_L}{(t_b - t_A)}, \tag{12.9}$$

defining the time intervals (see Figure 12.8)

$$\Delta t_1 = t_B - t_A, \quad \Delta t_2 = t_C - t_B, \quad \text{and} \quad \Delta t_3 = t_C - t_A = \Delta t_2 + \Delta t_1, \tag{12.10}$$

we have

$$\frac{R'_E - R_L}{\Delta t_2} = \frac{R_L}{\Delta t_1} = \frac{R'_E}{\Delta t_1 + \Delta t_2} \tag{12.11}$$

or

$$R'_E/R_M = k = \Delta t_3/\Delta t_1. \tag{12.12}$$

12.3.1 *Project 24. Determination of the size of the Moon and its distance from Earth — Hipparchus*

Recommended equipment: Real eclipse data similar to those shown in Figure 12.8, which can be found on the NASA website [1].

✓ Using the eclipse data illustrated in Figure 12.8 or similar, estimate the size of the Moon and its radius using the Hipparchus method. How do these results compare to those obtained with the Aristarchus method? Which method do you think is the most accurate and why?

✓ Compare your values to the accepted values for these quantities. What can you conclude from this analysis?

12.4 Earth-Sun distance

Aristarchus of Samos not only estimated the size of the Moon but also devised an ingenious scheme to measure the distance to the Sun. He reasoned that in its quarter phase, the Moon, the Earth, and the Sun would form exactly a right triangle. In Figure 12.9, the position of the Moon is represented in two complementary phases: first quarter

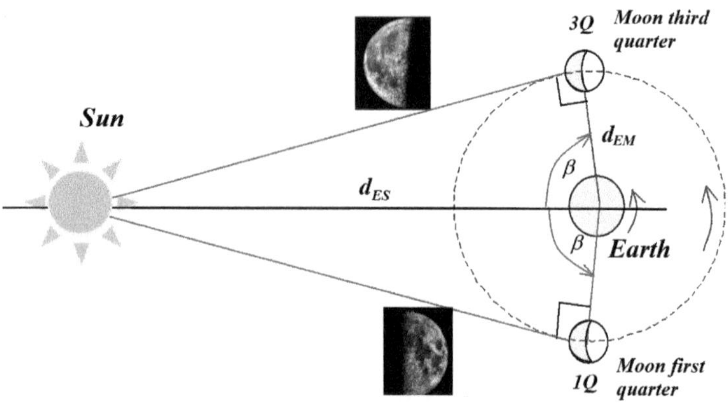

Figure 12.9 Moon phases. d_{ES} is the Sun-Earth distance and $2 \cdot \beta$ is the angle for which the Moon is in the first quarter (1Q) and the third quarter (3Q).

and last quarter. When the Moon is in these positions, the triangle formed by the Sun, the Moon, and the Earth is a right triangle, as shown in this diagram. This occurs in both the first quarter and third quarter [13].

If the Sun were infinitely far from the Earth ($d_{\mathrm{ES}} \gg d_{\mathrm{EM}}$), these two positions would be reached in the middle of the Moon's revolution period, T_{Moon}, around the Earth. The method proposed by Aristarchus consists of measuring the angle β in Figure 12.9, which relates to the time Tc between the complementary positions, i.e., the time between the third quarter and the first quarter. Since the modulus of the velocity of the Moon is approximately constant, it follows that

$$\frac{2\beta}{360^0} = \frac{T_C}{T_{\mathrm{Moon}}}. \tag{12.13}$$

Here, T_{Moon} represents the revolution period, more specifically, the synodic period of the Moon $T_{\mathrm{Moon}} = 29.53$ days. Using this technique, Aristarchus estimated $T_C = 14.25$ days, concluding that $\beta \approx 87°$. Hence, from the resolution of the corresponding triangles [13],

$$\cos\beta = d_{\mathrm{EM}}/d_{\mathrm{ES}} \Rightarrow \cos(87°) \approx 0.052 \approx 1/19, \tag{12.14}$$

that is,

$$d_{\mathrm{ES}} \approx 19 d_{\mathrm{EM}}. \tag{12.15}$$

Taking $Tc = 14.5$ days (note that to see the Moon in the first quarter and then in the third quarter, a whole number of days must pass, plus half a day, since the Moon is on opposite sides of the Earth in these two phases), $\beta \approx 88.4°$ and $d_{\mathrm{ES}}/d_{\mathrm{EM}} \approx 35$. Today we know that the relationship $d_{\mathrm{ES}}/d_{\mathrm{EM}} \approx 400$ and therefore $\beta \approx 89.85°$. The precision necessary to measure this angle is of the order of 0.02°, which is almost impossible to achieve with the Aristarchus method. What is interesting about this method is that it provided, for the first time, a way to estimate the distance from the Earth to the Sun and also managed to estimate roughly the order of magnitude of this distance. The limitation of this procedure is that the minimum angle variation that is possible to measure is 1/2 day [14], which means that the smallest possible angle difference that can be measured is

$\Delta\beta \approx (0.5d/29d) \cdot 360^0 \approx 7°$. On this basis, the most that can be said is that $d_{\mathrm{ES}}/d_{\mathrm{EM}} > 19$.

As mentioned above, the angles subtended by the Sun and the Moon from Earth are identical. It follows that $R_M/d_{\mathrm{EM}} = R_S/d_{\mathrm{ES}}$, from which

$$\text{Sun radius (km)} = R_s = (d_{\mathrm{ES}}/d_{\mathrm{EM}}) \cdot R_M \geq 33000\,\text{km}. \qquad (12.16)$$

The accepted value for the radius of the Sun is $R_s = 6.955 \times 10^5$ km.

12.4.1 *Project 25. Earth-Sun distance*

✓ On the basis that $Tc = 14.5$ days and using the Aristarchus method, estimate the value of the relation $d_{\mathrm{ES}}/d_{\mathrm{EM}}$.

✓ Discuss other methods available in the literature to improve the measurement of $d_{\mathrm{ES}}/d_{\mathrm{EM}}$.

12.5 Distances to other planets I

Venus is the second planet in the Solar System and, along with the Sun and the Moon, one of the three celestial bodies that can be seen by day. It is visible in the vicinity of the Sun, sometimes at sunrise and sometimes at sunset. The *elongation* of a planet is defined as the angular separation between the line of sight to the Sun and the line of sight to that planet. For Venus, maximum elongation occurs when the Sun, Venus, and the Earth form a right triangle like the one shown in Figure 12.10. An analogous situation occurs with Mercury, though it is not as easily visible as Venus.

The angle of maximum elongation can be easily obtained with a telescope, by observing the position of Venus over time. Actually, when Venus is observed through a telescope, it is easy to see its phases, similar to those of the Moon. In particular, in the position of maximum elongation, the phase of Venus is similar to a *waxing or waning quarter.*[c] To simplify the method and make it suitable for use in the classroom, we suggest using one of the many astronomy software applications that can be purchased on the market or downloaded from the Internet. The angle and time of maximum

[c]https://en.wikipedia.org/wiki/Lunar_phase

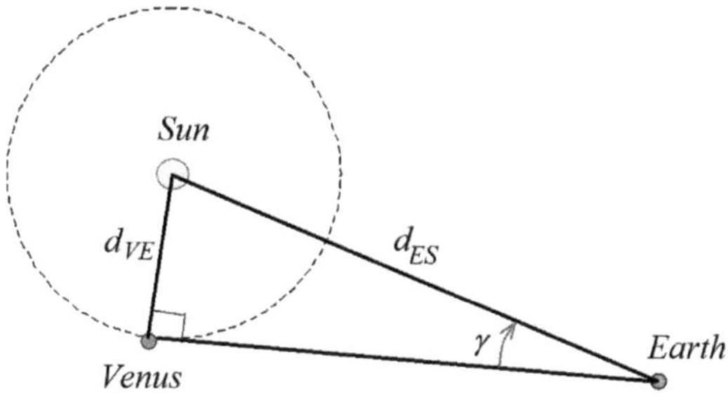

Figure 12.10 Position of Venus at its point of maximum elongation or angular separation with respect to the Sun. Observationally, this occurs when Venus is farthest from the Sun in the sky and appears stationary since it moves along a straight line with respect to the Earth. The triangle formed by the Sun, Venus, and the Earth is right-angled.

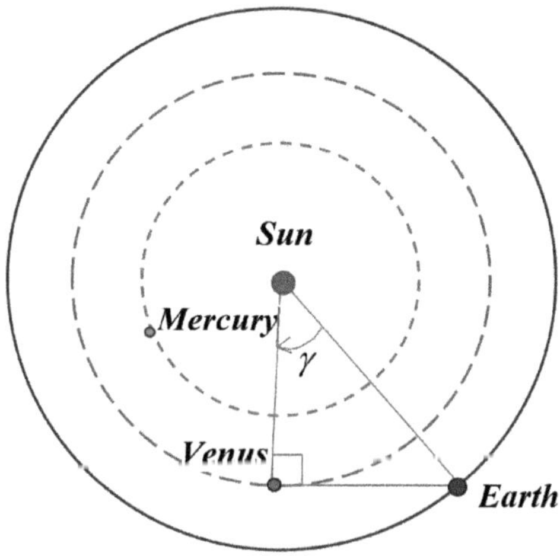

Figure 12.11 Diagram showing the positions of the Sun, Venus, and the Earth, obtained by simulation in the Winstars software. The positions correspond to January 14, 2009, and the maximum elongation angle γ can be measured on the screen or obtained through the specifications provided by the program for the planet to be studied on that date.

elongation can be determined with the software application Winstars [14] or similar. It is also possible to use data on the ephemeris of the planets that are available on the Internet [15]. These applications allow us to simulate the orbital motion of the planets (Figure 12.11) and determine the moment when Venus (or Mercury) is in its position of maximum elongation. It may also be useful to use simulations of the movement of these planets, available on various internet sites, in particular from Astronomy Education at the University of Nebraska [3]. They provide the date of occurrence and the angular positions of the Sun and the planets, from which the angle γ can be determined (see Figure 12.10). Venus was in this situation on January 14, 2009. The angle of maximum elongation is $\gamma = 46° \pm 2°$ (see Figure 12.11), which results in

$$d_{VS} = d_{ES}\sin\gamma. \tag{12.17}$$

Considering that the Earth-Sun distance, d_{ES}, also called AU (Astronomical Unit), is 149.6×10^6 km, we obtain $d_{VS} = (106 \pm 3) \times 10^6$km. This value is reasonable since the known distance is $(108{,}2 \pm 0,1) \times 10^6$km [3, 15].

The distance from Mercury to the Sun can be found in the same way. For April 26, 2009, Mercury's angle of maximum elongation was $\gamma = 19°$, therefore, $d_{\mathrm{SM}} = 50 \times 10^6$ km. This value is reasonable since the known distance is 57.9×10^6km.

Something remarkable is observed when a table of elongations for Venus and Mercury is analyzed for different occurrences of the maximum elongations because they vary constantly, in particular for Mercury. If the Copernican model of circular orbits centered on the Sun was correct, these elongations should always be the same. The fact that they are not clear indicates that the orbits of the planets are not well described by the Copernican model, and a different, more suitable model is required. Indeed, these variations of the maximum elongations can be used to trace the orbits of the planets and "discover" that they are elliptical, as established by the Kepler model.

12.5.1 *Project 26. Distance Venus-Sun and Mercury-Sun*

✓ Based on data from the ephemeris of the planets or a program to simulate the position of the planets, estimate the value of the angles of maximum elongation for Mercury and Venus.

✓ With these data, estimate the distance from the Sun to each of these planets.
✓ Compare your results to the accepted values for these parameters.
✓ Investigate some other method to measure these distances and compare it to the method discussed here.

12.6 Distances to other planets II

The method proposed in the previous section cannot be used for planets more external than Earth. However, it is possible to use a very simple geometric method, based on the observation of the times when the planets are in opposition (aligned on the same side of the Sun, so their elongation is 180°) and when they are in quadrature (the lines of sight to the Sun and the planet form an angle of 90°, i.e., the elongation of the planet is 90°).

These two situations for Earth and Mars are shown in Figure 12.12. From this figure, it is clear that

$$d_{\mathrm{MS}} = \mathrm{d}_{\mathrm{ES}}/\cos(\alpha - \beta). \tag{12.18}$$

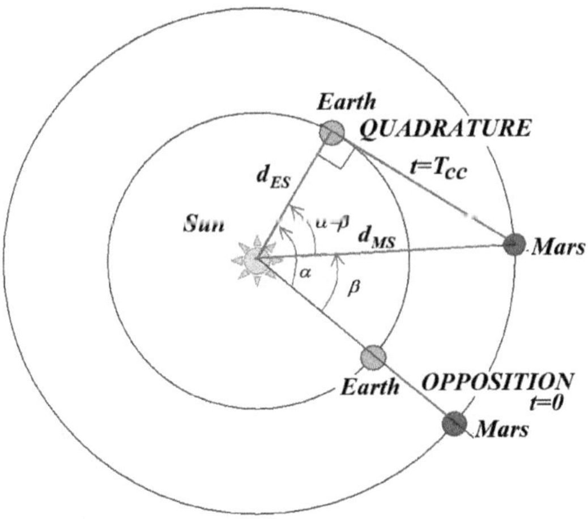

Figure 12.12 Positions of the Sun, the Earth, and an outer planet, for example, Mars at two times: in opposition and in quadrature. The time difference between these two consecutive events (quadrature and opposition) is denoted by T_{CC}.

To determine the angles α and β, measure the time T_{CC} between opposition and quadrature. If the sidereal periods of Earth and Mars are T_{Earth} and T_{Mars}, respectively, we have

$$\frac{\alpha}{360^0} = \frac{T_{\text{CC}}}{T_{\text{Earth}}} \quad \text{or} \quad \alpha = 360^\circ \times (T_{\text{CC}}/T_{\text{Earth}}) \tag{12.19}$$

and

$$\beta = 360^\circ \times (T_{\text{CC}}/T_{\text{Mars}}). \tag{12.20}$$

By combining these two expressions with Eq. (12.18), the distances to all the planets in the Solar System can be determined [3, 5]. The dates between the opposition and the quadrature of the planets can be found on the Internet [3] or from planetary motion simulation programs.

12.6.1 *Project 27. Distance to other planets*

Exercises

✓ Based on the dates of the opposition and the quadrature of Mars, obtained from the Internet [3] or from planetary motion simulation programs, estimate the values of the angles α and β (Ec.12.18).
✓ Use these data to estimate the distance from the Sun to Mars.
✓ Compare your results to the accepted values for this parameter.
✓ This method can be applied to estimate the distance to the other outer planets.
✓ Investigate some other method to measure these distances and compare it to the method discussed here.

12.6.2 *Project 28. Applications to astronomy and astrophysics — Kepler's laws and Hubble's law*

Exercises

✓ Using the data in Table 12.1, plot the length of the planetary year, T_p, as a function of its distance from the Sun, d_S. Using linear and logarithmic scales, describe the dependence of T_p on d_S.

Table 12.1 Average distance to the Sun of the different planets in the Solar System and their respective periods of rotation around the Sun (length of the year).

Planets	Distance to the Sun (km)	Orbit (days)
Mercury	5.79×10^7	87.97
Venus	1.08×10^8	224.70
Earth	1.50×10^8	365.26
Mars	2.28×10^8	686.98
Jupiter	7.78×10^8	4 328.90
Saturn	1.43×10^9	10 752.90
Uranus	2.87×10^9	30 663.65
Neptune	4.50×10^9	60 152.00
Pluto	5.91×10^9	90 717.10

✓ Kepler's third law[d] postulates that the square of the period T_p^2 is proportional to the cube of the distance d_S^3. Are your results consistent with this law?

✓ In many areas of science, there are conservation laws, such as the laws of conservation of energy or momentum. These laws establish that certain quantities do not vary before or after a given process, or that they are the same over time or when passing from one object to another in a system. In this sense, Kepler's third law establishes that the ratio T_p^2/d_S^3 is a constant for all the planets in the Solar System, i.e., this quantity is "conserved". From the data in Table 12.1, establish whether or not this statement is valid. Do your data support or refute this law?

✓ A hypothesis consistent with Newton's laws is to assume that if the planets move in circular or near-circular orbits, the Sun exerts a force on the planets that attracts them toward the center: a centripetal force. Suppose that this "gravitational" force depends on the distance r from the Sun in the form $F_{\text{grav}} = k/r^n$, where n is a real number, unknown for now. As we know, for an object to

[d]Johannes Kepler (1571–1630) was a German astronomer and mathematician known for his laws on the movement of the planets around the sun. He was a collaborator of Tycho Brahe, whom he replaced as Rudolf II's imperial mathematician.

move in a circular orbit of radius r and period T, the centripetal force necessary for it to stay in that orbit is

$$F_{\text{cent}} = F_{\text{grav}} \Rightarrow m \cdot \omega^2 r = m \cdot 4\pi^2 r/T^2 = k/r^n, \qquad (12.21)$$

where m represents the mass of the planet under study and r its distance from the Sun. From this relationship, it follows that

$$r^{n+1}/T^2 = 4\pi^2 k/m. \qquad (12.22)$$

Comparing this result with your graphical analysis, deduce the value of the exponent n of the gravitational force and how the constant k must depend on m so that Kepler's third law is fulfilled for all the planets in the Solar System.

Compare your results with what is known about the gravitational force. Can it be concluded that Kepler's laws lead to Newton's law of universal gravitation?

12.6.3 *Project 29. Expansion of the Universe and the Big Bang — How do we know this? When did it happen?*

Edwin Hubble[e] and his assistant Milton Humason[f] made one of the most remarkable discoveries in science. Using the 100" telescope at Mount Wilson, California, they discovered that the spectra of

---

[e] E. Hubble was born in Missouri in 1889. In his youth, he was an outstanding athlete and boxer, but in his early years of college, he became interested in mathematics and astronomy. However, he earned a scholarship to study law and became a lawyer. After practicing this profession for some time, he returned to the University of Chicago and earned a Ph.D. in astronomy in 1917. Before beginning his work as an astronomer at Mount Wilson, he served as an officer in World War I.

[f] M. Humason only went to school up to eighth grade. He worked as a muleteer during the construction of the Mount Wilson Observatory, in the days when the parts were transported by mules and donkeys. When the observatory was completed, he secured a job as a janitor. But he was curious, intelligent, and very interested in his work and always asked astronomers about what they were doing. In this way, he soon learned to use the telescope, and Hubble hired him as his collaborator. Humason's work contributed significantly to the discovery that the Universe is expanding.

light from distant galaxies were systematically "redshifted". This is a consequence of the Doppler effect [9, 10]. When a galaxy approaches us, the frequency of the light coming from it increases, i.e., the light appears bluish ("blueshift"). This is analogous to the fact that when a sound source approaches us, its frequency increases in proportion to its speed, while if the source moves away, the frequency of the sound becomes lower, or lower in proportion to its speed. In the same way, when a galaxy moves away from us, the frequency of the light coming from it decreases (it turns red, "redshift") in proportion to its recessional velocity.

The remarkable thing about Hubble and Humason's discovery is that they found that all the galaxies had a "redshift", i.e., they all seemed to move away from us at a speed (measured by the redshift of their spectrum) that was greater the farther away they were. A partial list of the Hubble data is shown in Table 12.2.

✓ Using these data, graph the variation in the recessional velocity of distant galaxies as a function of their distance. What kind of dependency do you see? Can these data be described by a linear relationship?

✓ If the dependence of the velocity v as a function of the distance d is linear, i.e., if $v = H\ d$, determine the value of H in units of km/s/Mpc and in 1/s.

✓ An interesting consequence of the relationship $v = H\ d$ is that, if we observe the universe, it is expanding. Therefore, in the past, it must have been smaller. If we "rewind the movie", we may wonder how long ago, T_{BB}, a given galaxy that today is at a distance d_0, was next to us. If the recessional velocity was constant, we could say that $T_{BB} = d_0/v = d_0/(H\ d_0) = 1/H$. Therefore, if $v = H\ d$, the time T_{BB} does not depend on the distance d_0 and is the same for all galaxies. In other words, T_{BB} measures the time that all the galaxies were together, including our own. From that moment, the universe was in expansion, as observed in the present. Thus, $T_{BB} = 1/H$ represents the time at which the universe began — the moment of the Big Bang! All the protons, neutrons, and electrons that make up our bodies, the Earth, the Sun, and the rest of the galaxies were formed at that moment. In other words,

Table 12.2 Receding speed of distant galaxies as a function of their distance. The speeds are in km/s and the distances in Mega Parsecs (Mpc),[g] which are equivalent to 3 million light years, i.e., 1 Mpc = 3.0857×10^{19} km. (Data taken from the original Hubble and Humason publication.)

Distance (MPc)	**Velocity (km/s)**	**Distance (MPc)**	**Velocity (km/s)**
2.05	308	27.6	2,002
2.07	96	28.2	992
4.66	467	28.7	780
4.68	255	30.7	1,789
8.3	732	31.7	2,373
9.9	519	34.9	1,788
13.0	731	39.0	3,221
13.5	943	45.7	3,857
15.6	1,314	46.1	4,707
16.1	942	48.9	2,900
16.6	1,580	50	3,856
19.3	941	58	3,216
20.8	1,791	82	4,273
24.4	1,790	84	5,122
25.5	1,365	86	8,682
26.4	2,746	90	5,121
26.5	2,321	111	6,922
27.6	2,002	158	9,619

what we are looking for is precisely the birthday of the universe.[h] Compare the time found from your graph with other times that provide information about the origin of the universe. Are your data consistent with what is known about when the Big Bang occurred?

12.7 Annex A: Path of a ray of light in the atmosphere

This annex provides a heuristic justification for the deflection of light in the atmosphere, assuming a flat Earth. Imagine that the

[g] https://en.wikipedia.org/wiki/Parsec
[h] https://en.wikipedia.org/wiki/Big_Bang

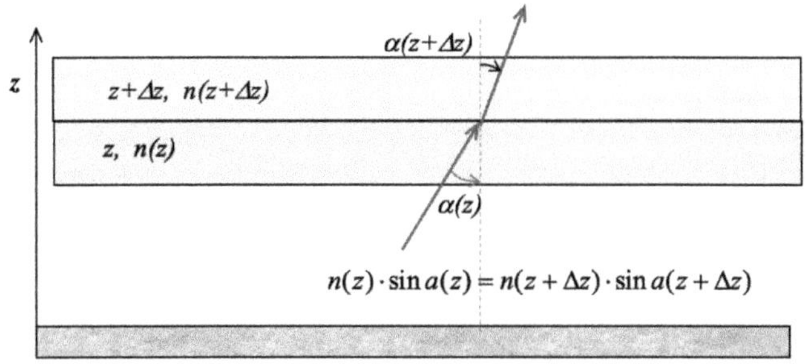

Figure 12.13 Schematic path of light rays when passing from an optimally denser medium to a less dense one.

atmosphere has layers of decreasing refractive indices $n\ (z)$, as shown schematically in Figure 12.13.

By Snell's law, we have

$$n(z) \cdot \sin\theta(z) = n(z + \Delta z) \cdot \sin\theta(z + \Delta z), \tag{12.23}$$

and in general,

$$n(z_1) \cdot \sin\alpha(z_1) = n(z_2) \cdot \sin\alpha(z_2), \tag{12.24}$$

where z_1 and z_2 are two different heights. In particular, if $z_1 = z_i$ $(=h)$ and $z_2 = z_f$ ($z_f = 0$ at sea level), according to (12.23), we have

$$\sin\alpha_f = \frac{n(z_i)}{n(z_f)} \cdot \sin\alpha_i = \left[\frac{1 + D \cdot \exp(-z_i/h_{00})}{1 + D \cdot \exp(-z_f/h_{00})}\right] \cdot \sin\alpha_i. \tag{12.25}$$

This expression contains the relevant factors of the expression (12.8) and explains the curvature of the light rays that start at an angle with respect to the vertical, $\alpha_i \neq 0$. Figure 12.14 shows the estimates of angles θ_i and δ expected using (12.8) and (12.5), respectively.

12.8 Annex B: Periods of the Moon

The sidereal period of the Moon (T_{Sid}) is the time that the Moon takes to travel 360° with respect to the distant stars. This period is 27.3216 days; see Figure 12.15. The *synodic period* ($T_{\text{Moon}} = 29.53$ días) is the time it takes for the Moon to reappear at the same point

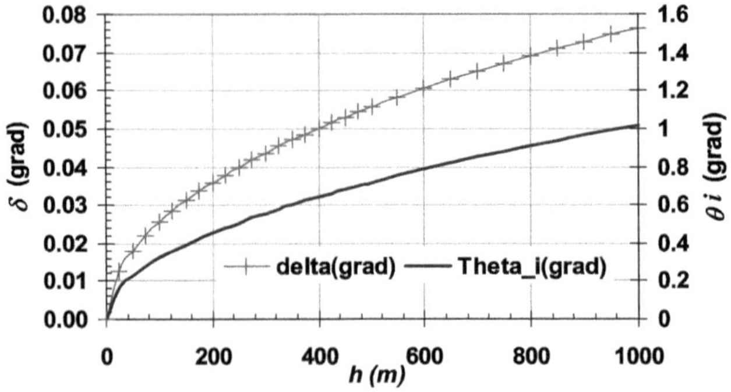

Figure 12.14 Correction angle, $\delta = \theta_f - \theta_i$, by effect of curvature of light rays, Eq. 12.8, crosses referred to the left vertical axis and angle θi as a function of the height h, referred to the right vertical axis.

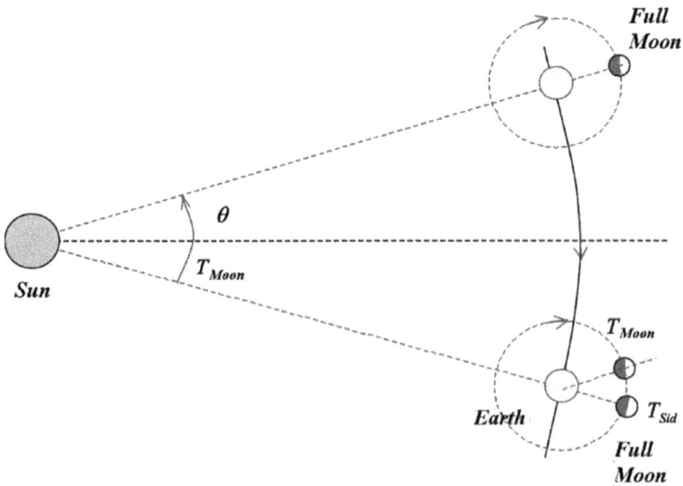

Figure 12.15 Diagram showing the respective positions of the Sun, Earth, and Moon.

in the sky relative to the Sun. From Figure 12.15, we see that

$$T_{\text{Moon}} = T_{\text{Sid}} + \frac{\theta}{360^0} T_{\text{Moon}}. \tag{12.26}$$

or

$$T_{\text{Moon}} = T_{\text{Sid}} \Big/ \left(1 - \frac{T_{\text{Sid}}}{T_{\text{Earth}}}\right). \tag{12.27}$$

Thus,

$$\frac{1}{T_{\text{Sid}}} = \frac{1}{T_{\text{Moon}}} + \frac{1}{T_{\text{Earth}}}, \tag{12.28}$$

where T_{Earth} is the Sidereal period of the Earth around the Sun (365 d).

It can be shown that the relation between the sidereal period (T_{Sid}) of a given planet and its synodal period (T_{Sin}) is given by

$$\frac{1}{T_{\text{Sid}}} = \frac{1}{T_{\text{Sin}}} - \frac{1}{T_{\text{Earth}}}, \tag{12.29}$$

if $T_{\text{Sid}} < T_{\text{Earth}}$, i.e., if the planet is inside the Earth's orbit, and

$$\frac{1}{T_{\text{Sid}}} = \frac{1}{T_{\text{Sin}}} + \frac{1}{T_{\text{Earth}}}, \tag{12.30}$$

if $T_{\text{Sid}} > T_{\text{Earth}}$, i.e., if the planet is outside the Earth's orbit.

References

[1] NASA, "Eclipses and transits," 2021. Available: https://www.nasa.gov/eclipse.

[2] NASA, "Moon in motion," 20121. Available: https://moon.nasa.gov/moon-in-motion/eclipses/.

[3] University of Nebraska-Lincoln, "Astronomy education," 2021. Available: http://astro.unl.edu/naap/ssm/modeling2.html.

[4] NASA, "The solar system exploration," 2021. Available: http://solarsystem.nasa.gov/index.cfm.

[5] P. Núñez, S. Carderón and S. Gil, "Astronomical measurement in a classroom setting (Mediciones astronómicas desde el aula)," *Lat. Am. J. Phys. Educ.*, vol. 3, no. 1, pp. 87–92, 2009.

[6] Hellenic Mathematical Society, Greece, "Eratosthenes experiment," 2021. Available: https://eratosthenes.ea.gr/.

[7] Department of Physics, UBA, Buenos Aires, Argentina, "Eratosthenes Project," 2021. Available: http://latierraesredonda.df.uba.ar/2020_en.html.

[8] D. Gangadharan, "Measuring the radius of the Earth from a Mountain top overlooking the Ocean," *Phys. Teach.*, vol. 47, no. 8, pp. 533–535, 2009.

[9] D. Giancoli, *Physics Principles with Applications*, 7th Ed., NY: Pearson, 2014.

[10] J. D. Wilson, *College Physics*, 7th Ed., NY: Pearson, 2009.

[11] J. J. Birriel and J. K. Adkin, "Estimating the size of Earth's umbral shadow using sky brightness light curves during a lunar eclipse," *Am. J. Phys.*, vol. 87, no. 12, pp. 994–996, 2019.
[12] NASA, "Lunar eclipses: Past and future," 2021. Available: https://eclipse.gsfc.nasa.gov/lunar.html.
[13] S. Derman, "Aristarchus at your service," *Phys. Teach.*, vol. 38, p. 574, 2000.
[14] WinStar, "WinStars is a planetarium running on computers, tablets and mobiles," 2021. [En línea]. Available: https://winstars.net/en/.
[15] NASA, "Twelve Year Planetary Ephemeris," 2021. Available: https://eclipse.gsfc.nasa.gov/TYPE/TYPE.html.
[16] S. Derman, "Correction: 'Aristarchus at your service'," *Phys. Teach.*, vol. 38, no. 3, p. 113, 2001.

Chapter 13

ELASTIC SYSTEMS — HOOKE'S LAW

This chapter studies the elastic behavior of springs and other elastic systems. In particular, it analyzes the behavior of coil springs, elastic bands, and some combinations of them.

Goals

- ✓ Behavior of springs and elastic bands
- ✓ Hooke's law
- ✓ Series and parallel springs

13.1 Introduction to linear elastic systems

In nature, there is a great variety of systems that have elastic properties with common properties, such as springs or elastic bands. A common characteristic of these systems is that, within certain limits, they present a linear response, i.e., their stretching is directly proportional to the applied force[1]. In about 1684, Robert Hooke, a contemporary of Newton, found that the stretch x undergone by a spring was approximately proportional to the applied force, F, that is,

$$F = -kx. \tag{13.1}$$

This relationship is known as *Hooke's law* and k is known as the spring constant. It is important to note that "Hooke's law" is not strictly a law, such as the law of conservation of energy or of linear momentum. "Hooke's law" simply describes a system or spring that obeys the relation (13.1). Perhaps it could be regarded as a definition of the *linear behavior* of certain elastic systems. Thus, when we speak of Hooke's "law", it is in a figurative sense, following the tradition.

13.1.1 *Project 30. Elastic properties of a spring*

Recommended equipment: A pair of helical springs, which stretch a distance of the order of a centimeter when a weight of about 100 g is hung from them. A ruler graduated in millimeters, a set of weights from 50 g to 500 g.

Hang different weights from a metal spring (coil spring) and study the dependence of the applied force, $F = mg$, with the elongation that it produces, $x = l - l_0$, where l_0 is the original length of the spring without load and l its loaded length. You can use an experimental setup such as the one shown in Figure 13.1. To facilitate reading the scale, it is convenient to attach a wire to the weight so that it acts as a pointer or indicator of its position relative to the scale and makes it easy to read the length of the spring. For some springs, it can be difficult to determine the value of l_0 because the spring in the absence of load can have deformations due to its own weight or because some springs have an original tension and require a minimum force F_0 to begin stretching. These effects are manifested in the appearance of an additive constant or a finite ordinate at the origin, in Eq. (13.1).

Suggested tasks:

✓ Graph the applied force $F = P$ as a function of elongation x. What relationship do you find between these quantities?

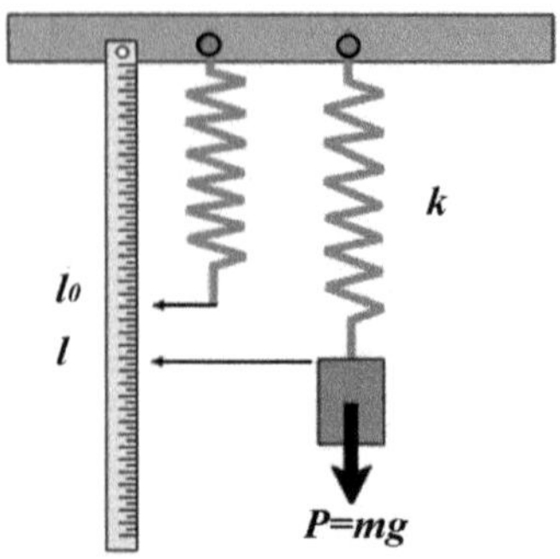

Figure 13.1 Diagram of a device for studying the elastic properties of a spring. It seeks to establish the relationship between force P and elongation x, ($x = l - l_0$).

✓ Do the springs that you studied obey "Hooke's law"? Do you get the same behavior when you unload the spring as when you load it? If so, determine the spring constant k. To do this, use the graph of F as a function of x and determine the line that best fits the experimental points, using the method of least squares (Chapter 7). The slope of the line determines the value of k. Pay attention to the units in which you express F and x. Discuss the physical meaning of this constant and what its units are.

✓ Using the techniques discussed in Chapter 7, determine the uncertainty or error in the value of k found.

Sometimes, when fitting a line to the points, a non-zero value of the ordinate at $x = 0$ is found, i.e., the line that best fits the data has the form $F = kx + b$. In general, the value of b is associated, as indicated above, with the fact that some springs have an original tension or compression, even without load or deformations. A finite value of the ordinate at the origin can also be a consequence of the choice that was made to define the origin of the x axis. If we write $F = k(x+b/k)$ and we define $z = (x+b/k)$, then $F = kz$, which recovers the original form of Hooke's law.

13.1.2 *Project 31. Elastic properties of an elastic band*

Recommended equipment: A pair of elastic bands. A ruler graduated in millimeters, a set of weights from 50 g to 500 g.

Using an elastic band (which can be clothing elastic or a rubber band) perform the same analysis as you did for the spring.

Suggested tasks:

✓ Graph the applied force F as a function of elongation x. In this first part, make sure that the stretched length l does not exceed the initial length (without load) l_0 by more than 50%. What relationship do you find between these quantities?

✓ Repeat the same graph of F as a function of x but this time load the system until the stretched length l exceeds twice the initial length, l_0. What relationship do you find between these quantities?

✓ Do the elastic bands studied obey "Hooke's law"? Discuss the differences and analogies between springs and rubber bands.

13.1.3 *Project 32. Series and parallel spring systems*

Recommended equipment: Two or more springs, similar to those used in the previous activities, or elastic bands. A ruler graduated in millimeters, a set of weights from 50 g to 500 g.

Consider two springs, one with an elastic constant k_1 and the other with k_2, which are connected in series, i.e., one after the other, as shown in Figure 13.2(a)), or in parallel, as shown in Figure 13.2(b)).

Series: In this case, the net stretch of the system is the sum of the stretches of each spring $x = x_1 + x_2$. Suppose that the weight $P = mg$ hung on them is much greater than the weight of either of the springs, which is usual in experiments with small springs. In this way, the force on each spring is the same and equal to P. Then,

$$x = x_1 + x_2 = \frac{P}{k_1} + \frac{P}{k_2}. \tag{13.2}$$

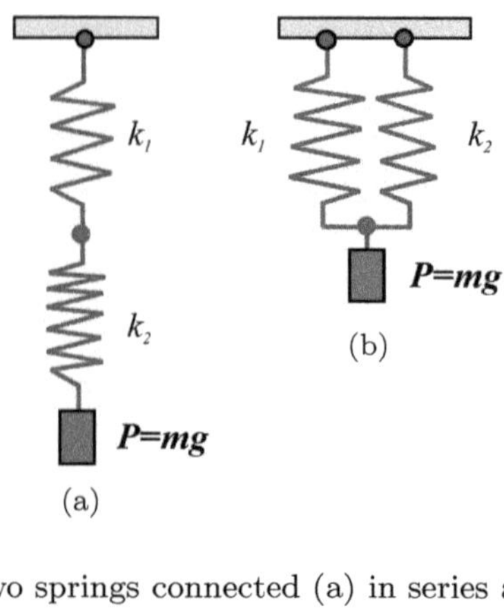

Figure 13.2 Two springs connected (a) in series and (b) in parallel.

The series spring can be considered as a new spring with an equivalent spring constant $k_s = P/x$, thus,

$$\frac{x}{P} = \frac{1}{k_s} = \frac{1}{k_1} + \frac{1}{k_2}, \tag{13.3}$$

or also,

$$k_S = \frac{k_1 k_2}{k_1 + k_2}. \tag{13.4}$$

The extension of this analysis for N springs in series is

$$\frac{1}{k_S} = \sum_{i=1}^{N} \frac{1}{k_i}. \tag{13.5}$$

Parallel: If the springs are connected in parallel (Figure 13.2(b)) and a force P is applied to them, the stretch of both will be identical ($x = \Delta l_1 = \Delta l_2$), although each spring will support different forces, F_1 and F_2. In equilibrium, it is true that

$$P = F_1 + F_2, \tag{13.6}$$

where $F_1 = k_1 \Delta l_1 = k_1 x$ and $F_2 = k_2 \Delta l_1 = k_2 x$. From (13.6), it follows that the equivalent spring constant of this system is $k_p = P/x$, that is,

$$k_P = k_1 + k_2 \tag{13.7}$$

For N springs in parallel, the previous result is generalized as follows:

$$k_P = \sum_{i=1}^{N} k_i \tag{13.8}$$

Suggested tasks:

- ✓ Choose two or more springs and measure the elastic constant k_i for each.
- ✓ Place the springs in a series configuration and measure the spring constant of the assembly, k_S.
- ✓ Do the same with the two springs arranged in parallel, i.e., determine k_P.
- ✓ Compare the measured values of the equivalent constants k_S and k_P with the expected values from the developed models, expressions (13.5) and (13.8).

13.2 Nonlinear elastic systems

13.2.1 *Project 33. Nonlinear elastic system [2, 3]*

Recommended equipment: Two equal light springs or a simple elastic band. A ruler graduated in millimeters, a set of weights from 50 g to 500 g.

Consider two springs of equal elastic constant k and natural length l_0 which are arranged in series, as shown in Figure 13.3(a). This time, the force is applied laterally and we wish to study the dependence of the applied force F_z with the lateral displacement z.

With reference to Figure 13.3(a), the force F_z can be written as follows:

$$F_z = 2F \cdot \cos(\theta) = 2k(l - l_0) \cdot \frac{z}{l}. \tag{13.9}$$

From the geometry of the system,

$$l = \sqrt{l_0^2 + z^2} = l_0\sqrt{1 + (z/l_0)^2} \approx l_0 \left(1 + \frac{z^2}{2l_0^2} - \frac{z^4}{8l_0^4} + \cdots\right), \tag{13.10}$$

from where

$$F_z \approx k\frac{z^3}{l_0^2}\left(1 - \frac{z^2}{4l_0^2}\right) \approx k\frac{z^3}{l_0^2}, \quad \text{for } z << l_0. \tag{13.11}$$

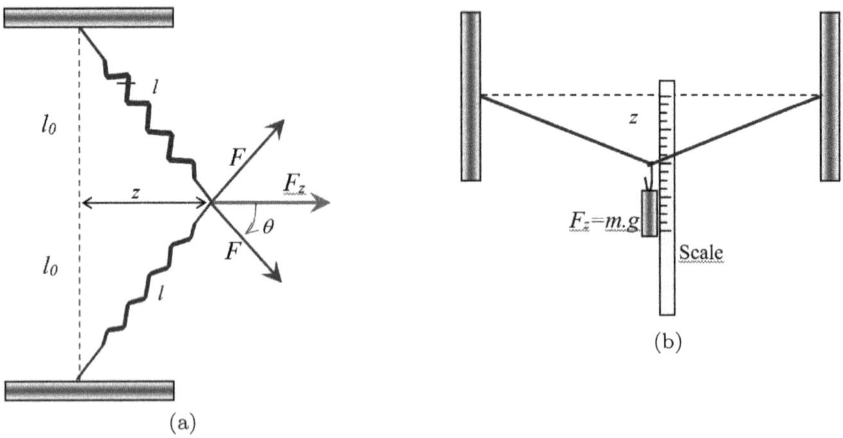

Figure 13.3 Nonlinear configurations of simple springs. (b) shows an experimental setup to study this system. Equal springs or an elastic band can be used.

This provides a simple way to achieve a *nonlinear* elastic force from *linear* springs. This type of nonlinear recovery force is very useful to build nonlinear oscillatory systems, which present very rich and interesting dynamics. Moreover, many characteristics of the chaos theory can be studied in them [2, 3]. These oscillators are very sensitive to variations in initial conditions. Small variations in the starting point can produce large differences in future behavior, complicating long-term prediction.

Suggested tasks:
Using two identical lightweight springs or a rubber band build a system similar to the one shown in Figure 13.3(b). Make sure that the initial stretch (without weight) of the springs or the elastic band is just enough so that the system is horizontal.

✓ Study experimentally the variation of the lateral elongation z as a function of the applied force $P = m.g$.
✓ Construct a graph of P as a function of z. Try to fit your results using a line through the origin and a cubic parabola as suggested by Eq. (13.11). Which one best reproduces your results?
✓ Discuss the validity of the expression (13.11) to reproduce your results.

References

[1] F. Sears, M. Zemansky, H. Young and R. Freedman, *College Physics*, NY: Pearson Ed., 2009.
[2] R. Khosropour and P. Millet, "Demonstrating the bent tuning curve," *Am. J. Phys.*, vol. 60, no. 5, p. 429, 1992.
[3] R. Dorner, L. Kowalski and M. Stein, "A nonlinear mechanical oscillator for physics laboratories," *Am. J. Phys.*, vol. 64, no. 5, p. 575, 1996.

Chapter 14

NEWTON'S LAWS AND FRICTION FORCE

The objective of this chapter is to carry out an experimental study of Newton's laws and the static and dynamic friction forces between a body and a dry surface. It proposes several ways to determine friction coefficients.

Goals

✓ Newton's laws
✓ Frictional force
✓ Determination of static and dynamic friction coefficients

14.1 Coulomb dry friction force

Friction is a frequent phenomenon in nature. *Dry friction* (also called *Coulomb friction*) [1, 2] often occurs between two surfaces in contact without lubrication. A classic example is a block resting on a table or an inclined plane. To set in motion a body that is resting on a table or other dry surface, it is often necessary to apply a force that exceeds the *maximum static friction* force, $f_r = f_s^{\max}$. Once the applied force reaches this value, the body begins to move. Everyday experience indicates that a somewhat smaller force is needed to keep the body moving [1, 3]. The minimum force that must be applied to the body to keep it moving is called the *dynamic friction* force f_d. Experience indicates that both the maximum static friction force and the minimun force necessary to keep the body in motion are proportional to the weight of the body, or more exactly, to the normal force that the support surface exerts on it, that is,

$$f_r = f_s^{\max} = \mu_s N \tag{14.1}$$

and

$$f_d = \mu_d N. \tag{14.2}$$

Here, N represents the normal force that the support surface exerts on the body and μ_s and μ_d are two constants, characteristic of the surfaces in contact, called *coefficients of static and dynamic friction*, respectively [1, 3]. From the above, $\mu_e > \mu_d$.

Equations (14.1) and (14.2) are phenomenological laws that generally have a limited range of validity. If a force F is applied to a mass such that $\mu_d mg < F < \mu_s mg$, and the body is disturbed to set it in motion, the friction force depends on the speed of the body [4]. This chapter proposes to study the validity of Eqs. (14.1) and (14.2).

Many suggestions on how to perform this type of experiment using a smartphone, which makes it easier to measure these effects, appeared in the literature in recent years [5–8].

14.1.1 *Project 34. Determination of the static friction coefficient*

Recommended equipment: A block of wood or plastic to which weights can be added, e.g., a wooden or plastic box. An inclined plane such as a board 50 to 100 cm long. A set of weights of masses from 50 g to 500 g. A millimeter ruler.

One way to test Eq. (14.1) and determine the coefficient of static friction, μ_e, between a body and a supporting surface is illustrated in Figure 14.1.

The experiment consists of determining the minimum angle θ_c that will set the body in motion. By tilting the plane, the mass M will remain at rest as long as the component of its weight parallel to the plane $F_p Mg\sin(\theta)$ does not exceed the maximum static friction force value $f_s^{\max}$. According to Eq. (14.1), the static friction force is given by $f_s^{\max} = \mu_s N = \mu_s Mg\cos\theta$, where N is the normal force or reaction that the plane exerts on the body. For the movement to

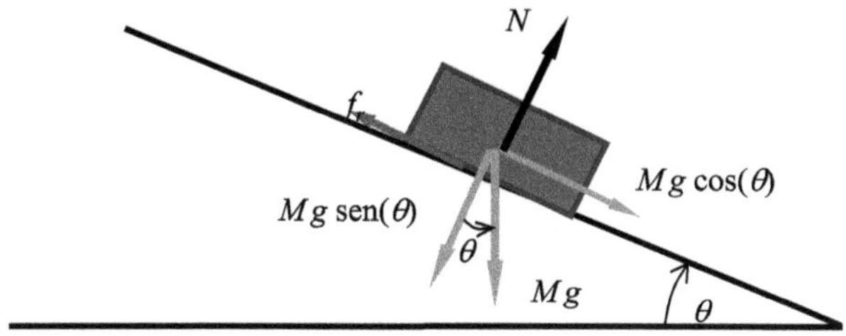

Figure 14.1 A body on an inclined plane remains at rest if the inclination θ is such that the component of its weight Mg sin(θ) does not exceed the maximum value that the static friction force can take on $f_s^{\max} = \mu_e N = Mg\cos(\theta)$.

start, the applied force F must satisfy $F \geq f_s^{\max}$, that is,

$$Mg\sin\theta \geq \mu_s Mg\cos\theta \quad \text{or} \quad \tan\theta \geq \mu_s. \tag{14.3}$$

Thus, the *minimum angle* θ_c at which the plane must be inclined to start the movement is related to the static friction coefficient as follows:

$$\mu_s = \tan\theta_c, \tag{14.4}$$

which makes it possible to determine μ_s by measuring θ_c. Note, however, that if Eq. (14.1) were of the form $f_r = f_s^{\max} = \mu_s N^\beta$, with β being an exponent $\neq 1$, from what was discussed above, it is clear that θ_c would depend on the mass of the body. Therefore, if θ_c is *independent* of body mass, this would be consistent with $\beta = 1$, and Eq. (14.1) is an adequate description of static friction.

An interesting alternative is to attach a smartphone to the top of the block in Figure 14.1 using double-sided adhesive tape so that the phone moves along with the block [7]. The smartphone can be used to detect the beginning of the motion using an accelerometer and even the angle [6].

Suggested tasks:

✓ Using wooden blocks or books for support, vary the tilt angle of your inclined plane, smoothly and controllably. If the length of the plane and the height of its highest part are known, the angle θ of inclination can be determined from the geometry.

✓ Determine the angle of inclination at which the body begins to move.
✓ Place an additional weight on top of the body but maintain the same contact surfaces. Determine in each case the angle θ_c at which the mass begins to move. Determine θ_c for at least five states of load or mass of the body. Within your measurement errors, does θ_c vary with total weight of the body?
✓ Find μ_s and estimate its uncertainty, which is associated with the determination of the angle θ_c.
✓ Expression (14.4) indicates that the coefficient of static friction does not depend on the mass (or total weight) of the body. Do your experiments verify this?
✓ What is the friction force when the plane is inclined at an angle less than θ_c? Before the body begins to move, what is the value of the friction force? In this case, is it true that $f_r = \mu_d N$? Explain carefully what happens in each case.

14.1.2 *Project 35. Determination of the coefficient of kinetic friction*

Recommended equipment: A block of wood (blackboard eraser) connected by an inextensible cotton string to a weight. A "smart pulley" connected to a photoswitch connected to a PC. A set of weights of masses from 50 g to 500 g. These measurements can also be taken by using a digital camera in video mode to study the movement of the system or an ultrasonic motion sensor to measure the position of the moving block as a function of time. Another alternative is to attach a smartphone to the horizontal block to measure its acceleration.

This activity intends to study the validity of Eq. (14.2) to describe dynamic friction force. A device like the one illustrated in Figure 14.2 can be used.

Mass M_1 could be a box (or block) that rests on a flat horizontal surface. This box can be easily loaded and unloaded by adding weights or a smartphone can be placed in it. The weights must remain

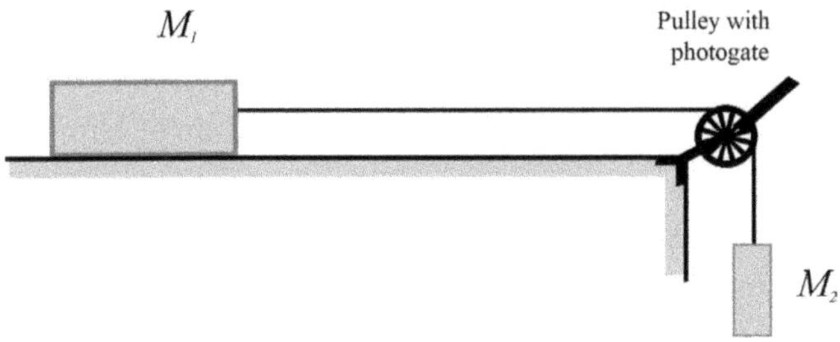

Figure 14.2 Experimental device to study the basic characteristics of the dynamic friction forces on dry surfaces. Use some device to prevent mass M_1 from colliding with the pulley.

fixed relative to the box during the movement. The movement is caused by releasing mass M_2, which is attached to mass M_1 by means of a string. A "smart" pulley with photogate [9, 10] is used to measure times and determine the speed of moving bodies, and a smartphone can be used to measure the acceleration [5]. Make sure the pulley is strong enough to support the weight of M_2 until the system begins to move. For this experiment, level the system so that the plane on which M_1 moves is horizontal. Use your hand or some other device to prevent mass M_1 from colliding with the pulley and breaking it.

The goal of the smart pulley associated with the photogate (see Chapter 9) is to determine the position of the system as a function of time. Of course, a digital camera (see Chapter 10), an ultrasonic motion detector, or a smartphone can also be used for this purpose.

Choose the masses so that movement is easily detected with the system used for measuring position according to time. The movement should be neither "too slow" nor "too fast". Also, start from a value of M_1 that is at least twice the mass of the unloaded box. Choose mass M_2 so that the system will move even for the largest value of M_1.

If the pulley mass M_p is small compared to the rest of the masses of the system (M_1 and M_2) and the dynamical friction force, f_d, is proportional to the value of the force N normal to the plane, that is, $f_d = \mu_d N$ (see Annex A), then according to Newton's second law, the acceleration can be written as follows:

$$a = g(M_2 - f_{\mathrm{d}}/g)/(M_2 + M_1). \tag{14.5}$$

So the acceleration is constant and we would expect linear dependence of the speed, v, of the system with respect to time t. The expression can also be written, introducing the pseudo-variable ξ, as follows:

$$\xi \equiv (M_2 + M_1)\frac{a}{g} = M_2 - f_{\mathrm{d}}/g = M_2 - \mu_d M_1, \tag{14.6}$$

which implies linear dependence of ξ with respect to M_1. Thus, we have two experimental indicators of the validity of the hypothesis (14.2): (a) the linear dependence of v on t or, equivalently, a constant value of the acceleration a and (b) the linear dependence of ξ on M_1.

Suggested tasks:

- ✓ Construct a graph of the velocity v of the system as a function of time for a given value of M_1. Alternatively, if you use a smartphone, plot a graph of the acceleration a of the system as a function of time. If the graph of v versus t is linear, determine the acceleration and its uncertainty for each value of M_1, see Chapter 7.
- ✓ Vary M_1, keeping M_2 fixed. Repeat this operation for at least five different values of M_1. In each case, construct a graph of v as a function of time and check whether it is linear or, alternatively, verify whether a is constant with t. If these graphs show a linear relationship, determine the acceleration of the system in each case, i.e., for the different values of M_1.
- ✓ Graph the pseudovariable $\xi \equiv a(M_1 + M_2)/g$ as a function of M_1. Annex A describes the movement of this system when the mass of the pulley is not negligible.
- ✓ Do your experimental results agree with Eq. (14.6)? To answer this question, analyze whether the pseudovariable graph $\xi \equiv a(M_1 + M_2)/g$ as a function of M_1 shows linear dependence.
- ✓ If the dependence of v on t and of ξ on M_1 is *linear* (or a is approximately constant over time), we can support the hypothesis expressed by Eq. (14.2). If so, we can find the coefficient μ_d from the slope of the graph of ξ with respect to M_1. Determine from your experimental results the coefficient of dynamic friction μ_d. Determine the uncertainty, see Chapter 7.

✓ Repeat the study carried out so far with another box (M_1) of the same material but in such a way that the contact area with the horizontal plane is approximately half that used previously. Do you note a significant variation in the value of μ_d with the area in contact?

14.2 Annex A: Study of the motion of the two-body system with dry friction

To study this system analytically, we apply Newton's laws to each of the bodies that make up the system (free-body diagram). We will assume that the dynamic friction force is proportional to the normal force, that is, $F_d = \mu_d N$.

Here, I_p is the moment of inertia of the pulley, R_p its radius, and $\alpha = a/R_p$ its angular acceleration. From the equations of motion of each of the bodies that make up the system, we obtain

$$(M_1 + M_2 + I_p/R_p^2)a = M_2 g - F_r. \tag{14.7}$$

If $F_d = \mu_d M_1 g$, we have

$$(M_1 + M_2 + I_p/R_p^2)\frac{a}{g} = M_2 - \mu_d M_1. \tag{14.8}$$

The parameters M_1, M_2, I_p, R_p, a, and g are accessible through direct measurements. Therefore, if we carry out an experiment keeping M_2 constant and by defining

$$\xi = (M_1 + M_2 + I_p/R_p^2) \tag{14.9}$$

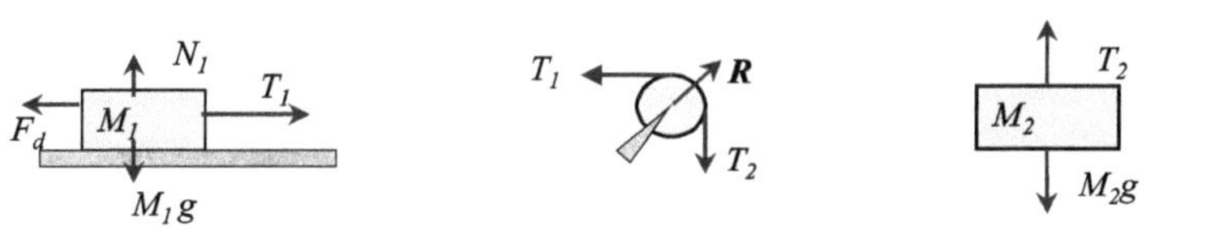

$$N_1 = M_1 g$$

$$M_1 a = T_1 - F_d$$

$$\vec{R} = \vec{T}_1 + \vec{T}_2$$

$$I_p \frac{a}{R_p} = R_p(T_2 - T_1)$$

$$M_2 a = M_2 g - T_2$$

Figure 14.3 Free-body diagrams of each part of the system in Figure 14.2. R_p is the radius of the pulley. T_1 and T_2 are the tensions in the string. R is the force applied to the pulley axis by the table. F_d is the same friction force.

and

$$x = M_1, \tag{14.10}$$

the expression (14.8) can be written as follows:

$$\xi = M_2 - \mu_d x. \tag{14.11}$$

Therefore, if the hypothesis $f_d = \mu N$ is satisfied (dynamic friction force $= F_r$), the dependence of the parameter ξ as a function of x is linear. If this hypothesis were not fulfilled, it is clear that the dependence of ξ with x would cease to be linear. In other words, whether or not the hypothesis $f_d = \mu N$ is valid is reduced to whether or not the dependence of the variable ξ on x is linear. On the other hand, if the dependence between these variables is indeed linear, the slope of the line expressed by (14.11) provides the value of μ_d and the y-intercept should match M_2.

Summary of important concepts

✓ Describe the dependence of the static and dynamic friction force on the area of the surfaces in contact.

✓ Does the dry friction force depend on the speed? How do you know this?

✓ How did you falsify the hypothesis that the dynamic friction force is proportional to the normal force?

✓ Are the values of the static friction coefficients μ_e and dynamic μ_d different? Which is greater?

References

[1] R. Halliday, D. Resnick and M. Krane, *Fundamentals of Physics*, NY: John Wiley & Sons, Inc., 2002.

[2] Wikipedia, "Friction," 2021. [Online]. Available: https://en.wikipedia.org/wiki/Friction.

[3] F. Sears, M. Zemansky, H. Young and R. Freedman, *College Physics*, NY: Pearson Ed., 2009.

[4] R. Morrow, A. Grant and D. P. Jackson, "A strange behavior of friction," *Phys. Teach.*, vol. 37, no. 7, p. 412, 1999.

[5] P. Vogt and J. Kuhn, "Analyzing free-fall with a smartphone acceleration sensor," *Phys. Teach.*, vol. 50, p. 182, 2012.

[6] S. Staacks, S. Hütz, H. Heinke and C. Stampfer, "Advanced tools for smartphone based experiments: phyphox," *Phys. Educ.*, vol. 53, p. 045009, 2018.
[7] A. Çoban and M. Erol, "Teaching and determination of kinetic friction coefficient using smartphones," *Phys.Educ.*, vol. 54, no. 2, p. 025019, 2019.
[8] J. Di Laccio and S. Gil, "Smartphone una herramienta de laboratorio y aprendizaje: laboratorios de bajo costo para el aprendizaje de las ciencias," *Am. J. Phys. Educ.*, vol. 11, no. 1, p. 1305, 2017.
[9] Pasco, CA, "Photogates, fences, and timers," 2021. [Online]. Available: www.pasco.com.
[10] Vernier Software & Technology, "Photogates," 2021. [Online]. Available: https://www.vernier.com/product/photogate/.
[11] Vernier Software, "Logger Pro and Vernier video analysis," 2021. [Online]. Available: www.vernier.com.

Chapter 15

HARMONIC OSCILLATORS

The harmonic oscillator is perhaps one of the most useful and frequent paradigms in physics. The basic characteristics of this phenomenon occur in many areas of physics and technology. The simplest examples are the mass-spring system and the pendulum. This chapter explores the motion of free and damped oscillating systems and the dependence of the oscillation frequency on different properties of the system, such as mass and friction. It considers viscous and turbulent friction forces, both dependent on speed.

Goals

✓ Oscillating systems
✓ Dynamics of a spring-mass system
✓ Free and damped oscillations
✓ Viscous and turbulent friction

15.1 Oscillatory systems

There are many experimental devices that make it possible to study free and damped oscillations. This chapter focuses on one of the simplest systems, which is a paradigm for studying this type of motion: the mass-spring system. The goals of the following experiments are to study the motion of these systems in different circumstances and to compare the results with the corresponding theoretical models.

15.2 Free oscillations

Consider a one-dimensional system, such as the one illustrated in Figure 15.1.

If the friction forces are negligible, the equation of motion obtained from the application of Newton's second law, which describes evolution over time of the position $x(t)$ of a mass m that oscillates connected to a linear spring of elastic constant k, is [1–3]

$$m\frac{d^2x}{dt^2} = -kx. \tag{15.1}$$

If we define the constant $\omega_0 = \sqrt{k/m}$, this equation can be written as follows:

$$\frac{d^2x(t)}{dt^2} = -\omega_0^2 x(t). \tag{15.2}$$

Therefore, the solution of this differential equation is a function $x(t)$ such that the second derivative is proportional to the original function, with the sign changed. The functions that have this property are the sine and cosine functions, therefore it is easy to prove by simple substitution into Eq. (15.2) that the function

$$x(t) = A_0 \sin(\omega_0 t + \phi) \tag{15.3}$$

is indeed a solution of the motion equation. The constants A_0 and φ depend on the initial conditions, i.e., on the value of the initial position, $x_0 = x(0)$, and the initial velocity, $v_0 = dx(0)/dt$, of the system at the time $t = 0$. In contrast, ω_0 is a dynamic property of the system, called *natural angular frequency*, which depends on the mass m and on the elastic constant k of the spring. Since the sine

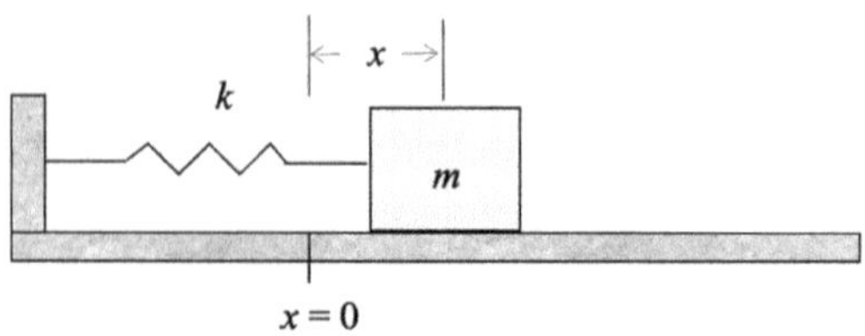

Figure 15.1 Schematic diagram of a one-dimensional mass-spring system.

function has period 2π, it is easy to show that the period T_0 of the system is given by $T_0 = 2\pi/\omega_0 = 2\pi\sqrt{m/k}$ or equivalently,

$$T_0^2 = \frac{1}{f_0^2} = \frac{4\pi^2}{k}m, \tag{15.4}$$

where f_0 is the *natural frequency*.

15.3 Damped oscillations

If the system described in Figure 15.1 is subject to a *viscous friction* force, that is, a force proportional to speed (see Annex A and B in Chapter 10), i.e.,

$$F_{\text{frict}} = -bv = -b\frac{dx}{dt}, \tag{15.5}$$

with b being a constant characteristic of the medium and of the shape of the oscillating mass, the equation of motion (15.2) becomes

$$m\frac{d^2x(t)}{dt^2} = -k \cdot x(t) - b\frac{dx(t)}{dt}. \tag{15.6}$$

If we define $\gamma = b/2m$, Eq. (15.6) can be written as follows [2–4]:

$$\frac{d^2x}{dt^2} + 2\gamma\frac{dx}{dt} + \omega_0^2 \cdot x = 0. \tag{15.7}$$

It is easy to check by simple substitution that a complete solution of this equation is given by

$$x(t) = A_0 \cdot e^{-\gamma \cdot t}\sin(\omega_p t + \phi), \tag{15.8}$$

with $\omega_p^2 = \omega_0^2 - \gamma^2$. Again, A_0 and φ depend on the initial conditions of the system. Figure 15.2 illustrates the behavior over time of a damped oscillator.

A useful parameter to characterize an oscillating system is the *logarithmic decrement* (LD), which is associated with the relative loss of energy per cycle. If at a given instant $t = t_1$ the sine term of (15.8) is equal to the unit, this term will also have this value for $t_2 = t_1 + nT$, where n is an integer and $T = 2\pi/\omega_p$. Therefore, for $t_2 = t_1 + nT$, the total energy of the system is elastic energy stored

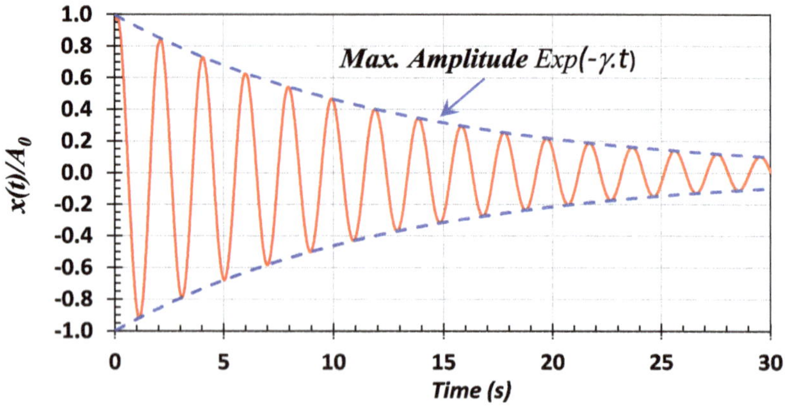

Figure 15.2 Response of a mass-spring system with viscous friction.

in the spring, since in this case the amplitude is maximum. The total energy is equal to the potential energy, which is $E(t) = kx^2(t)/2$. The variation of energy per cycle, according to (15.8), results in

$$DL = \ln\left[\frac{E(t_1)}{E(t_1+T)}\right] = \ln\left[\frac{(x(t_1))^2}{(x(t_1+T))^2}\right] = \ln\left[e^{2\gamma T}\right] = 2\gamma T = \frac{4\pi\gamma}{\omega_p}. \tag{15.9}$$

Another useful parameter of this system is the *quality factor*, defined as follows:

$$Q \equiv \frac{\omega_p}{2\gamma}, \quad \text{or} \quad Q = \frac{2\pi}{DL}. \tag{15.10}$$

Thus, the quality factor is related to the energy dissipation per cycle. A high value of the quality factor means that energy dissipation per cycle is small, and the opposite occurs if the quality factor is low.

15.3.1 *Project 36. Study of the oscillating system — Free oscillations*

Recommended equipment: A force sensor [5] connected to a PC. Several springs with elastic constants of the order of 100 N/m. Various masses (5 or more) from 20 to 200 g. A scale ranging from 1 to 500 g, with sensitivity of at least 1 g. Alternatively, a smartphone capable of measuring acceleration can be used [5–7].

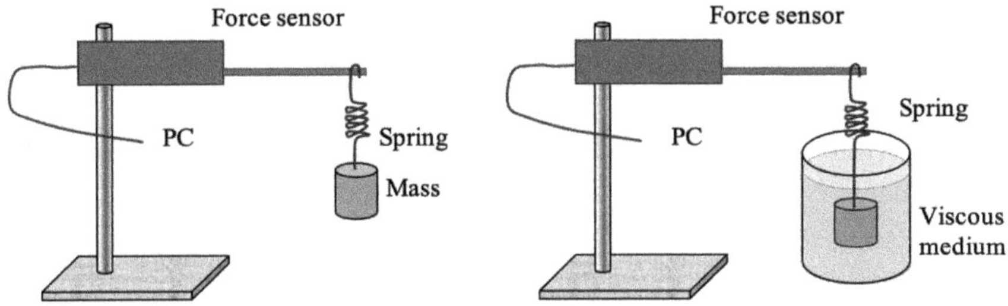

Figure 15.3 Experimental setup to study the dynamics of a mass-spring system. The force sensor connected to a computer enables real-time monitoring of the oscillations. On the right is a simple way to vary the friction in the system by means of a viscous liquid.

Figure 15.3 shows a suitable experimental setup for studying this system. It uses a force sensor [5] connected to a computer. For a linear spring (i.e., one that obeys "Hooke's law" $F = -kx$), the force that the spring applies to the sensor is proportional to the stretch x. Therefore, what the force sensor measures is a variable (force) that is proportional to the variation of the x coordinate as a function of time. To collect data with a smartphone, you can hang it at the end of the spring so that the phone acts as both a mass and measurement device [5–7]. The acceleration can easily be measured and stored in the smartphone. With this information, the variation of $v(t)$ and $x(t)$ can be found by simple numerical integration. See Chapter 10.

To study the dependence of the oscillation frequency on the mass in this system, you can use a spring whose elastic constant k is known (or measured, for example, by using the technique described in Chapter 13). The proposed experiment consists of determining the frequency of oscillation for different masses. Vary the value of the masses in the widest possible range, taking care not to cause permanent deformation of the spring.

Suggested tasks:

✓ Using the technique described in Chapter 13, determine the constant k of your spring and check its linearity over the range of forces to be used. Verify that even for the largest mass to be used, the spring returns to its original length once the mass is removed.

- ✓ Graph the dependence of the elongation of the spring as a function of the attached weights and determine the constant k and its uncertainty.
- ✓ Using this spring (for which you just measured k), study the oscillation of the system for the different selected masses. If you use a force sensor (which measures the force as a function of time, $F(t)$), you can find the frequency f for each mass by fitting this function with a sinusoidal function. If you use a smartphone, the measured parameter is acceleration as a function of time ($a(t)$). In this case, by fitting this function with a sinusoidal function, you can find the frequency f for each mass attached to the phone.
- ✓ Plot the period $T = 1/f$ and the frequency of oscillation f as a function of the mass m. Analyze the graphs on linear and logarithmic scales. What relationship do you find?
- ✓ Plot square of the period, $T^2 = 1/f^2$, as a function of mass m. Do you find a linear relationship between these variables? If so, from this graph, estimate the slope of the line that best fits your data.
- ✓ Using the theoretical model described above, Eq. (15.4), determine the value of the spring constant k and its uncertainty. Compare this value to the value of k found with the static method.
- ✓ Describe the characteristics of the friction forces involved. Are the friction forces viscous or not?

NOTE: There are several ways to determine frequency from the recorded force as a function of time $F(t)$, the acceleration $a(t)$, or the amplitude $x(t)$ of the oscillating system. The following are some alternatives:

Method 1. The simplest method is to estimate the times at which the signal of the oscillating function crosses zero. Make sure that the mean value of the signal is indeed zero, otherwise the oscillation signal will have an offset that will disturb the method. Because zero crossings occur regularly every half period ($\Delta t = T/2$), if you graph the successive zero crossing times as a function of the number of times n in which the crossing occurs, if the period is constant, you would expect linear dependence. The slope of the graph would be exactly equal to half a period. A departure from the linear behavior would be indicative of a variation in frequency over time, i.e., the occurrence of an "anharmonicity" in the signal under study [9].

Method 2. Another way to determine the period or frequency of an oscillating signal is by plotting the oscillating signal as a function of time and then fitting the measured signal with the sinusoidal function (15.3). This fitting can be done automatically using suitable programs or manually by varying the parameters of Eq. (15.3), i.e., by varying A_0, ϕ, and ω_0 until a good fit of the experimental data is achieved. To quantitatively measure the quality of the fit, it is helpful to define the parameter χ_T^2 which measures the degree of deviation between the measured signal and the theoretical one (see Chapter 7 and Appendix C):

$$\chi_T^2 = \sum\nolimits_{i=1} \left(X_i^{(\mathrm{exp})} - X_i^{(\mathrm{teo})} \right)^2, \qquad (15.11)$$

where $X^{(\mathrm{exp})}$ is the value of the experimental signal and $X^{(\mathrm{teo})}$ the corresponding value of the prediction given by Eq. (15.3). Vary the values of A_0, ϕ, and ω_0 to minimize the value of χ_T^2. The website www.fisicarecreativa.com (Physics Experiment Resources [10]) provides some examples of the use of this technique.

15.3.2 *Project 37. Damped oscillations — Viscous friction*

Use an appropriate container filled with some viscous liquid such as water or oil, as illustrated on the right side of Figure 15.3, so that the mass is totally submerged in the liquid but the spring is not. Study the resulting oscillatory motion experimentally. If you use a smartphone, you can hang a mass from it by means of a wire so that the mass is submerged in the liquid without affecting the smartphone.

Suggested tasks:

✓ For the chosen mass and spring, determine the frequency of free oscillation $\omega_0 = 2\pi f_0$ in air.

✓ Submerge the mass in the viscous medium and repeat the measurements. Plot the experimental results of x as a function of t. In the same graph, include the expected theoretical variation for this motion, Eq. (15.8). Vary the parameters of the theoretical model so that you can replicate the experimental results as closely as possible. When you achieve an adequate fit of the data, find the values of the characteristic parameters of the system described by Eq. (15.9): ω, A_0, φ, and γ.

✓ Fit a decaying exponential curve to the maximum experimental amplitudes. Use as many data as possible. From the best possible fit, determine the value of the constant γ that best describes the motion of the system and compare it to the one found previously. Is the relationship $\omega^2 = \omega_0^2 - \gamma^2$ verified?

✓ Determine the quality factor Q of your system and the logarithmic decrement (LD).

✓ What can you conclude about the variation of the friction force of your system with speed? Are the data compatible with a linear dependence of friction force on velocity?

15.3.3 *Project 38. ♣♣ Damped oscillations — Turbulent friction*

Another common condition is turbulent friction, where the friction force is quadratically dependent on the speed. As we have seen in Annex A of Ch. 10, this happens when the Reynolds number is large (of the order or greater than 3000) [11]. A simple way to achieve these conditions in air is to add a screen to the body to increase its effective surface interacting with the air, thereby increasing the friction force. In the case of the system oscillating in air, the condition can be simulated by enclosing the mass in a plastic or Styrofoam ball 5–10 cm in diameter. This envelope must be rigidly fixed to the mass so that during the oscillation it remains parallel and does not wobble. The frequency should be a few hertz. The idea is to create the conditions so that the Reynolds number is of the order of or greater than 3000 (see Annex A). Under these conditions, using the system described in Figure 15.3, study the variation in time of the oscillation amplitude of your system.

Suggested tasks:

✓ Plot your experimental results of x as a function of t. In the same graph, include the expected theoretical variation for this movement, described by Eq. (15.8). Does the amplitude of the signal vary exponentially?
✓ Perform the same analysis but this time testing with the amplitude expression described by the expression (15.24) discussed in Annex A. To do this, adjust the parameters A_0, ϕ, γ, and β in order to achieve the best possible fit to your experimental data.
✓ Is the model described by Eq. (15.24) adequate to explain your data? What can you conclude from this study?

15.4 Annex A: Friction forces dependent on the square of the velocity — Turbulent friction

For a sphere of diameter d, moving with velocity v in a viscous medium of dynamic viscosity μ, the drag force is given by the Stokes formula: [11]

$$F_{\text{drag}} = 3 \cdot \pi \cdot d \cdot v \cdot \mu. \tag{15.23}$$

This relationship is valid in the laminar regime, which occurs for Reynolds number values $Re < 1$:

$$Re = \frac{\rho \cdot d \cdot v}{\mu}. \tag{15.12}$$

Here, ρ is the density of the fluid media (air or water). For the case of high Reynolds numbers ($Re >> 1$), the drag force is given by [11]

$$F_{\text{drag}} = \frac{1}{2} \cdot C_d \cdot \rho \cdot A \cdot v^2, \tag{15.13}$$

where ρ is the density of the fluid, v the velocity, A the cross-sectional area of the body ($\pi d^2/4$, for a sphere), and C_d a numerical coefficient whose value depends on Re and the shape of the object [11].

The value of C_d is determined experimentally. For some simple geometries, C_d can be obtained from tables or graphs like the one in

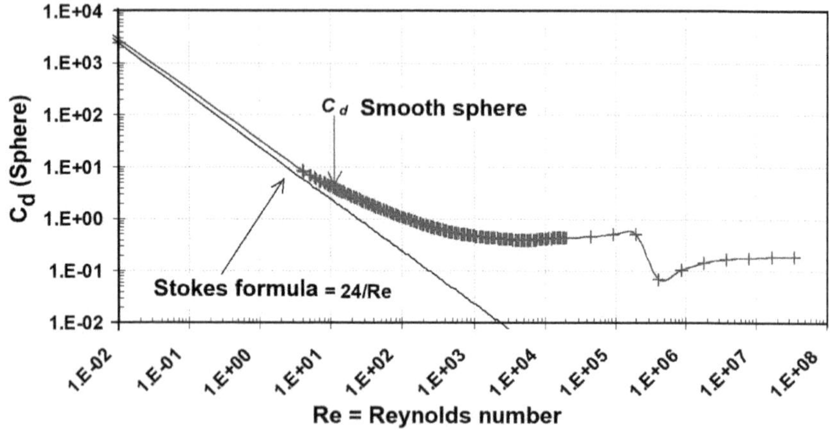

Figure 15.4 Variation of the drag coefficient C_d for a smooth sphere as a function of the Reynolds number.

Figure 15.4. The variation of C_d with Re can also be approximated with the semi-empirical formula $C_d(Re) = \frac{24}{Re} + \frac{1}{1+\sqrt{Re}} + 04$, valid in the range $0 < Re < 210^5$.

15.5 Annex A: Harmonic oscillator with turbulent friction force

The equation of motion for a harmonic oscillator with both a friction force proportional to velocity (laminar flow) and a friction force proportional to the square of velocity (turbulent flow) is

$$m\frac{d^2x}{dt^2} = -k\,x - b\frac{dx}{dt} - c\frac{dx}{dt}\left|\frac{dx}{dt}\right|. \tag{15.14}$$

An analytic solution cannot be found. However, it is possible to solve Eq. (15.4) by integrating it numerically with a program such as Mathematica or Matlab. Here we will try to find an approximate analytical solution, following the idea proposed by Nelson and Olsen [12]. Tentatively, we assume that

$$x = A(t)\cos(\omega t). \tag{15.15}$$

The total energy of the system can be written as follows:

$$E_T = \frac{1}{2}kA^2 \quad \text{and} \quad dE_T = kA dA. \tag{15.16}$$

The variation of total energy in a time equal to half a cycle ($\Delta t = \pi/\omega$) is

$$\Delta W = \int_0^{\Delta t} F_r v dt = -\int_0^{\Delta t} (bv^2 + cv^3) dt, \tag{15.17}$$

where $v = dx/dt$ is the speed. Combining Eq. (15.17) with Eq. (15.15) and assuming that $A(t)$ is practically constant for a period, we have

$$\Delta W = -\frac{\pi}{2} b\omega A^2 - \frac{4}{3} c\omega^2 A^3. \tag{15.18}$$

From energy conservation, we have $\Delta W = kA(dA/dt)\Delta t$, that is,

$$-\frac{\pi}{2} b\omega A^2 - \frac{4}{3} c\omega^2 A^3 = kA \frac{dA}{dt} \frac{\pi}{\omega} \tag{15.19}$$

or

$$\frac{dA}{dt} \approx -\frac{1}{2}\frac{b}{m} A - \frac{4}{3\pi}\frac{c}{m}\omega A^2. \tag{15.20}$$

If we define

$$\gamma = \frac{b}{2m} \quad \text{and} \quad \beta = \frac{4}{3\pi}\frac{c}{m}\omega, \tag{15.21}$$

Eq. (15.20) can be written as follows:

$$\frac{dA}{dt} \approx -\gamma A - \beta A^2. \tag{15.22}$$

This equation can be integrated, and it is easy to check that its solution is given by

$$A(t) = A_0 \frac{\gamma e^{-\gamma t}}{(\beta(1 - e^{-\gamma t}) + \gamma)}. \tag{15.23}$$

Here, A_0 is the initial oscillation amplitude. In this way, it is possible to write an approximate solution of Eq. (15.15) as follows:

$$x(t) = A(t)\cos(\omega_p t + \varphi) = A_0 \frac{\gamma e^{-\gamma t}}{(\beta(1 - e^{-\gamma t}) + \gamma)} \cos(\omega_p t + \varphi), \tag{15.24}$$

where A_0 and ϕ depend on the initial conditions of the problem and the parameters γ and β are given by (15.21). If $\beta < \gamma$, the value of ω_p can be calculated as follows:

$$\omega_p^2 \approx \omega_0^2 - \gamma^2. \tag{15.25}$$

Figure 15.5 shows how the shape of the signal decreases over time. Note that in this case, the amplitude does not decrease exponentially

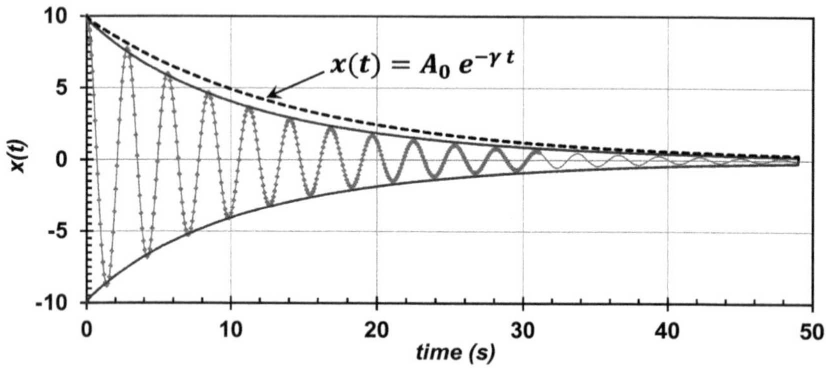

Figure 15.5 Response of an oscillatory system with friction force dependent on the square of the velocity. Note that in this case, the amplitude does not decrease exponentially.

but, especially at the beginning, it does so faster than a simple exponential.

References

[1] D. Giancoli, *Physics Principles with Applications*, 7th Ed. NY: Pearson, 2014.

[2] R. Halliday, D. Resnick and M. Krane, *Fundamentals of Physics*, NY: John Wiley & Sons, Inc., 2002.

[3] F. S. Crawford Jr., *Waves (Berkeley Physics Course, Vol. 3)*, Cambridge University Press, 2013. Vol. 3.

[4] F. Sears, *et al.*, *College Physics*. NY: Pearson Ed., 2009.

[5] Vernier Software, "Logger pro and Vernier video analysis," 2021. Available: www.vernier.com.

[6] P. Vogt and J. Kuhn, "Analyzing simple pendulum phenomena with a smartphone acceleration sensor," *Phys.Teach.*, vol. 50, p. 439, 2012.

[7] J. Di Laccio and S. Gil, "Smartphone una herramienta de laboratorio y aprendizaje: laboratorios de bajo costo para el aprendizaje de las ciencias," *Am. J. Phys. Educ.*, vol. 11, p. 1305, 2017.

[8] M. Santamaría, E. Cruz de Gracia and L. Dorneles, *Damped Oscillations — A Smartphone Approach,* (Preprint), 2021.

[9] S. Gil, E. Legarreta and D. E. Di Gregorio, "Anharmonicity in large amplitude pendulum," *Am. J. Phys.*, vol. 79, p. 843–847, 2008.

[10] S. Gil, "Fisica re-Creativa," *Phycics Experiments*, 2021. Available: https://www.fisicarecreativa.com/.
[11] B. R. Munson, D. F. Young and T. H. Okiishi, *Fundamentals of Fluid Mechanics.* New York: John Willey & Sons, Inc., 1994.
[12] R. A. Nelson and M. G. Olsson, "The pendulum- Rich physics from a simple system," *Am. J. Phys.*, vol. 54, p. 112–121, 1986.

Chapter 16

PENDULUMS

To continue with the study of oscillators, this chapter analyzes the behavior of different kinds of physical pendulums. It studies pendulums built of rings of different radii and a pendulum in which the mass distribution can be varied with respect to the suspension point while keeping the total mass constant, whose behavior shows several curious and counterintuitive characteristics. It also studies Kater's pendulum, by means of which g can be measured very precisely, and the behavior of a pendulum at large oscillation amplitudes and with different types of damping.

Goals

✓ Physical pendulum
✓ Oscillating rings
✓ Non-intuitive pendulum
✓ Kater's pendulum
✓ Determination of the value of g
✓ Large amplitude oscillations

16.1 Physical pendulums

A physical pendulum is a system composed of rigid parts that pivot from a suspension point that does not coincide with its center of mass [1, 2]. The only degree of freedom in this system is the angular displacement θ, see Figure 16.1.

The maximum angular displacement of each oscillation is called its amplitude, θ_0. If the pendulum is moved from its equilibrium position and released, assuming that the friction is negligible and the angular amplitudes of oscillation are small, $\theta_0 \leq 10°$, the system

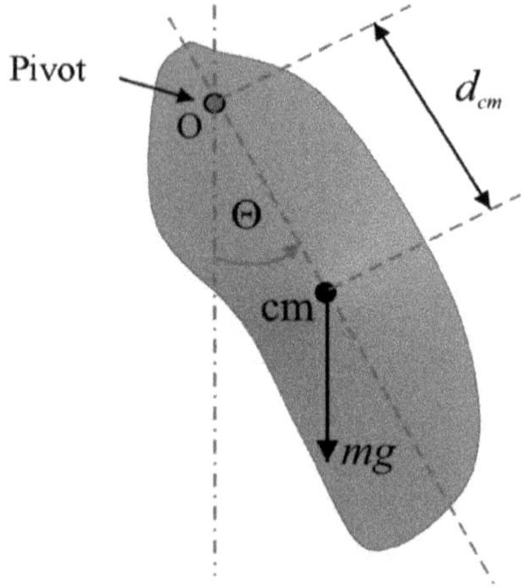

Figure 16.1 Physical pendulum. The point *cm* represents the center of mass of the system; d_{cm} is the distance from the suspension point (pivot) to *cm*.

describes a simple harmonic motion, similar to that of the mass-spring studied in the previous chapter. Figure 16.1 represents a physical pendulum consisting of a body of mass m suspended from a suspension point at a distance d_{cm} from its center of mass.

There are several means to measure the period of a pendulum. From a simple use of a clock to more elaborated photogates [3] connected to a PC or laptop. It is also possible to use a smartphone to measure the period of a pendulum [4, 5].

16.2 Period for small amplitudes

When the system is displaced from its equilibrium position, the weight of the body generates a torque around the pivot [1, 2]. Applying the laws of rotational motion to this system and neglecting the effects of friction, we have

$$-mgd_{\text{cm}}\sin(\theta) = I_P\alpha, \tag{16.1}$$

where I_P is the moment of inertia of the system with respect to the axis of rotation that passes through the pivot and α is the angular acceleration, $\alpha = d^2\theta/dt^2$. For *small oscillation amplitudes*, we can approximate $\sin(\theta) \approx \theta$ and Eq. (16.1) can be written as follows:

$$\frac{d^2\theta}{dt^2} = -\frac{mgd_{\mathrm{cm}}}{I_P}\theta. \tag{16.2}$$

This second-order differential equation represents a simple harmonic motion and, as shown in the previous chapter, its solution is

$$\theta(t) = \theta_0 \cos(\omega_0 t + \varphi) \tag{16.3}$$

with angular frequency $\omega_0 = 2\pi/T_0$ equal to

$$\omega_0 = \sqrt{\frac{mgd_{\mathrm{cm}}}{I_P}}. \tag{16.4}$$

Therefore, the period of oscillation T_0 for small oscillations is

$$T_0 = 2\pi\sqrt{\frac{I_P}{mgd_{\mathrm{cm}}}}, \tag{16.5}$$

which is independent of the amplitude θ_0. In this case, the oscillations are said to be *harmonic*.

It is important to note that this is only an approximate result for the case of small amplitudes, i.e., when $\sin(\theta) \approx \theta$, i.e., $\theta_0 < 10°$. In general, as discussed in Annex A, the amplitude of a pendulum depends on the amplitude θ_0. A better approximation to the variation of the period with the amplitude is given by

$$T(\theta_0) = T_0 \left(1 + \frac{k^2}{2^2} + \frac{3^2k^4}{2^24^2} + \frac{5^2k^6}{2^28^2} + \frac{35^2k^8}{2^264^2}\cdots\right), \tag{16.6}$$

where $k = \sin(\theta_0/2)$. When the oscillations depend on the amplitude, the oscillations are said to be *anharmonic*.

Other approaches are discussed in Annex A.

Preliminary exercises

✓ Using Eq. (16.6), construct a graph of $T(\theta_0)/T_0$ as a function of θ_0 between 0° and 90°. Examine how these figures vary if you take 1, 2, 4, 6, and 10 terms in the series. For $\theta_0 < 45°$, do you see any significant change in the graph?

✓ From the expression (16.6), discuss what the values of the amplitude θ_0 will be so that the expression (16.5) is valid to 1% precision. What should the amplitude be for Eq. (16.5) to be valid within 0.1%?

✓ How precisely should you measure the period of oscillation of a pendulum to obtain a significant variation between $T(\theta = 20°)$ and $T(\theta \to 0°)$? Suggest what instruments you could use to ensure such precision.

16.2.1 *Project 39. Study of an oscillating ring*

> **Recommended equipment**: A set of 5 or more rings of varying radii from approximately 4–20 cm. A photogate connected to a PC to measure the periods.

Consider a ring of uniform density which has a small, constant thickness compared to its radius. Call the outer radius R_e and the inner radius R_i.

Define the mean radius $R = (R_e + R_i)/2$; for a thin ring of radius R and mass m, the inertial momentum with respect to the suspension point is

$$I_{\mathrm{CM}} = \leftrightarrows mR^2. \tag{16.7}$$

For the setup shown in Figure 16.2, $d_{cm} \approx R$, since the center of mass is in the center of the ring. The period of oscillation for small amplitudes, according to Eq. (16.5), is

$$T = 2\pi\sqrt{\frac{2R}{g}}. \tag{16.8}$$

The objective of this project is to experimentally validate expression (16.8).

Suggested tasks:

✓ Using a device like the one suggested in Figure 16.2 and rings of different radii R, experimentally study the variation of the period of oscillation T as a function of the mean radius R of each ring.

✓ Plot the period T as a function of R and of T^2 as a function of R.

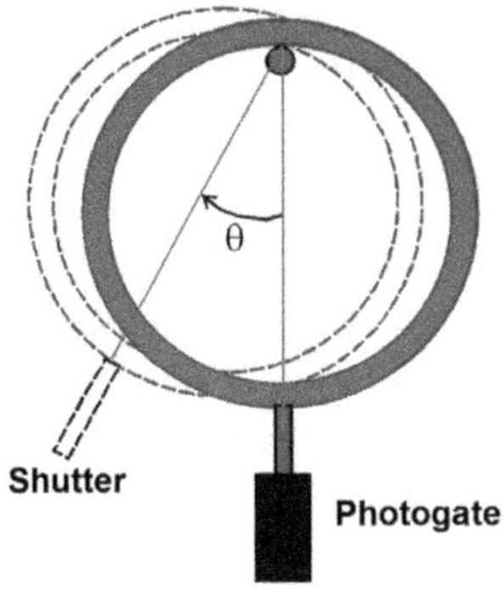

Figure 16.2 Oscillating ring or hoop. An opaque shutter, of negligible weight, is positioned to activate the photogate positioned at the bottom. The angle θ characterizes the angular displacement of the ring with respect to its equilibrium position.

✓ Compare the dependence of the experimental data with the theoretical prediction given by (16.4).

16.3 Counter-intuitive pendulum

The period of oscillation of a physical pendulum depends on the geometric shape of the suspended body, which affects the distance d_{cm}, and on its mass distribution around the axis of rotation, which determines the moment of inertia I_P. In this section, the aim is to analyze the behavior of a pendulum composed of a rigid rod and a mass whose position on the rod can be varied, as illustrated in Figure 16.3. The bar has mass m_b and length L_b and is suspended from one end. A mass m_a is attached to the bar by means of a screw at a distance x from the suspension point, and its position can be varied along the bar. The masses m_a and m_b should be approximately in a relation $m_a/m_b \approx 2$–4. The goal of this experiment is to investigate the variation of the period of oscillation T of this system for small amplitudes ($\theta_0 \leq 10°$) as a function of the x-position of the mass m_a. A photogate can be used to measure the period with a resolution of the order of 10 ms [3]. Alternatively, a smartphone can be attached to the mass m_b [6].

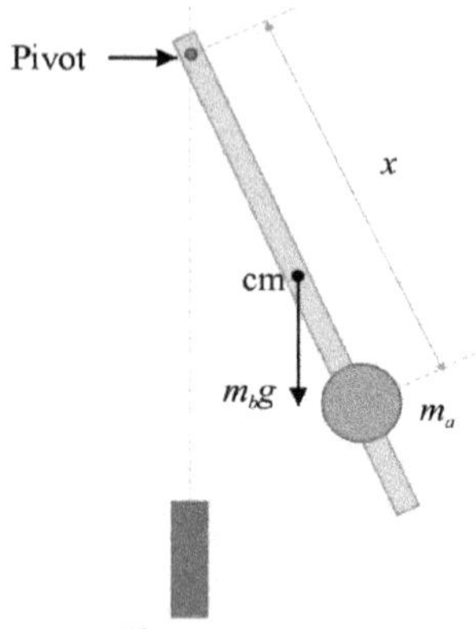

Figure 16.3 Physical pendulum with variable mass distribution. The bar has mass m_b and length L_b. An additional mass m_a is positioned at a distance x from the suspension point.

16.3.1 *Preliminary exercises*

✓ Show that for this system, the period for small oscillation amplitudes is

$$T(x) = 2\pi\sqrt{\frac{\left(\frac{1}{3}m_b L_b^2 + m_a x^2\right)}{g\left(\frac{L_b}{2}m_b + x m_a\right)}} = T_{00}\sqrt{\frac{1/3 + f(x/L_b)^2}{1/2 + f(x/L_b)}}, \quad (16.9)$$

with $f = m_a/m_b$ and $T_{00} = 2\pi\sqrt{L_b/g}$. A representation of $T(x)/T_{00}$ as a function of (x/L_b) for different values of f is shown in Figure (16.4).

16.3.2 *Project 40. Counter-intuitive pendulum*

Recommended equipment: A metal rod approximately 50 cm long, with a sliding mass comparable to that of the rod. An axis that passes through the bar at one end and allows it to oscillate. A photogate connected to a PC to measure the periods.

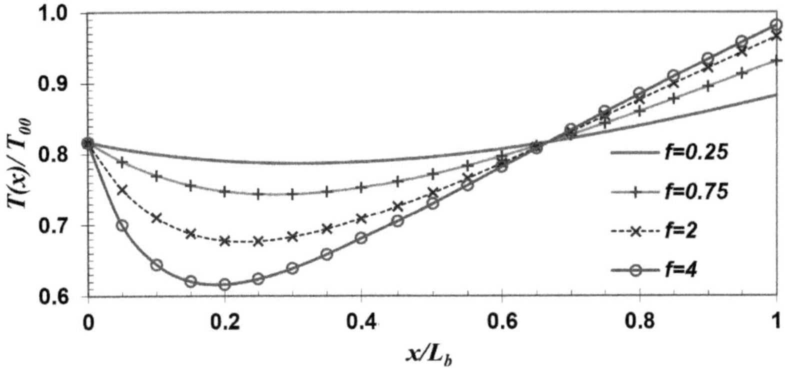

Figure 16.4 Variation of the reduced period $T(x)/T_0$ as a function of x/L_b for different values of $f = m_a/m_b$, according to Eq. (16.9).

Suggested tasks:

✓ Determine the period T_0 of the bar without the mass m_a.

✓ With the mass m_a in place, determine the period $T(x)$ for different positions x of the mass along the bar (at least 10 values of x distributed as evenly as possible over the length of the bar). Make sure that the mass is placed as close as possible to the oscillation axis and also to the other end of the bar.

✓ For each value of x, measure 10–20 oscillations and take the mean value and the standard deviation of the mean. See Chapter 3.

✓ On the same graph, represent the measured period $T(x)$ as a function of the distance x. Include its uncertainties as error bars and the theoretical expectations derived from Eq. (16.9). Use the measured values of m_b, L_b, and m_a. What can you conclude from this study?

16.4 Kater's reversible pendulum

In the early 19th century, Henry Kater, an English physicist and military man, presented the results of his measurements of the acceleration of gravity with great precision (of the order of 0.5%), using a "reversible" pendulum [7, 8]. Kater's reversible pendulum was subsequently refined, becoming the standard instrument for measuring Earth's gravity.

There are several embodiments of this type of pendulum [8, 9]. It typically consists of a rigid bar that can oscillate around two different suspension points O and O', as schematically illustrated in Figure 16.5.

The distance between the suspension points OO' is fixed and called L. The moment of inertia of the system with respect to the center of mass is designated as I_{cm} and the total mass of the system is M. The radius of gyration of the pendulum about its center of mass (c.m.) is called K_{cm}; $I_{\mathrm{cm}} = \rightleftarrows MK_{\mathrm{cm}}^2$. If a and a' are the distances from the center of mass to the suspension points O and O', respectively, the periods of the pendulum with respect to these two suspension points will be, respectively,

$$T = 2\pi\sqrt{\frac{K^2 + a^2}{g \cdot a}} \tag{16.10}$$

and

$$T' = 2\pi\sqrt{\frac{K^2 + a'^2}{g \cdot a'}}, \tag{16.11}$$

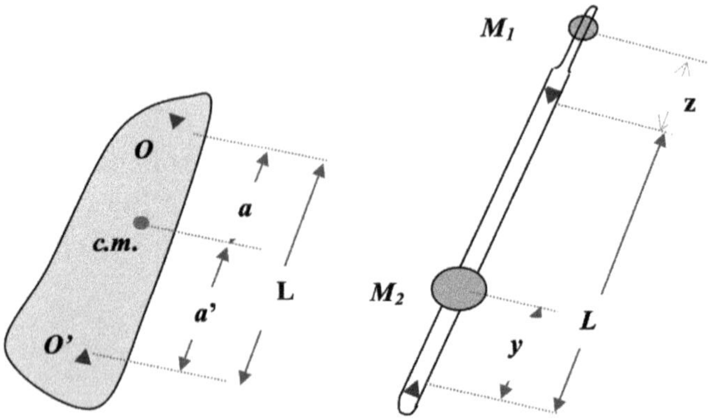

Figure 16.5 Kater's reversible pendulum. This pendulum can swing around either of the suspension points O and O'. The distance between these points, L, is fixed. It also has two masses, M_1 and M_2, with variable positions. The first mass allows a gross variation of the mass distribution, and therefore of the periods, with respect to O and O'. The second mass (the smaller one) is used to fine-tune the periods. The periods can be measured using a photogate.

with

$$L = a + a'. \tag{16.12}$$

If these two periods can be made equal by varying the mass distribution (location of M_1 and M_2), i.e., $T = T'$, then from (16.10) and (16.11), we have

$$K^2 = aa'. \tag{16.13}$$

Thus,

$$T_{\text{Kater}} = T = T' = 2\pi\sqrt{\frac{L}{g}}. \tag{16.14}$$

That is, when the two periods T and T' are equated, to determine g, the only dimension of the pendulum that needs to be accurately measured is L, i.e., the distance between the two suspension points.

16.4.1 *Project 41. Possible construction of Kater's pendulum: measuring g*

Recommended equipment: A Kater's pendulum. A photogate connected to a PC to measure the periods.

If you have a Kater's pendulum like the one described in Figure 16.5, you can study its behavior and attempt to determine the value of g with as little error as possible.

To measure the periods precisely, you could first represent in the same graph T and T' as a function of y, the position of the largest mass. The point where the curves intersect determines the optimal y value. Next, move the larger mass to this position and perform a new study of T and T' versus z, the position of the smaller mass. Move the smaller mass to the optimal position. Repeat the procedure until the two periods T and T' converge. It is not crucial to measure the value of $T_{\text{Kater}} = T = T'$. This value can also be inferred by extrapolating the plot of T and T' as a function of z and determining its value from the point where these curves intersect. When the measured data are closer to this point of intersection, there will be less uncertainty in the determination of the value T_{Kater}, see Appendix D. The periods can be measured using a photogate connected to a computer.

With the value of T_{Kater}, using the expression (16.14), you can find the value of g and its uncertainty.

Suggested tasks:

✓ Plot T and T' as a function of y, the position of the largest mass. Determine the value of y for which $T \approx T'$.
✓ For the value of y found in the previous point, plot T and T' again, both as a function of z, the position of the smaller mass. Determine the value of z for which $T \approx T'$.
✓ From this graph, find the best value of T_{Kate_r} and estimate its uncertainty.
✓ Find the value of g and estimate its uncertainty.

16.4.2 *Project 42. "Homemade" Kater's pendulum*

Recommended equipment: A wooden or metal bar approximately 60–100 cm long with a variable overload. A photogate connected to a PC to measure the periods.

A simple way to build a Kater's pendulum is to use a uniform bar of aluminum or other material [10] of length L_b (60–100 cm), as shown in Figure 16.6. The variable overload can be a medium binder paper clip that moves on a scale marked on the bar. The bar has two axes of oscillation or two perforations, through which a thin horizontal rod or blade can pass. The bar can oscillate around either of the two axes or holes. The distance between the suspension axes (OO') is denoted by L and δ is the distance from the upper end to the suspension point, of the order of 1 cm. With the letter y, we denote the distance of the overload to the point O'. If a and a' indicate the distances from c.m. of the system to the oscillation points O and O', respectively, when the condition $T = T'$ is achieved, we have [10]

$$a = \frac{L_b}{2} - \delta, \tag{16.15}$$

and according to Eq. (16.13) together with $K^2 = L_b^2/12$,

$$a' = \frac{L_b/6}{(1 - 2\delta/L_b)} \cong \frac{L_b}{6} + \frac{\delta}{3}. \tag{16.16}$$

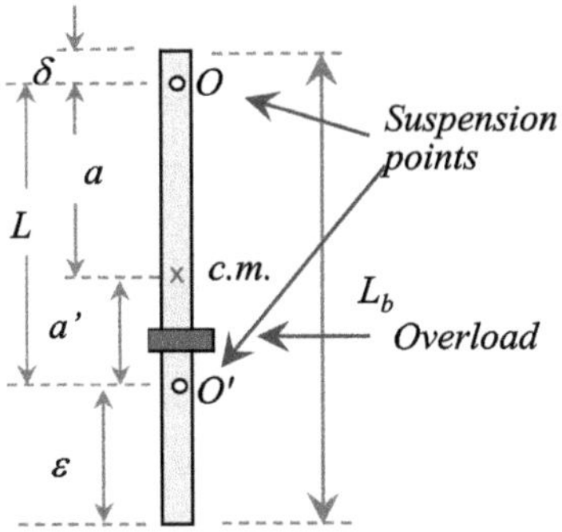

Figure 16.6 Setup of Kater's reversible pendulum. The bar of length L_b has two suspension points, separated from the center of mass (cm) by the distances a and a'; δ is the distance from the upper end to the suspension point. The overload can be varied along the bar. Its position y is measured from point O' or another point on the bar.

Since $\varepsilon = L_b - a'$, the distances from the other end of the bar to the oscillation point O', we have

$$\varepsilon \cong \frac{L_b - \delta}{3} \quad \text{and} \quad L \cong \frac{2}{3}(L_b - \delta). \tag{16.17}$$

Note that points O and O' *are not located symmetrically* with respect to the c.m.

Once the pendulum is prepared, using Eqs. (16.15) and (16.17) as a guide, it is advisable to measure L directly.

Suggested tasks:

✓ Show that in the case of a uniform bar, the periods will be equal when the distances to the center of mass are given by expressions (16.15) and (16.16).

✓ Using this pendulum, represent on the same plot the variation of T and T' as a function of y, where y measures the position of the overload with respect to point O'. Determine the value of y_0 for which $T \approx T'$.

✓ From the earlier plot, find the best value of the period T_{Kater} and estimate the error on its value.

✓ Find the value of g and estimate its uncertainty Δg.

16.5 Pendulum with friction

A pendulum coupled to a low-friction rotary sensor [11, 12] or shaft encoder can be used to set up a variety of interesting experiments in which you can take advantage of the measurements of the angle as a function of the time [13]. Many experiments can be performed using these types of devices. This section mentions just a few, using the commercial pendulum built by the firm LabTrek [14], but different experimental setups are feasible [12, 13]. For example, you could also use a simple aluminum disk of diameter 15–35 cm, connected to the rotary motion sensor, with a weight on one of the edges of the disk. Using an experimental setup like the one shown in Figure 16.7, the motion of the pendulum with two different kinds of damping can be studied: *dry friction* (the sliding friction between two solid surfaces in contact, which is almost constant over time) and *viscous friction* (as the friction of a body in a fluid, proportional to the velocity to some exponent, or the friction force caused by *eddy current*) [15].

(1) *Dry friction*: Use a rigid rod fixed to the rotary sensor axis (pivot) and a small mass m located at a distance L from the axis (in Figure 16.7, a perforated rubber ball is inserted on the rod). The dominant friction in this system is that associated with

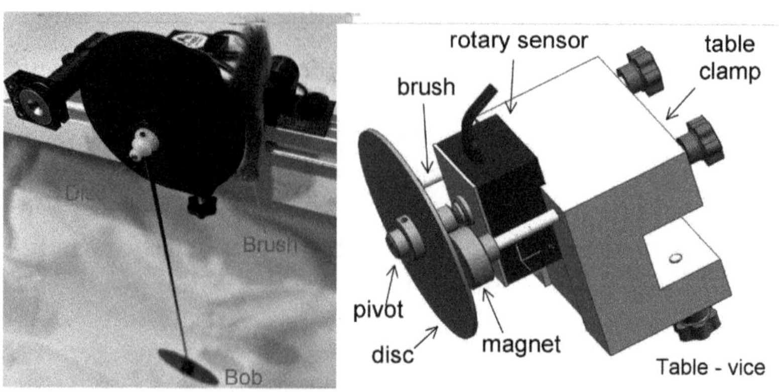

Figure 16.7 Pendulum with "dry" and "viscous" adjustable friction. It also enables large amplitude oscillations to be studied.

the bearing in the shaft encoder or rotatory sensor. Adjustable *dry* damping may be provided by a small, properly placed brush sweeping the disk.

(2) *Viscous friction*: Viscous damping can be achieved by a permanent magnet placed close to the *aluminum disk* fixed to the rotary sensor axis (the Foucault or eddy currents due to the magnet–disk interaction produce a restoring torque $\Gamma(\omega)$ proportional to the angular velocity ω).

16.6 Pendulum with friction — Small amplitudes

Equation (16.1) shows that for *small oscillation amplitudes* [approximating $\sin(\theta) \approx \theta$], and *negligible friction*, the pendulum motion Eq. (16.1) can be written as follows:

$$\frac{d^2\theta}{\mathrm{d}t^2} = -\frac{mgL}{I}\theta, \tag{16.18}$$

where L is the pendulum effective length and I is the moment of inertia of the system (ball + bar + disk)[a] with respect to the axis of rotation that passes through the pivot.

However, when friction is not negligible, Eq. (16.18) becomes

$$\frac{d^2\theta}{\mathrm{d}t^2} = -\frac{mgL}{I}\theta - \frac{\Gamma(\omega)}{I}. \tag{16.19}$$

Assuming only *viscous damping*, with torque proportional to the angular velocity ($\Gamma = -\gamma\omega$), the motion equation becomes

$$\frac{d^2\theta}{\mathrm{d}t^2} = -\frac{mgL}{I}\theta - \frac{\gamma}{I}\frac{\mathrm{d}\theta}{\mathrm{d}t}, \tag{16.20}$$

and with the positions $\delta = \gamma/2I$ and $\omega_o^2 = mgL/I$, it becomes

$$\frac{d^2\theta}{\mathrm{d}t^2} + 2\delta\frac{\mathrm{d}\theta}{\mathrm{d}t} + \omega^2\theta = 0. \tag{16.21}$$

[a]The momentum of inertia of the aluminum disk with radius r and mass m_d is $I_d = (2/3)m_d r^2$ and the momentum of inertia of the rod with mass m_r and length l is $I_r = (1/3)m_r l^2$ so that the total momentum is $I = mL + I_d + I_r$.

This equation, for small damping ($\delta << \omega_o$), has the analytical solution

$$\theta(t) = \theta_0 \left(e^{-\delta t}\right) \cos(\omega t), \quad \text{where} \quad \omega = \sqrt{\omega_0^2 - \delta^2}. \qquad (16.22)$$

The elongation amplitude should therefore decrease exponentially, i.e., the decrement during each half-period should decrease proportionally to the amplitude itself (see Figure 16.8).

Assuming only a *dry friction* torque with intensity C (with negative sign with respect to the sign of the angular velocity ω), we get $T = +C$ for $\omega > 0$, $T = -\mathrm{C}$ for $\omega < 0$, and $T = 0$ for $\omega = 0$, i.e., $T = \mathrm{C}\,\mathrm{sgn}(\omega)$,

$$\frac{d^2\theta}{dt^2} = -\frac{mgL}{I}\theta - \frac{C}{I}\mathrm{sgn}\frac{d\theta}{dt}. \qquad (16.23)$$

Let us analyze small angular oscillations starting from initial elongation θ_0 from rest ($\omega_0 = 0$). During the first half-oscillation, the

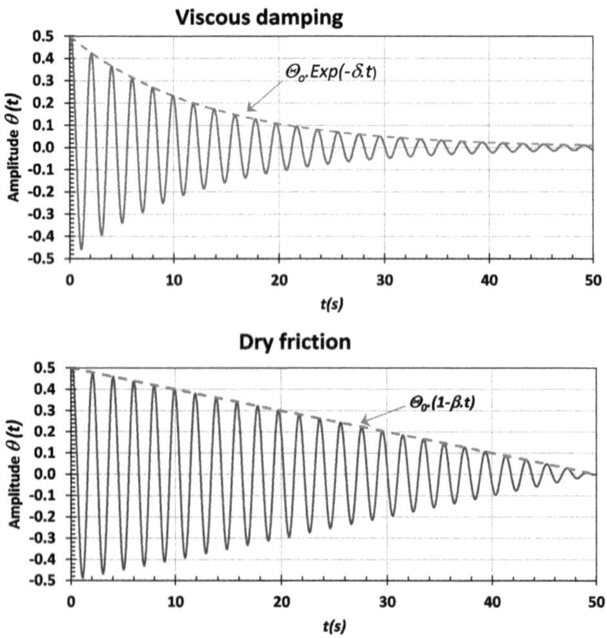

Figure 16.8 Oscillation damping with different types of friction. Top: Viscous damping, where the amplitude versus time follows an exponential function. Bottom: Dry friction, where the amplitude decreases linearly with time.

pendulum will sweep the total angle $\theta_0 + \theta_1$ and reach a situation $(\theta = \theta_1, \omega = 0)$, where we may calculate the energy balance.

The initial potential energy $mLg(1 - \cos\theta_0)$ equals the new potential energy $mLg(1 - \cos\theta_1)$ plus the energy lost due to the dry friction work $C(\theta_0 + \theta_1)$ or

$$mLg(\cos\theta_0 - \cos\theta) = C(\theta_1 + \theta_0). \tag{16.24}$$

By using the Werner formula for the trigonometric product, we have

$$\cos\theta_0 - \cos\theta_1 = 2\sin[(\theta_1 - \theta_0)/2]\sin[(\theta_1 + \theta_0)/2] \tag{16.25}$$

and the energy balance Eq. (16.24) gives for the elongation decrement $\Delta\theta = (\theta_0 - \theta_1)$ during one half-period: $\Delta\theta = 2C/mLg$. The calculation may be repeated for the second half-period, leading to the same result. To conclude, the elongation during each half-period decreases by the *constant* value $\Delta\theta = 2C/mLg$, giving an evaluation of the friction torque $C = mLg\Delta\theta/2$.

16.6.1 *Project 43. Pendulum with viscous and sliding friction*

Recommended equipment: Either a homemade or a commercial pendulum, mounted on a low friction rotatory sensor or shaft encoder connected to a PC to measure amplitudes as a function of time.

Using a shaft encoder or a low-friction rotary sensor connected to a PC, construct a pendulum like the one shown in Figure 16.7. A simple aluminum disk, with a weight on one of its ends, could be used for this experiment. Set up the data acquisition program to measure the angular amplitude $\theta(t)$ as a function of time.

(1) The pendulum with a lightweight, where the bolt bearing friction of the shaft encoder dominates, can generate a dry friction situation.
(2) If a permanent magnet can be placed close to the disk, a viscous friction situation could be generated.
(3) In all these instances, a large amplitude pendulum can easily be recreated.

Suggested tasks:

- ✓ Arrange your pendulum so that the dry friction dominates. Measure the angular amplitude $\theta(t)$ as a function of time.
- ✓ Compare your results to the expectation shown in the bottom panel of Figure 16.8.
- ✓ Arrange your pendulum so that the viscous friction dominates. Measure the angular amplitude $\theta(t)$ as a function of time.
- ✓ Compare your results to the expectation shown in the top panel of Figure 16.8.
- ✓ In a low friction situation, study the angular amplitude $\theta(t)$ as a function of time, for a large initial angle, i.e., $\Theta_0 \approx 90°$.
- ✓ In this situation, determine the variation of the period $T(\theta_0)$ as a function of the amplitude θ_0 of each oscillation.
- ✓ Compare your results to the prediction of the models discussed in Annex A, i.e., Eqs. (16.30), (16.31), and (16.32). See Ref. [15] for further analysis and ideas on this type of experiment.

16.7 Annex A: Large amplitude pendulum — Period

As we have seen in Annex A of Chapter 9, by applying Newton's laws to a simple pendulum *without friction*, we arrive at the equation of motion [2]:

$$\frac{d^2\theta}{dt^2} = -\frac{g}{L}\sin(\theta), \tag{16.26}$$

where the angle θ is measured from the vertical, g is the gravitational acceleration, and L is the length of the pendulum.

When the angle θ is *small*, $\sin(\theta) \approx \theta$ can be approximated, and the pendulum swings with simple harmonic motion around the equilibrium point ($\theta = 0$). The period is *independent* of the initial angle θ_0, i.e., the angle from which the pendulum is released:

$$T = 2\pi\sqrt{L}/g. \tag{16.27}$$

When the angle θ is *large*, the period T *depends* on the initial angle. We can find an expression for the period as a function of the angle

θ_0 starting from the equation of conservation of energy [16, 17]:

$$E = \frac{1}{2}mv^2 + mgL \cdot [1 - \cos(\theta)] = \frac{1}{2}mv_0^2 = mgL \cdot [1 - \cos(\theta_0)]. \tag{16.28}$$

Here, E is the energy of the pendulum, $v = Ld\theta/dt$ is the instantaneous velocity of the mass, v_0 is the maximum speed (which occurs when $\theta = 0$), and θ_0 is the maximum amplitude angle (which occurs when $v = 0$). From Eq. (16.20), the period can be found in an integral form:

$$T(\theta_0) = T_0 \cdot \frac{\sqrt{2}}{\pi} \int_0^{\theta_0} \frac{d\theta}{\sqrt{\cos(\theta) - \cos(\theta_0)}} \tag{16.29}$$

This equation has the following series expansion [18]:

$$T(\theta_0) = T_0 \cdot \left(1 + \frac{k^2}{2^2} + \frac{3^2 \cdot k^4}{2^2 \cdot 4^2} + \frac{5^2 \cdot k^6}{2^2 \cdot 8^2} + \frac{35^2 \cdot k^8}{2^2 \cdot 64^2} \cdots \right), \tag{16.30}$$

where $k = \sin(\theta_0/2)$ and $T_0 = \sqrt{L/g}$. Note that Eq. (16.30) reduces to Eq. (16.27) at the limit of small amplitudes ($k \approx 1$).

Sometimes it is useful to have approximate expressions to describe the variation of the period in a more compact form, for example [16, 17],

$$T(\theta_0) \approx T_0 \cdot \left(\frac{\theta_0}{sen\theta_0}\right)^{3/8} \tag{16.31}$$

and

$$T(\theta_0) \approx T_0 \cdot \left(1 - \left(\frac{\theta_0}{\pi}\right)^2\right)^{-\frac{\pi^2}{16}} \tag{16.32}$$

The reader can verify that the predictions of the period (16.30), (16.31), and (16.32), for amplitudes smaller than about 45°, are almost identical.

Numerical integration of the motion equation for large amplitude

The differential equations that describe the pendulum motion can be integrated numerically following various numerical techniques [19].

Some of these calculations can also be done using a spreadsheet, such as Microsoft Excel [15]. In the following example, $\alpha(t)$ denotes the angular acceleration as a function of time.

The initial relations to be used are

$$\omega(t_o + \Delta t) = \omega(t_o) + \alpha(t)\Delta t \quad \text{and} \tag{16.33}$$

$$\theta(t_o + \Delta t) = \theta(t) + \omega(t_o + \Delta t)\Delta t, \tag{16.34}$$

where we assume, within the time interval Δt, a constant acceleration equal to the mean value, as well as a constant velocity equal to the *mean velocity* in the same interval.

Repeating the process for the next time interval, we find the calculated values for $t = t_o + 2\Delta t$ and then for $t = t_o + 3\Delta t \ldots$ and so on.

There is one difficulty: the procedure described assumes that we know the *mean* value of acceleration, while in fact we know only the *initial* value. One solution (first suggested by Richard Feynman [20]) is to use the *initial acceleration* to calculate the velocity at the time $t_o + \Delta t/2$ and to take this value as the *mean value of the velocity* in the interval Δt and then to proceed with the recursive calculations that give positions at times $\Delta t, 2\Delta t, 3\Delta t, \ldots$ and velocities at the times $\Delta t + \Delta t/2, 2\Delta t + \Delta t/2, 3\Delta t + \Delta t/2, \ldots$ In order to simplify the procedure to be implemented in the electronic spreadsheet, we may use the *initial acceleration* to calculate the next velocity and the *final velocity* to calculate the next position. There are many good references the reader can consult, such as Refs. [15, 21].

For the simplified case of a pendulum *without friction* for which Eq. (16.1) simplifies into

$$\alpha(t) = (g/L)\sin\theta(t). \tag{16.35}$$

In case of *dry friction*, the acceleration becomes

$$\alpha(t) = -\frac{g}{L}\sin\theta(t) - A\,\mathrm{sgn}\,\omega(t) \tag{16.36}$$

and in case of *viscous friction*, the acceleration becomes

$$\alpha(t) = -\frac{g}{L}\sin\theta(t) - B\omega(t), \tag{16.37}$$

where A and B are two constants.

References

[1] F. Sears, M. Zemansky, H. Young and R. Freedman, *College Physics*, NY: Pearson Ed., 2009.
[2] R. Halliday, D. Resnick and M. Krane, *Fundamentals of Physics*, NY, John Wiley & Sons, Inc., 2002.
[3] Vernier Software & Technology, "Photogates," 2021. Available: https://www.vernier.com/product/photogate/.
[4] Phyphox, "Smartphone-experiment: Pendulum," phyphox App., 2021. Available: https://www.youtube.com/watch?v=xY3NFcDG3ZU.
[5] M. A. González and M. A. González, "Smartphones as experimental tools to measure acoustical and mechanical properties of vibrating rods," *Eur. J. Phys.*, vol. 37, no. 4, p. 045701, July 2016.
[6] P. Vogt and J. Kuhn, "Analyzing simple pendulum phenomena with a smartphone acceleration sensor," *The Phys.Teach.*, vol. 50, p. 439, 2012.
[7] H. Kater, "An account of experiments for determining the length of the pendulum vibrating seconds in the latitude of London," *Philos. Trans., R. Soc. Lond.*, vol. 104, no. 3, p. 109, 1818.
[8] AE Science Equipment and Services Co., Ltd., "Reversible Pendulum — Kater's pendulum," 2021. Available: https://en.aelab.com.vn.
[9] AmritaVirtual Lab Collaborative Platform, "Kater's pendulum," 2021. Available: https://vlab.amrita.edu/index.php?sub=1&brch=280&sim=518&cnt=1.
[10] R. D. Peters, "Student-friendly precision pendulum," *Phys. Teach.*, vol. 37, p. 390, 1999.
[11] USDIGITAL, "models Taiss- 360P/R, or Calt- GHS3806G," See for example the models Taiss- 360P/R, or Calt- GHS3806G or 2021. Available: https://www.usdigital.com/.
[12] Vernier Spoftware, "Rotary motion sensor," 2021. Available: https://www.vernier.com/product/rotary-motion-sensor/.
[13] S. Gil, E. Legarreta and D. E. Di Gregorio, "Anharmonicity in large amplitude pendulum," *Am. J. Phys.*, vol. 79, no. 9, pp. 843–847, 2008.
[14] LABTREK, "LABTREK," 2021. Available: http://www.labtrek.it/pendulum.
[15] G. Torzo and P. Peranzoni, "The real pendulum: Theory, simulation, experiment," *Lat. Am. J. Phys. Educ.*, vol. 3, no. 2, p. 1, 2009.
[16] M. I. Molina, "Simple linearization of the simple pendulum for any amplitude," *Phys. Teach.*, vol. 35, no. 35, p. 489, 1997.
[17] T. F. Zheng, M. Mears, D. Hall and D. Pushkin, "Teaching the nonlinear pendulum," *Phys. Teach.*, vol. 32, p. 248, 1994.

[18] R. A. Nelson and M. G. Olson, "The pendulum-rich physics from a simple system," *Am. J. Phys.*, vol. 54, p. 112, 1986.
[19] D. F. Griffiths and D. J. Higham, "Numerical methods for ordinary differential equations," NY: Springer, 2010.
[20] R. Feynman, *et al.*, *The Feynman Lectures on Physics, Boxed Set: The New Millennium Edition*, vol.1, Basic Books, 2011.
[21] S. Bilbao, *et al.*, "Teaching the numerical solution of ordinary differential equations using Excel 5.0."
[22] J. Head, "Faith in physics," *Phys. Teach.*, vol. 33, p. 10, 1995.
[23] H. D. Young, R. A. Freedman, *et al.*, *College Physics*, Vol. II, Addison-Wesley, 2009.

Chapter 17

CYCLOIDAL PENDULUM — BRACHISTOCHRONE AND TAUTOCHRONE

This chapter studies the behavior of pendulums whose bobs do not follow a circular path. In particular, it considers the case of a cycloidal trajectory, in which the pendulum has the property of being exactly isochronous for all its amplitudes. Around 1662, Christian Huygens, one of the most ingenious watchmakers in history, invented a pendulum clock that used this property and made it possible to determine the longitude of the location with great precision. This invention is an example of precision, technical ingenuity, and a clear theoretical understanding of the problem, all of which applied to Huygens. This chapter also studies other pendulums in which, in contrast to the simple pendulum, periods decrease with the amplitude.

Goals

✓ Cycloidal path
✓ Tautochrone and brachistochrone problem
✓ Evolutes and involutes
✓ Cycloidal pendulum
✓ Harmonic oscillations
✓ Pendulum perturbations, anharmonic oscillations

17.1 Cycloidal pendulum

A cycloid is the curve traced by a point on a circle that rolls along a straight line without slipping, Figure 17.1. Many mathematicians have studied the properties of cycloids, including Pascal, Galileo, Descartes, and Huygens, among others. The parametric equation of this curve is

$$\begin{aligned} x &= R(\theta - \sin\theta), \\ y &= R(1 + \cos\theta). \end{aligned} \tag{17.1}$$

If this curve is inverted with respect to the horizontal axis, as shown in Figure 17.2, and a particle moves along it without friction under the effect of gravity, the time of arrival at the lowest point is always the same, regardless of its starting point. This dynamic property is known as *tautochrone behavior*.[a] The equation of the

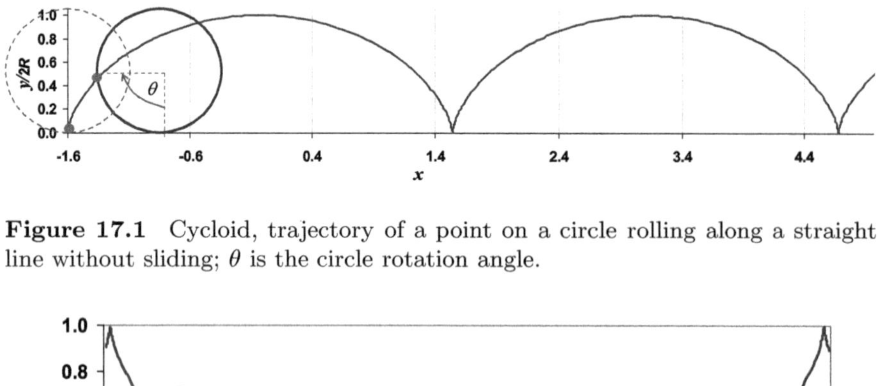

Figure 17.1 Cycloid, trajectory of a point on a circle rolling along a straight line without sliding; θ is the circle rotation angle.

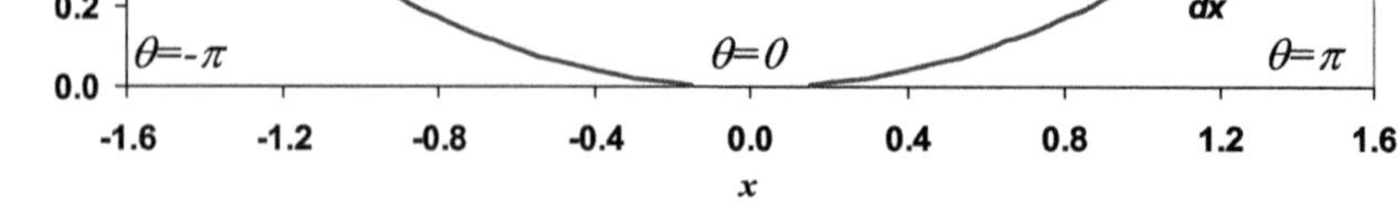

Figure 17.2 Particle falling smoothly along an inverted cycloid.

[a]A tautochrone curve (from Greek prefixes *tauto* meaning *same* and *chrono* meaning *time*) https://en.wikipedia.org/wiki/Tautochrone_curve

curve in Figure 17.2 is

$$\begin{array}{c} x = R(\theta + \sin\theta) \\ y = R(1 - \cos\theta) = 2\text{R} \cdot \sin^2(\theta/2) \end{array} \quad \text{and} \quad \begin{array}{c} dx = R(1 + \cos\theta)\text{d}\theta \\ dy = 2R \cdot \sin(\theta/2)\cos(\theta/2)\text{d}\theta. \end{array} \tag{17.2}$$

The length element of this curve is

$$ds^2 = dx^2 + dy^2 = \left[\left(\frac{dx}{d\theta}\right)^2 + \left(\frac{dy}{d\theta}\right)^2\right] d\theta^2 = 4R^2\sin^2(\theta/2)d\theta^2 \tag{17.3}$$

or also

$$ds = 2R\sin(\theta/2)d\theta \quad \text{and} \quad s(\theta) = -4R\cos(\theta/2). \tag{17.4}$$

Thus,

$$v = \frac{ds}{dt} \quad \text{and} \quad \frac{dy}{dt} = \frac{ds}{dt} \cdot \cos(\theta/2) = -\frac{ds}{dt} \cdot \frac{s}{4R}. \tag{17.5}$$

If a ball falls without friction along this curve, Eq. (17.2), the total energy must be conserved,
$E_0 =$ constant, therefore,

$$E_0 = E_p + E_k = mgy_0 = mg(y_0 - y) + \frac{1}{2}m\left(\frac{ds}{dt}\right)^2. \tag{17.6}$$

Taking the time derivative of this expression, we obtain the motion equation of the ball

$$0 = -g\frac{dy}{dt} + \left(\frac{ds}{dt}\right)\frac{d^2s}{dt^2} = +\frac{g \cdot s}{4R}\left(\frac{ds}{dt}\right) + \left(\frac{ds}{dt}\right)\frac{d^2s}{dt^2}, \tag{17.7}$$

that is,

$$\frac{d^2s}{dt^2} = -\frac{g}{4R}s = -\omega_0^2 s. \tag{17.8}$$

This differential equation is the same as for a harmonic oscillator of frequency ω_0 and period $T_0 = 2\pi/\omega_0 = 2\pi\sqrt{4R/g}$. Thus, a particle moving along a frictionless cycloidal path will perform a simple harmonic motion of period T_0, which *does not depend on the amplitude* of the oscillation. As in a simple harmonic oscillator, the oscillation is perfectly *isochronous*, that is, the oscillations are *harmonic*. Therefore, the time of fall of this particle, from any height to the lowest point, will always be the same. This means that if two particles are released in a cycloidal path, from different heights, they will reach the lowest point simultaneously after the time $T_0/4$. This is the aforementioned *tautochrone behavior*. On the other hand, the period of the particle in the cycloid described by Eq. (17.3) is equal to that of a simple pendulum (with small elongations) of length $L = 4R$. Another interesting property of the motion of a particle along a cycloid is that the fall time is the minimum of all possible trajectories. This property is known as a *brachistochrone* [1, 2]. When an oscillator moves with periods or frequencies that depend on the amplitude, we say that it performs *anharmonic* oscillations.

The motion of a particle moving along a frictionless path is almost impossible to implement experimentally. First, it is difficult to precisely shape the path or wire along a given curve. Second, it is even more difficult to prevent the movement along this path from being dominated by the forces of friction with the wire. Moreover, using spheres that rotate on predefined trajectories would introduce rotation as a new degree of freedom, thereby further complicating the problem [3].

This chapter proposes a simple experimental technique for building a pendulum whose bob moves along a predefined path. Consequently, the movement of this pendulum is equal to the originally proposed problem and avoids (largely) friction.

17.2 Involutes and evolutes

Consider a string of length R attached to point O of a curve (*evolute*), for example, the circle shown on the left side of Figure 17.3. If the string is held taut and tangent to the point of contact with the circle, the locus drawn by a pencil, attached to the end P of the string, is called the *involute* of the original curve.[b] This procedure is illustrated in Figure 17.4. In a strict sense, the evolute of a curve [4]

[b]https://en.wikipedia.org/wiki/Evolute

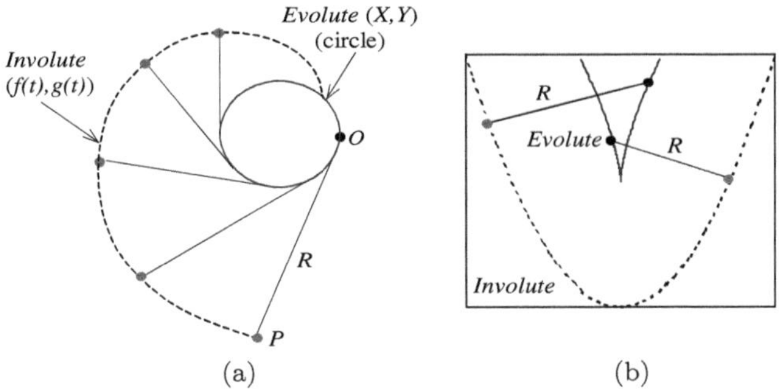

Figure 17.3 Two examples of evolute and involute pairs. (a) The end P of a chord of length R, attached to the point O of the blue circle (evolute), describes the spiral path (dashed line, involute). (b) The center of curvature of the involute (dashed line) describes the trajectory shown in thick blue lines (evolute).

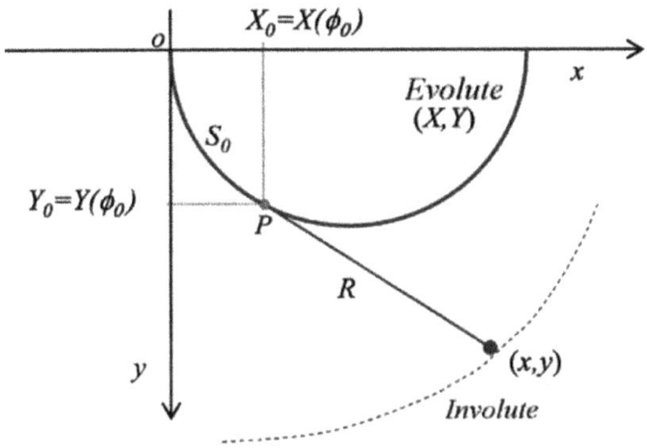

Figure 17.4 The bob of a pendulum, attached to point O through a wire of length $L = R + S_0$, describes the involute trajectory. The "cheek" S_0 is the length where the wire adheres to the evolute (X, Y). P is the center of curvature of the involute (x, y).

is determined by the centers of curvature of another curve that we call the involute.

There is a close relationship between the involutes and the evolutes. A single evolute corresponds to each involute. However, there can be an infinite number of involutes for each evolute, depending on the chosen *starting point* or the *length of the chord*, as in the example in Figure 17.3. This property of the pairs of evolutes–involutes can be

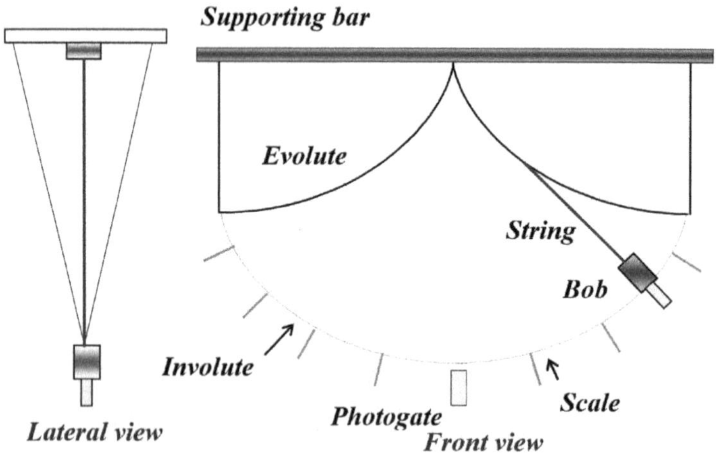

Figure 17.5 Example of a pendulum that follows the path described by the involute. The cheeks (evolute) and the trajectory form an evolute–involute pair. The side view illustrates one way to restrict the pendulum from moving in the same plane. The central thread or string is always in contact with the evolute (in the figure, the arc traced in red). A photogate placed at the bottom measures the period. The amplitude can be read on the lower scale, formed by a flexible ruler that has the approximate shape of the involute.

used to construct a pendulum whose bob travels a certain trajectory (involute). All you need to do is build a pair of suitable evolutes or *cheeks*, as shown schematically in Figure 17.5 [2, 4].

An interesting property of the cycloid curve, as represented by Eq. (17.2), is that if we use a thread of length $L = 4R$, joined to the vertex O of this curve (evolute), the other end of the thread also describes an equal cycloid. See Figure 17.4. This idea can be generalized, and with a suitable evolute, it is possible for the pendulum bob to describe any predetermined trajectory, as long as the appropriate evolute is chosen [2]. One of the objectives of the present investigation is to study the oscillatory motion of a particle in different types of potentials.

17.3 Experimental setup of a generalized pendulum

Figure 17.5 shows a schematic diagram of the experimental setup. Connect a photogate [5] to a personal computer and locate it at the lowest position of the pendulum. The photogate typically measures the period of oscillation with an uncertainty of the order of ms. Hang the bob from an inextensible cotton thread of length L. Place a flexible metric rod just below the trajectory of the bob to measure elongations. By visual inspection, the maximum amplitude S_{max} of each oscillation can be read directly from the scale.

It is easy to build an evolute. First, plot the evolute curve on paper on a 1:1 scale. Glue the evolute paper to a wooden board about 1 inch thick. Once the glue is dry, cut the wood to obtain the best evolute possible. You can use sandpaper to achieve a very good finish. Screw and glue the evolutes to a support bar, which supports the evolutes and the pendulum, as indicated in Figure 17.5. There are several interesting variations on these problems that the interested reader can investigate [6].

17.3.1 *Project 44. Simple pendulum — Variation of the period with the amplitude*

Recommended equipment: A bob attached to a thread of length L $\approx$ 1 m. A photogate connected to a PC. A scale to read the amplitudes of the pendulum.

Build a simple pendulum, like the one in Figure 17.5, with which you can measure its period for different amplitudes, starting with large amplitudes ($\theta_{\text{max}} \approx 90°$). An observer located near the pendulum dictates the maximum amplitudes of the pendulum, S_{max}, for each swing. Another observer writes down these amplitudes and the corresponding periods. With a little practice, this procedure is very simple to implement, since the period of a pendulum of $L \approx 1$ m is of the order of 2 s.

Suggested tasks:

- ✓ Measure the period T as a function of the amplitude S_{max}.
- ✓ Plot T as a function of S_{max}. Using the expression for period as a function of amplitude, described in Annex A of Chapter 16, try to describe your experimental results.
- ✓ Does the theoretical model adequately describe your experimental data?
- ✓ From the above analysis, determine the value of the period T_0 in the limit of small amplitudes ($S_{\text{max}} \to 0$). Construct a graph of T/T_0 as a function of the amplitude divided by L, that is, $\theta_0 = S_{\text{max}}/L$.
- ✓ What can you say about the variation of the period with respect to the amplitude? Does it increase or decrease?
- ✓ Compare your results from T/T_0 as a function of θ_0 to the theoretical expectations, Eqs. (16.22), (16.23), and (16.24) in Ch. 16.
- ✓ If two balls are dropped on the same trajectory of the bob (circle = involute) (without friction), which one reaches the lowest point first? The one that started from the farthest point or the one that started from the closest point to the lowest position? Note that the period measurements for each amplitude are precisely four times these decay times.

17.3.2 *Project 45. Cycloidal pendulum*

Recommended equipment: A pair of cycloidal evolutes with $R \approx 25$ cm. A photogate connected to a PC. A bob attached to a thread of length $L = 4R$. A scale to read the amplitudes of the pendulum.

Using an experimental setup like the one in Figure 17.5, measure the amplitude and period of the cycloidal pendulum simultaneously, starting with large amplitudes. To read the amplitude, you can use the same technique as in the previous experiment.

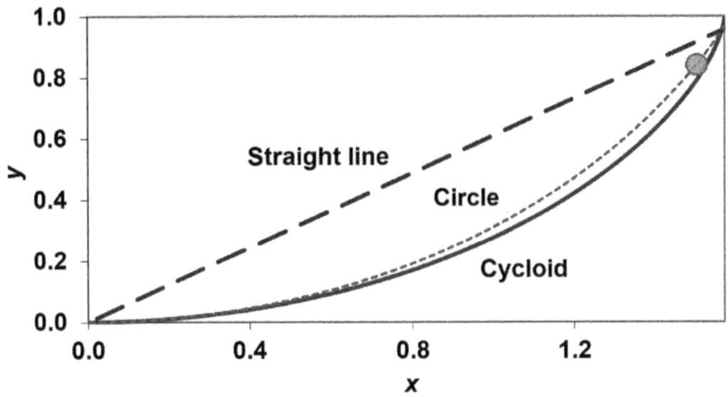

Figure 17.6 Comparison of straight, circular, and cycloidal paths that join the same two starting and ending points. See also "brachistochrone" at http://www.youtube.com.

Suggested Tasks:

✓ Measure the period T as a function of the amplitude S_{max}.

✓ Plot T as a function of S_{max}. Within the measurement errors, does the period vary as a function of the amplitude?

✓ Does the theoretical model previously developed describe your experimental data?

✓ If on the same trajectory of the bob (involute) two balls are dropped (without friction), which one reaches the lowest point first: the one that came from further away or the one that started from closer to the lowest position? Note that your measurements are proportional to these decay times.

17.3.3 *Project 46. Perturbed cycloidal pendulum — Anharmonic oscillations*

Recommended equipment: A pair of cycloidal evolutes with $R \approx 25$ cm. A photogate connected to a PC. A bob attached to a string of variable length L. A scale to read the amplitudes of the pendulum.

Use an experimental setup like the one in the previous project, only this time the thread is 15% longer and 15% shorter than $L_0 = 4R$. By varying the length of the pendulum with respect to L_0, the trajectory of the pendulum ceases to be a cycloid, and therefore it is to be expected that the oscillations will cease to be harmonic. To read the amplitude, you can use the same technique as in the previous experiment.

Suggested Tasks:

- ✓ For each of these pendulums, measure the period T as a function of the amplitude S_{max}.
- ✓ Plot T as a function of S_{max}.
- ✓ From the above analysis, determine the value of the period T_0 in the limit of small amplitudes ($S_{\text{max}} \to 0$). Build the graph of T/T_0 as a function of the amplitude divided by L, that is, S_{max}/L, and compare it to the graph corresponding to the simple pendulum and the cycloidal pendulum ($L = 4R$). What can you conclude from this analysis?
- ✓ If two beads are allowed to drop (without friction) along the same trajectories of the bob (involute) described by the modified cycloidal pendulums, which one will reach the lowest point first? The bead that started the farthest from this lowest point or the one that started the closest to it? Compare to your measurements.

17.3.4 *Project 47. Pendulum with semicubical evolute — Paradox of the race*

Recommended equipment: A pair of semicubical involutes with $L \approx 1$ m. A photogate connected to a PC. A scale to read the amplitudes of the pendulum.

The paradox of the race: Imagine two identical balls A and B on an inclined plane, both of which are released at the same time, starting from rest. If A starts from a point lower than B, it is clear that A will reach the base of the plane first. Here we will analyze a trajectory in which this intuitive notion is not fulfilled, i.e., the one that starts from the farthest distance arrives first.

Use an experimental arrangement like the one shown in Figure 17.5 but this time with a semicubical shape (or Niele parabola) of the form

$$Y(X) = L - \frac{3}{2}L^{1/3}X^{2/3}. \tag{17.9}$$

In this case, it can be proved that the involute described by the bob of the pendulum is [2]

$$\left.\begin{aligned} x(\varphi) &= \frac{2\varphi}{\sqrt{1+(\varphi/L)^2}} - \varphi \\ y(\varphi) &= 2L - \frac{2L}{\sqrt{1+(\varphi/L)^2}} - \frac{1}{2}\frac{\varphi^2}{L} \end{aligned}\right\}. \tag{17.10}$$

The interesting thing about this pendulum is that the pendulum period *decreases with amplitude*, contrary to what happens with the simple pendulum. In other words, if two balls are dropped on the trajectory of the pendulum, the one that is further from the equilibrium point arrives before the one that is closer to this point!

Suggested Tasks:

- ✓ Measure the period T as a function of the amplitude S_{max}.
- ✓ Plot T as a function of S_{max}. Within the measurement errors, does the period vary as a function of amplitude?
- ✓ From the above analysis, determine the value of the period T_0 at the limit of small amplitudes ($S_{\text{max}} \to 0$). Build the graph of T/T_0 as a function of the amplitude divided by L, that is, S_{max}/L, and compare it with the graph corresponding to the simple pendulum and the cycloidal pendulum. What can you conclude from this analysis?
- ✓ If two beads are allowed to drop (without friction) along this trajectory of the bob (involute) described by this pendulum, which one will reach the lowest point first? The bead that started the farthest from this lowest point or the one that started closer to it? Compare to your measurements.

References

[1] H. Goldstein, C. Poole and J. Safko, *Classical Mechanics*, 3rd Ed., Boston, MA: Addison-Wesley, 2001.

[2] S. Gil and D. E. Di Gregorio, "Nonisochronism in the interrupted pendulum," *Am. J. Phys.*, vol. 71, no. 11, pp. 1115–1120, 2003.

[3] J. Z. Villanueva, "Note on the rough cycloidal slide track," *Am. J. Phys.*, vol. 53, p. 490, 1985.

[4] A. Gray, *Modern Differential Geometry of Curves and Surfaces with Mathematica*, 2nd Ed., Boca de Ratón: CRC Press, 1997.

[5] Vernier Software & Technology, "Photogates," 2021. Available: https://www.vernier.com/product/photogate/.

Chapter 18

FORCED OSCILLATIONS

This chapter analyzes the motion of forced oscillation systems. In particular, it analyzes the oscillations induced by a periodic driving force and the phenomenon of resonance.

Goals

✓ Forced oscillations
✓ Resonance

18.1 Forced oscillations — Resonance

Forced oscillations: The equation of motion of a one-dimensional system of a mass m connected to a linear spring of elastic constant k, which is driven by a periodic external force of the form $F(t) = F_0 \cos(\omega t)$, assuming friction is proportional to speed, $f_r = -b\ v$, is [1, 2]

$$m\frac{d^2x}{dt^2} + b\frac{dx}{dt} + kx = F_0 \cos(\omega t), \tag{18.1}$$

dividing by m we get

$$\frac{d^2x}{dt^2} + 2\gamma\frac{dx}{dt} + \omega_0^2 x = B_0 \cos(\omega t). \tag{18.2}$$

Here, $B_0 = F_0/m$, with F_0 being the amplitude of the excitation force and ω the frequency of the external excitation, and γ and ω_0 having the same meaning as in Chapter 15, that is,

$$\gamma = b/2m \quad \text{and} \quad \omega_0^2 = k/m. \tag{18.3}$$

A *particular solution* of this differential equation, which can be verified by simple substitution, is [3]

$$x(t) = C_0(\omega)\cos(\omega t - \phi), \tag{18.4}$$

with

$$C_0(\omega) = \frac{F_0/m}{\sqrt{(\omega_0^2 - \omega^2)^2 + (2\gamma\omega)^2}} \tag{18.5}$$

and

$$\tan(\phi) = \frac{2\gamma\omega}{\omega_0^2 - \omega^2}. \tag{18.6}$$

The amplitude of the stationary signal, described by $C_0(\omega)$, presents a maximum for $\omega = \omega_R$, given by $(\partial C_0(\omega)/\partial\omega|_{\omega=\omega_R} = 0)$

$$\omega_R = \sqrt{\omega_0^2 - 2\gamma^2} = 2\pi f_R. \tag{18.7}$$

The frequency $f_R(\omega_R)$, for which the amplitude is maximum, is called the *resonance (angular) frequency*. Figure 18.1 shows the variation of the amplitude $C_0(\omega)$ and of the phase ϕ with reduced frequency ω/ω_R $(=f/f_R)$ described by (18.5) and (18.6), respectively. Here, ω_R is the value of the angular frequency for which the amplitude $C(\omega)$ reaches its maximum value. It is important to note that the frequency for which the amplitude of the speed is maximum does not coincide with this frequency.

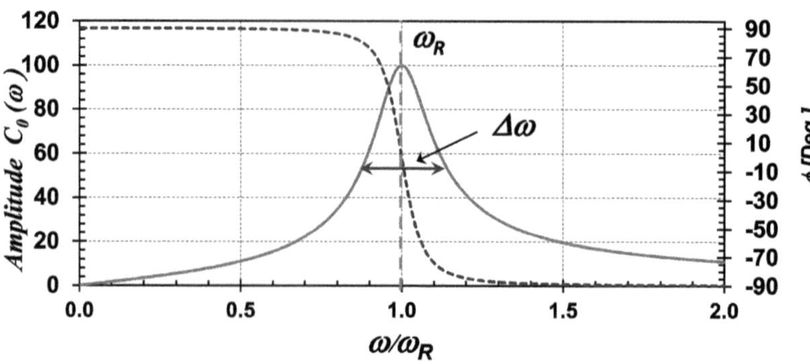

Figure 18.1 Variation of maximum amplitude and phase with frequency.

The part of the response described by (18.4) remains stable over time and is called the *stationary response* of the system.

The half-width of the resonance curve is defined as the difference between the frequencies for which the amplitude $C_0(\omega) = C_0(\omega_R)/\sqrt{2}$. Using Eq. (18.5), it is easy to prove that the half-width $\Delta\omega$ is related to the *quality factor* Q: [3, 4]

$$Q \cong \frac{\omega_0}{\Delta\omega} = \frac{\omega_R}{2\gamma} = \frac{2\pi}{DL}. \tag{18.8}$$

This ratio between the width of the resonance $\Delta\omega$ and the logarithmic decrement, DL, of energy, see Chapter 15, is very useful and prevalent in many physical problems. If the energy dissipation in an oscillatory system is large (large DL and small Q), the resonance curve will be wide, and it will be sharper if the friction is low. Figure 18.2 shows the effect of friction force (γ) on the resonance curve of an oscillating system.

The *complete solution* of the equation of motion (18.2) is given by a combination of Eq. (18.4) and the homogeneous solution discussed in Chapter 15, which is: [3]

$$x(t) = C_0(\omega)\cos(\omega_p t - \varphi) + A_0 \cdot e^{-\gamma \cdot t}\sin(\omega_p t + \phi). \tag{18.9}$$

Like before, A_0 and ϕ depend on the initial conditions of the system, while the parameters ω_p, γ, ϕ, and $C_0(\omega)$ depend on

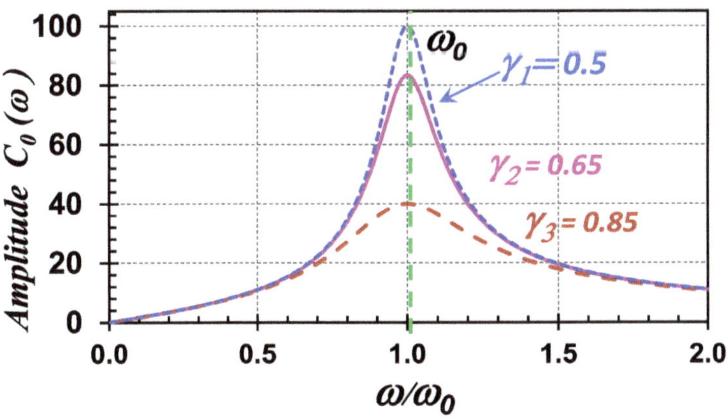

Figure 18.2 Resonance curve of an oscillatory system with three different values of damping. The red curve has the largest value of γ (lowest Q) and the dashed blue curve has the lowest value of γ (larger Q).

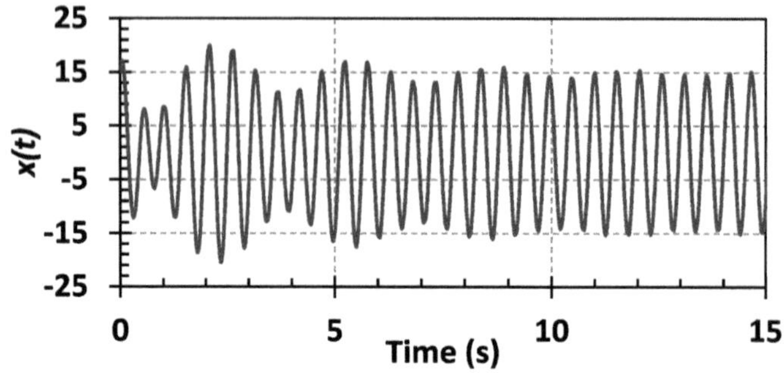

Figure 18.3 Complete response of a forced oscillatory system. At first, the beating effect is observed, due to the superposition of the transient and stationary signals.

the physical characteristics of the system. The second term on the left side of Eq. (18.9), which decreases exponentially with time, is called the *transient response* of the system, discussed in Chapter 15. The form of the complete solution (18.9) is shown schematically in Figure 18.3. If the frequencies ω and ω_p are close, at the beginning of the oscillation, a *beating effect* (superposition) between the transient and stationary signals can be observed.

18.2 Experimental setups

There are many experimental setups by means of which you can study the characteristics of a forced oscillator. Figures 18.4, 18.5, and 18.6 show three possible experimental arrangements. Other simple, interesting setups may also be used. Two experimental arrangements are particularly interesting. One uses a moving-coil analog ammeter, which is very easy to implement [5]. The other uses a rotating disk connected to an angle sensor [6, 7], providing a very suitable physical pendulum to carry out various studies.

For this setup, a disk is attached to a rotational motion sensor [8, 9] or shaft encoder.[a] The disk can be made of aluminum, about

[a] A shaft encoder or rotary encoder is an electromechanical or optical device used to convert the angular position of a shaft into an electrical, analog, or digital signal. These devices are used in many applications to determine an angular position.

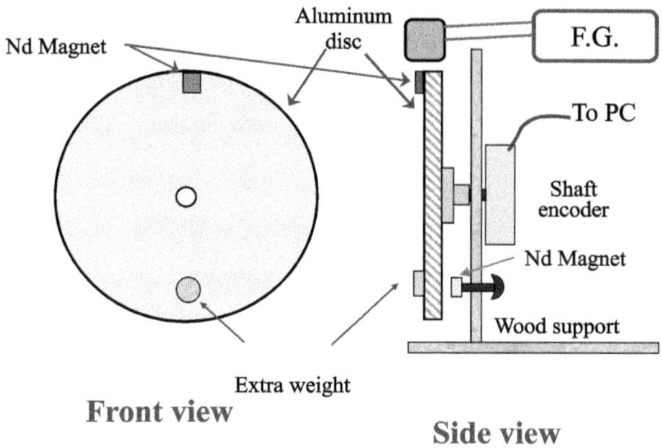

Figure 18.4 Experimental setup to study forced oscillations with a physical pendulum [7].

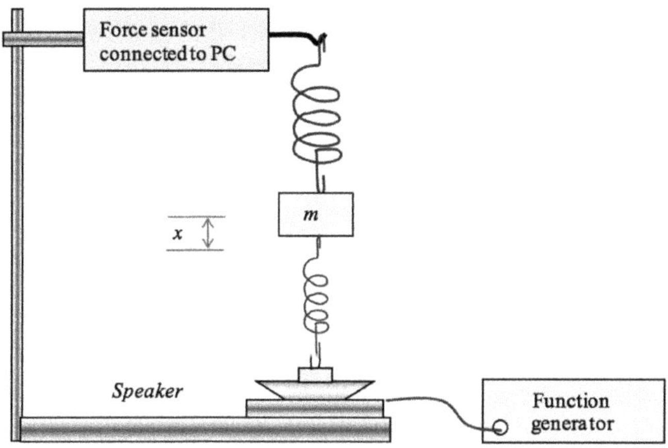

Figure 18.5 Diagram of a possible experimental arrangement to study a mass-spring system with viscous friction and driving force. An audio speaker connected to a function generator and linked to mass by a spring generates the periodic driving force at a frequency determined by the function generator.

20–30 cm in diameter and about 3 mm thick [7–10]. The aluminum disk is connected at its center to the axis of an optical shaft encoder. If desired, an extra weight can be placed close to the edge of the face of the disk to break its mass symmetry. A neodymium magnet can be used to vary the “viscous” friction coefficient of the system (γ).

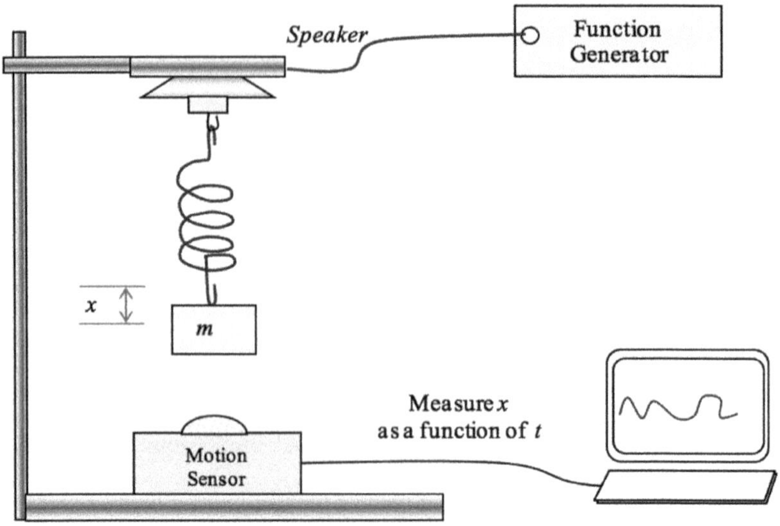

Figure 18.6 Diagram of another possible experimental arrangement to study a mass-spring system with impeller force. An audio speaker connected to the function generator and linked to ground by a spring generates the periodic driving force at a frequency determined by the function generator. The motion sensor measures the position of the mass x as a function of time.

The rotation sensor generates a digital signal that measures the turning angle and direction, in other words, it directly gives the angle value θ [8], which is the unique degree of freedom of the system (see Figure 18.4). The distance to the center of this overload and its magnitude can be used to regulate the period of the pendulum. A neodymium magnet can be positioned about 10 cm from the axis of rotation. Its distance to the disk can be adjusted with a screw. Eddy currents in the aluminum disk generate a *viscous* friction in the system, i.e., a friction force proportional to the speed of the disk ($d\theta/dt$). The magnitude of the friction force can be modified with the screw, which varies the distance from the magnet to the disk.

If a second lightweight magnet is fixed to the top edge of the disk, and a coil is connected to a function generator (FG), it is possible to produce a driving force whose frequency and amplitude can be varied at will by the FG. Furthermore, if a resistor is placed in series with the coil, the voltage drop across the resistor will be proportional to the current flowing through it. This electrical signal, connected to a computer data acquisition system, serves to monitor the magnitude of the driving force. In turn, the signal given by the rotatory sensor allows the rotation angle, θ, to be monitored. Thus, this system is very suitable for studying forced and free oscillations in various situations.

As noted above, there are many experimental setups that can be used to study the characteristics of a forced oscillator. Two other possible experimental arrangements are shown in Figures 18.5 and 18.6. In Figure 18.5, a force sensor is used to obtain a signal proportional to the displacement x of the mass m. An audio speaker or loudspeaker, with a jar cap glued to the cone and connected to a function generator, provides the driving force through a soft spring. In Figure 18.6, the mass hangs directly from the loudspeaker. In this case, an ultrasonic sensor is used to monitor the displacement x of the mass m.

18.2.1 *Project 48. Forced oscillations*

Using the experimental forced oscillatory setup available to you, study the variation in time of the displacement and velocity signal as a function of time.

Suggested tasks:

- ✓ First, turn off the driving force, and using the technique described in Chapter 15, determine the frequency of oscillation ω_p — close to natural frequency ω_0 if there is little friction — and the damping γ of your system.
- ✓ Turn on the driving force, working with a driving frequency close to ω_p, and try to observe the beating effect of the system.
- ✓ If you can measure $x(t)$, using a fast Fourier transform (FFT) [11] — which is available in many analysis programs such as Excel, Origin, Matlab, and Math — determine the frequencies present in the signal. Do these frequencies coincide with the frequency ω_p and the driving frequency as predicted by Eq. (18.9)?

✓ By varying the frequency of the driving signal, study the variation of the stationary amplitude with frequency. Plot your results. The range of angular frequencies of interest extends approximately between $\omega_p - 4\gamma$ and $\omega_p + 4\gamma$. On the same graph, superimpose the theoretical expectations described by Eq. (18.5), using the parameter F_0/m as variable but with the values of γ and ω_0 measured. Try to adjust the measured data. Discuss how adequately you can reproduce your data.

✓ Find the half-width $\Delta\omega$ of the resonance curve, and fitting expression (18.5), determine the values of γ and ω_0. What is the quality factor Q of your system? How do these values compare to those previously found with the free-oscillating system?

✓ By varying the frequency of the driving signal, study the phase variation ϕ between the driving signal and the stationary response $x(t)$, as a function of frequency. Plot your results. Compare your measurements with what is theoretically expected, Eq. (18.6).

References

[1] D. Giancoli, *Physics Principles with Applications*, 7th Ed., NY: Pearson, 2014.
[2] R. Halliday, D. Resnick and M. Krane, *Fundamentals of Physics*, NY: John Wiley & Sons, Inc., 2002.
[3] S. T. Thornton and J. B. Marion, *Classical Dynamics of Particles and Systems*, 5th Ed., Belmont, CA: Thomson Books/Cole, 2012.
[4] Wikipedia, "Q factor," 2021. Available: https://en.wikipedia.org/wiki/Q_factor.
[5] I. Moore, "Simple mechanical forced damped oscillator with electronic output," *Am. J. Phys.*, vol. 62, p. 140, 1993.
[6] Y. Kraftmakher, "Computerized physical pendulum for classroom demonstration," *Phys. Teach.*, vol. 43, no. 2, pp. 244–246, 2005.
[7] S. Gil, E. Legarreta and D. E. Di Gregorio, "Anharmonicity in large amplitude pendulum," *Am. J. Phys.*, vol. 79, no. 9, pp. 843–847, 2008.
[8] Vernier Spoftware, "Rotary motion sensor," 2021. [Online]. Available: https://www.vernier.com/product/rotary-motion-sensor/.
[9] US Digital, "US digital motion control products," 2021. Available: https://www.usdigital.com/.

[10] G. Torzo and P. Peranzoni, "The real pendulum: Theory, simulation, experiment," *Lat. Am. J. Phys. Educ.*, vol. 3, no. 2, p. 1, 2009.
[11] Wikipedia, "Fast Fourier transform," 2021.
[12] USDIGITAL, "models Taiss- 360P/R, or Calt- GHS3806G," See for example the models Taiss- 360P/R, or Calt- GHS3806G or 2021. Available: https://www.usdigital.com/.

Chapter 19

HANGING CHAINS

This chapter studies the shape of a flexible rope or chain supported at its ends, with and without additional weights, comparing it to theoretical expectations according to the laws of statics. Arches subject to simple compression are of great importance and interest in engineering architecture because they enable construction with traditional materials.

Goals

✓ Shape of hanging chains without load

✓ Shape of hanging chains with loads

19.1 Catenaries and parabolas

Ropes, flexible cables, and chains suspended from two points are often seen [1, 2]. An impressive, beautiful example is provided by the cables of suspension bridges, such as the Golden Gate in San Francisco. The problem of mathematically describing the shape of a chain suspended at its ends was first solved by Jakob Bernoulli in 1690 and also addressed by many prominent scientists, including Galileo, Leibniz, Huygens, and Euler [3].

Consider a rope or chain of length L_c and mass M_c suspended from its ends, as illustrated in Figure 19.1. Assume that the horizontal distance between the suspension points is L and the heights of the suspension points, measured with respect to the lowest point of the rope, which we take as the origin of coordinates, are h_1 and h_2. The shape taken on by the rope or chain (catenary) is described by the function $y(x)$, where x is the horizontal coordinate [2].

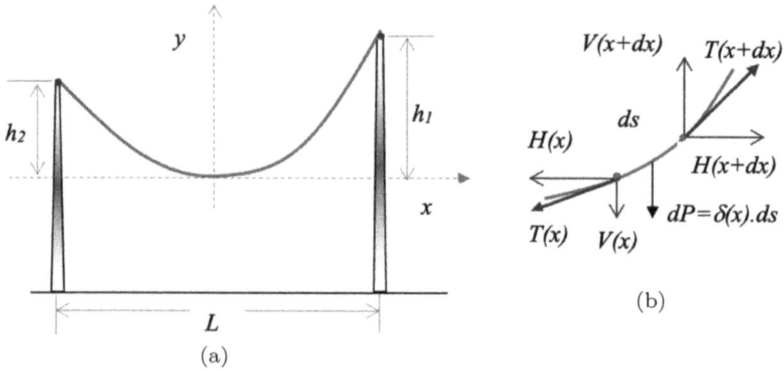

Figure 19.1 (a) Flexible chain or rope suspended at its ends from two fixed points. The coordinates of these points are $(-L_2, h_2)$ and (L_1, h_1), with $L_1 + L_2 = L$. (b) Forces acting on an infinitesimal segment of chord of length ds.

The weight of the infinitesimal element of length ds is $dP = \delta(x)gds$, where g is the value of the acceleration due to gravity and $\delta(x)$ the mass per unit length of the rope or chain. Here, we assume that the linear density of mass depends on the coordinate x. If we assume that the density is constant, then $\delta(x) = M_c/L_c.T(x)$ designates the value of the tension of the rope or chain at the x-coordinate point, pointing in the direction of the tangent to the curve $y(x)$. *It is precisely this property that defines a flexible chain or rope.* $V(x)$ and $H(x)$ designate the horizontal and vertical components of stress $T(x)$. From the physical requirement of equilibrium of forces in the x and y directions, we have

$$H(x + dx) = H(x) = H_0, \tag{19.1}$$

where H_0 represents the tension of the rope or chain at its vertex, where $dy/dx = 0$. Furthermore,

$$V(x + dx) - V(x) = dV = dP = \delta(x) \cdot g \cdot ds. \tag{19.2}$$

For a flexible rope or chain, the tension T always points in the direction of the tangent, therefore,

$$\frac{V(x)}{H(x)} = \frac{dy}{dx}. \tag{19.3}$$

Combining (19.1), (19.2), and (19.3), we get

$$dV = \frac{dV}{dx} \cdot dx = \frac{d^2y}{dx^2} \cdot H_0 \cdot dx = \delta(x) \cdot g \cdot ds. \tag{19.4}$$

Given that $ds = \sqrt{1 + (dy/dx)^2}$, Eq. (19.4) can be written as follows:

$$\frac{d^2y}{dx^2} = \frac{\delta(x) \cdot g}{H_0} \cdot \sqrt{1 + \left(\frac{dy}{dx}\right)^2}, \tag{19.5}$$

if we define

$$\lambda(x) = \delta(x) \cdot \frac{g}{H_0}, \tag{19.6}$$

the differential equation of the chain, Eq. (19.5), can be written as follows:

$$\left(\frac{d^2y}{dx^2}\right)^2 = \lambda^2(x) \cdot \left[1 + \left(\frac{dy}{dx}\right)^2\right]. \tag{19.7}$$

If we define $z(x) = dy/dx$, Eq. (19.7) can be easily integrated.

$$\int \frac{dz}{\sqrt{1 + z^2}} = \int \lambda(x) \cdot dx \Rightarrow z = \sinh(\lambda \cdot u(x)), \tag{19.8}$$

where

$$u(x) = \int \lambda(x) \cdot dx. \tag{19.9}$$

If the mass density were constant, $\delta(x) = M_c/L_c$ y $\lambda_0 = M_c \cdot g/(H_0 \cdot L_c)$. The above expressions would lead to

$$z(x) = \frac{dy(x)}{dx} = \sinh(\lambda_0 \cdot x) + c_1. \tag{19.10}$$

By choosing a coordinate axis system such that the origin coincides with the lowest point of the chain, where the tangent ($z = dy/dx$) is zero, $c_1 = 0$. By integrating once more, we obtain the equation of the chain or *catenary* $y(x)$ being sought:

$$y(x) = \frac{1}{\lambda_0} \cosh(\lambda_0 \cdot x) + c. \tag{19.11}$$

The constants c and λ_0 can be determined by enforcing the boundary conditions: for $x = L_1$, $y = h_1$ and for $x = -L_2$, $y = h_2$.

For simplicity, in the following we will assume that $h_1 = h_2 = h$ and $L_1 = L_2 = L/2$. Under these conditions, we have

$$h = \frac{1}{\lambda_0}\cosh(\lambda_0 \cdot L/2) + c, \tag{19.12}$$

as $y(x = 0) = 0$,

$$0 = \frac{1}{\lambda}\cosh(0) + c \rightarrow c = \lambda_0. \tag{19.13}$$

The resulting catenary equation is

$$y(x) = \frac{1}{\lambda_0} \cdot (\cosh(\lambda_0 \cdot x) - 1). \tag{19.14}$$

From this expression, it follows that the length of the chain can be calculated as follows:

$$L_C = 2 \cdot \int_0^{L/2} \sqrt{1 + (dy/dx)^2} \cdot dx = \frac{2}{\lambda_0} \cdot \sinh(\lambda_0 \cdot L/2). \tag{19.15}$$

By combining (19.12) with (19.15), we obtain the following condition:

$$\frac{\lambda_0 \cdot L_c}{2 \cdot (\lambda_0 \cdot h + 1)} = \tanh(\lambda_0 \cdot L/2). \tag{19.16}$$

This expression indicates that the parameters $L, L_c, h,$ and λ_0 are related. If L, L_c, and h are known, it is possible to determine the value of λ_0 by solving the transcendent Eq. (19.16). A simple method for this is to make a graph of the values of the first and second members of (19.16) as a function of λ_0. The λ_0 values for which the curves intersect provide the roots of Eq. (19.16).

Sometimes it is useful to have the expression of the catenary taking as the origin of the coordinates in the upper left corner from which the chain hangs. In this case, it is easy to demonstrate that expression (19.14) becomes

$$y(x) = \pm\frac{1}{\lambda_0} \cdot \left(\cosh\left[\frac{\lambda_0 \cdot L}{2}\right] - \cosh\left[\lambda_0 \cdot \left(x - \frac{L}{2}\right)\right]\right), \tag{19.17}$$

where the + or − sign corresponds to whether the y axis is adopted pointing downwards or upwards, respectively.

The mass per unit horizontal length is $dm/dx = \delta(x) \cdot ds/dx$. If this density is constant, $dm/dx = \beta$, note that $dm/dx =$ constant, it implies that the horizontal load is constant, which is not the same as saying that $\lambda_0 =$ constant, from Eq. (19.4), we have

$$\frac{d^2y}{dx^2} = \frac{g}{H_0}\beta = a, \quad \text{that is,} \quad y(x) = a \cdot x^2 + b \cdot x + c. \tag{19.18}$$

Therefore, in this case, which corresponds to the case of a suspension bridge, where most of the mass is on the platform or deck hanging from it, we would expect the main cables to take the shape of a parabola. In engineering and architecture, arches subjected only to simple traction, such as the catenary or the parabola, are known as *funicular forms* [4].

19.2 Loaded and unloaded chains

✓ Show that expressions (19.14) and (19.17) are equivalent, as stated above.

✓ For a rope of $Lc = 2\,\text{m}$ and $L = 1\,\text{m}$ and $h = 0.8\,\text{m}$, by solving Eq. (19.14), show that $\lambda_0 = 2.030$ and 4.86. Discuss the meaning of having two roots.

19.2.1 *Project 49. Simple chain supported at its ends*

Recommended equipment: A digital camera (smartphone) with a resolution of at least 460×600 pixels. A chain 1–3 m long, supported at its ends.

Using a digital camera, acquire images of the chain attached at its ends. The separation between the suspension points is L and they are at the same height h. Likewise, assume the length of the chain Lc is known. It is suggested to place two graduated rulers, one vertical and the other horizontal, to be able to reconstruct the real scales in the image (see Figure 19.3). Use the recommendations made in Chapter 10.

Another way to "calibrate" the image to real dimensions is to have a background grid of known dimensions, as shown in Figure 19.2.

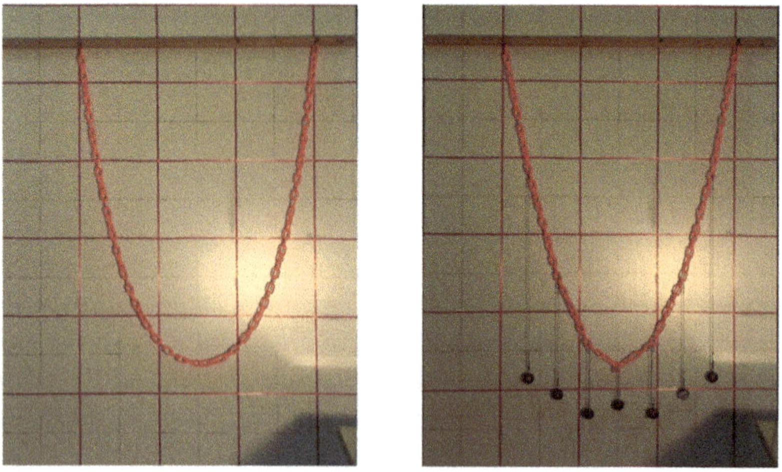

Figure 19.2 Hanging chain with a gridded background of known dimension, 20 cm × 20 cm in this case. On the left is a chain without load and on the right is a chain with loads evenly distributed along the horizontal axis.

A useful element for this experiment consists of having a plumb line at the bottom of the table and a well-leveled horizontal line. Try to level the camera such that the vertical plumb line and horizontal line are as parallel as possible to the margins of the frame. If you use a background grid, it suffices to simply level it well [2].

Suggested tasks:

✓ Acquire several images of the unloaded chain and measure the values of L and Lc if necessary.

✓ Using an image viewing program (see Chapter 10), find the coordinates of each point where the mouse is located in pixel units. Choose the coordinate origin in the way most convenient for you. A convenient point to choose for the origin is the lowest point of the chain.

✓ Transfer these points to a spreadsheet and compare the experimental results directly with the theoretical expressions discussed above, Eq. (19.14) or (19.17).

✓ An alternative way to make this comparison is to incorporate the digital image, Figure 19.2, in the drawing area of a graph

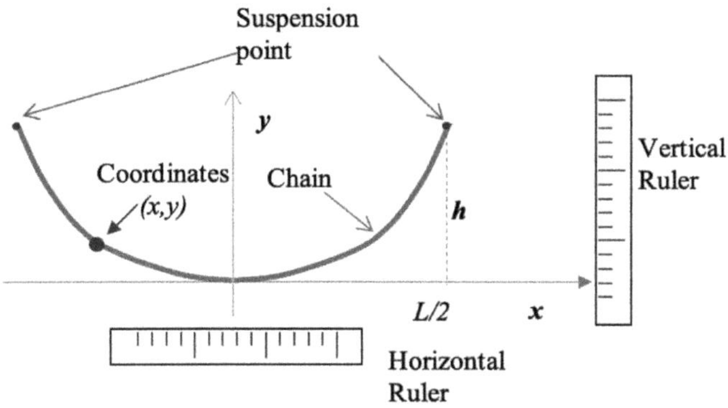

Figure 19.3 Flexible chain or rope suspended at its ends from two fixed points. Graduated rulers can be used to define an absolute scale to measure distances. It is advisable to place the camera on an axis perpendicular to the plane of the catenary that passes close to the origin of the coordinates. Make sure the rulers are well placed in vertical and horizontal positions, respectively.

and superimpose the theoretical curves on it, following a method similar to those described in Chapter 10.

19.2.2 *Project 50. Chain with loads*

> **Recommended equipment**: A digital camera or smartphone with a resolution of at least 460 × 600 pixels. A chain 1–3 m long, supported at its ends.

Study the case of a chain with loads placed uniformly along the horizontal direction, as illustrated on the right-hand side of Figure 19.2. Analyze this case with the same technique as used in the previous project [2].

Suggested tasks:

- ✓ Acquire multiple images of the loaded chain.
- ✓ Using the same techniques as discussed in the previous activity, compare the experimental results directly to the theoretical expressions discussed above, Eq. (19.18).
- ✓ Discuss how adequate the proposed model is to explain your data.

References

[1] J. D. Wilson, *College Physics*, 7th Ed., NY: Pearson, 2009.

[2] S. Gil, H. D. Reisin and E. Rodríguez, "Using a digital camera as a measuring device," *Am. J. Phys.*, vol. 74, p. 768, 2006.

[3] University of St Andrews Scotland, "Famous curves index: Catenary diagram, history, and connections to other curves," 2021. [Online]. Available: https://mathshistory.st-andrews.ac.uk/Curves/.

[4] S. P. Timoshenko and J. M. Gere, *Theory of Elastic Stability*, NY: McGraw- Hill, 1985.

Chapter 20

ELASTIC PROPERTIES OF MATERIALS — MODULUS OF RIGIDITY

This chapter looks at the elastic properties of materials and discusses various techniques to determine the modulus of elasticity or Young's modulus of different samples. It also analyzes the shapes taken on by cantilevered bars when subjected to loads or under their own weight. Finally, it studies the vibrations of bars and their normal modes of oscillation. These experiments illustrate some of the modern non-destructive testing techniques currently used in many industrial applications.

Goals

✓ Traction and compression
✓ Modulus of rigidity
✓ Bending a bar
✓ Determination of Young's modulus
✓ Vibrations of a bar

20.1 Mechanical properties of materials

If a material is subjected to traction, i.e., if it is stressed from its ends in opposite directions, Figure 20.1, its length increases and eventually, if the applied force is large, the material may break. In this section, we will study the connection between the effects of forces and the deformations that they produce in a material.

Consider a cylindrical material sample of cross-section A and initial length L_0 which is subjected to traction. As a consequence of the force F acting along its axis, the sample will undergo a stretch ΔL. For small deformations, when $\Delta L/L_0 \ll 1$, it is found

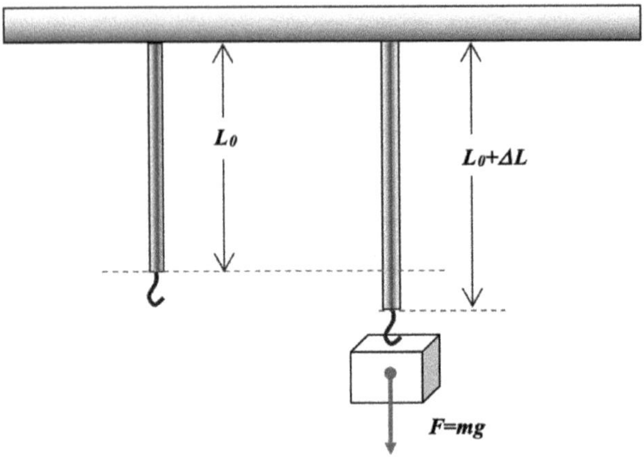

Figure 20.1 Cylindrical bar of original length L_0, subjected to a traction tension.

experimentally that ΔL is proportional to the applied force (F) and to its original length (L_0) and inversely proportional to its cross-sectional area (A), that is,

$$\frac{\Delta \mathrm{L}}{L_0} = \left(\frac{1}{E}\right)\frac{F}{A}. \tag{20.1}$$

Here, E is a characteristic constant of the material, called *Young's modulus* or *modulus of elasticity*. This modulus is usually designated interchangeably with the letters Y or E. The phenomenological expression (20.1) is known as *Hooke's law*. Although this expression is valid for a great variety of materials, it is not universal. Its validity is limited to the so-called *proportionality zone*, see Figure 20.2.

The quotient F/A is called *stress*, denoted by the letter σ and expressed in pressure units (Pa$=$N/m^2). The quotient $\Delta L/L_0$ is called *strain*, denoted by the letter ε, and dimensionless. With this notation, the expression (20.1) can be written as follows:

$$\sigma = E \cdot \varepsilon = Y \cdot \varepsilon. \tag{20.2}$$

Figure 20.2 shows a typical curve for the strain as a function of stress, commonly found in many materials [1, 2].

At the beginning of the stretch, the deformation is proportional to the force, within the range of validity of Hooke's law. This occurs

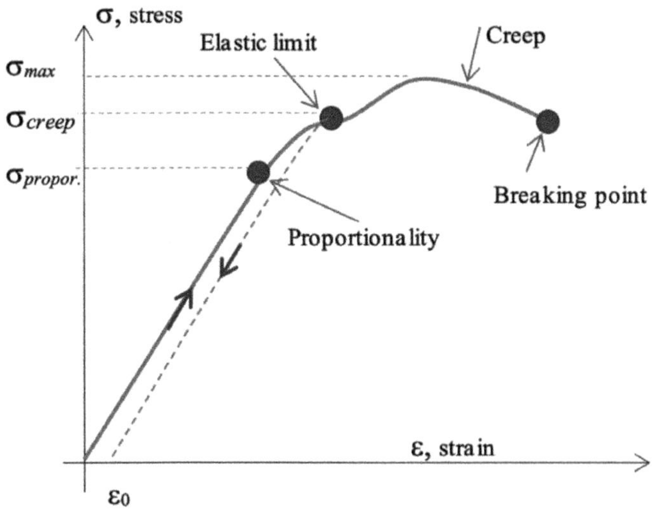

Figure 20.2 Relationship between applied stress σ and strain ε. Hooke's law is strictly valid up to the limit of proportionality. When the elastic limit is exceeded and the applied stress is suppressed, the material is permanently deformed. This is indicated in the graph by the arrows. The value ε_0 indicates the magnitude of permanent deformation. Up to the limit of proportionality, $\varepsilon_0 < 10^{-4}$. The yield point is defined as the intersection of a parallel to the line of the elastic zone that passes through the permanent deformation point $\varepsilon_f = 0.002$ (0.2%).

until the applied force reaches a value called *limit of proportionality* (σ_{propor}). Up to the limit of proportionality, the relationship between strain and stress is linear, i.e., expression (20.2) is valid. If the material is subjected to stress up to this value, when the stress is removed, the material generally returns to its original shape without undergoing permanent deformation. Beyond the limit of proportionality, the graph of ε as a function of σ deviates from the line and there is no simple relationship between σ and ε. However, up to the elastic limit, the object will return to its original length if the applied force is removed, i.e., the applied forces produce negligible permanent deformation (characterized by the residual deformation value ε_0) on the material. More quantitatively, the *elastic limit* is defined at the point of the curve $\sigma - \varepsilon$ where, by removing the applied stress, the remaining deformation is $\varepsilon_0 \approx 2.10^{-3}$. The zone from the origin to the elastic limit is called the *elastic zone*. For most metals, the limit of proportionality differs little from the elastic limit. If the object is subjected to a stress beyond the elastic limit, it enters the *plastic region* or *creep zone* [3] and will not return to its original length when the applied force is removed and will be permanently deformed. If the stress continues to increase beyond the elastic limit, rupture may be reached. Between the elastic limit and the breaking point, there is often a *yield zone*, where the material deforms easily, without the need to increase the stress (or with little effort, flat region of the curve).

Depending on the type of material, this creep region may or may not exist. If the creep region is small or non-existent, the material is *brittle*, and if this region is wide, the material is *ductile*. If we do not want a sample to break, it is important not to exceed the stress associated with the *yield strength* σ_Y (or *tensile strength*). The yield point or elastic limit is defined as the point of intersection of the curve $\sigma - \varepsilon$ and the curve parallel to the line of the elastic zone that passes through the permanent deformation point $\varepsilon_f = 0.002$ (0.2%).

When a cylindrical sample is subjected to traction, in addition to stretching in the direction of the applied force, its transversal dimensions become narrower. If we call the diameter of a cylindrical sample d, it will decrease by a magnitude Δd due to the applied stress. We define the transverse deformation $\varepsilon_T = \Delta d/d$. In general, ε_T is proportional to ε, that is,

$$\varepsilon_T = -\mu \cdot \varepsilon. \tag{20.3}$$

The constant μ is called the *coefficient of transverse deformation* or *Poisson coefficient*. For an incompressible sample, the value would be $\mu = 0.5$ and this is the case for rubber, for example [4–6]. Experimentally, the value of μ usually ranges from 0.2 to 0.5, typically being 0.3 for many materials (see Annex A). An interesting case is cork, for which $\mu \approx 0$, making it ideal for bottle stoppers.

Incompressible material: Imagine a cylindrical bar of length L and diameter d. If it is subjected to traction tension, the length L increases by ΔL and the diameter decreases by Δd. The variation of the volume of the bar will be

$$\frac{\Delta V}{V} \approx \frac{\Delta L}{L} + 2\frac{\Delta d}{d} \approx \varepsilon - 2\mu \cdot \varepsilon = \varepsilon(1 - 2\mu). \qquad (20.4)$$

If the volume does not vary, as in an incompressible solid, we have

$$\frac{\Delta V}{V} \approx \varepsilon(1 - 2\mu) = 0, \quad \text{that is,} \quad \mu = \frac{1}{2}. \qquad (20.5)$$

Therefore, for an incompressible solid, $\mu = 0.5$.

20.1.1 *Project 51. Measuring Young's modulus for copper or steel wires, etc. by the loading and unloading methods*

Recommended equipment: Enameled copper wire of diameter 0.20–1 mm. A goniometer and weights of 0.25–7 kg.

A possible experimental arrangement is suggested in Figure 20.3. Since commercial copper wires are very malleable and available in well-defined diameters and high purity, they are a good sample to test. Other possible samples are aluminum or iron wires.

In this version of the experiment, a goniometer is firmly attached to a horizontal axis of diameter d in the order of 5–10 mm. This

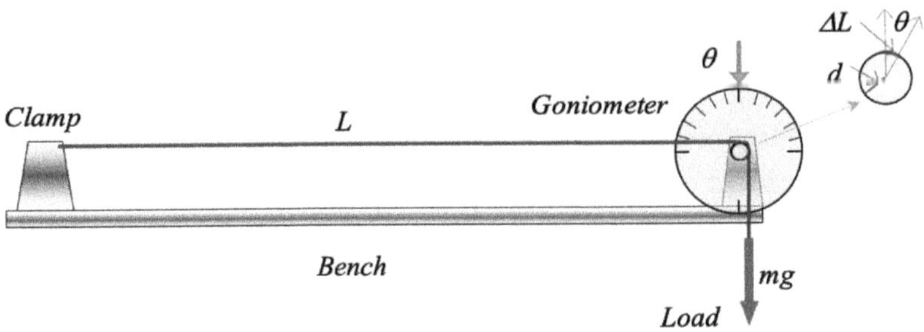

Figure 20.3 Experimental arrangement to study the dependence of the stretch on the applied force. A wire of length L is supported by a clamp (bracket) that is rigidly attached to the test bench. At the other end of the bench, the wire rests, without sliding, on a horizontal axis, of diameter d, attached to a goniometer. The turning angle θ of the goniometer enables the stretch ΔL of the sample to be measured. Force F can be applied by hanging weights from the end of the wire.

horizontal axis can rotate freely, supported by two bearings that keep it attached to one end of the bench, as shown in Figure 20.3. The sample wire of length L is attached at the other end of the bench by means of a clamp that holds it firmly, without pinching or damaging. The wire rests without sliding on the horizontal axis. The wire would be able to wrap completely around the axis before connecting to a container with weights placed at the end of the wire that hangs vertically, from which we apply force $F = mg$. The parameter that is measured is the angle of rotation θ of the goniometer as a function of the mass m that hangs from the wire. Thus, Δl can be measured as a function of the applied force F.

We assume that the diameter of the wire φ (which can be measured directly) and the length L are known. The relationship between the angle of rotation θ is given by

$$\theta = \frac{\Delta L}{d/2} = 2 \cdot \frac{L}{d} \cdot \frac{\Delta L}{L} = 2 \cdot \frac{L}{d} \cdot \varepsilon \to \theta$$
$$= 2 \cdot \frac{L}{d} \cdot \frac{\sigma}{E} = \frac{1}{E} \cdot \left(\frac{8}{\pi} \frac{L}{d} \frac{g}{\phi^2} \right) \cdot m \qquad (20.6)$$

Thus, by measuring θ as a function of the load m, we can determine the value of E and study the relationship between the strain of the wire $\varepsilon = \Delta L/L$ and the stress $\sigma = mg/(\pi\varphi^2/4)$. Note that the shaft diameter, d, acts as a magnification factor for the measured angle: the smaller the d, the greater the sensitivity of the measurements. However, a balance must be sought so that the horizontal axis is sufficiently rigid and does not bend and its value can be measured with reasonable precision. A 5–8 mm steel shaft, supported by two bearings approximately 1 cm apart, can be used for this experiment.

Suggested tasks:

✓ Using an experimental device similar to the one in Figure 20.3, study the relationship between ε as a function of stress σ for various diameters φ of wires of the same material. For each diameter, plot σ versus ε.

✓ For each sample, determine the maximum load for which, upon removal, the wire returns to its original length. In other words, try to avoid having the maximum weight exceed the elastic limit. It is advisable to take as original length the value which, when supporting a certain weight, makes the wire remain stretched, without deformation. Use this weight as the base value from which stretch measurements are made.

✓ If you have wires of different diameters φ of the same material, combine all the results of σ as a function of ε on the same graph. What can you conclude from this study? What is the connection between θ versus m? Is this relationship linear? Plot σ versus ε Is their dependance linear?

✓ Find the value of the slope of the graph of σ as a function of ε and its error (see Chapter 7). How does the modulus of elasticity E found in your study compare to the known values E for these materials?

✓ For one of your wires, study the relationship of σ as a function of ε when you make the wire exceed the yield point. Plot the relationship between σ and ε for the loading process and the unloading stage of the same sample. Are the loading and unloading paths identical in this case? Do you note permanent deformation after pulling the material with forces that exceed the yield point?

20.2 ♣ Beam deflection — Euler–Bernoulli theory

Consider a section of a bar undergoing a bending process, as illustrated schematically in Figure 20.4. While the upper part of the bar is subjected to compression stresses, its lower part is subjected to tensile stresses. There will also be a surface of the bar, close to its central zone, which will not be subjected to tension or compression. This surface is known as the neutral zone of the bar.

For a bar to flex, it is necessary to apply a bending moment or *torque* M. The bending of a bar is associated with the occurrence of a non-zero second derivative in the function that describes its shape (of the neutral line). If we apply Hooke's law to an infinitesimal

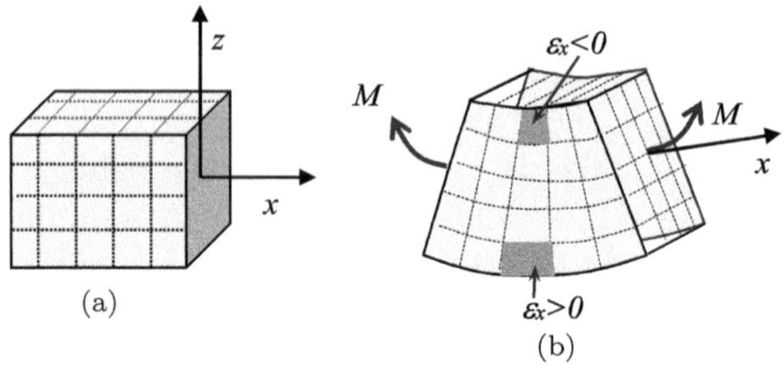

Figure 20.4 (a) Segment of a bar in equilibrium. (b) The same bar segment subjected to bending by a torque M. Note that the infinitesimal elements of the upper part of the bar are subjected to compression, while the elements of the lower part are subjected to traction.

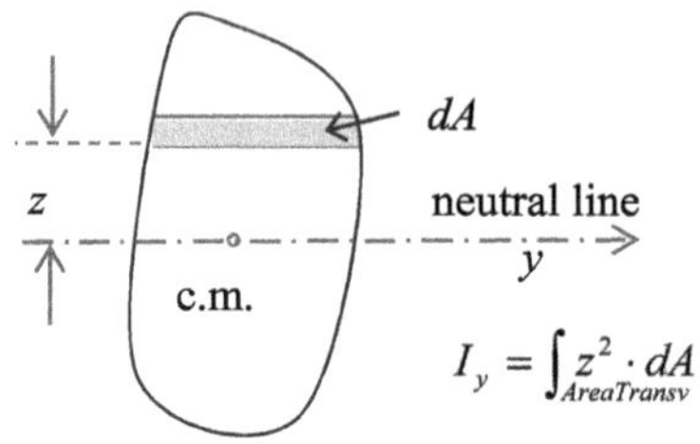

Figure 20.5 Definition of the areal moment of a bar of uniform section. The neutral line (y-axis) passes through the center of gravity of the figure (c.m.). The integral is applied to the entire cross-section.

element of the bar, it is not difficult to show that [4–7]

$$M = \frac{E \cdot I_y}{R}, \tag{20.7}$$

where we have defined the *areal moment* I_y or *moment of inertia* of the cross-section of the bar, see Figure 20.5, as

$$I_y = \int_{Area} z^2 \cdot dA. \tag{20.8}$$

In Eq. (20.7), E represents the modulus of rigidity or Young's modulus of the material. The product $I_{;y}E$ is known as *stiffness* or *flexural rigidity* to bending of the bar. In Eq. (20.7), R represents the radius of curvature of the bar, that is [7–10],

$$\frac{1}{R} = \frac{d^2z/dx^2}{[1 + (dz/dx)^2]^{3/2}} \approx \frac{d^2z}{dx^2}. \tag{20.9}$$

This approximation of the radius of curvature is valid for small deformations of the bar under study.

The relationship between the moment or bending couple and the deformation in the case of small deformations is given by the following relationship:

$$M \approx E \cdot I \frac{\partial^2 z}{\partial x^2}. \tag{20.10}$$

20.2.1 ♣ *Beam deflection*

A common difficulty in various problems involving bars or beams is to determine the value of the bending moment $M(x)$ at a point of coordinate x. To do this, imagine a bar fixed at one end, as illustrated in Figure 20.6(a). If we cut the bar at the point of coordinates x (line mn), to keep the section to the right of the line mn in equilibrium, a vertical shear force $V(x)$ and a bending moment $M(x)$ will be necessary, which act on the section of the beam to the left of the line mn. The equilibrium conditions ($\sum F_i = 0 y \sum M_i = 0$) can help provide the values of $M(x)$ and $V(x)$ [2, 7, 11].

20.2.2 *Light cantilever supporting a weight*

This situation is illustrated in Figure (20.6(a)). If we consider the section of the beam to the right of mn, from the equilibrium conditions, we obtain

$$V(x) = -P \tag{20.11}$$

and

$$M(x) = -P \cdot (L - x). \tag{20.12}$$

From expression (20.10), assuming small deformations of the bar, we have [7]

$$M(x) = -P \cdot (L - x) \approx E \cdot I_y \cdot \frac{\partial^2 z}{\partial x^2}. \tag{20.13}$$

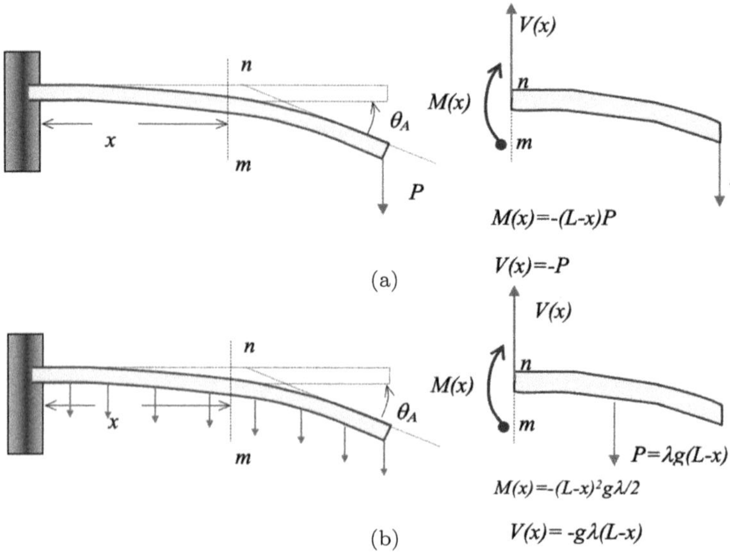

Figure 20.6 (a) Light cantilever beam of length L, subjected to bending by a force P applied at its free end. (b) Cantilever beam of length L, subjected to bending by its own weight. The distributed force along its length is given by the force per unit length: $g\lambda$. The value of the bending moment $M(x)$ and the force $V(x)$, at a point of coordinate x, is obtained from the equilibrium conditions ($\sum F_i = 0$ and $\sum M_i = 0$). The angle θ_A formed by the tangents to the bar at its two ends at a point of coordinate x is obtained from the equilibrium conditions and is known as the *angle of rotation.*

By integrating this expression, we obtain

$$z(x) \approx -\frac{P}{E \cdot I_y} \cdot \left(\frac{Lx^2}{2} - \frac{x^3}{6}\right) + C_1 x + C_2. \tag{20.14}$$

The constants of integration (C_1 and C_2) can be determined from the boundary conditions for the bar, that is, $z(0) = 0$ and $\partial z(0)/\partial x = 0$, therefore [7, 9],

$$z(x) \approx -\frac{1}{6} \cdot \frac{P \cdot L}{E \cdot I_y} \cdot x^2 \cdot \left(3 - \frac{x}{L}\right) \quad \text{and} \quad f = z(L) \approx \frac{1}{3} \cdot \frac{P \cdot L^3}{E \cdot I_y}. \tag{20.15}$$

These expressions can be compared to the experimental results to test the arguments supporting them.

20.2.3 *Cantilever subjected to bending by its own weight*

This situation is illustrated in Figure (20.6(b)). If we consider the section of the beam to the right of mn, from the equilibrium conditions, it is easy to obtain

$$V(x) = -g \cdot \lambda \cdot (L - x), \tag{20.16}$$

where $g\lambda$ is the load per unit length, $\lambda = \rho \mathrm{A}$ linear mass density, ρ density (mass per unit volume), A the cross-sectional area of the bar, and g the acceleration of gravity. We also have

$$M(x) = P \cdot \frac{(L - x)}{2} = -\frac{g \cdot \rho \cdot A}{2} \cdot (L - x)^2. \tag{20.17}$$

From expression (20.10), for small deformations of the bar and with the boundary conditions $z(0) = 0$ and $\partial z(0)/\partial x = 0$, we obtain

$$z(x) \approx -\frac{1}{24} \cdot \frac{g.m \cdot L}{E \cdot I_y} \cdot x^2 \cdot \left(6L - 4x + \frac{x^2}{L}\right) \quad \text{and}$$

$$f \approx \frac{1}{8} \cdot \frac{m.g \cdot L^3}{E \cdot I_y}. \tag{20.18}$$

These expressions can be tested experimentally.

20.3 ♣ Vibrations of a cantilever beam

The transverse vibrations of a cantilever beam are very important in many applications. To describe these vibrations, consider an infinitesimal element of the bar. Equations (20.12) and (20.17) indicate that

$$V(x) = \frac{\partial M(x)}{\partial x} = \text{transverse shear force} \tag{20.19}$$

and

$$w(x) = -\frac{dV(x)}{dx} = -\frac{\partial^2 M(x)}{\partial x^2} = \text{vertical load per unit length.} \tag{20.20}$$

It can be shown that these relationships between the moments $M(x)$ on a section of the x-coordinate bar and the vertical shear force $V(x)$, and the load per unit length $w(x)$ have general validity for a bar [7, 8].

If λ denotes the mass per unit length of the bar, from Eqs. (20.10) and (20.20) combined with Newton's second law, we have

$$EI \cdot \frac{\partial^4 z}{\partial x^4} \approx -\lambda \frac{\partial^2 z}{\partial t^2}. \tag{20.21}$$

This is the motion equation of a bar vibrating transversely with small amplitudes, assuming there are no frictional forces or other external forces. If the bar vibrates, as is usual in a viscous medium (air, water, etc.) or due to internal friction forces of the material, Eq. (20.21) is modified to

$$E \cdot I_y \cdot \frac{\partial^4 z}{\partial x^4} = -\rho \cdot A \cdot \frac{\partial^2 z}{\partial t^2} - b\frac{\partial z}{\partial t}, \tag{20.22}$$

where we have introduced *viscous friction* whose magnitude is described by the parameter b. Defining

$$c_0^2 = \frac{E \cdot I_y}{A \cdot \rho} \quad \text{and} \quad \gamma = \frac{b}{E \cdot I_y}, \tag{20.23}$$

Eq. (20.22) becomes

$$\frac{\partial^4 z}{\partial z^4} + \frac{1}{c_0^2}\frac{\partial^2 z}{\partial t^2} + \gamma\frac{\partial z}{\partial t} = 0. \tag{20.24}$$

If there were an external driving force described by the function $f(x, t)$, the equation of motion would become

$$\frac{\partial^4 z}{\partial x^4} + \frac{1}{c_0^2}\frac{\partial^2 z}{\partial t^2} + \gamma\frac{\partial z}{\partial t} = f(xt). \tag{20.25}$$

More details of these equations of motion can be found in Refs. [4] and [7].

20.3.1 ♣ *Vibrations of a cantilever beam*

If a cantilever beam (without overload) vibrates, as illustrated in Figure (20.7(b)), its motion can be described by Eq. (20.22).

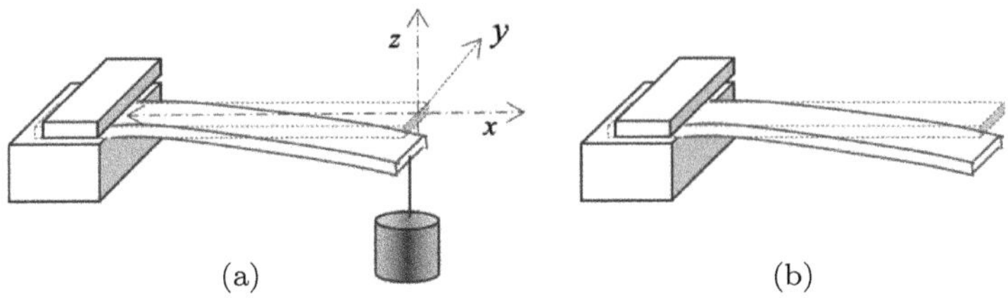

Figure 20.7 Bar of length L, with one end supported by a vice and the other end free. On the left, with an overload at its free end, and on the right, vibrating without overload.

In the first approximation, we will ignore friction ($\gamma = 0$). The vibration frequency is found by solving Eq. (20.22) with the boundary conditions $\{z(x = 0) = 0,\ \partial z/\partial x|_{x=0} = 0;\ \partial^2 z/\partial x^2|_{x=L} = 0$; and $\partial^3 z/\partial x^3|_{x=L} = 0\}$. The solution of this differential equation is easily obtained by separating variables [7–9]. The fundamental frequency is given by the following expression [10–12]:

$$f_1 = \frac{0.28}{L^2} \cdot \sqrt{\frac{E \cdot I_y}{A \cdot \rho}} = \frac{0.28}{L^2} \cdot c_0, \tag{20.26}$$

with c_0 given by Eq. (20.23). More details of the solution can be found in Refs. [7–9].

20.3.2 ♣ *Vibrations of a bar with both ends free*

For a bar with both ends free, the solution of Eq. (20.24) leads to the following natural frequencies [2, 7–9]:

$$f_1 = \frac{1.7908}{L^2} \cdot c_0 \quad \text{if } n = 1 \tag{20.27}$$

and

$$f_n = \left(n + \frac{1}{2}\right)^2 \cdot \frac{\pi}{2} \cdot \frac{1}{L^2} \cdot c_0 \quad \text{if } n > 1. \tag{20.28}$$

If there is friction, using Eq. (20.21), the modified natural frequencies are obtained. If $\omega_{0k} = 2\pi . f_{0k}$ are the natural frequencies without friction, Eqs. (20.27) and (20.28), the corresponding frequencies ω_k

including friction are

$$\omega_k^2 = \omega_{0k}^2 + \delta^2 \quad \text{with} \quad \delta = \frac{I_y \cdot E}{2 \cdot A \cdot \rho} \cdot b = \frac{c_0^2}{2} \cdot b. \tag{20.29}$$

In this equation, δ represents the attenuation factor that determines how the amplitude of oscillation decreases over time, i.e., for a given point of the bar, the oscillation over time can be written as follows:

$$y(x_0, t) = B(x_0) \cdot \exp(-\delta \cdot t) \cdot \sin(\omega_k t + \phi(x_0)), \tag{20.30}$$

where x_0 indicates the coordinate of the bar where the oscillation is observed.

20.3.3 *Project 52. Measuring Young's modulus of bars by the static method — Beam deflection*

Recommended equipment: Aluminum, bronze, iron, or steel bars about 1 m long with uniform cross-section. The dimensions and shape may vary, but bars of rectangular sections of approximately 12×2 mm may be suitable. Bars of circular sections with diameters of 4–12 mm can also be used. A millimeter ruler or, optionally, a micrometric comparator and weights between 50 g and 1 kg.

A possible arrangement for this experiment is shown in Figure 20.8(a)). The bar without load should not have significant deflection. The cross-section should be as uniform as possible, since the areal moment I_y (see Annex B) depends on high powers of the lateral dimensions (radius or thickness and width), therefore small uncertainties in these dimensions imply large errors in I_y and consequently in the value of Young's modulus, E.

Using two pieces of wood or a metal block and a clamp, fix the bar to a table horizontally and determine the free length L of the bar. Attach a wire to the free end — preferably light but stiff wire — so that the weights can be hung easily. Place a vertical ruler or comparator near the free end to measure the various arrows f, where

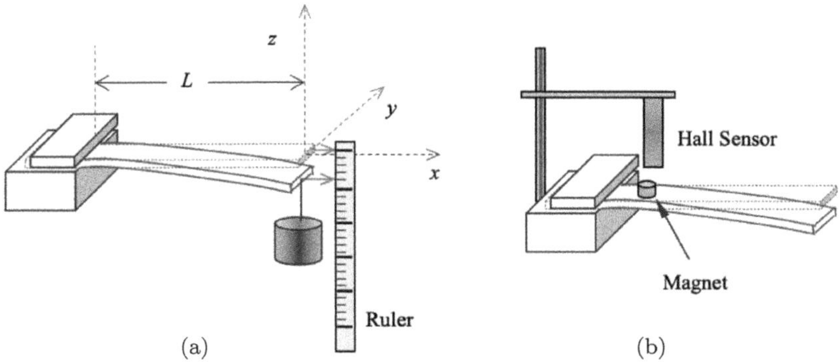

Figure 20.8 Bar of length L, with one end supported by a clamp and the other end free. On the left, with an overload at its end, and on the right, vibrating without overload.

f is the deflection of the free end of the bar generated by different weights $P = mg$. Arrange the setup so that you can measure the defection on a ruler as you load the bar. Make sure, however, that the maximum weight used does not create permanent deflection in the bar [13].

Suggested tasks:

✓ For each value of L used, measure the f arrows produced by at least 5 weights.

✓ Vary the value of L (do this for at least three values of L) and repeat the measurement of f as a function of m.

✓ Construct a graph of f as a function of the weight m for each value of L used. Is the relationship between these quantities linear?

✓ Using all your data, construct a graph of f as a function of PL^3. Is the dependency shown by these data linear? How do these results compare with those expected from the model developed above, Eq. (20.15)?

✓ If the model described by Eq. (20.15) describes your data well, find the best value of Young's modulus for your bar and its uncertainty. Compare the value of E found experimentally to the tabulated values. Discuss the agreement or discrepancy between these values.

20.3.4 *Project 53. Beam deflection — Determining its shape by means of digital photographs*

Recommended equipment: Aluminum, bronze, iron, or steel bars about 1 m long with uniform cross section, similar to those used in the previous project, and a digital camera or smartphone to obtain a digitized image of the bar.

Using the precautions outlined in the previous project, place the bar horizontally, with a background such that the shape of the bar is clearly visible and has good contrast for shooting. You can optimize this process by varying the lighting and doing some preliminary testing. Place a ruler with clearly visible divisions behind the bar so that it can be used as a scale for your photogram, as described in Chapter 10. Position the camera at a distance greater than three times the size of the ruler.

Suggested tasks:

- ✓ Using a load m that makes the deflection of the bar clearly visible, obtain a photogram for each value of L and m.
- ✓ Using some of the techniques discussed in Chapter 10, compare the shape observed in your frames to the shapes predicted by the theoretical model, Eq. (20.15).
- ✓ Discuss the degree of agreement or disagreement found. What can you conclude about the theoretical model discussed to describe your experimental data?
- ✓ If the theoretical model adequately describes your data, find the value of E and compare it to those expected from tables.

20.3.5 *Project 54. Cantilever deflection by its own weight — Using digital photographs*

Recommended equipment: Aluminum, bronze, iron or steel bars about 1 m long with uniform cross section, similar to those used in previous projects, but thinner. A digital camera or smartphone to obtain a digitized image of the bar.

This time we want to study the deflection of a cantilever by its own weight. To do so, it is convenient to use a bar such as a long metal ruler that often has a significant deflection due to its own weight. This means that its arrow or maximin deflection is greater than about 3 cm. Position the bar horizontally, with a background that makes the shape of the bar clearly visible and has good contrast for shooting. Place a ruler with a clearly visible scale behind the bar so that it serves as a scale for each photogram, as described in Chapter 10. Position the camera at a distance greater than three times the size of the ruler.

Suggested tasks:

✓ For different lengths of the same bar, obtain a photogram for each value of L.
✓ Using some of the techniques discussed in Chapter 10, compare the shape observed in the photograms to the shapes predicted by the theoretical model, Eq. (20.18).
✓ Discuss the degree of agreement or disagreement found. What can you conclude about the theoretical model that should describe your experimental data?
✓ If the theoretical model adequately describes your data, find the value of E and compare it to those expected from tables.

20.3.6 *Project 55. ♣ Measuring Young's modulus of bars by the dynamic method*

Recommended equipment: Aluminum, bronze, iron, or steel bars about 1 m long with uniform cross section, similar to those used in the previous project, and a photogate or a Hall sensor connected to a computer. Two small magnets.

We wish to study the vibration frequencies of a cantilever beam like the one in Figure (20.8(b)), which illustrates a possible arrangement for this experiment. It is desirable that the bar should not have significant deflection due to its weight. A steel ruler may be suitable. To measure the frequency of oscillation of the bar, you can use a photogate connected to a PC near the free end of the bar (when the bar vibrates, it interrupts the passage of light, see Chapter 9). Make the bar oscillate slowly by hand and make sure that the photogate triggers once or twice per each oscillation cycle. This information is crucial to determine the vibration frequency correctly. If the light blockage is not well defined, you can tape a piece of opaque paper to the end of the bar to improve the interruption of the light beam.

Another alternative for measuring the frequency of vibration is to place a small magnet near the fixed end of the bar. A Hall sensor (see Chapter 33) can be used to monitor the vibration frequency. If the bar is not ferromagnetic, you may place two small magnets, on opposite sides of the bar (or use glue). The idea is that the magnets with small mass should not affect the mass of the bar too much. Magnets like the ones found on earphones or audio headphones may be suitable.

Suggested tasks:

✓ For different lengths of the same bar, measure the vibration frequency as a function of L.

✓ Construct a graph of the frequency f as a function of L and another of f as a function of $1/L^2$. Do either of these graphs show a linear trend?

✓ Using all of your data, construct a graph of $L^2 f$ as a function of L. Is the dependence shown by these data linear? How do these results compare to those expected from the model developed above, Eq. (20.25)?

✓ Discuss the degree of agreement or disagreement found between your measurements and the theoretical predictions, Eq. (20.25). What can you conclude about the theoretical model discussed to describe your experimental data?

✓ If the theoretical model adequately describes your data, find the value of E and compare it to those expected from tables.

✓ For the bars used in this experiment, compare the results found for Young's modulus with the dynamic method to one of the static methods described above. Discuss the advantages and disadvantages of each technique.

20.3.7 *Project 56. ♣♣ Measurement of Young's modulus from the sound emitted when a sample is struck*

Recommended equipment: Hollow tubes made of bronze, steel, or aluminum, about 0.1–2 mm thick and 10–90 cm long. A microphone connected to a computer or a smartphone or microphone connected to a computer data acquisition system. A wooden rod to strike the pipes.

We want to study the oscillation frequencies of a hollow tube that can vibrate without restrictions (see Figure 20.9). For this experiment, it is advisable to choose a thin-walled tube that will emit a clearly audible sound when struck. This property of pipes and other objects is used in wind chimes and bells.

Thin tubes made of aluminum, bronze, steel, etc. may be suitable for this experiment. Cut several tubes, from the same sample, with different lengths L. For example, five tubes of length L between 10 and 50 cm may be suitable. Secure them with a thin thread, tied at approximately 22% of their length from one end. The fundamental mode of transverse vibration of a bar has a node near

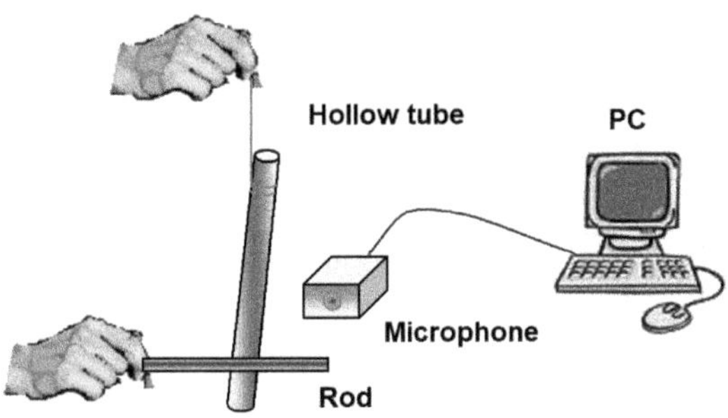

Figure 20.9 Tube of length L, supported by a thin wire. The sound produced by striking the tube with the rod is recorded by the microphone and stored in the PC.

this point [13, 14], therefore the thread does not interfere with the vibration of the tube.

The theoretical model that describes these transverse oscillations of a bar was discussed above and the natural frequencies predicted by this model are given by Eqs. (20.26) and (20.27). Our goal in this experiment is to experimentally test these predictions and use them to obtain Young's modulus of the tube materials.

When the tube is struck with the wooden rod, various oscillation modes are excited. To analyze the frequencies present in the sound emitted by the tube, it is necessary to perform a Fourier analysis of the recorded signal [8, 9, 13, 14]. The discrete frequencies present in the Fourier spectrum, i.e., the peaks of the spectrum, constitute the natural frequencies of the tube. Fortunately, there are many programs that can do this mathematical operation and return the Fourier spectrum of the recorded signal. For a smartphone, FFT plot [15–17] provides the fast Fourier spectra of the sound registered by the microphone. Also, by recording the sound, it is possible to obtain the Fourier spectrum using Microsoft Excel®, Math-Works Matlab®, Wolfram Research, Inc. Mathematica®, OriginLab Origin®, etc. To achieve signals with clear spectra, it is advisable to do some preliminary tests until you have some experience in striking the tube. Since the sound emitted only lasts a fraction of a second, just recording the sound for its duration immediately after striking it is enough to achieve a clear spectrum.

Suggested tasks:

- ✓ Choose a set of tubes of different lengths cut from the same sample. Record the sound emitted after striking each length.
- ✓ Find the natural frequencies present in each tube by conducting a Fourier analysis.
- ✓ In each case, identify the fundamental frequency and all the harmonics that you can identify. Build a table where for each length, L, the different natural frequencies are indicated, f_n.

✓ From the table, construct a graph of the different frequencies f_n, as a function of L, and another graph of the different f_n as a function of $1/L^2$. If the theoretical model discussed above adequately describes the data found, Eqs. (20.26) and (20.27), we would expect the second graph to show a linear trend. Do your experimental results show a linear trend?

✓ What can you conclude about the theoretical model discussed to describe your experimental data?

✓ If the theoretical model adequately describes your data, find the value of E and its uncertainty. Compare these results to table values for these materials.

✓ Repeat the analysis for all the materials you have. In each case, compare the results found for E to the tabulated values.

✓ Compare the results found for Young's modulus using the dynamic method suggested here to one of the static methods described above. Discuss the advantages and disadvantages of each technique.

20.4 Annex A: Young's modulus for some materials

Table 20.1 Modulus of elasticity (units 1 Kg/mm^2 = 9.8 MPa, 1 Kg/cm^2 = 0.098 Mpa, 100 PSI = 0.689 MPa, 10 dyn/cm^2 = 1 Pa) [6, 18].

Material	**Young's modulus** E **[GPa]**	**Yield stress** σ_F **[Mpa]**
Iron	150–170	180
Stainless steel	195	550
Steel	200	250
Aluminum	70–72	88–96
Copper 99.9%	70	220
Bronze	90.2	70–800
Glass	48–78	65
Nylon	45	75

20.5 Annex B: Areal moments for different cross-sections

Cross-section	**Moments**	**Distance to c.m.**
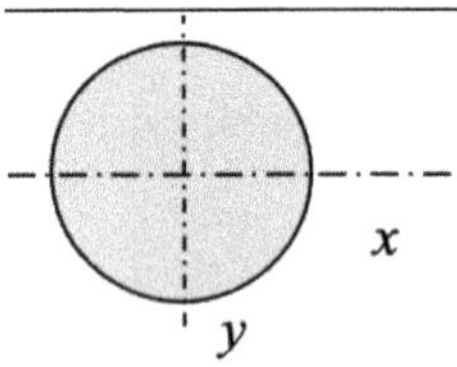	Radius circle R $I_x = I_y = \frac{\pi}{4} \cdot R^4$	
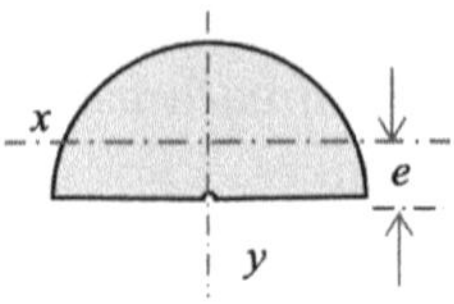	Radius semicircle R $I_x = \left(\frac{\pi}{8} - \frac{8}{9\pi}\right) \cdot R^4 I_y = \frac{\pi}{8} \cdot R^4$	$e = \frac{4}{3\pi} \cdot R$
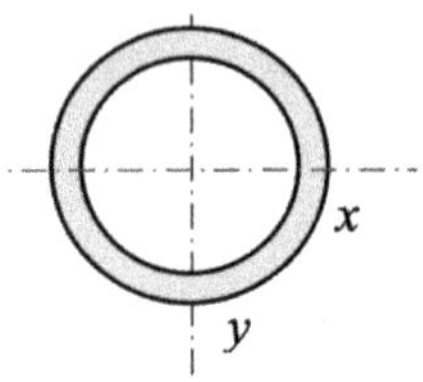	Tube radius R (outer) and r (inner) $I_x = I_y = \frac{\pi}{4} \cdot (R^4 - r^4)$	
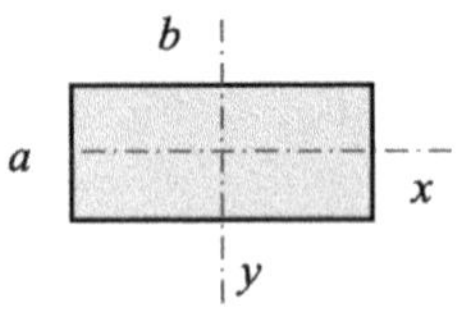	Rectangle with sides a and b $I_x = \frac{b \cdot a^3}{12} I_y = \frac{b^3 \cdot a}{12}$	

Here c.m. means center of mass. Refs: [7, 11]

References

[1] Wikipedia, "Yield (engineering)," 2021. Available: https://en.wikipedia.org/wiki/Yield_(engineering).

[2] R. Feynman, *et al.*, *The Feynman Lectures on Physics, Boxed Set: The New Millennium Edition*, Vol. 1, Basic Books, 2011.

[3] Wikipedia, "Dislocation and creep," 2021. Available: https://en.wikipedia.org/wiki/Dislocation_creep.

[4] S. P. Timoshenko, *History of Strength of Materials*, NY: Dover, 1983, pp. 28–40.

[5] P. A. Laura and M. J. Maurizi, *Introduction to the Mechanics of Solids*, Buenos Aires: Eudeba, 1979, pp. 249–256.
[6] Wikipedia, "Tensile strength and yield from Wikipedia," 2021. Available: http://en.wikipedia.org/wiki/Tensile_strength or https://en.wikipedia.org/wiki/Yield_(engineering).
[7] W. A. Nash and M. C. Potter, *Strength of Materials*, 2nd Ed., NY: Mc Graw-Hill, 2011.
[8] C. R. Wylie, *Higher Mathematics for Engineering*, 2nd Ed., NY: Mc Graw-Hill, 1996.
[9] L. A. Pipes, *Higher Mathematics for Engineers and Scientists*, NY: Mc Graw-Hill, 1970.
[10] W. W. Seto, *Mechanical Vibrations, Schaum's Outline Series*, NY: McGraw-Hill, INC, 1970.
[11] S. P. Timoshenko and J. M. Gere, *Theory of Elastic Stability*, NY: McGraw-Hill, 1985.
[12] M. Liebl, "Saw blades and resonance," *Phys. Teach.*, vol. 43, p. 282, 2005.
[13] G. W. Baxter and K. M. Hagenbuch, "A student project on wind chimes," *Phys. Teach.*, vol. 36, pp. 204, 209, 1998.
[14] M. A. González and M. A. González, "Smartphones as experimental tools to measure acoustical and mechanical properties of vibrating rods," *Eur. J. Phys.*, vol. 37, no. 4, p. 045701, July 2016.
[15] "FFT Plot, sound frequency analyzer RTA," Available: https://apps.apple.com/us/app/fft-plot/id569468015#:~:text=Seller-,ONYX%203,-Size, 2021; https://apps.apple.com/us/app/fft-plot/id569468015.
[16] Faber Acoustical, LLC, "SignalScope X," *Dynamic Signal Analysis Tools*, 2021. https://apps.apple.com/it/app/signalscope-x/id1304775893?l=en.
[17] ONYX Apps, "FFT Plot, Sound frequency analyzer RTA," Available: https://apps.apple.com/us/app/fft-plot/id569468015#:~:text=Seller.
[18] The Engineering ToolBox, "Engineering ToolBox, (2008). Engineering materials," 2022. Available: https://www.engineeringtoolbox.com/engineering-materials-properties-d_1225.html.

Chapter 21

DYNAMICS OF A MOVING CHAIN

This chapter studies the dynamics of systems of chains that fall in different ways, such as sliding along the edge of a table or attached to a pivot at one end with the other end free, simulating a bungee jumper. Because the driving forces are not constant in these systems, their movement is far from intuitive, and describing it poses interesting challenges. These experiments can be conducted with very low-cost equipment.

Goals

✓ Movement of hanging chains
✓ Simulation of a bungee jumper

21.1 Chain hanging from the edge of a table

The purpose of this section is to describe the equation of motion of a chain of length L and mass M, which initially lies on a table and has a section of length y_0 that hangs vertically over the edge of the table, as illustrated in Figure 21.1. We assume that the horizontal end of the chain is attached to a thread of negligible mass, which passes through a pulley at the other edge of the table and has a small mass m_a hanging from the other end. The proposed experimental arrangement is illustrated in Figure 21.1. This problem has attracted the attention of many researchers, and there is rich and varied bibliography on the subject [1–6].

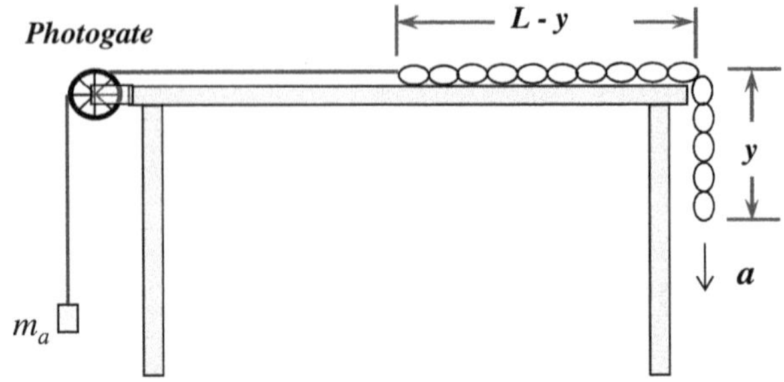

Figure 21.1 Experimental arrangement of a chain falling over the edge of a table or through an L-shaped tube.

The pulley is associated with a photogate (smart pulley) [7] that can be used to measure the y coordinate and its speed dy/dt. To conduct this experiment under controlled conditions, a plastic tube with a 90° elbow (e.g., an inexpensive L-shaped PVC drainpipe) can be used as a guide for the chain. This guide restricts the problem to a system with a single degree of freedom, y. The pulley's mass, radius, and moment of inertia are designated by the notations m_p, R_p, and I_p, respectively. The tensions in the rope are designated by T_1 (horizontal) and T_2 (vertical). Using free body diagrams for chain, pulley, and mass m_a, we have

$$Ma = \frac{y}{L}Mg - T_1 - \mu_d\frac{(L-y)}{L}Mg - f_r, \tag{21.1}$$

$$I_p\frac{a}{R_p} = T_1R_p - T_2R_p, \tag{21.2}$$

$$m_a a = T_1 - m_a g. \tag{21.3}$$

Here, μ_d represents the coefficient of dynamic friction between the chain and the tube (or the table), $M(L–y)/L$ represents the mass of the part of the chain that is horizontally supported, and f_r is the value of the friction force exerted by the 90° elbow on the chain. Each link in the chain collides with that elbow, generally inelastically, losing some of its kinetic energy. For the moment, we will assume that the value f_r is constant. If we divide Eq. (21.2) by R_p and add it to Eqs. (21.1) and (21.3), we obtain

$$M_Ta = \frac{y}{L}Mg - \mu_d\frac{(L-y)}{L}Mg - f_r - m_a g, \tag{21.4}$$

where we have used $M_T = M + m_a + I_p/R_p^2$. If the pulley is shaped like a ring with spokes, its moment of inertia is $I_p = \alpha M_p R_p^2$, where α is a geometry-dependent constant [7].

Thus, $M_T = M + m_a + \alpha M_p$. The equation of motion (21.4) can be rewritten as follows:

$$a = \frac{d^2y}{dt^2} = \frac{y}{L}\frac{M}{M_T}g - \mu_d\frac{(L-y)}{L}\frac{M}{M_T}g - \frac{f_r}{M_T} - \frac{m_a}{M_T}g \tag{21.5}$$

or

$$\frac{d^2y}{dt^2} - \left(\frac{M}{M_T}\right)\frac{g}{L}(\mu_d+1)y = -g\left(\frac{\mu_d + m_a/M_T}{\mu_d+1}\right) - \frac{f_r}{M_T}. \tag{21.6}$$

As we have assumed that f_r is constant, for simplicity, we define the constant β as $f_r = \beta M_T g$. It is convenient to define the following parameters:

$$\lambda^2 = \left(\frac{M}{M_T}\right)\frac{g}{L}(\mu_d+1) \tag{21.7}$$

and

$$y_d \cdot \lambda^2 \equiv g\left(\frac{\mu_d + m_a/M_T}{\mu_d+1}\right) + \frac{f_r}{M_T} = g\left(\frac{\mu_d + m_a/M_T}{\mu_d+1}\right) + \beta g \tag{21.8}$$

or

$$y_d = L\left(\frac{M_T}{M}\right)\left(\frac{\mu_d + m_a/M_T}{(\mu_d+1)^2}\right)\left[1 + \beta\left(\frac{\mu_d+1}{\mu_d + m_a/M_T}\right)\right]. \tag{21.9}$$

With these definitions, the differential Eq. (21.6) can be written as follows:

$$\frac{d^2y}{dt^2} - \lambda^2\, y = -\lambda^2\, y_d, \tag{21.10}$$

whose solution is immediate:

$$y(t) = A\,\cosh(\lambda\, t + \varphi) + y_d, \tag{21.11}$$

where A and ϕ are two constants that depend on the initial values of $y(t)$ and dy/dt. In particular, if for $t = 0, dy/dt = 0$ and $y(0) = y_0$, then (21.11) becomes

$$y(t) = (y_0 - y_d)\,\cosh(\lambda t) + y_d, \tag{21.12}$$

and the velocity is

$$v(t) = \frac{dy(t)}{dt} = (y_0 - y_d)\,\lambda\sin h(\lambda \mathrm{t}). \tag{21.13}$$

The form of these functions is shown in Figure 21.2.

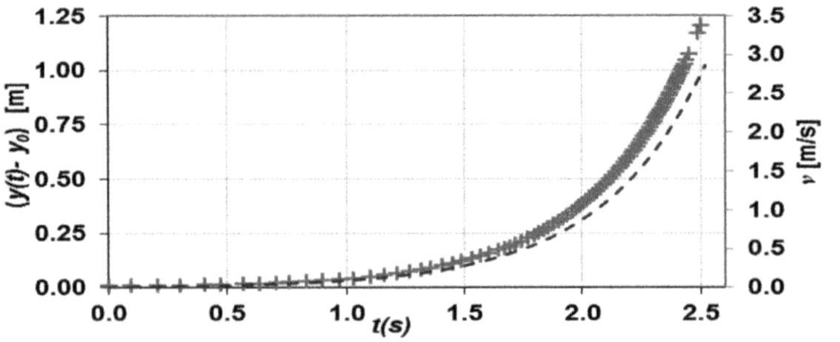

Figure 21.2 Prediction of the behavior of position (red crosses, referred to the left vertical axis) and velocity (dotted blue line, referred to the right vertical axis) as a function of time for a chain moving as shown in Figure 21.1, using the expressions (21.12) and (21.13).

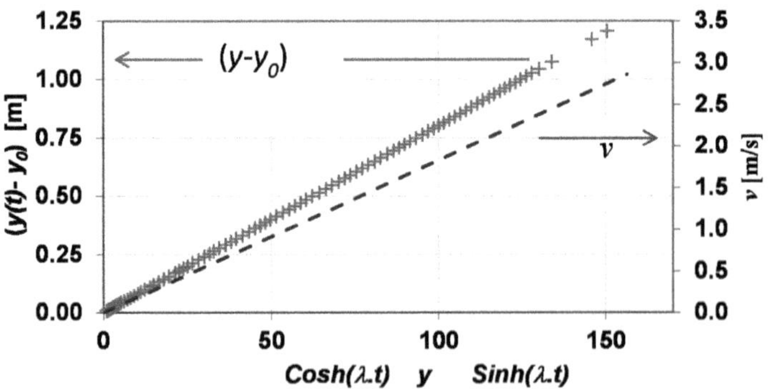

Figure 21.3 Prediction of the behavior of position (crosses, referred to the left vertical axis) and speed (dotted line, referred to the right vertical axis) as a function of time, using expressions (21.12) and (21.13). The horizontal axes were modified to approximately linearize the expressions (21.12) and (21.13), as described in the text.

Figure 21.3 presents the same data, using as vertical axis $(y–y_0)$ and taking as horizontal axis $[\cos(\lambda t)]$. The speed $v(t) = dy/dt$ is also represented on the same graph as a function of $\sinh(\lambda t)$. With this parametrization, the solutions (21.12) and (21.13) are linearized. The values of the constants β and μ_d are found from the fits of the experimental data. It is generally expected that the order of magnitude of $\mu_d \approx 0.3$ and $\beta \approx 0.02$.

21.1.1 *Project 57. Chain hanging from the edge of a table or through a tube*

Recommended equipment: A metal or plastic chain 1.5–2 m long. A PVC pipe with a right-angle elbow, through which the chain can slide smoothly. A "smart pulley" connected to a photogate or photoswitch. These measurements can also be made using a digital camera or smartphone in video mode to study the movement of the lower end of the chain.

Suggested tasks:

- ✓ Using an experimental device similar to the one illustrated in Figure 21.1, study the falling motion of a chain for different initial lengths y_0.
- ✓ For each case, determine the variation of $y(t)$ and $v(t)$ as a function of time.
- ✓ Represent the experimental results and theoretical expression of $y(t)$, Eq. (21.11), in the same figure.
- ✓ Represent the experimental and theoretical values, expression (21.13), of $v(t)$ in the same figure. Adjust the values of the parameters μ_d and β to obtain the best possible agreement between the data and the model.
- ✓ Discuss the degree of agreement between your measurements and the proposed theoretical model.

Note: Since the initial movement of the chain is very slow at the beginning and very fast at the end, small inaccuracies in defining the initial moment of the fall can have appreciable effects at the end. Therefore, consider defining the initial instant as one where the chain already has a well-defined motion. In this case, extract the initial values from your data y_0 and v_0 and modify Eqs. (21.11) and (21.13) to take into account these specific initial conditions.

21.2 Vertical drop chain — Bungee jumper

This section intends to study the motion of a falling bungee jumper, namely someone who engages in the risky sport of jumping off bridges

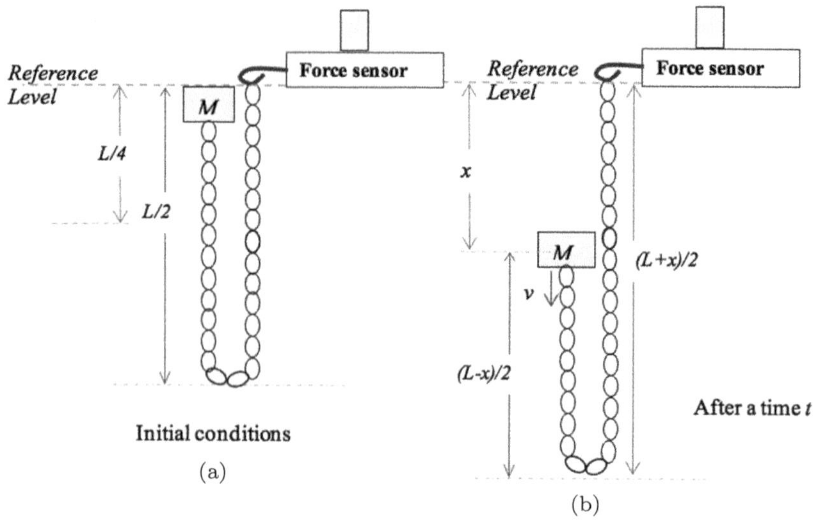

Figure 21.4 Experimental arrangement. A mass M attached to a chain of mass m moves in free fall, simulating a bungee jumper. The force sensor measures the vertical force exerted by the chain in its fall. A digital camera can record the falling motion of mass M.

while tied to an elastic rope. One of the remarkable characteristics of bungee jumper movement is that its acceleration exceeds that of free fall (if the rope mass is not negligible). This can be easily verified, without needing to take the risk of bungee jumpers, by using an arrangement like the one shown in Figure 21.4. If two falling masses M — one attached to a chain and another not — are released simultaneously and filmed, it can be seen that the one attached to the chain falls faster than the one in free fall [8–10].

To analyze this problem, we will assume that the energy of the system is conserved at least until the chain is fully extended [3, 11–15]. We will assume that the chain has a length L and its mass m is of the order of mass M and one of its ends is tied to a fixed point (in the experiment, a force sensor, and in the case of the bungee jumper, the bridge). We take the reference level at the chain attachment point, Fig. 21.4. If x is the distance traveled by the mass M in its fall, its total mechanical energy, E_T, equals the initial potential energy:

$$E_T = E_P = -\frac{1}{4}mgL. \tag{21.14}$$

In this expression, we have taken as $E_p = 0$ the chain attachment point, that is, when $x = 0$. See Figure 21.4.

Ignoring the small energy losses in the movement of the chain, from the conservation of mechanical energy, E_T, we have

$$E_T = \frac{1}{2}\left[\frac{m}{L}(L-x)+M\right]\dot{x}^2 - \frac{m}{2L}(L-x)\mathrm{g}\left(x+\frac{L-x}{4}\right)$$
$$-m\left(\frac{L+x}{2L}\right)\mathrm{g}\left(\frac{L+x}{4}\right) - Mgx. \qquad (21.15)$$

By clearing $v = dx/dt$, we get [13]

$$v^2 \equiv \left(\frac{dx}{dt}\right)^2 = g\,x\,\frac{(4ML+2mL-mx)}{(mL+2ML-mx)}$$
$$= g\,L\,y\,\frac{(4+2\alpha-\alpha\,y)}{(2+\alpha-\alpha\,y)}, \qquad (21.16)$$

where $\alpha = m/M$ and $y = x/L$. Differentiating Eq. (21.16),

$$a = g\left[1+\frac{mx(4ML+2mL-mx)}{2\,(mL+2ML-mx)^2}\right]$$
$$= g\left[1+\frac{\alpha\,y\,(4+\alpha/2-\alpha\,y)}{2\,(\alpha+2-\alpha\,y)}\right]. \qquad (21.17)$$

The expression (21.16) can be integrated analytically to find $x(t)$ or alternatively $t(x)$ as follows:

$$dt = \frac{dx}{v(x)} \Rightarrow t(x) = \int_0^x \frac{dx}{v(x)}, \qquad (21.18)$$

which provides an implicit relationship of $x(t)$.

However, to find a theoretical expression of $x(t)$[a] that can be compared to the experimental results, it is more convenient to integrate expression (21.18) numerically. To do so, a spreadsheet can be organized with the variable x, varying it from 0 to L in about 100 or more steps, as illustrated schematically in Table 21.1.

[a]The solution of Eq. (21.18) can be written as $\sqrt{g/[2L(1+2M/m)]}t(x) = \sqrt{2}\int_0^{\varphi}[\cos^2\varphi/\sqrt{1+\cos^2\varphi}]d\varphi$, where $\sin\varphi = \sqrt{x/L}$; this integral can be expressed in terms of elliptic integrals [3, 12].

Table 21.1 Example of numerical integration of Eqs. (21.17) and (21.18) to find the theoretical expression of $x(t)$.

$x[m]$	$y = x/L$	$a(x)$	$v(x)$	$t(x)$
0	0.00	10		0
0.01	0.01	10		0.05
0.02	0.02	10		0.06
x_i	x_i/L	$a(x_i) =$ Ec(21.17)	$v(x_i) =$ Ec(21.18)	$t_i = t_{i-1} + dx/v$
$x_{i+1} = x_i + dx$	x_{i+1}/L	$a(x_{i+1})$	$v(x_{i+1})$	$t_{i+1} = t_i + dx/v$
0.05	0.05	10		0.10
$x_{max} = L$	$y_{max} = 1$	a_{max}	v_{max}	t_{max}

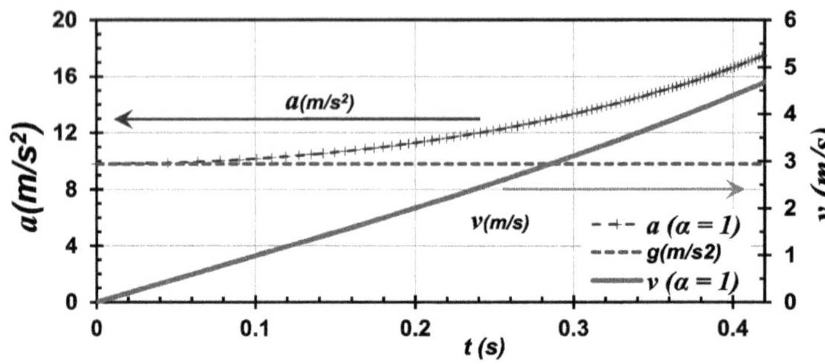

Figure 21.5 Variation of the acceleration (crosses, referred to the left vertical axis) and the velocity (solid line, referred to the right vertical axis) of a mass M attached to a chain of mass m moving in free fall, simulating a bungee jumper. The horizontal, green, dotted line is the value of the acceleration of gravity $g = 9.8\,\text{m/s}^2$, $\alpha = 1$, and $L = 1$ m.

One way to efficiently integrate Eq. (21.18) is to use the following trapezoidal expression [12, 16]:

$$t_{i+1} = t(x_{i+1}) = t_i + \frac{2\,dx}{(v_i + v_{i+1})} = t_i + \frac{2\,(x_{i+1} - x_i)}{(v_i + v_{i+1})}. \qquad (21.19)$$

Using this technique, we can obtain a graph of $a(t)$ and $v(t)$ like the one shown in Figure 21.5. We see that in all cases the acceleration of the mass M (equivalent to the bungee jumper) is greater than g [8, 9, 13].

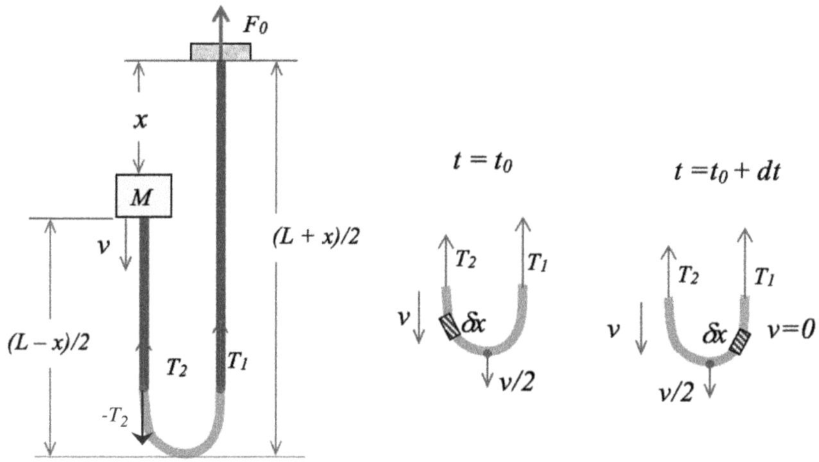

Figure 21.6 Dynamics of the lower curve of the chain. As the time passes from t_0 to t_0+dt, a fraction of the chain goes from having a velocity v downwards to being stationary. F_0 is the force that supports the chain from the pivot (force sensor) and T_1 and T_2 are the tensions at the lower end of the chain on each side. The reaction to the force T_2 acting on the vertical branch is $-T_2$, which is an extra force acting on the left side of the chain.

The reason for this surprising result, acceleration greater than g, is related to what happens in the lower curve of the chain, see Figure 21.6.

When the mass M falls a distance dx, at the lower end, the portion $\delta x \approx dx/2 = v \cdot dt/2$ of the chain passes from the left side (with speed v) to the right side (with zero speed). Thus, in a time interval dt, the portion δx of the chain goes from having a velocity v downwards to being static at the right branch, Figure 21.6. This implies that the tensions of the two branches (T_1 and T_2) must provide this force in order to stop this section of the chain. The momentum change, dp, of this mass fraction is

$$dp = \delta x(m/L)v = \frac{1}{2}vdt(m/L)v, \qquad (21.20)$$

thus,

$$F = dp/dt = \frac{1}{2}v^2(m/L) \approx T_1 + T_2. \qquad (21.21)$$

In this expression, we have assumed that the weight of the horizontal portion of the chain is negligible. If we assume that the tension in the rope or chain varies continuously along the lower span, we have $T_1 \approx T_2$. Thus, the mass M is subject to a force that is the sum of its weight plus that of the section of chain hanging from it, $-T_2$. Therefore, the force acting on the mass M and the left section of the chain, $(L-x)/2$, is its weight plus T_2. Consequently, its acceleration will be greater than g and will increase as v increases, since

$$T_1 \approx T_2 \approx \frac{1}{4} v^2 (m/L). \tag{21.22}$$

As can be seen in Figure 21.5, the force F_0 on the pivot that supports the chain — or the bridge in the case of a bungee jumper — will be the sum of T_1 plus the weight of the part of the chain that is not moving (in Figure 21.6, the one on the right), that is,

$$F_0 = T_1 + \frac{1}{2}\left(\frac{L+x}{2L}\right) mg = \frac{1}{4} v^2 (m/L) + \frac{1}{2}\left(\frac{L+x}{L}\right) mg. \tag{21.23}$$

The objective of this experiment is to falsify the proposed model. To do so, we propose to build an experimental system of the type illustrated in Figure 21.4 and measure, if possible, simultaneously, the falling movement of the weight M (bungee jumper) and the force exerted on the force sensor as a function of time.

21.2.1 *Project 58. Hanging chain in vertical fall — Bungee jumper*

> **Recommended equipment**: A metal or plastic chain 1–1.5 m long. A weight of mass comparable to that of the chain. A force sensor connected to a computer. These measurements can be complemented by using a digital camera or smartphone in video mode to study the movement of the end of the chain attached to the weight, M.

Using a chain of length L of 1–1.5 m and a force sensor connected to a computer as schematically illustrated in Figure 21.4, determine the experimental value of $F_0(t)$ with the force sensor. With a digital camera in video mode, record the falling motion of the mass M.

To synchronize the two measurements (force and position), place the PC screen that records the force on one side of the system to record the time readings on the screen and the mass M in the same video frame. These data enable you to synchronize the two measurements, although it is not essential to do so because you can also measure force and position sequentially or separately. If you only measure the force with a sensor connected to a PC, you can compare these results to Eq. (21.23), and if you only measure the drop in mass M, that is, $x(t)$ with a digital camera, you can compare these results to Eqs. (21.18) and (21.19).

Suggested Activities

✓ A *preliminary experimental test* that can provide a clue of the motion of the mass M hanging from a chain, and that of a similar mass, released simultaneously, but in free fall, as illustrated in Refs. [8, 9]. Differences should be observable at first glance. However, if these falls are filmed with a digital camera, the differences can be seen more clearly. Further details of this technique are described in Ref. [10].

✓ On the same graph, compare the measured results of x as a function of time to the values predicted by the model, Eqs. (21.18) and (21.19).

✓ If possible, use the data recorded with the camera to determine the speed of the lower end of the chain. On the same graph, draw the data for the velocity of the mass M and the lower end of the chain. How do these two speeds compare?

✓ Do any of the models proposed here adequately explain your experimental results?

✓ Determine from your data the time the chain takes to fall $t(x = L)$ and estimate the uncertainties of this determination.

✓ On the same graph, compare the measured results of F_0 as a function of time to the values predicted by the proposed models.

✓ Does the proposed model adequately explain your experimental results?

References

[1] A. Sommerfeld, *Lectures on Theoretical Physics, Mechanics*, Vol. I, New York: Academic Press, 1952.

[2] M. Spiegel, *Theoretical Mechanics*, Shaum's Outline Series, NY: McGraw-Hill, 1967.

[3] M. G. Calkin and R. H. March, "The dynamics of falling chain: I," *Am. J. Phys.*, vol. 57, p. 154, 1989.

[4] D. Prato and R. J. Gleiser, "Another look at the uniform rope sliding over the edge of a smooth table," *Am. J. Phys.*, vol. 50, p. 532, 1982.

[5] J. R. Sanmartín and M. A. Vallejo, "Comment on" Another look at the uniform rope sliding over the edge of a smooth table," *Am. J. Phys.*, vol. 52, p. 585, 1983.

[6] M. Denny, "A uniform explanation of all falling chain phenomena," *Am. J. Phys.*, vol. 88, pp. 94–101, 2020.

[7] Vernier Software, "There are several suppliers of commercial equipment that manufacture this type of device, 'Smart Pulley' among them: Pasco and Vernier Software," 2021. Available: www.vernier.com.

[8] Veritasium an Element of Truth, "Chain drop," 2021. Available: https://www.youtube.com/watch?v=X-QFAB0gEtE.

[9] P. Hewitt, "Chain Drop Re-visited," *Phys. Teach.*, vol. 54, p. 439, 2016.

[10] S. Calderón and S. Gil, "Experiments with falling objects with acceleration greater than g," *Lat. Am. J. Phys. Educ.*, vol. 5, no. 2, pp. 501–507, 2011.

[11] D. Kagan and A. Kott, "The greater-than-g acceleration of a bungee jumper," *Phys. Teach.*, vol. 34, p. 368, 1996.

[12] G. Arfken, *Mathematical Methods for Physicist*, NY: Academic Press, 1995.

[13] H. Biezeveld, "The bungee jumper: A comparison of predicted and measured values," *Phys. Teach.*, vol. 41, p. 238, 2003.

[14] W. Tomaszewskia, P. Pieranski and J. C. Geminard, "The motion of a freely falling chain tip," *Am. J. Phys.*, vol. 74, no. 6, pp. 776–783, 2006.

[15] C. E. Mungan, "Newtonian analysis of a folded chain," *Phys. Teach.*, vol. 56, p. 278, 2018.

[16] M. Spiegel, *Theory and Problems of Advanced Mathematics for Scientists and Engineers*, Schaum's Outline, NY: McGraw-Hill Book Co., 1971.

Chapter 22

VARIABLE MASS SYSTEMS — GRANULAR MATERIALS

This chapter explores the flow of granular materials through an orifice. To do so, it looks at the general characteristics of the flow of sand through a hole and compares it to the flow of liquids. After discussing the properties of granular flow, it continues with the design and study of physical systems of variable mass. In particular, it analyzes the behavior of a variable mass Atwood machine and the characteristics of a simple variable mass oscillator.

Goals

- ✓ Sand flow through a hole
- ✓ Atwood — fixed mass machine
- ✓ Atwood — variable mass machine
- ✓ Variable mass oscillator

22.1 Granular materials

Granular materials[a] are conglomerates of solid particles. The particles are large enough (diameter greater than 5 μm) so that their movement due to thermal fluctuations is negligible. Examples of granular materials are sand, salt, a conglomerate of seeds such as wheat, corn, rice, and nuts, a large set of balls, etc. Depending on the average kinetic energy of the individual grains, granular materials can behave as solids, liquids, or gases. If the average kinetic energy is small, the particles remain at rest. This is why we can build

[a]https://en.wikipedia.org/wiki/Granular_material

sandcastles or sand piles. Under these conditions, the system can withstand shear stresses, that is, the conglomerate behaves similarly to a solid. If the energy of the individual grains is increased, for example, by shaking the sample, these materials begin to flow, as in a liquid. By increasing the energy further, the particles behave like a dust with properties similar to those of a gas, showing turbulent behavior. Granular materials can be homogeneous or heterogeneous, isotropic or anisotropic. In addition, they present the characteristic phenomenon of *avalanches*. These materials tend to dissipate energy quickly [1, 2].

The importance of granular materials is that they are very prevalent in nature and in industry. The phenomena of avalanche and fluidification of soil are of great practical relevance. To understand these phenomena, it is necessary to develop models of the general behavior of granular materials. The behavior of these materials is also important to understand the dynamics of systems, such as icebergs, the asteroid belts of the Solar System, and the rings of Saturn. In industry, after water, the presence of granular materials is very preponderant.

Granular materials can flow through holes. Unlike what happens with a liquid, where the flow of a tank depends on the height of the liquid that fills it, in the case of granular materials, the flow is constant and independent of how full the tank is. This property of constant flow is used in hourglasses to measure time. It is precisely this peculiar property of granular media that we will study in this chapter, taking advantage of it to build different dynamic systems of variable mass.

22.2 Granular material flow

To generate a model of how sand (or any other granular medium) flows through a hole, we try to identify the relevant parameters that may determine the magnitude of the flow. We will conjecture that the flow depends on each of these parameters raised to a certain power. Using a dimensional analysis [3, 4], we determine these exponents. In all that follows, we will assume that the diameter d of the hole is much larger than the characteristic dimensions δ of the particles of the granular medium. We will also provisionally assume that the flow does not depend on the filling height of the container.

We propose that the mass flow depends on the area A of the hole, the mean density ρ of the granular medium, and the value of the acceleration of gravity, g, since if g were zero, there would be no flow. Thus, we conjecture that

$$\frac{dm}{dt} = C\rho^k g^n A^p, \tag{22.1}$$

where C is a dimensionless proportionality constant and k, n, and p are exponents to be determined through dimensional analysis. This analysis consists of replacing each variable in Eq. (22.1) with the units in which each of them is expressed, looking for the conditions to obtain "dimensional consistency" in both members of the equation, that is, for what values of the exponents both sides of the equation are equal. Using MKS units, we have

$$[kg/s] = [C][kg/m^3]^k[m/s^2]^n[m^2]^p = \frac{kg^k}{m^{3k-n-2p}s^{2n}}. \tag{22.2}$$

Therefore, assuming that C is a dimensionless constant, $k = 1$, $n = 1/2$, and $p = 5/4$. In other words, our working hypothesis is that the flow of matter is independent of the mass or volume of the filling or its height in the container, i.e. [3],

$$\frac{dm}{dt} = C\rho\, g^{1/2} A^{5/4}. \tag{22.3}$$

To test this hypothesis, we propose an experiment that enables us to measure the flow of matter (dm/dt) and study its dependence on the variables: filling height h, A, ρ, and g. We will first attempt to test the hypothesis that the sand flow does not depend on the filling height of the container and compare it to the case where the container is filled with a liquid.

Note on sand: It is important to conduct these studies on a granular medium that is *homogeneous* and *dry*. We know by simple observation that wet or damp sand does not flow, so it is advisable to use dry sand. Just leave the sand in a tray in the sun for a few hours to eliminate moisture. Then, use a sieve to remove grains that exceed a certain size.

22.2.1 *Project 59. Experimental study of water and sand flows*

Recommended basic equipment: A force sensor connected to a computer. A plastic bottle with a screw cap. Several caps, with holes of different sizes. 2–3 kg of dry, sifted sand.

For this experiment, use a transparent plastic beverage bottle (such as a mineral water, soda, or soft drink bottle), with a uniform cross-section of at least 10 cm in diameter, and several screw caps. Drill a hole of known size in each cap (size determined by the bit used). It is advisable to have holes ranging from a couple of millimeters to about 8 or 10 millimeters in diameter. Cut off the base of the bottle. Use strings to hang the bottle upside down from the force sensor, as shown in Figure 22.1. Choose one cap with a perforation of known diameter at a time, and fill the bottle with water or sand, covering the hole with your finger or adhesive tape until the start of the measurement. It is advisable to use two different bottles

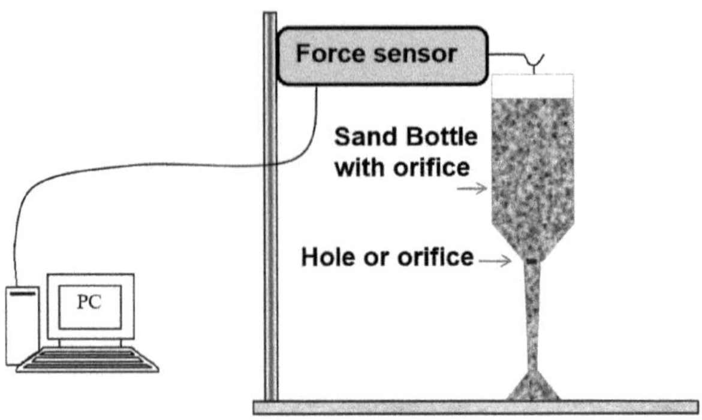

Figure 22.1 Experimental setup of a possible system to measure the variation of the mass of a bottle with an orifice. The force sensor records mass as a function of time [5].

(of the same type) for the water and the sand, so as not to wet the sand with the water. The computer records the variation of the mass or weight of the bottle as a function of time while it empties. In order to prevent changes in the section of the bottle from affecting the measurement, in this first stage, it is recommended to use bottles with constant cross-section over most of their main body.

Suggested tasks:

✓ Take as references the weight of the bottle when filled up to the maximum height, $m_{\max}$, and its weight when filled up to the minimum height, $m_{\min}$; between these masses, it is assumed that the cross-section of the bottle is uniform. Measure the distance or height h_0 between the maximum and minimum heights. With these data, $(m_{\min}, 0)$ and $(m_{\max}, h_0)$, draw a calibration line that gives the filling height h as a function of the mass measured with the force sensor. Do this manually, i.e., unload the bottle gradually and measure the height h and its mass m. Use these data for the "calibration" to find the dependence of m on h. Of course, this calibration will be different for sand and water.

✓ Once the system has been calibrated, hang the bottle of water or sand from the force sensor and start the measurement, using the data acquisition system, of m as a function of t. Determine $m(t)$ for at least two different hole diameters each for water and sand. Plot your results.

✓ The slope of the graph of m vs. t is the flow dm/dt. Construct a plot of the flow as a function of time or alternatively as a function of the height h. If these graphs are consistent with a horizontal line, it is indicative that dm/dt is independent of time and filling height.

✓ Discuss the differences and similarities you find between the emptying behavior of sand and water. A linear dependence between mass as a function of time is indicative that the flow remains constant. Do you find evidence that the flow is constant for either of the cases studied? Do you find a difference between the behavior of water and sand?

22.2.2 *Project 60. Influence of container shape on the flow of sand*

Recommended basic equipment: A force sensor connected to a computer. A plastic bottle with a screw cap. Several caps, with holes of different sizes. 2–3 kg of dry, sifted sand.

The aim of this experiment is to study how the flow of sand varies with the shape of the bottle that contains it. The previous activity studied the flow of sand from a bottle with a constant cross-sectional area. Now, we want to find out what happens if the diameter of the opening is fixed but the shape of the bottle changes, having variable section, e.g., a classic Coca-Cola bottle.

Suggested tasks:

✓ Use a plastic bottle of non-uniform cross-section, such as a classic Coca-Cola, and caps with holes similar to those used in Project 59 to determine the dependence of m as a function of t for the sand. Also, graph the flow dm/dt as a function of time. For the same outlet orifice size, is there a significant difference between the flow of sand in a vessel with constant section and one with variable section?

✓ Analyze experimentally whether for the same hole, the dependence of m and dm/dt depending on t for water varies with the shape of the container. What can you conclude about the flows of sand or water regarding their behavior according to the shape of the container?

22.2.3 *Project 61. Dependence of sand flow on the area of the exit hole*

Recommended basic equipment: A force sensor connected to a computer. A plastic bottle with a screw cap. Several caps, with holes of different sizes. 2–3 kg of dry, sifted sand.

The aim of this experiment is to study the variation of the flow, dm/dt, with the area of the exit hole. In other words, it is an experimental test of expression (22.3). We can use the same experimental setup as in Project 60, taking the same precautions to keep the sand dry and sifted. Use at least five caps with different holes of known areas, ranging from about 2 mm to 10 mm in diameter.

Suggested tasks:

- ✓ For each cap (with its corresponding orifice of area A), determine the flow (dm/dt) by measuring the slope of the graph of m as a function of t.
- ✓ Plot dm/dt as a function of A on linear scale and a log-log scale. What can you conclude about the dependence of flow on area?
- ✓ One way to test expression (22.3) is to graph dm/dt as a function of $A^{5/4}$. If the graph shows a linear trend, it is indicative that the hypothesis indicated by Eq. (22.3) is indeed fulfilled. What can you conclude about this hypothesis, using your experimental results?

22.2.4 *Project 62. Determining the moment of inertia of a pulley*

> **Recommended basic equipment:** A smart pulley with photogate connected to a PC. Two or more weights of the order of 100 g.

The objective of this preliminary experiment is to determine the moment of inertia I_p of a "smart pulley" associated with a photogate [6]. Figure 22.2 shows a possible setup for this activity.

In this case, a mass m hangs from a thread about 50 cm long of negligible mass, wound on the pulley, so that it detaches from the pulley when it unwinds completely.

While the thread attached to the mass is wound around the pulley, it exerts a force that accelerates the pulley with an angular acceleration α^+. When the thread attached to the mass is released, the smart pulley continues to rotate, but the friction force of the pulley decelerates it with angular acceleration α^-. Thus, the

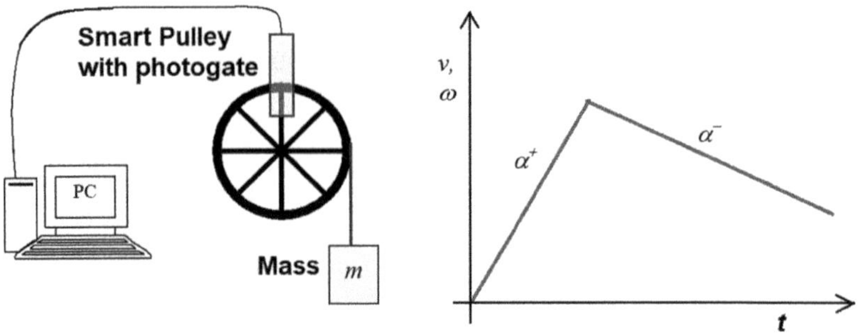

Figure 22.2 Experimental setup for determining the moment of inertia of a pulley associated with a photogate.

movement of the pulley looks something like the diagram on the right in Figure 22.2.

When the mass is attached to the pulley, using a free-body diagram, we obtain the equations of motion of the hanging mass m:

$$ma = mR_P\alpha^+ = mg - T, \tag{22.4}$$

where T is the tension of the thread and R_p the radius of the pulley. If f_R represents the friction force on the pulley, the equation of motion of the pulley is

$$I_p\alpha^+ = (T - f_R)R_p. \tag{22.5}$$

Eliminating T from (22.4) and (22.5), we have

$$\left(\frac{I_p}{R_p^2} + m\right) R_p\alpha^+ = \left(\frac{I_p}{R_p^2} + m\right) a^+ = mg - f_R. \tag{22.6}$$

Once the thread loses contact with the pulley, the movement is decelerated with angular acceleration α^- given by

$$\frac{I_p}{R_p^2} R_p\alpha^- = \frac{I_p}{R_p^2} a^- = -f_R. \tag{22.7}$$

Since f_R is the same in Eqs. (22.6) and (22.7), we have

$$I_p = mR_p^2 \frac{g - a^+}{a^+ - a^-}, \tag{22.8}$$

where $a^+ = R_p\alpha^+$ and $a^- = R_p\alpha^-$ represents linear accelerations. Sometimes it helps to write $I_P = k \cdot M_pR_p^2$, where M_p is the mass of the pulley and R_p its radius. The constant k is generally of the order of a unit.

Suggested tasks:

- ✓ Build a system like the one in Figure 22.2. Choose the mass (of the order of 100 g) so that the system will move when released.
- ✓ Determine $x(t)$ and $v(t) = dx/dt$ as a function of time. To do so, count the number of rays N_R of the pulley wheel and measure its inner diameter d_p (i.e., the inner diameter of the pulley groove, where the thread passes). Thus, one ray occupies the position of the previous one each time the thread travels a distance $dx = \pi d_p/N_p$. As in principle, the photogate can measure the time interval between two interruptions of the light beam, that is, the time dt between the passage of two consecutive rays, we can determine $t, x(t)$ and $v(t) = dx/dt$.
- ✓ Plot $x(t)$ and $v(t)$ as a function of time. If $v(t)$ as a function of t has a linear trend in both sections (see Figure 22.2), from the slopes, find the accelerations a^+ and a^- and their respective uncertainties.
- ✓ Using Eq. (22.8), find the value of the moment of inertia I_p and its uncertainty.

22.2.5 *Project 63. Atwood machine with constant mass*

Recommended basic equipment: A smart pulley with photogate connected to a PC. Two or more weights of the order of 100 g. Alternatively, a smartphone can also be used to measure acceleration, instead of the smart pulley with photogate.

The aim of this experiment is to study the dynamics of an Atwood machine, Figures 22.3 and 22.6, with fixed masses. We want to study this system first and determine the friction of the pulley. This parameter will be useful when we study an Atwood machine with variable mass. Annex A briefly describes the theoretical treatment

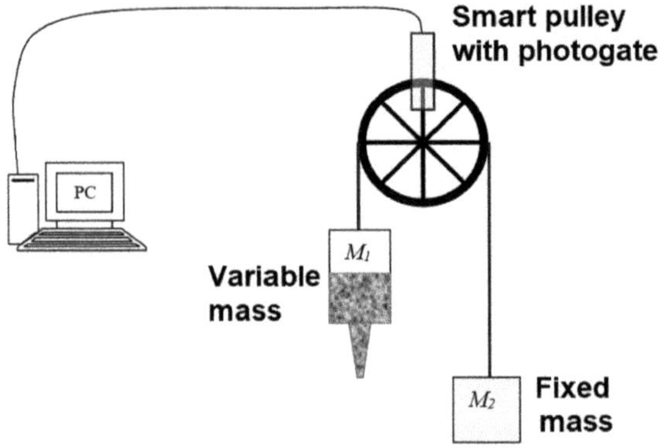

Figure 22.3 Experimental setup of a variable mass Atwood machine. A smart pulley with a photogate connected to a PC is used to determine the position of the masses as a function of time [5]. Alternatively, a smartphone can also be used to record the motion of this system [7, 8].

of this system. Another possible setup for doing this experiment is to use a smartphone that hangs as one of the masses of the system, to measure acceleration, instead of the smart pulley with photogate. In Refs. [7, 8], the details of this mode of carrying out this experiment can be found.

Suggested tasks:

✓ Build a system like the one in Figure 22.3, with fixed masses M_1 and M_2, of the order of 100 g each. The masses are different, so the system moves when released. Make sure you can vary the combination of masses M_1 and M_2 to achieve at least five different combinations of the difference $M_2 - M_1$.

✓ For each combination of masses, determine $x(t)$ and $v(t) = dx/dt$ as a function of time.

✓ Plot $x(t)$ and $v(t)$ as a function of time. If $v(t)$ is a linear function of t, from the slope you can find the acceleration a and its uncertainty Δa.

✓ From the set of all the data obtained with the different combinations of masses M_1, M_2, and a, graph the variables or pseudovariables $y = M_T a = (M_1 + M_2 + kM_p)a$ as a function of $z = (M_2 - M_1)$. Discuss the linearity or lack of linearity of these graphs (see Annex A).

✓ From this study, determine the value of the friction force f_R of its pulley and discuss the suitability of the model proposed in Annex A, to describe the present experimental system.

22.2.6 *Project 64.* ♣♣ *Variable mass Atwood machine*

> **Recommended basic equipment**: A smart pulley with photogate connected to a PC. A plastic bottle with a screw cap. Several caps, with holes of different sizes. 2–3 kg of dry, sifted sand. Force sensor connected to a PC. Alternatively, the dynamics of the system can be studied using a digital camera or smartphone in video mode, which replaces the photogate and PC.

The aim of this experiment is to study the dynamics of an Atwood machine with variable mass [5]. Annex B discusses the theoretical model proposed to describe this system. The proposed experimental system is schematically illustrated in Figure 22.3. Again, as indicated in the previous projects, use a smartphone that hangs as the fixed mass of the system, to measure acceleration, instead of the smart pulley with photogate. See Refs. [7, 8] for more details.

Suggested tasks:

✓ To achieve a variable M_1 mass, we can use a small bottle or aluminum tube (such as a medicine or vitamin tube), with a hole at the bottom through which the sand can empty in a few seconds. This means that, if at the beginning of the movement $M_1 > M_2$, at some point during the movement of the system, this relationship is reversed. It is convenient to have two or three tubes or bottles with different openings.

✓ For each tube or bottle, with different openings, using the experimental setup suggested in Figure 22.1, determine the value of the mass flow. To do this, find the slope of the graph of m as a function of t in static mode, as is described in Project 61.

- ✓ Determine the value of the initial masses for each system you use. Using the photogate associated with a Smatpuley [6], determine $x(t)$ and $v(t) = dx/dt$ as a function of time. Remember that in general, ordinary photogates only measure the time between two consecutive interruptions, therefore the direction of movement (i.e., the sign of the speed) must be entered manually. This is easy to do because when you plot velocity as a function of time, at the instant the system reverses its movement, the graph $v(t)$ shows a conspicuous V-shape. Just by changing the sign of dx from this point on, it will be seen that both $v(t)$ and $x(t)$ take the form of a continuous and differentiable function. If a digital camera or smartphone is used to study motion, the sign change is evident when analyzing the video frame by frame to determine $x(t)$.
- ✓ In the same graph, and for each case studied, represent the measured values of $x(t)$, $v(t)$ and the corresponding theoretical predictions using the model discussed in Annex B, Eqs. (22.29) and (22.30).
- ✓ Discuss the adequacy of the model proposed in Annex B to describe this experiment.

22.3 Just for fun: Cup and key experiment

A delightful demonstration based on Atwood's machine is the *cup and key* experiment. This experiment is intended to amuse and surprise children from 3 to 95 years old: a party trick that shows that physics can be fun.

Take a heavy object such as a mug and tie it to a thin rope or cotton string of about 1–1.2 m long. Choose a flexible string that is strong enough to support the weight of the mug. Tie a key or other light object to the other end of the string. Take a pencil and hold it firmly horizontally with one hand. With the other hand, hold the key so that the cup is hanging by the thread that passes over the pencil. Make sure the cup is just over half a meter from the ground and that the string is flexible. Cotton string is best. Figure 22.4 illustrates the procedure.

Figure 22.4 Cup and key experiment.

Can you predict what happens when the key is released? Will the cup break on the floor? After rehearsing a few times, you'll be ready to amuse your friends. This experiment, also known as the swinging Atwood's machine, has been widely discussed in the literature [9–11]. Despite being a system with only two degrees of freedom, in general, its equations of motion cannot be integrated, and it has generated many theoretical studies in recent times.

22.3.1 *Project 65. ♣♣ Variable mass oscillator*

Recommended basic equipment: Force sensor connected to a PC. A plastic bottle with a screw cap. Several caps, with holes of different sizes. 2–3 kg of dry, sifted sand. A spring enables the system with the bottle filled with sand to oscillate at a frequency of approximately 1 Hz.

The aim of this experiment is to study the dynamics of a simple variable mass oscillator [12, 13]. Annex C discusses the theoretical model proposed to describe this system. The proposed experimental system is illustrated in Figure 22.5. Since the force exerted by the spring on the force sensor is proportional to its stretch x (Hooke's law), the signal registered by the force sensor as a function of time has a magnitude proportional to $x(t)$.

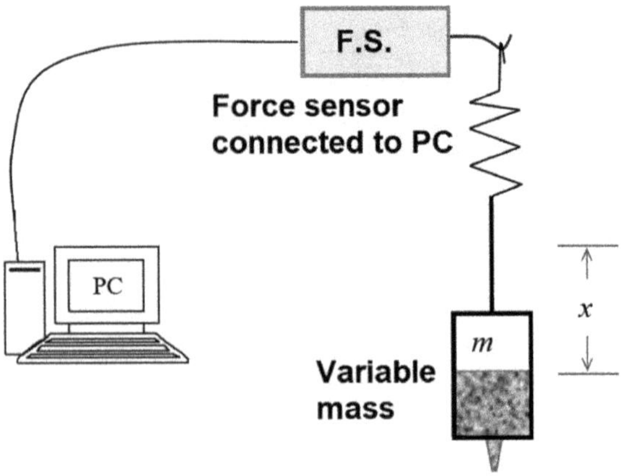

Figure 22.5 Experimental setup of a variable mass oscillator [12].

To achieve a variable mass m, a small bottle of sand with caps with holes of different sizes can be used. This allows the flow of sand to be easily varied. Different behaviors of the system can thus be explored, depending on how the energy loss of the system due to mass loss is compared with the decrease in energy due to the characteristic friction of the oscillator with the air. For this reason, as a preliminary experiment, it is useful to characterize these variables independently (friction and mass flow). To work comfortably, select a spring that produces free oscillations at a frequency of the order of 1 Hz when the bottle is full.

Suggested tasks:

✓ For each bottle cap, determine the value of the mass flows, $c = dm/dt$ and $\lambda = c/m_0$, where m_0 is the initial mass of the bottle with sand.

✓ Use the bottle of sand with a cap without a hole to study the free movement of the system and determine the damping coefficient γ for your system. To do this, determine the variation of x as a function of time. Using a simple viscous damping model,

$$x(t) = A_0 \exp(-\gamma t) \sin(\omega t + \phi) \tag{22.9}$$

adjust parameters A_0, γ, ω, and ϕ in order to reproduce the experimental data, as discussed in Chapter 15.

✓ Using a cap with a hole which allows you to measure at least about 40 oscillations before the bottle of sand empties, measure x as a function of time. Since the mass varies continuously, the oscillation curve will be superimposed on this variation. To analyze the oscillatory movement more comfortably, subtract the continuous variation given by $x_c(t) = X_0(1 - \lambda t)$ from the $x_{\text{med}}(t)$ measured by the force sensor. Here, λ is the parameter that determines the flow for the orifice used and X_0 an initial amplitude, chosen such that the resulting function, $x(t)$, oscillates around zero:

$$x(t) = x_{\text{med}}(t) - X_0(1 - \lambda t). \tag{22.10}$$

✓ Plot your results and observe the variation in frequency over time.

✓ Using the technique of determining the period of anharmonic oscillations, and calculating the zero crossings of the signal $x(t)$, as discussed in Chapter 15, determine the variation of the period T of the signal $x(t)$ with time. Plot T as a function of time and mass. Also plot T^2 as a function of $m(t)$, the instantaneous mass of the system.

✓ Analyze whether the experimental data are consistent with (see Annex C)

$$\omega = \sqrt{\frac{k}{m(t)}} \quad \text{and} \quad T^2 = \frac{4\pi^2}{k} m(t). \tag{22.11}$$

✓ Experimentally, study the variation of $x(t)$ for the bottle with different types of mass flows (see Annex C):

- ➢ when $\lambda \ll \gamma$, that is, when the energy loss of the oscillator is dominated by the viscous friction force,
- ➢ when $\lambda \gg \gamma$, that is, when the loss of energy of the oscillator is dominated by the loss of mass of the system,
- ➢ when $\lambda \approx 2.\ \gamma$, that is, when the loss of energy due to loss of mass is comparable to that of the viscous friction force.

✓ In the same graph, and for each case studied, represent the measured values of x as a function of time and the corresponding theoretical predictions, using the model discussed in Annex C.

✓ Discuss the adequacy of the model proposed in Annex C to describe your experimental results.

22.4 Annex A: Atwood machine with constant masses

The Atwood machine is a classic example of the application of Newton's second law. This machine consists of two masses (M_1 and M_2) connected by an inextensible string of negligible mass, which passes through a pulley that is fixed, as illustrated in Figure 22.6. The pulley has a mass M_P and radius R_p. If the masses are fixed, the only degree of freedom of the system is the position x of the mass M_2 with respect to the center level of the pulley (Figure 22.6).

If we assume that the moment of inertia of the pulley with respect to its center is $I_p = kM_pR_p^2$, where k is a constant $0 < k < 1$, and that the friction of the pulley generates a torque R_pf_r, the equations of motion, obtained using a free-body diagram (Figure 22.6), are

$$M_1\frac{d^2x}{dt^2} = +T_1 - M_1g, \tag{22.12}$$

$$M_2\frac{d^2x}{dt^2} = -T_2 + M_2g, \tag{22.13}$$

$$I_p\alpha = R_p(T_2 - T_1 - f_r). \tag{22.14}$$

And, therefore,

$$I_p\frac{1}{R_p}\frac{d^2x}{dt^2} = R_p(T_2 - T_1 - f_r) \quad \text{or} \quad kM_p\frac{d^2x}{dt^2} = T_2 - T_1 - f_r. \tag{22.15}$$

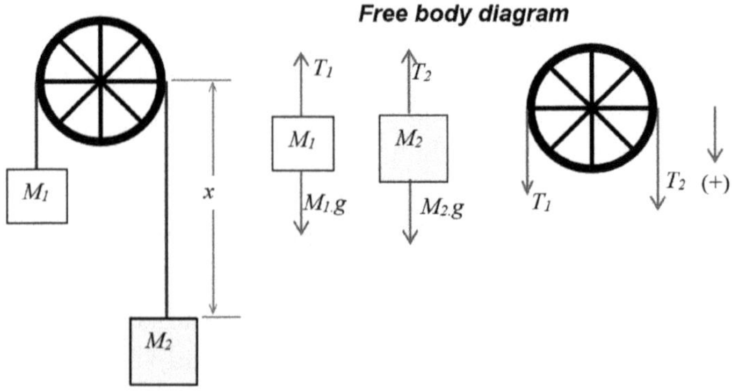

Figure 22.6 Experimental setup of an Atwood machine with constant masses.

Here, we have assumed that the positive direction of x is that of the downward movement of M_2. If the sign resulting from the acceleration is negative, it means that the mass M_2 is rising. Combining these last three equations, we have

$$(M_1 + M_2 + kM_p)\frac{d^2x}{dt^2} = (M_2 - M_1)g - f_r. \tag{22.16}$$

Since the parameters M_1, M_2, M_p, and k are easy to determine, a simple way to falsify this result is to measure the acceleration of the system, $a = d^2x/dt^2$, for various combinations of M_1 and M_2. If we define $y = M_T a = (M_1 + M_2 + k\ Mp)a$ and $z = (M_2 - M_1)$, according to Eq. (22.16), we have

$$y = gz - f_r. \tag{22.17}$$

Therefore, if a is constant over time for each pair M_1 and M_2, the dependence of y as a function of z should be linear, if the model just described, Eq.(22.16), is correct. In such case, the slope should be equal to g. If these predictions are supported by the experimental results, it shows that Eq. (22.16) is indeed a good description of the physical system under study. Also, from the intercept on the graph $y(z)$, we can find the value of f_r.

22.5 Annex B: Atwood machine with variable mass

For a variable mass system, Newton's second law can be written as follows [5, 14, 15]:

$$\frac{dp}{dt} = F_T, \tag{22.18}$$

where p is the total momentum of the system and F_T is the net force acting on it. Since the mass of the system varies over time, we must be very careful when referring to the variation of p in Eq.(22.18) because it includes the impulse associated with the mass that leaves the system [5, 14]. Let us consider the variation of the moment between t and $t + dt$. At time t, the mass of the system is $M(t)$ and it moves with a velocity $v(t)$, therefore its moment is $p(t) = M(t)v(t)$. After a time dt, the system has lost a mass dM (< 0) and its speed has changed by dv. If the ejected mass dM leaves the system with a velocity u relative to the container, the total moment in $t + dt$ will be [14, 15]

$$p(t + dt) = (M - dM)(v + dv) + (v - u)dM. \qquad (22.19)$$

Thus,

$$\frac{dp(t)}{dt} = M\frac{dv}{dt} - u\frac{dM}{dt}. \qquad (22.20)$$

If the escape velocity of the ejected mass (sand) with respect to the container is zero ($u = 0$), i.e., there is no *thrust*, then

$$\frac{dp(t)}{dt} = M(t)\frac{dv}{dt}, \qquad (22.21)$$

which is the same equation as for a constant mass system.

Going back to the Atwood machine, if we now assume that M_1 varies with time, but the sand leaves with zero velocity with respect to the container ($u = 0$), there is no thrust, and the equations of motion, Eqs. (22.12)–(22.16), are the same and the result of Eq. (22.16) can be written as follows:

$$M_T \cdot a = (M_2 - M_1(t))g - f_r, \qquad (22.22)$$

with $M_T(t) = (M_1(t) + M_2 + kM_p)$.

Since the geometry of the variable mass system and the constant mass system are equal, it is reasonable to assume that the friction force f_r is the same in both cases. Therefore, we can use the value given by Eq. (22.17). To estimate the outflow of the sand, since the mass M_1 is accelerating, according to Eq. (22.3), this acceleration affects the flow. Therefore, we will assume that

$$\frac{dM_1(t)}{dt} = c(a) = c_0(1 + a/g)^{1/2}, \qquad (22.23)$$

where c is the flow to acceleration a and c_0 is the mass flow value (dm/dt) when the mass M_1 is at rest. This dependence of the flow with the acceleration is obtained from Eq. (22.3). If the mass were in free fall, that is, if $a = -g$, the flow would be zero, which is consistent with what we expect from a system like this.

If $a \ll g$, we can use the approximation

$$c(a) \approx c_0 \left(1 + \frac{a}{2g}\right). \tag{22.24}$$

Defining the parameters

$$m_{10} = M_1(t = 0), \quad m_{12}^0 = M_{10} - M_2 \tag{22.25}$$

and

$$M_0 = M(t = 0) = m_{10} + M_2 + 1/2M_p, \tag{22.26}$$

the equation of motion, obtained from Eqs. (22.22)–(22.24), is [5]

$$a(t) = \frac{dv}{dt} = \frac{\left(m_{12}^0 - c_0\, t\right) g - f_r}{\left(M_0 - c_0 \eta t\right)}, \tag{22.27}$$

where we have defined

$$\eta = \left(1 - \frac{1}{2}\left(1 - \frac{a}{g}\right)\right). \tag{22.28}$$

We thus see that η depends linearly on a, thereby introducing a second-order dependence on the acceleration. As the quotient a/g can be approximated in the first order by Eq. (22.27), taking its mean value, i.e., replacing M_1 by its average value $\langle M_1(t)\rangle \approx m_{10}/2$ and ignoring the friction force, and using the previously introduced parameters M_0 and m_{10}, we have

$$\eta \approx 1 - \frac{1}{2}\left(1 - \frac{\frac{1}{2}m_{10} - M_2}{M_0 - \frac{1}{2}m_{10}}\right) = 1 - \frac{1}{2}\left(\frac{2M_2 + \frac{1}{2}M_p}{M_0 - \frac{1}{2}m_{10}}\right). \tag{22.28}$$

The equation of motion (22.27) can now be integrated to obtain the velocity and position of the masses:

$$v(t) = v(t = 0) + \frac{gt}{\eta} + \frac{\left(M_0 - \eta m_{12}^0\right) g + \eta f_r}{c_0 \eta^2} \ln\left(1 - \frac{c_0 \eta t}{M_0}\right) \tag{22.29}$$

and

$$x(t) = x(t = 0) + v(t = 0)\, t + \frac{g\, t^2}{2\eta} + M_0 \frac{\left(M_0 - \eta m_{12}^0\right) g + \eta f_r}{c_0^2 \eta^3}$$
$$\times \left(1 - \frac{c_0 \eta t}{M_0}\right)\left[\ln\left(1 - \frac{c_0 \eta t}{M_0}\right) - 1\right]. \tag{22.30}$$

These equations can be compared to the experimental results of $x(t)$ and $v(t)$.

22.6 Annex C: Variable mass oscillator

As described in Annex B, the equation of motion of a variable mass system when the lost mass leaves the container with velocity $u = 0$, i.e., there is no thrust effect, is the following:

$$\frac{dp}{dt} = m(t)\frac{dv}{dt}. \tag{22.31}$$

The friction force f_r of air for an object at low velocities is proportional to the velocity, $f_r = -bv$, where the constant of proportionality b depends on the geometry of the object and the viscosity of the fluid. Therefore, the equation of motion of a variable mass, $m(t)$, attached to a linear spring of constant k, as illustrated in Figure (22.5) can be written as follows [12]:

$$m(t)\frac{d^2x}{dt^2} + b\frac{dx}{dt} + kx = 0. \tag{22.32}$$

If we multiply this expression by $\frac{dx}{dt}$, we obtain

$$m(t)\frac{d^2x}{dt^2}\frac{dx}{dt} + b\frac{dx}{dt}\frac{dx}{dt} + kx\frac{dx}{dt} = 0. \tag{22.33}$$

The derivative of the kinetic energy of a system of variable mass can be written as follows:

$$\frac{dE_k}{dt} = \frac{1}{2}\frac{dm(t)}{dt}\left(\frac{dx}{dt}\right)^2 + m(t)\frac{dx}{dt}\frac{d^2x}{dt^2}. \tag{22.34}$$

Since the potential energy of the system is $E_p = 1/2k \cdot x^2$, Eq. (22.33) can be written as follows:

$$\frac{d(E_k + E_p)}{dt} - \frac{1}{2}\frac{dm(t)}{dt}\left(\frac{dx}{dt}\right)^2 + b\left(\frac{dx}{dt}\right)^2 = 0. \tag{22.35}$$

In our case, we assume that

$$m(t) = m_0 - ct = m_0(1 - \lambda t) \text{ and } \frac{dm(t)}{dt} = -c = -m_0\lambda. \tag{22.36}$$

Hence, Eq. (22.35) becomes

$$\frac{d(E_k + E_p)}{dt} = -\left(b + \frac{c}{2}\right) \cdot \left(\frac{dx}{dt}\right)^2. \tag{22.37}$$

This expression indicates that the mechanical energy loss of our variable mass oscillator is made up of two terms: the friction energy loss and the mass-loss energy decrease. To determine the variation of the amplitude of the oscillations with time, we assume

$$A(t) = A_0 f(t). \tag{22.38}$$

It is important to emphasize that $A(t)$ is the maximum amplitude of each oscillation. Total mechanical energy can be written as $E_T = (1/2)kA^2(t)$. Thus, if we consider two consecutive oscillations, we have

$$\frac{d(E_k + E_p)}{dt} = \frac{dE_T}{dt} = \frac{d}{dt}\left(\frac{1}{2}kA^2(t)\right) \approx -\left(b + \frac{c}{2}\right)\langle v^2 \rangle, \tag{22.39}$$

where $\langle v^2 \rangle$ represents the root mean square velocity over an entire period of oscillation and we can estimate it as $\langle v^2 \rangle \approx A^2\omega^2/2$, where ω is the angular frequency of the oscillation associated with the period of oscillation that we are considering. If we also make the approximation $\omega^2 \approx k/m(t)$, which we can verify experimentally, from (22.11), we have

$$kA(t)\frac{dA(t)}{dt} \approx -\left(b + \frac{c}{2}\right)\frac{1}{2}\frac{k}{m(t)}A^2(t). \tag{22.40}$$

Thus,

$$\frac{dA}{dt} \approx -\frac{1}{2}\left(b + \frac{c}{2}\right)\frac{A}{m(t)}. \tag{22.41}$$

This expression can be interpreted as the differential equation that describes the amplitude of the oscillations, which can be easily integrated to give

$$A(t) = A_0 f(t) = A_0\left(1 - \frac{ct}{m_0}\right)^{\varepsilon} = A_0(1 - \lambda t)^{\varepsilon}, \tag{22.42}$$

where $\varepsilon \equiv b/2c + 1/4 = \gamma\lambda + 1/4$. Here, we have introduced the parameter $\gamma \equiv b/2m_0$. The parameter ε measures the ratio of the frictional energy losses to the mass loss of the system and determines the concavity of the envelope of the oscillations. If $\varepsilon > 1$, we have $A''(t) > 0$, therefore the envelope of the oscillations is concave. Otherwise, if $\varepsilon < 1$, the envelope of the oscillations is convex. For $\varepsilon = 1$, $A''(t) = 0$.

The following particular cases are notable:

(a) If $(b/2c) \to \infty (\text{or} \gamma/\lambda \to \infty)$, that is, we are assuming that the loss of energy due to the variation in mass is negligible compared to viscous friction, from Eq. (22.42), we have

$$\lim_{\gamma/\lambda \to \infty} A(t) = A_0 \exp\left(-\frac{b}{2m_0}t\right) = A_0 \exp\left(-\gamma t\right), \qquad (22.43)$$

which matches the standard expression for the amplitude of a damped oscillator. This case is illustrated in Figure (22.7(a)).

(b) If $\varepsilon = 1$, or if $(b/2c) = \gamma/\lambda = 3/4$, we get

$$A(t) = A_0 \left(1 - \frac{ct}{m_0}\right). \qquad (22.44)$$

In this case, the amplitude of the oscillations decreases linearly in time as illustrated in Figure (22.7(b)).

(c) *If* $\varepsilon = 1/4(b/2c \ll 1 \text{ or } \gamma/\lambda \ll 1)$, that is, the loss of mass completely dominates the energy variation of the system,

$$A(t) = A_0 \left(1 - \frac{ct}{m_0}\right)^{1/4}. \qquad (22.45)$$

This case is illustrated in Figure (22.7(c)).

Now that we have obtained an adequate description of the amplitude, we can try to find an approximate solution to Eq. (22.32). To do so, we propose a solution of the type

$$x(t) = A_0 \left(1 - \frac{ct}{m_0}\right)^{\varepsilon} \sin\left[h_0(t) + \varphi\right], \qquad (22.46)$$

where, as usual, the constants A_0 and ϕ are determined by the initial conditions of the problem. The replacement of Eq. (22.46) in Eq. (22.32) enables us to obtain a differential equation for $h_0(t)$, whose solution is [12]

$$h_0(\tau) = \frac{2\beta}{c}\left[\arctan\left(\sqrt{\frac{1-\alpha-\tau}{\alpha}}\right) - \left(\sqrt{\frac{1-\alpha-\tau}{\alpha}}\right)\right], \qquad (22.47)$$

where

$$\beta = \frac{1}{2}\sqrt{\left(b + \frac{1}{2}c\right)\left(b + \frac{3}{2}c\right)}, \alpha = \frac{\beta^2}{m_0 k}, \quad \text{and} \quad \tau = \frac{ct}{m_0} = \lambda \cdot t. \qquad (22.48)$$

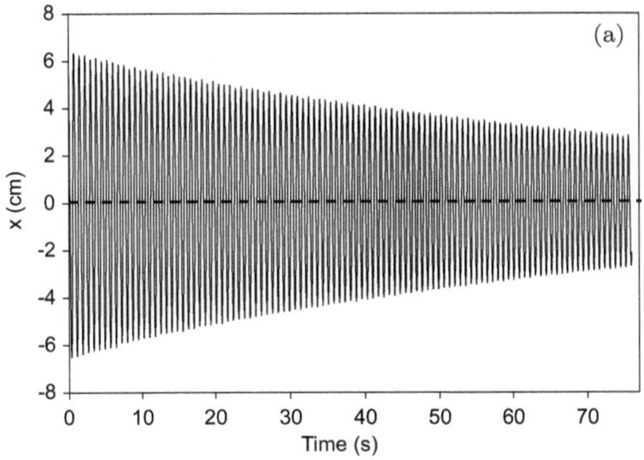

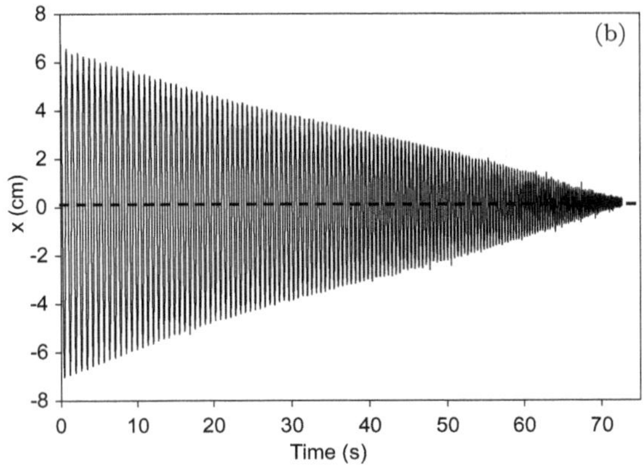

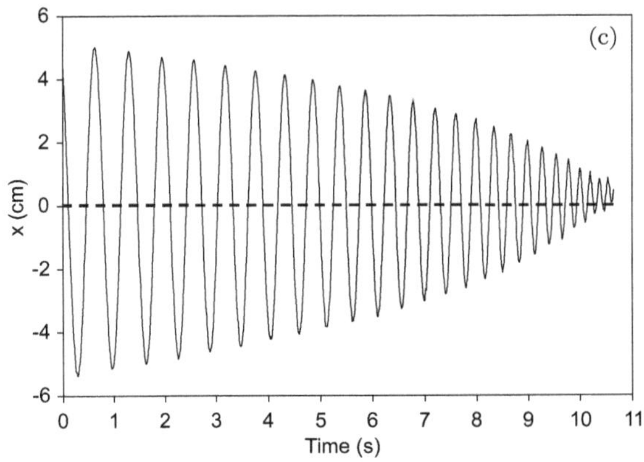

Figure 22.7 Variation of the position, $x(t)$, as a function of time, for the three types of oscillations. (a) Typical behavior of the underdamped oscillations corresponding to the small flow case $\varepsilon > 1$. (b) Case of $\varepsilon \approx 1$. (c) Case where $\varepsilon < 1$, that is, when the loss of mass dominates the decrease in energy.

When $(c \to 0)$, Eq. (22.47) is reduced to the familiar equation:

$$h_0(t) \to \omega t = \left(\sqrt{\frac{k}{m_0}} - \gamma^2\right) t. \tag{22.49}$$

Another consequence of solution (22.46) is that the angular frequency and period of the system can be written as follows: [12, 16]

$$T^2(t) \equiv \frac{4\pi^2}{\omega^2(t)} \approx 4\pi^2 \frac{m(t)}{k}. \tag{22.50}$$

The expression (22.46) is an approximate solution of (22.32) as long as the following condition is met:

$$\frac{(b + c/2)(b + 3c/2)}{4k} \ll m(t), \tag{22.51}$$

therefore, we can expect that our model will no longer be adequate to describe the system when the bottle is close to running out of sand. The situation when the friction force is not proportional to v is discussed in Ref. [13].

References

[1] H. M. Jaeger, *et al.*, "The physics of granular materials," *Phys. Today*, vol. 49, no. 4, pp. 32–38, 1996.
[2] L. A. Pugnalon, "Azúcar, pimienta y sal," *Ciencia Hoy*, vol. 14, no. 79, pp. 36–44, 2004.
[3] M. Yersel, "The flow of sand," *Phys. Tech.*, vol. 38, no. 5, pp. 290–291, 2000.
[4] B. Munson, D. Young and T. Okiishi, *Fundamentals of Fluid Mechanics*, New York: Wiley, 1994.
[5] J. Flores, G. Solovey and S. Gil, "Flow of sand and a variable mass Atwood machine," *Am. J. Phys.*, vol. 71, no. 7, pp. 715–720, 2003.
[6] Vernier Software & Technology, "Photogates," 2021. Available: https://www.vernier.com/product/photogate/.
[7] M. Monteiro, C. Stari, C. Cabeza and A. Marti, "The Atwood machine revisited using smartphones," *The Phys. Teacher,* vol. 53, no. pp. 373–374, 2015.
[8] D. Lopez, *et al.*, "Study of a variable mass Atwood's machine using a smartphone," *The Phys. Tech.*, vol. 56, p. 182, 2018.
[9] A. R. Marlow, "A surprising mechanics demonstration," *Am. J. Phys.*, vol. 59, no. 10, pp. 951–952, 1991.

[10] D. J. Griffiths and T. A. Abbott, "Comments on: A surprising mechanics demonstration," *Am. J. Phys.*, vol. 60, no. 10, pp. 951–953, 1992.
[11] Wikipedia, "Swinging Atwood's machine," Wikipedia, the free encyclopedia, 2021. Available: http://en.wikipedia.org/wiki/Swinging_Atwood's_machine.
[12] J. Flores, G. Solovey and S. Gil, "Variable mass oscillator," *Am. J. Phys.*, vol. 71, no. 7, pp. 721–725, 2003.
[13] R. M. Digilova, M. Reiner and Z. Weizman, "Damping in a variable mass on a spring pendulum," *Am. J. Phys.*, vol. 73, p. 901, 2005.
[14] A. Sommerfeld, *Lectures on Theoretical Physics, Mechanics*, Vol. I, New York: Academic Press, 1952.
[15] J. G. Roederer, *Mecánica Elemental*, Buenos Aires: Ed. EUDEBA, 2002.
[16] A. Franco García, "Oscilador amortiguado de masa variable," Curso Interactivo de Física en Internet — Universidad del País Vasco, 2021. Available: http://www.sc.ehu.es/sbweb/fisica3/oscilaciones/variable/variable.html.

Chapter 23

ROTATION OF SOLIDS — A FUN PARADOX

The goal of this chapter is to study the stability of a rotating system using a simple, fun device that shows some amazing results regarding the rotation of a rigid body. This system can be used for a quantitative experimental study of the stability properties of a symmetrical spinning top or — why not? — to entertain friends at a party and share the pleasure of doing physics. This system also illustrates the supercritical bifurcation phenomenon.

Goals

- ✓ Rigid bodies in rotation
- ✓ Reference system fixed to the body
- ✓ Stability of rotating bodies
- ✓ Supercritical fork

23.1 Rotational stability

It is not always easy to share with relatives and friends the satisfaction and enthusiasm that science lovers tend to feel about studying natural phenomena. Observing curious or paradoxical effects often requires sophisticated equipment, or these effects may be hidden in formalities that are not accessible to the uninitiated.

Here is a simple device that can be used to entertain friends at a party and illustrate the kinds of problems that physicists enjoy studying. It can also be used in the laboratory to perform quantitative measurements that enable us to explore the properties of rotating bodies. More specifically, we propose to study the rotational stability of a rigid body. The experiment consists of rotating a pencil-sized metal rod or a ring attached to a wire, as illustrated in Figure 23.1.

At rest, the bar or ring will hang with its center of mass at its lowest position. When we begin to rotate the wire with our fingers, the bar or ring rotates with its center of mass in the same position. But if the frequency of rotation of the wire increases above a critical value, the center of mass begins to rise gradually as the speed of rotation increases. Finally, the bar or ring ends up rotating on a horizontal plane. Here is the paradox: Why does the center of gravity of the rotating body rise? Isn't a system more stable when its center of gravity is at the lowest position? At least, this is the situation when we hang any object from a wire or string.

The objective of this experiment is to analyze this phenomenon. This system illustrates the phenomenon of *supercritical bifurcation* [1]. A bifurcation occurs when there is a sudden qualitative change in the behavior of a system (bar or ring) due to the infinitesimal variation in one of its parameters (angular velocity) [1, 2].

The physics associated with the movement of this type of system (Figures 23.1 and 23.2) has important applications in various areas of physics, such as classical and molecular mechanics, nuclear physics,

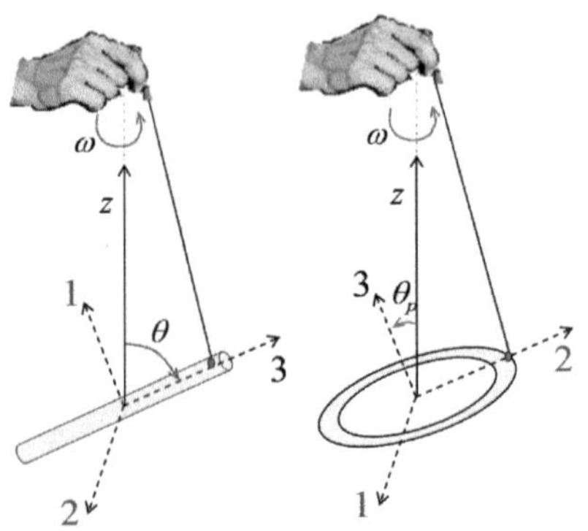

Figure 23.1 The image on the left shows a bar attached to a blue wire that is rotated by hand. The figure on the right shows a ring supported by a blue wire, also rotated by hand. In both cases, 1, 2, and 3 denote the main axis of symmetry of the body.

and astrophysics. The theory of the rotation of a rigid body is discussed in most introductory and intermediate mechanics texts [3–5]. This experiment proposes a simple model to explain the behavior of this system. A more refined model to describe this system is outlined in Annex A [6]. Other amusing behaviors of rotating bodies can be seen in several videos [7] and in Refs. [8–10].

23.2 Naive model — Rotating rod

It is possible to develop an extremely simplified, semi-quantitative model to explain the phenomenon depicted in Figure 23.2, which we will call the "naive model", which contains the basic ingredients to understand this phenomenon. We will make the following simplifying hypotheses: a rod of length h and mass m hangs from a very long, flexible string of length L. That is, $L \gg h$. This system is shown schematically in Figure 23.2. An improved model of the same system is described in Annex A.

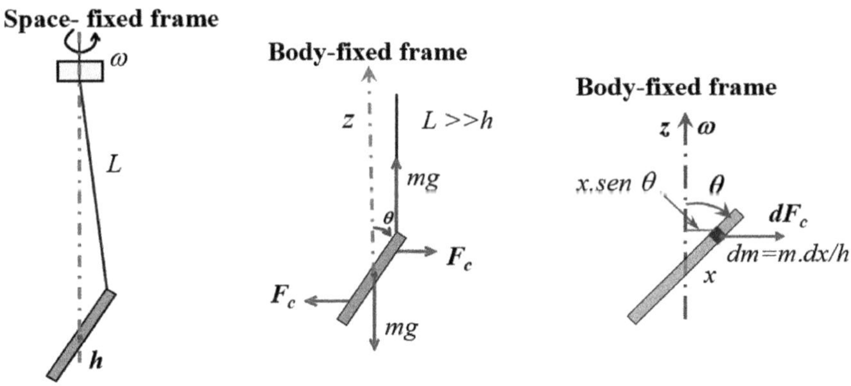

Figure 23.2 Schematic representation of a rod, of length h, rotating around a vertical axis (z) with an angular frequency ω, hanging from a string of length L. The figure on the left shows the rod in the laboratory's reference system. The figures in the center and on the right show the same bar in the reference system fixed to the body (reference system that rotates with the bar). Here, we assume that $L \gg h$. Therefore, the cable is almost parallel to the vertical direction (naive model). At the far right, an element of the bar of length dx is shown, at a distance x from the center of the bar.

The angular frequency of rotation along the z axis is ω. In the frame of reference fixed to the body, i.e., in the system that rotates with frequency ω around the z axis, the bar is stationary, i.e., it does not rotate. The rotating system fixed to the body is not an inertial system [3–5], therefore, there is a centrifugal force, in addition to the external forces on the body, that must be taken into account. More specifically, in the rotating system, two pairs of forces (torques) act on the bar in opposite directions. The first torque is due to the weight (τ_w) and the string tension (τ_c). This torque tends to restore the bar to the vertical direction. The second torque, due to the centrifugal force, tends to bring the bar to a horizontal position.

To calculate the torque due to the centrifugal force, imagine a bar element, of length dx, at a distance x from its center. The centrifugal force on this infinitesimal element is

$$dF_c = dm\,\omega^2 x \sin\theta = \frac{m}{h}\omega^2 x \sin\theta dx, \tag{23.1}$$

and the corresponding infinitesimal torque is

$$d\tau_c = \frac{m}{h}\omega^2 x \sin\theta dx x \cos\theta = \frac{m}{h}\omega^2 x^2 \sin\theta \cos\theta dx. \tag{23.2}$$

By integrating over x from $-h/2$ to $h/2$, we obtain the resulting centrifugal torque:

$$\tau_c = \frac{m}{h}\omega^2 \sin\theta \cdot \cos\theta \cdot 2. \int_{-h/2}^{h/2} x^2 dx = \frac{m}{12} h^2 \sin\theta \cdot \cos\theta. \tag{23.3}$$

At equilibrium, $\tau_c = \tau_w = mg(h/2)\sin\theta$. Thus,

$$m \cdot g \cdot \frac{h}{2} \cdot \sin\theta = m \cdot \omega^2 \frac{h^2}{12} \cdot \sin\theta \cdot \cos\theta. \tag{23.4}$$

From Eq. (23.4), it follows that there are two possible solutions for θ: a trivial one corresponding to $\sin(\theta) = 0$ (i.e., $\theta = 0$) and the other equilibrium position is determined by

$$\cos\theta = \frac{6g}{h\omega}^2 = \left(\frac{\omega_{c0}}{\omega}\right). \tag{23.5}$$

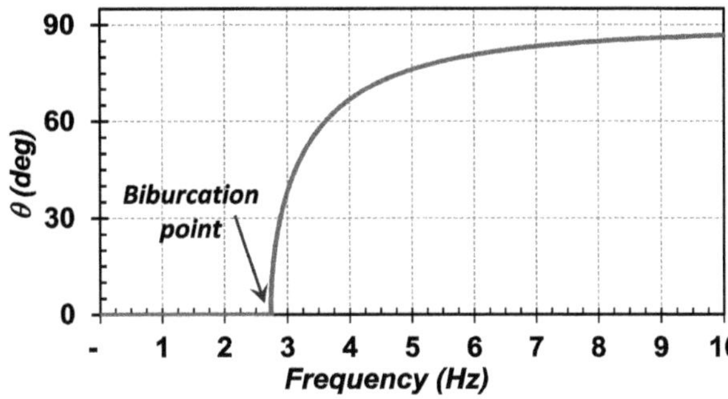

Figure 23.3 Dependence of the equilibrium angle θ on the frequency f described by Eq. (23.5). The critical frequency $f_c = \omega_{c0}/2\pi$ represents the bifurcation point.

As $\cos\theta \leq 1$, this solution is only possible if $\omega_{c0} < \omega$, where

$$\omega_{c0} = \sqrt{\frac{6g}{h}}. \tag{23.6}$$

If $\omega_{c0} > \omega$, the only feasible solution is the trivial one, $\theta = 0$. Consequently, we expect that at low rotational frequencies, $\omega < \omega_{c0}$, the bar rotates in vertical position ($\theta = 0$). For $\omega > \omega_{c0}$, the angle θ of equilibrium will increase as we increase ω, according to Eq. (23.5). In Figure 23.3, the expected dependence of the equilibrium angle θ on the frequency f is shown. When by varying a parameter, the system changes its structure or its stability condition, we say that the system experiences a *bifurcation.* The value of the parameter in which this transformation occurs is called the *critical value.* In this case, $\omega_{c0} = 2\pi \cdot f_c$ represents the critical value of the bifurcation.

The model we have just discussed provides a simplified explanation of the phenomenon. It is possible to develop more general models that take into account the finite length of the string that holds the bar and that consider other shapes of the rotating solid [6]. The derivation of these models can be found in the bibliography; here we will only describe the simplest model in Annex A, which takes into account the finite length of the cable.

23.3 Experimental arrangement

A possible experimental arrangement, suitable for studying quantitatively the behavior of a rotating bar, is illustrated in Figure 23.4. The device consists of a direct current motor (with mechanical reduction) powered by a variable DC voltage source. The voltage regulates the speed of the motor, and the mechanical reduction enables it to work at low frequencies, $f < 10\,\text{Hz}$, applying moderate torques. A plastic shutter attached to the motor shaft is used to interrupt the light beam of a photogate [10] connected to a PC. This photogate measures the frequency of rotation of the motor shaft. The bar can be an aluminum tube of about 8–14 mm outer diameter, about 1 mm thick, of various lengths, between about 15 cm and 40 cm. A small hole is drilled near one end of each tube to connect a thin wire (of the kind usually used in electronics, 0.2–0.5 mm in diameter or AWG 24–28) of length L between 20 and 50 cm. This cable connects the motor shaft to the bar.

The experiment consists of measuring at each rotation frequency f the orientation θ of the bar with respect to the vertical direction. This can be accomplished using a webcam, digital camera, or smartphone. An advantage of low-cost webcams is that as the lighting decreases, the camera's shutter speed is automatically reduced, meaning that the exposure time of each frame increases. Thus, by controlling the illumination, it is possible to obtain a photograph of the bar or a ring that is the average of several positions, as schematically illustrated in Figure 23.5. Otherwise, by filming the rotating object, it is possible to determine the angle θ by analyzing the movie frame by frame [11].

23.3.1 *Project 66. Study of a rotating bar*

Recommended equipment: A few bars of lengths between 15 and 40 cm. A variable speed motor and a rotational frequency sensor, for example, a photogate. A digital camera, webcam, or smartphone to determine the orientation of the bar.

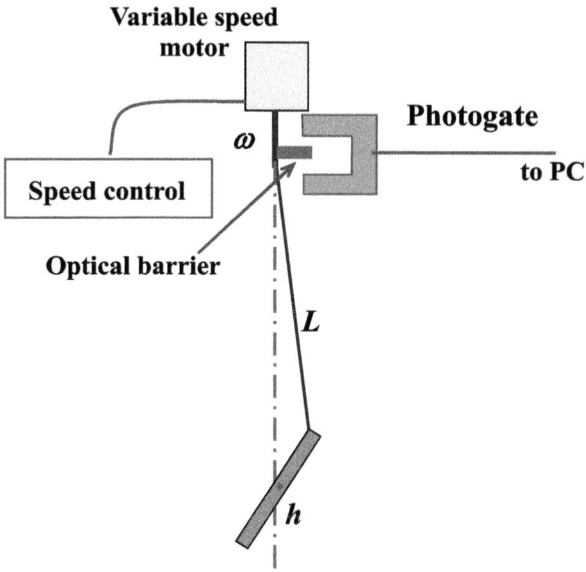

Figure 23.4 Experimental arrangement to quantitatively study the dynamics of a rotating bar. The variable speed motor allows the rotation frequency to be varied in a controllable way. The photoswitch connected to a PC monitors the rotation frequency. A plastic (opaque) object connected to the axis of rotation acts as an optical switch.

Having selected your experimental device, equivalent to the one suggested in Figure 23.4, study the dependence of the orientation angle θ of the bar as a function of the frequency of rotation f. Cover a frequency range in which the bar rotates vertically until it is approximately horizontal. In particular, carefully study the behavior of the bar in the vicinity of the *critical frequency* (f_c), when the bar changes its dynamic equilibrium.

Suggested tasks:

Using the chosen experimental arrangement, for each bar, study the following:

✓ Variation of θ as a function of the rotation frequency f or $\omega = 2\pi f$.

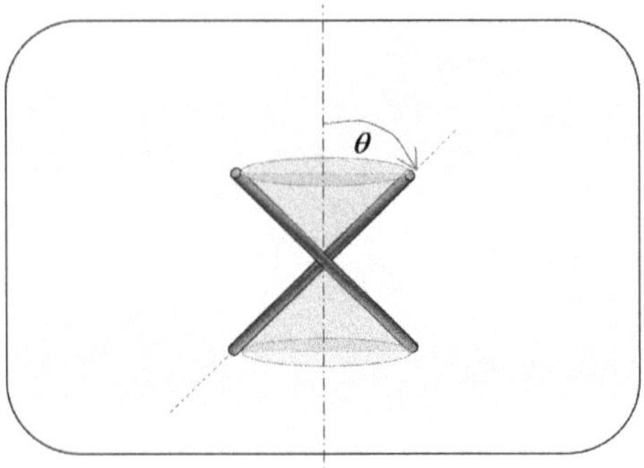

Figure 23.5 Schematic diagram of a photograph of a rotating bar. By varying the lighting, it is possible to increase the exposure time, which automatically averages the positions of the bar. A weight hanging from a thread in the foreground can be used to define the vertical axis.

✓ In the same figure, plot the experimental results of θ as a function of ω and the theoretical expectations obtained with the "naive model" developed previously, Eqs. (23.5) and (23.6).
✓ What can you conclude from this graph? Does the naive model provide an adequate description of the phenomenon under study?
✓ For the bars analyzed, extract from your experimental data the best value of ω_{co}. Plot the values of $6g/\omega_{c0}^2$ as a function of h and compare them to the prediction of (23.6). What can you conclude from this analysis? Is the naive model an adequate description of your data?
✓ Compare your experimental results for each bar analyzed to the predictions of the improved model, Eqs. (23.17) and (23.18). What can you conclude from this analysis regarding both models to describe the position of the bar?

23.4 Annex A: ♣♣ Theory of a rotating bar

This annex extends the naive model of Section 23.2 to take into account the finite length of the string or wire and the fact that it is tied to the bar at a distance δ from its end [6]. Consider a bar of length h hung from a string of length L at one end, as illustrated in Figure 23.6. At a certain instant, the rotation frequency is $\omega =$

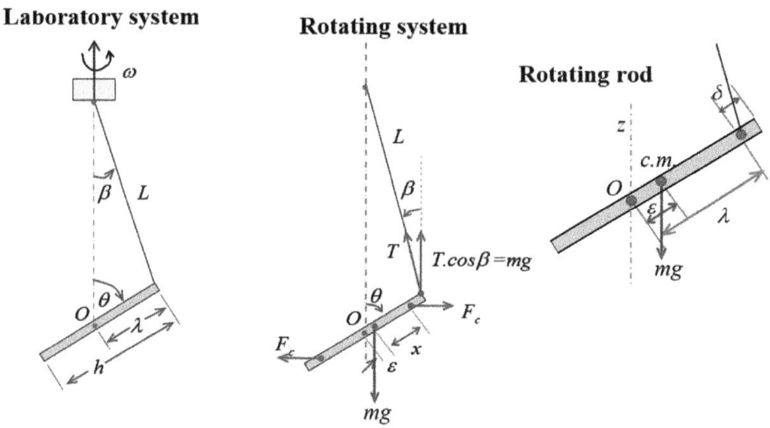

Figure 23.6 Representation of a rod, of length h, rotating around a vertical axis (z) with an angular frequency ω, by means of a string of length L. On the center, we see the rod in the laboratory's reference system. On the right, we see the same bar, in the reference system fixed to the body.

$2\pi f$, and we assume that the bar forms an angle θ with the vertical (axis z). In the frame of reference that rotates with the body, that is to say, in the system fixed to the body, the centrifugal force tends to bring the bar to a horizontal position, while its weight tends to restore the bar to the vertical position. Throughout our analysis, we will assume that the string is longer than the bar, that is, $L > h$. The balance of vertical and horizontal forces, see Figure 23.6, implies that

$$T\cos\beta = mg \quad \text{and} \quad T\sin\beta = F_c^{(cm)} = m\omega^2\varepsilon\sin\theta, \tag{23.7}$$

where T is the tension of the string and $F_c^{(cm)}$ is the centrifugal force acting on the center of mass rotating around the vertical passing through the suspension point, that is, $F_c^{(cm)}$ is the net centrifugal force.

Here, ε is the distance from the center of mass of the rod to point O (Figure 23.6) where the rod intersects the vertical passing through the point of suspension of the string. From the geometry of our system, we have

$$\lambda\sin\theta = L\sin\beta. \tag{23.8}$$

where λ is the distance from O to the point on the bar where it is tied to the string, z axis. From this relationship, we have

$$\tan\beta = \frac{\sin\beta}{\sqrt{1+\sin^2\beta}} = \frac{\lambda}{L}\frac{\sin\theta}{\sqrt{1-(\lambda/L)^2\sin^2\theta}}. \qquad (23.9)$$

Also, from the geometry of the problem, we see that $\lambda = h/2 - \delta + \varepsilon$, where δ is the distance from the point of union of the thread with the bar to the end of it. From Eqs. (23.7) and (23.9), we have

$$\varepsilon(\omega,\theta) = \frac{g}{\omega^2}\frac{\lambda}{L}\frac{1}{\sqrt{1-(\lambda/L)^2\sin^2\theta}}. \qquad (23.10)$$

Note that $\varepsilon > 0$. As ε depends on the relationship $\lambda/L \approx h/2L$, we see that $\varepsilon \to 0$ for $L \gg h$.

We will use point O to calculate torques. The torque due to the centrifugal force on an infinitesimal element dx of the bar at a distance x from its center of mass is

$$d\tau_c = (x+\varepsilon)\cos\theta\,\mathrm{dF}_c = (x+\varepsilon)^2\cos\theta\,dm\,\omega^2\sin\theta. \qquad (23.11)$$

Here, ω is the angular velocity of the bar and $dm = (m/h)dx$ is the mass of the infinitesimal element of the bar. As $\lambda \approx h/2$, the resulting centrifugal torque will be

$$\tau_c \approx \frac{m}{h}\cos\theta\sin\theta\omega^2\int_{-h/2}^{h/2}(x+\varepsilon)^2dx = \frac{m}{12}h^2\omega^2\cos\theta\sin\theta[1+12\varepsilon^2/h^2]. \qquad (23.12)$$

The torque due to the weight of the bar and the tension of the thread, which tends to bring the bar to its vertical position, is

$$\tau_w = mg\lambda\sin\theta + mg\frac{\sin\beta}{\cos\beta}\lambda\cos\theta - mg\varepsilon\sin\theta \qquad (23.13)$$

or also,

$$\tau_w = mg(\lambda-\varepsilon)\sin\theta\left[1\frac{\lambda}{(\lambda-\varepsilon)}\frac{\lambda}{L}\frac{\cos\theta}{\sqrt{1-(\varepsilon/L)^2\sin^2\theta}}\right]. \qquad (23.14)$$

At equilibrium, $\tau_w = \tau_c$, from which

$$\omega^2 = \frac{6g}{h}\left[\frac{1}{\cos\theta} + \frac{\lambda}{(\lambda-\varepsilon)}\frac{\lambda}{L}\frac{1}{\sqrt{1-(\lambda/L)^2\sin^2\theta}}\right] \times \frac{2(\lambda-\varepsilon)/h}{(1+12\varepsilon^2/h^2)} \quad \text{for}\, \omega \geq \omega_{\text{crit}}, \tag{23.15}$$

with

$$\omega_{\text{crit}}^2 = \frac{6g}{h}\left[1 + \frac{h}{2L}\frac{1}{(1-2\delta/h)}\left(1 - \frac{2\delta}{h} + \frac{2\varepsilon_c}{h}\right)\right]\frac{(1-2\delta/h)}{(1+12\varepsilon_c^2/h^2)}. \tag{23.16}$$

As $\lambda - \varepsilon = h/2 - \delta$, expanding the previous expression to first order in δ/h and ε/h, we get

$$\omega_{\text{crit}}^2 \approx \frac{6g}{h}\left[1 - \frac{2\delta}{h} + \frac{h}{L}\left(1 - \frac{4\delta}{h} + \frac{4\varepsilon_c}{h}\right)\right] \tag{23.17}$$

or also

$$\omega_{\text{crit}}^2 \approx \omega_{c0}^2\left[1 + \frac{h}{2L} - \frac{2\delta}{h} - \frac{2\delta}{L} + \frac{2\varepsilon_c}{L}\right], \tag{23.18}$$

with $\varepsilon_c \equiv \varepsilon(\omega_{\text{crit}}, \theta = 0) \approx (g/\omega_{c0}^2)\lambda/L \approx (1/12)(h^2/L)$, which is the value of ε, Eq. (23.10), at critical frequency ω_{crit}. As in general ε_c and δ are small compared to L and h, Eq. (23.18) can be written as follows:

$$\omega_{\text{crit}}^2 \approx \omega_{c0}^2\left[1 + \frac{h}{2L} - \frac{2\delta}{h} - \frac{2\delta}{L} + \frac{h^2}{6L}^2\right], \tag{23.19}$$

where ω_{c0} is given by Eq. (23.6). Therefore, the effect of taking into account a finite length for L results in the value of ω_{crit} increasing with respect to the value obtained with the naive model, Eq. (23.6). However, the fact that the contact point of the thread with the bar does not match the end of the bar ($\delta > 0$) leads to ω_{crit} decrease with respect to the case $\delta = 0$.

References

[1] Wikipedia, "Bifurcation theory," 2021. Available: http://en.wikipedia.org/wiki/Bifurcation_theory.

[2] F. Moisy, "Supercritical bifurcation of a spinning hoop," *Am. J. Phys.*, vol. 71, no. 10, pp. 999–1004, 2003.
[3] H. Goldstein, C. Poole and J. Safko, *Classical Mechanics*, 3rd Ed., Boston, MA: Addison-Wesley, 2001.
[4] S. T. Thornton and J. B. Marion, *Classical Dynamics of Particles and Systems*, 5th Ed., Belmont, CA: Thomson Books/Cole, 2012.
[5] A. Sommerfeld, *Lectures on Theoretical Physics: Mechanics*, Vol. I, New York: Academic Press, 1952.
[6] C. M. Sendra, F. Della Pica and S. Gil, "Rotational stability, an amusing physical paradox," *Eur. J. Phys.*, vol. 28, pp. 845–857, 2007.
[7] Veritasium, "The Dzhanibekov effect or Tennis Racket theorem," 2021. Available: https://www.youtube.com/watch?v=1VPfZ_XzisU.
[8] P. Müller, A. Sack and T. Pöschel, "Misconceptions about gyroscopic stabilization," *Am. J. Phys.*, vol. 88, p. 175, 2020.
[9] M. Danny, "Why does a spinning egg rise?" *Eur. J. Phys.*, vol. 39, p. 025002, 2018.
[10] European Space Agency, "Gyroscopes in space," 2021. Available: https://www.esa.int/ESA_Multimedia/Videos/2016/03/Gyroscopes_in_space.
[11] Vernier Software & Technology, "Photogates," 2021. Available: https://www.vernier.com/product/photogate/.
[12] Vernier Software, "Logger pro and Vernier video analysis," 2021. Available: www.vernier.com.

MODULE III

EXPERIMENTS OF ELECTROMAGNETISM

Chapter 24

SIMPLE CURRENT CIRCUITS — OHM'S LAW

This chapter investigates the relation between current and voltage applied to various electrical devices, metallic and carbon film resistors, electric light bulbs, diodes, etc., and analyzes in which cases these devices follow Ohm's law. It also looks at different methods of measuring electrical resistance and the use of ammeters, voltmeters, and ohmmeters. It shows how to determine internal resistances of voltmeters and ammeters. Finally, it suggests an experimental study to measure the internal resistance of a voltage source.

Goals

- ✓ Voltage and current measurement
- ✓ Dependence of current on voltage
- ✓ Ohm's law
- ✓ Internal resistance
- ✓ Equivalent circuit of a source
- ✓ Thévenin and Norton's theorems

24.1 Current dependence on voltage — Ohm's Law

For an electric current I to flow through a material, an electric field (a potential difference V) needs to be applied between two points of the material and there must be *free* charges capable of moving in the material (*free charges* can be electrons or ions). This occurs in metals, semiconductors, electrolytic solutions, etc.

By varying the applied voltage V, the current I will generally also change. This variation depends on the type of material being used.

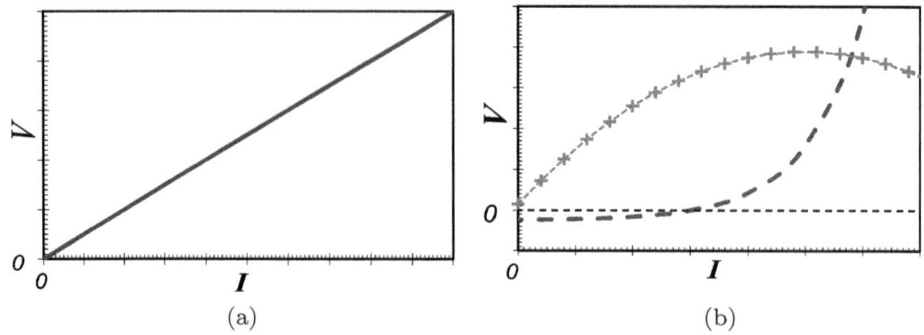

Figure 24.1 (a) Example of an ohmic system. (b) Non-ohmic system.

If the relationship between V and I is linear, as illustrated in Figure 24.1(a), that is, if we can write [1–3]

$$V = R \cdot I \quad \text{with} \quad R = \text{constant}, \tag{24.1}$$

we say that we are in the presence of an *ohmic* material [4] (or element) and the relationship (24.1) that describes this behavior is known as *Ohm's law* [1–3].

The constant R is the resistance of the sample, and its unit is Ohms (Ω =Volt/Ampere).

If the relation between V and I is not linear, a value $R = V/I$ can still be defined, but in this case, R will vary with V and I and the element under study will not obey Ohm's law. Figure 24.1(b) illustrates the behavior of non-ohmic systems.

It is important to note that the relationship (24.1) *is not universal*, i.e., it does not apply to all materials but rather to a restricted set of materials, mainly metals, semiconductors, and some electrolytes under special conditions (e.g., constant temperature). The expression (24.1) can be considered the definition of an ohmic material or element. It is a phenomenological relationship, similar to Hooke's law.

For the case of a cylindrical conductive sample, of constant cross-section A and length l, as shown in Figure 24.2, the resistance of the sample can be written as follows [1–3]:

$$R = \rho \frac{l}{A}, \tag{24.2}$$

where ρ is an intrinsic property of the material called *resistivity*.

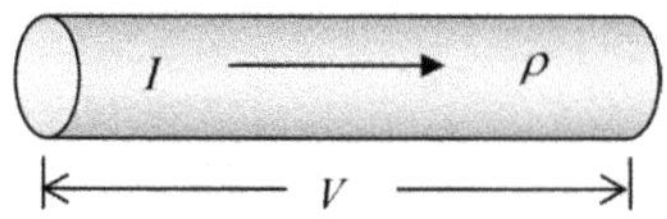

Figure 24.2 Metal wire of length l, cross-section A, and resistivity ρ.

When a current flows through a resistor (ohmic element), heat is generated by the *Joule effect*, and the power dissipated is [1–3]

$$P = I \cdot V = \frac{V^2}{R} = I^2 R. \tag{24.3}$$

When designing a circuit, it is important to ensure that each element used can effectively dissipate the heat generated by the Joule effect. Otherwise, an accident may occur that could potentially damage the experimenter and the circuit and equipment in use. A general recommendation, when designing an electrical circuit, is to estimate the currents that will pass through each of its branches, to calculate the dissipated powers using Eq. (24.3), and to be sure that all elements used can effectively dissipate these powers. If this condition is not met, some element will burn or suffer irreversible damage.

Instruments that measure voltages are generally called *voltmeters*; those which measure small ranges of voltages are called millivoltmeters, microvoltmeters, etc. Instruments that measure currents are called *ammeters* (there are also milliammeters, microammeters, etc.) and those that measure resistance are called *ohmmeters*. Currently, *multimeters*, which are instruments that can measure currents, voltages, resistances, and frequencies, are very common.

Other very useful elements in laboratories are variable resistors, for example, *potentiometers*, *rheostats*, and standard resistors or *resistance boxes*. Figure 24.3 illustrates these elements.

24.2 Construction of a voltage divider

Variable voltage sources are needed in many practical applications. These sources are common devices in almost all current laboratories and are available in a wide range of models suitable to the most varied requirements [5]. However, it is very useful to be able to build a variable voltage source from a fixed voltage source.

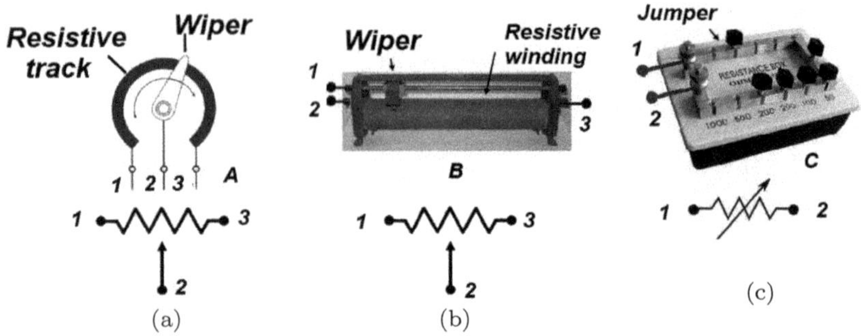

Figure 24.3 (a) Potentiometer, (b) rheostat, and (c) resistance box. Symbols commonly used to represent these elements are shown at the bottom.

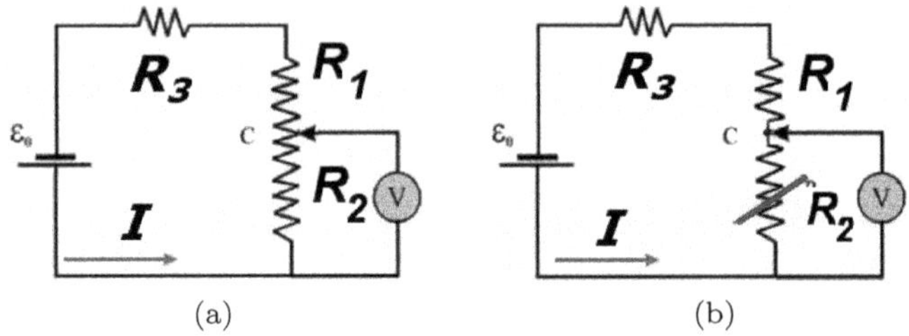

Figure 24.4 Two possible versions of a resistive voltage divider. ε_0 is the constant voltage source. On the left, R_1 and R_2 are parts of the same rheostat or potentiometer. The total resistance of the rheostat is $R_{\text{rheo}} = R_1 + R_2$. Point C can be moved continuously to define the values of R_1 and R_2. The resistance R_0 is a current-limiting resistor. On the right is a similar circuit, with a variable resistor R_2 (resistor box) and a fixed resistor R_1.

Two possible setups are illustrated in Figure 24.4, based on a resistive voltage divider, constructed with a rheostat, variable resistor, or potentiometer. The *constant voltage* source (ε_0) can be, for example, a 9 V battery. R_1 and R_2 are parts of the same rheostat. The moving point C (wiper) divides the rheostat and defines the values of R_1 and R_2 through the relation $R_{\text{rheo}} = R_1 + R_2$. The resistance R_0 is a resistor that limits (the *limiting resistor*) the intensity of the current in the circuit (when R_0 is not present, it is replaced by the internal resistance of the constant voltage source), and the *variable voltage* output is V.

As noted above, before connecting a resistor to a circuit, it is necessary to verify whether it will be able to dissipate the power generated by the current flowing through it. If a limiting resistor is not used ($R_0 = 0\Omega$), the rheostat must be capable of dissipating power:

$$P = \varepsilon_0^2/(R_1 + R_2). \tag{24.4}$$

If, for example, $R_{\text{rheo}} = R_1 + R_2 \approx 100\Omega$ and the source voltage is $\varepsilon_0 \approx 10\,\text{V}$, then $P \approx 1\,\text{W}$. To determine whether the potentiometer or rheostat in use can effectively dissipate this power, it is necessary to check their specifications. If a suitable value of R_0, a limiting resistor, is used, it can help to decrease the current in the circuit and consequently decrease the powers that the different elements have to dissipate.

Using Kirchhoff's law of meshes [1–3], it is easy to show that if we include a limiting resistor in the circuit ($R_0 > 0$), the voltage measured by the voltmeter will be

$$V = \frac{R_2}{R_0 + R_1 + R_2} \cdot \varepsilon_0 = \frac{\varepsilon_0}{(R_0 + R_{\text{rheo}})} \cdot R_2, \tag{24.5}$$

and the power dissipated in the rheostat will be

$$P_{\text{rheo}} = \frac{\varepsilon_0^2}{(R_0 + R_{\text{rheo}})^2} \cdot R_{\text{rheo}}. \tag{24.6}$$

Preliminary exercise: Using a circuit like the one shown in Figure 24.4, verify that the voltage measured by the voltmeter actually changes by varying the position of the wiper C or by varying the value of R_2.

24.2.1 *Project 67. Determining the voltage–current characteristics of a metallic conductor — Ohm's law*

Recommended basic equipment: Two multimeters (or a voltmeter and an ammeter). A source of direct voltage or battery of 5–10 V. A variable resistor of approximately 100 Ω (rheostat of 100 Ω and 1 W). Some commercial resistors (metal or carbon film) of approximately 50 Ω and 2 W.

The objective of this experiment is to study the voltage–current characteristic (curve V–I) of a commercial metal (or carbon film) resistor R. To do so, we propose to investigate the dependence of the current I_R that passes through the resistor with the voltage difference V between its terminals, using ammeters and voltmeters to measure the corresponding magnitudes. It is proposed to use a circuit like the one shown in Figure 24.5. A commercial resistor R, with value between 50 and 300 Ω that can dissipate at least 2 W, can be used. With the actual values of ε_0 and R that you will use, estimate the power generated on R and make sure that the resistor can dissipate this power. The variable voltage source can be a voltage divider like the one described above (Figure 24.4) or a variable voltage source, between 0 V and 10 V.

In the circuit shown in Figure 24.5, the current measured by the ammeter, I_t, is not exactly the current I_R that passes through resistance R. However, as the internal resistances of voltmeters are usually very high, generally greater than 1MΩ, the value of i_v is very small, if R is much less than 1MΩ. Thus, $I_R \approx I_t$. If the condition $R \ll r_v$ is not met, the circuit suggested in Figure 24.5 must be changed. This case and the methods for determining the internal resistance of voltmeters and ammeters are discussed in Annex A.

Suggested tasks:

✓ Using a circuit like the one illustrated in Figure 24.5, vary the applied voltage between 0 and 10 V. For each value of the source voltage, record the values of V and I_t. Plot V as a function of I_t.

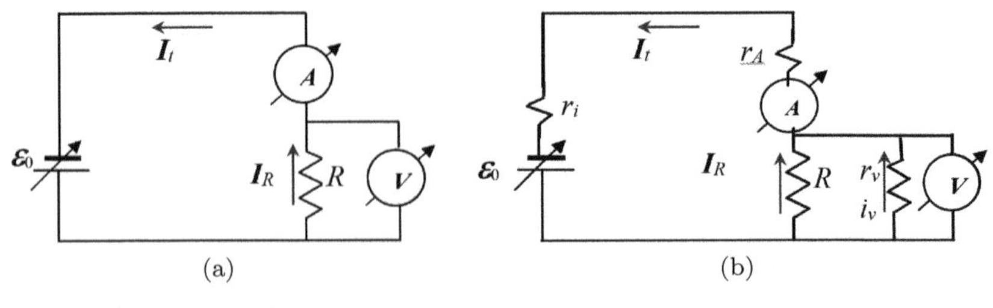

Figure 24.5 (a) Basic circuit for measuring voltage difference, V, and current, I_t. The current through resistor R is I_R. If the internal resistance of the voltmeter, r_v, is much greater than R, then $I_R \approx I_t$. The voltage provided by the voltage source is assumed to be variable. (b) Equivalent diagram of the circuit on the left.

✓ If the relation between V and I_t is linear, we say that R obeys Ohm's law [1–3]:

$$V = R \cdot I_t, \tag{24.7}$$

and the resistance is given by the slope of the graph $V(I_t)$. Find the value of resistance R using the least squares method.

✓ Determine the best value of R and estimate its uncertainty ΔR, see Chapter 7.

✓ The value of R can also be found by using a multimeter in ohmmeter mode. This technique for determining R is called *two-point method* which is used to determine the value of R. Compare the R values found using each method.

Many current multimeters can be used to directly measure the value of the resistance of an element in ohmmeter mode. These instruments usually have a current source (they generate constant current, independently of the connected load) and measure the voltage drop for that current value. However, if the element is not ohmic (I does not vary linearly with V), these instruments *do not allow* direct detection of this effect. The V–I feature provides more detailed and reliable information on an unknown element. When a component is suspected to be non-ohmic, the V–I curve can be obtained using a circuit like the one in Figure 24.5.

24.2.2 *Project 68. Resistors in series and in parallel — Using an ohmmeter*

Recommended basic equipment: A multimeter capable of measuring resistance (ohmmeter). A few commercially available resistors of approximately 1 kΩ, 4.7 kΩ, and 10 kΩ.

Using two different resistors, R_1 and R_2, of the same order of magnitude, determine the resistance value of each combination: (a) in series and (b) in parallel. Remember that for the series combination, we expect $R_S = R_1 + R_2$, and for the parallel combination, $R_P = R_1R_2/(R_1 + R_2)$ [1–3]. For this experiment, you could use the *two-point method*, that is, use an ohmmeter or multimeter connected to the ends of the resistor or group of resistors.

Suggested tasks:

✓ First, using an ohmmeter, apply the two-lead or *two-point method* to measure the value of the resistances R_1 and R_2. Estimate the errors.

✓ Connect these resistors in series and in parallel and determine their values with the ohmmeter.

✓ Compare the measured values for the series and parallel configuration and compare these values with the theoretically predicted value.

24.3 V–I curve using an acquisition system connected to a PC

When an element is suspected to be non-ohmic, it is necessary to obtain the curve V–I. An alternative to the circuit shown in Figure 24.5 is to use a data acquisition system connected to a PC, as illustrated in Figure 24.6. The acquisition system measures the values of the voltages V_A and V_B with respect to ground (GND). We implicitly assume that this acquisition system works in *common mode* [5], in other words, it measures the voltages always referred to the same ground (GND). The element whose curve V–I we wish to study is the Z element. The ohmic resistance R_0, of known value ($R_0 \approx 50\Omega$), is used to monitor the current flowing through the

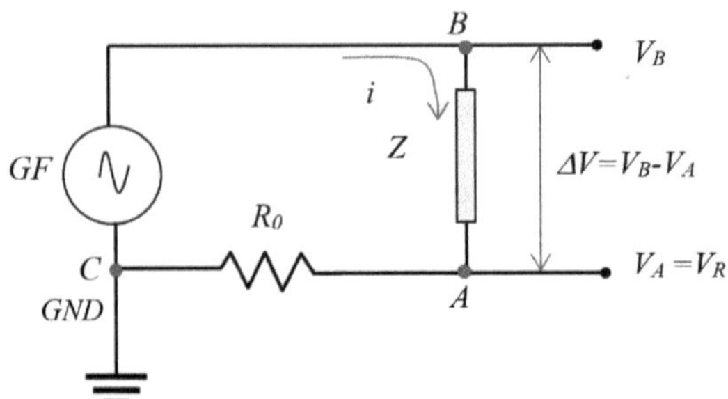

Figure 24.6 Circuit to determine the voltage–current characteristic through a Z element. The ohmic resistance R_0 is of known value, and the voltage drop ($V_R = V_A$) signal is proportional to current, i. The voltage drop across element Z is ΔV. The FG generates a sinusoidal signal and GND is the common ground of the whole circuit.

circuit, since $I = V_A/R_0 = V_R/R_0$. The voltage drop ΔV across Z is measured indirectly from V_B and V_A ($\Delta V = V_B - V_A$). When working in *common mode*, all the grounds of the different channels are the same as the ground of the function generator (FG) or power supply. This common point in this case is designated GND (*ground*). Therefore, we assume that the voltages V_B and V_A are measured.

The function generator (FG) provides a variable voltage, whose frequency, amplitude, and shape can be varied within a wide range. When using an external power source, such as an FG, one has to be careful that the applied voltage applied to the input of the data acquisition system never exceeds the maximum allowed voltage. Always check the maximum allowed value for the system in use. Choose the amplitude of the FG signal so that the current is tolerable for all elements of the circuit (see Figure 24.6). The frequencies to be used must be selected carefully, especially if there are inductive or capacitive elements. For example, if we want to study the curve $I - V$, a frequency of a few Hertz may be adequate. Moreover, the data collection rate of the acquisition system must be greater than the frequency used. As a practical rule, the acquisition rate should be at least 10 times higher than the frequency used. With these precautions, you can proceed to collect data.

24.3.1 *Project 69. Determining the voltage–current characteristics of a resistor and a diode*

Recommended basic equipment: A data acquisition system connected to a PC with two input channels. An alternating voltage source or a function generator. A variable resistor of approximately 50 Ω (rheostat 100 Ω and 1 W). Some ordinary resistors of approximately 50 Ω and 1 W. A common diode of 1 A.

✓ The aim of this experiment is to study the voltage–current characteristics (curve V–I) of a generic electrical element Z. To do so, we propose to use a circuit like the one shown in Figure 24.6. For this experiment, choose a frequency for the FG of about 10 Hz, and set the data acquisition sampling rate to about 100 hz.

The aim is to measure the voltage drop between the points A and B:V_A and V_B. With this information, knowing the value of R_0, we can obtain $i = V_A/R_0$ and $\Delta V = V_B - V_A$.

Suggested tasks:

V–I ratio for a metallic or carbon film resistor

✓ Using a resistor R like the one used in Project 67, plot ΔV as a function of i: the V–I curve.

✓ The circuit is illustrated in Figure 24.6, where the resistor R under investigation replaces the element Z.

✓ If the dependence between ΔV and i is linear, determine the value of R and its uncertainty.

V–I ratio for a diode

✓ The circuit is illustrated in Figure 24.6, where the diode replaces the element Z.

✓ Determine the V–I curve for a silicon diode. Remember that a diode is an element that allows current to flow in only one direction. Do your data support this characteristic of the diode? The diode is an example of a nonlinear element [5].

24.4 ♣ Common mode and differential inputs

Electronic instruments and devices (amplifiers, measuring instruments, etc.) have inputs that can operate in *common mode* or *differential mode*. In *common mode* (which is the usual case for many data acquisition systems), the *ground* is the same for all channels. Consequently, it is not possible to measure simultaneously and independently the voltage drops of two elements that are connected in series. For example, in the circuit shown in Figure 24.6, if we want to measure the voltage drop between the points A and B, $\Delta V = V_{AB}$, we must measure the voltages of A (V_A) with respect to C (GND) and of B (V_B) with respect to C. Then, based on the difference, we can calculate the voltage sought, namely $V_{AB} = V_A - V_B$.

On the contrary, in the case of *differential mode*, [5, 6] the voltage between the two input leads is "floating", meaning that it has no reference to ground. The measurement is taken as the voltage difference between these two wires. An additional benefit of a differential measurement is noise rejection because the noise is added to both wires and can be more easily filtered out than in the common mode configuration. Of course, there are some data acquisition systems that do possess differential inputs [6, 7]. Therefore, with a differential input data acquisition system, the voltage drop across the element Z, V_{AB}, could be measured directly by placing the inputs between A and B. Battery-operated multimeters are a typical example of an instrument with differential inputs. Some PC data acquisition systems can operate in differential mode using a special interface: a differential amplifier [7]. To know whether or not a measuring instrument (data collection system, oscilloscope, multimeter, etc.) has differential inputs, check the specifications.

24.4.1 *Project 70. Do incandescent light bulbs obey Ohm's law?*

Recommended basic equipment: Two multimeters (or a voltmeter and an ammeter). A variable voltage source from 0 to 10 V. A light bulb of about 6–12 V and power of the order of 10 W.

The traditional incandescent light bulb, invented by Thomas Edison around 1880, is now being rapidly replaced by the modern LED lamp, which is more efficient and has a longer lifespan. Nonetheless, incandescent lightbulbs are wonderful artifacts that can readily be used in many physics labs for a number of interesting experiments. As resistors, their temperatures can be varied in a wide range [8]. Incandescent lightbulbs are also simple, low-cost blackbody radiation sources [9–11].

Using an incandescent light bulb of approximately 10 W @ 12 V[a] instead of the element R in the circuit of Figure 24.5, we propose to study the voltage–current relationship for this device.

[a]This notation, 10 W @ 12 V, means that when the lamp is connected to 12 V, it dissipates a power of 10 W, that is, its nominal operating resistance is $(12\text{V})^2/10\,\text{W} \approx 15\Omega$.

Suggested tasks:

✓ First, choose an amplitude for the input signal of the power supply so that the light bulb does not glow. With this setting, study the variation of ΔV as a function of i. Plot ΔV versus i. What relationship do you find between V and i while the filament shows no incandescence?

✓ Next, increase the amplitude of the input signal of the power supply so that the light bulb does glow. Under this condition, plot ΔV as a function of i. What relationship do you find between V and I, from when $i = 0$ to when the filament is incandescent?

✓ If you look at the $I - V$ curve for an incandescent light bulb, you will surely observe nonlinear behavior. This is because an important parameter in the problem — the temperature of the filament — varies significantly [8, 10]. This time, the nonlinearity is a consequence of the fact that when the current varies through the filament, its temperature varies considerably, and consequently, the resistance also changes.

In this experiment, it is important to consider that the temperature of the filament varies significantly. The temperature of an incandescent filament ranges from 1000°C to 2700°C, although at about 500°C it already begins to have a reddish glow. If the filament is made of tungsten, its coefficient of variation of resistance with temperatures [1, 5] is $\alpha \approx 0.0045\text{K}^{-1}$, so for a temperature variation $\Delta T \approx 222\,\text{K}$, the resistance of the filament is expected to double. In other words, in this experiment, the resistance is expected to vary significantly, that is, it will have very different temperatures, as I varies.

24.4.2 *Project 71. Determining the voltage–current characteristics of an incandescent light bulb using a data acquisition system*

Recommended basic equipment: A data acquisition system connected to a PC with two input channels. An alternating voltage source or a function generator. A variable resistance of approximately 50 Ω (rheostat 50 Ω and 1 W). An incandescent light bulb of about 5 W @ 12 V.

The aim of this experiment is to study the voltage–current characteristics (V–I curve) of an incandescent light bulb, in a similar way as in the previous project, but using a data acquisition system (DAS) to enable you to see how the light bulb behaves, by varying the frequency of the functions generated. This provides some information about the thermal inertia of the light bulb. For this, we propose to use a circuit like the one shown in Figure 24.6.

Suggested tasks:

V-*I* ratio for an incandescent light bulb

✓ For this first part of the experiment, choose a relatively low frequency for the FG — say 1–3 Hz — and the sampling rate of the DAS of about 50 or 100 Hz. You want to measure the voltage drop between the points A and B, *i.e.*, V_A and V_B. With this information, knowing the value of R_0, you can find $i = V_A/R_0$ and $\Delta V = V_B - V_A$.

✓ Perform this study for an incandescent light bulb, using the circuit shown in Figure 24.6. Construct a plot of ΔV versus i, making sure that the filament does not reach incandescence. By varying the amplitude of the FG voltage, it is possible to control the maximum current that passes through the filament of the light bulb.

✓ Using the same frequency, study the V–I curve for the incandescent light bulb but increase the amplitude of the FG so that the filament does glow now. Compare the V–I curves of these two exercises. Do you notice any difference? Can you explain the difference qualitatively?

✓ Next, increase the frequency of the FG to about $f \approx 70$ Hz, increase the sampling rate of the DAS to about 700 Hz, and repeat the same study just made at low frequencies. Determine the V–I curve for this case. Is the V–I curve the same as before?

✓ What differences do you see in the graphs of low and high frequencies? How can you explain your results? Remember that at high frequencies, thermal inertia makes the temperature of the light bulb remain constant. At low frequencies, if the filament cooling time is comparable to the signal period, its temperature varies continuously throughout the cycle [11].

24.5 Equivalent circuits of a source — Thévenin and Norton's theorems

It is often useful to model a real complex circuit using an equivalent fictitious one, which, seen from its terminals, behaves similarly to the real one. Thévenin and Norton's theorems [5, 12] enable us to find equivalent circuits of voltage and current sources, respectively, which combined with a resistor (or impedance) can model the behavior of complex circuits. In this section, we propose to conduct an experimental study of these theorems applied to voltage sources and function generators.

Thévenin's theorem states that a real voltage source can be modeled by an *ideal voltage source* (without internal resistance) and a resistance (or impedance) in series with it. Similarly, *Norton's theorem* states that any current source can be modeled by means of an *ideal current source* and an impedance in parallel. These models for real sources are shown schematically in Fig. 24.7. The objectives of these experiments are to study the validity of these theorems for some sources that are available in the laboratory and to determine the parameters of the model, namely source voltage (*emf*ε) and its internal impedance (r_i). The power source will also be studied. The sources can be Direct Current (DC) or Alternating Current (AC). These theorems apply equally well to power sources, function generators, or batteries.

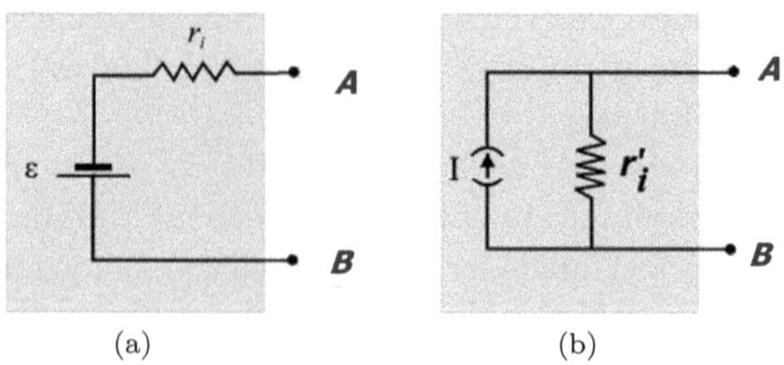

Figure 24.7 Equivalent circuits for a real voltage source. (a) Thévenin equivalent circuit; r_i is its equivalent resistance. (b) Norton equivalent circuit.

24.5.1 *Project 72. Model of a source*

> **Recommended basic equipment**: A DC voltage source or 9 V battery, two multimeters (or a voltmeter and an ammeter). A variable resistance of approximately 200 Ω (rheostat 200 Ω and 1 W).

Build a circuit like the one depicted schematically in Fig. 24.8. Before connecting the source, make sure that the resistance R can dissipate the electrical power that will be generated when applying the maximum voltage to be used. To do this, estimate the maximum current that will pass through it and the maximum power that will be dissipated. The worst case corresponds to $r_i = 0$. Check the specifications of the resistors and the source to make sure that they can dissipate this power. In general, these types of precautions must be considered every time you build a circuit.

Under the hypothesis that Thévenin's theorem is valid, applying Kirchhoff's laws [1] to the circuit in Fig. 24.8, we expect the relationship between V (voltage measured by the voltmeter between A and B) and I (current measured by the ammeter) to be

$$V = \varepsilon - r_i \cdot i, \tag{24.8}$$

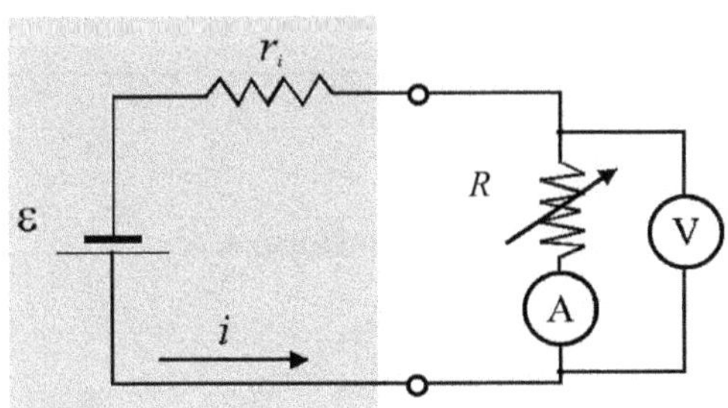

Figure 24.8 Circuit to determine internal resistance r_i from a voltage source. It is advisable for the ammeter to be connected as shown in the figure so that its internal resistance, r_A, does not affect the measurement of r_i.

where r_i is the internal resistance of the source (momentarily unknown), while R is a variable resistance or rheostat that can be varied, causing i and V to vary.

Thus, if the voltage source obeys Thévenin's theorem, the relationship of V with i will be linear, as R changes, and the slope of the graph is the internal resistance of the source.

In this project, we assume that the ammeter is close to *ideal* (negligible internal resistance r_A or $r_A \ll R$) and that the voltmeter is close to *ideal* (i.e., with infinite internal resistance r_V *or* $r_V \gg R$).

Suggested tasks:

✓ Using the circuit shown in Figure 24.8, vary the value of R and construct a table with the measured values V and i.

✓ Plot the value of voltage V, measured by the voltmeter, as a function of current i (as in Figure 24.9).

✓ Using the measured values of V as a function of i, discuss the validity of Thévenin's theorem, Eq. (24.8), to describe your experimental results. If possible, determine the values of r_i and ε for the voltage source used.

✓ Compare the determination of ε with a measurement of the source emf made with a voltmeter when the source is not connected to the external circuit ("open circuit" measurement, i.e., removing R in the circuit shown in Figure 24.8).

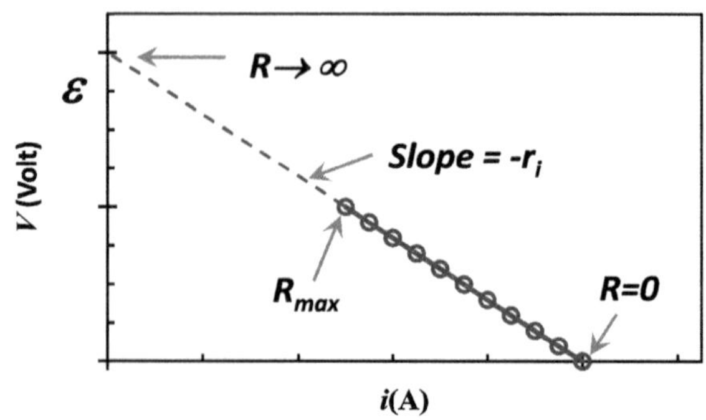

Figure 24.9 Schematic representation of V versus i for the circuit shown in Figure 24.7

24.6 Annex A: ♣ Internal resistance of voltmeters and ammeters

♣ Systematic error introduced by voltmeters: Look again at the circuit in Figure 24.5. V is the value of the voltage measured by the voltmeter and I_t is the value measured by the ammeter. The quotient V/I_t is not necessarily equal to the resistance R, since the current that passes through the resistance is I_R and in general differs from I_t. The value of the resistance we wish to detect is $R = V/I_R$. The problem is that we do not know I_R.

From Kirchhoff's law of nodes [1–13] (conservation of charge), we have

$$I_t = I_R + i_v, \tag{24.9}$$

where i_v is the current flowing through the voltmeter, whose resistance we call r_V. From Ohm's law, we have

$$R = \frac{V}{I_R} = \frac{V}{I_t - i_v} = \frac{V}{I_t - V/r_v} = \frac{V}{I_t} \cdot \frac{1}{(1 - V/I_t r_v)}. \tag{24.10}$$

If the condition $R \ll r_v$ is satisfied, then $I_t \gg i_v$ and $V/I_t \approx R$. Doing a Taylor series expansion, in the first approximation, we have

$$R \approx \frac{V}{I_t}\left(1 + \frac{R}{r_v}\right) \quad \text{and also} \quad \frac{\Delta R}{R} \approx \frac{R}{r_V}. \tag{24.11}$$

Here $\Delta R/R$ designates the relative systematic error in R, which results from the finite value of r_v. Thus, as indicated above, the experimental arrangement illustrated in Figure 24.5 is adequate to measure the value of R as long as $R \ll r_v$. If we call the quotient between the voltage and the current actually measured *equivalent resistance*, R_{eq}, that is,

$$R_{\text{eq}} = \frac{V}{I_t}, \tag{24.12}$$

then, according to (24.10), we have

$$\frac{1}{R} = \frac{1}{R_{\text{eq}}} - \frac{1}{r_v} \quad \text{or} \quad R \approx R_{\text{eq}}(1 + \frac{R_{\text{eq}}}{r_v}). \tag{24.13}$$

Typically, $r_v \geq 1\ \text{M}\Omega$, but it can also be measured directly. To do this, we could use a *two-point* measurement (or *two-lead method*) with an ohmmeter. To do so, connect the ohmmeter to the terminals of the voltmeter, making sure that the voltmeter is on the same scale that is going to be used for voltage measurement. The value indicated by the ohmmeter is the value of r_v.

Note that a finite value of r_v introduces a systematic error in the measurement of R, since it always affects the determination of R in the same sense. In this case, since it is a systematic error, that can be corrected, the error signs in Eqs. (24.11) and (24.13) are relevant and must be considered.

♣ **Systematic error introduced by ammeters:** If the condition r_v is not fulfilled, the circuit in Figure 24.5 is not suitable for measuring R. In this case, an arrangement such as shown in Figure 24.9 must be considered.

In the circuit in Figure 24.10, V is the value of the voltage measured by the voltmeter and I_R is the current measured by the ammeter. In this case, we again define the *equivalent resistance as* $R^*_{\text{eq}} = V/I_R$, which is not equal to $R = V_R/I_R$, which we wish to measure. From the circuit analysis, we have

$$\boldsymbol{V = V_R + I_R \cdot r_A}, \tag{24.14}$$

where r_A is the internal resistance of the ammeter. The value of R in this case is

$$R = \frac{V_R}{I_R} = \frac{V}{I_R} - r_A = R^*_{eq} - r_A. \tag{24.15}$$

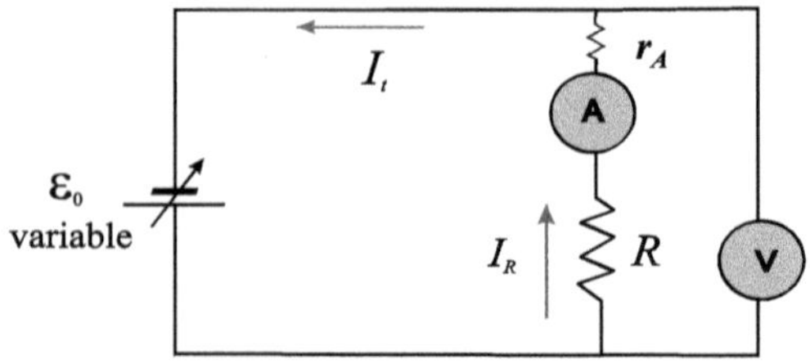

Figure 24.10 Alternative circuit for measuring voltage, V, and current, I, through a resistor, R.

Thus, if $R \gg r_A$ and $R \approx V/I_R$, the relative systematic error in R will be

$$\frac{\Delta R}{R} \approx -\frac{r_A}{R}. \tag{24.16}$$

The circuit in Figure 24.9 is suitable for measuring a resistance R as long as the internal resistance of the ammeter (r_A) is small compared to $R(r_A \ll R)$. However, if this condition is not met, it is possible to correct the systematic error that introduces a finite value of r_A using Eq. (24.15).

By combining the measurements of voltages and currents of the circuits in Figure 24.5 and Figure 24.10, it is possible to minimize the systematic errors introduced by the non-ideal characteristics of the ammeters and voltmeters. However, as we will see later, when the resistances are small, more specifically, less than about 10 Ohm, it is preferable to use other experimental techniques to measure their value, for example, the *four-lead method*, which is discussed later in this book.

♣ **Determining the internal resistance of ammeters and voltmeters** As stated above, it is sometimes necessary to know the values of the internal resistances of these devices. As indicated above, the simplest way is to use an ohmmeter to measure these resistances (*two-lead method*). There are several other possibilities, such as the circuits illustrated in Figure 24.11. In circuit 24.10(a), R_x is an external variable resistor or potentiometer with the maximum value in the range of 1–100 MΩ. As in general $r_A \ll r_v$, in the first

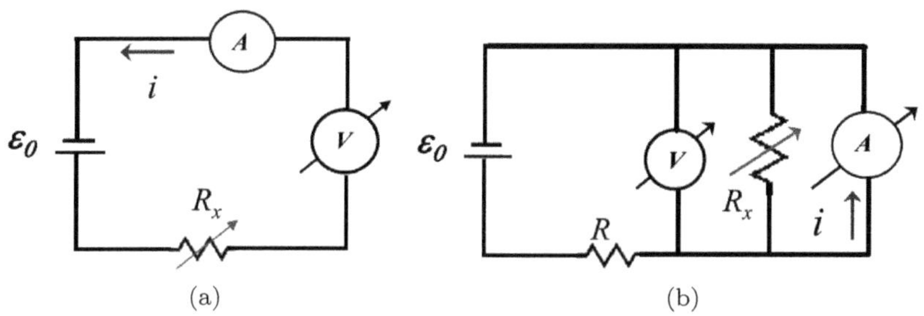

Figure 24.11 Circuits to determine the internal resistance of voltmeters and ammeters.

instance, we set the value of the variable resistance to $R_x = 0$. In these conditions, we have

$$r_v \approx \frac{V}{i}, \tag{24.17}$$

where V is the value measured by the voltmeter and i the value measured by a milliammeter. Next, we increase the value of R_x until the current i drops to half its original value. In these conditions, $R_x = r_v$. This value should coincide with the one originally found in Eq. (24.17) and serves to verify the consistency of our method.

Circuit 24.11(b) can be used to measure the internal resistance of an ammeter, r_A. This time it is advisable to use a potentiometer R_x that varies between 0 and 10 Ω, approximately. We start by doing $R_x = \infty$ (open circuit, i.e., the potentiometer R_x is disconnected). The limiting resistance R is chosen so that the current through the ammeter does not exceed the maximum value allowed on its scale. When $R_x = \infty$, then $i = V/r_A$, where V is the value of the voltage measured by the voltmeter. Then we connect R_x and vary it until the current in the ammeter drops to half its original value, monitoring that the voltage V remains constant during this process. In these conditions, $R_x = r_A$; since in general $r_A \approx 1$, the variable resistance R_x should vary between zero and a few Ohms.

Resistance color code: Most commercial resistors have a *nominal value* indicated by the manufacturer through a label or some code [14]. Its color code is very useful for quickly knowing its approximate value, therefore it is useful to learn these color codes by heart. The actual capacity may generally differ from this value by a few percent, so if the actual value of a resistor is important, it must be measured carefully.

References

[1] R. Halliday, D. Resnick and M. Krane, *Fundamentals of Physics*, NY: John Wiley & Sons, Inc., 2002.

[2] F. Sears, M. Zemansky, H. Young and R. Freedman, *College Physics*, NY: Pearson Ed., 2009.

[3] E. M. Purcell and D. J. Morin, *Electricity and Magnetism, Berkeley Physics Course*, 3rd Ed., NY: Cambridge University Press, 2014, p. 839.
[4] Wikipedia, "Electrical elements," 2021. Available: https://en.wikipedia.org/wiki/Electrical_element.
[5] P. Horowitz and W. Hill, *The Art of Electronics*, 3rd Ed., NY: Cambridge University Press, 2015, p. 1220.
[6] Wikipedia, "Differential amplifier," 2021. Available: https://en.wikipedia.org/wiki/Differential_amplifier.
[7] Vernier, "Instrumentation amplifier," Vernier Software, 2021. Available: https://www.vernier.com/product/instrumentation-amplifier/.
[8] H. S. Leff, "Illuminating physics with light bulbs," *Phys. Teach*, vol. 28, pp. 30–35, 1990.
[9] J. Dryzek and K. Ruebenbauer, "Planck's constant determination from blackbody radiation," *Am.J.Phys.*, vol. 60, p. 251, 1992.
[10] G. Brizuela and A. Juan, "Planck's constant determination using a light bulb," *Am. J. Phys.*, vol. 64, pp. 819–821, 1996.
[11] G. Torzo, M. D'Anna and B. Pecori, "Playing with a bulb lamp: RTL measurements and modelling," *Phys. Educ.*, vol. 51, no. 5, p. 055004, 2016.
[12] Wikipedia, "Thévenin's theorem," 2021. Available: https://en.wikipedia.org/wiki/Th%C3%A9venin%27s_theorem.
[13] R. Feynman, *et al.*, *The Feynman Lectures on Physics, boxed set: The New Millennium Edition*, Vol.1, Basic Books, 2011.
[14] University of south Dakota, "Resistor colour codes," 2021. Available: http://apps.usd.edu/coglab/psyc770/resistors/r4beginner.html.
[15] B. Denardo, "Temperature of a lightbulb filament," *The Phys. Teach.*, vol. 40, p. 101, 2002.

Chapter 25

RESISTOR NETWORKS

This chapter studies resistor networks in one and two dimensions. Different methods for estimating the equivalent resistances of a circuit are discussed, for example, the superposition method.

Goals

✓ Resistance networks
✓ Networks in one and two dimensions

25.1 Resistor arrays

It is often useful to consider systems that involve large numbers of resistances in various geometries, sometimes of infinite dimension. There are many practical applications that use this type of network, for example, geophysical exploration and water and oil exploration. In engineering, it is also very useful to know how networks operate, e.g., the electrical, water, or gas distribution networks in a city. The objective here is to present a series of simple network models that can be studied in the laboratory.

25.2 Resistor networks in 1D — Fibonacci ratio

A simple, interesting system to analyze is the so-called *ladder network* illustrated in Figure 25.1. This network consists of n loops, each consisting of a pair of identical resistors of value R_0. To calculate the equivalent resistance of this system, the following recursive procedure can be used [1–3]. We call the equivalent resistance of the first loop R_1, the next R_2, and so on.

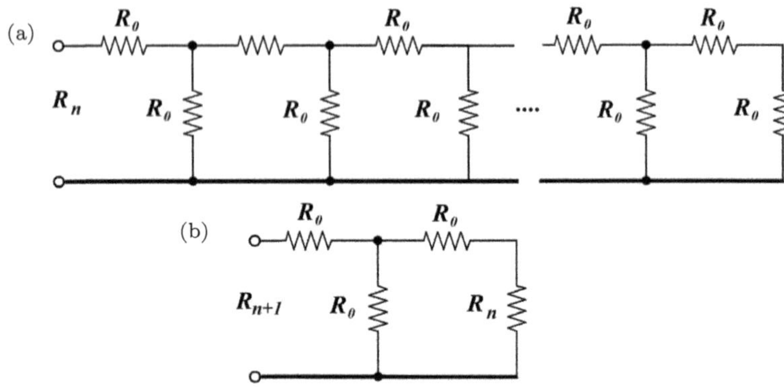

Figure 25.1 Mesh or net in 1D or ladder. (a) 1D resistive network consisting of n loops of pairs of resistors of value R_0. (b) Diagram for determining the R_{n+1} in terms of R_n and R_0.

The resistance of R_{n+1} can be calculated in terms of R_n as shown in Figure 25.1(b), this is,

$$R_{n+1} = R_0 + \frac{R_0 \cdot R_n}{R_0 + R_n}, \quad \text{with} \quad R_1 = 2R_0. \tag{25.1}$$

This recursive expression enables us to calculate the resistance R_n for any value of n. It is very easy to do this calculation in a spreadsheet. In particular, if we apply this relation to an infinite network, $(n \to \infty)$, so $R_{n+1} = R_n \equiv R_\infty$. From Eq. (25.1), we obtain

$$R_\infty = R_0 + \frac{R_0 \cdot R_\infty}{R_0 + R_\infty}, \quad \text{and} \quad R_\infty = R_0 \left(\frac{1+\sqrt{5}}{2} \right) = R_0 \cdot 1,60938. \tag{25.2}$$

The number $\Phi = (1+\sqrt{5})/2 = 1,60938\cdots$ is known as the *golden number* or *golden ratio*[4]. This number is the asymptotic value to which the ratio of the consecutive terms a_n/a_{n+1} of a *Fibonacci sequence* tends,[a] i.e., if a_n are the terms of a sequence of this type, $a_n/a_{n+1} \to \Phi$ for $n \to \infty$.

[a]A Fibonacci sequence meets the condition $a_{n+1} = a_n + a_{n-1}$, where a_0 is an arbitrary real number.

Table 25.1 Theoretical values of equivalent resistances for a ladder network of n loops.

n	R_n/R_0	$(R_n - R_\infty)/R_0$
1	2	3.82×10^{-1}
2	5/3	4.86×10^{-2}
3	13/8	6.97×10^{-3}
4	34/21	1.01×10^{-3}
5	89/55	1.48×10^{-4}
6	233/144	2.16×10^{-5}
7	610/377	3.15×10^{-6}
∞	Φ	0

Exercise 1: Demonstrate through Eq. (25.1) the values shown in Table 25.1.

25.2.1 *Project 73. 1D resistor networks*

Recommended basic equipment: A multimeter capable of measuring resistance (ohmmeter), preferably $4^1/_2$ or more digits. A network of 1D resistors of $n \geq 10$.

Build a network of resistors as illustrated in Figure 25.1, with approximately $n \geq 10$ loops. One possibility is to use a set of resistors R_0 from 1 kΩ to 10 kΩ. Choose a set in which, as far as possible, the resistors differ from each other by no more than 0.2%. To build the network, you can use a *breadboard*.

Suggested tasks:

✓ Using a multimeter in ohmmeter mode, preferably $4^1/_2$ or more digits, measure the *equivalent resistance* R_n for n varying between 1 and 10 approximately.

✓ On a graph, represent how the relationship R_n/R_0 varies as a function of n. On the same graph, draw the measured values with their errors and the theoretical predictions from Table 25.1.

✓ What can you conclude about the proposed model?

25.2.2 *Project 74. Resistance of graphite lines*

Recommended basic equipment: A multimeter capable of measuring resistance (ohmmeter). A sheet of paper with uniform lines of different widths, made using an inkjet printer or a graphite pencil.

Print or draw several horizontal lines of different widths on the paper, as shown in Figure 25.2, using an inkjet printer or a graphite pencil. It is recommendable for this experiment to use blunt tips for the multimeter, such as a banana plug connector, to avoid perforating the sheet of paper on which the lines are drawn. It is recommendable to draw equidistant dotted lines, separated by a distance d, for reference.

Black pencil leads are often made from a mixture of graphite, clay, and wax, so lines drawn with a regular pencil do not always conduct electricity or their resistance varies erratically. On the other hand, graphite itself and the kind of graphite pencils used for artwork do conduct electricity, so a pure graphite pencil (purchased at an art supply store) or a B6 pencil (very soft lead) can be used to draw well-defined resistance lines. In this experiment, we propose to characterize the resistance of graphite lines as a function of length and then use this technique to construct several types of networks of resistors using graphite lines. To ensure that the graphite strokes are even, it is advisable to use smooth-textured paper and draw the lines using a ruler, taking care to go over the same mark at least

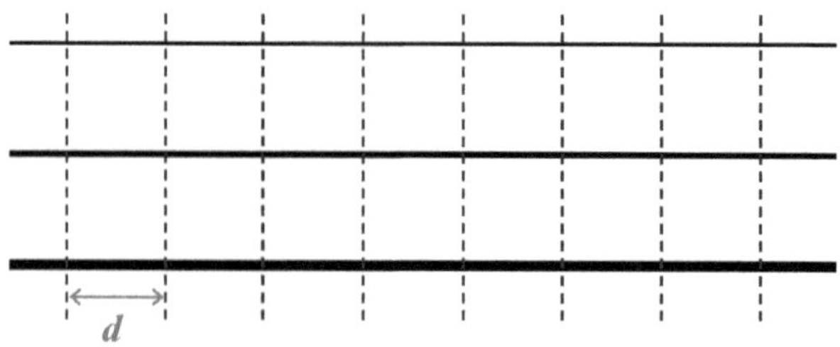

Figure 25.2 Schematic diagram of lines of different widths, printed with an inkjet printer or drawn with a graphite pencil. The equidistant dotted lines, spaced at a known distance d, are used as reference to measure the resistance of each line as a function of its length.

four times until a completely even, black line is formed. With a little practice, you will be able to draw graphite lines with well-defined, uniform resistance along their length.

Another alternative is to use an inkjet printer. The inks used in this type of printer are electrically conductive, although conductivity can vary greatly among brands. To find out if a given printer produces conductive lines that would be useful for this type of experiment, it is best to perform a test with a smooth sheet of paper. Print several lines with different thicknesses, as shown in Figure 25.2, and use a multimeter or ohmmeter to test whether the resistance is measurable with the available instruments.

Suggested tasks:

- ✓ Taking segments of equal length on a line ($d \sim 1$ or $2\,\text{cm}$) but in different positions along the line, make a histogram to characterize the uniformity of the resistances per unit length for the line studied. Estimate the mean value of this resistance per unit length and its dispersion.
- ✓ Over a line of given width, measure the resistance $R(x)$ as a function of the segment length x between the ends of the ohmmeter leads.
- ✓ Plot R as a function of x. If the relationship is linear, estimate the value of the resistance per unit length $R(x)/x$, that is, the slope of this graph and its uncertainty.
- ✓ How does this value compare with the result of the first measurement?

25.3 ♣ 2D resistor networks

An example of a 2D resistor network is schematically illustrated in Figure 25.3. It can easily be constructed with the technique employed in the previous project or using commercial carbon film resistors, hand-soldered into a prototype board or printed circuit board (PCB).

To make a network of $n \times n$ cells, a total of $2n(n+1)$ individual resistors are required. Before proceeding to solder the resistors, it

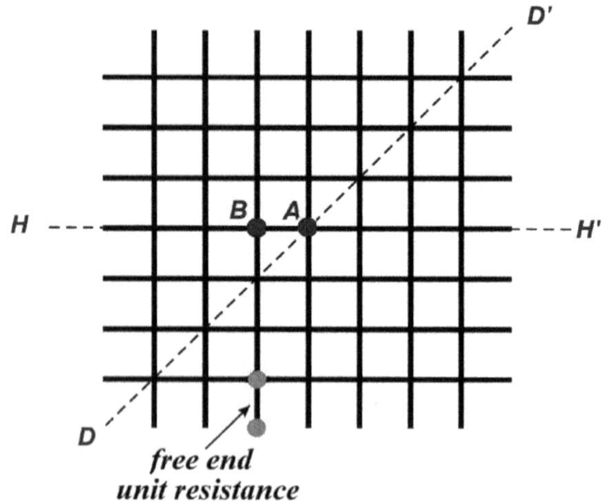

Figure 25.3 Graphite mesh or net. This network can be built using the technique employed in the previous project or using commercial resistors.

is advisable to measure each resistor to be used and select those that are within a predetermined range of variation, for example, 0.5%. One possibility is to use 10 kΩ resistors, since they are easy to measure with good precision with most standard instruments. In other words, to construct this network, from a set of 10 kΩ resistors, select only units that fall within the range 9950 Ω–10050 Ω. Once soldered to a PCB Universal Experiment Matrix Circuit Board [5], the resistors can be attached to a non-conductive base for measurements.

Another possibility is to build the network using graphite lines drawn with a pencil or printed by an inkjet printer. In this case, it is important to verify that the different segments have similar resistance. This can be done by measuring the resistances of the loose ends on the edge of the network, as shown in Figure 25.3. The value of the unit resistance, R_0, can be estimated as the mean value of these resistances using some of the techniques discussed in the previous projects.

A limitation of a network made with graphite pencils or inkjet printers is that the individual resistors often have more dispersion than preselected commercial carbon film resistors do.

25.4 ♣ 2D resistor networks — Theoretical model

Imagine that we have a very large network of resistors, with a geometry like the one shown in Figure 25.3, made with identical resistors whose unit value is R_0. Our objective is to determine the value of the resistance between two adjacent points such as points A and B in Figure 25.3. To do so, we use the *principle of superposition* [1–3, 5].

Imagine that a current I is injected through point A of this circuit, which we assume to be of infinite extension. By symmetry, the current will flow in a similar way through the four branches that start from A. In particular, through the section AB, a current $I/4$ will flow in the direction from A to B. This current will extend in a decreasing way to the limits of the network, at infinity. If in a second mental exercise, we assume that a current I is drawn through point B, this time, A remains isolated. Again, by symmetry, through section AB a current $I/4$ will circulate in the direction from A to B. If we now superimpose these two cases simultaneously, that is, through A we inject a current I and by B we extract the same current, by overlapping [3] the two previous schemes, a current $I/2$ will flow through section AB in the direction from A to B. Therefore, the voltage drop V_{AB} will be $R_0 \cdot I/2$. That is, the equivalent resistance between A and B, $R_{AB} = V_{AB}/I = R_0/2$.

Note, by the way, that this way of estimating the equivalent resistance of this network is much simpler than the infinite combinations of meshes that would have to be performed to calculate this value.

References [2, 3, 6] and [7] provide suitable formalisms to study resistance networks such as the one illustrated in Figure 25.3. In particular, it can be shown that the resistance equivalent to a set of nearby points, for a resistance network such as those illustrated in Figure 25.3, has values that are summarized in Table 25.2 [3, 5, 6].

Table 25.2 Theoretical values of equivalent resistances [3, 7, 8] for a 2D network as illustrated in Fig. 25.3.

Relationship	R_{eq}/R_0	*Points*
Adjacent side points	1-Feb	AB o EF
Diagonal adjacent points	$2/\pi$	*AF o FK*
Non-adjacent lateral points	$2-4/\pi$	*AC o EG*
Non-adjacent diagonal points	$8/3\pi$	*AK o BI*
Chess horse move	$4/\pi-1/2$	*AG o EC*

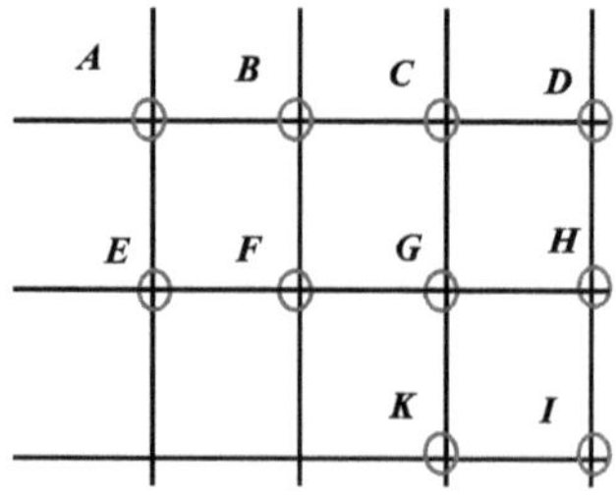

25.4.1 *Project 75. 2D resistor networks*

Recommended basic equipment: A multimeter capable of measuring resistance (ohmmeter). A network of resistances. A network of commercial carbon resistors is best, but an inkjet printer or a graphite pencil can also be used to construct the network.

Build a resistor network of approximately 10×10 elements. One possibility is to use resistors of about 10 kΩ that are easy to measure. Make sure beforehand that the resistors used have values that do not disperse from the nominal more than 0.5%.

Suggested tasks:

- ✓ Using a multimeter in ohmmeter mode, determine the unit resistance R_0 of the network.
- ✓ Verify that the resistance of two adjacent points of the network (points A and B, for example) is equivalent to $R_0/2$.
- ✓ Plot the resistance values as a function of distance for points along a central line (such as HH'), taking the central point as fixed (A) and varying the other along the HH' line.
- ✓ Perform the same study for points along a diagonal (line DD').
- ✓ Measure the equivalent resistance value for points such as those provided in Table 25.2 and compare the measured values of R_{eq}/R_0 to those calculated theoretically, as indicated in said table.
- ✓ ♣ (Optional) Interpret your results theoretically. References [6] and [7] provide adequate formalisms to study these cases.

References

[1] E. M. Purcell and D. J. Morin, *Electricity and Magnetism, Berkeley Physics Course*, 3rd. Ed., NY: Cambridge University Press, 2014, p. 839.
[2] H. A. Mavromatis, "Infinite and polygonal capacitor networks: Comparison with analogous, Fibonacci sequence related, resistor networks," *Am. J. Phys.*, vol. 63, no. 11, pp. 981–986, 1999.
[3] B. Denardo, J. Earwood and V. Sazonova, "Experiments with electrical resistive network," *Am. J. Phys.*, vol. 63, no. 1, pp. 85-88, 1995.
[4] Wikipedia, "Golden number Wikipedia," 2021. Available: https://en.wikipedia.org/wiki/Golden_ratio.
[5] Amazon, "Prototype Universal experiment circuit," 2021. Available: https://www.amazon.com/100Pcs-Prototype-Universal-Experiment-Circuit/dp/B07VJJRCGQ.
[6] P. Horowitz and W. Hill, *The Art of Electronics*, 3rd Ed., NY: Cambridge University Press, 2015, p. 1220.
[7] L. D. Woolf and H. H. Streckert, "Graphite pencil line for exploring resistance," *Phys. Teach.*, vol. 34, p. 440, 1997.
[8] D. Atkinson and F. J. Van Steenwijk, "Infinite resistive lattices," *Am. J. Phys*, vol. 67, no. 6, p. 486, 1999.

Chapter 26

WHEATSTONE BRIDGE

This chapter deals with the Wheatstone bridge and the wire bridge to measure resistances and small variations in resistances.

Goals

✓ Measurement of resistances
✓ Wheatstone Bridge
✓ Wire bridge

26.1 Introduction to resistance bridge

It is often useful to be able to detect small *variations* in the value of a resistor, rather than its absolute value. The bridge (devised by Samuel H. Christie and later improved by Sir Charles Wheatstone [1] in 1833) is very useful for this purpose. It is schematically illustrated in Figure 26.1. This circuit is known as a *Wheatstone bridge* [2–4], and with some modifications, it is also the basis of the *slide wire bridge*, which is discussed in this chapter. Appropriately modified, this experimental arrangement can be used to measure impedances, capacitances, and inductances. This bridge is also widely used in electronic instrumentation [4, 5].

Assume that between points C and D there is a *differential voltmeter* that measures *the differential potential* V_{CD} or a current measuring instrument (galvanometer or ammeter) whose internal resistance is designated R_v. Without loss of generality, assume that the resistance $R_1 \equiv R_x$ may vary, while the other resistors (R_2, R_3, R_4, R_v, and r) are constant.

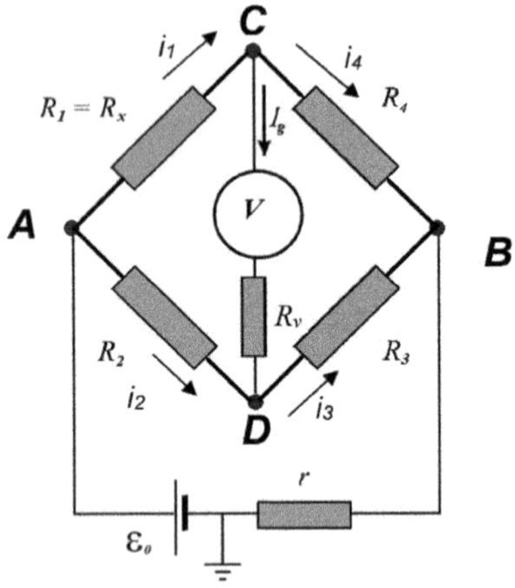

Figure 26.1 Wheatstone bridge; r is a resistor that limits current through the circuit and ε_0 is the bridge supply voltage. The voltmeter connected between C and D measures the potential difference between these points, V_{CD}, and its internal resistance, R_v.

If a galvanometer or milliammeter is connected between nodes C and D to measure current I_g, when I_g is null, it is said that *the bridge is balanced.* Alternatively, if a differential voltmeter is placed between points C and D, the bridge is balanced if the voltage difference between C and D is zero. In short, if *the bridge is balanced*, the electric potentials of points C and D are equal, that is, $V_{CD} = 0$. It is easy to see that in such a case, $i_1 = i_4$ and $i_2 = i_3$, and that the voltage drops V_{CA} and V_{AD} are the same, and therefore,

$$i_1 R_x = i_2 R_2 \tag{26.1}$$

and

$$i_1 R_4 = i_2 R_3. \tag{26.2}$$

Dividing member by member the last two expressions, we get

$$R_x = \frac{R_2}{R_3} R_4 \equiv R_1^0, \tag{26.3}$$

which is the *balance condition* of the bridge. Therefore, if the values of R_2, R_3, and R_4 are known, it is possible to calculate the value of R_x. The notation R_1^0 is used to designate the value of $R_1 \equiv R_x$ (*equilibrium*) that balances the bridge.

The important thing about this circuit is that it enables determination of small variations in the value of one of the resistors (R_x) if the others remain constant. Note that the equilibrium condition (26.3) does not depend on the value of the applied voltage ε_0, whether the applied voltage is direct (DC) or alternating (AC). Moreover, as the equilibrium condition depends on a *null value determination*, it does not depend on the absolute calibration of the instrument scale. This type of circuit is very useful for measuring *resistance variations.*

Usually, the voltage difference between points C and D is measured with a very high resistance voltmeter, e.g., $R_v \geq 1\,\mathrm{M\Omega}$, and in general, the following relation must be satisfied:

$$R_v \gg \mathrm{Max}\,(R_1, R_2, R_3, R_4). \tag{26.4}$$

If this condition is met, we can assume that R_1 and R_4 are in parallel with the combination of R_2 and R_4. Furthermore, this condition implies that $I_g = 0$. Under these conditions, the voltage difference between points C and D as a function of the voltage difference between A and B is

$$V_{CD} = \frac{V_{AB}}{(R_1 + R_4)} R_1 - \frac{V_{AB}}{(R_2 + R_3)} R_2 = V_{AB} \frac{R_1 R_3 - R_2 R_4}{(R_1 + R_4)(R_2 + R_3)}. \tag{26.5}$$

Including the effect of limiting resistance, r, using Kirchhoff's laws [2–4], it can be shown that the value of the voltage difference between points C and D, V_{CD}, is generally given by [5]

$$\varepsilon = \frac{V_{CD}}{\varepsilon_0} = \frac{R_1 R_3 - R_2 R_4}{\Delta}, \tag{26.6}$$

where

$$\Delta = r \cdot (R_1 + R_2 + R_3 + R_4) + (R_1 + R_4)(R_2 + R_3). \tag{26.7}$$

It is clear that Eq. (26.6) reduces to Eq. (26.5) if $r \to 0$. In any case, when the equilibrium condition (26.3) is fulfilled, the voltage drop between C and D, $V_{CD} = 0$. Equation (26.6) can also be written as follows:

$$\varepsilon(x) = \frac{V_{CD}}{\varepsilon_0} = \frac{R_1^0 \cdot R_3}{r \cdot (R_1(x) + R_2 + R_3 + R_4) + (R_1(x) + R_4) \cdot (R_2 + R_3)} \cdot x, \tag{26.8}$$

where we have introduced the dimensionless parameter x, defined by

$$x \equiv \left(R_1/R_1^0\right) - 1 \quad \text{or also} \quad R_1(x) = R_1^0(1 + x). \tag{26.9}$$

As previously stated, R_1^0 is the value of R_1 when the bridge is balanced ($x = 0$).

Figure 26.2 illustrates the dependence of $\varepsilon(x)$ as a function of x, described by Eq. (26.8). There are two interesting properties of the Wheatstone bridge:

(A) The value of the voltage measured between points C and D depends only on how much R_1 deviates from its equilibrium value, R_1^0, through the variable x.

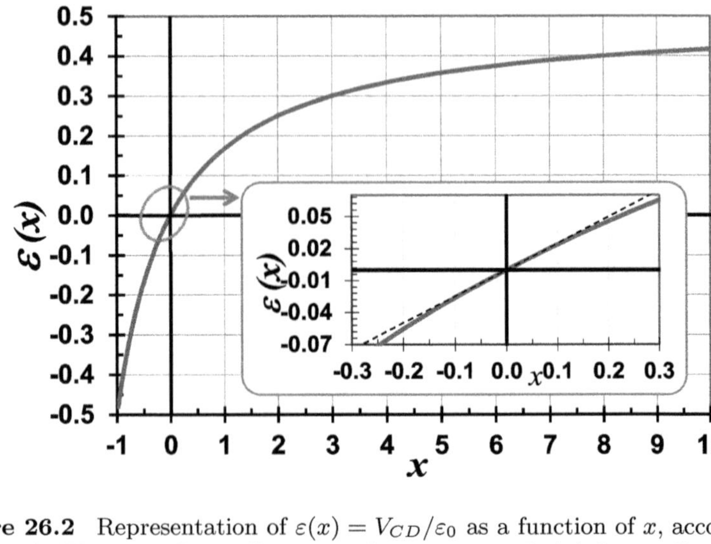

Figure 26.2 Representation of $\varepsilon(x) = V_{CD}/\varepsilon_0$ as a function of x, according to Eq. (26.8). The inset figure illustrates the variation of $\varepsilon(x)$ for small variations of x.

(B) If the deviation of R_1 with respect to R_1^0 is small ($|x| \leq 0.2$), $\varepsilon(x)$ varies approximately linearly with x. On the contrary, if the variation of R_1 relative to its equilibrium value is large, the dependence of $\varepsilon(x)$ on x is not linear, as illustrated in Figure 26.2.

In particular, if $r \ll M\acute{i}n(R_1, R_2, R_3, R_4)$, the expression (26.8) is equivalent to Eq. (26.6), and $\varepsilon(x)$ can be written as follows:

$$\varepsilon(x) = \frac{R_1^0 R_3}{(R_2 + R_3)\left(R_1^0 + R_1^0 x + R_4\right)} x = S_R(\lambda, x) \cdot x, \qquad (26.10)$$

where

$$S_R(\lambda, x) = \frac{R_1^0 R_3}{(R_2 + R_3)(R_1^0 + R_1^0 x + R_4)} = \frac{\lambda^{\square}}{(\lambda + 1) \cdot (\lambda + \lambda \cdot x + 1)} \qquad (26.11)$$

with

$$\lambda \equiv \frac{R_1^0}{R_4} = \frac{R_2}{R_3}. \qquad (26.12)$$

An important property of the Wheatstone bridge is its sensitivity, that is, the value of $\varepsilon(x)$, for small changes of R_1 or x. According to (26.10), this sensitivity depends on the function $S_R(\lambda, x)$. For $x \approx 0, S_R(\lambda, 0)$ has a maximum for $\lambda = 1$. In other words, the *maximum sensitivity* of the Wheatstone bridge is obtained when it operates near equilibrium conditions, i.e., when the following requirements are fulfilled: $R_1^0 \approx R_4$ and $R_3 \approx R_2$. This property makes this circuit very useful for making very precise measurements of *resistance variation* (R_1) regardless of its absolute value.

Note that when using a Wheatstone bridge, the instrument that measures the unbalance between C and D must be a device that operates in *differential mode*, see Section 24.4. That is, the instrument must have a *floating ground*, for example, a battery-powered multimeter. In general, standard data acquisition systems (A/D interfaces) connected to computers operate in *common mode* and therefore cannot be used directly to measure the voltage difference between C and D. Note that if the common ground is connected to point C or D, it would make the resulting circuit completely different from the one in Figure 26.2 that we are analyzing, since it would mean that one of the points C or D is connected to ground.

To monitor DC voltage variations with a standard data acquisition system, two channels of the interface can be used: one to measure the voltage of C with respect to the ground and the other to measure the voltage of D with respect to the ground. Then, the difference between these two signals is found using software, and hence, V_{CD} may be calculated. Another alternative, which would allow the use of a data acquisition system, would be to supply the bridge with a floating ground battery, in which case it would be possible to measure V_{CD} directly with a single channel of the acquisition system. Of course, this problem can also be avoided by using a differential input data acquisition system.

26.1.1 *Project 76. Experimental study of the Wheatstone bridge*

Recommended basic equipment: Carbon resistors from 1 to 10 kΩ. A 10 kΩ potentiometer or a variable resistance box. A DC voltage source or battery. A millivoltmeter.

Build a Wheatstone bridge using four similar resistors of resistance between 1 and 10 kΩ, with one of them (R_1) being variable. For R_1, you can use a potentiometer or a resistor box. If you use a potentiometer as R_1, it should have a lap counter or dial (scale reading) to indicate the resistance value. If you use a resistor box, its resistance value R_1 can be read easily. Choose a limiting resistor r so that the currents in the various branches of the bridge do not exceed the maximum allowable currents.

Suggested tasks:

✓ If the resistance R_1 is an uncalibrated potentiometer, use an ohmmeter to calibrate the resistance dial R_1 so that the value of this resistance can be ascertained by reading the dial. If you use a resistor box, check its calibration with the ohmmeter. Also measure each of the resistors, R_2, R_3, R_4, and r.

✓ Using this bridge, test expressions (26.6) and (26.8). Plot the measured values of V_{CD} and $\varepsilon(x)$ as a function of x together with the theoretical predictions, that is, Eq. (26.8). Choose the variation range of R_1 so that x varies between about 0.2 and 15. What can you say about the sensitivity of the method for detecting variations in $R_1(x)$ and the linearity of the function $\varepsilon(x)$ with respect to x?

✓ Discuss the possibility of using a computer to record V_{CD} variation. For this purpose, it is advisable to use a battery, e.g., a 9 V battery, to power the bridge and avoid possible grounding problems.

26.2 Slide wire bridge

A variant of the Wheatstone bridge is the *slide wire bridge* illustrated in Figure 26.3. Note that the circuit topology is the same as that described in Figure 26.1, but R_2 and R_3 are now the resistances of the sections AD and DB of the *same wire* that joins points A and B. The wire is usually approximately one meter long and made of metallic alloy. Nichrome, for example, has high resistivity compared to pure metals and good mechanical properties. It is also desirable for its cross-section to be as uniform as possible.

The metal wire is often held tightly on top of a plastic or wooden ruler or meter stick. The ends of the wire are screwed onto the ruler. The cursor or moving contact (MC) can move along the wire, thereby changing the length of each section and, consequently, the

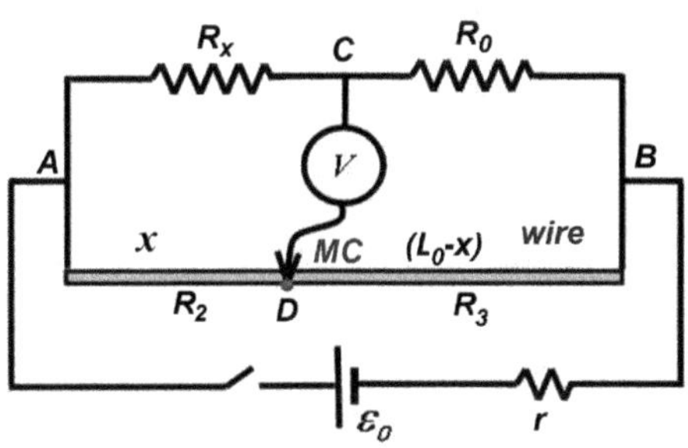

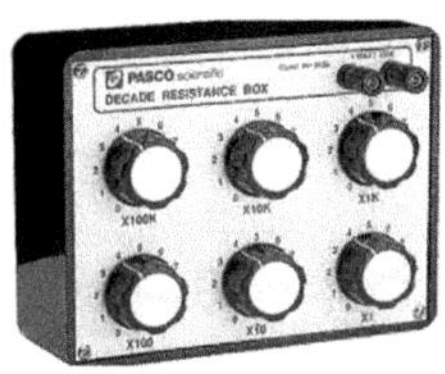

Figure 26.3 On the left, a wire bridge. On the right, a resistor box.

values of R_2 and R_3. The position x of the moving contact can easily be measured on the scale of the ruler. Under these conditions, the *differential voltmeter* is connected between the fixed point C and the moving point D [6].

For a wire of length L_0, uniform cross-sectional area A, and resistivity ρ, according to Eq. (24.2), its resistance is proportional to its length ($R = \rho L_0/A$),

$$R_2 = R_{23}\frac{x}{L_0} \quad \text{and} \quad R_3 = R_{23}\frac{L_0 - x}{L_0} \text{ with } R_{23} = \rho\frac{L_0}{A}. \tag{26.13}$$

The balance condition of the bridge (26.3) is expressed as follows:

$$R_x = \frac{R_2}{R_3}R_0 = \frac{x}{L_0 - x}R_0, \tag{26.14}$$

which enables R_x to be found from the known value of R_0 and measurement of the lengths x and L_0.

26.2.1 ♣ *Accuracy of a wire bridge*

Let's analyze the precision with which we measure R_x with the bridge. If ΔR_x is the absolute uncertainty of the determination of R_x, the precision of the determination is $\Delta R_x/R_x$. To estimate the uncertainty in determining R_x, we start from (26.14):

$$\begin{aligned}\frac{\Delta R_x}{R_x} &\approx \frac{\Delta R_0}{R_0} + \frac{\Delta x}{x} + \frac{\Delta(L_0 - x)}{L_0 - x}\\ &= \frac{\Delta R_0}{R_0} + \Delta x\left(\frac{1}{x} + \frac{1}{L_0 - x}\right) = \frac{\Delta R_0}{R_0} + \Delta x \cdot g(x),\end{aligned} \tag{26.15}$$

where we assumed $\Delta L_0 = 0$, since the length of the thread is a constant. To measure R_x, the value of R_0 must be known as precisely as possible. For this reason, the resistance R_0 must be a precision resistor of well-known value. If we do not have a well-calibrated resistor box, we can use a resistor of known value or that we can measure carefully. The variation of $\Delta R_x/R_x$ as a function of x can be analyzed using Eq. (26.15). This function has an extreme when

$$\frac{d(\Delta R_x/R_x)}{dx} = 0. \tag{26.15}$$

From the expression (26.15), it follows that

$$\frac{d(\Delta R_x/R_x)}{dx} \approx -\frac{\Delta x}{x^2} + \frac{\Delta x}{(L_0 - x)^2} = 0, \tag{26.16}$$

hence it follows that the minimum relative error in R_x is achieved when $x \approx L_0/2$. This result indicates that the bridge will provide more accurate values if the balance is reached with the cursor located at the midpoint of the wire. Under these conditions, from Eq. (26.11), we see that $R_x \approx R_0$.

26.2.2 ♣ *Uncertainties in slide wire bridge measurements*

To find ΔR_x using (26.12), it is necessary to estimate the uncertainties of the lengths $L_1 = x$ and $L_2 = L_0 - x$. Since the uncertainties ΔL_1 and ΔL_2 are related, it is reasonable to propose that $\Delta x \equiv \Delta L_2 = -\Delta L_1$. On the other hand, the uncertainty x has two sources:

a) appreciation of the length measuring instrument x, Δx_{ap},
b) the determination of the cursor position when balancing the bridge, Δx_{det}, which depends on the sensitivity of the voltmeter and the size of the contact between the cursor and the wire.

To determine the first component, we define

$$\Delta x_{\mathrm{ap}} = \frac{1}{2}\ (\text{nominal appreciation of the ruler}).$$

To determine Δx_{det}, we will consider the sensitivity of the voltmeter. To determine this parameter, one possibility is to position the voltmeter on the most sensitive scale and find the position x that balances the bridge. Then, move the cursor (*increase* x) slowly until the voltmeter increases by a value equal to the appreciation of the instrument (the least variation that can be clearly detected on the differential voltmeter) and measure the new position $x + d_1$. Next, *decrease* the value of x, until the voltmeter again increases by a value equal to the appreciation, and record the new position $x + d_2$:

$$\Delta x \frac{d_1 + d_2}{2}_{det}. \tag{26.17}$$

The combined uncertainty of x is

$$\Delta x = \sqrt{\Delta x_{\mathrm{ap}}^2 + \Delta x_{det}^2}. \tag{26.18}$$

This value can be used in expression 26.15 to calculate ΔR_{x}.

26.2.3 *Project 77. Determining the value of an unknown resistance using a slide wire bridge*

Recommended basic equipment: Voltage source from 0 to 10 V. Slide wire bridge, voltmeter, resistance of known value, and an unknown resistance whose value is sought.

Suggested tasks:

- ✓ Using a slide wire bridge, determine the value of an unknown resistance R_x. To do this, it is proposed to use the circuit in Figure 26.3. If possible, find the approximate value of the unknown resistance with an ohmmeter and choose the resistance R_0 so that its value is close to that of R_x, in order to minimize errors in determining R_x.
- ✓ Connect the unknown resistances R_x and R_0 on the corresponding branch of the bridge using short connecting cables. It is advisable to use a resistor box as R_0.
- ✓ Connect the voltmeter between points C and D (moving cursor on the wire) and initially operate the voltmeter on a coarse scale. This is necessary because you will not know in advance what the magnitude of V_{AB} will be when the bridge is out of balance. Once the balance is achieved, you can change the voltmeter to a more sensitive scale.
- ✓ Determine the best value of R_x and its corresponding uncertainty.

✓ Use other unknown resistances, preferably of known values, to obtain additional verification of the experimental method used. Express your results in the form $R_x \pm \Delta R_x$.

Note: Gently slide the cursor over the wire and do not overtighten to make good contact. If the wire becomes uneven or deformed due to strong forces applied, its resistance will no longer be uniform due to the local decrease in wire section.

Self-evaluation questions

✓ Discuss the effect of source voltage (ε_0) on the precision of the Wheatstone bridge. Is it reasonable to think that doubling the voltage ε_0 would double the precision of the measurement?
✓ Discuss the effect of voltmeter sensitivity on bridge accuracy. Is it reasonable to think that twice the accuracy will be achieved with a twice more sensitive voltmeter?

References

[1] Wikipedia, "Charles Wheatstone," 2021. Available: https://en.wikipedia.org/wiki/Charles_Wheatstone.
[2] R. Halliday, D. Resnick and M. Krane, *Fundamentals of Physics*, NY: John Wiley & Sons, Inc., 2002.
[3] E. M. Purcell and D. J. Morin, *Electricity and Magnetism, Berkeley Physics Course*, 3rd. Ed., NY: Cambridge University Press, 2014, p. 839.
[4] P. Horowitz and W. Hill, *The Art of Electronics*, 3rd Ed., NY: Cambridge University Press, 2015, p. 1220.
[5] R. Pallás-Areny and J. G. Webter, *Sensors and Signal Conditioning*, NY: John Wiley & Sons, 2012, p. 608.
[6] Blender IITB, "Meter bridge experiment assembly," 2021. Available: https://www.youtube.com/watch?v=ab7_MUMSQ1I.
[7] Wikipedia, "Wheatstone bridge," 2021. Available: https://en.wikipedia.org/wiki/Wheatstone_bridge.

Chapter 27

FOUR-POINT RESISTANCE MEASUREMENT OR KELVIN METHOD

This chapter looks at the four-lead method for measuring resistance. The key advantage of this technique is that it eliminates the contribution of wiring resistance and contact potentials. This procedure, also known as the Kelvin method, is very useful for measuring very low-value resistors. It is widely used in measurement laboratories and geophysical prospecting. It is suitable for measuring resistivity of samples of various shapes or geometries.

Goals

✓ Low-value resistance measurement
✓ Four-point method
✓ Resistivity of a flat sample
✓ Trans-resistance

27.1 Determination of low-value resistors

Determining the resistivity or conductivity of a sample is very useful in many experiments and industrial applications. The four-point technique or Kelvin method is one of the most common and useful procedures for this purpose. Originally developed by Lord Kelvin, it was later perfected in the early 1900s by Frank Wenner [1], who used it to measure the resistivity of soil samples [1–3]. In geophysics, it is known as the *Wenner method.* It is also widely used in the semiconductor industry to control the production process.

To measure a resistance of intermediate values (between a few tens of Ohms (Ω) a few MΩ) perhaps the simplest is to use the

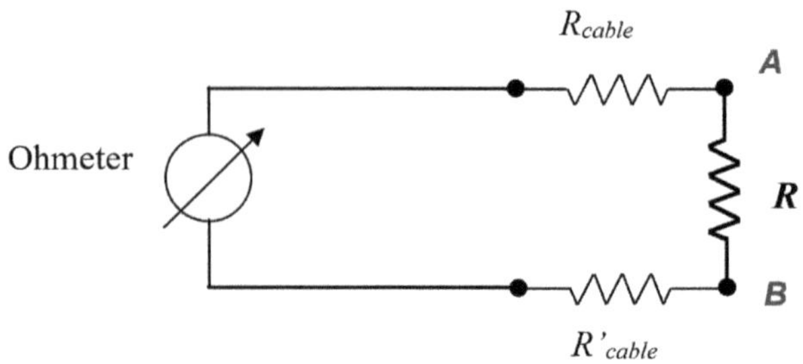

Figure 27.1 Determination of the resistance of a sample using an ohmmeter or multimeter or the *two-wire test* configuration. The resistance of interest is R, however what the ohmmeter measures is $R + R'_{\text{cable}} + R_{\text{cable}}$.

two-lead method or *two-wire test* configuration, using a multimeter (ohmmeter), as shown in Figure 27.1 and discussed in previous chapters. The resistance of interest is R, but what the ohmmeter measures is the sum of: $R + R'_{\text{cable}} + R_{\text{cable}}$. The measured value will be very close to R only if $R \gg R'_{\text{cable}} + R_{\text{cable}}$. For resistance of small magnitude, $R < 10\Omega$, this condition is almost never satisfied. Therefore, to measure a small resistance (less than about 10 Ω), it is necessary to take into account both the resistances of the cables and the *contact potentials* that may be present when interconnecting two different metals. These contact potentials are common in the junctions of different metals and can vary with temperature (Peltier–Seebeck effect [1]). Partly due to these effects, the effective resistance of the system may depend on the polarity of the source, that is, the resistances are not necessarily the same if the current flows in one direction or another. The resistance measurement method described in the following solves some of the aforementioned problems of the *two-lead method* and is particularly useful for measuring low-value resistors.

27.2 Four-point method or Kelvin method

In Kelvin's method (schematically shown in Figure 27.2), the current flows mainly in the external loop. As modern voltmeters generally have high internal resistances, greater than 10 MΩ, practically no current flows in the inner loop. In this case, the voltage measured by the voltmeter will be

$$V^+ = \varepsilon_A + I^+ R - \varepsilon_B, \tag{27.1}$$

where ε_A and ε_B represent the contact potentials at each junction. The limiting resistance R_{ext} is chosen such that the current in the circuit does not damage the source or the other elements of the circuit. The superscript (+) indicates that the current flows as shown in Figure 27.2. We use the superscript (−) when the direction of the current changes by inverting the polarity of the voltage source (ε_{ext}) but without altering the rest of the circuit.

If the polarity of the voltage source is reversed, the voltage measured by the voltmeter will be

$$-\boldsymbol{V}^- = \boldsymbol{\varepsilon_A} - \boldsymbol{I}^-\boldsymbol{R} - \boldsymbol{\varepsilon_B}. \tag{27.2}$$

The voltage and current values indicated in Eqs. (27.1) and (27.2) are the absolute values of what the instruments indicate. Subtracting these equations from each other gives

$$V^+ + V^- = (I^+ + I^-)R. \tag{27.3}$$

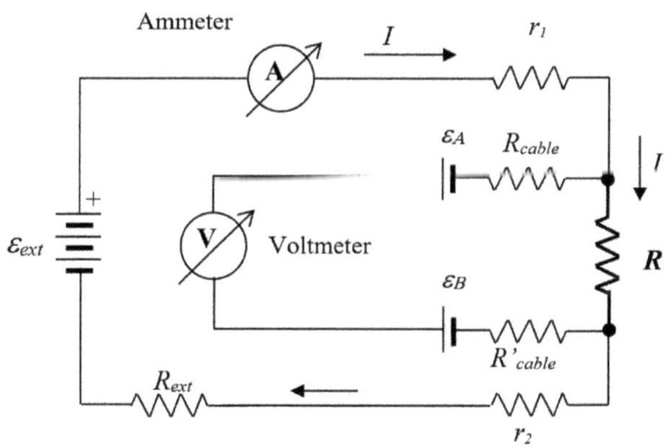

Figure 27.2 Determination of the resistance R using the *four-point method*. Note that as most voltmeters have high internal resistance ($R_{\text{voltmeter}} > 10$ MΩ), practically all the current flows through the external loop and there is no voltage drop across the voltmeter connecting cables (r_2 and r_3). ε_{ext} is the external voltage source and ε_A and ε_B are the contact potentials. R_{ext} is the current limiting resistor.

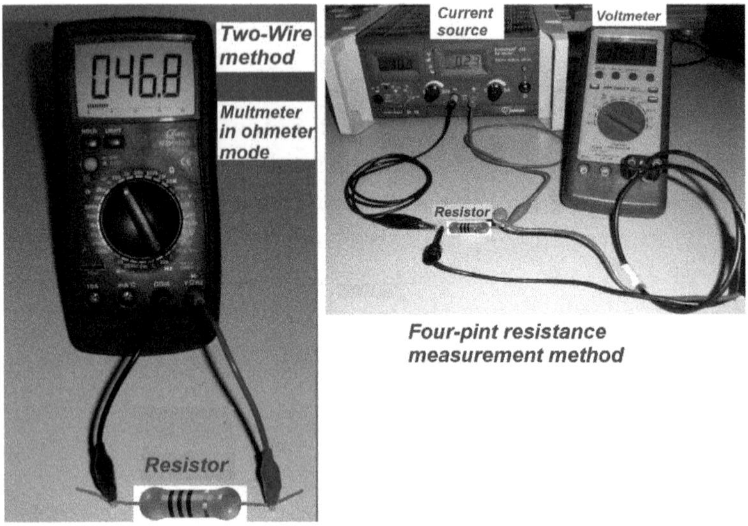

Figure 27.3 Two-point (left) and four-point (right) resistance measurement methods. Some special instruments have an arrangement to measure four points directly. However, it is always possible to design an array with conventional instruments, as illustrated in the right panel, to perform the four-point measurement.

Therefore, by reversing the direction of the flow of the current and taking the difference of the measured potentials, we can cancel the effect of the contact potentials. More specifically, we have

$$R = \frac{V^+ + V^-}{(I^+ + I^-)} = \frac{|V^+| + |V^-|}{|I^+| + |I^-|}. \tag{27.4}$$

When applying expression (27.4) to a specific case, it is advisable to carefully analyze the signs that you use for $I^\pm$ and $V^\pm$. The second manner of writing Eq. (27.4) resolves this possible ambiguity.

We thus see that the four-point method enables us to simultaneously eliminate the effect of cable resistances and contact potentials, as well as to evaluate the magnitude of these potentials. At first glance, it may seem surprising that the current intensity through the circuit varies if the polarity of the external source is reversed: the values of I^+ and I^- are different when there are oxides at the contacts, which cause the resistance values to be different if the current flows in one direction or the other (as in the case of a diode). Furthermore, the value of the effective voltage applied to the resistance R, formed by the external source and the contact potentials, varies by changing the polarity of the external source.

When using an alternating voltage source (AC), it is advisable to measure the voltage with an instrument that filters the direct components (DC). Many instruments provide the option to activate this measurement mode, e.g., oscilloscopes, multimeters, and *lock-in* amplifiers. If the voltage is measured in AC mode, Eq. (27.1) becomes

$$V^{\mathrm{AC}} = I^{\mathrm{AC}} R, \tag{27.5}$$

since in this mode the contact potentials (DC) are automatically filtered by the measuring instrument. Therefore, in this case, the four-point measurement method is simplified.

Finally, it is interesting to note that nowadays, many multimeters have four ports — two for current input and output and two to measure voltage — which can provide the result in ohms directly [5, 6].

27.3 Measurement of the resistivity of a sample of simple geometry – 1D case

Imagine a sample in the shape of a cylindrical wire of diameter φ and cross-sectional area $A(=\pi\varphi^2/4)$. The potential difference between two points separated by a distance L will be [7]

$$\Delta V = I\,R = I\,\rho\,\frac{L}{A}. \tag{27.6}$$

Using Eqs. (27.4) and/or (27.5), we have

$$\rho = (A/L)(\Delta V/I) = (A/L)\left\{\frac{|V^+|+|V^-|}{|I^+|+|I^-|}\right\} = (A/L)\frac{V^{\mathrm{AC}}}{I^{\mathrm{AC}}}, \tag{27.7}$$

depending on whether a DC ($^{+\,-}$) or AC source is used, respectively. In any case, it is important that the geometry of the wire should be well known, i.e., that the values of A and L can be measured with small uncertainties. Therefore, the wire diameter cannot be very small, since the uncertainty in the measurement of the diameter would greatly limit the determination of the resistivity. By using larger diameters, the resistance of the sample decreases, and therefore the use of the four-prong method is crucial for this type of measurement.

27.3.1 *Project 78. Measuring the resistivity of a wire by the four-point method*

Recommended basic equipment: Copper, aluminum, silver, or iron wire, 1–3 mm in diameter and about 1 m long. Two multimeters to measure current and voltage (millivolts). A DC or AC voltage source of about 12 V @ 2 A.

This project requires samples of some pure metals ($\approx$ 99% purity) in order to easily compare the measured values with the tabulated ones for each material. Build a circuit like the one in Figure 27.2 to use the Kelvin method to measure resistances.

Suggested tasks:

✓ Select a set of pure samples of known materials, for example, Cu, Al, Fe, and Ag. It is important that the geometry of the sample can be well characterized, so it is suggested to use wires with a diameter ϕ of 1–3 mm and a length L of approximately 1 m, in order to enable ϕ and L to be measured with precisions of the order of 1%. It is important to remember that the distance between the voltmeter leads determines the value of the length L of the wire between points A and B (see Figure 27.2). The current connectors may be joined (or soldered) to those same points (A and B) or they can also be connected to points outside the wire segment A–B. Discuss and justify this procedure for connecting the voltage and current leads.

✓ The leads carrying the current must have a cross-section that is sufficient to withstand the maximum current (a few amperes) without overheating. The leads connected to the voltmeter can be thinner, since practically no current will flow through them.

✓ Use a DC voltage source with a current limiting resistor (R_{ext}) in series, in the range of 10–50 Ω and capable of withstanding currents of a few amperes, or a source that has current regulation or limiting current. If the current through the circuit is 1 A, for a sample resistance of some mΩ, we expect to measure voltages of the order of a few mV. Estimate the expected value of voltages for your specific case and choose the appropriate range on your multimeter to measure these voltages.

✓ Vary the polarity of the external voltage source and investigate whether the magnitudes of the measured voltage and currents change significantly.

✓ Perform various voltage measurements with different current values. To keep the temperature constant, you may submerge the coil in a glass of water and monitor the temperature with a thermometer. Plot the measured voltage (V^{AC} or $\mathrm{V}^{+} + V^{-}$) as a function of current (I^{AC} or $I^{+} + I^{-}$). Using expression (27.4) or (27.7), find the best value of R and its corresponding error.

✓ Knowing the diameter and the length of the wire (distance between the points of contact with the voltmeter connectors), determine the value of the resistivity ρ of the material and estimate its error.

✓ Discuss the degree of agreement found with the values in the corresponding tables.

27.4 Determination of the resistivity of a large two-dimensional sample

Imagine a flat conducting sample, of infinite extension, with thickness t and resistivity ρ, as shown in Figure 27.4. Suppose that a current I is *injected* at a point O of the sample and *drained* at infinite distance. By symmetry, we can imagine that the current is distributed *uniformly* in all directions in the sample. Thus, the potential difference between two points (placed on two circles, centered on O, and separated by a small distance dr, at the distance r from O) will be

$$dV' = I \cdot \delta R = I \cdot \frac{\rho}{t \cdot 2\pi} \cdot \frac{dr}{r}. \tag{27.8}$$

Here, δR represents the resistance of a ring of thickness t times dr and length $2\pi r$. The current flows *radially* through the area section $2\pi r t$ for a distance dr.

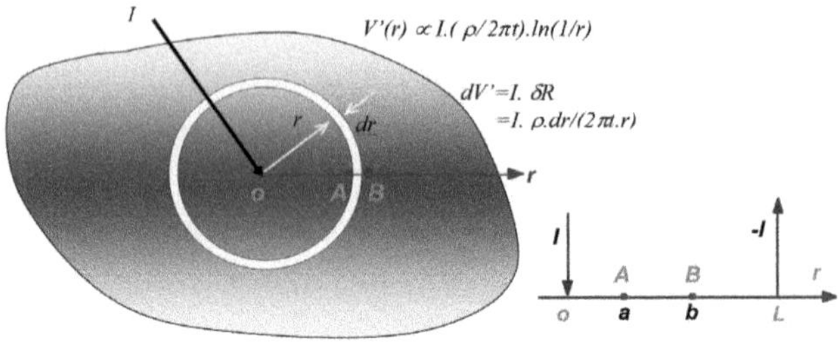

Figure 27.4 Flat sample of infinite extension and thickness t, with a current I injected through the point O. dV' represents the potential difference between two points separated by a distance dr, due only to the injected current I. $V'(r)$ is the potential generated only by this injected current. On the right, a schematic diagram of the position of the current injection point O and drain point O', and of the voltage measuring points A and B.

We can thus associate a potential to this current I in the form

$$V'(r) = I \cdot \frac{\rho}{t \cdot 2\pi} \cdot \ln(r) + C, \tag{27.9}$$

where C is a constant.

The potential difference between two points A and B that are at distances a and b, respectively, from the injection point O will be

$$\Delta V'(AB) = I \cdot \frac{\rho}{t \cdot 2\pi} \cdot \ln\left(\frac{b}{a}\right). \tag{27.10}$$

If through another point P, at a distance L from the first point, we drain a current $-I$, the potential difference between points A and B will be

$$\Delta V'(AB) = I \cdot \frac{\rho}{t \cdot 2\pi} \cdot \ln\left(\frac{L-b}{L-a}\right). \tag{27.11}$$

Implicitly, we are assuming that the four points in question (O, A, B, and P) are aligned. If we now imagine that we have the injection and extraction acting simultaneously, by *superposition* of the two previous cases (see Chapter 25), the potential difference between the previous points will be

$$\Delta V(AB) = I \cdot \frac{\rho}{t \cdot 2\pi} \cdot \ln\left[\frac{b}{a} \cdot \left(\frac{L-a}{L-b}\right)\right]. \tag{27.12}$$

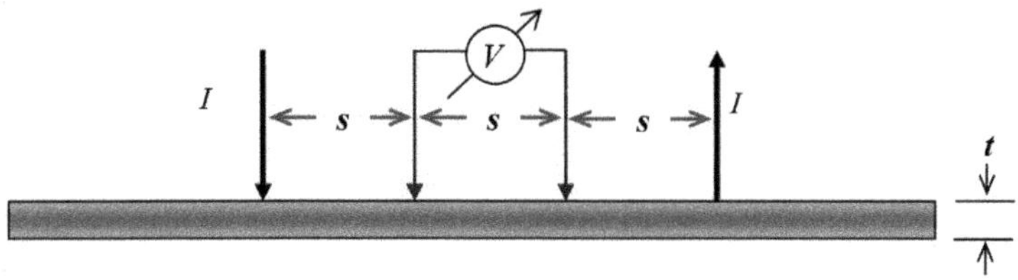

Figure 27.5 Four electrodes separated by the same distance s on a flat sample of thickness t, with $s \gg t$.

If the points O, A, B, and P are equally spaced, as shown in Figure 27.5, that is, if $a = s$, $b = 2s$, and $L = 3s$, where s is the distance between the contact electrodes, then $b/a = 2$ and $(L - a)/(L - b) = 2$, we have

$$\Delta V(AB) = I \cdot \frac{\rho}{t \cdot \pi} \cdot \ln(2) \quad or \quad \rho = \frac{\Delta V}{I} \cdot \frac{t \cdot \pi}{\ln(2)}. \tag{27.13}$$

Therefore, in a plane geometry and with electrodes equidistant and separated by a distance $s \gg t$, as illustrated in Figure 27.5, the resistivity of the sample can be extracted from the measurement of the injection current I and the measurement of the potential difference ΔV, as indicated by Eq. (27.13). Note that the distance s does not enter in the calculation of ρ, although it is required that $s \gg t$ for the plane geometry assumption to hold. Another implicit condition in this method is that the characteristic dimensions d of the flat plate are much greater than the distance between the electrodes (s). If the condition $d \gg s$ is not met, a finite dimension correction coefficient f_1 must be used [3, 8, 9]. In this case, the resistivity is calculated by

$$\rho = f_1 \cdot \frac{\pi}{\ln(2)} \cdot t \cdot \frac{\Delta V}{I}, \tag{27.14}$$

with the correction coefficient f_1 given in Figure 27.6 [7, 10]. Similarly, if the sample is not very thin, that is, if the condition $s \gg t$ is not valid, it is necessary to introduce an analogous correction [3, 7, 11].

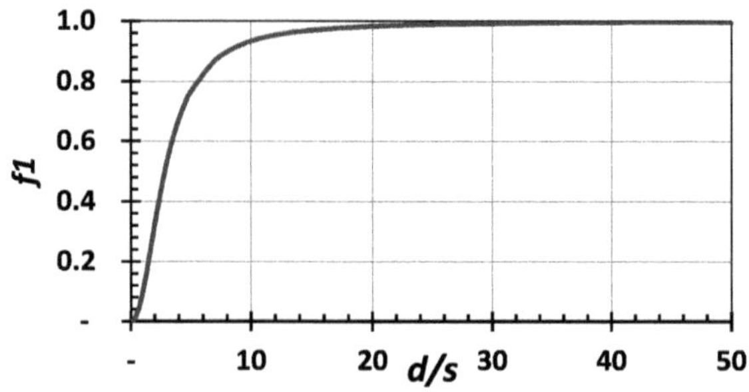

Figure 27.6 Finite sample correction coefficient, f_1 as a function of the quotient d/s, with d being the characteristic dimension of the flat sample and s the distance between the electrodes.

27.4.1 *Project 79. Determining the resistivity of a flat sample*

Recommended equipment: Flat metal samples of Cu, Al, or some other pure metal, 0.25–1 mm thick. Two multimeters, a DC or AC voltage or current source. A system of four evenly spaced electrodes.

This project requires samples of some pure metals (99% purity) in order to easily compare the measured values to those tabulated for the same material. Build a circuit like the one in Figure 27.5. One way to ensure that the four electrodes are evenly spaced and make good contact is to mount four copper or bronze screws at equal distances along a straight line on an acrylic bar approximately 1 cm thick. The ends of the bar are supported so that it rests firmly on the specimen. The bar is supported on the plate with clips. The screws must have pointed tips so that their position is well defined. By tightening the screws slightly, good contact with the metal plate is achieved. Another alternative, if a copper or bronze plate is used, is to tin-solder the electrodes. There are also commercial four-point systems [12] available in different configurations, with well-determined tip geometries. These devices are useful for high-precision measurements.

Suggested tasks:

✓ Cut the metal sheets carefully into flat shapes that fulfill the condition $d \gg t$.

✓ Connect the current injection and voltage measurement electrodes so that they are aligned and equidistant.

✓ Use a DC or AC source with a series current limiting resistor; 50 Ω @ 5 W may be adequate for a current of about 100 mA to flow. If you use a current-regulated power source, the limiting resistor is not necessary. Assuming that resistance between the measurement points is a few $m\Omega$, voltages of the order of 0.1 mV are expected. Therefore, choose the appropriate range on your multimeter to measure these voltages and the corresponding currents.

✓ If you are using a DC source, vary the direction of the current and investigate whether the measured voltage changes significantly.

✓ Knowing the thickness of the sample, determine the value of the resistivity of the material and estimate its errors.

✓ Discuss whether your sample and measurement system meet the conditions $d \gg s \gg t$ and whether it is necessary to make corrections for these characteristics.

✓ Discuss the degree of agreement between your resistivity values and the values provided in tables for the same materials.

27.5 ♣♣ Van der Pauw method: Trans-resistances – Flat sample

In many cases of practical interest, it is not useful or not possible to use an equidistant electrode arrangement as discussed above, Figure 27.4, e.g., when the sample is very small. In these cases, the van der Pauw method [13] can be very useful, since both the current input and output points, as well as the voltage measurement points, can be arbitrarily located on the edge of the sample. In this method, the effects due to size and spacing are irrelevant [13]. The only requirements are that the thickness of the sample is uniform and that the sample has no holes and is homogeneous and isotropic. Here, we will only describe the procedure and transcribe the results. For a justification, see Ref. [13].

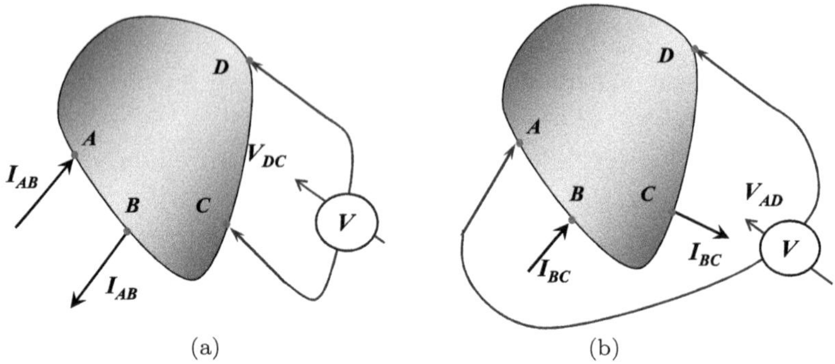

Figure 27.7 Van der Pauw method to measure trans-resistances. (a) Configuration to determine $R_{AB,CD} = V_{DC}/I_{AB}$. (b) Configuration to determine $R_{BC,AD} = V_{AD}/I_{BC}$.

Imagine that the current I_{AB} flows in the sample through peripheral points A and B, as shown in Figure 27.7(a), and the voltage difference is measured between points D and C, V_{DC}. *Trans-resistance* [14] is defined as $R_{AB,CD} = V_{DC}/I_{AB}$. If the current input-output points and voltage difference measurement points are altered as shown in Figure 27.7(b), another trans-resistance can be defined as $R_{BC,AD} = V_{AD}/I_{BC}$. Note that unlike **resistance**, which is obtained by dividing the voltage difference by the current between two well-defined points, **trans-resistance** [14] is the voltage difference between two points divided by the current between two other different points.

It can be shown that [13]

$$\exp[-\pi \cdot t \cdot R_{AB,CD}/\rho] + \exp[-\pi \cdot t \cdot R_{BC,DA}/\rho] = 1, \qquad (27.15)$$

where ρ is the resistivity of the sample and its thickness is t. This expression makes it possible to implicitly obtain the resistivity of the sample by measuring the trans-resistances [14] $R_{AB,CD}$ and $R_{BC,AD}$ and the thickness of the sample. Equation (27.14) cannot be solved analytically, but it can be solved numerically or graphically. A simple way to solve it is to graph the following functions:

$$y_1(\rho) = \exp[-\pi \cdot t \cdot R_{AB,CD}/\rho] \quad \text{and}$$
$$y_2(\rho) = 1 - \exp[-\pi \cdot t \cdot R_{BC,DA}\rho], \qquad (27.16)$$

as a function of ρ. The value of ρ where the functions $y_1(\rho)$ and $y_2(\rho)$ intersect gives us the solution.

27.5.1 *Project 80. Determining the resistivity of a small, flat sample*

> **Recommended equipment:** Flat metal samples of Cu, Al, bronze, or another pure metal of interest, 0.25–1 mm thick, in the shape of a circle or quadrilateral of approximately 3 cm. The shape is not important. Ideally, it should be a piece of the same material that was used in the previous activity. Two multimeters, a DC or AC voltage or current source.

For this project, it is advisable to use a piece of a sample that was used in the previous project, whose resistivity was previously measured. Cu and bronze samples have the advantage that they can be soldered with tin. Build a circuit like the one in Figure 27.7.

Suggested tasks:

- ✓ Use a DC or AC source with a 50 Ω @ 5 limiting resistor. This will allow a current of about 100 mA to flow. If you use a current-regulated power source, the limiting resistor is not necessary. Assuming a resistance between the measurement points of a few mΩ, choose the appropriate range on your multimeter to measure these voltages and the corresponding currents.
- ✓ Connect the current input and output electrodes to points A and B. Simultaneously measure the voltage difference between D and C. Plot V_{DC} as a function of I_{AB}. From the slope of this graph, find the value of the trans-resistance $R_{AB,CD}$.
- ✓ Reverse the current input and output points so that the trans-resistance can be measured $R_{BC,AD}$.
- ✓ If you are using a DC source, vary the direction of the current and investigate whether the measured voltage changes significantly.
- ✓ Knowing the thickness t of the sample, determine the value of the resistivity of the material and estimate its errors.

✓ Discuss whether your sample and measurement system meet the conditions $d \gg s \gg t$ and whether it is necessary to make corrections for these characteristics.

✓ Discuss the degree of agreement between your resistivity values and the values provided in tables for the same materials.

✓ If the sample is of the same material that you used in any of the previous projects, compare the resistivities obtained with this technique and the one used previously.

27.6 ♣ Large three-dimensional sample Wenner method

The four-point method can also be used to estimate the resistivity of a large three-dimensional sample. In this case, due to the size of the sample, the separation between the electrodes is much greater. An example would be the measurement of the conductivity of a region of the soil [15]. To justify the expressions to be used, we first consider the case of a current I that is injected into a three-dimensional sample, similar to the case illustrated in Figure 27.8. Under these conditions, given the three-dimensional nature of the problem, the potential difference between two adjacent points separated by an infinitesimal distance dr will be

$$dV' = I \cdot \delta R = I \cdot \frac{\rho}{2\pi} \cdot \frac{dr}{r^2}. \tag{27.17}$$

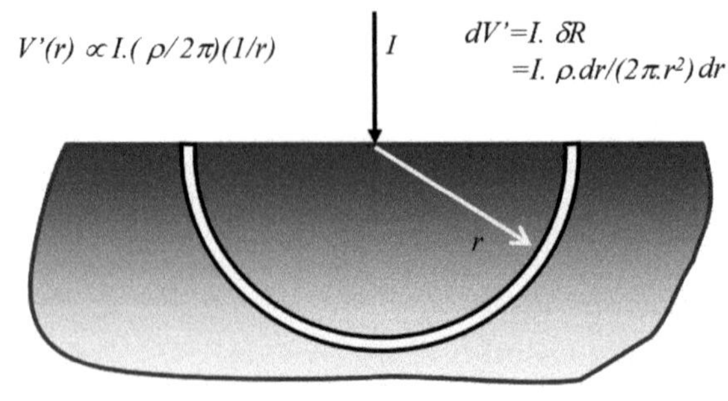

Figure 27.8 Variation of the potential in a semi-infinite three-dimensional sample, in which a current I is injected at a point on its surface. dV' represents the potential difference at two points separated by a distance dr, due only to the injected current I.

δR represents the resistance of the spherical shell of radius r and thickness dr, and again the dV' indicates the potential difference due only to the injected current.

The potential difference between two electrodes at distances $a = s$ and $b = 2s$ from the injection point, similar to the case shown in Figure 27.5, is found by integrating Equation (27.19):

$$\Delta V'(AB) = I \cdot \frac{\rho}{2\pi} \cdot \left(\frac{1}{a} - \frac{1}{b}\right) = I \cdot \frac{\rho}{4\pi} \cdot \frac{1}{s}. \tag{27.18}$$

If again we use for the electrodes a geometry similar to that shown in Figure 27.5, that is, the four electrodes aligned and separated by a distance s, using superposition, we have

$$\Delta V = I \cdot \frac{\rho}{2\pi} \cdot \frac{1}{s}, \quad \text{or} \quad \rho = 2\pi \cdot s \cdot (\Delta V/I) \tag{27.19}$$

This arrangement for measuring resistivities is also known as Wenner's four-electrode method [16]. This type of method is used in geophysical prospecting to measure the resistivity of the Earth and to determine the depth of a layer of different composition or conductivity, such as water or oil.

27.6.1 *Summary of important concepts and review questions*

Discuss some of the possible applications and implications of the previous experiments. For example,

(1) Why does the two-wire method, Figure 27.1, have difficulty measuring resistances of less than about 10 Ω?
(2) If you want to determine the resistivity of a wire, copper, for example, why not use a very thin and long wire so that it has high resistivity? By doing so, the two-wire technique could be used, which is simpler than the four-point technique. Analyze the errors of the various quantities you need to measure in this case, Eq. (27.7). In particular, discuss how the relative error of the wire diameter influences your resistivity measurement.

(3) What are contact potentials?
(4) Why can the current in the circuit shown in Figure 27.2 change in magnitude if the polarity of the source is reversed?

References

[1] F. Werner, "A method of measuring earth resistivity," *Bur. Stand. U.S. Bull.*, vol. 12, p. 469–478, 1915.
[2] Wikipedia, "Four-terminal sensing," 2021. Available: http://en.wikipedia.org/wiki/Four-terminal_sensing.
[3] A. P. Schuetze, *et al.*, "A laboratory on the four-point probe technique," *Am. J. Phys.*, vol. 72, no. 2, pp. 149–153, 2004.
[4] Wikipedia, "Seebeck effect," 2021. Available: https://en.wikipedia.org/wiki/Thermoelectric_effect.
[5] "Bridge technology," 2021. Available:http://four-point-probes.com/.
[6] Four Dimension Inc., "Bridge technology Inc.," 2021. Available: http://www.4dimensions.com/.
[7] D. Henry, "Resistance of a wire as a function of temperature," *Phys. Teach.*, vol. 33, p. 96, 1995.
[8] D. E. Vaughan, "Four-probe resistivity measurements on small circular specimens," *J. Appl. Phys.*, vol. 12, pp. 414–416 , 1961.
[9] R. A. Wellera, "An algorithm for computing linear four-point probe thickness correction factors," *Rev. Sci. Instrum.*, vol. 72, no. 9, pp. 3580–3586, 2001.
[10] F. M. Smits, "Measurement of sheets resistivities with a four-point probe," *Bell Syst. Tech. J.*,pp. 711–718, 1958.
[11] J. Shia and Y. Sun, "New method of calculating the correction factors for the measurement of sheet resistivity of a square sample with a square four-point probe," *Rev. Sci. Instrum.* vol. 68, no. 4, pp. 1814–1817, 1997.
[12] D. W. Koon and C. J. Knickerbocker, "What do you measure when you measure resistivity?" *Rev. Sci. Instrum.*, vol. 63, no. 1, pp. 207–210, 1992.
[13] L. J. van der Pauw, "A method for measuring specific resistivity and Hall effect of discs of arbitrary shape," *Phillips Res. Rep.*, vol. 13, no. 1, 1958.
[14] Wikipedia, "Transconductance," 2021. Available: https://en.wikipedia.org/wiki/Transconductance.
[15] Wikipedia, "Soil resistivity," 2021. Available: https://en.wikipedia.org/wiki/Soil_resistivity.

[16] B. Avants, D. Soodak and G. Ruppeinera, "Measuring the electrical conductivity of the earth," *Am. J. Phys.*, vol. 67, no. 7, pp. 593–598, 1999.

[17] M. R. Spiegel, *et al.*, *Complex Variables, Schaum's Outline Series*, 2nd Ed., NY: Mc Graw Hill, 2009.

Chapter 28

VARIATION OF ELECTRICAL RESISTANCE WITH TEMPERATURE

This chapter studies the effect of temperature on the electrical conduction of some solids (a pure conductor, an alloy, and a semiconductor) and some common electronic devices (carbon resistors and thermistors). We will try to explain the experimental results using simple microscopic conduction models for each case. Different electrical resistance measurement methods are used.

Goals

- ✓ Variation of resistance with temperature
- ✓ Metallic resistances and alloys
- ✓ Thermal behavior of a thermistor

28.1 Simplified model of electric conduction

The behavior of solids regarding transport of electric current can be characterized by measuring electrical resistance $R = V/I$ (Ohm's law). From this measurement, it is possible to obtain the macroscopically defined resistivity ρ for a homogeneous conductor of length L and section S, assuming a prismatic sample of known geometry and constant cross-section, $R = \rho L/S$, or the *conductivity* $\sigma = 1/\rho$.

Conductivity can also be expressed as $\sigma = J/E$, i.e., the ratio between current density J and the intensity of electric field E.

A solid is electrically conductive if it contains mobile charges (*carriers*) capable of conducting electricity. In metals, the moving charges are electrons; in semiconductors, they may be electrons and *holes*.[a]

To explain Ohm's law, a microscopic model can be used for the motion of the charge carriers. If we think of electric current as a flow of charged particles with concentration n (i.e., number of charge carriers per unit of volume), all with the same charge q and the same *drift velocity* v_d, then the current density (i.e., the amount of charge that passes through the unit section in the unit of time in the conductor) can be written as $J = qnv_d$. This means that in one second, all the charge carriers contained in the volume of the cylinder of section S and height v_d pass through the unit surface S.

Consider a sample of cross-section S and length l of a solid through which an electric current i flows, as shown in Figure 28.1.

It is possible to relate this current in a very general way to the average speed v_d (drift velocity) of charge carriers, to the number n of carriers per unit volume, and to the charge transported by each carrier (e is the elementary charge) [1, 2].

$$i = \ neSv_d. \tag{28.1}$$

Microscopically, the motion of the charge carriers is the superposition of an ensemble motion in the direction of the applied electric field and the disordered motion of thermal agitation with mean square velocity which, in the classical model, can be estimated as equal to $v_m = \sqrt{3k_BT}/m_e$, where k_B is the Boltzmann constant, T is the absolute temperature, and m_e is the charge carrier mass. At room temperature, this relationship gives $v_m \approx 10^5$ m/s. For comparison, the speed of sound in the air is about 350 m/s.

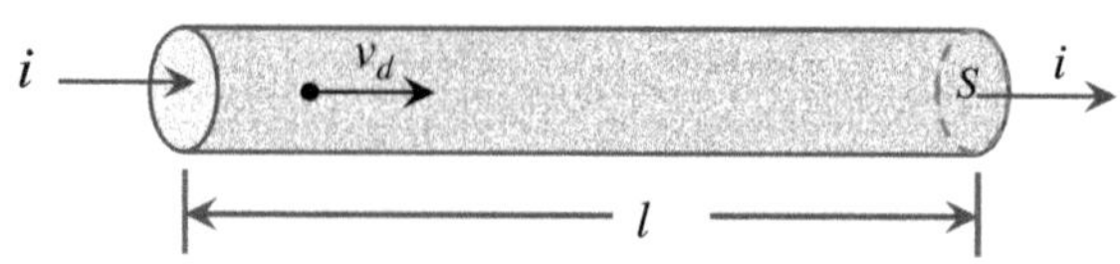

Figure 28.1 A current i flows through a conductor of length L and section S of known geometry and constant cross-section. $i = n.S.v_d.q$ and $J = i/S = n.v_d.q$.

[a] A hole is the absence of an electron from a full valence band. https://en.wikipedia.org/wiki/Electron_hole

In addition to the force due to the electric field, there is also the resistance due to collisions with the inevitable impurities present and with the crystal defects and vibrations of the lattice (phonons). This makes the motion of the carriers along the direction of the applied electric field occur at constant average speed v_d (*drift velocity*). This is something like what happens when a marble falls down a staircase: it accelerates between steps, but on average, it moves at constant speed. The energy lost by the carriers in these collisions increases the temperature of the material (Joule effect).

The drift velocity v_d is small. It is a *resultant component* of the velocity in the direction of the applied electric field. The thermal agitation, having completely random directions, has a *null average value*.

If the material in question obeys Ohm's law, the dependence of the voltage V on the current i is linear ($V = iR$). The electrical resistance R of the (cylindrical) sample under consideration is given by

$$R = \rho l/S, \tag{28.2}$$

where ρ is the resistivity of the material. If we assume that the electric field across the cylinder is uniform, we can write $E = V/l$, and from (28.1) and (28.2), we have

$$\rho = RS/l = (V/i)(S/l) = (El/neSv_d)(S/l) = E/nev_d. \tag{28.3}$$

If we denote the ratio $\mu(= v_d/E)$ as the *drift mobility*, then we may write the resistivity as

$$\rho = 1/ne.\mu \tag{28.4}$$

and its reciprocal $1/\rho = \sigma$, which is the *conductivity* $\sigma = J/E$ of the material, becomes

$$\sigma = qn\mu. \tag{28.5}$$

In conclusion, the conductivity σ is proportional

(a) to the *concentration* n of the electric charge carriers,
(b) to their *mobility*, i.e., the parameter that takes into account the resisting forces they encounter in their movement.

In our model, the drift velocity is proportional to E ($v_d = \mu E$). The order of magnitude of the mobility at room temperature is

10 or 1,000 $cm^2V^{-1}s^{-1}$, respectively, for metals and semiconductors, and therefore, even with a fairly intense applied electric field (E $\approx$ 10 V/cm), the drift velocity is *much smaller* than the velocity due to thermal agitation.

The drift mobility ($\mu = v_d/E$) in microscopic terms can be expressed as follows:

$$\mu = e\tau/m, \tag{28.6}$$

where m is the effective mass of the carrier and τ is the average time between two collisions of the carrier against the crystal lattice or impurities in the crystal. This expression allows us to understand why the increase in temperature decreases σ (or increases ρ). As the temperature increases, the effective area of the ions in the crystal increases due to vibration, therefore, τ and μ decrease, increasing the value of ρ.

This relationship can also be obtained by assuming that in each collision, the charge carrier loses all the energy acquired by the electric field over time τ until the next collision. Being subject to acceleration $a = eE/m$, it acquires the average speed

$$v_d = a\tau = eE\tau/m. \tag{28.7}$$

The Drude model [3] predicts $\rho \propto T$ for the dependence of the resistivity on the temperature in metal conductors, where the concentration of charge carriers can be assumed to be *constant*, and therefore the *only parameter* that is affected by the variations in temperature in the relation $\rho = 1/(qn\mu)$ is the *mobility*.

In semiconductors, things are very different. The concentration of charge carriers n is *no longer constant* as the temperature changes: in the case of "pure" semiconductors, n increases exponentially with T, while in the case of "doped" semiconductors, n can be approximately constant (*extrinsic* zone at low temperatures) or it may grow exponentially with T (*intrinsic* zone at high temperatures).

To summarize: In a metallic material (n = constant), as the temperature increases, the ions that form the crystal vibrate more around their equilibrium positions (presenting a greater probability of collision), which results in an increase in effective section σ_{ef} and a consequent decrease in mobility and therefore increase in resistivity with temperature. For temperature ranges that are not very large, the variation of R with T in a metallic conductor is linear of the form

$$\boldsymbol{R(T)/R_0 = [1 + \alpha\,(T - T_0)]}, \qquad (28.8)$$

where R_0 is the resistance at temperature T_0 and α is the coefficient of variation of resistance with temperature, which is a characteristic parameter of each material.

In semiconductors, in the intrinsic zone (thermistors type NTC = Negative Temperature Coefficient), it is often found that

$$R(T) = R(T_0)\exp[\beta(1/T - 1/T_0)], \qquad (28.9)$$

where $R(T_0)$ is the value of its resistance at temperature T_0 and β is a characteristic parameter of the thermistor used. Annex A in this chapter discusses the origin of this expression. The following experiments are intended to study the validity of these predictions.

28.2 Measurement of resistance as a function of temperature

A simple device for measuring the variation of resistance with temperature consists of a container of water or oil, as illustrated in Figure 28.2, with a thermometer to measure its temperature. The resistance whose properties are to be studied is immersed in the liquid so that it acquires that temperature. Warmer or cooler liquid is added to vary the temperature. Since ordinary water conducts electricity, care must be taken so that this does not affect the resistance to be measured. However, as the resistivity of water is high, if resistances of less than about 500 Ω are measured, water generally does not affect their value. Otherwise, we can use distilled water or oil, which have much higher resistivities.

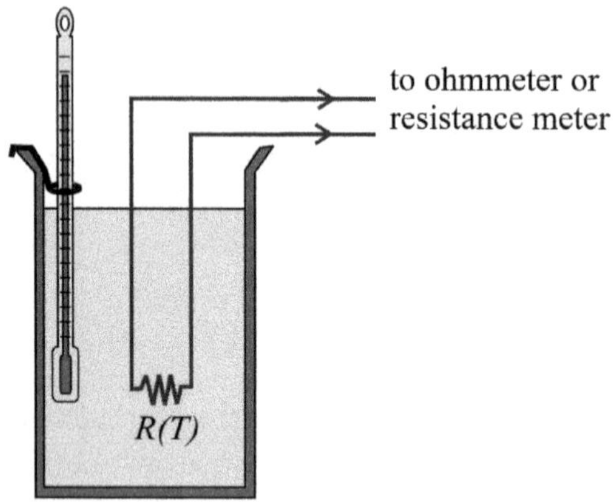

Figure 28.2 Experimental device to measure the variation of resistance with temperature. The temperature control bath consists of a container filled with a liquid of low electrical conductivity. It uses a thermometer to record temperature and some method to measure resistance R.

28.2.1 *Project 81. Measurement of the variation of resistance with temperature of a metallic wire by the four-point method*

Recommended basic equipment: Copper, aluminum, or iron wire. Two multimeters to measure currents and voltage (millivolts). A DC or AC voltage source of about 12 V @ 2 A. A thermometer.

In this project, you will need some samples of pure metals (~99% purity) so that you can easily compare the values you measure to tabulated values for the same material. Build a circuit like the one in Figure 27.2 to use the four-lead method or Kelvin method to measure resistance. See Chapter 27 for more details on this measurement technique.

Select wires of known materials, e.g., Cu or Al, of diameter 0.1–1 mm and length of about 2 or 3 m. To make it easier to place the sample in the water, it is helpful to wind the wire into a small coil, making sure that the turns do not touch each other. For copper, this is easy to do by using enameled wire. For aluminum, you can either use anodized wires, which are insulated, or construct the coil very carefully by winding the wire onto a nonconductive cylinder without the turns touching. Then use temperature-resistant epoxy to seal the arrangement. The coil material must be able to withstand a temperature cycle ranging from 0° to 100° C without deteriorating.

Suggested tasks:

✓ If you use a circuit like the one in Figure 27.2, which uses a source without current regulation, include a current limiting resistor (R_{ext}) in series of about 10–50 Ω and capable of withstanding a few amps. If the source is current-limited, this resistor is not essential.

✓ Immerse the sample (coil) in the container of water (or oil), as shown in Figure 28.2. Determine the value of R for each temperature T, varying the temperature between approximately 0° C and 100° C.

✓ Plot R as a function of T, and investigate the functional dependence between these variables. Can the data be adequately described with a linear relationship?

✓ If the relationship between R and T is linear, take one of the temperatures used as a reference, T_0, for example, the ambient temperature of the water, and as R_0 the resistance at this temperature. From the graph of R/R_0 as a function of T, determine the slope and its error. From these data, determine the coefficient α of variation of resistance with temperature and its corresponding error.

✓ Compare your results to the table values for the coefficient of variation of resistance with temperature. Discuss the degree of agreement or disagreement with the table values. Remember that commercial materials are rarely pure, therefore some variations with tabulated values are to be expected.

28.2.2 *Project 82. Variation of resistance with temperature of a metallic alloy*

Recommended basic equipment: Nichrome, manganin, or constantan wire a few meters long and diameter such as to achieve a resistance of the order of about 50 Ω or greater. A multimeter that can measure electrical resistance in the range $1-10^4\Omega$. A thermometer.

Many alloys are used to build resistors. They usually have high electrical resistivities and are resistant to high temperatures. For example, constantan is a copper and nickel alloy, nichrome is a nickel and chromium alloy, and manganin is a copper, manganese, and nickel alloy. These materials are inexpensive and can be purchased in the form of wires of different diameters. Many of them are sold with insulators. If they are not insulated, when you wind them, take care to electrically isolate the different turns from each other. Use any of them to build resistors of the order of about 50 Ω or greater. This kind of resistor makes it possible to use an ohmmeter to measure the resistance, i.e., the two-point method.

With an experimental device like the one shown in Figure 28.1, study experimentally the variation of R with T.

- ✓ Immerse the coil in the container of water and determine the value of R for each temperature T, varying the temperature between approximately 0°C and 100°C.
- ✓ Plot R as a function of T, and investigate the functional dependence between them. Can the data be described adequately with a linear relationship?
- ✓ If possible, find the coefficient of variation of R with temperature, α, and compare its value to those in the table.

28.2.3 *Project 83. Variation of resistance with temperature of a thermistor*

Recommended basic equipment: A Negative Resistance Temperature (NRT) thermistor of resistance less than 1 kΩ at room temperature and a multimeter that can measure electrical resistance in the range $1–10^4\Omega$. A thermometer and a multimeter in ohmmeter mode.

Thermistors are semiconductor devices widely used to measure temperatures, due to their low cost and great sensitivity. Their thermometric property is electrical resistance, so it is important to know precisely the variation of R with T. However, the dependence on temperature is not simple. As discussed earlier, Eq. (28.7) and

Annex A, an expression that could describe the behavior of these components is

$$R(T) = R(T_0)\exp\left(\beta\left[\frac{1}{T}-\frac{1}{T_0}\right]\right) = R_0\exp\left(\frac{E_g}{k_B}\left[\frac{1}{T}-\frac{1}{T_0}\right]\right), \tag{28.10}$$

where T and T_0 are absolute temperatures and $R(T_0) = R_0$ is the resistance at T_0, a temperature taken as a reference. $\beta(=Eg/k_B)$ and E_g are constants to be determined experimentally and $k_B = 1.38065 \times 10^{-23}\,\mathrm{J}/k = 8.61734 \times 10^{-5}\ eV/k$ is Boltzmann's constant.

Choose a low-resistance thermistor ($R(T_0) < 1\ \mathrm{k}\Omega$) so that when it is immersed in water, its value is not affected by the resistance of the water. If the resistance of the thermistor is less than about 5 kΩ, you can use tap water that will not affect its resistance significantly. If you want to study high-resistance thermistors, you can substitute the water in the thermal bath for some non-flammable, non-polluting oil, such as silicone oil. If you use a thermistor with a resistance of the order of 0.5–5 kW, a multimeter in ohmmeter mode can be use to measure the resistance.

Suggested tasks:

✓ Plot the thermistor resistance as a function of temperature and the quotient $R(T)$ as a function of $1/T$ on a semi-logarithmic scale. What kind of dependency between R and T does this graph suggest?

✓ Choose a temperature as a reference, for example, T_0, ambient temperature. Call $R_0 = R(T_0)$. Graph the quotient $R(T)/R_0$ as a function of $1/T$ on a semi-logarithmic scale.

✓ Say whether expression (28.11) adequately accounts for your experimental results. If possible, find the coefficient β.

✓ If expression (28.11) adequately describes your results, find the value of E_g. For a wide variety of semiconductors $E_g < 5\,\mathrm{eV}$, in the case of silicon, its value is $E_g \approx 1.12\,\mathrm{eV}$.

✓ (Optional) What can you conclude about Boltzmann's statistics, Eq. (28.9), to describe the behavior of electrons in a semiconductor from this study?

Questions

Discuss some of the possible applications and implications of the previous experiments. For example:

(a) If you want to make a "standard resistor" with which to compare other resistors, what parameters are important to consider for a good choice of material? Would you choose a pure metal or an alloy to build said resistance?
(b) If you want to use a resistor as a temperature sensor, what criteria would you use for the choice of material? Would you choose a pure metal or an alloy to build said resistance?
(c) Using the ideas discussed in this activity, consider the coefficient of variation of the resistance with the temperature of tungsten and discuss how you would estimate the temperature of the filament by measuring its resistance. Suggest a setup to perform this experiment.

28.3 Annex A: ♣ Simple model of conduction in semiconductors

In semiconductors, the ions vibrate more with increasing temperature, but there is also a significant increase in the number of charge carriers n, with increasing temperature, which has a greater impact on conduction. Therefore, in semiconductors, resistivity generally decreases as temperature increases.

A very simplified model to estimate the number of charge carriers (electrons) in the conduction band of a semiconductor consists of assuming that in order for them to pass from the valence band to the conduction band, they must jump a potential barrier gap E_g, which is usually referred to as *band gap.* Within this model, only the electrons in the conduction band conduct electricity, while the ones in the valence band are bound to the ions and unable to conduct electricity.

An important result of statistical mechanics is that if we have a set of N_0 particles at a temperature T, which can be at two energy levels, one lower, E_1, and another higher, E_2 (with $E_g = E_2 - E_1$),

Chapter 29

ELECTRICAL CONDUCTION IN LIQUIDS — ELECTRON CHARGE ESTIMATION

This chapter studies electrical conduction in liquids and how its resistance varies with temperature, and also explains introduction to the phenomenon of electrolysis and estimation of the value of the electric charge of the electron.

Goals

✓ Electrical conduction in liquids
✓ Electrolysis
✓ Estimation of electron charge

29.1 Simple model of conduction in liquids — Electrolysis

As discussed in previous chapters, for a material medium to conduct electrical current, it must contain mobile charges capable of carrying electricity. In metals, the mobile charges are the electrons of the outermost shells of the atoms (conduction electrons) [1]. In some liquids, such as water, dissolved salts, acids, or bases dissociate into positive and negative ions that can move through the liquid, making electrical conduction possible. It is the migration of positive and negative ions through the liquid that carries the electric current in them. This conduction is particularly sensitive to the types of ions present and to temperature.

An important difference between electrical conduction in liquids and solids is that in liquids, it is accompanied by the transport of matter. For simplicity, assume that in the liquid there is only one type of ion, whose charge is ve (e is the charge of the electron and v the charge number per ion). The mass of the ion is denoted by μ_{ion}. The charge Q transported through the liquid due to the passage of N ions is

$$Q = N \cdot \nu \cdot e. \tag{29.1}$$

On the other hand, this number N of ions transported is related to the mass transfer Δ m *to* the electrode:

$$\Delta m = N \cdot \mu_{\text{ion}} = N \cdot \frac{\text{Mol}_{\text{ion}}}{N_A}, \tag{29.2}$$

where Mol_{ion} is the molar mass of the ions and N_A is Avogadro's number. Clearly, $\mu_{\text{ion}} = \text{Mol}_{\text{ion}} / N_A$.

29.2 Faraday's law of electrolysis

Combining Eqs. (29.1) and (29.2), we have [1, 2]

$$\Delta m = \frac{1}{v.e} \mu_{\text{ion}} \cdot Q = \frac{1}{v.e} \left(\frac{\text{Mol}_{\text{ion}}}{N_A} \right) \cdot Q = \left[\frac{1}{F} \frac{\text{Mol}_{\text{ion}}}{\nu} \right] \cdot Q. \tag{29.3}$$

This expression relates linearly the transported mass, Δm (deposited on the electrodes), with the charge Q transferred in the process. Equation (29.3) is known as *Faraday's law of electrolysis* [4]. The constant $F = N_A.e = 96500C$ is the *Faraday constant* [3] and the quotient ($E = \text{Mol}_{\text{ion}}/v$) is known as the *chemical equivalent* or *equivalent weight* of the ion that carries the current. The phenomenon of transport and deposition of elements or ions by means of electricity is known as electrolysis [6], and Eq. (29.3) is sometimes called Faraday's law of electrolysis.

An important aspect of electrical conduction in liquids is that due to the transport and deposition of matter (ions) on the electrodes, the characteristics of the liquid medium (*electrolyte*) change as current flows through it. This means that if we want to study the properties of the conducting medium under stationary conditions, we must use alternating currents so that the load carried is on average zero during the experiment. If we use DC current, the test must only last a short time, to prevent major changes in the composition of the electrolyte.

As discussed in Chapter 28, for a medium to obey Ohm's law, the average (constant) speed of the charge carriers must depend linearly on the magnitude of the applied field. In a liquid, it is reasonable to assume that ions move in a viscous medium and suffer a viscous force $F \alpha V_{\text{ion}}$(V_{ion} =*drift velocity of the ions*) that must balance the force applied by the electric field E. Therefore, in equilibrium, we can write

$$F = -veE = -6\pi a\eta . V_{\text{ion}}. \tag{29.4}$$

The third member of this expression is the value of the viscous force (Stokes' law [5]) for a sphere of radius a moving with velocity v in a medium of viscosity (dynamic) η, ve is the ion charge, and E is the value of the electric field. Therefore, based on the arguments discussed in Chapter 28, we can write for the resistivity ρ, Eq. (29.3),

$$\rho = \left(\frac{E}{V_{\text{ion}}}\right) \cdot \frac{1}{nve} = \left(\frac{6\pi a\eta}{ve}\right) \cdot \frac{1}{nve} \propto \frac{\eta}{n}. \tag{29.5}$$

In this expression, ρ is the resistivity of the medium and n is the number of charge carriers (ions) per unit volume. The number of charge carriers n depends on the concentration of the ions. Since the viscosity η of liquids generally decreases with temperature, Eq. (29.5) provides a model that allows us to understand qualitatively the dependence of resistivity on temperature. It is expected that for liquids, resistivity should decrease with the concentration of ions and with temperature.

Figure 29.1 schematically shows a device that can be used to investigate electrical conduction in liquids. The liquid is contained in a vessel made of insulating material (with high resistivity, it may be plastic or glass). An alternating voltage source, a function generator, with an output range of 0–10 V is connected to the two electrodes immersed in the liquid. An ammeter — connected in series with the source and the liquid — is used to measure the current flow and a voltmeter — connected in parallel — to measure the voltage drop across the electrodes.

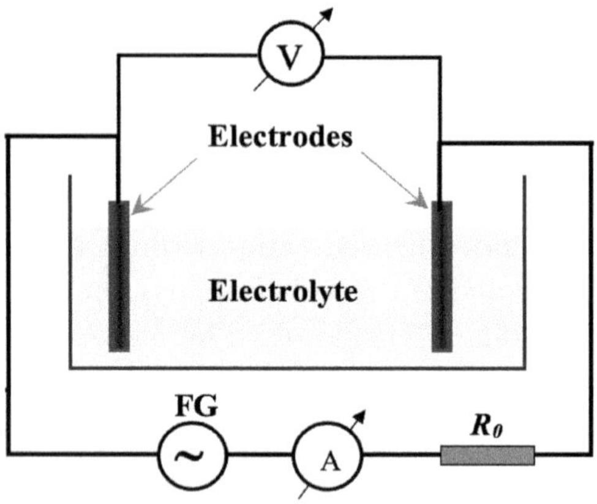

Figure 29.1 Tank with electrodes and electrolyte. The voltage source can be a function generator (FG) or a variable alternating voltage source. V is the voltmeter and A is the ammeter, both operating in AC mode. This circuit may include a current-limiting resistor R_0.

29.2.1 *Project 84. Conductivity of a liquid — Semi-quantitative study*

Recommended equipment: A glass vessel or beaker of approximately 1/2 l with two electrodes. A source of alternating voltage of 0–10 V. A voltmeter and an ammeter.

Start by pouring into the tank the purest distilled water you can get, and apply a voltage of about 10 V. Measure the current with the ammeter. It is important to verify that the ammeter is set to measure alternating current (AC mode). In a separate vessel, prepare a solution of 1 g of common salt, NaCl, in 1/2 l of water. This solution provides the ions necessary for electrical conduction.

Suggested Tasks:

✓ Starting with distilled water, add the NaCl solution to the vessel one drop at a time, and observe how the current varies. Keep the applied voltage stable and measure the current I that flows through the electrodes.

✓ Plot I as a function of the number of drops. How does the number of charge carriers (ions) affect the conductivity of the electrolytic medium?
✓ What conclusion can you draw from this experiment?

29.2.2 *Project 85. Conductivity of a liquid — Voltage–Current Relationship*

Recommended equipment: A vessel or beaker of approximately 1/2 l with two electrodes. A 0–12 V alternating voltage source. A voltmeter and an ammeter.

The aim of this experiment is to investigate the dependence of voltage V on the current I for a solution of defined concentration. It is proposed to use the same experimental arrangement as in the previous project. Using a voltage divider or a variable voltage source, study the variation of the current I that flows through the liquid when the voltage V applied to it is varied.

Suggested tasks:

✓ Plot Ias a function of V. If the relationship between V and I is linear, the conducting medium is said to obey Ohm's law. How can you describe the observed dependency?
✓ If your liquid obeys Ohm's law, find the value of R.

29.2.3 *Project 86. Conductivity of a liquid — Effect of temperature*

Recommended equipment: A vessel or beaker of approximately 1/2 l with two electrodes. A 5–10 V alternating voltage source. A voltmeter and an ammeter. A thermometer.

Suggested Tasks:

✓ Using the device illustrated in Figure 29.1, use a thermometer to measure the temperature of the liquid and study the variation of resistance with temperature. For example, you can use a heater to heat the liquid. Before taking measurements, make sure that the temperature of the liquid is stable. Stir the liquid with the thermometer and wait for thermal equilibrium before taking measurements. Take care not to alter the geometry of the plates (electrodes) when performing these manipulations. Determine the resistance of the system for different temperatures.

✓ Plot R as a function of temperature T.

✓ Consult the literature to see how the viscosity of water varies with temperature. Using these viscosity data for the various temperatures, plot R as a function of temperature T. Next, with the same data, plot the quotient R/η as a function of temperature T.

✓ Do you find any correlation or link between the quantities involved in this analysis? What can you say about the validity of the model proposed above for charge transport in liquids?

29.2.4 *Project 87. Electrolysis and electron charge*

> **Recommended equipment**: A vessel or beaker of approximately $1/2$ l with two light copper or bronze electrodes. A direct voltage source of 5–12 V @ 2A. A voltmeter and an ammeter. A scale that can weigh the electrodes with a precision ≈ 0.01 g or better. SO_4Cu solution.

Use the device shown in Figure 29.1 to investigate the phenomenon of electrolysis. Before beginning this study, label each electrode clearly and weigh them carefully. This time, the voltage source is DC and the voltage and current meters must operate in DC mode. Apply a voltage in the range 2–10 V. Place a copper sulfate solution of known concentration, for example, 50 g/l, in the vessel. Adjust the current (by setting the proper voltage) between 0.25 A and 2 A and wait for about half an hour. Observe what happens to each of the electrodes as the current flows. Keep the current as stable as possible and measure the time Δt during which the current circulates. This will let you know how much charge ($Q = I \cdot \Delta t$) has been transferred. At the end of this test, remove the electrodes, let them air-dry or place them carefully on filter paper or a towel to drain. Once they are dry, weigh them carefully to find for each electrode the change in mass Δm, i.e., the mass of copper deposited on the cathode.

Semi-quantitative study

✓ What happened to the mass of each electrode: did it increase or decrease?

✓ Repeat this experiment with other values of current flowing through the tank, in order to obtain a table of values of Δm and Q in as wide a range of Q as possible.

✓ With the data you have recorded, graph Δm as a function of Q.

Quantitative study

✓ If the relationship between Δm and Q is linear, determine the value of the slope and estimate its error. If the solution is SO_4Cu, $\mu_{\text{ion}} = 63.546g/N_A$, and for the cation Cu++, $v = 2$, use Eq. (29.3) and the value of the slope to estimate the value of the charge on the electron e, and estimate the error of this determination.

✓ Compare the result found for e with the accepted value for the charge of the electron. What can you conclude from this method of measuring e?

References

[1] Wikipedia, "Electronic band structure," 2021. Available: https://en.wikipedia.org/wiki/Electronic_band_structure.

[2] E. M. Purcell and D. J. Morin, *Electricity and Magnetism, Berkeley Physics Course*, 3rd. Ed., NY: Cambridge University Press, 2014, p. 839.

[3] B. H. Mahan, *University Chemistry*, 3rd Ed., NY: Narosa Publishing House, 1986.

[4] Wikipedia, "Faraday's laws of electrolysis," 2023. Available: https://en.wikipedia.org/wiki/Faraday%27s_laws_of_electrolysis.

[5] Wikipedia, "Faraday constant," 2021. Available: https://en.wikipedia.org/wiki/Faraday_constant.

[6] Wikipedia, "Faraday's laws of electrolysis," 2021. Available: https://en.wikipedia.org/wiki/Faraday%27s laws of electrolysis.

[7] Wikipedia, "Stokes law," 2021. Available: https://en.wikipedia.org/wiki/Stokes%27 law.

Chapter 30

CAPACITORS AND DIELECTRICS

The objective of this chapter is to study the basic properties of capacitors, in particular, the dependence of capacitance on geometry and the effect of the presence of dielectrics. We will also study some techniques to measure capacity and analyze systems of capacitors in series and in parallel.

Goals

✓ Capacity measurement
✓ Capacity dependence on the geometry of a flat capacitor
✓ Dielectrics

30.1 Capacitors

A *capacitor* is a system formed by two conductors in close proximity, of arbitrary shape, which have equal charges Q of different signs, as illustrated in Figure 30.1. This system of conductors is called a capacitor. The conductors are called capacitor plates and Q is the capacitor charge [1–3].

An interesting fact is that in general, the potential difference between plates is proportional to the charge Q. The *capacity* C is defined as the quotient between the absolute value of the charge Q of the plates and the potential difference V between them [1, 2, 4, 5]:

$$C = \frac{Q}{V}. \tag{30.1}$$

If we imagine that the capacitor is charged by carrying infinitesimal charges from one plate to another until the value Q is reached, work will have been done, and stored in the electric field that was created

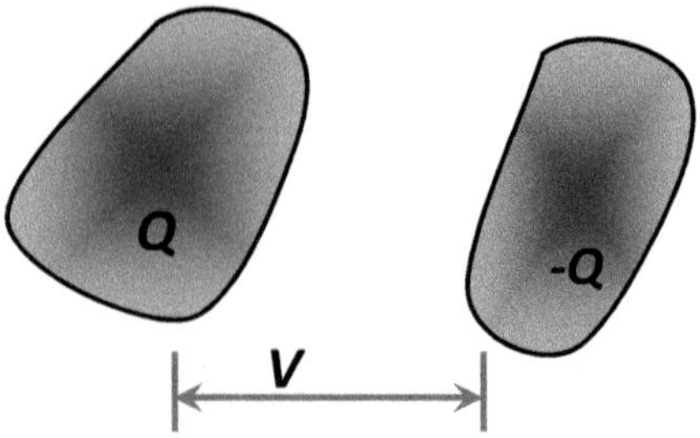

Figure 30.1 Capacitor formed by two conductors. Both conductors usually have the same charge, Q, but with opposite signs. The potential difference between the two conductors is V.

as it was being charged. The work done is equal to the energy stored in the capacitor. Its value is simple to calculate [1, 2, 5]:

$$W(Q) = \int_{q=0}^{Q} V(q) \cdot dq = \int_{q=0}^{Q} \frac{q}{C} \cdot dq = \frac{1}{2}\frac{Q^2}{C} = \frac{1}{2}CV^2. \qquad (30.2)$$

A capacitor is a device that stores energy, and the *capacity* measures *how much energy the capacitor can accumulate*, for a difference in voltage of $\sqrt{2}$ volts between the two plates.

The capacity of a system depends on the geometric shape of the capacitor and the type of dielectric material placed between the plates. In particular, for a flat capacitor, i.e., a capacitor with parallel flat plates of equal areas A, separated by a distance d, when there is vacuum between the plates, the capacity is [1, 5]

$$C_0 = \varepsilon_0 \frac{A}{d}, \qquad (30.3)$$

where $\varepsilon_0 = 8.85 \times 10^{-12}\ \mathrm{C^2/Nm^2}$ is the *permittivity of vacuum*, which can also be expressed as $\varepsilon_0 = 8.85\ \mathrm{pF/m}$.

An insulator or dielectric is a material that does not conduct electricity. If a *dielectric* fills the entire space between the plates of a flat capacitor, the capacitance is [1, 2, 5]

$$C = k\varepsilon_0 \frac{A}{d} = kC_0, \qquad (30.4)$$

that is, k times greater than the capacity C_0 of an identical condenser but with vacuum between its plates. So, $k = C/C_0$ is the *dielectric constant* of the inserted medium and depends on the nature of the material and on the temperature.

30.2 Capacitance meter

Capacitance meters are used to measure the capacity of a discrete capacitor. They come in a wide spectrum of sophistication and prices. Currently, there is low-cost equipment on the market with an approximate measuring range from 2 nF to 2,000 μF. Before using a *capacitance meter*, you should inspect it carefully and read the instructions manual. Find out about its mode of operation, precautions for use, measurement range, sensitivity, and calibration controls. For each measurement you make, estimate the measurement uncertainty according to the specifications provided by the manufacturer. Many commercial multimeters include capacity measurement among their functions.

Commercial capacitors have a *nominal value* indicated by the manufacturer by means of a label or a code [6], such as a color code or alphanumeric characters. The actual capacity often differs from this value by a few percentage points.

The main objective of the activities proposed in this chapter is to study the basic properties of a parallel plate capacitor and analyze the dependence of the capacitance with the geometry and with the presence of a dielectric material between the plates. We will also study some methods to determine capacitance and analyze systems of capacitors in series and in parallel.

30.2.1 *Project 88. Serial and parallel capacitors — Basic measurements*

Recommended basic equipment: Capacitance meter with a range spanning from 200 pF to 100 μF. Two capacitors of different values in the order of 1 μF.

Important: Before connecting the capacitor to a capacitance meter, check that the capacitor is discharged, otherwise the instrument may be damaged. To check this, measure the potential difference between the capacitor terminals with a voltmeter. If the measurement is different from zero, keep the voltmeter connected so that the capacitor discharges through it and wait until a null reading is obtained. You can also place a resistance of a few tenths of an Ohm between its terminals if you want a faster discharge.

Suggested tasks:

✓ To become familiar with the instrument, start by measuring the capacitance of known capacitors. Compare each measured value with the capacity rating of each capacitor you use. Observe any possible discrepancies that may exist and estimate any measurement error.

✓ Using two capacitors of different values, but of the same order of magnitude, determine the value of the capacitance of each one and estimate the uncertainty of their measurements.

✓ **Serial configuration:** Using two previously measured capacitors, connect them in series, Figure 30.2(a), and measure the capacity of the combination. Compare the experimental result to the value of the equivalent capacity C_S of two capacitors C_1 and C_2 in series, given by the following expression [1, 2, 5]:

$$C_S = \frac{C_1 C_2}{C_1 + C_2}. \tag{30.5}$$

✓ Are the measured values consistent with the calculated values? Is there discrepancy between these data? What criteria do you use to decide this?

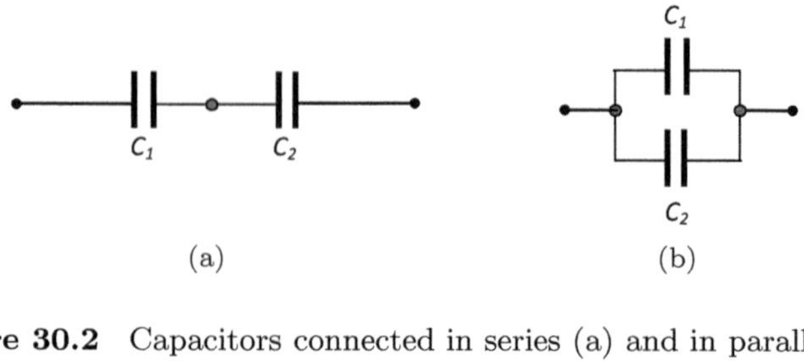

Figure 30.2 Capacitors connected in series (a) and in parallel (b).

✓ **Parallel configuration:** Connect the same capacitors from the previous exercise in parallel and measure the capacity of the combination. Compare the experimental result to the expected value of the equivalent capacity according to the expression of the equivalent capacity C_P for two capacitors in parallel:

$$C_P = C_1 + C_2. \tag{30.6}$$

✓ Are the measured and calculated values comparable?

In all cases, carefully analyze the measurement errors and consider the uncertainty propagation formula to express the uncertainties of the measurement of C_S and C_P.

30.2.2 *Project 89. Parallel flat plate condenser — Variation of capacity with geometry*

Recommended basic equipment: Capacitance meter with a range spanning from 200 pF to 100 nF. Two flat metal plates with an area comparable to that of a letter-size sheet of paper (or A4) and some thin dielectric plates, such as glass, acrylic, mylar, or paper of comparable size.

Two metal plates of area A, facing each other and separated by a distance d, form a parallel plates capacitor. To study the basic properties of a capacitor of this type, the plates must be sufficiently rigid to prevent them from deforming during the experiment. Aluminum, copper, or bronze plates 2–4 mm thick are suitable, and of course, other metals can be used. Aluminum foil can be attached to a rigid wooden board. One way to keep the plates separated at a fixed distance d during the experiment is to interpose spacers of known heights between them, which can be placed near the corners of the plates. For example, spacers can be made of square pieces of paper of about 1 cm^2, placed at the corners of the plates. By varying the number of papers placed at the corners, you can change the distance between the plates. When a rigid dielectric material is placed between the plates, the thickness of the material will define the gap between the plates.

When using paper as dielectric, the thickness of the sheets can be calculated by measuring the height of a large number of these sheets. As the height of the stack of sheets depends on the pressure to which it is subjected, it is useful to place a flat wooden board with a weight on the top sheet. This weight should be the same throughout the experiment. A possible embodiment of the experiment is schematically illustrated in Figure 30.3. The connecting cables from the plates to the capacity meter can be soldered or screwed to the plates. If you choose to use screws, make sure that they do not protrude from the surfaces of the plates that face each other (flat-head screws are suitable). Otherwise, you can simply use thin aluminum or copper foil in contact with each plate [7, 8].

Using two flat metal plates, study the variation of the capacity with the distance d between the plates and the area of the plates. Perform the experiment with air between the plates and determine the capacity value for at least five different separations d. One way to obtain different values of areas of the conductive plates is to have

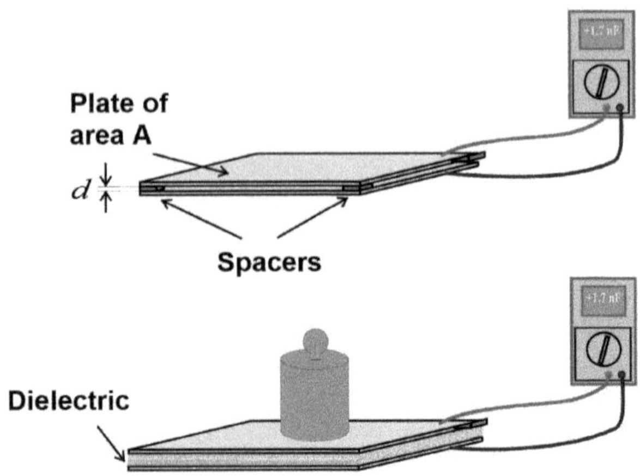

Figure 30.3 Schematic diagram of the experimental device proposed to study the variation of the capacity with the distance d between plates and the nature of the dielectric between plates. In the lower diagram, sheets of dielectric (paper, glass, acrylic, nylon, mylar foils, etc.) are *sandwiched* between the conducting plates.

several pairs of plates of different areas, A, and then combine them in parallel, preserving the same separations between them. Another way of varying the area is to move one of the plates so that the overlapping area changes.

Suggested tasks:

✓ Plot C as a function of d (gap between plates) and C as a function of $1/d$.

Note: The graphical representation of expression (30.3) can be *linearized* if the capacity is plotted as a function of the inverse of the separation between the plates. In this *linearized* plot, the regression line should pass through the origin. Look at your experimental data and analyze whether they agree with this expected behavior. If the fitting line in the graph of C as a function of $1/d$ does not cross the origin, give a physical meaning to intercept at $1/d = 0$ as predicted by Eq. (30.3). A possible reason may be related to the fact that the capacitance meter measures an *equivalent capacity* (C_{eq}) which includes the capacity of the connection cables, C_{cabl}, which can be significant in relation to the system under examination. That is, what the capacitance meter measures is C_{eq}:

$$C_{\text{eq}} \approx C_{\text{cabl}} + C = C_{\text{cabl}} + k\varepsilon_0 \frac{A}{d}. \tag{30.7}$$

✓ Plot C_{eq} as a function of A (area of the plates), if you can make the areas vary.

✓ Plot C_{eq} as a function of A/d for all the data available (in this graph, you can combine data taken by your colleagues) with the same dielectric between the plates.

✓ What can you conclude from these graphs about the variation of capacity with geometry?

✓ Discuss the degree of agreement between your results and the standard expression for the capacity of a flat plate capacitor, Eq. (30.3).

✓ If the plot of C_{eq} as a function of A/d is linear, for the case that the space between the plates is basically air, determine the slope and its error. Find the experimental value of the vacuum permittivity

ε_0 and compare it to the value provided in the literature. For this comparison to be consistent with the system of units you use to measure C_{eq} and A/d, in general, it is suggested to work with the international system (SI) units.

30.2.3 *Project 90. Variation of capacity with dielectric medium*

Recommended basic equipment: Capacitance meter with a range spanning from 200 pF to 100 nF. Two flat metal plates of equal areas. Some thin dielectric plates, such as glass, acrylic, mylar, or paper.

Using insulating or dielectric media that fill the entire volume between the plates, study the variation of the capacitance with the type of insulating media (dielectric). As a dielectric, you can use glass, acrylic, mylar, or paper. For a capacitor, the ratio of its capacity with and without dielectric, C_0 and C_k, respectively, determines the value of the dielectric constant k of the medium (as long as the geometry is maintained, i.e., equal area and distance between the plates) [1, 7, 8]:

$$k = C_k/C_0. \tag{30.8}$$

One way to find the value of k is to study the variation of C with A/d for the capacitor, in air and when it has a dielectric. The quotient of the slopes of each graph gives the relationship between the respective dielectric constants.

Suggested tasks:

- ✓ Plot C as a function of A/d for each dielectric material available.
- ✓ Determine the slopes of these graphs and their respective uncertainties.
- ✓ Determine the value of the dielectric constant of the materials used and estimate their uncertainties.
- ✓ Compare your results to the tabulated values of k. Discuss the degree of agreement or discrepancy between the values.

References

[1] R. Halliday, D. Resnick and M. Krane, *Fundamentals of Physics*, NY: John Wiley & Sons, Inc., 2002.

[2] E. M. Purcell and D. J. Morin, *Electricity and Magnetism, Berkeley Physics Course*, 3rd Ed., NY: Cambridge University Press, 2014, p. 839.

[3] R. Feynman, *et al.*, *The Feynman Lectures on Physics, Boxed Set: The New Millennium Edition*, Vol. 1, Basic Books, 2011.

[4] J. R. Reitz, F. J. Milford and R. W. Christy, *Fundamentals of Electromagnetic Theory*, NY: Addison-Wesley, 1996.

[5] F. Sears, M. Zemansky, H. Young and R. Freedman, *College Physics*, NY: Pearson Ed., 2009.

[6] Aspencore — Electronic Tutorial, "Capacitor colour codes," 2021. [Online]. Available: https://www.electronics-tutorials.ws/capacitor/cap_5.html.

[7] T. T. Grove, M. F. Masters and R. E. Miers, "Determining dielectric constants using a parallel plate capacitor," *Am. J. Phys.*, vol. 73, no. 1, pp. 52–56, 2005.

[8] B. B. Wells, *et al.*, "An adjustable parallel-plate capacitor instrument — Test of the theoretical capacitance formula," *Am. J. Phys.*, vol. 84, pp. 723–726, 2016.

[9] M. N. Sadiku, *Elements of Electromagnetics* (The Oxford Series in Electrical and Computer Engineering) 4th Edn., Oxford: Oxford University Press, 2006.

their distribution between these two levels is given by *Boltzmann's statistics* [4, 5]:

$$\frac{\text{Number of particles in } E_2}{\text{Number of particles in } E_1} = \frac{n_2}{n_1} = \exp(-E_g/k_B T), \tag{28.11}$$

where $n_1 + n_2 = N_0$ and k_B (=1.3806503 × 10^{-23} J·K^{-1}) is Boltzmann's constant. According to this result, if $n = n_2 \ll n_1 \approx N_0$, the number of electrons in the conduction band at absolute temperature T will be

$$n \approx N_0 \exp(-E_g/k_B T), \tag{28.12}$$

where N_0 is a constant. Therefore, according to Eq. (28.5) for a semiconductor, we would expect that

$$\rho \propto \frac{\sigma_{\text{ef}}}{n} \propto \sigma_{\text{ef}} \exp(E_g/k_B T) \tag{28.13}$$

or also,

$$\rho(T) = \rho(T_0) \exp[(E_g/k_B)(1/T - 1/T_0)], \tag{28.14}$$

where $\rho(T_0)$ is the resistivity of the semiconductor at the reference temperature T_0.

References

[1] D. Halliday, R. Resnick and J. Walker, *Fundamentals of Physics: Extended*, 11th Ed., John Willey and Sons Inc., 2018.
[2] H. Young, R. Freedman and F. Sears, *Sears and Zemansky's University*, NY: Pearson Pub., 2004.
[3] Wikipedia, "Drude model," 2021. Available: https://en.wikipedia.org/wiki/Drude_model.
[4] Wikipedia, "Statistical mechanics," 2021. Available: http://en.wikipedia.org/wiki/Statistical_mechanics.
[5] Physics with computer Interactive Physics Course on the Internet, "Boltzmann distribution function," 2021. Available: http://www.sc.ehu.es/sbweb/fisica/estadistica/boltzmann/formula/formula.htm.

Chapter 31

RC CIRCUIT

The goal of this chapter is to study the charging and discharging processes of a capacitor and determine the characteristic time constants of the circuit. It analyzes the basic properties of an RC circuit and the response of an RC circuit to a square and sinusoidal signal. It also considers techniques for determining the phase difference of sinusoidal signals.

Goals

- ✓ Charging and discharging a capacitor
- ✓ Determining the time constant of an RC circuit
- ✓ Phase difference between two signals

31.1 RC circuit

A capacitor and resistor connected in series with a voltage source make up what is called a series *RC circuit* (Figure 31.1).

By Kirchhoff's first law, [1, 2] the voltage V_0 applied to the circuit is equal to the sum of voltage drops across the resistor and the capacitor [3–5]:

$$V_0 = iR + \frac{q}{C}, \tag{31.1}$$

where q is the instantaneous charge of the capacitor and i is the instantaneous current of the circuit. As $i = dq/dt$, by replacing in Eq. (31.1), the differential equation of the circuit is obtained:

$$\frac{dq}{dt} + \frac{1}{RC}q = \frac{V_0}{R}. \tag{31.2}$$

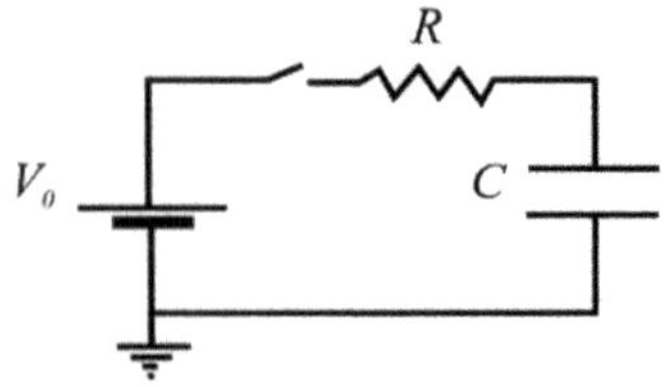

Figure 31.1 Series RC circuit with source.

If at instant $t = 0$, when the circuit is first connected to the source, the capacitor is discharged ($q = 0$), the solution to Eq. (31.2) is [1, 2]

$$q(t) = Q_0(1 - e^{-t/\tau}), \tag{31.3}$$

where $Q_0 = CV_0$ is the maximum charge that the capacitor will acquire and $\tau = RC$ is the circuit *time constant* in the "charging process". During charging processes, the voltage drop across the capacitor, $V_C(t) = q(t)/C$, grows according to

$$V_C(t) = V_0(1 - e^{-t/\tau}). \tag{31.4}$$

Equation (31.4) states that, after a time τ from the instant the source is connected, the capacitor has a potential difference between its plates $V_C(\tau) = V_0\ (1 - e^{-1}) = 0.63\ V_0$, that is, 63% of its final maximum voltage V_0. The capacitor is fully charged when $t \to \infty$, but in practice it is usual to consider that "the capacitor is charged" when a time $t > 5\tau$ has elapsed, when the charge has reached 99% of its maximum value Q_0.

During the charging process, the current $i(t)$ decreases exponentially in time from its initial maximum value $I_0 = V_0/R$ as follows:

$$i(t) = \frac{dq}{dt} = I_0\ e^{-t/\tau}. \tag{31.5}$$

Since the voltage drop across the resistor is $V_R(t) = i(t)\ R$, we have

$$V_R(t) = V_0\ e^{-t/\tau}. \tag{31.6}$$

Once the capacitor is charged, it can be disconnected from the power supply and will remain charged unless the charge is drained through some element. One way to achieve controlled discharge is to connect

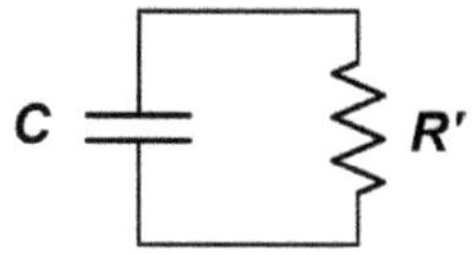

Figure 31.2 Serial RC circuit without a source. In this case, the previously charged capacitor is connected to a resistor R'.

it to a resistor R', as shown in Figure 31.2. This kind of circuit is called RC circuit with no source.

The equation of this circuit is identical to Eq. (31.2), which is the same as before, but with $V_0 = 0$. In this case, the solution of the equation shows that the charge of the capacitor decreases exponentially over time, from the moment it is connected to the resistor, according to

$$q(t) = Q_0 \, e^{-t/\tau'}, \tag{31.7}$$

and the potential drop across the capacitor is

$$V_C(t) = V_0 \, e^{-t/\tau'}. \tag{31.8}$$

Here, Q_0 and V_0 are the initial values of the charge and voltage in the capacitor, respectively, at the start of the discharge process, $t = 0$, and $\tau' = R'C$ is the discharge time constant. This time constant represents the time that must elapse from the start of discharge for the capacitor charge to decrease to 36% of the initial value Q_0.

31.1.1 *Project 91. Charging and discharging a capacitor using a data acquisition system connected to a PC*

Recommended basic equipment: Computer data collection system with a sampling frequency of at least 3 Hz. We can use a standard data collection system or a multimeter connected to a PC. DC voltage source or 9 V battery. Commercial resistor and capacitors ($C \approx 50$ or $100\ \mu$F and $R \approx 100$ kΩ).

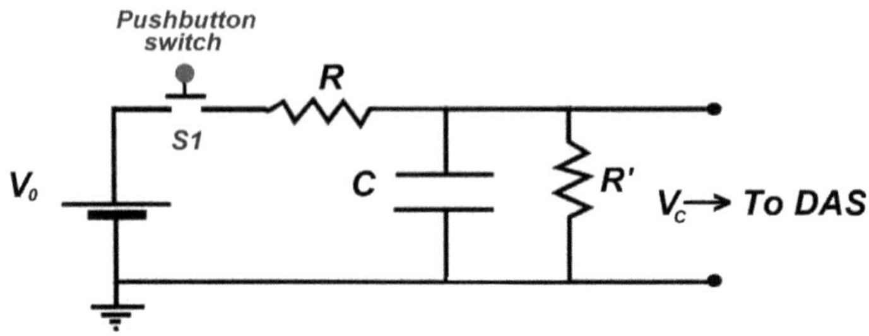

Figure 31.3 Circuit to study the discharge of a capacitor using a Data Acquistion System (DAS) connected to a PC.

Using the circuit shown in Figure 31.3, you can study the discharge of a capacitor connected to a resistance (RC circuit). For this experiment, use a capacitor C and a resistor R such that the product RC is of the order of 5 or 10 s. Connect the terminals of the data collection system to the capacitor terminals as shown in Figure 31.3. It is also useful (though not essential) to have a pushbutton switch. Make sure the value of the voltage source V_0 is lower than the maximum admissible by the data collection system (generally a 9 V battery is adequate for this study). Choose a sampling rate (amount of data per second that the acquisition system records) in the order of $100/(RC)$. Use the push button to charge the capacitor.

Suggested tasks:

✓ Plot the voltage drop across the capacitor V_C as a function of time during the charging process (choose time $t = 0$ when the pushbutton is switched off). Graph the experimental values of V_C as a function of t on a linear scale and on a semilogarithmic scale (see Chapter 2). In the same graph, plot the values $V_C(t)$ given by Eq. (31.8) and vary the value of parameter τ until you find the best fit of the theoretical curve to the experimental data. Evaluate whether the theoretical discharge description in Eq. (31.8) adequately describes this process. Determine the best value of τ from your graphs.

✓ In order to linearize the plot, construct a graph of $dV_C(t)/dt$ as a function of $V_C(t)$. (Suggestion: Read Chapter 2, which discusses the fundamentals of graphic analysis.) From these graphs, determine the time constant τ characteristic of the RC circuit. Compare this value to the one found in the previous point. Remember that both voltmeters and acquisition systems have an internal resistance r_V (see Annex A of Chapter 24). Therefore, the resistance R of the circuit in Figure 31.3 or Eq. (31.8) represents the equivalent resistance of the parallel connection between the selected resistance and r_V.

✓ Repeat this analysis for another set of R and C values and, for each combination, determine the value of τ from the fit of the data of $V_C(t)$ as a function of t. In each case, measure the values of C and R using a capacitor and an ohmmeter.

✓ For all the cases studied, plot the values of τ (obtained from the fits of the data of $V_C(t)$ as a function of t) as a function of the product RC, measured with the ohmmeter and capacitor, for each case.

✓ What can you conclude from this study about the dependence of the characteristic time τ with respect to the values of RC?

31.1.2 *Project 92. Determining the internal resistance of a voltmeter or data acquisition system*

Recommended basic equipment: A voltmeter and/or a computer data collection system with a sampling frequency of at least 10 Hz. DC voltage source or 9 V battery. A capacitor of known value in the order of 10 or 100 μF.

Annex A of Chapter 24 discusses a way to determine the internal resistance of a voltmeter, r_V. This section presents a complementary method that can also be used to measure the internal resistance of a data acquisition system connected to a PC. The technique consists of measuring the time constant τ of the capacitor of known value C when it is connected to the input of a voltmeter or acquisition system, using an arrangement like the one described in Figure 31.3, without including the external resistance R. That is, the capacitor connects directly to the input of the voltmeter or acquisition system. In other words, in this case, the resistance R is the internal resistance of the voltmeter or acquisition system.

For the system you want to study, build a circuit like the one in Figure 31.3, connecting the capacitor to the input of the voltmeter or data collection system. If you use a voltmeter that does not have a connection to a PC to collect data, you should read the voltage on the screen (or "display") every one or two seconds to build a table of V_C as a function of time. To make the process slower and easier to measure, you may want to use a capacitor of about 100 μF, in which case, even with an internal resistance of 1 MΩ, it would have a time constant of about 100 s, which would allow you to measure by direct reading simply and comfortably.

Suggested tasks:

✓ Plot the voltage drop across the capacitor, V_C, as a function of time in the capacitor discharge process. Choose the instant $t = 0$ when the pushbutton is switched off. Plot the experimental values of V_C as a function of t on a semi-logarithmic scale. In the same figure, report the values $V_C(t)$ given by Eq. (31.8). Vary the parameter value τ until you find the best fit of the theoretical curve to the experimental data. Determine the best value of t from your data.

✓ Estimate the value of the internal resistance $r_V = \tau/C$. Compare this technique to the one discussed in Annex A of Chapter 24.

31.1.3 *Project 93. RC circuit — Square excitation signal*

Recommended basic equipment: Two-channel oscilloscope (10 MHz or faster) or a data collection system with a sampling frequency greater than 300 Hz. Function generator with square output. Commercial resistor and capacitors ($C \approx$ 10 nF and $R \approx$ 500 kΩ).

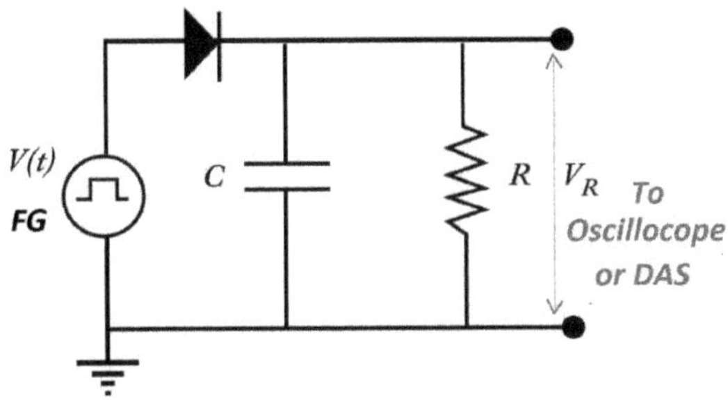

Figure 31.4 Setup for studying the response of an RC circuit to a square signal using an oscilloscope. The diode prevents the capacitor from discharging through the internal resistance of the function generator.

An RC circuit responds in different ways when excited by voltages of different shapes, e.g., square or sinusoidal voltages. The following experiments can be performed to study the response (current) of an RC circuit when it is connected to different types of excitations.

The circuit in Figure 31.4 may be suitable to study this behavior when the excitation voltage $V(t)$ is applied using a function generator. The configuration proposed in this figure is related to the fact that we assume that we are working with instruments operating in common ground. The diode prevents the capacitor from discharging through the internal resistance of the FG. An oscilloscope (or data acquisition system DAS) is used to monitor the voltage signals in each element of the circuit.

Note: Not all oscilloscopes and data acquisition systems work in *differential mode*, that is, with independent input grounds. Check the specifications of the system you are using. If your equipment works in *common mode* – which is often the case and should be assumed if you are not sure – the signals can only be measured referring to the *real ground* of the circuit. This requires carefully choosing a point in the circuit where the ground is to be connected and also connecting the *ground of the function generator* at that point. Also, remember that both the oscilloscope and the function generator have internal resistances. Oscilloscopes generally have a very high impedance input, in the order of or greater than 10 MΩ. Function generators usually have an internal resistance $R_i \approx 50\ \Omega$.

For this part of the experiment, work with a capacitor C and a resistor R such that the product $\tau = RC$ is of the order of 20 ms. Set the source output to apply a square signal at a frequency f in the order of 5 Hz, that is, with a period $T = 1/f \approx 0.2$ s. In this way, T will be of the order of 5τ, and in each cycle, you will be able to see the complete loading and unloading processes. To do this, connect the square signal, $V(t)$, of the FG to the capacitor through a diode. The diode prevents the capacitor from discharging through the internal resistance of the function generator. In this configuration, the only discharge path for the capacitor is the resistance R (and the internal resistance r_V of the DAS I parallel with R). Measure the value of R and the value of C used with a multimeter. Connect V_R to a channel of the oscilloscope or to an input channel of your data acquisition system. Compare the shape of the function generator output signal to the signals observed for V_R.

Figure 31.5 shows a plot used to estimate the value of τ.

Suggested tasks:

✓ Qualitatively study the behavior of $Vc(t)$ as a function of time. Does the discharge present the qualitative behavior indicated by Eq. (31.8) or in Figure 31.5?

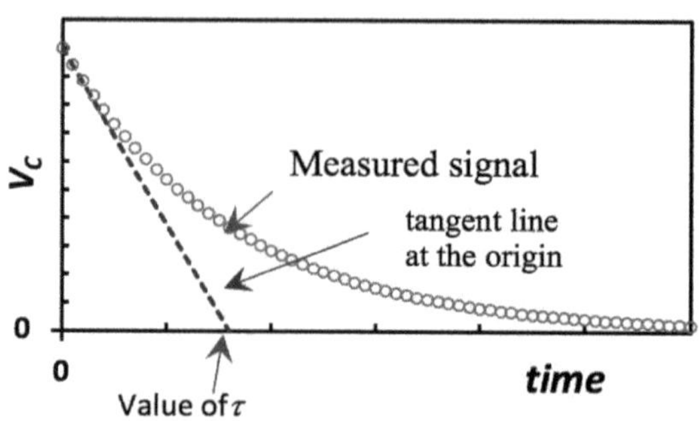

Figure 31.5 Graphical procedure to find the time constant using an oscilloscope. The example corresponds to the discharge of the capacitor in an RC circuit. The line is drawn tangent to the experimental curve at the origin ($t = 0$, when the discharge begins). The value of the abscissa where the line intersects the horizontal axis gives the value of τ.

✓ Determine the value of τ with the method suggested in Figure 31.5. Is this value consistent with the value of RC obtained from the measurements of C and R?

Optional if you use a digital oscilloscope or data acquisition system

✓ If you are using a *digital* oscilloscope (i.e., an oscilloscope that can store the numerical values of V_C as a function of t) or a data acquisition system (DAS), plot the experimental values of V_C as a function of t on a linear scale and on a semi-logarithmic scale. In the same graph, include the theoretical expectation of $V_C(t)$ given by Eq. (31.8) and vary the value of the parameter τ, until you find the best fit of the theoretical curve to the experimental data. To simplify the fit process, use a semi-log scale to plot V_C as a function of t.

✓ Evaluate whether the theoretical discharge description, Eq. (31.8), describes this process adequately. Determine the best value of τ from your plot.

✓ Repeat this analysis for another set of R and C values, and for each combination, determine the value of τ from the fit of the data of $V_C(t)$ as a function of t. In each case, measure the values of C and R using a capacitance meter and an ohmmeter.

✓ For all the cases studied, plot the values of τ (obtained from the adjustments of the data of $V_C(t)$ as a function of t) as a function of the product RC, measured with the ohmmeter and capacitance meter.

✓ What can you conclude from this study about the characteristic time dependence τ with respect to the values of RC?

31.2 Excited RC circuit — Forced response

If the circuit under study includes an alternating voltage source or a function generator (FG), we know – from Thevenin's theorem, Chapter 24 – that such a source can be modeled as an ideal voltage source $\varepsilon(t)$ and a certain internal resistance in series r_{int}. The circuit under consideration is illustrated in Figure 31.6. R is used

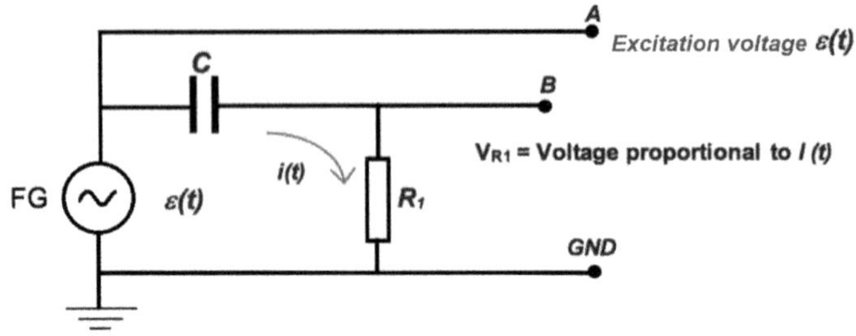

Figure 31.6 Forced RC circuit, R represents total resistance $(R_1 + r_{\text{int}})$. The voltage measured at A is the input voltage; at point B, a voltage proportional to the current in the circuit is measured. GND represents the circuit ground.

to designate the total or equivalent resistance of the circuit, that is, $R = R_1 + r_{\text{int}}$. Here, R_1 is the external resistor connected to the circuit. Applying Kirchhoff's mesh law, we have [4, 5]

$$R\frac{dq}{dt} + \frac{1}{C}q = \varepsilon(t). \tag{31.9}$$

Assume that the function $\varepsilon(t)$ is sinusoidal: $\varepsilon(t) = V_0 \sin(\omega t)$. A particular solution of the inhomogeneous differential equation (31.9) is

$$q(t) = -Q_0 \cos(\omega t + \beta) \tag{31.10}$$

and

$$i(t) = I_0 \sin(\omega t + \beta) = \omega\, Q_0 \sin(\omega t + \beta) \tag{31.11}$$

with

$$I_0 = \frac{V_0}{\sqrt{R^2 + (1/C\,\omega)^2}} = \frac{V_0}{R}\frac{\tau \cdot \omega}{\sqrt{1 + (\tau \cdot \omega)^2}} \tag{31.12}$$

and

$$\tan\beta = \frac{1/C\omega}{R} = \frac{1}{\tau \cdot \omega}. \tag{31.13}$$

On the other hand,

$$I_0 = \frac{V_0}{\sqrt{R^2 + (1/C\omega)^2}} = \frac{V_0}{R}\frac{\tau \cdot \omega}{\sqrt{1 + (\tau \cdot \omega)^2}}. \tag{31.14}$$

Equation (31.12) shows that for $\omega \to 0$, the current is zero, and for $\omega \to \infty$, the capacitor behaves like a short circuit. For intermediate frequencies, there is a phase shift between the applied voltage $\varepsilon(t)$ and the current b given by Eq. (31.13). The following project studies the behavior of this circuit experimentally.

31.2.1 *Project 94. RC circuit: Stationary response — Sinusoidal excitation signal*

Recommended basic equipment: Two-channel oscilloscope (10 MHz or faster) or a data collection system with a sampling frequency greater than 300 Hz. Function generator with sinusoidal output. Commercial resistor and capacitors ($C \approx 100$ nF and $R \approx 100$ kΩ).

Stationary response: Using the circuit shown in Figure 31.6, with the FG in sinusoidal mode, study the amplitudes of the voltages at points A and B as a function of $\omega = 2\pi f$. Call the signals measured with respect to the common ground (GND) V_A and V_B. In this case, $V_{R1} = V_B$, $V_C = V_A - V_B$, and $\varepsilon = V_A$. β denotes the phase difference between the applied voltage and the current, that is, between V_B and V_A. Annex A describes a method to determine β using an oscilloscope.

Suggested tasks:

- ✓ Plot the ratio of the measured amplitudes V_{R1}/V_0 as a function of ω. On the same graph, include the predictions of Eq. (31.14).
- ✓ Plot the ratio of the measured amplitudes V_C/V_0 as a function of ω.
- ✓ Study the phase shift β of V_{R1} with respect to V_A. On the same graph, plot the measured values and the theoretical values of β, Eq. (31.13), as a function of ω.
- ✓ Adjust the value of τ in your graphs to find the best possible fit for your data. Compare the value of τ found from your fit to the value of RC.
- ✓ What can you conclude about the proposed model?

31.3 Annex A: Determination of the phase difference between two signals

Consider two sinusoidal signals of the same frequency but different amplitudes and with a phase difference between them (*phase shift*) that we call ϕ. Consider the two signals $x(t)$ and $y(t)$:

$$x(t) = A \cdot \sin(\omega \cdot t), \tag{31.15}$$

$$y(t) = B \cdot \sin(\omega \cdot t + \varphi). \tag{31.16}$$

This section presents two complementary ways of determining the phase shift between two signals. If each of these signals is sent to channels 1 and 2 of a dual-trace oscilloscope, they will appear as shown in Figure 31.7.

The phase shift ϕ of the waves is related to the difference Δt (see Figure 31.7) by

$$\varphi = 2\pi \frac{\Delta t}{T}, \tag{31.17}$$

where $T = \frac{2\pi}{\omega}$. Both Δt and T can be measured on the graph, and these values determine ϕ.

If the same signals are analyzed with an oscilloscope operating in "XY mode", we obtain a graph like the one shown in Figure 31.8, that is, $y(t)$ as a function of $x(t)$, i.e., as a parametric plot of

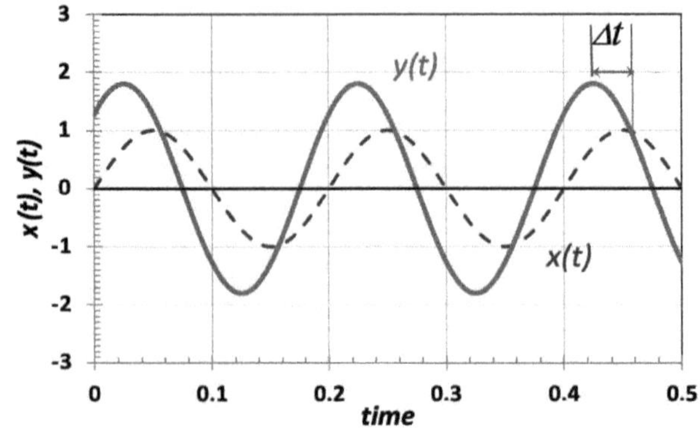

Figure 31.7 View of two sinusoidal signals of equal frequency with a phase difference ϕ between them, as a function of time.

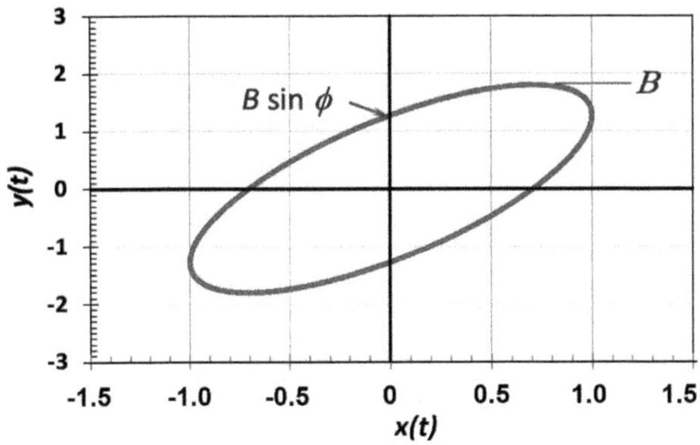

Figure 31.8 Signals $x(t)$ and $y(t)$ shown in XY mode.

Eqs. (31.15) and (31.16). The resulting locus is an ellipse, with its axes rotated by an angle ϕ relative to the horizontal axis of the oscilloscope.

It is easy to see in Eqs. (31.15) and (31.16) that for $t = 0$, $x(0) = 0$ and $y(0) = B\ \sin(\phi)$; therefore the quotient between the value of y, where the ellipse intersects the vertical axis, and the maximum value of the signal y ($y_{\text{máx}} = B$), gives the value $\sin(\phi)$. A similar analysis applies to the horizontal axis.

If the signals are acquired with a data collection system connected to a PC, the two waves $x(t)$ and $y(t)$ must be measured for at least one period and then plotted in a parametric representation.

If the signals $x(t)$ and $y(t)$ are in phase (i.e., if $\phi = 0$), the graph of y as a function of x is a line. On the other hand, if $\phi = \pi/2$, the graph of y as a function of x is an ellipse with its axes parallel to the coordinate axes.

References

[1] R. Halliday, D. Resnick and M. Krane, *Fundamentals of Physics*, NY: John Wiley & Sons, Inc., 2002.
[2] F. Sears, M. Zemansky, H. Young and R. Freedman, *College Physics*, NY: Pearson Ed., 2009.
[3] R. Feynman, *et al.*, *The Feynman Lectures on Physics, Boxed Set: The New Millennium Edition*, Vol. 1, Basic Books, 2011.

[4] E. M. Purcell and D. J. Morin, *Electricity and Magnetism, Berkeley Physics Course*, 3rd. Ed., NY: Cambridge University Press, 2014, p. 839.
[5] J. R. Reitz, F. J. Milford and R. W. Christy, *Fundamentals of Electromagnetic Theory*, NY: Ed. Addison-Wesley, 1996.

Chapter 32

LORENTZ FORCE AND AMPERE'S LAW

This chapter analyzes the properties of magnetic forces. More specifically, it proposes to study the variation of the force between two flat, parallel circular loops as a function of the common current that passes through them and of the distance between their planes.

Goals

✓ Lorentz force
✓ Force between two circular loops with equal number of turns

32.1 Force between two circular loops

When a current i circulates through a straight conducting wire of length l which is immersed in a magnetic field $\vec{B}$, a force $\vec{F}$, perpendicular to the field $\vec{B}$, is exerted on the wire, given by [1–3]

$$\vec{F} = i\vec{l} \times \vec{B}. \tag{32.1}$$

This expression is known as the *Lorentz force*. On the other hand, if a current I flows through a wire, according to *Ampere's law* [1–4], a magnetic field $\vec{B}$ is generated around the wire, given by

$$\oint_C \vec{B} \cdot d\vec{l} = \mu_0 I. \tag{32.2}$$

In particular, through a long wire of length L, at a distance $d \ll L$, the field intensity is [1–4]

$$B = \frac{\mu_0}{2\pi}\frac{I}{d}. \tag{32.3}$$

Therefore, the attractive (or repulsive) force F between two parallel wires, of length L, through which the currents I and i flow in the same (or opposite) direction, separated by a distance d, is given by [3, 5, 6]

$$F = iLB = \pm\frac{\mu_0}{2\pi}\frac{iI}{d}L. \tag{32.4}$$

The + sign indicates attraction (if both currents flow in the same direction) and the − sign indicates repulsion (when the currents flow in opposite directions).

One way to test expression (32.4) experimentally and falsify Eqs. (32.1) and (32.2) indirectly consists of studying the force between two flat, parallel circular loops (or circular coils) of radius R, with N_1 and N_2 turns each, through which a common current I circulates. If the distance between the planes of the coils is d, as long as $d \ll R$, we can assume that the force between them is similar to that of two parallel cables, and from (32.4), we have

$$F = \pm\mu_0\frac{N_1 \cdot N_2 I^2}{d}R. \tag{32.5}$$

To test this expression experimentally, one possibility is to use a device such as the one illustrated in Figure 32.1.

The lower loop rests on a 10 cm thick slab of expanded polystyrene (EPS), which is light and rigid [7]. This material separates the turns from the balance, thereby minimizing the possible interactions of the ferromagnetic parts in the balance with the magnetic field of the turns. The electronic balance or digital scale, can be replaced by a force sensor connected to a PC. In this case, the upper circular coil hangs from the force sensor and the variation of its weight is measured as a function of the common current in the coils and of the distance between them.

To minimize the interference of the connecting cables with the measured circular coil weight, it is advisable to use thin, flexible cables that are capable of withstanding a couple of amperes. If they are suspended from a support, their impact on the measurement can be made negligible.

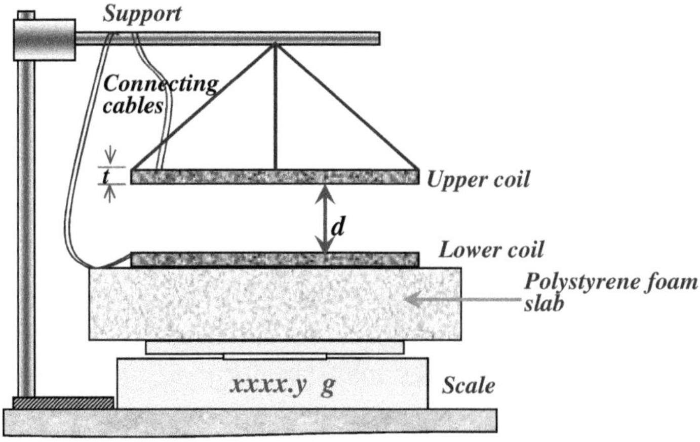

Figure 32.1 Experimental arrangement to measure the force of attraction or repulsion between two flat, parallel coils. The balance measures the weight changes of the lower coil caused by the magnetic force between them. The mean distance between the turns is d and the thickness of the turns is t.

32.1.1 *Project 95. Study of the magnetic force between two circular loops*

Recommended equipment: Two circular coils of equal radius R between 10 and 15 cm, with 200–500 turns each, made of wires suitable for currents of the order of 2 A. An electronic balance with sensitivity of at least 0.1 g and range greater than the mass of the heavier circular loop. Instead of a balance, you could use a force sensor connected to a PC. A variable DC source with an output current of the order of 2 A (when connected to both turns in series) and a multimeter or ammeter to measure the current through the turns.

Use an experimental setup like the one shown in Figure 32.1, with distances d between turns greater than the thickness t of the coils, but less than approximately $R/3$, i.e., $t < d < R/3$. If d_i is the distance between the closest faces of the coils, the average coil separation d is $d \approx d_i + t$. When the two coils are connected in series, the same current flows through both. By changing the polarity of the connection of one of them, you can reverse the force from attractive to repulsive.

Suggested tasks:

✓ For each distance d, measure the force of attraction or repulsion for at least five values of current through the turns.
✓ Plot the force as a function of I and I^2 for each value of d.
✓ For the same current, plot the force as a function of d and $1/d$.
✓ Using all the data obtained, but considering that the condition $t < d < R/3$ is always met, plot the force as a function of the parameter $y = I^2/d$.
✓ What can you conclude regarding the validity of Eq. (32.5) to explain your data?
✓ Discuss the linearity of the plot of the force as a function of the parameter: $z = N_1 N_2 R I^2/d$.
✓ Note that Eq. (32.5) assumes that the wires are in a vacuum, whereas the experiment was carried out in air. In this case, the equations are modified by introducing the relative permeability of air μ_r as a factor:

$$F = \pm\mu_0\mu_r \frac{N_1 N_2 I^2}{d} R = \pm\mu_0\mu_r \cdot z. \tag{32.6}$$

If the plot of the force as a function of z is linear, the slope of the line allows estimation of the value of the product $\mu_0\mu_r$. If these hypotheses are fulfilled, find the best value of the product $\mu.\mu_0$ and estimate its uncertainty. Remember that the permeability of vacuum μ_0, is defined in the International System of Units exactly as $\mu_0 = 4\pi 10^{-7}\mathrm{T}\cdot\mathrm{m}\cdot\mathrm{A}^{-1}$, while μ_r is dimensionless.

✓ Compare the value of μ_r for the air found in your experiment with the known value for air under standard conditions of temperature and pressure, $\mu_r(\text{air}) \approx 1$.
✓ What can you conclude from this experiment? Do your data agree with the predictions of Ampere's law and Lorentz force?
✓ What fundamental laws come into play to explain your results?

Optional

✓ For the case where the distance $d \approx R$, measure the force of attraction and repulsion for at least five current values in the turns. Include this data set in the above graph of force as a function of $z = N_1 N_2 R I^2 / d$. Does this data set fall within the same trend line as the previous ones? Propose an explanation of the causes. In this case, is it reasonable to approximate the forces between the coils to the force of two long, parallel wires? See Eqs. (32.3) and (32.4).

References

[1] R. Halliday, D. Resnick and M. Krane, *Fundamentals of Physics*, NY: John Wiley & Sons, Inc., 2002.

[2] F. Sears, M. Zemansky, H. Young and R. Freedman, *College Physics*, NY: Pearson Ed., 2009.

[3] E. M. Purcell and D. J. Morin, *Electricity and Magnetism, Berkeley Physics Course*, 3rd Ed., NY: Cambridge University Press, 2014, p. 839.

[4] R. Feynman, *et al.*, *The Feynman Lectures on Physics, Boxed Set: The New Millennium Edition*, Vol. 1, Basic Books, 2011.

[5] J. R. Reitz, F. J. Milford and R. W. Christy, *Fundamentals of Electromagnetic Theory*, NY: Ed. Addison-Wesley, 1996.

[6] D. J. Griffiths, *Introduction to Electrodynamics*, 4th Ed., MA: Cambridge University Press, 2017, p. 620.

[7] Wikipedia, "Expanded polystyrene (EPS)," 2022. Available: https://en.wikipedia.org/wiki/Polystyrene#Expanded_polystyrene_(EPS).

Chapter 33

MAGNETIC FIELDS

This chapter studies the magnetic field generated by turns and coils. An estimate of the Earth's magnetic field will be made in the laboratory. Use of Hall effect sensors to measure magnetic field is described.

Goals

✓ Ampere and Biot–Savart's law

✓ Earth's magnetic field

33.1 Ampère and Biot–Savart's law

Ampère's law is one of the fundamental equations of electromagnetism [1–3]. It expresses that the line integral of the magnetic field B along a closed loop C is proportional to the intensity of the current I that passes through the loop C: [1–3].

$$\oint_C \vec{B} \cdot d\vec{l} = \mu_0 I. \tag{33.1}$$

This law is closely related to Biot–Savart's law, which also establishes a relationship between the magnetic field and the currents that generate it [1–3]:

$$B = \frac{\mu_0}{4\pi} \int I \frac{d\vec{l} \times \vec{r}}{r_3}. \tag{33.2}$$

Here, the integral extends along the cables that carry the current. It is usual to apply these laws to calculate the magnetic field produced by current distributions, such as straight wires, coils, and solenoids. [1–3].

The following projects compare the theoretical expectations obtained from these laws to the results of our measurements. Then, some experiments are proposed to measure the Earth's magnetic field in the laboratory and the magnetic field of coils and permanent magnets. A widely used tool for measuring magnetic fields is the Hall effect sensor. For semi-quantitative determinations, a compass can be used.

33.1.1 *Project 96. Earth's magnetic field (Using a magnetic compass)*

Recommended basic equipment: Direct voltage source (DC) with a variable voltage between 0 and 12 V, capable of supplying a current of the order of 1 A. A circular coil of a few hundred turns and a diameter of 15–30 cm. A magnetic compass and a milliammeter.

This experiment suggests a simple technique to measure the intensity of the Earth's magnetic field using a magnetic compass, a voltage or current source, an ammeter, and a coil of a few hundred turns. One possible circuit is shown in Figure 33.1. The goal is to apply and measure a variable current ($I < 1$ A) through the coil to produce a known magnetic field at the position of the magnetic compass, which

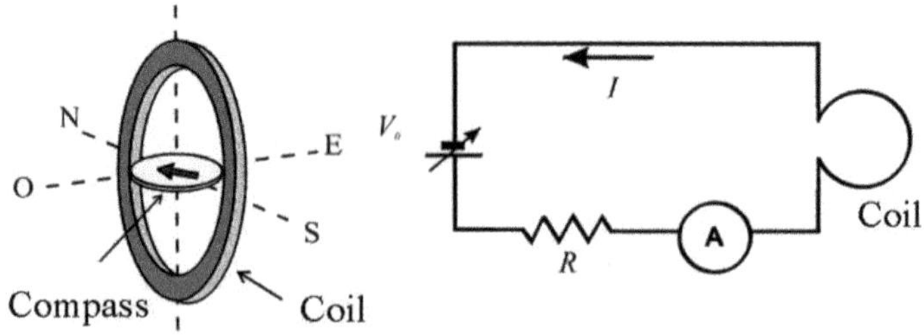

Figure 33.1 Diagram of the experimental setup to measure the Earth's magnetic field. The figure on the left shows the geometry, and the figure on the right shows the circuit. The circular coil must be oriented with its axis in east-west direction so that the field generated by it is perpendicular to the Earth's magnetic field.

is placed in the center of the coil. Use a compact winding circular coil of N turns, which is built so that its thickness is much less than the coil radius.

Design the experimental setup so that it allows you to hold the coil with its axis oriented in east-west direction (the needle of the compass is in the same plane as the coil). The experimental device must be assembled far from any ferromagnetic materials, such as metal supports, iron columns, tables, and chairs. It is also useful to have a ruler made of a non-magnetic material (wood or plastic) placed on the axis of the coil to measure the position of the compass on the axis (z-axis) at different distances from the coil center.

For this part, we will assume that based on the geometry of the coil, we know the magnetic field in the center of the coil of radius R, $\boldsymbol{Be}$, with N turns and current I flowing through it. The z component of the magnetic field $\boldsymbol{B_e^0}$ at the center of the coil is in the direction of its axis (z) and its value is [1–3]

$$B_z(z=0) = B_e^0 = \frac{\mu_0}{2}\frac{IN}{R}, \tag{33.3}$$

where $\mu_0 = 4\pi 10^{-7}\text{Tesla}\cdot\text{m}\cdot\text{A}^{-1}$ is the permittivity of vacuum. Here, the z axis coincides with the axis of the coil.

The proposed method consists, first, of determining the direction of the Earth's magnetic field $\boldsymbol{B_T}$ with the compass. Then, place the coil so that its plane is in the direction of the Earth's magnetic field, i.e., its axis (z axis) is in east-west direction. Place the compass in the center of the coil and slowly increase the current through it. The compass needle will be oriented in the direction of the magnetic field $\boldsymbol{B_B}$ that results from combining the magnetic fields $\boldsymbol{B_e}$(coil) and $\boldsymbol{B_T}$ (Earth). Take the north-south direction to measure the angles, i.e., take $\theta = 0°$ as the orientation of the compass needle when the current through the coil is zero.

Suggested tasks:

✓ In the position of the compass, the Earth's magnetic field $\boldsymbol{B_T}$ and the magnetic field generated by the coils $\boldsymbol{B_e^0}$ combine to produce the resulting magnetic field $\boldsymbol{B_B}$. Using the principle of

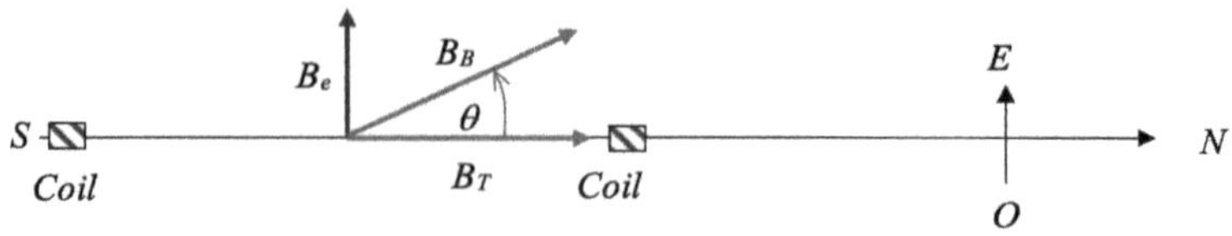

Figure 33.2 Side view of the coil. The magnetic compass is aligned in the plane of the coil. $\boldsymbol{Be}$ is the magnetic field of the coil at its center. $\boldsymbol{B_T}$ is the Earth's magnetic field. $\boldsymbol{B_B}$ is the resultant magnetic field.

superposition, the orientation angle θ of the compass needle is related to the quotient of $B_e^0(I)$ and B_T (Figure 33.2) through

$$\tan(\theta) = \frac{B_e^0(I)}{B_T}. \tag{33.4}$$

✓ When $\theta = 45°$, the magnetic field of the coil (known) is equal to the Earth's magnetic field. Adjust the current through the coil so that this condition is met and determine the B_T value.

✓ A more accurate way to determine the value of B_T is to plot the value of $B_e(I)$ as a function of $\tan(\theta)$. According to (33.3) and (33.4), we have

$$B_e^0(I) = B_T \cdot \tan(\theta). \tag{33.5}$$

Note that a plot of $\boldsymbol{B_e^0}(I)$ as a function of $\tan(\theta)$ enables you to find $\boldsymbol{B_T}$ using a linear regression. In fact, if our hypotheses are correct, we expect the dependence between $B_e^0(I)$ and $\tan(\theta)$ to be linear with slope equal to B_T. Discuss the validity of these hypotheses and find B_T and its uncertainty.

✓ Note that this experiment only enables you to determine the horizontal component of the magnetic field (on the plane of the compass).

✓ Compare the experimental value you have found in this test to the value of the magnetic field B_T in your area provided by a reliable source (scientific database, manual) [4, 5]. Remember that in a building with furniture, there are generally many sources that disturb the value of the Earth's magnetic field. It is therefore very likely that only the orders of magnitude of your measurements will be comparable to the table values.

Note: In an experiment of this type, in which a set of circular coils has to be aligned with respect to the direction indicated by the needle of a compass (usually small), there are often significant alignment errors. One way to assess whether the alignment has been adequate is to analyze the graph obtained from $\tan(\theta)$ as a function of I. According to Eq. (33.5), a straight line is expected, but there may be appreciable deviations from this behavior. A possible cause of the deviation may be the error in the definition of the east-west direction from which you measure the angle θ. One way to correct the deviation from linearity is to modify the values of the measured angles θ by adding to or subtracting from them a systematic correction term $\Delta\theta$ so that the ordinate at the origin of the graph of B_e as a function of $\tan(\theta)$ is zero. To achieve this, it is advisable to measure $\tan(\theta)$ for positive and negative values of I. From this graph, determine $\Delta\theta$, making it stop at $I = 0$, $\theta = 0$. If you cannot correct the nonlinearity of the graph with this procedure, there are likely other sources of error that have not been detected. In this case, carefully review the environment where you are conducting your experiment and try to remove as much ferromagnetic material as possible.

33.1.2 *Project 97. The compass as a magnetometer – Axial magnetic field of a coil*

Recommended basic equipment: A coil of a few hundred turns and a diameter between 15 and 30 cm. A compass. DC source from 0 to 12 V, with a current output of up to 1 A. An ammeter.

Use the same geometry and orientation as in the previous project and place a wooden or plastic ruler along the z-axis of the coil so that the compass can be moved in parallel along the z-axis (axis of the coil). As you move the compass along the z axis away from the coil, the tilt angle θ of the needle will change. Determine the angle θ for different z-positions of the compass.

From measurements of θ as a function of z, determine the value of the magnetic field of the coil along its axis, $B_z(z)$. To do this, use Eq. (33.5) and the value you found for the Earth's magnetic field B_T in the previous experiment. For this experiment, it is also important that there are no ferromagnetic materials in the vicinity of the experimental device.

Suggested tasks:

✓ Plot the experimental value of $B_z(z)$ as a function of z. [1, 3] On the same graph, plot the function

$$f(z) = \frac{\mu_0}{2} \cdot \frac{I \cdot N}{R} \cdot \frac{R^3}{(R^2 + z^2)^{3/2}} = B_e^0 \frac{R^3}{(R^2 + z^2)^{3/2}}, \qquad (33.6)$$

which describes the axial magnetic field of a coil of radius R, thickness $d \ll R$, with N turns, through which a current I flows.

✓ How do the experimental values compare to the theoretical expectations in (33.6)? Discuss your results.

33.2 Hall effect sensor

The "Hall probes" are magnetic field sensors based on the Hall effect [1, 6–8] that enable very precise measurement of the component of the field perpendicular to its work plane. When a current I_H is supplied to the sensor, it develops a voltage $V_H = V_{ab}$ (Hall voltage) in the direction perpendicular to I_H, which is proportional to the value of the magnetic field component perpendicular to the plane of the probe, $B_\perp$, and the intensity of the current I_H (see Figure 33.3).

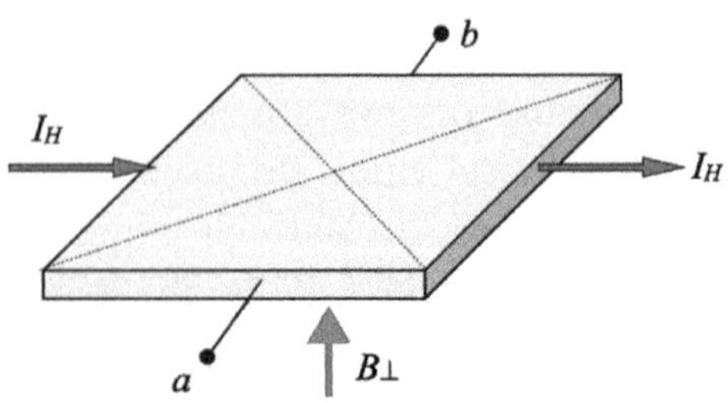

Figure 33.3 Hall probe. The current I_H is fed to the sensor. The Hall voltage V_H is measured between two points in the transverse direction: from a to b.

For a given value of the current I_H, the measurement of $V_H = V_a - V_b$ enables determination of B if the sensor calibration curve is known. These probes are generally thin films of a semiconductor material and operate with a stable current. Miniature Hall probes (sensitive area 0.1 mm × 0.1 mm) are currently available. They develop a Hall voltage of the order of 100 mV when a field is applied to them. $B \approx 1000$ Gauss and current $I_H \approx 5$ mA.

An alternative is to use a low-cost commercial sensor (which costs about US$ 2), such as UGN3503U [9–11] (Figure 33.4). This sensor has linear voltage response for magnetic fields $B < 900$ G, sensitivity 1.30 mV/G when it is supplied with direct voltage V_{CC} = 5V [6], and has a specially labeled face, which is the surface that must be located perpendicular to the direction of the magnetic field to be measured. Another series of similar sensors are the DRV5055, which are available in different sensitivities (from 1.25 to 10 mV/G) [12], or OMH3150 (3.5 mV/G) [11], TL173C (mV/G) [7], and SS49E (1.4 mV/G) [13].

The sensor can be mounted on a non-magnetic rod (acrylic, for example) for manipulation. It is suggested to use a regulated power

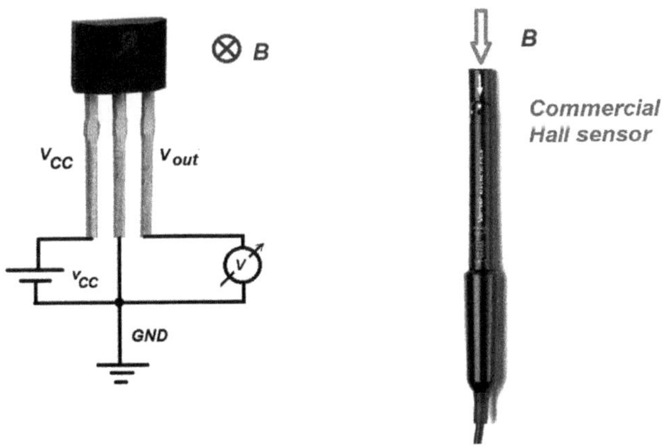

Figure 33.4 Left: Electrical circuit of a commercial Hall IC; the DC voltage source is V_{CC} and the voltmeter measures the generated voltage. Right: Commercial hall sensor. In general, these sensors are connected directly to an interface connected to a PC [8].

supply that polarizes the probe with $V_{cc} \approx 5$V with respect to ground (GND). To measure the output voltage, a voltmeter is connected to measure the potential difference between V_{OUT} and GND. Moreover, many manufacturers make Hall sensors [8] that can be connected to different data acquisition systems, which are pre-calibrated or can be calibrated, and measure magnetic fields in different ranges (see Figure 33.4). Nowadays, there are some low-cost development platforms, *Arduino*, based on a free hardware electronic board that incorporates a programmable microcontroller. These devices can establish connections between the microcontroller and the different sensors and PC [9, 14].

In order to use a Hall effect sensor, it is important to know that the plane of the sensor must be oriented perpendicular to the field. This can be done by placing the sensor in an area where the direction of the magnetic field is known, for example, in the center of a coil or solenoid, or facing the magnetic pole of a permanent magnet, which by symmetry allows you to know the direction of its field. When the sensor is rotated, two opposite orientations are found where the Hall potential value passes through a maximum and minimum, respectively. This direction defines the axis perpendicular to the plane of the Hall sensor. Some commercial sensors provide indication of the direction of the field (Figure 33.4).

33.2.1 *Project 98. Measuring magnetic fields using a Hall effect sensor*

Recommended basic equipment: A coil of a few hundred turns and a diameter between 15 and 30 cm. A Hall sensor connected to a millivoltmeter or a magnetic field sensor connected to a computer data acquisition system (DAS). DC source from 0 to 12 V, capable of applying a current of up to 1 A. A milliammeter.

Imagine that you want to measure the magnetic field of a circular coil of radius R, from its center to a distance of approximately $2R$, along the axis of the coil, which you can take as the z-axis. To do this, it is necessary to design a mechanism (that does not contain ferromagnetic materials) that allows you to move the Hall probe from the center of the coil, along the z axis. The mechanism must move the probe (or the coil) so that the orientation of the probe with respect to the coil is always constant. A plastic cable duct or cable trunking with cover can be used for this purpose. Orient the Hall probe to measure the z component of the coil's magnetic field, vary its position along the coil axis (z), and measure the Hall voltage value, V_H, as a function of the distance z.

Suggested tasks:

✓ Plot the value of the Hall voltage $V_H(z)/V_H(0)$, measured at the distance z from the origin (center of the coil). Here, $V_H(0) = V_H(z = 0)$ represents the value of the Hall voltage at the point $z = 0$. Vary the distance z between 0 and $2R$, approximately.

✓ On the same graph, plot the value of the function

$$g(z) = \frac{R^3}{(R^2 + z^2)^{3/2}}, \tag{33.7}$$

which predicts the axial magnetic field of the coil at the distance z [1, 2]. Equation (33.7) is derived from theoretical expression (33.7). Discuss the suitability of expression (33.7) to describe the observed data.

✓ The value $B_z(0)$ of the magnetic field at the coil center can easily be calculated from Eq. (33.3). By measuring $V_H(0)$, you can calibrate the Hall probe, i.e., you can find the relationship between V_H and $B_\perp$ (component of B perpendicular to the sensor plane). Once calibrated, this sensor can be used to measure the magnetic field.

✓ Explore what happens when the Hall sensor is rotated 180° so that the magnetic field enters the opposite plane of the sensor.

✓ Explore what happens if an ordinary piece of iron is brought close to the measuring device.

33.2.2 *Project 99. Magnetic field of a permanent magnet*

Recommended basic equipment: A Hall sensor connected to a voltmeter and preferably calibrated. A cylindrical permanent magnet. Rare earth magnets are particularly suitable for this experiment.

Using a Hall sensor, study the variation of the magnetic field of a cylindrical permanent magnet. Rare earth magnets in the shape of short cylinders (coin-shaped) are particularly suitable for this experiment. Study in particular the variation of $B_z(z)(z =$ cylinder axis) and $B_z(x)$ (x: axis perpendicular to z and passing through the center of the cylinder).

Suggested tasks:

✓ The goal of this experiment is to model the magnetic field of the permanent magnet in the shape of a coin of radius R by the field of a coil of equal radius. To do this, use the Hall probe to measure the component z of the magnetic field as a function of z for values of $z \approx R$ to $z \approx$ 5R. Likewise, measure the z component of the field along a radial axis, x, i.e., perpendicular to the z axis for values of x between $x \approx 2R$ and $x \approx 5R$. Here, we have assumed that the origin of z is the geometric center of the permanent magnet.

✓ On the same graph, draw the following:

(a) the experimental values $B_z(z)$ (found with the Hall probe) as a function of z,

(b) the theoretical values of the field $B_z(z)$ of a coil of radius equal to the magnet radius, given by Eq. (33.7), with $N = 1$. Vary the value of I in the plotted function so as to achieve the best possible fit to your experimental data.

✓ On the same graph, draw the following:

(a) a possible model to test, assuming that the magnetic field of the cylindrical magnet is similar to that of a magnetic dipole [3],

(b) the experimental values $B_{\mathrm{z}}(\mathrm{x})$ as a function of x,

(c) the theoretical values of the field $B_z(x)$ as a function of x for a magnetic coil [1, 3]:

$$B_z(\mathrm{x}) = -\frac{\mu_0 R^2}{4x^3} I \quad \text{with} \quad x > 2R. \tag{33.8}$$

For this graph, use the same value of I that you used to represent the theoretical value of $B_z(z)$ as a function of z.

✓ Based on your graphs, discuss whether the model of a coil of diameter equal to that of your magnet represents a suitable current distribution for modeling your permanent magnet. If this model is adequate, elaborate on the physical reasons that justify it.

✓ Discuss the value of the intensity or magnitude of current I that you found from the settings. Compare this value to a real physical system through which a current of this magnitude can flow.

33.2.3 *Project 100. Study of the magnetic field of a Helmholtz coil pair*

> **Recommended basic equipment:** Two coils of the same radius, between 15 and 30 cm, and the same number of turns, between 300 and 400. A Hall sensor connected to a millivoltmeter. DC source from 0 to 12 V, capable of applying a current of up to 1 A. An ammeter.

A very common physical system in many practical and laboratory applications consists of the Helmholtz coils [15], which are two identical parallel turns of radius R, separated from each other by a distance $d = R$ (see Figure 33.5). A very interesting property of these coils is that the field in their *central zone* is extremely constant. This property can be analyzed using a Hall effect sensor. This experiment consists of measuring the magnetic field along the axis of symmetry z of a pair of identical flat coils separated by a distance d, which can be varied. The experimental setup is shown in Figure 33.5. The same current circulates through the coils and produces a magnetic field in the same direction on the axis of symmetry of each one. The coils can

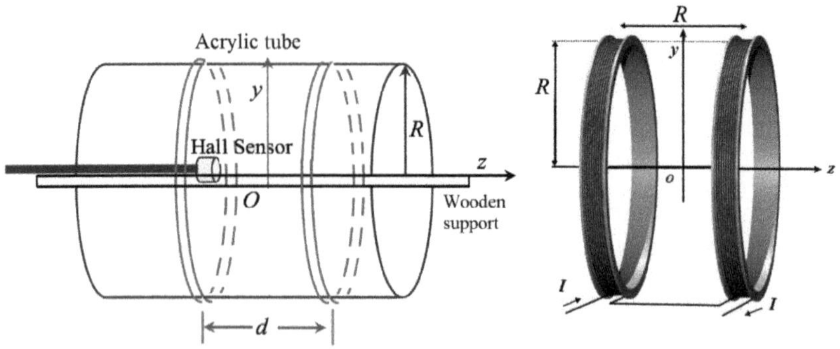

Figure 33.5 Measurement of the axial magnetic field of a pair of coils, composed of two identical flat coils of radius R separated by a distance d. The sensor moves along the axis of symmetry (z) of the assembly. When $d = R$, as in the figure to the right, the system is called *Helmholtz coil.*

be moved on the acrylic tube. When the separation between coils is $d = R$, we have what is called a *Helmholtz coil* [3, 15].

You can use a wooden ruler placed in the direction of the axis of symmetry, z-axis, on which the sensor rests and moves (if the ruler is a few millimeters below the axis, the center of the sensor will move approximately on the axis). You can also use a ruler to measure the position of the probe. Take the origin of coordinates at the midpoint between the coils, point O in Figure 33.5. The distance z along the axis of symmetry is measured from this point. Call the axis perpendicular to the z axis "y".

Suggested tasks:

✓ Set the current through the coils at a fraction of an Ampere, compatible with the thickness of your coil wires. Make sure the magnetic field of each coil is oriented in the same direction on the axis of symmetry of the pair. To do this, it may be helpful to explore the field direction of each coil, at its center, using a compass. Make sure that the series connection of both coils generates a field that is the sum (in the same direction and sense) of the field of each of them separately.

✓ Record the magnetic field at regular distance intervals, measuring the field along the axis of symmetry and passing from one side of the coil pair to the other. Take enough measurements to clearly define the spatial dependence of the field.

✓ To analyze the acquired data, compare your results to the Biot–Savart's law for the magnetic field produced by the pair of coils, separated by a distance d, along the axis of symmetry [2, 3]. The resulting magnetic field is the superposition of the magnetic fields generated independently by each coil of radius R and N turns of wire and is represented by [2, 3]

$$B(z) = \frac{\mu_0 N I R^2}{2\left[R^2 + (z - d/2)^2\right]^{3/2}} + \frac{\mu_0 N I R^2}{2\left[R^2 + (z + d/2)^2\right]^{3/2}}. \quad (33.9)$$

✓ Discuss the degree of agreement of (33.9) with your experimental data.

✓ Evaluate the size of the area where the magnetic field can be considered constant in a Helmholtz pair.

References

[1] R. Halliday, D. Resnick and M. Krane, *Fundamentals of Physics*, NY, John Wiley & Sons, Inc., 2002.
[2] F. Sears, M. Zemansky, H. Young and R. Freedman, *College Physics*, NY: Pearson Ed., 2009.
[3] J. R. Reitz, F. J. Milford and R. W. Christy, *Fundamentals of Electromagnetic Theory*, NY Ed. Addison-Wesley, 1996.
[4] Wikipedia, "Earth's magnetic field," 2021. Available: https://en.wikipedia.org/wiki/Earth%27s_magnetic_field.
[5] British Geological Survey, "The Earth's magnetic field: An overview-Geomagnetism," 2021. Available: http://www.geomag.bgs.ac.uk/education/earthmag.html.
[6] E. Ramsden, *Hall-Effect Sensors: Theory and Application*, NY: Elsevier, 2006.
[7] Texas Instruments, "Linear hall-effect sensors," 2021. Available: https://www.bristolwatch.com/hall_effect/tl173c.pdf.
[8] Vernier Software, "Magnetic field sensor," Vernier software & technology, 2021. Available: https://www.vernier.com.
[9] Amazon, "Linear Hall effect sensors," 2021. Available: https://www.amazon.com/s?k=linear+hall-effect+sensor&rh=n%3A306929011&ref=nb_sb_noss.

[10] All data Sheet.com, "Electronic components datasheet search," 2021. Available: https://www.alldatasheet.com/view.jsp?Searchword=Ugn3503ua&gclid=Cj0KCQiAoY-PBhCNARIsABcz770l2qC1u4NCoxN5hFiUg-FuKhdichX3DjposBrAnSHspyhCMki_D5waAiK6EALw_wcB.

[11] TT Electronics, "Ratiometric linear Hall-effect sensor," 2021. Available: https://www.ttelectronics.com/TTElectronics/media/ProductFiles/Optoelectronics/Datasheets/OMH3150.pdf.

[12] Texas Instruments, "DRV5055 ratiometric linear hall effect sensor," 2021. Available: https://www.ti.com/lit/ds/symlink/drv5055.pdf?ts=1616759616048&ref_url=https%253A%252F%252Fwww.google.com%252F.

[13] Honeywell Co., "Magnetic sensors," 2021. Available: https://sps.honeywell.com/us/en/products/sensing-and-iot/sensors/magnetic-sensors.

[14] Wikipedia, "Arduino," 2021. Available: https://en.wikipedia.org/wiki/Arduino.

[15] Wikipedia, "Helmholtz coil," 2021. Available: https://en.wikipedia.org/wiki/Helmholtz_coil.

[16] Lucio Sciamanna Electronics, "Comparison of several linear ratiometric Hall sensors," 2021. Available: https://www.sciamannalucio.it/10-best-linear-hall-sensors/.

[17] Magnet Man, "Experiments with electronics, build your own Gaussmeter," 2021. Available: Available: http://www.coolmagnetman.com/magmeter.htm.

[18] Lewis Loflin, "Using ratiometric hall effect sensors," 2021. Available: https://www.bristolwatch.com/hall_effect/ratiometric_hall_effect.htm.

[19] J. I. Antorino, *et al.*, "Experimentos con un sensor de efecto Hall," *Lat. Am. J. Phys. Educ.*, vol. 3, no. 3, p. 606, 2009.

Chapter 34

ELECTROMAGNETIC INDUCTION — MUTUAL INDUCTION

This chapter studies Faraday's law of induction. It begins by exploring the Electromotive Force (emf) induced in a secondary circuit by the variation of magnetic flux generated by a primary circuit. Then it explores the basic features of electromagnetic induction and suggests several activities in which this law is used to study the characteristics of magnetic fields produced by different systems of currents. In particular, it studies the magnetic field of a loop using a lock-in amplifier and discusses some paradoxes associated with Faraday's law of induction.

Goals

✓ Faraday's law of induction
✓ Mutual induction
✓ Using a lock-in amplifier
✓ Electromagnetic paradox

34.1 Faraday's law

Faraday's law of induction is one of the fundamental equations of electromagnetism [1–4]. It was discovered almost independently by Michael Faraday and Joseph Henry in the early 19th century. It is the principle upon which the operation of electrical generators, transformers, and many other electrical devices is based. Faraday's law states that the voltage induced ε (emf) in a closed circuit C is directly proportional to how fast the magnetic flux Φ_B changes through the surface S, which has the curve C as border. That is to say,

$$\varepsilon \equiv \oint_C \vec{E} \cdot d\vec{l} = -\frac{d\Phi_B}{dt}. \tag{34.1}$$

This chapter proposes to investigate some implications of this law experimentally.

34.1.1 *Project 101. Faraday's law I — Qualitative analysis*

> **Recommended equipment:** Two coils of a few hundred turns. A data acquisition system connected to a computer or oscilloscope. A function generator (FG).

A possible setup to explore Faraday's law of induction is shown in Figure 34.1. The resistance R $\approx$ 30 Ω is introduced to limit the current in the primary coil.

Suggested tasks:

✓ Apply a sinusoidal voltage of a few volts to the primary with frequency $f \approx 130$ Hz. View the signal generated on the secondary, using a data acquisition system (DAS) or an oscilloscope. If you

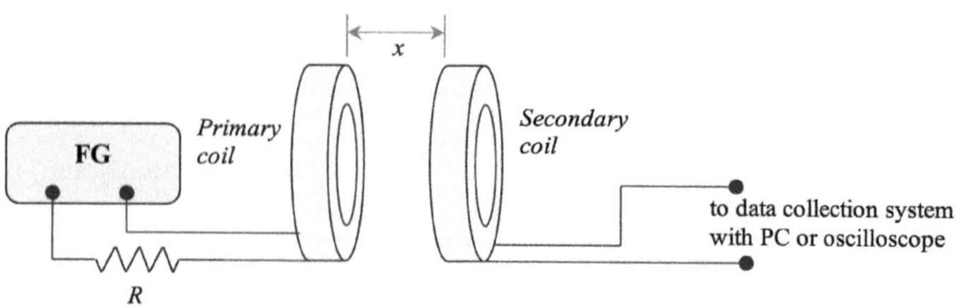

Figure 34.1 Experimental arrangement to generate an emf in a secondary coil. The function generator (FG) supplies the primary coil. The resistance R serves to limit the current in the primary circuit. The distance between the coils is called x.

use a DAS, choose sampling frequency as $10 \times f$ so that you can see the signal in the secondary clearly.

✓ Analyze qualitatively what happens to the signal in the secondary:

- if distance x between the coils varies,
- if the voltage in the primary rises,
- if the frequency of the signal applied to the primary varies,
- if you insert a ferromagnetic cylinder, for example, a screw or thick nail, into the common axis of the two coils,
- if you rotate one of the coils so that their axes are perpendicular,
- if the signal applied to the primary changes from sinusoidal to triangular. How does the signal change in the secondary?

✓ Explain how these results can be interpreted. This system can be regarded as a simplified model of a transformer.

34.1.2 *Project 102. Faraday's law II — Quantitative analysis*

Recommended equipment: A coil of a few hundred turns and a diameter between 10 and 25 cm. Another coil of a few hundred turns, smaller than the primary so that it can be placed within the field generated by the primary. Wooden wedges with 30° and 45° angles. A data acquisition system connected to a computer or an oscilloscope. A function generator.

A possible setup for this activity is shown in Figure 34.2. It consists of a primary coil, with a number of turns of the order of 300 or more. It is connected to a function generator through the resistor R of a few tens of Ohm. This serves to limit the current in the coil and to measure the current I through the voltage drop V_R ($I = V_R/R$). In general, low impedance elements ($Z < 20\Omega$) should not be directly connected to a voltage source (the function generator in this case) to avoid either damaging the source or burning out the coil.

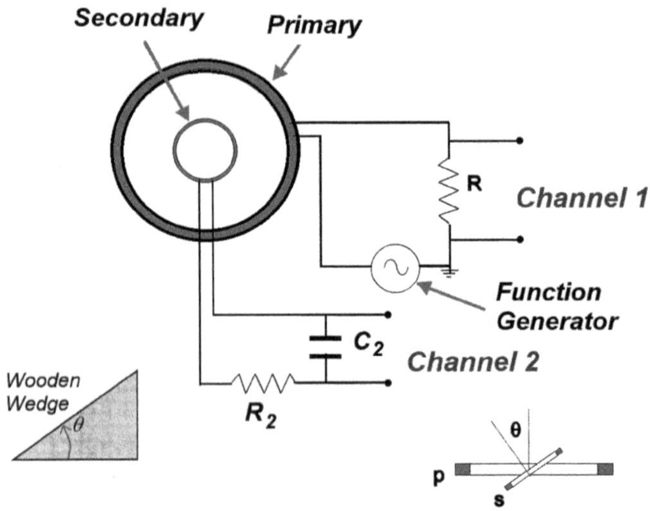

Figure 34.2 Experimental arrangement to generate an *emf* in a secondary coil (**s**). The function generator (FG) supplies the current to the primary coil (**P**). The resistance R serves to limit the current in the primary circuit. The angle θ, which forms the axes of the primary and secondary coils, can be varied using wooden wedges whose angle θ is known.

The voltage drop V_R across the resistor R (proportional to the current in the primary) may be measured by the oscilloscope or by a data collection system. It is advisable to connect the resistance to the common ground, the same ground as the function generator (FG), and the data collection system or oscilloscope.

A second coil with a diameter smaller than the primary coil and with a number of turns of the order of 500 is used as secondary and connected to the other channel of the oscilloscope or data collection system.

Note: If there is a lot of noise in the secondary signal, it may be advisable to connect a capacitor C_2 in parallel with the secondary coil and a resistor R_2, with a time constant $\tau_2 \equiv R_2 \cdot C_2 \ll 1/f$, where f is the frequency applied by the FG. The connections to Channel 1 and Channel 2 refer to the channels of the oscilloscope or the data collection system.

Suggested tasks:

When *triangular voltage* [5] is applied to the primary circuit shown in Figure 34.2 with a frequency of the order of 100 or 200 Hz, if you are using a data collection system, remember to set the sampling frequency at around *10xf*.

- ✓ Describe *qualitatively* the current signals in the primary and the voltages $V_2(t)$ (efm) in the secondary. Remember that the derivative of a triangular signal is a square one. See Figure 34.3.
- ✓ How does the amplitude of the voltage in the secondary vary with the *frequency* of the input signal or that of the FG? How does the amplitude of the voltage in the secondary vary with the *amplitude* of the FG signal fed to the primary?
- ✓ Varying the amplitude V_0 and frequency f of the function generator in the primary, verify that the amplitude in the secondary is proportional to the product of $V_0 \times f$.
- ✓ Investigate the dependence of the amplitude of the secondary voltage V_2^0 on the angle θ between the two coil axes. If possible, investigate the case for the angles $\theta = 0°, 30°, 45°, 60°$, and $90°$. Plot V_2^0 as a function of $\cos(\theta)$. What can you conclude from your graphs about the dependence of V_2^0 on the magnetic flux Φ_B generated by the primary coil?
- ✓ Using all the data found in the previous studies, attempt to answer whether or not the magnitude of the voltage V_2 in the secondary is proportional to $-d\Phi_B/dt$.

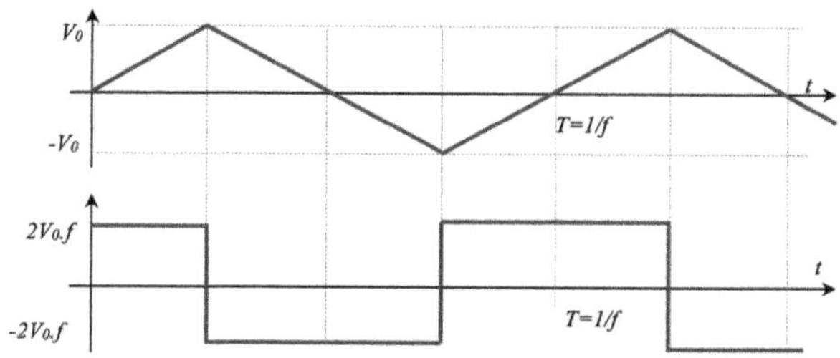

Figure 34.3 Triangular signal (top) amplitude V_0 and frequency $f = 1/T$ and its derivative (bottom). The amplitude of the derivative of the triangular signal (bottom) is $2V_0 \cdot f$ and its frequency is f.

34.1.3 *Project 103. Faraday's law III — Variation of the number of turns*

> **Recommended equipment:** Some coils of a few hundred turns, smaller than the primary coil, so that they can be placed within the field generated by the primary. A data acquisition system connected to a computer or oscilloscope. A function generator.

This activity is similar to the previous one, but here, the aim is to study how the voltage in the secondary varies when the number of turns of the secondary varies. The secondary coils must be geometrically similar to each other and have a number of N_2 turns of the order of 200, 500, 1,000, 2,000, etc.

Suggested tasks:

Using secondary coils of different numbers of turns N_2, but keeping the conditions of the primary constant in amplitude and frequency, investigate the dependence of the amplitude V_2^0 as a function of N_2. Apply *sinusoidal* voltage to the primary circuit with a frequency of the order of 300 or 500 Hz. If you are using a PC data collection system, remember to set the sampling frequency to values at around $10 \times f$.

✓ Describe *qualitatively* the current signals in the primary and the voltages $V_2(t)$ in the secondary. Are these signals in phase? As discussed in the Annex of Chapter 31, if the signals under study are observed in XY mode of an oscilloscope or DAS, it is possible to measure the phase shift ϕ between the two signals.
✓ How does the amplitude of the voltage in the secondary vary when the number N_2 of turns in the secondary is changed?
✓ Is it true that the amplitude of the signal V_2^0 is proportional to N_2?
✓ Justify your results theoretically.

34.1.4 *Project 104. Magnetic field of a loop along its axis, using Faraday's law*

Recommended equipment: A loop of a few hundred turns and a diameter between 15 and 30 cm. A smaller coil with a few hundred turns is used as a secondary. A wooden or plastic bar along which the distance z between the coils can be varied and measured. A data acquisition system connected to a computer or an oscilloscope. A function generator.

The aim of this activity is to study how the magnetic field of a loop varies along its axis (see Project 97), but this time, using Faraday's law to determine its magnitude. In this activity, *sinusoidal* voltage is applied to the primary circuit with a frequency of the order of kHz. Avoid using a frequency that is a multiple of the line frequency (50 Hz or 60 Hz) to reduce interference with these signals, which are generally always present (line pick-up noise). A second, smaller coil with approximately 300–2,000 turns is used as a secondary in the z position, where the magnetic field is to be measured. Of course, the secondary coil must be moved along the axis common to both coils, which is designated as the z-axis, as illustrated in Figure 34.4.

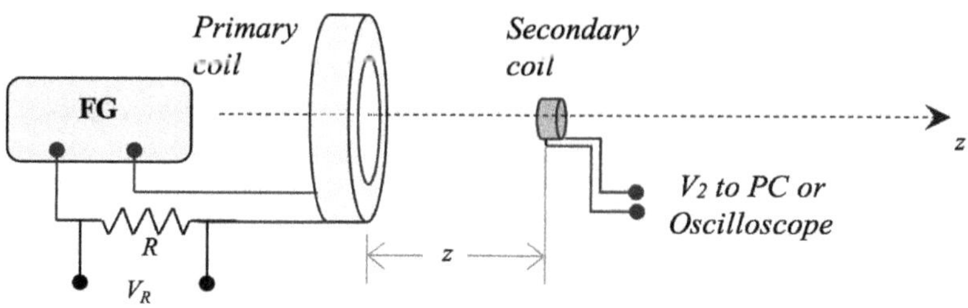

Figure 34.4 Experimental arrangement to study the magnetic field of a loop using Faraday's law. The function generator (FG) supplies the primary with a frequency f. V_R is the voltage drop across R, proportional to the current through the primary coil.

The signal amplitude induced in the secondary coil V_2^0 according to Eq. (34.1) is

$$V_2^0(z) = -N_2 \cdot A_2 \cdot 2\pi \cdot f \cdot B_1(z) = k \cdot B_1(z). \tag{34.2}$$

Here, N_2 and A_2 are the number of turns and area of the secondary coil, respectively. At the fixed frequency f, the amplitude V_2^0 in the secondary coil is proportional to $B_1(z)$, which is the z-component of the magnetic field generated by the primary loop that is to be determined, and k is the constant of proportionality. The aim is to determine $B_1(z)$ by measuring $V_2^0(z)$. Clearly, this setup for measuring the magnetic field can also be used for a system that produces an AC magnetic field, for example, the Helmholtz coils studied in Project 100. If you are using a DAS, remember to set the sampling rate at around $10 \times f$. As you can see when taking your measurements, the signal detected in the secondary coil consists of a signal of frequency f associated with the emf induced by the primary, plus a noise signal, generated by all the *stray fields* present in the laboratory, that is,

$$V_2^0(zt) = V_2^0(z) \cdot \sin(2\pi f \cdot t + \varphi) + V_2^{(\text{noise})}(t). \tag{34.3}$$

Let ϕ be the phase difference between the *emf* induced in the secondary and the signal generated by the FG. The component of ambient noise $V_2^{\text{noise}}(t)$ determines the sensitivity of the method, since the minimum *measurable* amplitude $V_2^0(z)$ is of the order of magnitude of the noise amplitude. For this experiment, avoid the presence of ferromagnetic materials near your system.

Suggested tasks:

Using an experimental setup like the one shown in Figure 34.4, keeping the frequency of the FG fixed, determine the following:

✓ The amplitude $V_2^0(z)$ from $z = 0$ to the maximum distance $z_{\max}$ at which you can still distinguish the emf induced in the secondary from the ambient noise.

✓ Plot the experimental value of $V_2^0(z)/V_2^0(0)$ as a function of z. On the same graph, plot the following function:

$$B_z(z)/B_z(0) = \frac{R^3}{(R^2 + z^2)^{3/2}}, \tag{34.4}$$

which describes the axial magnetic field of a coil of radius R, see Eq. (33.6) [1, 2, 4].

✓ How do the experimental values compare to the theoretical expectations of Eq. (34.4)? Discuss your results.

34.1.5 *Project 105 ♣♣ Magnetic field of a loop along its axis using a lock-in amplifier*

> **Recommended equipment:** A loop of a few hundred turns and a diameter between 15 and 30 cm. A second, smaller coil, with a few hundred turns, is used as a secondary. A wooden or plastic bar along which the distance z between the coils can be varied and measured. A lock-in amplifier, or a 20 Mhz or faster 2-channel digital oscilloscope in conjunction with a function generator.

As studied in the previous project, the limit to the sensitivity of the method used is determined by the *noise level.* Measuring small signals in the presence of high noise is a permanent challenge in many fields of science and technology: the main goal of this experiment is to illustrate the use of a very powerful technique for measuring small signals (of well-known frequency), whose amplitude can be much smaller than the noise signals. The amplitude of the signal that we are looking for in the secondary is the one associated with the frequency f of the FG and the primary. On the other hand, noise, which can be important, does not necessarily have the same frequency. This characteristic of the problem makes it possible to separate the signal sought from the noise, even if the noise is louder than the signal sought. This technique is usually used in *Lock-in Amplifiers* (LIA) also called *Phase Sensitive Detectors* (PSD), which are very useful instruments in modern laboratories. See Appendix E for more details on how the lock-in works [6]. A commercial lock-in or a digital oscilloscope in combination with a function generator can be used for this experiment. In the latter case, the digital oscilloscope is used to detect the input signals and the reference signal to study the variation of the magnetic field of a loop along its axis. The algorithms for extracting the signal of interest will be carried out offline, using the LIA algorithm, as described in Appendix D. The experimental setup is like the one illustrated in Figure 34.4, except that an LIA procedure is used to determine the amplitude of the signal $V_2^0(z)$.

The frequency should be on the order of kHz. If you use a digital oscilloscope or data collection system, the acquisition rate of the system is at least 10 times faster than the frequency f of the input signal. Avoid using a frequency in the primary that is a multiple of the line frequency (50 Hz or 60 Hz) to avoid interference with these signals, which are generally strong. Also avoid the presence of ferromagnetic materials near your system.

Suggested tasks:

✓ If you use a lock-in amplifier, study the variation of $V_2^0(z)$ (peak voltage in the secondary) as a function of the distance z.

✓ Plot the experimental value of $V_2^0(z)/V_2^0(0)$ as a function of z. On the same graph, plot the function given by Eq. (34.4), which describes the variation of the axial magnetic field of a coil of radius R, see Eq. (33.6).

✓ How do the experimental values compare to the theoretical expectations of Eq. (34.4)?

Note.

✓ If you are using a digital oscilloscope or data collection system, record the signals from Channels 1 and 2. Channel 1 measures the voltage V_R which is the reference signal (PSD) and Channel 2 measures the voltage $V_2(z,t)$.

✓ For each measurement (each value of z), using the technique described in Appendix E: find the value of $V_2^0(z)$ for each value of z used. To do this, perform multiplication (PSD) and software filtering operations using a spreadsheet or an *ad hoc* program.

Plot V_2^0 as a function of z and $B_z(z)$ as a function of z. Compare the curves obtained with the theoretical expectation predicted by Eq. (34.4).

34.2 What do voltmeters measure?

34.2.1 *Project 106. Faraday's law — Electromagnetic paradox or what do voltmeters measure?*

Recommended equipment: A coil of about 500 or more turns, with a core iron bar. A variable autotransformer or variac. Two or more resistances of 50 and 100 Ω approximately. Two multimeters or a two-channel oscilloscope.

The objective of this experiment is to understand what voltmeters actually measure when they are connected to an alternating current circuit, when this current is generated in it by an alternating magnetic field. In this study, a series of paradoxes [8–11] emerge, which we hope will serve to illustrate how measuring instruments such as voltmeters and oscilloscopes work, and in a broader sense, the meaning of Faraday's law of induction.

Use a primary coil of about 500 or more turns, made from a wire at least 0.25 mm in diameter, with a total resistance of the order of 50 or 100 Ω. An external resistor in series with the coil can be used to achieve this resistance. The primary coil is wound around an iron rod of about 2–3 cm in diameter, which protrudes from it by about 5–10 cm. A torus-shaped iron core can also be used, as illustrated in Figure 34.5. The primary coil is connected to the output of a variac, which supplies an alternating voltage of variable amplitude. Two resistors of different values R_1 and R_2 (for example, $R_1 = 50\,\Omega$ and $R_2 = 100\,\Omega$) are used, connected as shown on the right of Figure 34.6.

The question is as follows: Are the voltages V_1 and V_2 in Figure 34.6 the same or not? Suppose that $R_1 \neq R_2$ and that the Fe core of the primary coil is the one that induces the current in the circuit under study.

(i) On the one hand, since both voltmeters are connected to exactly the same points (A and B), we would tend to answer that $V_1 = V_2$.

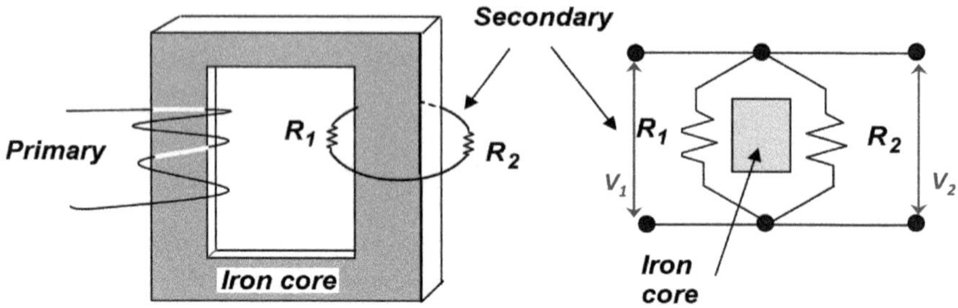

Figure 34.5 Experimental arrangement to generate an alternating magnetic field in the circuit on the right. The primary is connected to an alternating voltage source. You want to measure the voltage difference between A and B of the secondary circuit.

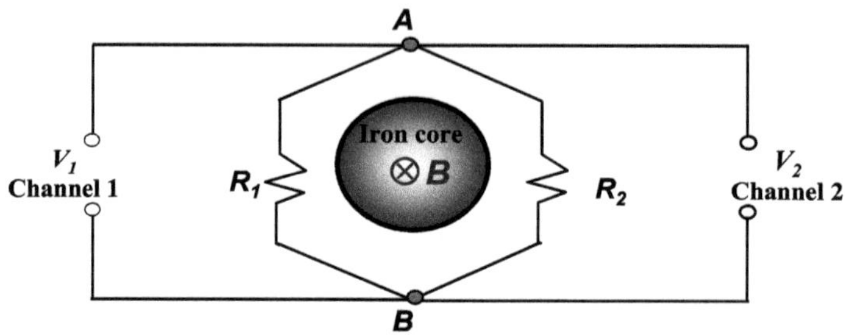

Figure 34.6 The aim is to determine the voltage difference between points A and B using two voltmeters connected to the same points A and B. That is, one voltmeter measures voltage V_1, and the other measures V_2. The question is as follows: Are voltages V_1 and V_2 the same or not? Here, assume that $R_1 \neq R_2$ and that the Fe core of the primary coil is the one that induces the current in the circuit under study.

(ii) On the other hand, as the impedance of the voltmeters is much greater than the resistances involved, the common current flowing through the circuit is the same and equal to a certain value $i(t) = I_0 \sin(2\pi f t)$. Thus $V_1(t) = R_1 I_0 \sin(2\pi ft)$ and $V_2(t) = R_2 I_0 \sin(2\pi f t)$. As $R_1 \neq R_2$, we conclude that $V_1 \neq V_2$.

What is the correct answer and why? This is the paradox to analyze.

Suggested tasks:

✓ Study the voltages measured by voltmeters 1 and 2, as indicated in Figure 34.6. An oscilloscope can also be used by connecting Channels 1 and 2 to the same points on the secondary. Analyze your results in detail. Are the tensions V_1 and V_2 the same or not?

✓ Repeat the experiment bringing voltmeter 1 to the same side as voltmeter 2. Are the tensions V_1 and V_2 the same or not?

✓ What can you conclude from these studies regarding the uniqueness of the voltage in a circuit when it encloses a variable magnetic flux? What do voltmeters measure in these systems? Explain your results. An interesting and instructive discussion of this system can be found in Refs. [8, 10, 12].

References

[1] R. Halliday, D. Resnick and M. Krane, *Fundamentals of Physics*, NY: John Wiley & Sons, Inc., 2002.

[2] F. Sears, M. Zemansky, H. Young and R. Freedman, *College Physics*, NY: Pearson Ed., 2009.

[3] E. M. Purcell and D. J. Morin, *Electricity and Magnetism, Berkeley Physics Course*, 3rd Ed., NY: Cambridge University Press, 2014, p. 839.

[4] J. R. Reitz, F. J. Milford and R. W. Christy, *Fundamentals of Electromagnetic Theory*, NY: Ed. Addison-Wesley, 1996.

[5] R. Duffy and U. Haber-Schal, "Establishing V=L (dI/dt) directly from experiment," *Am. J. Phys.*, vol. 45, p. 170, 1977.

[6] P. Kromer, R. Robinett, R. Bengtson and C. Hay, "PC-based digital lock-in detection of small signals in the presence of noise," Department of Physics, University of Texas at Austin, 2021. Available: http://citeseerx.ist.psu.edu/viewdoc/summary?doi=10.1.1.594.2679.

[7] R. Romer, "What do volmeters 'measure'?: Faraday's law in multiple connected region," *Am. J. Phys.*, vol. 50, p. 1089, 1982.

[8] F. Reif, "Generalized Ohm's law, potential difference, and voltage measurements," *Am. J. Phys.*, vol. 50, no. 11, pp. 1048-9, 1982.

[9] P. Peters, "The role of induced emf's in simple circuits," *Am. J. Phys.*, vol. 52, p. 208, 1984.

[10] E. Lanzara and R. Zangara, "Potential difference measurements in the presence of a varying magnetic field," *Phys. Educ.*, vol. 30, no. 2, pp. 85–89, 1995.

[11] A. Bossavit, "What do voltmeters measure?" *COMPEL Int. J. Comput. Math. Electr.*, vol. 27, no. 1, pp. 9–12, 2008.

[12] H. Nicholson, "What does the voltmeter read," *Am. J. Phys.*, vol. 73, pp. 1194–1196, 2005.

[13] Department of Physics, University of Texas at Austin, "PC-based digital lock-in detection of small signals in the presence of noise," 2021. Available: http://mrflip.com/papers/LIA/AAPTPaper/aaptpaper.html.

[14] G. Ballarà Madrid and J. Asensi Lopez, "Lock-in amplifier based on virtual instrumentation," Facultat de Física, Universitat de Barcelona, 2016. Available: http://diposit.ub.edu/dspace/handle/2445/102753.

Chapter 35

SELF-INDUCTION AND RL CIRCUIT

In this chapter, we will study some properties of self-inductances of different devices and the basic characteristics of the RL circuits. We will discuss some techniques to determine the value of self-inductance. We plan to study the transient and stationary response of this circuit. We also plan to analyze the response of this circuit to square and sinusoidal excitation signals.

Goals

✓ Determination of self-inductance
✓ RL circuits
✓ Transient and stationary response

35.1 Inductance

Due to Faraday's law of induction, whenever a current changes through a coil, it induces an electromotive force (emf); the *inductance* is the tendency of a coil to resist these changes in current. The *voltage–current characteristic* of an element in an electric circuit is the relationship between the voltage drop across this element and the current flowing through it. For example, Ohm's law, $V = R.i$, describes the voltage–current characteristic of a resistor. The goal of the following experiment is to find the equivalent relationship to Ohm's law for an inductance using a circuit similar to that of Figure 35.1 using a data acquisition system (DAS) or a digital oscilloscope.

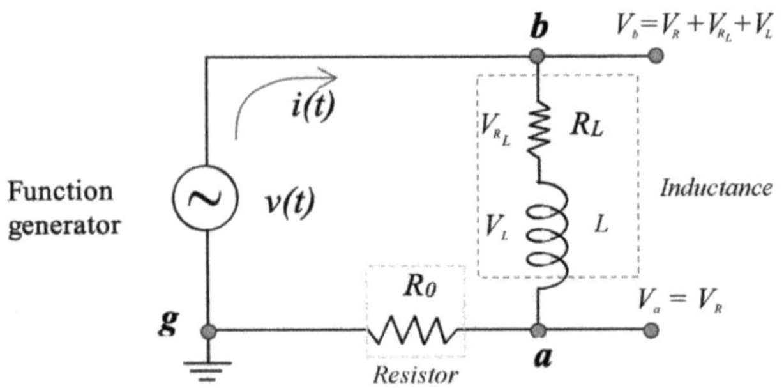

Figure 35.1 Circuit to study the $V - I$ characteristic of an inductance.

35.2 RL circuit

When a variable current circulates through a coil, the magnetic flux in its interior also varies; according to Faraday's law of induction, a voltage or emf, ε_L, is produced at its ends given by

$$\varepsilon_L(t) = -\frac{d\Phi_B}{dt} = -\frac{d\Phi_B}{di}\frac{di}{dt} = -L\frac{di}{dt}, \tag{35.1}$$

where $L \equiv d\Phi_B/di$ is the *inductance*[1] of the system under consideration or self-inductance, since emf is produced by the change in the current that circulates through this element.

When we study circuits that contain an inductance L, we must remember that these devices, in general, are made of wound wires, which also have a certain electrical resistance, which we call R_L or *intrinsic resistance*. Its value can be easily measured with an ohmmeter. Therefore, the total or equivalent resistance between points **g** and **b** of the circuit in Figure 35.1 is $R_t = R_0 + R_L$.

In general, data acquisition systems and oscilloscopes have more than one analog input, so it is possible to monitor simultaneously two or more variables at the same time. However, usually the inputs of these devices work in *common mode*, that is, the ground of each channel is the same (only some versions have *differential inputs*, it is necessary to analyze the specifications of each instrument to verify if they work in differential or common mode, see Chapter 24).

In the circuit of Figure 35.1, we want to monitor two signals simultaneously. These are as follows: the signal between points **g** and **a** and the signal between points **g** and **b**, which are supposed to be connected to the inputs of two channels of the data acquisition system (DAS) or digital oscilloscope. We designate these signals as V_a and V_b, respectively. We assume that the function generator produces a sinusoidal signal of approximately 100 Hz (this frequency can be varied in wide ranges, as long as the acquisition system can take at least 10 samples for each period, approximately). The measured voltages V_a and V_b can be written in terms of the circuit components as follows:

$$V_a = V_R = R_0 \cdot i \quad \text{and} \quad V_b = R_1 \cdot i + V_L, \tag{35.2}$$

where i is the current and $R_1(= R_L + R_0)$ is the total ohmic resistance of branch **g-a-b**. The values of R_0 and R_L are obtained using an ohmmeter. The value of V_a allows us to know the current i, and from V_b, we can know the voltage drop V_L in the inductance:

$$V_L = V_b - R_1 \cdot i = V_b - V_a \cdot \left(\frac{R_1}{R_0}\right). \tag{35.3}$$

The value of V_L can be compared with the value predicted by Eq. (35.1). If it is verified that when plotting V_L as a function of $di/dt = (dV_a/dt)/R_0$, we obtain a linear relationship, that is, if it is verified that

$$V_L = -L\frac{di}{dt}, \tag{35.4}$$

it means that our results are consistent with expression (35.1), and from the slope of this graph, we can obtain the value of L. Note that with this technique, we are analyzing the behavior of actual inductance stripped of its intrinsic resistance.

The interesting thing about this experiment is that it allows us, on the one hand, to falsify the expression (35.1), which is derived from Faraday's law, and obtain the value of the inductance [2]. In fact, Eq. (35.4) represents the *voltage–current characteristic* of the inductance.

35.2.1 *Project 107. Voltage–current characteristic of an inductance*

Recommended equipment: Coil (inductance of a few tens of mH) and resistances of a few tens of Ohms. A data acquisition system connected to a computer or digital oscilloscope. A function generator.

For this experiment, it is convenient to use one or more inductances of a few tens of mH. These can be constructed using enameled copper wire with thicknesses of the order of 0.2 ± 0.1 mm in diameter, winding a few hundred turns on a 1 or 2 cm spool. An approximate expression for calculating the inductance of a coil is given in Annex A. The suggested circuit is illustrated in Figure 35.1.

Suggested tasks:

✓ For your circuit, determine the values of R_0, R_L, and its uncertainties, using an ohmmeter.

✓ Connect the terminals of channels A and B from your measurement system (DAS or oscilloscope) to points a and b, respectively, and the common ground of the two channels is point g. Choose a frequency f in the function generator (FG) between 50 and 100 Hz. Set the sampling frequency of the data collection system so that it is greater than $10 \times f$. The amplitude of the input signal must not exceed the maximum admissible voltage for the data collection system. In this way, V_a and V_b are functions of time.

✓ On the same graph, plot $V_L(t)$ and $i(t)$ as functions of time for at least two complete periods. On another graph plot, plot $V_L(t)$ as a function of i. The output of the function generator is assumed to be *sinusoidal.* Do you see a simple relationship between $V_L(t)$ and $i(t)$?

✓ Using some numerical derivation technique, obtain $di(t)/dt$. Using these values, plot V_L versus $di(t)/dt$.

✓ If the variation V_L with $di(t)/dt$ is linear, determine L (the slope) and its uncertainty (see Chapter 7).

✓ Vary the frequency of the FG (between 50% and 300% of the initial one) and check if the relationship between V_L and $di(t)/dt$ is still linear. In this case, determine L and its uncertainty. Does the value of L vary with frequency?

✓ What can you conclude regarding the validity of Eq. (35.1) and Faraday's law in this case?

✓ Using the same circuit, vary the FG signal to a triangular and square signal of the same frequency as the original sinusoidal. For this case, repeat the previous analysis. What can you say about the relationship between V_L and di/dt?

35.3 RL circuit — Transient response

If in the circuit of Figure 35.1, we choose in the function generator a signal of square shape of amplitude V_{00}, the current through the circuit will not vary in the same way (in general, it will not be square). This change in form of the current is due to the fact that the energy in self-inductance is [3–5]

$$E_L = \frac{1}{2} L \cdot i^2. \tag{35.5}$$

Since energy cannot jump, because this would imply infinite power, the current through the inductor cannot have sudden jumps and must vary smoothly; $i(t)$ cannot have infinite derivative with respect to time. Therefore, the current in an inductor must always be a continuously differentiable function of time. To know the response or current $i(t)$ in the circuit, through which no current was flowing and at $t = 0$, we apply a constant voltage V_{00}. We must apply Kirchhoff's law of meshes, that is [3–5],

$$L\frac{di}{dt} + Ri = V_{00} \tag{35.6}$$

or

$$\frac{di}{dt} + \frac{1}{\tau} i = \frac{V_{00}}{L} \quad \text{with} \quad \tau = L/R. \tag{35.7}$$

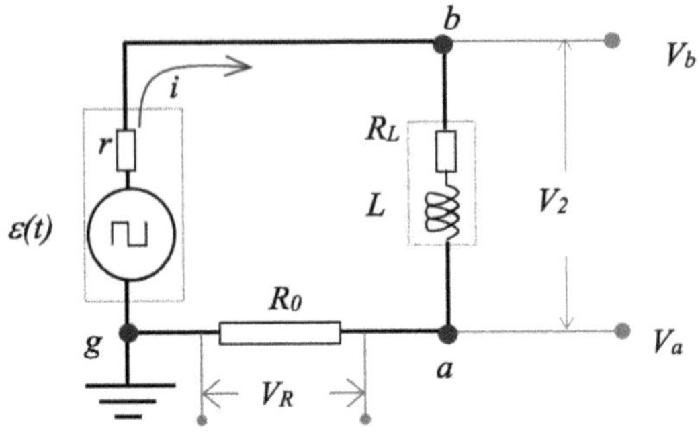

Figure 35.2 *RL* circuit connected to a function generator with internal resistance r.

It's easy to show [3–5] that the solution of this differential equation is

$$i(t) = \frac{V_{00}}{R}\left[1 - e^{-t/\tau}\right], \tag{35.8}$$

which satisfies the initial condition $i(t) = 0$. Parameter $\tau = L/R$; it is known as time constant or characteristic time of the RL circuit. In the following activity, we will attempt to study the behavior of an RL circuit when we apply a square voltage. In this case, we must remember that the resistance R in Eqs. (35.6)–(35.8) is the total resistance, which includes the external resistance R_0, the intrinsic resistance R_L of the inductance, and the internal resistance r of the FG. See Figure 35.2.

35.3.1 *Project 108. Characteristic time of the RL circuit*

Recommended equipment: A coil of a few tens of mH and resistors of a few tens of Ohms. A two-channel oscilloscope or data acquisition system connected to a computer. A function generator.

Using the circuit in Figure 35.2 and an oscilloscope or DAS, experimentally study the variation of the voltages V_a and V_b when in the FG we apply a square voltage of amplitude V_{00}. Measure the value of R_L and choose R_0 so that $\tau \approx 10\,\mu$s. Choose the frequency of FG $f_0 \approx 0.2/\tau$ so that during each half cycle, the current reaches a value very close to that of $t = \infty$. If possible, for R_0 use a resistor box or a variable resistor of known value. Likewise, the amplitude V_{00} must be less than the maximum voltage that the data collection system can accept.

Suggested tasks:

- ✓ Using a digital oscilloscope or DAS, record the values of $V_a(t)$ and $V_b(t)$, for at least two full cycles.
- ✓ Fitting the experimental data, of i versus t, similar to Figure 31.5, determine the experimental value of τ.
- ✓ Vary the value of R_0 and determine the new experimental value of τ. Repeat this operation for at least four values of R_0, making sure to cover an approximate range of $(R_0/2 - 2R_0)$.
- ✓ Plot τ as a function of $R(= R_0 + R_L + r)$ and then plot τ as a function of the quotient R/L.
- ✓ What conclusions do you draw from this study regarding the model discussed in the previous section?

35.4 RL circuit — Stationary response

Let's consider the circuit in Figure 35.2 which includes the internal resistance r of the FG. Usually, this resistance is of the order of $50\,\Omega$ and in general it may be found in the instrument datasheet or determined using some of the techniques discussed in Chapter 24. Thus $R = R_t = R_0 + R_L + r$. If $v(t)$ is the output voltage of the FG, which we assume sinusoidal, that is,

$$v(t) = V_{00}\sin(\omega t) \quad \text{with} \quad \omega = 2\pi \cdot f, \tag{35.9}$$

applying Kirchhoff's mesh law, we have [5, 6]

$$L\frac{\mathrm{di}}{\mathrm{dt}} + \mathrm{Ri} = V_0 \sin(\omega t). \tag{35.10}$$

The solution of this differential equation is [3–5]

$$i(t) = I_0 \sin(\omega t - \phi) \tag{35.11}$$

with

$$I_0 = \frac{V_0}{\sqrt{R^2 + (L\omega)^2}} \quad \text{and} \quad \tan\phi = \frac{L\omega}{R}. \tag{35.12}$$

In this circuit, the measurement of V_a allows us to know the current $i(= V_a/R_0)$. On the other hand, $V_b = v(t) - r \cdot i(t)$, which is what we can measure directly. It can be written as follows:

$$V_b(t) = V_0 \sin(\omega t + \varphi_0). \tag{35.13}$$

And according to Eq. (35.11), the current $i(t)$ should be

$$i(t) = I_0 \sin(\omega t - \phi + \varphi_0), \tag{35.14}$$

with

$$I_0 = \frac{V_0}{\sqrt{(R_0 + R_L)^2 + (L\omega)^2}} \quad \text{and} \quad tan\,\phi = \frac{L\omega}{R_0 + R_L}. \tag{35.15}$$

Thus, the phase difference between V_a and V_b is precisely ϕ. For each frequency, this phase difference can be determined using a parametric plot (XY) of V_a and V_b (see Annex A of Chapter 31). On the other hand, V_0 is the amplitude of the signal V_b and I_0 is the amplitude of (V_a/R_0).

35.4.1 *Project 109. Response of the RL circuit in frequency*

Recommended equipment: Inductances (of a few tens or hundreds of mH) and resistances of a few tens of Ohms. A data acquisition system connected to a computer or digital oscilloscope. A function generator.

For this experiment, it is convenient to use one or more inductances of a few tens or hundreds of mH. Remember to use a sampling rate in the DAS of the order of $10 \times f_m$, f_m being the maximum frequency to be used. In order to study the frequency dependence of the RL, it is convenient to use a frequency range between $f_r/3$ and $3f_r$, with $f_r = (R_0 + R_L)/(2\pi \cdot L)$. Use the *sinusoidal* output on the FG.

Suggested tasks:

Using the circuit in Figure 35.2, study the shape of the signals V_a and V_b as a function of time:

✓ Analyze the phase difference ϕ between V_a and V_b for $f \approx f_r$.
✓ If V_0 and I_0 are the amplitudes of the measured signals $V_b(t)$ and $I(t) = V_a(t)/R_0$, respectively, plot I_o and V_0 as a function of ω.
✓ Construct a plot of $(V_0/I_0)^2$ as a function of ω^2. According to the proposed model, Eq. (35.15), it is expected that in this representation we would obtain a straight line. Do your results match this theoretical expectation? If so, obtain the slope and ordinate to the origin and compare these values with those expected according to Eq. (35.15).
✓ Determine the phase difference ϕ between V_a and V_b. Build a plot of $\tan(\phi_0)$ as a function of ω. On the same graph, include the results of your theoretical expectations, Eq. (35.15). What do you conclude about the theoretical model in this comparison?

35.5 Annex A. Estimation of the self-inductance of a coil

In many practical situations, it is useful to have approximate expressions to estimate the value of the inductance of a coil. The inductance L of a linear solenoid of length h, cross-section S, and diameter d, having N turns can be calculated as follows [6–9]:

$$L = k\frac{\mu_0 \mu_r S N^2}{h}, \tag{35.16}$$

where k is a constant that depends on the h/d ratio of the particular solenoid according to Table 35.1. μ_r is the relative permittivity of the coil core and μ_0 is the permittivity of the vacuum. In usual units, Eq. (35.16) can be written as follows [7]:

$$L[\mu H] = 0.01257\, k\frac{\mu_r S[\text{cm}^2] N^2}{h[\text{cm}]}. \tag{35.17}$$

Table 35.1

Relation h/d	0.1	0.5	1	5	10
k	0.2	0.5	0.6	0.9	1.0

The inductance of a cylindrical coil, with an air core, can also be estimated using the following semi-empirical expression:

$$L(mH) = \frac{N^2 d^2}{18d^2 + 40h}. \tag{35.18}$$

$d = 2R$ is the diameter of the winding (coil) and h is the total length of the coil, both expressed in inches; N is the number of total turns of the coil.

35.6 Estimation of the inductance of a circular loop

It is useful to have an expression to calculate the inductance L of a coil of radius a, with N turns of wire of radius r whose relative permittivity is 1. According to Rosas [8, 9], we obtain

$$L(H) \approx \mu_0 N^2 a \left[\left(1 + \frac{r^2}{8a^2}\right) \ln \frac{8a}{r} + \frac{r^2}{24a^2} - 1,75 \right] + O[(r/a)^4]. \tag{35.19}$$

If $a \gg r$, we get

$$L(H) \approx \mu_0 N^2 a \left[\ln\left(\frac{8a}{r}\right) - 1.75 \right]. \tag{35.20}$$

References

[1] Wikipedia, "Inductance," 2021. Available: http://en.wikipedia.org/wiki/Inductance.

[2] R. Duffy and U. Haber- Schai, "Establishing V=L (dI/dt) directly from experiment," *Am. J. Phys.*, vol. 45, p. 170, 1977.

[3] R. Halliday, D. Resnick and M. Krane, *Fundamentals of Physics*, NY: John Wiley & Sons, Inc., 2002.

[4] F. Sears, M. Zemansky, H. Young and R. Freedman, *College Physics*, NY: Pearson Ed., 2009.

[5] E. M. Purcell and D. J. Morin, *Electricity and Magnetism, Berkeley Physics Course*, 3rd Ed., NY: Cambridge University Press, 2014, p. 839.

[6] J. R. Reitz, F. J. Milford and R. W. Christy, *Fundamentals of Electromagnetic Theory*, NY: Ed. Addison-Wesley, 1996.
[7] N. Koshkin and M. Shirkevich, *Physics Handbook*, Moscu: MIR, 1975.
[8] E. Rosa, "The self and mutual inductances of linear conductors," *Bull. Bur. Stand.*, vol. 4 no. 2, pp. 301–344, 1908. Available: https://nvlpubs.nist.gov/nistpubs/bulletin/04/nbsbulletinv4n2p301_a2b.pdf.
[9] E. Rosa and L. Cohen, "Formulae and tables for the calculation of mutual and self-inductance," 1908. Available: https://www.nist.gov/publications.

Chapter 36

PHYSICS OF A MAGNET FALLING THROUGH A TUBE

This chapter proposes an experimental study of the passage of a permanent magnet through a loop and in free fall inside a conductive tube. The goal is to characterize the motion of a magnet inside the tube and to study the physics of this process. An attempt is made to explain the results using the laws of electromagnetism.

Goals

✓ Magnetic moment of a permanent magnet
✓ Passage of a magnet through a loop
✓ Movement of a permanent magnet inside a metal tube

36.1 Magnet falling through a metal tube

A very striking and interesting demonstration is the fall of a strong magnet through a thick-walled copper or aluminum tube. Although there is no magnetic attraction between the magnet and these metals, the fall of the magnet inside the tube is significantly delayed by the force caused by the interaction between the currents induced in the tube and the magnet itself. This experiment is a remarkable demonstration of the effect of *Lenz's law*. The purpose of these experiments is to carry out a quantitative study of the effects that occur in this process.

Given the importance of the magnetic properties of the magnet being used, the first part of this chapter is devoted to determining the magnetic moment of the magnet. Then, it studies the shape of the pulses induced by the magnet when it passes through a coil and finally, the fall of the magnet through the interior of a metal tube.

36.2 Oscillation of a permanent magnet in a uniform field

Consider the case of a cylindrical permanent magnet, magnetized along its axis, immersed in a uniform magnetic field **B**, as shown in Figure 36.1. The magnet is supposed to hang from its middle point (horizontally) from a soft, light thread that does not exert any appreciable torque when twisted (e.g., a cotton sewing thread). If the magnet is rotated from its equilibrium position (Figure 36.1), it will begin to oscillate.

The external magnetic field **B** will exert a restoring torque given by [1–3]

$$\tau = -m_B B \sin(\theta), \tag{36.1}$$

where m_B is the magnetic moment. If the effect of friction is negligible, the equation of motion of the magnet is

$$-B \cdot m_B \sin(\theta) = I \frac{d^2\theta}{dt^2}, \tag{36.2}$$

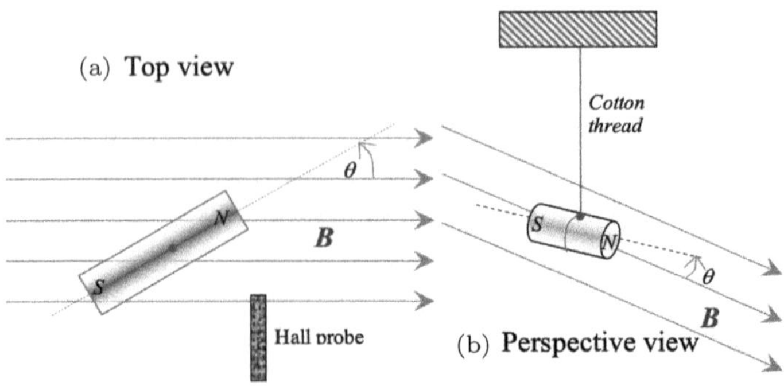

Figure 36.1 Magnet in a uniform magnetic field B: (a) top view and (b) perspective view. The Hall probe detects the magnet oscillations. The axis of the cylindrical magnet is indicated by the dotted blue line.

where I is the moment of inertia with respect to the vertical axis that passes through the magnet's center of mass and is perpendicular to the axis of the cylindrical magnet. The magnetic field B could be the Earth's magnetic field or the field produced by Helmholtz coils.

In what follows, it is assumed that the oscillations are small so that $\sin(\theta)$ can be approximated to θ. Therefore, Eq. (36.2) becomes

$$\frac{d^2\theta}{\mathrm{dt}^2} \approx -\frac{B \cdot m_B}{I}\theta, \tag{36.3}$$

which is the equation of motion of a system that oscillates with simple harmonic motion with angular frequency ω_0, given by

$$\omega_0^2 = \left(\frac{2\pi}{T}\right)^2 = \frac{B \cdot m_B}{I}. \tag{36.4}$$

The period of oscillation T can be measured directly by measuring the time corresponding to a number n of oscillations or using a device such as a photogate, a pick-up coil, or a Hall probe connected to a computer data acquisition system. Care must be taken that the measuring devices (photogate, coil, Hall probe, etc.) do not contain ferromagnetic parts.

By measuring the period T (or the frequency, ω_0), knowing the value of field B and the moment of inertia I, m_B can be calculated from Eq. (36.4).

36.3 Determination of the magnetic moment of a permanent magnet

There are several methods to determine the magnetic moment of a permanent magnet. As discussed in the previous section, its oscillation can be studied within a uniform magnetic field, e.g., within a Helmholtz coil, but also, as we will see later, by measuring its magnetic field at different distances from its center.

(a) **Oscillation of the magnet in the center of a Helmholtz coil:** As discussed in Chapter 33, a Helmholtz coil consists of two circular coils of radius R, separated by a distance equal to their radius. If both coils have N turns and the same current, I, circulates through both coils (in the same direction), the magnetic field in the central region between the coils, within a volume of radius R, is extremely constant. The value of this field B_H, along the axis of the coil, is given by [3, 4]

$$B_H(z) = \mu_0 \frac{8}{\sqrt{125}} \frac{NI}{R}. \tag{36.5}$$

If a permanent magnet is hung by a thread in the center of this Helmholtz pair, as shown in Figure 36.1, it will orient itself, like the arrow on a compass, in the direction of the total magnetic field $\vec{B}_T$ resulting from the coil and the Earth's magnetic field $\vec{B}_{\text{Earth}}$, *i.e.*,: $\vec{B}_T = \vec{B}_H + \vec{B}_{\text{Earth}}$.

A practical way to proceed is to orient the axis of the Helmholtz coil perpendicular to the north-south direction. With the compass in the center of the coil, the field is increased until the needle is oriented along the axis of the coil. When this happens, we know that the Earth's magnetic field is negligible compared to that of the coil, therefore the field value B_H is used for Eq. (36.4). The moment of inertia I can be determined by knowing the mass and geometry of the magnet.[a] The mass of the magnet can be determined with a sensitive balance. To avoid spurious influences or magnetizing the balance, it is advisable to place the magnet on top of a light plastic or expanded polystyrene (Styrofoam) cup placed upside-down on the balance plate so that the magnet field interacts as little as possible with the balance.

In a plot of $I\omega_0^2$ as a function of B_H, the slope of the linear fit provides the value of m_B [5]. Once m_B is determined, if the current through the coil stops, the frequency of oscillation of the magnet gives the value of B_{Earth} through Eq. (36.4).

(b) **Using a Gaussmeter or Hall probe:** Another technique to determine m_B is to measure the axial field of the magnet in

[a]A thin rod rotating on an axis that goes through the center of the rod (perpendicular to its length), with mass M and length L, has a moment of $I = ML^2/12$. See also https://en.wikipedia.org/wiki/List_of_moments_of_inertia

the direction of the z axis (take the z axis coincident with the direction of $\vec{m}_B$) and in a direction perpendicular to it (x axis). Remember that for a dipole, at large distances compared to its dimensions ($r \gg L$), its magnetic field is given by [3, 4, 6]

$$B_r(r,\theta) = \frac{\mu}{4\pi} 2m_B \frac{\cos}{\theta} r^3 \tag{36.6}$$

and

$$B_\theta(r,\theta) = \frac{\mu}{4\pi} m_B \frac{\sin\theta}{r^3}, \tag{36.7}$$

the value of the axial field will be $B_z(z) = B_r(\theta = 0,\ r = z)$ and the perpendicular component $B_\perp(x) = B_\theta(\theta = \pi/2, r = x)$, that is,

$$B_z(z) = \frac{\mu}{2\pi} m_B \frac{1}{z^3} \tag{36.8}$$

and

$$B_\perp(x) = B_z(x) = \frac{\mu}{4\pi} m_B \frac{1}{x^3}. \tag{36.9}$$

By plotting the measured values of $2\pi B_z(z)/\mu_0$ as a function of $(1/z^3)$, or of $4\pi B_\perp(x)/\mu_0$ as a function of $(1/x^3)$, the value of m_B can be found from the respective slopes of the linear fit to these data. For a magnet of length L, the expression (36.8) can be improved by the following expression [7, 8]:

$$B_z(z) = \frac{\mu_0}{4\pi} \frac{m_B}{L} \left[\frac{1}{(z - L/2)^2} - \frac{1}{(z + L/2)^2} \right]. \tag{36.10}$$

This expression reduces to (36.8) for $L \to 0$.

36.3.1 *Project 110. Determining the magnetic moment of a magnet inside a Helmholtz coil*

Recommended equipment: A rare earth magnet, a Helmholtz coil, a DC power supply, and a Hall probe, preferably connected to a data acquisition system connected to a PC. Optional: A photogate and a data acquisition system associated with a PC.

As discussed previously, the magnetic moment m_B of a magnet is very important to understand quantitatively the physics associated with the effects that a permanent magnet produces under various circumstances. It is therefore necessary to know the value of this parameter [1, 2]. Of course, there are many experimental methods for measuring m_B, but we will focus only on the two procedures discussed above. Begin by simply hanging a cylindrical magnet from a string, as illustrated in Figure 36.1. Another alternative, if two identical rare earth magnets are used (of the kind that resemble button-type batteries), consists of joining the two identical magnets by their opposite poles and trapping the cotton thread between them. Of course, what will be measured in this case is the moment resulting from the combination of the two magnets.

The second experimental technique suggested to determine the magnetic moment m_B of a permanent magnet is to measure with a Hall probe the magnetic field along the axis of the magnetic moment m_B and perpendicular to it, as indicated by Eqs. (36.8) and (36.9), using an experimental setup like the one suggested in Project 99 of Chapter 33.

Suggested tasks:

✓ Determine m_B for the chosen magnet using the method that is most appropriate for you and estimate its uncertainty.

✓ If possible, use the two techniques discussed above and compare their advantages and disadvantages.

36.4 Pulses induced by a magnet passing through a coil

Consider a permanent magnet of length L, diameter d, and magnetic moment m_B that moves with velocity v along an axis (z), passing through a circular coil of radius R and N turns, whose axis coincides with the z axis, as illustrated in Figure 36.2. The problem we want to analyze is the shape of the pulse induced in the coil. Assume that $d < 2R$ and that the thickness of the coil is much less than L.

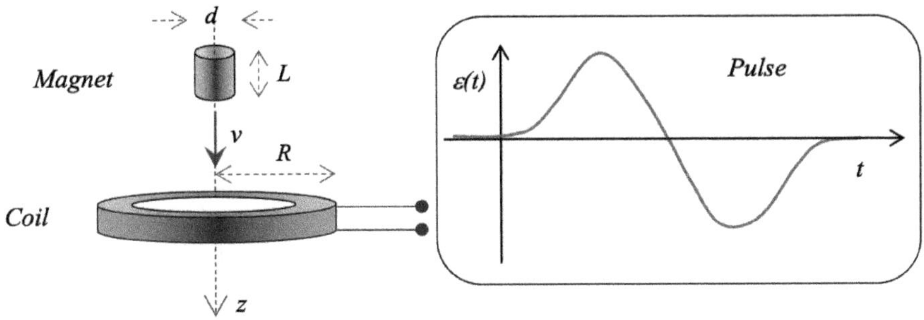

Figure 36.2 Permanent magnet of length L and magnetic moment m_B that passes through a coil of radius R with speed v. The z axis coincides with the direction of movement of the magnet and the axis of the coil.

Using Faraday's law of induction in combination with a model of the magnetic field of a magnet, Eq. (36.10), we will try to obtain an approximate expression of the voltage generated in the coil. This expression is deduced in detail in Refs. [7] and [8]. Here is a brief summary of these arguments.

An approximate way of obtaining an order of magnitude of the pulses is to assume that the flux in the coil is approximately $\Phi_B \approx \pi R^2 B_z(z)$. If we assume that the z coordinate of the magnet is given by $z = v \cdot t$, the induced voltage is $\varepsilon(t) = -(d\Phi_B/dz) \cdot (dz/dt)$. Using Eq. (36.10) for B, we have

$$\begin{aligned}\varepsilon(t) &\approx \pm\pi R^2 N \frac{dB_z(z)}{dz}\frac{dz}{dt} \\ &\approx \pm v N m_B \frac{R^2}{2L}\mu_0 \left\{\frac{1}{(z(t)+L/2)^3} - \frac{1}{(z(t)-L/2)^3}\right\}. \end{aligned} \tag{36.11}$$

The sign of the induced voltage depends on the orientation of the magnet with respect to the coil.

In this model, we assumed that the field over the entire area of the coil is equal to the field along the z-axis of the magnet, which may be approximately true if the radius of the coil is small. If this is not the case, it is necessary to make a better approximation. Reference [7] provides an improved model to consider the case of a finite value of R, whose result is

$$\varepsilon(z) = \pm v N m_B \frac{R^2}{2L}\mu_0 \left\{ \frac{1}{[(z+L/2)^2+R^2]^{3/2}} - \frac{1}{[(z-L/2)^2+R^2]^{3/2}} \right\}, \tag{36.12}$$

where it is assumed that $z = vt$. Therefore, if we take as the origin of time, t, the point at which the magnet is in the center of the coil (when the induced voltage passes through zero), and assume that during the passage of the magnet, the speed is constant and equal to v, we have that the induced voltage will vary in time as follows:

$$\varepsilon(t) = \pm v N m_B \frac{R^2}{2L}\mu_0 \left\{ \frac{1}{[(vt+L/2)^2+R^2]^{3/2}} - \frac{1}{[(vt-L/2)^2+R^2]^{3/2}} \right\}. \tag{36.13}$$

This result can be tested (falsified) by recording the shape of the induced pulse, as indicated in the following activities.

Using Eq. (36.12) or (36.13), the value of the integral of $\varepsilon(t)$ between zero and infinity, that is, the area of a half pulse, results in [7]

$$I_\varepsilon = \text{Area}_{\text{half pulse}} = \int_0^\infty \varepsilon(t)dt = \left[\frac{N\mu_0}{(L^2+4R^2)^{1/2}} \right] \cdot m_B. \tag{36.14}$$

It can be seen that this area of the pulse is independent of velocity. On the other hand, as N, L, and R are easily measurable parameters, the value of I_ε enables us to determine the value of m_B, from the measured pulse. The m_B value obtained with this method can be compared to those obtained by the methods discussed in Section 36.3.

36.4.1 *Project 111. Pulses induced by a magnet passing through a coil in free fall*

Recommended equipment: A rare earth magnet, a glass or acrylic tube, and a coil of about 20–100 turns. A data acquisition system associated with a PC.

This experiment studies the shape of the voltage pulse induced in a coil when the magnet passes through it at different speeds. For this project, it is useful to have a non-conductive tube, preferably transparent (acrylic or glass), with an internal diameter greater than the magnet so that the magnet can fall through it without touching the walls. It is advisable to have a coil of about 20–100 turns on the tube, which can slide along it, or several coils of this type distributed approximately every 10 cm, all connected in series. The coils can be made using enameled copper wire 0.1–0.2 mm in diameter.

The tube is hung as shown in Figure 36.3. A weight may be hung from the lower end of the tube to keep it vertical. It is also useful to have a container to catch the magnet, as rare earth magnets are fragile. A practical suggestion is to surround the magnet with a paper tube attached with adhesive tape to prevent the magnet from tilting inside the tube. This makes the fall more stable.

The coil (or the coils in series) are connected to the input of a computer data acquisition system that can record the pulses generated as the magnet passes through the various coils. If you have an instrumentation amplifier (differential) [9], you can use it between the coil and the input of the data collection system. The

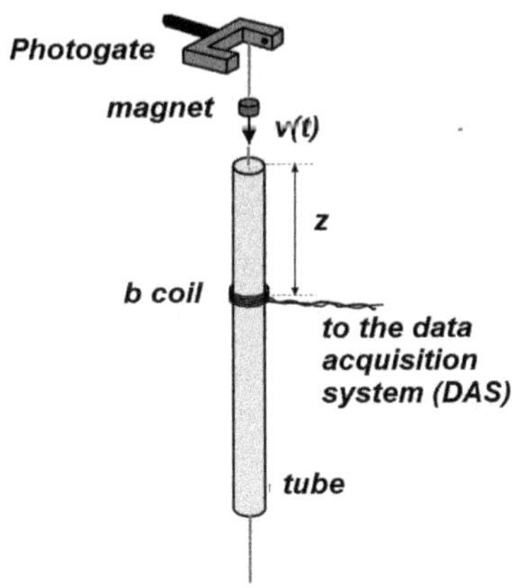

Figure 36.3 Diagram of the suggested experimental device to measure the shape of the pulse generated by a permanent magnet passing through a coil in free fall.

pulse amplitude can be amplified by increasing the number of turns of each coil or by using a magnet with a stronger field strength.

If you have a photogate, place it at the top and connect it to another input of the data collection system. This photogate generates a signal when the magnet begins to fall, which can be used to trigger data collection. Thus, the time interval t between the photogate signal and the zero crossing of the loop signal gives the magnet fall time. These data may be useful but is not crucial to carry out this experiment. The distance z between the tube inlet (position of the photogate) and the position of the coil defines the speed v of the magnet passing through the coil. Assuming that the fall takes place with constant acceleration, the speed v of the magnet as it passes through the coil will be

$$v = \sqrt{2g \cdot z}, \tag{36.15}$$

where g is the gravitational acceleration.

The goal of this project is to experimentally test the expressions (36.12) and (36.13), assuming that the parameters N, R, v, L, and m_B are known or have been previously measured.

Suggested tasks:

- ✓ Perform the experiment for different values of z and record the voltage pulses induced in the coil by the passage of the magnet.
- ✓ From your measurements and using Eqs. (36.12) and (36.13), compare the measured pulses to the calculated pulses. Discuss the degree of agreement or disagreement.
- ✓ Given that in this accelerated motion, the speed of the magnet is not constant, to estimate the area of half a pulse, calculate the area of the absolute value of the entire pulse and divide by two, that is,

$$I_\varepsilon = \text{Area}_{\text{half pulse}} = \frac{1}{2}\int_{-\infty}^{\infty} |\varepsilon(t)|dt, \tag{36.16}$$

where the integral must be calculated using a numerical area calculation procedure and using the measured values of $\varepsilon(t)$.

In this way, the value obtained is closer to the estimate given by Eq. (36.14). Calculate the experimental value of I_ε by performing a numerical integration of the pulses measured as a function of time. Is the value of I_ε independent of speed and is it the same for all measured pulses?

✓ If the values of I_ε are approximately constant, from the mean value of I_ε, find the value of m_B using Eq. (36.14) and compare it to the results found for m_B of the magnet used with other experimental techniques. See Section 36.3.

✓ Discuss how good the model described by Eqs. (36.12), (36.13), and (36.14) is.

36.5 Magnet falling through a conductive tube

This section considers a magnet of magnetic moment m_B falling through a *metal* tube. We will assume that the tube has a mean radius (in the center of the wall) a, the conductivity of the material is σ, and the wall thickness is δ. We will also assume that the magnet has a mass m and length L. Let the coil (or several coils) that measure the pulse voltage, $\varepsilon_{\mathrm{M}}(t)$, have N turns and radius $a_{\mathrm{coil}} \approx a$. All these parameters are known or measurable in principle.

The problem of a magnet falling through a tube has been studied by several authors [7, 8, 10]. The magnetic drag force, F_M, on a strong magnet falling inside a vertical conductive tube is a consequence of the current induced on each ring that constitutes the tube. Due to Lenz's law, each of these currents tends to oppose the motion of the magnet, thereby generating the magnetic drag or break. This phenomenon is analyzed in detail in Refs. [10–14]. Here, we only summarize some of the basic results. Since the magnetic drag force is proportional to the magnet speed v, that is, $F_M = -Kv$, with K being a constant, the magnet will accelerate until its velocity makes $F_M = -mg$, thus the net force will be zero and the magnet will fall with a *constant* speed, v_∞, given by

$$v_\infty = \frac{m \cdot g}{K}. \tag{36.17}$$

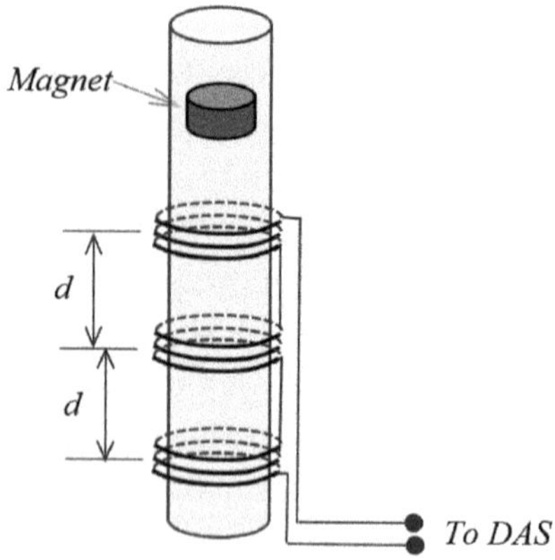

Figure 36.4 Diagram of the suggested setup to study the fall of a magnet through a conductive tube. A set of coils spaced at regular distances d from each other and connected in series to a data acquisition system (DAS) is used to register the passage of the magnet through each coil and measure its velocity.

To better analyze this problem, it is advisable to use a setup like the one suggested in Figure 36.4.

By plotting the positions $z_i = i \cdot d$ of the coils along the tube versus the times t_i at which the voltage pulses $\varepsilon_M(t)$ pass through zero, we can study the magnet motion (v_i) along the tube. Here, i is the index of the coil wound on the tube. Once the magnet reaches a constant speed, the relationship between z_i and t_i is linear, and from the slope, we can calculate $v_\infty = dz/dt$.

The azimuthal currents $\delta i_\theta(z)$ induced in each section (infinitesimal ring of length dz) of the metal tube, are related to the induced emf, $\varepsilon(z)$, by [10]:

$$\delta i_\theta(z) = \frac{\varepsilon(z)}{2\pi a\rho/(\delta dz)} = \frac{\sigma\delta\varepsilon(z)}{2\pi a}dz. \tag{36.18}$$

Here, $\sigma(= 1/\rho)$ represents the conductivity of the material of which the tube is made, δ is its thickness, a is its mean radius, and z is the distance of the infinitesimal element of a tube (of height dz) to the center of the magnet. By conservation of energy, when the magnet reaches the limit speed v_∞, the gravitational energy difference of the magnet is converted into tube heating (Joule effect), that is [10, 11],

$$P = v_\infty mg = \int \varepsilon(z)\delta i_\theta dz = \frac{\sigma\delta}{2\pi a} v_\infty \int_{-\infty}^{\infty} \varepsilon^2(t)dt. \tag{36.19}$$

Since $\varepsilon(t)$ is proportional to m_B, defining $K_0 = \int_{-\infty}^{\infty} \varepsilon^2(t)dt/m_B^2$, we have

$$v_\infty = K_0 \frac{\sigma \cdot \delta}{a \cdot m} m_B^2, \tag{36.20}$$

from where we get an expression that relates the magnetic force with some characteristics of the tube. The voltage induced in the coils (with N turns)is: $\varepsilon_M(t) = N\varepsilon(t)$, therefore:

$$F_M = mg = \frac{\sigma\delta}{2\pi a N^2} \int_{-\infty}^{\infty} \varepsilon_M^2(t)dt = \frac{\sigma\delta}{2\pi a N^2} \cdot m_B^2 K_0. \tag{36.21}$$

Since according to Eq. (36.13), $\varepsilon(t)$ is proportional to m_B, from Eq. (36.21), we have $F_M \propto v_\infty m_B^2$. Therefore, Eq. (36.17) can be written as follows:

$$v_\infty = \frac{m \cdot g}{k_a \cdot m_B^2}, \tag{36.22}$$

where k_a is a constant of proportionality. Thus, we see that if we form a magnet by joining n equal magnets, we expect that the mass of this compound magnet will be $m_n = n\ m$ and its magnetic moment $m_{nB} = n \cdot m_B$. Therefore, the rate of fall of this compound magnet will be

$$v_\infty^{(n)} = \frac{n \cdot m \cdot g}{k_a \cdot n^2 \cdot m_B^2} = \frac{mg}{k_a \cdot m_B^2} \frac{1}{n} = \frac{v_\infty^{(1)}}{n}, \tag{36.23}$$

where $v_\infty^{(n)}$ represents the terminal or final velocity of the combination of n magnets and $v_\infty^{(1)}$ the terminal velocity of a single magnet. This relationship allows us to validate the model of a magnet drop in a conductive tube in a simple way, that is, by measuring the terminal or final velocity of a combination of identical magnets.

36.5.1 *Project 112. Fall of a permanent magnet through a conductive tube I*

Recommended equipment: 4–6 identical rare earth magnets and an aluminum or copper tube with a wall thickness δ from 2–4 mm. Several coils (about 30 turns) evenly distributed along the tube. A data acquisition system.

The goal of this experiment is to study the fall kinematics of a permanent magnet through a metal tube. To do so, record the voltage pulses $\varepsilon_M(t)$ induced in the turns when the magnet passes through them and measure the limiting speed v_∞ that the magnet acquires as it passes through the tube. After analyzing the case of a falling magnet, consider the same situation with n identical stacked magnets, [15] with $n = 1 - 5$. From these measurements, compare the results with the theoretical model of Eq. (36.23).

For this project, an experimental setup like the one shown in Figure 36.4 can be used, except that this time, the tube is made of a conductive, non-ferromagnetic material. One possibility is to use copper or aluminum tubes about 2–4 mm thick. Several evenly spaced coils of equal number $N(\approx 30)$ of turns should be wound onto the tube and connected in series.

Suggested tasks:

- ✓ Using a single magnet, determine the measured voltage $\varepsilon_M(t)$ which is recorded when the magnet falls through the tube.
- ✓ If the times, t_i, at which voltage pulses, $\varepsilon_M(t)$, pass through zero are plotted vs. the position z_i of the turns along the tube, you can study the movement of the magnet. Once the magnet reaches a constant speed, the relationship between z_i and t_i is linear, and from the slope, you can calculate v_∞.
- ✓ Determine the value of the limiting velocity v_∞ for $1, 2, \ldots, n$ magnets.
- ✓ Construct a graph of $v_\infty^{(n)}$ as a function of $1/n$. If this relationship is linear, find the value d from the slope $v_\infty^{(1)}$ and compare it to the value found for a single magnet. What can you conclude from the model made to obtain Eq. (36.23)?

36.5.2 *Project 113. ♣♣ Fall of a permanent magnet through a conductive tube II*

Recommended equipment: A rare earth magnet and an aluminum or copper tube with wall thickness $\delta = 2 - 4\,\text{mm}$. A coil of about 100 slidable turns or, better, several coils (of approximately 30 turns each) evenly distributed along the tube. An associated data acquisition system.

The objective of this experiment is to study the shape of the voltage pulses $\varepsilon_{\text{M}}(t)$ induced in the coils when the magnet passes through them. Compare the results of these measurements to the expectations of the theoretical model discussed above.

For this project, you can use the experimental arrangement shown in Figure 36.4, but this time, the tube is made of a conductive, non-ferromagnetic material. Wind along the tube several evenly spaced fixed coils of equal number of turns ($N \approx 30$) connected in series.

Suggested tasks:

✓ Using a single magnet, determine the measured voltage $\varepsilon_{\text{M}}(t)$ which is recorded when the magnet falls through the tube.
✓ Determine the value of the limiting speed v_∞.
✓ Compare the shape of the measured pulses, $\varepsilon_{\text{M}}(t)$, to those predicted by the model Eq. (36.13).
✓ Using Eq. (36.20), determine the value of the magnetic force F_M from the numerical integration of the measured pulses $\varepsilon_{\text{M}}(t)$, once the speed of the magnet reaches its limit value v_∞.
✓ Using Eqs. (36.20) and (36.21), calculate the theoretical value of v_∞ and compare these results to the measured values. What can you conclude?

References

[1] R. Halliday, D. Resnick and M. Krane, *Fundamentals of Physics*, NY: John Wiley & Sons, Inc., 2002.

[2] F. Sears, M. Zemansky, H. Young and R. Freedman, *College Physics*, NY: Pearson Ed., 2009.
[3] E. M. Purcell and D. J. Morin, *Electricity and Magnetism, Berkeley Physics Course*, 3rd Ed., NY: Cambridge University Press, 2014, p. 839.
[4] J. R. Reitz, F. J. Milford and R. W. Christy, *Fundamentals of Electromagnetic Theory*, NY: Ed. Addison-Wesley, 1996.
[5] E. Bisquert, S. Hurtado and J. Pina Mafé, "Oscillation of a dipole in a magnetic field: An experiment," *Am. J. Phys.*, vol. 58, p. 567 and 838, 1990.
[6] J. Jackson, *Classical Electrodynamics*, New York: Wiley and Sons, 1991.
[7] J. Manzanares, *et al.*, "An experiment on magnetic induction pulses," *Am. J. Phys.*, vol. 62, p. 702, 1994.
[8] N. Derby and S. Olbert, "Cylindrical magnets and ideal solenoids," *Am. J. Phys.*, vol. 78, p. 229, 2010.
[9] Vernier, 2022. Available: https://www.vernier.com/product/instrumentation-amplifier/; https://www.vernier.com/product/instrumentation-amplifier/.
[10] G. Donoso, C. Ladera and P. Martín, "Damped fall of magnets inside a conducting pipe," *Am. J. Phys.*, vol. 79, p. 193, 2011.
[11] K. Hahn, *et al.*, "Eddy current damping of a magnet moving through a pipe," *Am. J. Phys.*, vol. 66, p. 1066 , 1998.
[12] J. Iniguez, V. Raposo and A. Hernandez-Lopez, "Study of the conductivity of a metallic tube by analysing the damped fall of a magnet," *Eur. J. Phys.*, vol. 25, p. 593–604, 2004.
[13] G. Ireson and J. Twidle, "Magnetic braking revisited: Activities for the undergraduate laboratory," *Eur. J. Phys.*, vol. 29, p. 745–751, 2008.
[14] M. Roy, M. Harbola and H. Verma, "Demonstration of Lenz's law: Analysis of a magnet falling through a conducting tube," *Am. J. Phys.*, vol. 75, p. 728, 2007.
[15] C. Sawicki, "A Lenz's law experiment revisited," *Phys. Teach.*, vol. 38, p. 438, 2000.
[16] R. Nicklin, "Faraday's law — Quantitative experiments," *Am. J. Phys.*, vol. 54, pp. 422–428, 1986.
[17] J. Clack and T. Toepker, "Magnetic induction experiment," *Phys. Teach.*, vol. 28, p. 236, 1990.
[18] J. J. Priest and B. Wade, "A Lenz law experiment," *Phys. Teach.*, vol. 30, no. 2, p. 106, 1992.
[19] C. MacLatchy, P. Backman and L. Bogan, "A quantitative magnetic breaking experiment," *Am. J. Phys.*, vol. 61, pp. 1096–1101, 1993.

[20] P. Carpena, "Velocity measurements through magnetic induction," *Am. J. Phys.*, vol. 65, pp. 65, 135–140, 1997.
[21] C. MacLatchy, P. Backman and L. Bogan, "A quantitative magnetic breaking experiment," *Am. J. Phys.*, vol. 61, p. 1096–1101, 1993.
[22] J. Palesko, M. Cesky and S. Huertas, "Lenz's law and dimensional analysis," *Am. J. Phys.*, vol. 73, no. 1, p. 37–39, 2005.
[23] B. Berls and M. Ruiz, "Lenz's law with aluminum foil and a lengthwise slit," *Phys. Educ.*, vol. 53, p. 4, 2018.
[24] L. Wood, R. Rottmann and R. Barrera, "Faraday's law, Lenz's law, and conservation of energy," *Am. J. Phys.*, vol. 72, p. 3, 2004.

Chapter 37

ELECTROSTATIC FIELDS AND POTENTIALS — LAPLACE EQUATION

This chapter studies the electric potentials for diverse field configurations with different boundary conditions in two dimensions. It presents a technique to determine equipotential surfaces and electric field lines for different electrode configurations, immersed in a conductive liquid medium and connected to a low-voltage source. The experimental results are compared to the corresponding theoretical predictions, which result from solving the Laplace equation in two dimensions. The relaxation method is introduced using a spreadsheet to solve the Laplace equation. We also discuss how to deal with the different boundary conditions, particularly those of Neumann and Dirichlet.

Goals

✓ Equipotential surfaces and field lines
✓ Boundary conditions
✓ Laplace's law
✓ Relaxation method

37.1 Introduction to electrostatics

The *electric field* at a given point in space is related to the electric force that is exerted on a positive test charge q. If at the coordinate point (x, y) there is an electric field $\vec{E}(x, y)$, a force is exerted on the test charge placed at that point $\vec{F}(x, y)$. According to the definition of electric field, we have [1–4]

$$\vec{F}(x, y) = q \cdot \vec{E}(x, y). \tag{37.1}$$

The force $\vec{F}$ is a vector and the charge q is a scalar, therefore $\vec{E}$ is also a vector. On the other hand, the electric potential, V, is related to the work W required to move a charge from one point to another; more precisely, the work necessary to move the charge q between two points 1 and 2 will be $W(1,2) = q \cdot \Delta V(1,2)$. Here, $W(1,2)$ is the work required to move the charge q from point 1 to point 2 and $\Delta V(1,2)$ the *electric potential* difference between 1 and 2. Since the work is a scalar quantity, so is the potential. More specifically, the potential variation between two points separated by a distance dl is

$$dV = \frac{dW}{q} = -\frac{1}{q}\vec{F}(x, y) \cdot d\vec{l} = -\vec{E} \cdot d\vec{l}. \tag{37.2}$$

Therefore, the components of the electric field can be expressed as a function of the electric potential [1–3]:

$$E_x = -\frac{dV}{dx}, \quad E_y = -\frac{dV}{dy}, \quad \text{and} \quad E_z = -\frac{dV}{dz} \tag{37.3}$$

or, more generally,

$$E = -\left[\frac{\mathrm{dV}}{dl}\right]_{\max}. \tag{37.4}$$

This expression means that the modulus of $\vec{E}$ is equal to the derivative of the electric potential with respect to the displacement, in the direction in which this derivative is maximum. Furthermore, this direction is the direction of the field $\vec{E}$. This is written more precisely using the *gradient* notation $\vec{\nabla}$ (readers unfamiliar with this notation may ignore this equation for the time being) as follows [1, 2, 4]:

$$\vec{E} = -\vec{\nabla} \cdot V. \tag{37.5}$$

Moreover, when $dV = 0$, as occurs on an *equipotential line* or *surface*, the component of $\vec{E}$ along this line is zero. In other words, $\vec{E}$ *is always perpendicular to equipotential lines (or surfaces)*. The central idea of this experiment consists of establishing experimentally the potentials determined by a given configuration of electrodes. Equipotential lines or surfaces are obtained by determining *the locus of points of equal potential*. From these equipotential surfaces, the *electric field lines* can be found by tracing the orthogonal trajectories to the equipotential surfaces. The electric field values can be obtained from the potential values around a point using Eq. (37.4) or (37.5).

Gauss's law [1–4] (physically equivalent to Coulomb's law) relates fields to charges and can be expressed in two equivalent manners. In integral form, it is expressed as follows:

$$\oiint \vec{E} \cdot d\vec{S} = \frac{q_{\text{net}}}{\varepsilon_0}, \tag{37.6}$$

where the integral is over a closed surface and q_{net} is the net charge inside said surface, or in differential form, Gauss's law is written as follows:

$$\vec{\nabla} \cdot \vec{E} = \frac{\rho}{\varepsilon_0}, \tag{37.7}$$

where the first member represents the *divergence* of the electric field and ρ is the bulk density charge. Combining Eqs. (37.5) and (37.7) provides *Poisson's equation* [3, 4], which relates the potentials to the charges:

$$\nabla^2 \cdot V = -\frac{\rho}{\varepsilon_0}. \tag{37.8}$$

When the charge density is zero, i.e., in areas where there is no net charge, this equation reduces to *Laplace's equation* [3–5]:

$$\nabla^2 \cdot V = 0, \tag{37.9}$$

which is a partial differential equation. This is the equation that must satisfy the electric potential in a region free of net charge, under static conditions or with low-frequency alternating fields, for which the associated wavelengths are much larger than the dimensions of the system [6]. For a frequency of 50 Hz, the associated wavelength is 6,000 km in air. Thus, for systems with dimensions of the order of the meter, the field and potential can be described by means of a static model up to frequencies of several kHz.

Typically, the conductivity, σ, of the water ranged between 10^{-4} and $10^{-3}\Omega^{-1} \cdot \mathrm{m}^{-1}$ to prevent electrolysis, which would alter the chemical properties of the media with time. Therefore, alternating current may be used, with frequency, f, in the range of 50 hz to 1Khz. In this manner, the effective potential distribution does not change with time, making the problem stationary. At these frequencies, the wavelength associated with the electromagnetic field, λ, is several hundred kilometers ($\lambda = c/n.f \approx 10^6$m, where c is the speed of light in vacuum and $n^2 = k$ the dielectric constant of the liquid, at the corresponding frequency). Since the wavelength is much larger than the dimension of the experimental setup, the electric field configuration is essentially quasi-static and equivalent to the electrostatic case. It is possible to use a DC source as well, but caution must be taken so that electrolysis does not change the properties of the media. For this, we must try to do all the measurements for a given configuration as short as possible, let's say less than 30 min, and then turn off the power supply.

Of course, there are several manners to carry out these experiments; it is possible to buy in the market conductive paper with grid [7]; here the electrodes are marked with conductive silver paint on the conductive paper that already has a reference grid. Of course, when using this version of the setup, a DC power supply can be used without restriction [8].

37.1.1 *Project 114. Semi-quantitative analysis*

> **Recommended basic equipment:** A transparent glass or acrylic tray, approximately 30 cm × 20 cm × 4 cm. A source of alternating voltage of 5–12 V. A voltmeter. Metal plates (copper, bronze, and aluminum) to use as electrodes.

Use the experimental device like the one illustrated in Figure 37.1. It consists of a transparent glass or plastic tray containing water with low conductivity, e.g., clean tap water or tap water mixed in equal proportions with distilled water. Under the tray, place a sheet of

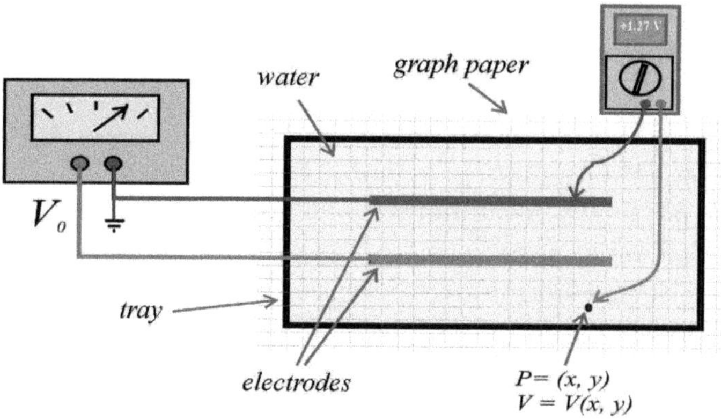

Figure 37.1 A transparent tray made of insulating material contains distilled water. The thick lines represent metal electrodes. At the coordinate point (x, y), the value of the electric potential $V(x, y)$ is measured with the voltmeter.

paper printed with a grid of known dimensions (e.g., 1 cm × 1 cm) or graph paper. An alternating voltage source V_0 of about 5 V feeds the electrodes. This source can be the output of a transformer with a secondary isolated from the primary or a function generator. A voltmeter, in AC mode, with ground connected to the source ground, is used to measure the voltages at each point on the tray. The coordinates x, y of these points are measured on the reference grid.

Suggested tasks:

✓ Determine the equipotential lines and the field lines. For each equipotential line, measure at least six points approximately 1 cm apart. Then, on graph paper or a sheet of paper like the one used as a grid, mark these points of equal potential and connect them with solid lines. These are the equipotential lines corresponding to the electrode geometry selected. Determine at least 10 equipotential lines, some of them close to the electrodes and others in the central areas. In each case, find the areas where the electric field is highest.

✓ For the configuration indicated in Figure 37.1, the horizontal equipotential line that goes through the middle (center) of the two electrodes is at the effective potential $V_0/2$. By symmetry, this equipotential line extends to infinity. Therefore, if we want our potential to satisfy the usual convention of $V = 0$ at infinity, we must re-normalize our measured values, V_{meas}, and take a new $V(x, y) = V_{\text{meas}}(x, y) = V_0/2$. It should be emphasized that what is physically meaningful is the potential difference. But the interesting thing of re-normalizing the potential in this manner is that the new re-normalized potential resembles more closely the usual dipole potential found in most introductory books.

✓ Once the equipotential lines are characterized, use any program of your preference to draw a 3-D plot of the potential and the equipotential lines and the electric field lines as well. There are several spreadsheet programs that allow this type of visualization [9].

✓ **Perturbation of the potential configuration**: For the same setup as above, Figure 37.1, place a conductor between the electrodes, for example, a metal cylinder that fits between the two electrodes as illustrated in Figure 37.2, and determine how the equipotential lines are modified. In particular, study the equipotential lines around the conductor. How are the field lines and the field itself on the surface of the conductor modified? Plot the result of the new potential field.

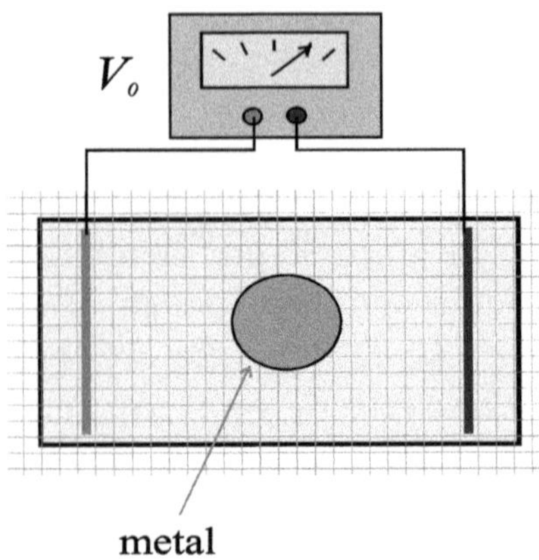

Figure 37.2 The tray contains a sample of a conductor or an insulator between the electrodes.

37.2 Numerical solution of Laplace's equation — Relaxation method

There are many methods for solving Laplace's equation analytically [3, 4], which are particularly suitable for systems with simple geometries. However, when the geometry of the electrodes is complex or the boundary conditions are diverse, such as those generally encountered in practical situations, it is necessary to use numerical methods, which can be easily implemented using spreadsheets in a computer. A simple way to solve Laplace's equation numerically is to use the *relaxation method* [3, 6, 10, 11]. Here, we will briefly describe the two-dimensional case.

If the problem is two-dimensional, Eq. (37.9) can be written as follows:

$$\frac{\partial^2 V}{\partial x^2} + \frac{\partial^2 V}{\partial y^2} = 0. \tag{37.10}$$

If we discretize the $x - y$ plane so that it forms a two-dimensional mesh, as illustrated in Figure 37.3, the coordinates (x, y) are replaced by the indices i, j. Remembering the basic expressions of numerical derivation, we have

$$\frac{\partial V}{\partial x} \approx \frac{V_{i+1,j} - \cdot V_{i,j}}{h}, \quad \frac{\partial V}{\partial y} \approx \frac{V_{i,j+1} - \cdot V_{i,j}}{h}, \tag{37.11}$$

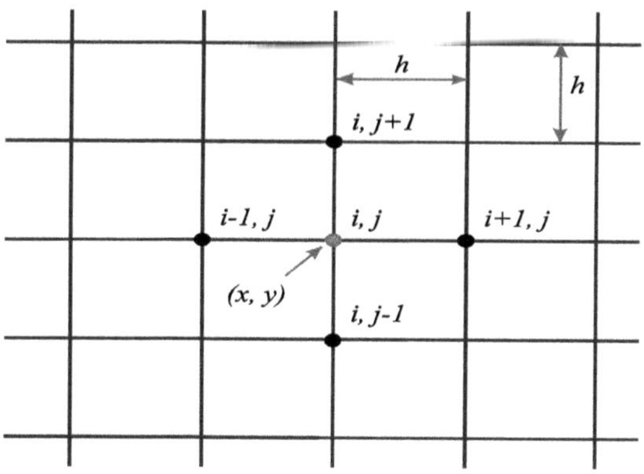

Figure 37.3 Discretization of the plane. Square cells have sides of length h. Coordinate point (x, y) is represented by the node (i, j), with $x = ih$ and $y = jh$.

and

$$\frac{\partial^2 V}{\partial x^2} \approx \frac{V_{i+1,j} - 2 \cdot V_{i,j} + V_{i-1,j}}{h^2}, \quad \frac{\partial^2 V}{\partial y^2} \approx \frac{V_{i,j+1} - 2 \cdot V_{i,j} + V_{i,j-1}}{h^2}. \tag{37.12}$$

Using these relationships, expression (37.10) can be approximated as follows:

$$\frac{V_{i+1,j} - 2 \cdot V_{i,j} + V_{i-1,j}}{h^2} + \frac{V_{i,j+1} - 2 \cdot V_{i,j} + V_{i,j-1}}{h^2} = 0, \tag{37.13}$$

where h is the size of the discretization or unit cell, as illustrated in Figure 37.3. By solving (37.13) for $V_{i,j}$ [$\approx V(x, y)$], we get

$$V_{i,j} = \frac{1}{4} \cdot (V_{i+1,j} + V_{i-1,j} + V_{i,j+1} + V_{i,j-1}). \tag{37.14}$$

This means that if the function $V(x, y)$ satisfies Laplace's equation in two dimensions, the value of the potential $V(i, j)$ at a given point on the plane (i, j) is equal to the average of the potential value at the *four nearest neighboring points*.

The relaxation method uses this property of Laplace's equation. In a spreadsheet, the value of a given cell is the average of its closest neighbors, except for those points that have a fixed potential (coinciding with the potential of the electrodes or other conditions) whose values are established by the boundary conditions. Applying the method involves iterating until the cell values do not change or their variation is less than a preset value, e.g., variations smaller than 0.1%.[a]

37.3 Dirichlet and Neumann boundary conditions

There are two basic types of boundary conditions in the experimental setup described in Figures 37.2, 37.3, and 37.4. On one hand, there are potential values determined by the electrodes whose values are constant. In these areas, what is known as the *Dirichlet*

[a]An example of the application of this procedure can be found in the relax.xls template, which can be obtained from the website www.fisicarecreativa.com.

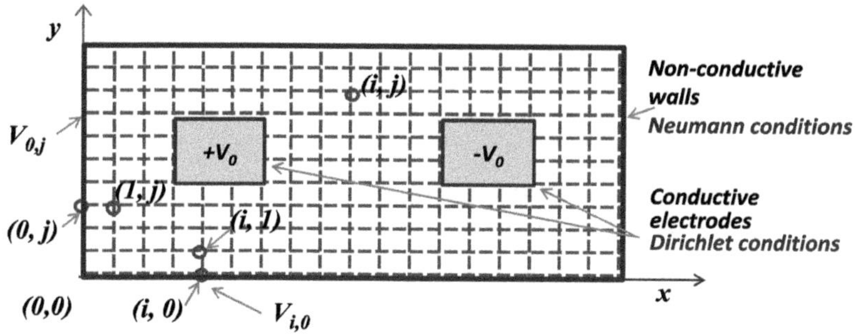

Figure 37.4 The tray made of insulating material (its edges shown by the thick rectangular outline) contains a low-conductive medium, with two metal electrodes at potential $+V_0$ and $-V_0$, respectively. On the electrodes, the Dirichlet edge conditions apply, $V_{i,j} = \pm V_0$, while at the edges of the container, the Neumann edge conditions are satisfied. For example, on the left edge, $V_{0,j} = V_{1,j}$ and on the lower edge, $V_{i,0} = V_{i,1}$.

condition [4, 12, 13] is applied as boundary condition, $V =$ constant. Operationally, this is achieved by imposing that the potential values in the cells that define these edges are constant. For these points, $V_{i,j} = V$.

On the other hand, there are boundary conditions at the container walls, which are non-conductive. At the walls, the electric current ($\vec{j} = \sigma \cdot \vec{E}$) in the conductive medium (liquid) can only have components parallel to the non-conducting walls. In other words, at the walls of the tank, the perpendicular component of the electric field is zero, that is,

$$E_\perp = 0 \quad \text{or} \quad \frac{\partial V}{\partial n} = 0, \quad \text{where } n \text{ is the normal to the wall.} \tag{37.15}$$

This is called *Neumann boundary condition* [4, 12, 14]. Operationally, this condition is achieved by making the values of the cells that define the edges of the container equal to the values of the contiguous interior cells, perpendicular to the wall. For example, if the left wall of the container coincides with the y-axis, Figure 37.4, the cells representing this wall are characterized by the indices ($i = 0, j$), the value of the potential on this wall fulfills $V_{i=0,j} = V_{i=1,j}$, with which the condition of Eq. (37.15) is satisfied. Similarly, for the lower face of the container, coinciding with the x-axis, we have $V_{i,j=0} = V_{i,j=1}$ [6].

37.3.1 *Project 115. Quantitative analysis — Relaxation method I*

> **Recommended equipment:** A transparent glass or acrylic tray, approximately 30 cm × 20 cm × 4 cm. An alternating voltage source of 5–12 V. A voltmeter. Metal plates (copper, bronze, and aluminum) to be used as electrodes.

Use an experimental setup like the one illustrated in Figure 37.1. It consists of a transparent glass or plastic tray containing water of low conductivity, for example, clean tap water or mixed in equal proportions with distilled water. On a sheet of paper, covering the whole area of the bottom of the tray, a grid of known dimensions is printed, for example, 1 cm × 1 cm. Graph paper can also be used, taped or glued on the external part of the tray. An alternating voltage source, about 5 V, feeds the circuit. This source can be the output of a transformer with a secondary isolated from the primary or a function generator. A voltmeter, in AC mode, with its ground connected to the ground of the source, as shown in Figure 37.1. The other probe measures the voltages at each point of the medium, whose coordinates are determined by the reference grid.

Suggested tasks:

- ✓ Determine the potential for all points on the tray using a grid with a pitch of approximately 1 cm. Represent your results on a two-dimensional matrix and make a three-dimensional graphical representation of your results, that is, (x, y, V) [6, 8].
- ✓ For the chosen geometry, calculate the potential by the relaxation method. Represent the numerical results using the same graphing criteria as the ones you chose to represent the experimental

results. Compare your measurements to the numerical solution [6, 8].

37.3.2 *Project 116. Quantitative analysis — Relaxation method II*

Recommended equipment: A transparent glass or acrylic tray, approximately 30 cm × 20 cm × 4 cm. An alternating voltage source of 5–12 V. A voltmeter. Metal plates (copper, bronze, and aluminum) to be used as electrodes.

Using one of the experimental arrangements illustrated in Figure 37.5, perform the same study as in the previous project.

Suggested tasks:

✓ For the chosen geometry, determine the point-to-point experimental potential and graph your results.

✓ Use the relaxation method and represent the numerical results using the same graphing criteria as the ones you chose to represent the experimental results. Compare your measurements to the numerical solution.

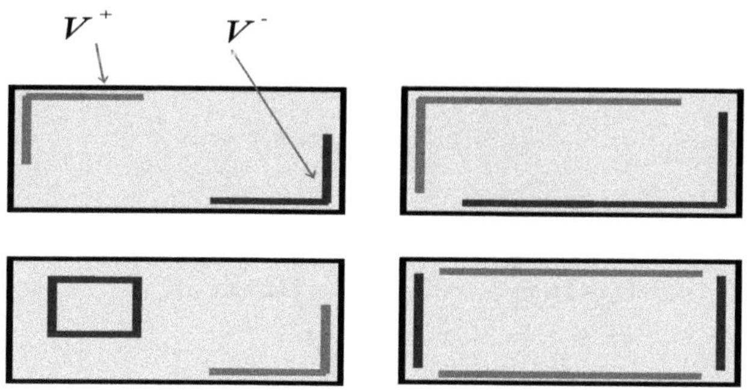

Figure 37.5 Examples of possible electrode configurations to study the distribution of potentials.

37.3.3 *Project 117. ♣♣ Estimation of the electric field vector*

Recommended equipment: A transparent glass or acrylic tray, approximately 30 cm × 20 cm × 4 cm. An alternating voltage source of 5–12 V. A voltmeter. Metal plates (copper, bronze, and aluminum) to be used as electrodes.

The objective of this project is to estimate the electric field at some points. For a given electrode configuration, select points on some of the equipotential lines found in one of the previous projects. Measure the potential around the chosen point (Figure 37.6). These measurements can be used to estimate the value of $\vec{E}$ from its components E_x and E_y by approximating Eq. (37.3) by finite difference quotients:

$$E_x \approx -\frac{\Delta V}{\Delta x} = -\frac{V(x+\delta x, y) - V(x-\delta x, y)}{2\delta x} \quad \text{or}$$

$$E_x(i,j) \approx \frac{1}{2h}(V_{i-1,j} - V_{i+1,j}), \tag{37.16}$$

$$E_y \approx -\frac{\Delta V}{\Delta y} = -\frac{V(x, y+\delta y) - V(x, y-\delta y)}{2\delta y} \quad \text{or}$$

$$E_y(i,j) \approx \frac{1}{2h}(V_{i,j-1} - V_{i,j+1}). \tag{37.17}$$

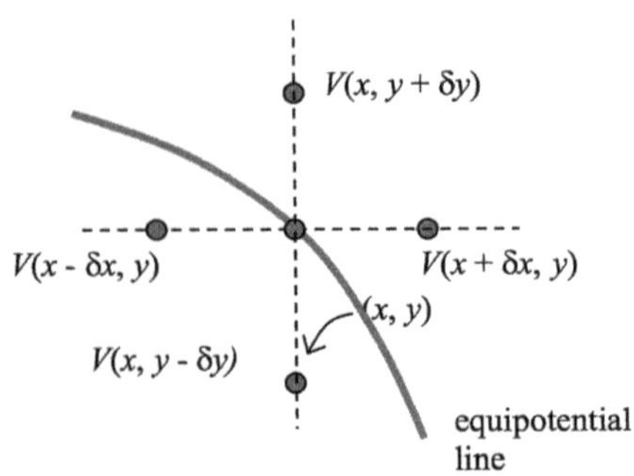

Figure 37.6 Measurement of V in the vicinity of a point (x, y) to estimate the field at the coordinate point (x, y) equivalent to (i, j).

From these components, we can define the intensity and direction of the field:

✓ Intensity:

$$E = \sqrt{E_x^2 + E_y^2} \quad \text{or} \quad E(i,j) = \sqrt{E_x^2(i,j) + E_y^2(i,j)}. \qquad (37.18)$$

✓ Orientation with respect to the axis x:

$$\theta = \arctan\left(\frac{E_y}{E_x}\right) \quad \text{or} \quad \theta(i,j) = \arctan(E_y(i,j)/E_x(i,j)). \qquad (37.19)$$

Suggested tasks:

✓ Select points (x, y) over at least four equipotential trajectories and estimate the electric field vector $\vec{E}(x, y)$ in each case. On the equipotential map, draw the calculated vectors at each chosen point; use appropriate scales to represent the vectors.

✓ What can you conclude about the orientation of the electric field with respect to an equipotential line?

37.4 Self-evaluation questions

(1) What do you see if you measure the potential difference between two points on the same electrode with the voltmeter? What can you conclude about the potential variation on the metal electrode? Explain your results.

(2) The walls of the tray are not conductive, and this imposes boundary conditions on the electric field and potential. From the observation of the maps recorded:

   (a) What can you say about the potential gradient on the wall of the tray?

   (b) What is the electric field like in the vicinity of the wall? In particular, what is the component of the field normal to the wall?

(3) In what units is the experimentally obtained electric field expressed in Eqs. (37.16) and (37.17)? Show the equivalence of

these units to those arising from the definition of electric field according to Eq. (37.1).

(4) If the experiments were to be repeated by connecting the electrodes to a source that provides half the voltage that you used, predict what would change in the results. Consider possible changes in the electric field, in the configuration of equipotential lines, and so on. Explain your reasoning.

References

[1] R. Halliday, D. Resnick and M. Krane, *Fundamentals of Physics*, NY: John Wiley & Sons, Inc., 2002.
[2] F. Sears, M. Zemansky, H. Young and R. Freedman, *College Physics*, NY: Pearson Ed., 2009.
[3] E. M. Purcell and D. J. Morin, *Electricity and Magnetism, Berkeley Physics Course*, 3rd Ed., NY: Cambridge University Press, 2014, p. 839.
[4] J. R. Reitz, F. J. Milford and R. W. Christy, *Fundamentals of Electromagnetic Theory*, NY: Ed. Addison-Wesley, 1996.
[5] Wikipedia, "Laplace Equation," 2022. Available: https://en.wikipedia.org/wiki/Laplace%27s_equation.
[6] S. Gil, M. Saleta and D. Tobia, "Experimental study of the Neumann and Dirichlet boundary conditions in 2D electrostatic problem," *Am. J. Phys.*, vol. 70, no. 12, p. 1208, 2002.
[7] PASCO, "Conductive paper with grid," Field Mapper Kit · PK-9023, 2022. Available: https://www.pasco.com/products/lab-apparatus/electricity-and-magnetism/electrostatics-and-electric-fields/pk-9023.
[8] S. Johnston, "Equipotential lines, electric and magnetic field mapping," 2021. Available: https://www.youtube.com/watch?v=WcSSWN4Tnoo.
[9] R. Beichner, "Visualizing potential surfaces with a spreadsheet," *Phys. Teach.*, vol. 35, no. 2, pp. 95–97, 1997.
[10] M. Di Stacio and W. McHarris, "Electrostatic problems? Relax!" *Am. J. Phys.*, vol. 49, p. 440, 1979.
[11] Wikipedia, "Relaxation method," 2022. https://en.wikipedia.org/wiki/Relaxation_(iterative_method).

[12] J. Jackson, *Classical Electrodynamics*, 3rd Ed., NY: John Willey and Sons, 1998.
[13] Wikipedia, "Dirichlet boundary condition," 2022. Available: https://en.wikipedia.org/wiki/Dirichlet_boundary_condition.
[14] Wikipedia, "Neumann boundary condition," 2022. Available: https://en.wikipedia.org/wiki/Neumann_boundary_condition.

Chapter 38

FREE AND FORCED OSCILLATIONS OF AN RLC CIRCUIT

This chapter studies the characteristics of series and parallel RCL circuits and coupled oscillations. The first part analyzes the free response of a series RLC circuit, without an external driving voltage. The second part studies the response of an RLC circuit in series and in parallel when connected to a sinusoidal voltage input, focusing on the response of the circuit to variation in amplitude and frequency of the excitation input voltage. This leads to an introduction to the concepts of admittance and impedance and to the phenomenon of resonance. Next, it analyzes the response of the RLC circuit to square and triangular driving voltages and explores the validity of the superposition principle. The following chapter proposes to study the behavior of two or more coupled RLC circuits. These devices are an interesting and instructive example of coupled oscillating systems, which have classical and quantum analogs in various areas of physics.

Goals

✓ Free RLC circuit
✓ Electrical oscillations
✓ Oscillations of a forced RLC circuit
✓ Oscillations of a parallel RLC circuit

38.1 Free oscillations of an RLC circuit

When studying circuits that contain an inductance L, we must remember that the inductances always have an intrinsic resistance R_L, whose value may be measured with an ohmmeter, see Chapter 35. The total or equivalent resistance of the circuit in Figure 38.1 is $R = R_1 + R_L$.

If $q(t)$ represents the charge on the capacitor at an instant t, using Kirchhoff's mesh law, we may write [1–3]

$$L\frac{d^2q}{\mathrm{dt}^2} + R\frac{\mathrm{dq}}{\mathrm{dt}} + \frac{1}{C}q = 0. \tag{38.1}$$

We may define

$$\omega_0^2 = \frac{1}{LC}, \quad \gamma = \frac{R}{2L}, \tag{38.2}$$

where $f_0 = \omega_0/2\pi$ is the *natural frequency* and γ is the *damping coefficient* [1] of the circuit. It is also useful to define

$$\omega_1^2 = \omega_0^2 - \gamma^2 = \frac{1}{LC} - \left(\frac{R}{2L}\right)^2. \tag{38.3}$$

It is easy to verify that Eq. (38.1) can be written as a homogeneous, linear, second-order differential equation:

$$\frac{d^2q}{\mathrm{dt}^2} + 2\gamma\frac{\mathrm{dq}}{\mathrm{dt}} + \omega_0^2 q = 0, \tag{38.4}$$

whose solution depends on the sign of the term $\omega_0^2 - \gamma^2$.

✓ **Underdamped case:** If $R < R_{\text{crit}} = 2\sqrt{L/C}$, so that $(\omega_0^2 - \gamma^2 > 0)$,

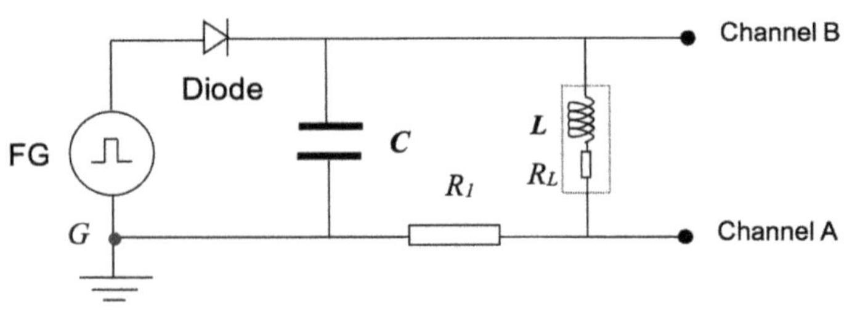

Figure 38.1 Series RLC circuit: R_L is the internal resistance of the inductance. The diode prevents the capacitor from discharging through function generator, FG, and ensures that the input signal is always positive.

the solution of Eq. (38.4) is

$$q(t) = q_0 \exp(-\gamma \cdot t) \cdot \sin(\omega_1 t + \varphi), \tag{38.5}$$

where q_0 and φ are two constants whose values depend on the initial conditions, that is, the values of $q(t)$ and $i(t)$ for $t=0$. In this case, the system oscillates with decreasing amplitudes over time.

✓ **Overdamped case**: If $R > R_{\text{crit}}$, so that $(\gamma^2 - \omega_0^2) > 0$,

$$q(t) = q_{10}\exp(-\lambda_1 t) + q_{20}\exp(-\lambda_2 t), \tag{38.6}$$

where q_{10} and q_{20} are two constants whose values depend on the initial conditions and the values of λ_1 and λ_2 are the solutions of the quadratic equation:

$$\lambda^2 + 2\gamma\lambda + \omega_0^2 = 0 \rightarrow \lambda_{1,2} = -\gamma \pm \sqrt{\gamma^2 - \omega_0^2}. \tag{38.7}$$

✓ **Critical damping case:** If $R = R_{\text{crit}}$, so that $(\omega_0^2 - \gamma^2) = 0$.

$$q(t) = q_{10}\exp(-\gamma t)(1 + Bt), \tag{38.8}$$

where q_{10} and B are two constants whose values depend on the initial conditions. These three solutions to the homogeneous differential equation (38.4) are often referred to as a *transitory solution*, since they all decay over time.

38.2 Phase diagram

A *phase diagram* is very instructive in many physical systems. For mechanical systems with a single degree of freedom, the parameters that specify the state of the system are the coordinate $x(t)$, and the velocity $v = dx/dt$ (*conjugate momentum*). The *phase diagram* is constructed by plotting $v = dx/dt$ as a function of $x(t)$. In other words, at each instant of time, the system is represented by a point in the *phase space*[a] of the position velocity (x, v). As the system evolves in time, the representative point of the state of the system describes a trajectory in the phase space.

In electrical systems like the RLC circuit, the *generalized coordinate* that specifies the state of the system is the capacitor charge $q(t)$,

[a]https://en.wikipedia.org/wiki/Phase_space

and the conjugate momentum is the electric current $i = dq/dt$. Therefore, the phase diagram in this case is constructed by plotting $i = dq/dt$ as a function of $q(t)$.

Exercises:

✓ Show that for both a frictionless one-dimensional harmonic (mechanical) oscillator and for an ideal circuit LC $(R = 0)$, the phase diagram is an ellipse.

✓ For the circuit RLC, in the underdamped case ($R < R_{\text{crit}}$), show that the phase diagram resembles that schematically illustrated in Figure 38.2.

38.2.1 *Project 118. Underdamped free RLC circuit response*

Recommended equipment: Coils (inductances of a few tens of mH), two or more capacitors with capacitance between 10 nF and 500 nF, and resistors of 50 Ω, 100 Ω, 500 Ω, 1 kΩ, and 10 kΩ. A data acquisition system connected to a computer (DAS). A function generator. The acquisition rate should be at least 10 times greater than the characteristic frequencies to be measured. If a DAS is not available, a 20 MHz or faster two-channel oscilloscope can be used.

Measure the values of L and R_L (see Chapter 35). Choose a capacitor C so that the natural frequency, defined in Eq. (38.2), $f_0 = \omega_0/2\pi$,

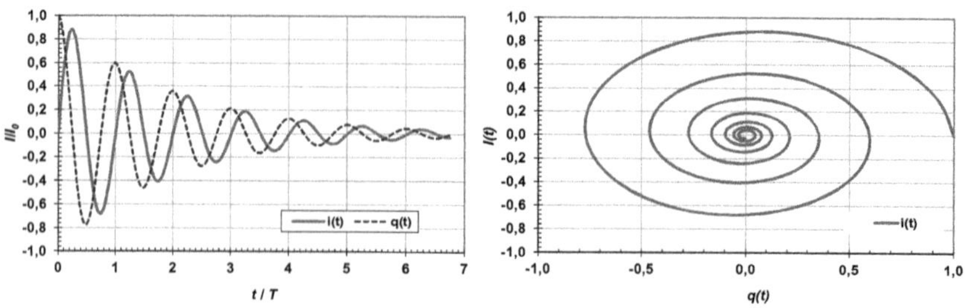

Figure 38.2 Oscillation of an underdamped free RLC circuit. The diagram on the left shows the variation of charge and current as a function of t/T, where T is the period of oscillation. The diagram on the right is the corresponding phase diagram.

is less than about 3 or 4 kHz. Make sure the acquisition system has an acquisition rate of about 10 times this frequency or greater. Calculate the value of $R_{\text{crit}} = 2\sqrt{L/C}$ for your circuit and make sure that $R_{\text{crit}} > R_1 + R_L$. This condition is crucial to ensure that the oscillations are underdamped. In a circuit like the one in Figure 38.1, the function generator (FG) charges the capacitor through a diode. The diode prevents the capacitor from discharging through the FG.

Connect the terminals of two channels of your DAS to the points indicated as channels A and B in the circuit shown in Figure 38.1. Note that for both channels (A and B), the point G is the common ground (GND). In fact, it is also the ground of FG. The FG should be set to provide a *square signal* of very low frequency; its purpose is only to charge the capacitor. With this arrangement, study experimentally the variation of the voltage in R, L, and C when the capacitor discharges.

Suggested tasks:

- ✓ Plot your experimental results for $i(t)$ and $q(t)$ as functions of time.
- ✓ On the graph of $q(t)$, include the theoretical expectation given by expression (38.5). Vary the model parameters (q_0 and φ) to achieve the best possible fit to the experimental results. From the fit of the theoretical curve to the data, find the values of f_0 and γ. Estimate the errors in obtaining these parameters, see Chapter 15.
- ✓ Compare the values of $f_0 = \omega_0/2\pi$ and γ obtained from the fit of $q(t)$ and $i(t)$ as a function of time, to the values predicted by the theory, Eqs. (38.2), (38.3), and (38.4) using the measured values of L, C, R_L and R_1. What can you conclude about the proposed model to explain the data you recorded in this experiment?
- ✓ In the underdamped case ($R < R_{\text{crit}}$), plot your experimental results of $i(t)$ as a function of $q(t)$ (phase diagram). Interpret the results found.
- ✓ Vary the value of R_1 so that $R > R_{\text{crit}}$. Study the variation of the current as a function of time and compare it to the results predicted by the model. What can you conclude from this comparison?

Note: Remember that standard oscilloscopes and computer-based acquisition systems typically operate in "common mode," meaning that the input channels have the same ground connection or G-point. See Chapter 24.

38.3 Forced oscillations of an RLC circuit

If the RLC circuit is connected to a voltage source, we know from Thevenin's theorem (see Chapter 24) that this source can be modeled as an ideal voltage source $\varepsilon(t)$ in series with an internal resistance r_{int}. This circuit is illustrated in Figure 38.2. If R designates the total or equivalent resistance of the circuit, that is, $R = R_1 + R_L + r_{\text{int}}$, by applying Kirchhoff's mesh law, we have [3, 4]

$$L\frac{d^2q}{\mathrm{dt}^2} + R\frac{\mathrm{dq}}{\mathrm{dt}} + \frac{1}{C}q = \varepsilon(t). \tag{38.9}$$

Without loss of generality, we can assume that the function $\varepsilon(t)$ is sinusoidal: $\varepsilon(t) = V_0\sin(\omega t)$, since according to Fourier's theorem, any function can be expressed as a linear combination of sinusoidal functions. A particular solution of the inhomogeneous differential equation Eq. (38.9) is

$$q(t) = -Q_0\cos(\omega t + \beta) \tag{38.10}$$

with an amplitude

$$Q_0 = \frac{V_0}{\omega\sqrt{R^2 + (L\omega - 1/C\omega)^2}} \tag{38.11}$$

and

$$\tan\beta = -\frac{L\omega - 1/C\omega}{R}. \tag{38.12}$$

Taking the derivative of (38.10), we have

$$i(t) = I_0\sin(\omega t - \varphi), \tag{38.13}$$

where

$$I_0 = \frac{V_0}{Z(\omega)} = \frac{V_0}{\sqrt{R^2 + (L\omega - 1/C\omega)^2}} \tag{38.14}$$

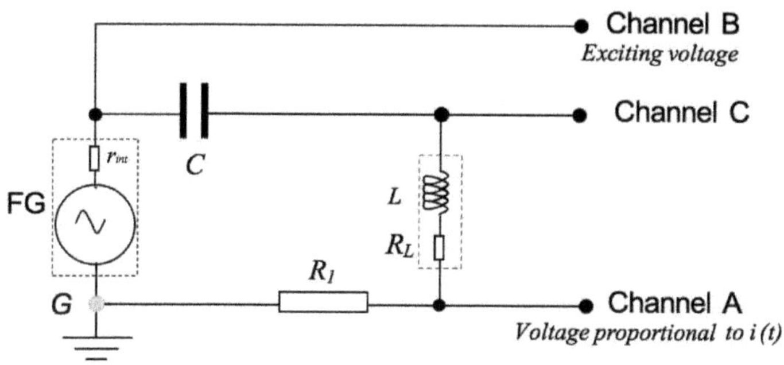

Figure 38.3 Forced series RLC circuit. R_L represents the internal resistance of the inductance. The voltage measured in channel A is proportional to the current $i(t)$, while the driving voltage $\varepsilon(t)$ is measured in channel B; G is the common GND of the system.

and

$$\tan\varphi = -\tan\beta = \frac{L\omega - 1/C\omega}{R} = \frac{2\gamma\omega}{\omega_0^2 - \omega^2}. \qquad (38.15)$$

The parameter $Z(\omega)$ is known as the impedance of the circuit at the angular frequency $\omega = 2\pi f$ and is a generalization of the concept of resistance for DC circuits, while φ is the phase angle.

The *particular solutions* of the differential equation [Eqs. (38.10) and (38.13)] of the circuit constitute the *stationary response* of the system. If we want to obtain the complete solution to Eq. (38.9), we must add to the particular solutions [Eqs. (38.10) and (38.13)] the *general solutions* of the homogeneous differential equation. This means that the complete solution to Eq. (38.9) is found by adding Eq. (38.5) or (38.6) or (38.8) to expression (38.10), depending on the value of R with respect to R_{crit}. The initial conditions, that is, the values of $q(0)$ and $i(0)$, determine the value of the arbitrary constants q_0 and φ.

The frequency for which the phase difference between the applied current and voltage is zero is called the *resonant frequency* f_{res}. In Figure 38.4, the current amplitude $I_0 = V_0/|Z(\omega)|$ is normalized to its maximum value $I_{\max}$ as a function of the normalized frequency (f/f_{res}). Note that $I_0(\omega)$ has a maximum for $f = f_{\text{res}} = \omega_0/2\pi$ and

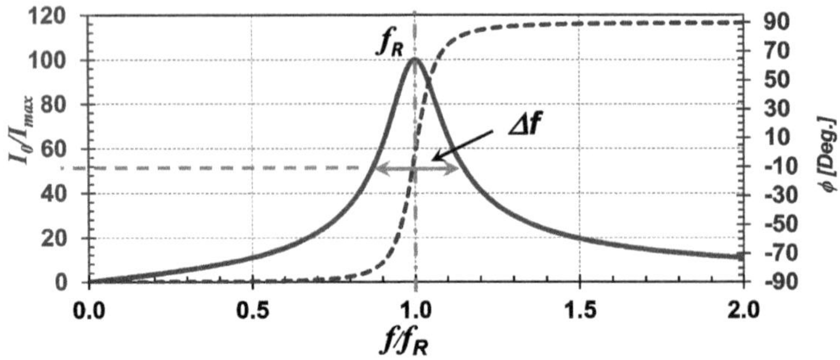

Figure 38.4 Frequency response of a forced RLC circuit. The red curve represents the value of the normalized amplitude $I_0(f)/I_{\max}$. The green dashed curve, referred to the right scale, represents the phase difference (φ) between the applied voltage and the current.

that the phase difference φ passes through zero. In some systems, there may be several resonant frequencies.

38.4 Complex reactances and impedances

If we work with *complex notation*, using the imaginary unit $j \equiv \sqrt{-1}$, the input voltage can be considered the complex part of $V_0 \cdot \exp(j\omega t)$. We define *capacitive reactance* as $X_C = 1/(jC\omega)$ and *inductive reactance* $X_L = jL\omega$. The *ohmic reactance* is $X_R = R$.[b] With this notation, we combine the reactances as if they were resistors but preserving their complex character. Applying this formalism to the series RLC circuit shown in Figure 38.3, we obtain for the total *impedance* Z[c]:

$$Z(\omega) = R + j(L\omega - 1/C\omega). \tag{38.16}$$

With this notation, we can write

$$i(t) = \frac{V_0 \exp(j\omega t)}{Z(\omega)} = \frac{V_0}{|Z(\omega)|} \exp(j\omega t - \varphi). \tag{38.17}$$

[b]https://en.wikipedia.org/wiki/Electrical_reactance
[c]https://en.wikipedia.org/wiki/Electrical_impedance

These two expressions are completely equivalent to Eqs. (38.13), (38.14), and (38.15), with the difference that the method that uses complex reactances (impedances) can be applied to any combination (series, parallel) of elements.

The *quality factor* Q_M is defined as the ratio of the resonant frequency and the difference between the frequencies that determine the half-width of the resonant curve (see Figure 38.4), that is [3–5],

$$Q_M = \frac{\omega_{\text{res}}}{\omega_1 - \omega_2} = \frac{f_{\text{res}}}{\Delta f}. \tag{38.18}$$

While for f_{res} the current reaches its maximum value I_{max}, for frequencies ω_1 and ω_2, the current $I(\omega_{1,2}) = I_{\text{max}}/\sqrt{2}$. For a series RLC circuit, it is easy to prove[d] that

$$Q_M = \frac{\omega_0 L}{R}. \tag{38.19}$$

It is often useful to use the *complex admittance*, which is the reciprocal of impedance $Y(\omega) = 1/Z(\omega)$.

38.4.1 *Project 119. Response of the forced RLC circuit*

> **Recommended equipment:** Coil (inductance of a few tens of mH), two or more capacitors between 10 nF and 500 nF, and resistors of 50 Ω, 100 Ω, 500 Ω, 1 kΩ, and 10 kΩ. A data acquisition system connected to a computer and/or a two-channel digital oscilloscope, 20 MHz or faster. A function generator. It is advisable for the acquisition rate of the data logger to be at least 10 times greater than the characteristic frequencies to be measured.

Measure the values of L and R_L. Choose a capacitor C so that the natural frequency $f_0 = \omega_0/2\pi$ is less than about 3 or 4 kHz. If you have done Project 118, you may use the same elements but this time using the circuit shown in Figure 38.3. Make sure the acquisition system has an acquisition rate about 10 times higher

---

[d]https://en.wikipedia.org/wiki/RLC_circuit#Series_circuit

than the maximum working frequency. Connect the terminals of your acquisition system or oscilloscope to the points indicated as channels A and B so that the signal of channel A is proportional to $i(t)$ and the signal of channel B measures the driving voltage.

Preliminary calculations: For the circuit shown in Figure 38.3, assuming a sinusoidal signal, with the values of R, C, and L selected, determine the complex impedance $Z(\omega)$ of the circuit. Plot the module of $Z(\omega)$ as a function of frequency. On the same graph, using a second vertical axis, plot the values of the phase φ of $Z(\omega)$.

Suggested tasks:

✓ Using the circuit shown in Figure 38.3 with a sinusoidal voltage source, from the measurement of V_A, find $i(t)$ as a function of time. Plot $i(t)$ and $V(t)(= V_B)$ as functions of time. What shapes do these signals have?

✓ Determine the phase difference between $i(t)$ and $V(t)$ using the procedure described in Annex A of Chapter 31 to measure the phase shift between two signals.

✓ If $I_0(f)$ and $V_0(f)$ denote the amplitudes of the current $i(t)$ and voltage $V(t)(= V_B)$ signals, and $\varphi(f)$ denotes the phase difference between these signals at frequency f, determine the frequency for which $\varphi(f_{\text{res}}) = 0$ and the frequency for which the relationship $I_0(f)/V_0(f)$ is maximum. Do these two frequencies match?

✓ For frequencies in the vicinity of f_{res}, determine the phase difference $\varphi(f)$ as a function of the driving frequency f. To do so, it is useful to plot $i(t)$ (channel A) as a function of $V(t)$ (channel B) (e.g., using the oscilloscope in XY mode). This method is discussed in more detail in Annex A of Chapter 31.

✓ Compare the value obtained experimentally for the frequency f_{res}, for which $\varphi(f_{\text{res}}) = 0$, to the value calculated from expressions (38.14) and (38.15).

✓ Vary the frequency around f_{res} and construct an experimental graph of $I_0(f)/V_0(f)$ and $\phi_{(f)}$ as a function of f_{res}. On the same graph, draw the theoretical expectation that is obtained from the measured values of R, L, and C using expressions (38.14) and (38.15). What can you conclude from this comparison?

✓ From your experimental graph, find the quality factor Q_M of the circuit using the expression (38.18) and compare it to the theoretical value (38.19).

38.4.2 *Project 120 ♣♣ Parallel RLC circuit response — Resonance*

Recommended equipment: Coil (inductance of a few tens of mH). If a high-quality factor coil is desired, a few turns coil with a ferrite core can be used. A capacitor between 10 nF and 500 nF and resistance R_0 between 1 and 10 kΩ. A function generator. A 20 MHz or faster two-channel oscilloscope.

It is often useful to define the transfer function $T(\omega)$. For a linear circuit, like the one in Figure 38.5 with sinusoidal excitation, $T(\omega)$ is defined as the quotient between the (complex) voltage of the output $V_B(t)$ and the complex sinusoidal input voltage $V_A(t)$ [3, 6]:

$$V_{\text{out}}(t) = T(\omega)V_{\text{in}}^0 \exp(j\omega t). \tag{38.20}$$

Note that the transfer function is generally a complex function of ω. If the input is not sinusoidal, the transfer function is defined as the ratio of the Laplace transform of the output signal to the Laplace transform of the input [6].

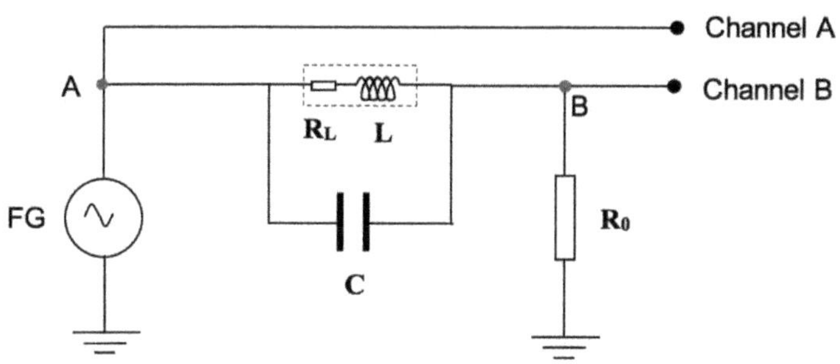

Figure 38.5 Parallel LC circuit: R_L represents the internal resistance of the inductance and R_0 is an auxiliary resistance that enables you to measure the current by measuring the voltage in channel B.

✓ Measure the values of L and R_L. Choose a capacitor C so that the natural frequency defined by Eq. (38.2) $f_0 = \omega_0/2\pi$ is of the order of 2 kHz–5 kHz. Build a circuit like the one shown in Figure 38.5. Connect the oscilloscope test leads to the points indicated as channels A and B. The resistance R_0 should be between 50 Ω and 200 Ω.

Preliminary calculations: For the circuit in Figure 38.5, assuming a sinusoidal input, calculate the complex impedance $Z(\omega)$ of the part of the circuit between points A and B using the chosen values of R, C, and L. Also calculate the transfer function $T(\omega)$ between output V_B and input V_A. Plot the modules of $Z(\omega)$ and $T(\omega)$ as a function of frequency. On the same graph, using a secondary vertical axis, plot the phase of $T(\omega)$ as a function of the frequency.

Suggested tasks:

✓ Using a sinusoidal voltage source, study the relationship between $i(t)$ ($= V_B/R_0$ measured on channel B) and $V(t)$ (measured on channel A) as a function of time.

✓ Determine the phase difference between $i(t)$ and $V(t)$ using the procedure described in Annex A of Chapter 31.

✓ If $I_0(f)$ and $V_0(f)$ are the amplitudes of current and voltage, respectively, and if $\phi(f)$ is the phase difference between $I_0(f)$ and $V_0(f)$, determine the frequency f_{res} for which the relation $I_0(f)/V_0(f)$ is maximum and determine the phase difference $\phi(f_{\text{res}})$ at this frequency. Compare the value f_{res} found experimentally to the value calculated theoretically.

✓ Vary the frequency around f_{res} and plot $I_0(f)/V_0(f)$ and $\phi(f)$ as functions of f. On the same graph, draw the theoretical expectation obtained from the measured values of R_0, R_L, L, and C. What can you conclude from this comparison?

38.5 ♣♣ Linear systems

One of the most remarkable and interesting properties of linear systems is that they obey the *superposition principle* [6, 7]. If a linear system, governed by a linear differential equation, which we denote symbolically with $\hat{L}$[e], has a response $i_1(t)$ for an excitation $v_1(t)$, that is to say,

$$\hat{L}\{i_1(t)\} = v_1(t), \tag{38.21}$$

similarly, if the system has the response $i_2(t)$ for an excitation $v_2(t)$,

$$\hat{L}\{i_2(t)\} = v_2(t), \tag{38.22}$$

the response of the linear system for a combined excitation $[a\,v_1(t) + b\,v_2(t)]$ will be $[ai_1(t) + bi_2(t)]$. That is, for a linear system, the following condition applies [7]:

$$\hat{L}\{ai_1(t) + bi_2(t)\} = av_1(t) + bv_2(t), \tag{38.23}$$

where a and b are two real constants. This relationship is very important in circuit theory and constitutes a necessary and sufficient condition for a system to be considered linear. It implies that if in steady state, the input signal $v(t)$ is doubled, then the output signal $i(t)$ also doubles with respect to the original value and without modifying the signal shape.

From Fourier's theorem [8], we know that any periodic function $f(t)$, of period T, can be expressed as a linear combination of sinusoidal functions, that is,

$$f(t) = \frac{a_0}{2} + \sum_{n=1}^{\infty}\left(a_n\cos\left(n\frac{2\pi}{T}t\right) + a_n\sin\left(n\frac{2\pi}{T}t\right)\right). \tag{38.24}$$

If we call $\omega_0 = 2\pi/T$, in complex notation, Fourier's theorem can be written as follows:

$$f(t) = \sum_{n=-\infty}^{\infty}\left(c_n\exp\left(jn\frac{2\pi}{T}t\right)\right) = \sum_{n=-\infty}^{\infty}(c_n\exp(jn\omega_0 t)), \tag{38.25}$$

[e]Referring to equation (38.1), the operator $\hat{L}$ would be $\hat{L} = Ld^2/dt^2 + Rd/dt + 1/C$. With this notation, the differential equation (38.1) would be written as $\hat{L}\{q(t)\} = 0$.

where the constants c_n are determined from the form of the function $f(t)$ over a period T. More explicitly [8],

$$a_n = \frac{2}{T}\int_{t_0}^{t_0+T} f(t)\cos(n\omega_0 t)dt, \quad b_n = \frac{2}{T}\int_{t_0}^{t_0+T} f(t)\sin(n\omega_0 t)dt, \tag{38.26}$$

and

$$c_n = \frac{1}{T}\int_{t_0}^{t_0+T} f(t)\exp(-jn\omega_0 t)dt, \tag{38.27}$$

where t_0 is an arbitrary time indicating the start of a period of $f(t)$. Therefore, if we know that the response of the linear system for a sinusoidal excitation $v(t) = V_0\exp(j\omega t)$ is $i(t) = V_0\exp(j\omega t)/Z(\omega)$, that is to say,

$$\hat{L}\{i(t)\} = \hat{L}\left\{\frac{V_0}{Z(\omega)}\exp(j\omega t)\right\} = V_0\exp(j\omega t), \tag{38.28}$$

then we can calculate the response $i(t)$ for any periodic excitation by means of the superposition theorem. In particular, if the input excitation $v(t) = f(t)$ is described by Eq. (38.25), according to the superposition theorem, we have

$$i(t) = \sum_{n=-\infty}^{\infty}\left(\frac{c_n}{Z(n\omega_0)}\exp(jn\omega_0 t)\right), \tag{38.29}$$

where $Z(n\,\omega_0)$ is the complex impedance of the system at frequency $\omega_n = n\omega_0$.

In particular, if

$$f(t) = \sum_{n=1}^{\infty}(a_n\sin(n\omega_0 t)), \tag{38.30}$$

then

$$i(t) = \sum_{n=1}^{\infty}\left(\frac{a_n}{|Z(n\omega_0)|}\sin(jn\omega_0 t + \phi(n\omega_0))\right). \tag{38.31}$$

It is often useful to present the constants c_n of the expression (38.25) in the form of a histogram or the constants a_n of the expression (38.30) as a function of the frequencies $\omega_n = n\,\omega_0$. This diagram is known as the *Fourier spectrum* of the function (38.25) or (38.30). There are many programs that perform Fourier analysis and enable the corresponding spectrum of a given function $f(t)$ to be found. In many programs, this function is designated by the abbreviation FFT (*Fast Fourier Transform*). Origin, Excel, Mathematica, and MatLab, among others, provide this tool. It is easy to see from Eqs. (38.29) and (38.31) that the Fourier spectrum of the current $i(t)$ for an excitation $f(t)$ is the product of the Fourier spectrum of the excitation $f(t)$ multiplied by the corresponding admittance $Y(\omega)$. It is therefore possible to analyze a system in the time domain, that is, for each excitation function $f(t)$, to find its corresponding response $i(t)$. It is often also useful to analyze the system in the frequency domain by performing a Fourier transformation.

38.5.1 *Project 121. ♣♣ Response of the RLC circuit forced to a square and triangular excitation*

> **Recommended equipment:** Coil (inductance of a few tens of mH), a capacitor between 10 nF and 500 nF, and resistors between 50 and 500 Ω. A data acquisition system connected to a computer or digital oscilloscope. A function generator.

The aim of this project is to test an application of the Fourier theorem for an RLC circuit excited by square and triangular signals. You can use the same circuit as in Project 119, using the setup shown in Figure 38.3.

Preliminary calculations: For the circuit shown in Figure 38.3, assuming a sinusoidal input, with the values of R, C, and L selected, determine the complex impedance $Z(\omega)$ of the circuit and its resonant frequency f_{res}.

Suggested tasks:

✓ Using a square input signal, of frequency f of approximately $f_{\text{res}}/10$, record the input signal $V(t)$ and the output signal $i(t)$. Next, double the input signal or decrease it by a factor of 2 and determine the output signal again. Check whether the linearity condition is met, Equation (38.22), for your system. Physically, this means that if the input is doubled ($V(t)$), the output also doubles, without altering its shape.

✓ Acquire the input and output data for several periods. Using an appropriate program, perform a Fourier analysis (FFT) of both signals: input ($V(t)$) and output ($i(t)$) containing frequencies that are multiples of the frequency f of the function generator. Discuss your results.

✓ Using Fourier's theorem, determine the coefficients of expansion (a_n, b_n, and c_n) of the input signal ($V(t)$). Using expression (38.25) or (38.26), find the theoretical form of $i(t)$ and compare the shape of the signal obtained experimentally. Plot the measured signals $i(t)$ and those obtained theoretically through Eq. (38.31). Perform the sum up to a value of n (n-maximum) such that it causes the error in the calculated signal to be of the order of 1%.

✓ Using a triangular input signal, of frequency f of approximately $f_{\text{res}}/10$, perform the same experimental study.

✓ What can you conclude from your studies regarding the superposition theorem for your system?

Note: The Fourier expansion for a square signal of amplitude V_0 and period $T(=\pi/\omega_0)$, like the one shown in Figure 38.6, is [8]

$$f(t) = \sum_{n=1}^{\infty}(a_n \cdot \sin(n\omega_0 \cdot t)) = -\frac{4 \cdot V_0}{\pi} \cdot \left[\frac{\sin(\omega_0 t)}{1} + \frac{\sin(3\omega_0 t)}{3} + \frac{\sin(5\omega_0 t)}{5} + \frac{\sin(7\omega_0 t)}{7} + \cdots\right], \tag{38.32}$$

and for a triangular signal, like the one shown in Figure 38.6, we have [8]

$$f(t) = \sum_{n=1}^{\infty}(a_n \sin(n\omega_0 t)) = +\frac{4 \cdot V_0}{\pi} \times \left[\frac{\sin(\omega_0 t)}{1^2} + \frac{\sin(3\omega_0 t)}{3^2} + \frac{\sin(5\omega_0 t)}{5^2} + \frac{\sin(7\omega_0 t)}{7^2} + \cdots\right]. \tag{38.33}$$

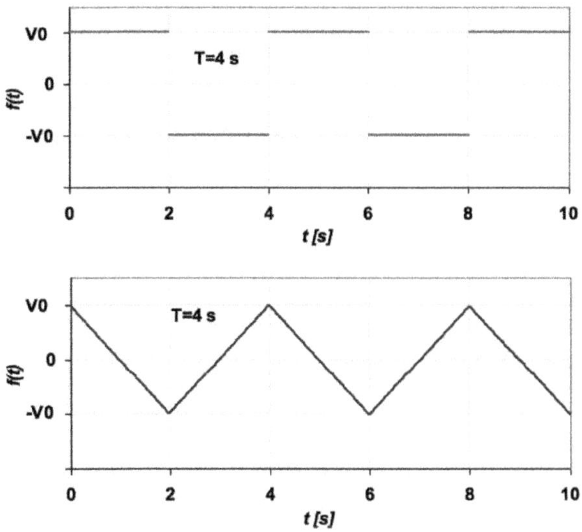

Figure 38.6 Square and triangular signals with amplitude V_0 and period T.

References

[1] R. Halliday, D. Resnick and M. Krane, *Fundamentals of Physics*, NY: John Wiley & Sons, Inc., 2002.

[2] F. Sears, M. Zemansky, H. Young and R. Freedman, *College Physics*, NY: Pearson Ed., 2009.

[3] J. R. Reitz, F. J. Milford and R. W. Christy, *Fundamentals of Electromagnetic Theory*, NY: Ed. Addison-Wesley, 1996.

[4] R. Feynman, *et al.*, *The Feynman Lectures on Physics, Boxed Set: The New Millennium Edition*, Vol. 1, Basic Books, 2011.

[5] E. M. Purcell and D. J. Morin, *Electricity and Magnetism, Berkeley Physics Course* 3rd. Ed., NY, Cambridge University Press, 2014, p. 839.

[6] P. Horowitz and W. Hill, *The Art of Electronics*, 3rd Ed., NY: Cambridge University Press, 2015, p. 1220.

[7] J. DiStefano III, A. Stubberud and I. Williams, *Schaum's Outline of Feedback and Control Systems*, 2nd Ed., NY: McGraw-Hill, 2011, p. 512.

[8] M. Spiegel, *Theory and Problems of Advanced Mathematics for Scientists and Engineers*, Schaum's Outline, NY: McGraw-Hill Book Co., 1971.

[9] R. A. Wellera, "An algorithm for computing linear four-point probe thickness correction factors," *Rev. Sci. Instrum.*, vol. 72, no. 9, pp. 3580–3586, 2001.
[10] D. E. Vaughan, "Four-probe resistivity measurements on small circular specimens," *J. Appl. Phys.*, vol. 12, pp. 414–416 , 1961.
[11] F. M. Smits, "Measurement of sheets resistivities with a four-point probe," *Bell Syst. Tech. J.*, pp. 711–718, 1958.
[12] J. Shia and Y. Sun, "New method of calculating the correction factors for the measurement of sheet resistivity of a square sample with a square four-point probe," *Rev. Sci. Instrum.*, vol. 68, no. 4, pp. 1814–1817, 1997.
[13] A. P. Schuetze, *et al.*, "A laboratory on the four-point probe technique," *Am. J. Phys.*, vol. 72, no. 2, pp. 149–153, 2004.
[14] Wikipedia, "Four-terminal sensing," 2021. Available: http://en.wikipedia.org/wiki/Four-terminal_sensing.
[15] D. W. Koon and C. J. Knickerbocker, "What do you measure when you measure resistivity?" *Rev. Sci. Instrum.*, vol. 63, no. 1, pp. 207–210, 1992.
[16] D. Henry, "Resistance of a wire as a function of temperature," *Phys. Teach.*, vol. 33, p. 96, 1995.
[17] F. Werner, "A method of measuring earth resistivity," *Bur. Stand. U.S. Bull.*, vol. 12, p. 469–478, 1915.
[18] Four Dimension Inc., "Bridge Technology Inc.," 2021. Available: http://www.4dimensions.com/.
[19] "Bridge Technology," 2021. Available:http://four-point-probes.com/.
[20] L. J. van der Pauw, "A method for measuring specific resistivity and Hall effect of discs of arbitrary shape," *Phillips Res. Rep.*, vol. 13, no. 1, 1958.
[21] M. R. Spiegel, *et al.*, *Complex Variables*, Schaum's Outline Series 2nd Ed., NY: Mc Graw Hill, 2009.
[22] B. Avants, D. Soodak and G. Ruppeinera, "Measuring the electrical conductivity of the earth," *Am. J. Phys.*, vol. 67, no. 7, pp. 593–598, 1999.
[23] Wikipedia, "Seebeck effect," 2021. https://en.wikipedia.org/wiki/Thermoelectric_effect.
[24] Wikipedia, "Transconductance," 2021. Available: https://en.wikipedia.org/wiki/Transconductance.
[25] Wikipedia, "Soil Resistivity," 2021. Available: https://en.wikipedia.org/wiki/Soil_ resistivity.

Chapter 39

COUPLED CIRCUITS

This chapter discusses different methods for determining the admittances and impedances of electrical circuits. It introduces two very useful experimental techniques to study complex impedances or admittances in a physical system. The first, which is simpler, uses a standard data acquisition system connected to a computer, and the second uses a *lock-in amplifier*. The second technique will be used to study the behavior of two or more coupled RLC circuits. These devices are an interesting and instructive example of coupled oscillating systems, which have classical and quantum analogs of great interest and application in various areas of Physics.

Goals

- ✓ Techniques for determining the resonance curves of different circuits
- ✓ Coupled RLC circuits
- ✓ Coupled electrical oscillations

39.1 ♣♣ Coupled LC circuits

Coupled oscillating systems are of great interest in Physics, since many of their properties are applicable to a wide variety of phenomena and have several analogies that range from classical to quantum physics [1, 2]. Therefore, we can say that *coupled oscillator* systems are an interesting and instructive paradigm of Physics. This section analyzes the free (not forced) response of two LC circuits coupled through their mutual inductance, M, which we will assume depends on the separation x between them. We will assume $M(x)$ as schematically illustrated in Figure 39.1. Here, the magnetic fields of the coils generate the coupling between them, which foreseeably depends on the distance x.

Kirchhoff's equations [3–6] for this circuit are

$$\frac{q_1}{C_1} + L_1\frac{di_1}{dt} \pm M\frac{di_2}{dt} = 0 \qquad (39.1)$$

and

$$\frac{q_2}{C_2} + L_1\frac{di_2}{dt} \pm M\frac{di_1}{dt} = 0. \qquad (39.2)$$

In these equations, $q_k(t)$ and $i_k(t) = dq_k(t)/dt$ refer to the charges (in the capacitors) and currents in the circuits $k = 1, 2$ (where 1 refers to the left circuit and 2 to the one on the right of Figure 39.1). Expressions (39.1) and (39.2) constitute a system of differential equations coupled through the coefficient of mutual inductance M. The sign in M reflects the two possible relative orientations of the coil windings (inductances L_1 and L_2). In addition, we have the condition [2, 6] $M^2 \leq M_{\text{max}}^2 \equiv L_1 \cdot L_2$. If we assume that

$$q_k(t) = A_k \exp(j\omega t), \qquad (39.3)$$

replacing into Eqs. (39.1) and (39.2), we get

$$\frac{A_1}{C_1} - L_1A_1\omega^2 \mp MA_2\omega^2 = 0 \qquad (39.4)$$

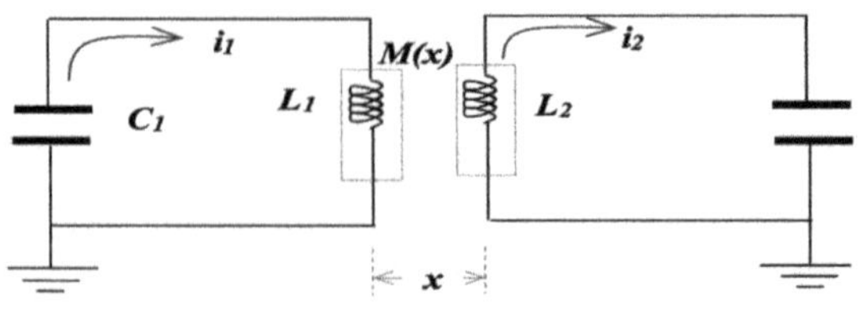

Figure 39.1 Coupled LC circuits. Coupling occurs through mutual inductance $M(x)$ between the two coils, whose centers we assume to be separated by a distance x. With 1, we refer to the left circuit and with 2, to the one on the right.

and

$$\frac{A_2}{C_2} - L_2 A_2 \omega^2 \mp M A_1 \omega^2 = 0 \tag{39.5}$$

with the definitions

$$\omega_{10}^2 = \frac{1}{L_1 C_1} \quad \text{and} \quad \omega_{20}^2 = \frac{1}{L_2 C_2}. \tag{39.6}$$

This system of algebraic Eqs. (39.4) and (39.5) can be written in matrix form as follows:

$$\begin{pmatrix} (\omega_{10}^2 - \omega^2) & \mp \frac{M}{L_1}\omega^2 \\ \mp \frac{M}{L_2}\omega^2 & (\omega_{20}^2 - \omega^2) \end{pmatrix} \begin{pmatrix} A_1 \\ A_2 \end{pmatrix} = 0. \tag{39.7}$$

This system of equations has a solution other than the trivial one ($A_1 = A_2 = 0$) only if the determinant of the matrix is equal to zero, that is,

$$\begin{vmatrix} (\omega_{10}^2 - \omega^2) & \mp \frac{M}{L_1} \cdot \omega^2 \\ \mp \frac{M}{L_2} \cdot \omega^2 & (\omega_{20}^2 - \omega^2) \end{vmatrix} = (\omega_{10}^2 - \omega^2)(\omega_{20}^2 - \omega^2) - \frac{M^2}{L_1 L_2}\omega^4 = 0. \tag{39.8}$$

Defining the coupling μ:

$$\mu = \frac{1}{1 - [M^2/L_1 \cdot L_1]} \geq 1, \tag{39.9}$$

the solution of Eq. (39.7) can be written as follows:

$$\tilde{\omega}^2 = \frac{\mu}{2}(\omega_{10}^2 + \omega_{20}^2) \pm \sqrt{\frac{\mu^2}{4}(\omega_{10}^2 + \omega_{20}^2)^2 - \mu(\omega_{10}^2 \omega_{20}^2)}. \tag{39.10}$$

This equation enables calculation of the natural frequencies in the presence of coupling μ (if M $= 0 \rightarrow \mu = 1$, and for $M \neq 0 \rightarrow \mu > 1$) as a function of the undisturbed natural frequencies (ω_{10} and ω_{20}). Note that since $\mu(\approx 1 + [M^2/L_1 \cdot L_1] + \cdots)$ is always greater than 1, if $M \neq 0$, regardless of the sign of M, from Eq. (39.10), it can be shown that

$$|\tilde{\omega}_1^2 - \tilde{\omega}_2^2| = 2\sqrt{\frac{\mu^2}{4}(\omega_{10}^2 + \omega_{20}^2)^2 - \mu(\omega_{10}^2 \cdot \omega_{20}^2)} \geq |\omega_{10}^2 - \omega_{20}^2|. \tag{39.11}$$

This indicates that as the coupling increases ($|M(x)| > 0$), *the difference between the resonance frequencies of the two circuits increases*, i.e., the resonance frequencies tend to separate. This effect is also observed in quantum systems: when there is coupling between two energy levels, it always generates a separation between the energy levels [2, 7, 8]. This phenomenon of frequency repulsion is also known as the *Wigner–von Neumann effect* or *avoided crossing* [9].

To quantitively test the validity of Eq. (39.10) applied to the case of coupled LC circuits, it is necessary to study experimentally the variation of M as a function of x. This may be done by placing two cylindrical coils on a common axis on which the distance between them can be varied without changing their orientations, and simultaneously measuring $M(x)$ and the distance x between the coils [10].

39.2 Determination of mutual inductances M

As indicated above, the value of the mutual inductance M of two coils, with self-inductance L_1 and L_2, has an upper limit given by $M_{\text{max}}^2 \equiv L_1 \cdot L_2$ [6]. In all the circuits discussed above, the value of the mutual inductance $M(x)$ is fundamental to describe the behavior of coupled systems. The self-inductance of a coil can be determined with any of the techniques discussed in Chapter 35. The value of M can be found using a setup like the one shown in Figure 39.2.

39.2.1 *Project 122. Determining the mutual inductance M(x) as a function of the separation of the coils*

Recommended equipment: Coils (inductances of a few tens of mH) and resistances from 20 Ω to 50 Ω. A data acquisition system connected to a computer or a two-channel oscilloscope. A function generator.

The value of M as a function of the distance x between two coils can be easily studied using a device like the one shown in Figure 39.5. The device consists of two coils of a few hundred mH, which can be moved

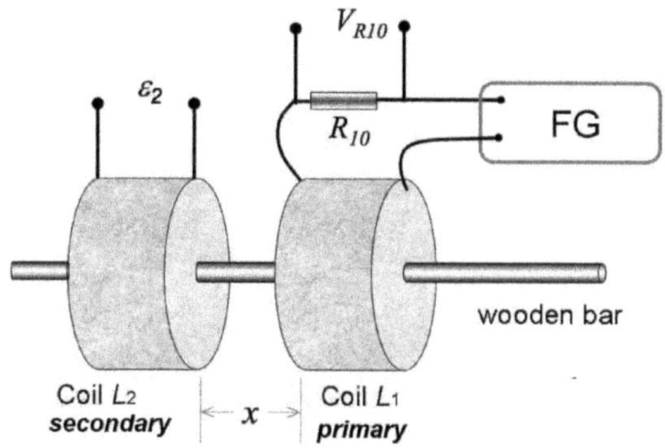

Figure 39.2 Coupled LC circuits. Coupling occurs through mutual inductance $M(x)$ between the two coils, whose centers we assume to be separated by a distance x.

along a common axial bar made of a non-ferromagnetic material, such as wood. Let x be the separation between the centers of the coils. The primary coil is fed by the function generator (FG) at a certain frequency f $(= \omega/2\pi)$ through a limiting resistor $R_{10}(\approx 30\,\Omega)$. By measuring the voltage drops in the secondary and in the resistor R_{10}, it is possible to determine the amplitudes of the voltages $V_{20}(x)$ and $V_{R10}(x)$ for each separation x.

Suggested tasks:

- ✓ Determine the value of each self-inductance using any of the techniques discussed in Chapter 35.
- ✓ Using the setup shown in Figure 39.2, for each value of x, determine $V_{20}(x)$ and $V_{R10}(x)$. The value of $M(x) = R_{10}\,V_{20}(x)/\omega\,V_{R10}(x)$.
- ✓ Plot your experimental results of $M(x)$ as a function of x.
- ✓ If possible, adjust your results with the best function that describes this behavior [10, 11].

39.3 ♣♣ Forced coupled RLC circuits — Wigner–von Neumann effect

A practical and useful way to study the response of coupled RLC circuits is to use an alternating power source in one of them

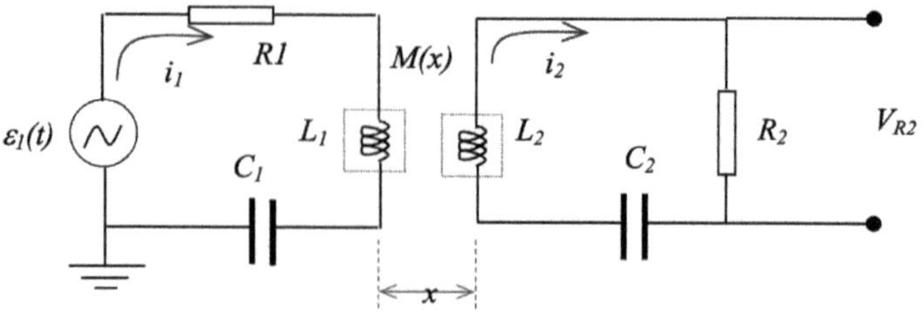

Figure 39.3 Coupled RLC circuits. The coupling occurs through the mutual inductance $M(x)$ between the two coils, whose centers we assume separated by a distance x.

and measure the induced signal in the second one. This section analyzes the response of two RLC circuits coupled through their mutual inductance, M, when one of them is excited by an external source. We assume that M depends on the separation x between the coils, that is, $M(x)$. Figure 39.3 illustrates the characteristics of this system [11, 12]. In this case, we assume, without loss of generality, that the driving signal of the primary circuit is $v_1(t) = V_{10}$ $\exp(-j\omega t - \phi_1)$. Therefore, the circuit equations in this case are the same as those described by Eqs. (39.1) and (39.2), except that we have incorporated a resistor in each circuit and the second member of Eq. (39.1) is now equal to $v_1(t)$.

To solve this circuit, given its linearity, we will assume that the current in the primary and secondary are of the following form:

$$i_1(t) = I_{10}\exp(j\omega t) \quad \text{and} \quad i_2(t) = I_{20}\exp(j\omega t + \phi_2). \tag{39.12}$$

We define the impedances of the primary and secondary as follows:

$$Z_1(\omega) = R_1 + j(L_1\omega - 1/\omega C_1) = R_1 + jX_1(\omega) \tag{39.13}$$

and

$$Z_2(\omega) = R_2 + j(L_2\omega - 1/\omega C_2) = R_2 + jX_2(\omega). \tag{39.14}$$

Kirchhoff's equations for this system, Eqs. (39.1) and (39.2), lead to

$$V_{10}\exp(j\omega t - \varphi_1) = Z_1(\omega)i_1(t) \pm M(x)j\omega i_2(t) \tag{39.15}$$

and

$$Z_2(\omega)i_2(t) \pm M(x)j\omega i_1(t) = 0. \tag{39.16}$$

Removing $i_2(t)$ from these equations, they result in [10, 12]

$$V_{10}\exp(j\omega\, t - \varphi_1) = i_1(t)[Z_1(\omega) + \omega^2 M^2/Z_2(\omega)] = i_1(t)Z_1^*(\omega), \tag{39.17}$$

with

$$Z_1^*(\omega) = \left(R_1 + \frac{\omega^2 M^2 R_2}{R_2^2 + X_2^2}\right) + j\left(X_1 - \frac{\omega^2 M^2 X_2}{R_2^2 + X_2^2}\right) = R_e + jX_e(\omega). \tag{39.18}$$

By solving for i_1 and i_2, we get

$$i_1(t) = I_{10}\exp(j\omega t) = V_{10}\exp(j\omega t - \varphi_1)[Z_2(\omega)(Z_1(\omega)\, Z_2(\omega) + \omega^2 M^2)] \tag{39.19}$$

and

$$\begin{aligned} i_2(t) = I_{20}\exp(j\omega\, t + \phi_2) &= j\, V_{10}\exp(j\omega\, t - \varphi_1) \\ &\times [M\omega(Z_1(\omega)\, Z_2(\omega) + \omega^2 M^2)]. \end{aligned} \tag{39.20}$$

On the other hand, the maximum values I_{10} and I_{20} can be written as follows:

$$I_{10} = V_{10}\left(\frac{\sqrt{R_2^2 + X_2^2}}{D(\omega)}\right) \quad \text{and} \quad I_{20} = V_{10}\left(\frac{\omega M}{D(\omega)}\right), \tag{39.21}$$

with

$$D^2 = \{(R_1R_2 - X_1X_2 + \omega^2M^2)^2 + (R_1X_2 + R_2X_1)^2\} \tag{39.22}$$

or also

$$\begin{aligned} D^2(\omega) = (R_1R_2)^2 \cdot \{&[1 - \lambda_1\lambda_2 \cdot \omega^2(1 - (\omega_{10}/\omega)^2) \\ &\cdot (1 - (\omega_{20}/\omega)^2) + \omega^2\lambda_M^2]^2 + [\lambda_1 \cdot \omega(1 - (\omega_{10}/\omega)^2) \\ &+ \lambda_2 \cdot \omega(1 - (\omega_{20}/\omega)^2)]^2\}, \end{aligned} \tag{39.23}$$

with

$$\lambda_1 = \frac{L_1}{R_1}, \quad \lambda_1 = \frac{L_1}{R_1}, \quad \text{and} \quad \lambda_M^2 = \frac{M^2}{R_1 \cdot R_2}. \tag{39.24}$$

Here, ω_{10} and ω_{20} are defined by Eq. (39.6). Equation (39.21) enables I_{20} (maximum of the current in the secondary) to be found as a function of the frequency. Since M always has a maximum value given by $M_{\max} = \sqrt{L_1 \cdot L_2}$, it is useful to define the parameter $M(x) = \kappa \cdot M_{\max}$. Therefore κ determines the percentage of coupling of the system, with respect to the maximum coupling possible. Typical behavior of a coupled RLC circuit is illustrated in Figure 39.4, which shows typical variation of I_{20} and I_{10} as a function of the excitation frequency. The dotted green curve corresponds to the case of $\kappa = 0$ and the solid red curve to $\kappa = 0.5$ (50% of the maximum possible). The coupling always separates the levels [2, 9].

Figure 39.5 shows the frequency shifts corresponding to the maxima of I_{20} as a function of coupling $\kappa = M/\sqrt{\mathbf{L_1 L_2}}$. It is interesting to note that the position of the maxima with coupling is described by Eq. (39.10). Also note that the effect of coupling is always to separate the maxima, as discussed above [1, 2].

This effect has a highly relevant analogous effect in quantum mechanics [7, 9]. If a quantum system without interaction (unperturbed system) has two energy levels E_{10} and E_{20}, then, when an

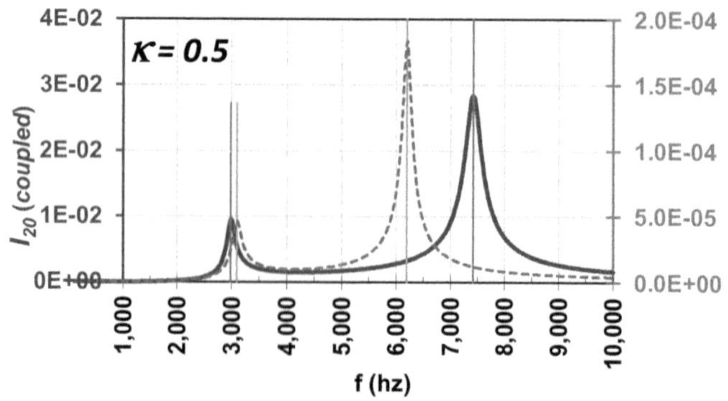

Figure 39.4 Current in the secondary (I_{20}) of a coupled RLC system. The dotted line represents the response with negligible coupling, and the vertical lines represent the position of the maximums. The red curve corresponds to the case with the coupling of 50% of the maximum possible ($\kappa = 0.5$). The maximums of the quasi-decoupled system coincide with the frequency values ω_{10} and ω_{20} given by Eq. (39.6). The maxima of the coupled system coincide with the frequencies given by Eq. (39.10). Note that the coupling always separates the levels.

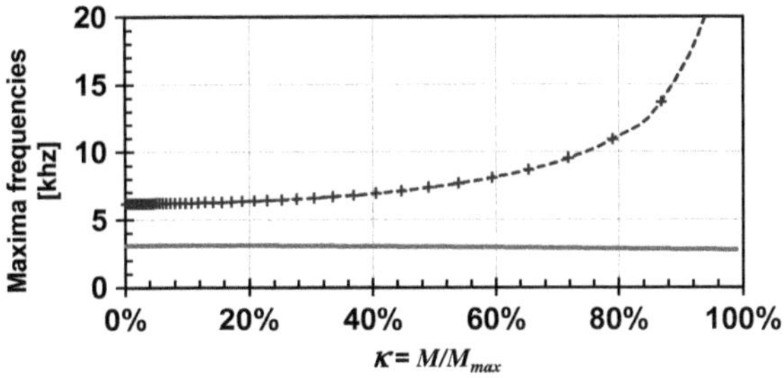

Figure 39.5 Prediction of resonance frequencies (maxima of I_{20}) as a function of coupling $\kappa = M/M_{\max}$, with $M_{\max} = \sqrt{\mathbf{L_1 L_2}}$. Note that the effect of coupling is always to separate the maxima.

interaction or coupling (perturbation) is introduced, the position of these levels will vary in general. If the perturbation is U, the new energy states will be E_1 and E_2, with [1, 7]

$$E_{1,2} = \frac{1}{2}[(E_{10} + E_{20}) \pm \sqrt{(E_{10} - E_{20})^2 + 4\,U^2}], \tag{39.25}$$

with the property

$$|E_1 - E_2| \geq |E_{10} - E_{20}|, \tag{39.26}$$

which is analogous to expression (39.12).

39.3.1 *Project 123. Characterization of the resonance curve using a data acquisition system*

> **Recommended equipment:** Coil (inductance of a few tens of mH). Capacitors between 10 nF and 500 nF and resistors of 30 Ω, 100 Ω, 500 Ω, and 1 kΩ. A data acquisition system connected to a computer. A VCO (function generator with the frequency controlled by a variable DC input signal). A variable DC voltage source.

Figure 39.6 shows a very useful setup to determine or characterize the resonance curve of a series RLC circuit. Choose the values for C and L so that the resonant frequency is of the order of about

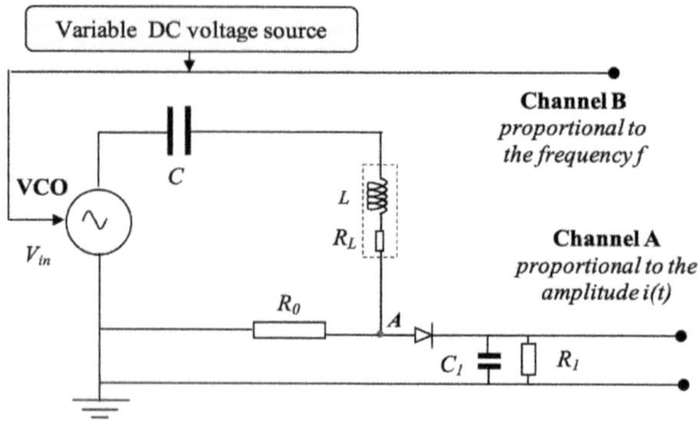

Figure 39.6 Forced series RLC circuit: R_L represents the internal resistance of the inductance. The variable voltage source DC controls the frequency of the FG or VCO; the measurement of V_{in} on channel B is proportional to the frequency. The rectifier circuit, using a diode and a low-pass filter, gives a DC signal proportional to the amplitude of the current.

3–4 kHz. In the first part of this study, use a resistor $R \approx 30\ \Omega$. Measure each of the components R, L, C, and R_L (coil resistance) as accurately as possible. The technique discussed here can be used in a wide variety of resonant circuits. The idea is to use the Variable Controlled Frequency (VCF) input of the Function Generator (FG) or Voltage Controlled Oscillator (VCO) [13], which, using a DC signal between −10 V and 10 V, enable the FG to produce a signal with a frequency proportional to the input voltage (V_{in}). Before using either of these techniques, it is advisable to review the FG or VCO manuals. With this circuit, by monitoring the voltage on channel B, you can measure the frequency f.

The rectifier made of a diode, a capacitor C_1, and a resistor R_1 enables achievement of a DC signal proportional to the amplitude that is connected to channel A. The low-pass filter time constant $\tau_1 = C_1 R_1$ must be much greater than the period $(1/f)$ of the driving signal provided by the VCO but less than the sweep time from one frequency to another. In this way, the signal measured on channel A is proportional to the amplitude of the sinusoidal signal at point A (channel A). A value of $\tau_1 \approx 0,5$ s is generally adequate if the frequencies studied are greater than 300 Hz ($f \cdot \tau_1 \gg 1$).

Suggested tasks:

✓ Frequency calibration: For this experiment, it is advisable to know the relationship between the voltage applied to the FG or VCO input and the frequency f produced by the FG or VCO. To do this, build a table of the value of the DC voltage applied to the input (V_{in}) and the frequency (f) produced by the VCO. With these data, plot f as a function of V_{in}. Fit a function that adequately describes your data.

✓ Once channel B has been calibrated in frequency, you are ready to study the RLC circuit shown in Figure 39.6. By measuring the signal at point A, make sure that channel A indeed detects a signal proportional to the amplitude of the current, I_0 in the circuit. Also check that when the frequency varies, the amplitude V_0 of the signal produced by the VCO is constant.

✓ Using this experimental setup and the data acquisition system, plot the current amplitude I_0 as a function of the frequency f. Determine the resonance frequency, f_{res}, for which $I_0(f)$ is maximum. Compare the experimental value for the frequency f_{res} to the theoretical value using the expression ($f_{\text{res}} = \sqrt{1/LC}/2\pi$) and to the previously measured values of L and C.

✓ In the same plot, compare the resonance curve obtained experimentally to the one predicted theoretically, see also Chapter 38.

✓ If you use an oscilloscope, find the resonance curve for this system with the conventional procedure, i.e., by measuring the amplitude of the current for each frequency.

✓ Discuss the advantages and disadvantages of each of the methods proposed above. Can you suggest other circuits that would enable you to characterize the resonance curve?

✓ Optional: Repeat the study characterizing the resonance curve of the previous system, but insert a ferrite core in the coil. Estimate the variation of the coil inductance with and without the ferrite core.

Note: There are many commercially available function generator models that can be programmed to obtain a frequency sweep within two limits set by the operator in a given time interval. In this case, the frequency produced by the VCO is proportional to time so that by measuring the current amplitude as indicated in Figure 39.6, it is possible to determine the portion of the resonance curve that falls between the preset frequency limits. In this case, it is not necessary to use the input V_{in} of the VCO. By measuring the signal proportional to amplitude (Channel A in Figure 39.6) *as a function of time*, the resonance curve can be obtained.

Finally, if you have an FG or VCO with connection to a PC, using programs like LabView® [14], it is possible to develop an algorithm to perform these measurements simply and automatically.

39.3.2 *Project 124. ♣ Characterization of the resonance curve using a lock-in amplifier*

> **Recommended equipment:** Coil (inductance of a few tens of mH). Capacitors between 10 nF and 500 nF and a resistance of the order of 30 Ω. A lock-in amplifier connected to a computer for data acquisition.

Appendix E of this book discusses the basic operating characteristics of a *lock-in amplifier* [15, 16]. This instrument is particularly suitable when very small signals are being measured in the presence of large noises and also when it is necessary to measure the admittances and impedances of physical systems. This section presents an activity that illustrates how to use the lock-in amplifier to determine the admittance and complex impedance of a series RLC circuit. The experimental technique can easily be generalized to study the admittance and complex impedance of any other circuit.

A schematic diagram of the proposed experimental setup is illustrated in Figure 39.7.

In this circuit, the sinusoidal drive of frequency f to the circuit is provided by the output reference of the *lock-in* (Ref. out), and the voltage drop of the resistor R (proportional to the current i) is used as the input signal to the lock-in. The lock-in control computer is programmed to generate a sinusoidal reference signal with a frequency sweep that generously covers the resonance frequency f_{res} of the system under study. For example, $f_{\text{min}} \approx f_{\text{res}}/2$ and

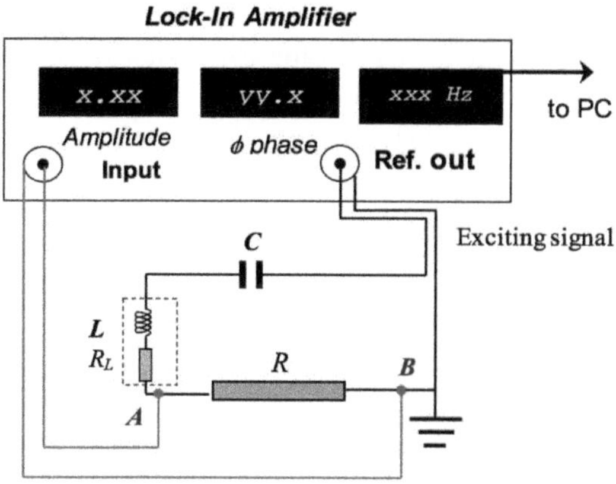

Figure 39.7 Schematic diagram of an experimental arrangement to study the complex admittance of a series RLC circuit using a lock-in amplifier. The driving signal of the circuit is provided by the output reference (Ref. out) of the lock-in, and the voltage drop of the resistance R (proportional to the current i) is used as an input signal to the lock-in.

$f_{\max} \approx 2f_{\text{res}}$. If the value of f_{res} is not known, a preliminary quick scan can be done to detect its value.

If you are working with a two-channel lock-in, the output will consist of a list of data that includes frequency f, signal in phase with excitation V_X, and signal out of phase 90° with excitation V_Y (see Appendix E). From these signals, and knowing the amplitude V_0 of the input (reference signal), the real and imaginary parts of the admittance of the circuit can be found. As the current component in phase ($\text{Re}\{i\}$) and the 90° phase shift with respect to the excitation ($\text{Im}\{i\}$) are, respectively,

$$\text{Re}\{i(f)\} = V_x(f)/R \quad \text{and} \quad \text{Im}\{i(f)\} = V_Y(f)/R, \qquad (39.27)$$

the real and imaginary parts of admittance $Y(f)$ of the circuit will be

$$\text{Re}\{Y(f)\} = \frac{\text{Re}\{i(f)\}}{V_0} = \frac{V_x(f)}{R \cdot V_0} \quad \text{and}$$
$$\text{Im}\{Y(f)\} = \frac{\text{Im}\{i(f)\}}{V_0} = \frac{V_Y(f)}{R \cdot V_0}. \qquad (39.28)$$

With these values, both the real and imaginary parts of the admittance $Y(f)$ and impedance $Z(f) = 1/Y(f)$ can be obtained as a function of f, as well as the module and phase of the impedance and/or admittance, that is, the polar form of the impedance.

$$|Y(f)|^2 = \frac{1}{|Z(f)|} = (\mathrm{Re}\{Y(f)\})^2 + (\mathrm{Im}\{Y(f)\})^2 \quad \text{and}$$
$$\tan\phi = (\mathrm{Re}\{Y(f)\})/(\mathrm{Im}\{Y(f)\}). \tag{39.29}$$

Suggested tasks:

✓ Using the setup shown in Figure 39.7, plot the amplitudes, admittance $Y(f)$, impedance $Z(f)$, and phase ϕ as a function of frequency f for the RLC circuit under consideration. Determine the frequency f_{res} for which $Y(f)$ is maximum or $\phi = 0$. Compare the experimentally obtained value for the frequency f_{res} to the theoretical value. See Chapter 38.

✓ Superimpose on the experimental graph that you have just obtained, the theoretical curve of $Y(f)$, $Z(f)$, and ϕ as a function of frequency f. Keep in mind that the total or effective resistance of the system in Figure 39.7 is $R_{\mathrm{total}} = R + R_L + R_G$, where R_G is the internal resistance of the lock-in reference output, usually $50\,\Omega$, but check the specifications of your instrument. R_L is the resistance associated with the autoinductance wires, which can be measured with a standard ohmmeter, see Chapter 35.

✓ From the half-width of the resonance curve Δf, find the quality factor $Q = f_{\mathrm{res}}/\Delta f$ of the circuit and compare it to the theoretical value [6, 17] $Q = \sqrt{L}/C/R$.

✓ If you have an oscilloscope, compare the curve obtained using the acquisition system to the one obtained with the oscilloscope. Discuss the advantages and disadvantages of each method.

39.3.3 *Project 125. Response of the RLC-C circuit*

Recommended equipment: Coil (inductance of a few tens of mH). A capacitor between 10 nF and 500 nF and resistance $R_0 \approx 30\,\Omega$. A function generator. For data collection, you can use a two-channel oscilloscope, a data acquisition system (DAS) connected to a PC, or a *lock-in amplifier.*

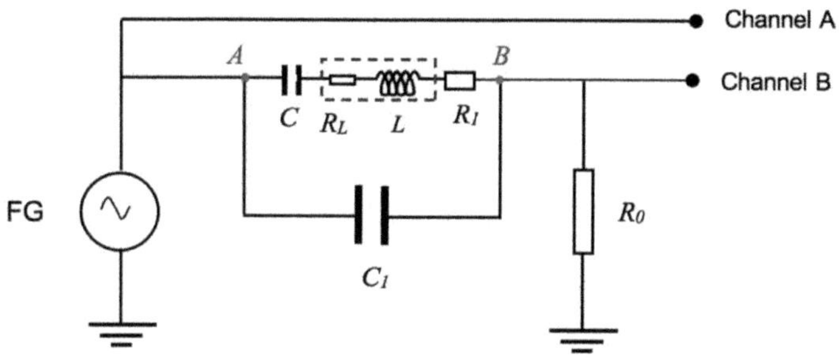

Figure 39.8 Parallel LC circuit: R_L represents the internal resistance of the inductance and R_0 is an auxiliary resistor that enables you to measure the current by measuring the voltage in channel B.

This activity can be performed using any of the experimental techniques described above, an oscilloscope combined with an FG, a DAS, or a *lock-in amplifier*. The proposed circuit illustrated in Figure 39.8 consists of a series RLC circuit in parallel with a capacitor C_1. The goal of this activity is to characterize the admittance or impedance of this circuit and compare it to the theoretical expectations.

Suggested tasks:

✓ Using the complex reactance technique, theoretically study the physical characteristics of the circuit shown in Figure 39.8. Show that the effective impedance of the part of the circuit between points A and B is

$$Z^{-1}(\omega) = Y(\omega) = \frac{R - j\left(\omega L - \frac{1}{\omega C}\right)}{R^2 + \left(L\omega - \frac{1}{C\omega}\right)^2} + j\omega C_1, \tag{39.30}$$

with $R = R_1 + R_L$. From Eq. (39.30), you can find both the impedance module and the phase difference between the applied voltage (channel A) and the current (proportional to the voltage in channel B) as a function of f. Also show that

$$|Z(\omega)|^2 = \frac{R^2 + \left(\omega L - \frac{1}{\omega C}\right)^2}{C_1^2\omega^2 R^2 + \left(1 - \omega C_1\left[L\omega - \frac{1}{C\omega}\right]\right)^2}. \tag{39.31}$$

✓ From your theoretical analysis, make a plot of the modulus of admittance, $|Y(f)|$, and the phase difference $\phi(f)$ as a function of the frequency f. How many maxima for $Y(\omega)$ and resonance frequencies $\phi(f_{\text{res}}) = 0$ do you foresee?

✓ Using your chosen experimental technique, determine the frequency (or frequencies) for which the modulus of admittance, $Y(\omega)$, is maximum, and the phase difference $\phi(f) = 0$. Compare the experimentally obtained value to the frequencies predicted by the theoretical model.

✓ Compare the experimental values of $Y(\omega)$ and $\phi(\omega)$ as a function of ω to the ones predicted theoretically by Eqs. (39.30) and (39.31).

39.3.4 *Project 126 ♣♣ Coupled RLC Circuits. Wigner–von Neumann effect of frequency repulsion*

Recommended equipment: Two coils (a few tens of mH), two capacitors between 10 nF and 500 nF, and resistors from 20 Ω to 50 Ω. For data collection, you can use a two-channel oscilloscope, a data acquisition system connected to a PC, or preferably a *lock-in amplifier.*

This project studies the response of two or more coupled oscillating systems, similar to the system discussed in Section 39.3 of this chapter. This activity can be performed using any of the experimental techniques described above: an oscilloscope combined with an FG, a computer acquisition system, or preferably a *lock-in amplifier.* An example of this type of system could be the system shown in Figure 39.9, which consists of two RLC circuits, close together, coupled through mutual inductance $M(x)$, which depends on the separation x between the centers of its self-inductances. We assume that the separation of the inductance is done preserving its relative orientations.

For this experiment, it is advisable to have determined the variation of the mutual inductance $M(x)$ as a function of the spacing x between the coils, as described in Project 122 [10, 18], [11]. Also,

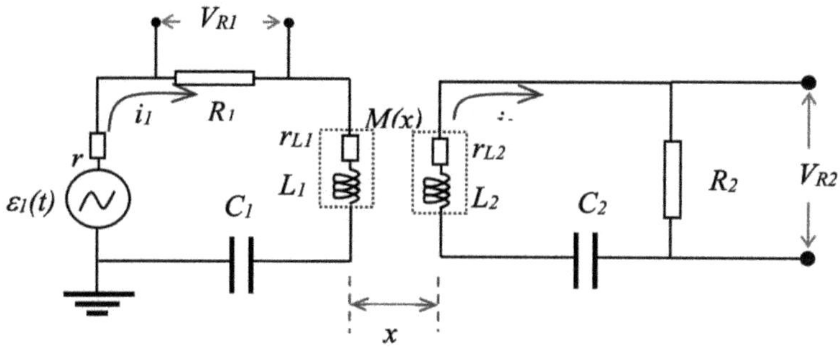

Figure 39.9 Coupled series RLC circuits: the coupling occurs through the mutual inductance $M(x)$ between the two coils, whose centers are assumed to be separated by a distance x. This circuit enables the forced response to be determined. The voltage signals across the resistors R_1 and R_2 (V_{R1} and V_{R2}) can be fed to the inputs of a lock-in amplifier. The drive can be provided by a function generator or the reference signal of a lock-in amplifier.

using the method that is most easily available to you, measure the values of L_i, R_i, R_{Li}, and C_i, $i = 1, 2$, for the two circuits to be used.

Suggested tasks:

✓ For each RLC circuit (primary and secondary) separately, ($x \to \infty$), using a technique like those discussed in Project 124 or 125, find the natural frequencies of oscillation for each circuit individually.

✓ Using the available experimental method (see Projects 124 and 125, Chapter 38), determine for each separation x between the coils, the resonance curve of the primary circuit, by measuring the variation of the amplitude of the signal V_{R1} as a function of frequency. Plot your results for the various x separations used. Make sure that x varies between the smallest possible value and $x \approx 1\,\mathrm{m}$.

✓ For each spacing x between the coils, determine the resonance curve of the primary circuit by measuring the variation of the amplitude of the signal V_{R1} as a function of frequency. Verify that for large separations, the resonance curve is unimodal, but as the coils get closer, the resonance curve becomes bimodal [2, 10, 11]. Plot your results.

✓ For each spacing x between the coils, determine the resonance curve of the secondary circuit, by measuring the variation of the amplitude of the signal V_{R2} as a function of frequency. Verify that for large separations, the resonance curve is unimodal, but as the coils get closer, the resonance curve becomes bimodal [10, 11]. Plot your results. Compare your results with the theoretical expectation in Figure 39.4. From both families of resonance curves, find the natural frequencies of oscillation of both circuits as a function of coil separation x.

✓ Plot the variation of the resonance frequencies of both circuits as a function of x and compare your results to the theoretical expectations, Eq. (39.11) and Figure 39.5.

✓ Optional: Compare your results to those reported in References [10] and [11].

✓ Discuss qualitatively how your results would change if instead of two coupled RLC circuits, you had three coupled RLC circuits. See Ref. [18].

References

[1] R. Feynman, *et al.*, *The Feynman Lectures on Physics, Boxed Set: The New Millennium Edition*, vol. 1, Basic Books, 2011.

[2] W. Frank and P. Von Brentano, "Classical analogy to quantum mechanical level repulsion," *Am. J. Phys.*, vol. 62, p. 706, 1994.

[3] R. Halliday, D. Resnick and M. Krane, *Fundamentals of Physics*, NY: John Wiley & Sons, Inc., 2002.

[4] F. Sears, M. Zemansky, H. Young and R. Freedman, *College Physics*, NY: Pearson Ed., 2009.

[5] E. M. Purcell and D. J. Morin, *Electricity and Magnetism, Berkeley Physics Course*, 3rd Ed., NY: Cambridge University Press, 2014, p. 839.

[6] J. R. Reitz, F. J. Milford and R. W. Christy, *Fundamentals of Electromagnetic Theory*, NY: Ed. Addison-Wesley, 1996.

[7] C. Cohen-Tannoudji, B. Diu and F. Laloe, *Quantum Mechanics*, vol. 1, Chap. 4, NY: John Wiley & Sons, 1977.

[8] C. Garrido Alzar, M. A. G. Martinez and P. Nu, "Classical analog of electromagnetically induced transparency," *Am. J. Phys.*, vol. 70, no. 1, p. 37, 2002.

[9] Wikipedia, "Avoided crossing," 2021. Available: https://en.wikipedia.org/wiki/Avoided_crossing.

[10] M. J. Schauber, *et al.*, "Measurement of mutual inductance from the frequency dependence of impedance of AC coupled circuits using a digital dual-phase lock-in amplifier," *Am. J. Phys.*, vol. 76, p. 129, 2008.

[11] R. Gamarra, M. Josebachuili, P. Zurita and S. Gil, "Experimental study of the frequency repulsion effect," *Am. J. Phys.*, vol. 75, pp. 1073–1077, 2007.

[12] B. Kurrelmayer and W. Mais, *Electricity and Magnetism*, Van Nostrand, 1967.

[13] Wikipedia, "Voltage controlled oscillator," 2021. Available: https://en.wikipedia.org/wiki/Voltage-controlled_oscillator.

[14] Wikipedia, "LabVIEW," visual programming language from National Instruments, 2021. Available: https://en.wikipedia.org/wiki/LabVIEW.

[15] Wikipedia, "Lock-in amplifier," 2021. Available: https://en.wikipedia.org/wiki/Lock-in_amplifier.

[16] J. H. Scofield, "Frecuency-domain description of a lock-in amplifier," *Am. J. Phys.*, vol. 62, pp. 129–133, 1994.

[17] P. Horowitz and W. Hill, *The Art of Electronics*, 3rd Ed., NY: Cambridge University Press, 2015, p. 1220.

[18] H. Atwater, "Laboratory exercises in classical electromagnetic field theory," *Am. J. Phys.*, vol. 36, p. 672, 1968.

[19] Stanford Research Instruments, "Instruction manuals of lock-in amplifiers," 2021. Available: http://www.thinksrs.com.

Chapter 40

QUASI-STATIONARY FIELDS — EDDY CURRENTS

This chapter studies eddy currents induced by time-varying electromagnetic fields on conductors. It also studies the important topic of electromagnetic shielding and the skin effect which occurs when moderate-frequency alternating currents circulate through conductors. These experimental techniques provide a way to measure the conductivity or resistivity of samples without making physical contact with them, which is very useful for studying archaeological artifacts and conducting non-destructive tests of materials

Goals

- ✓ Eddy currents
- ✓ Electromagnetic shielding
- ✓ Skin effect
- ✓ Measurement of conductivity and resistivity

40.1 Eddy currents

Eddy currents [1–5] are currents that circulate inside conductors as a consequence of the presence of electromagnetic fields that vary over time. The Joule power dissipated in a conductor as a result of this effect varies, approximately, proportionally to the square of the frequency of the applied field [2, 4]. This effect is used, for example, in the so-called induction furnaces, which are very useful in industry, and operate at high frequencies ($f \approx 400\,\text{kHz}$), currents, and power [5], as well as induction cooktops [6]. The reader may have noted that the cores or armatures of electrical transformers are made of insulated sheets, which considerably reduces the power losses that occur as a result of eddy currents. These currents are also the basis of the operation of the *metal detectors* used at airports, as well as the metal detectors that are used on beaches, in mining explorations, or in the construction industry, to detect hidden pipes or iron bars inside walls and other structures.

Eddy currents enable the operation of magnetic levitation systems, such as MAGLEV trains [5, 7, 8]. They also have applications in numerous non-destructive tests for inspecting the physical integrity of metal parts and welds. Eddy currents are also very effective in shielding high-frequency electromagnetic fields. When a high-frequency current circulates through a conductor, it tends to circulate preferentially on its surface or periphery. This phenomenon is called *skin effect* [5, 9]. As a result of the skin effect, when high frequencies are used, it is advantageous to use hollow conductors or conductors made of multiple, very fine, insulated, and stranded wires known as *Litz wires* [10], which are very common in audio systems, induction cookers, antennas, etc.

The following experimental projects study some basic characteristics of the effects of electromagnetic fields that vary with time, in the presence of conductors. These studies are of great theoretical and practical interest. These experiments are also intended to help the reader to become familiar with the *lock-in amplifier*, a useful instrument in modern laboratories (see Appendix E). A lock-in amplifier can be used to make highly precise and sensitive measurements of signals whose variation over time is known. It can also measure the phase variation of an input signal with respect to its output. This is particularly useful for determining the variation in frequency of a complex admittance of a system.

40.2 Quasi-stationary electromagnetic fields in conductors

Maxwell's equations for a material medium in the International System (SI) of units are [1–5]

$$\vec{\nabla} \cdot \vec{E} = \frac{\rho}{K\varepsilon_0} \tag{40.1}$$

$$\vec{\nabla} \times \vec{E} = -\frac{\partial \vec{B}}{\partial t} \tag{40.2}$$

$$\vec{\nabla} \cdot \vec{B} = 0 \tag{40.3}$$

$$\vec{\nabla} \times \vec{B} = +\mu_0 \mu_r \vec{J} + \mu_0 \mu_r K \varepsilon_0 \frac{\partial \vec{E}}{\partial t}. \tag{40.4}$$

Here, ε_0 and μ_0 are the permittivity and permeability of the vacuum, respectively, while K is the dielectric constant of the medium and μ_r its relative magnetic permittivity. These equations, together with the Lorentz force

$$\vec{F} = q\,(\vec{E} + \vec{v} \times \vec{B}) \tag{40.5}$$

and the equation

$$\vec{J} = \sigma\,\vec{E}, \tag{40.6}$$

where $\boldsymbol{J}$ is the current density and σ the conductivity of the medium, enable us to find the electromagnetic fields in the presence of conductive media.[a] By applying the operator $\vec{\nabla}\times$ to Eqs. (40.2) and (40.4), it is easy to obtain the following expressions for the vector fields $\boldsymbol{E}$ and $\boldsymbol{B}$ [3–5]:

$$\nabla^2 \vec{E} - \sigma \mu_0 \mu_r \cdot \frac{\partial \vec{E}}{\partial t} + \mu_0 \mu_r K \varepsilon_0 \frac{\partial^2 \vec{E}}{\partial t^2} = 0 \tag{40.7}$$

and

$$\nabla^2 \vec{B} - \sigma \mu_0 \mu_r \cdot \frac{\partial \vec{B}}{\partial t} + \mu_0 \mu_r K \varepsilon_0 \frac{\partial^2 \vec{B}}{\partial t^2} = 0. \tag{40.8}$$

As

$$\frac{1}{c^2} = \mu_0 \varepsilon_0, \quad \text{with } c = \text{speed of light in vacuum}, \tag{40.9}$$

and

$$\omega = 2\pi \frac{c}{\lambda} \quad \text{or} \quad f = \frac{c}{\lambda}, \tag{40.10}$$

[a]Strictly speaking, expression (40.6) is only valid in equilibrium, but in most conductors, the relaxation time (i.e., the time it takes for the system to reach equilibrium) is of the order of $\tau \approx K\varepsilon_0/\sigma \approx 10^{-14}$ s [8]. Therefore, in typical conductors, this relationship is valid up to frequencies close to infrared [26].

for the case of alternate fields of the form,

$$\vec{E}(rt) = \vec{E}(r)\exp(-j\omega t), \tag{40.11}$$

where $j \equiv i \equiv \sqrt{-1}$[b] is the imaginary unit, the expressions (40.7) and (40.8) reduce to

$$\nabla^2 \vec{E}(r) + \left(\frac{4\pi^2 \mu_r K}{\lambda^2} + j\frac{2}{\delta^2}\right)\vec{E}(r) = 0 \tag{40.12}$$

$$\nabla^2 \vec{B}(r) + \left(\frac{4\pi^2 \mu_r K}{\lambda^2} + j\frac{2}{\delta^2}\right)\vec{B}(r) = 0 \tag{40.13}$$

with

$$\delta^2 = \frac{2}{\sigma\mu_0\mu_r\omega} \quad \text{in SI system and} \quad \delta^2 = \frac{c^2}{2\pi\sigma\mu_r\omega} \quad \text{in cgs system.} \tag{40.14}$$

The parameter δ has dimensions of length and is known as the *penetration length*. In Eqs. (40.12) and (40.13), the first term within the parentheses is a consequence of the *displacement* current, while the second is related to the *conduction* current. If we consider the *quasi-stationary* case, that is, those cases in which the dimensions of the system of interest are such that its characteristic length l is much shorter than the wavelength λ of the electromagnetic field, that is,

$$\lambda = \frac{2\pi c}{\omega} \gg l of \ll= \frac{c}{l}, \tag{40.15}$$

we have that the first term within the parentheses of Eqs. (40.12) and (40.13) associated with the displacement current is negligible compared to the term associated with the conduction current.

[b]In electromagnetism it is preferable to use j to designate the imaginary unit, to avoid possible confusion with the symbol i, which is used here for the electric current.

Therefore the expressions (40.12) and (40.13) are reduced to

$$\nabla^2 \vec{E}(r) + j\frac{2}{\delta^2} \cdot \vec{E}(r) = 0 \tag{40.16}$$

and

$$\nabla^2 \vec{B}(r) + j\frac{2}{\delta^2} \cdot \vec{B}(r) = 0. \tag{40.17}$$

Using (40.6), we also have that

$$\nabla^2 \vec{J}(r) + j\frac{2}{\delta^2} \cdot \vec{J}(r) = 0. \tag{40.18}$$

These equations, along with the boundary conditions

$$B_{n1} = B_{n2}, \quad B_{t1}/\mu_{r1} = B_{t2}/\mu_{r2} \tag{40.19}$$

and

$$E_{n,\text{conductor}} = \frac{J_{n,\text{conductor}}}{\sigma_{\text{conductor}}} = 0, \tag{40.20}$$

where the subscripts t and n represent the tangential and normal components to the metal surface (2) and air or vacuum (1), respectively, are the starting point to find the electromagnetic fields in many cases of interest. Expression (40.20) is associated with the fact that the currents, by continuity, cannot have a component perpendicular to the conductor–vacuum interface, since the surrounding environment is non-conductive (air or vacuum), $\sigma_1 = 0$.

For electromagnetic fields with frequencies of $1 \approx$ MHz, the wavelength is $\lambda \approx 300\,\text{m}$. Therefore, for systems whose dimensions are of the order of one meter, the quasi-stationary approximation is generally adequate for fields with frequencies equal to or less than a few MHz.

Physically, the *quasi-stationary approximation* is valid to the extent that we can consider negligible the effects of displacement currents in relation to conduction currents. This criterion is equivalent to considering negligible all the effects linked to the finite speed of propagation of electromagnetic fields. It is clear that this will be possible as long as the dimensions of the system are small with respect to λ.

40.3 Electromagnetic shielding — Cylindrical symmetry

Consider the system shown schematically in Figure 40.1, consisting of a pair of cylindrical coils placed concentrically along a common axis. The outer coil (longer and with a larger diameter) is the primary. The second, smaller coil, placed inside the first, is the secondary. A metal cylinder (shield) can be inserted between the two. Our goal is to calculate the effect that the introduction of the metal cylinder, which acts as a shield, will have on the induced voltage in the secondary coil.

In this case, it is possible to solve the problem using Eqs. (40.16) and (40.17) to find the electromagnetic fields, but it is simpler and physically more transparent to use the scheme developed by S. Fahy *et al.* [11, 12], starting from the same assumptions that were used to deduce Eqs. (40.16) and (40.17).

Assume that the metal cylinder used as a shield has radius R, length l, and thickness d. Assume also that $d \gg \delta \gg R$, where δ is the penetration length defined in Eq. (40.14). Let B_1 be the magnetic field generated inside the primary solenoid. Let B_p be the magnetic field generated by the shield itself inside the shield cylinder, where the secondary solenoid is located. Thus, the resulting magnetic field

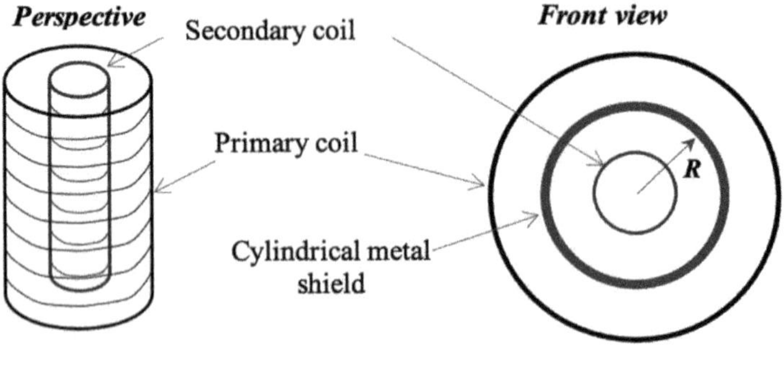

Figure 40.1 Experimental device to study electromagnetic shielding in cylindrical geometry. The outer (larger) coil is the primary, and the inner (smaller) coil is the secondary. A metal cylinder of radius R and thickness d is placed between the two coils.

B_2 in the inside area of the metal shield will be

$$B_2 = B_1 + B_p, \tag{40.21}$$

with

$$B_1(t) = B_{10}\, e^{j\omega t}, \tag{40.22}$$

where $\omega = 2\pi f$ is the angular frequency of the field applied to the primary. B_P is the field generated by the cylindrical shield *due to eddy currents* induced by the variable magnetic field B_2.

By Faraday's law, the emf ε generated at the surface of the metal cylinder is

$$\varepsilon = -\frac{d\varphi_B}{dt} = (\pi R^2) j\omega B_2(t). \tag{40.23}$$

This emf generates an azimuthal current on the surface of the cylinder. The current per unit length is given by the following expression:

$$J_\phi = \frac{\varepsilon}{\delta R} = \frac{\varepsilon}{2\pi R\rho/d} = j\frac{Rd\omega}{2\rho} B_2(t). \tag{40.24}$$

Here, $\delta\mathfrak{R}$ represents the resistance of a strip of the cylinder of unit length and $\rho = 1/\sigma$ is the resistivity of the material that the shield is made of. This azimuth current in the cylinder generates a magnetic field B_p similar to that of a solenoid of similar dimensions to the metal cylinder. By Ampère's law, this field is [1, 2]

$$B_p = \mu_0 J_\phi K_{\text{geom}} = j\mu_0 \frac{\pi R d f}{\rho} K_{\text{geom}} B_2(t). \tag{40.25}$$

Here, K_{geom} is a geometric correction coefficient, or geometric factor, that takes into account the fact that the shield has a finite length.[c] If the solenoid is long, this coefficient is the unit; otherwise, its value

[c]By analogy with a finite solenoid of length l and radius R, the magnetic field in its center, B_o, can be expressed as the field corresponding to an infinite solenoid B_8 (i.e., $l \gg R$) multiplied by a form factor $K_{\text{geom}} = 1/\sqrt{1 + (2R/l)^2}$, that is, $B_o = B_8 \cdot K_{\text{geom}}$ [1, 2, 15].

will depend on the geometry and the relationship R/l but will be constant over time. By replacing (40.25) in (40.21), we have

$$B_2 = B_1(t) + j\mu_0 \frac{\pi\, \mathrm{Rd\,f}}{\rho} K_{\text{geom}} B_2(t) \tag{40.26}$$

Thus,

$$B_2(t) = \frac{B_1(t)}{\left(1 - j\frac{R\,d\,\mu_0\,\omega}{2\rho} K_{\text{geom}}\right)} = \frac{B_1(t)}{\left(1 - j\,\frac{R\,d}{\delta^2}\,K_{\text{geom}}\right)}. \tag{40.27}$$

By measuring in the secondary circuit the induced voltage $V_2(t)$ proportional to $\omega B_2(t)$ with shield (*WS*) and comparing it to the voltage induced in the secondary, with no shield (*NS*), which is proportional to $\omega B_1(t)$, we have

$$\frac{V_2^{(WS)}(t)}{V_2^{(NS)}(t)} = \frac{\omega \cdot B_2(t)}{\omega \cdot B_1(t)} = \left(1 - j\frac{Rd}{\delta^2} \cdot K_{\text{geom}}\right)^{-1} \tag{40.28}$$

or also for the peaks or effective values (rms), we have

$$\frac{V_2^{(WS)}}{V_2^{(NS)}} = \frac{B_2}{B_1} = (1 - j\,\tan\varphi)^{-1} = \cos\varphi \cdot e^{-j\varphi} \tag{40.29}$$

where the angle of phase shift ϕ between the voltage with and without shield is given by

$$\tan\varphi = \frac{Rd}{\delta^2} \cdot K_{\text{geom}} = \mu_0 \pi\, R\,d\,f. \tag{40.30}$$

The quotient of the peak or rms values is known as the *shielding or screening factor* ($SF \equiv |\cos\varphi|$) and will be proportional to the modulus of expression (40.29), i.e.,

$$SF = \left.\frac{V_2^{(WS)}}{V_2^{(NS)}}\right]_{\text{peak}} = \frac{1}{\sqrt{1 + \left(\frac{R.d}{\delta^2} K_{\text{geom}}\right)^2}} \frac{1}{f} = |cos\,\varphi|. \tag{40.31}$$

According to the initial hypothesis, $d \ll \delta \ll R$, expression (40.31) is valid in the frequency range $f_{\min} \equiv 1/(\pi\sigma\mu_0 \cdot R^2) \ll f \ll f_{\max} = 1/(\pi\sigma\mu_0 \cdot d^2)$. Note that for $f = f_{\min}$, $\delta(f_{\min}) = R$ and for $f = f_{\max}$, $\delta(f_{\max}) = d$. In particular, if $R\delta K_{\text{geom}} > d^2$, or if $f \gg f_m \equiv 1/(\pi\sigma\mu_0 R\, d\, K_{\text{geom}})$, Eq. (40.31) simplifies to

$$\frac{V_2^{(WS)}}{V_2^{(NS)}}\Big|_{\text{peak}} \cong \frac{\delta^2}{RdK_{\text{geom}}} = \left[\frac{1}{\mu_0\pi Rd\cdot K_{\text{geom}}}\right]\frac{1}{f}, \quad \text{if } fm \ll f \ll f_{\max} \tag{40.32}$$

Another useful way to present this result is to take logarithm in both members, that is,

$$\ln\left[\frac{V_2^{(WS)}}{V_2^{(NS)}}\Big|_{\text{peak}}\right] \cong \ln(\mu_0\pi Rd\cdot K_{\text{geom}}) - \ln(f), \quad \text{if} \quad fm \ll f \ll f_{\max}. \tag{40.33}$$

Figure 40.2 shows the generic form of the shielding factor *SF* dependence on frequency.

The expressions (40.29), (40.30), (40.31), and (40.33) can be used to validate the model experimentally. Measuring the dependence of the shielding factor *SF* and of the phase shift ϕ on the frequency enables these results to be compared to the predictions (40.29), (40.30), and (40.31), thereby ascertaining the extent to which the proposed model is adequate for explaining the data.

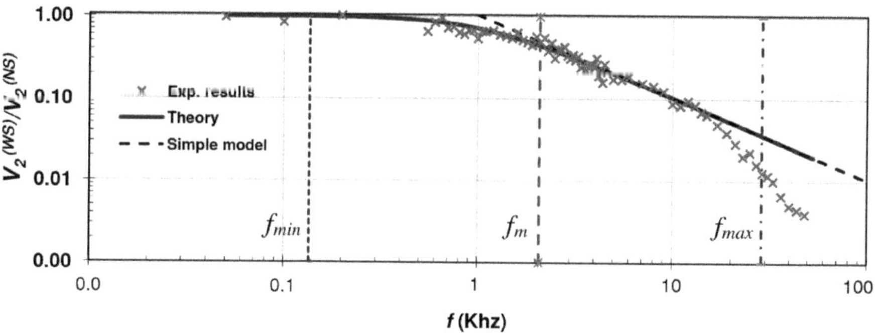

Figure 40.2 Dependence of *SF* on frequency: thick blue line Eq. (40.31). Frequencies $f_{\min}, f_m$, and $f_{\max}$ are also indicated. The black dashed line is the approximation given by the model in Eqs. (40.32–40.33). The crosses represent a typical set of experimental results. For $f > f_{\max}$, Eq. (40.31) does not adequately describe the data.

Table 40.1 Resistivities for some common metals [12]. Note that the resistivity of commercial metals may differ from these values, which are for pure metals.

Material	ρ (Ωm) at 20°C	σ (S/m) at 20°C	Cond. Relative to Cu
Silver	1.59×10^{-8}	6.30×10^{7}	105.7
Copper	1.68×10^{-8}	5.96×10^{7}	100.0
Annealed copper	1.79×10^{-8}	5.60×10^{7}	94.0
Aluminum	2.86×10^{-8}	3.5×10^{7}	58.7
Commercial Aluminum	3.8×10^{-8}	2.6×10^{7}	43.6
Iron	1.0×10^{-7}	1.00×10^{7}	16.8
Tin	1.09×10^{-7}	9.17×10^{6}	16.3
Bronze#	7.7×10^{-7}	1.3×10^{7}	21.8
Nickel	6.99×10^{-8}	1.43×10^{7}	24.0

Note: #Note that bronze is an alloy of variable composition.

Table 40.1 lists the resistivities and conductivities of some common materials that can be used as shields.

40.3.1 *Project 127. Electromagnetic shielding I*

Recommended equipment: Two concentric coils of 300–1,000 turns each, forming a kind of transformer with an outer primary and an inner secondary. Tubes made of aluminum, copper, or some known conductor, of different thicknesses between 0.2 mm and 2 mm. It must be possible to place these metal tubes between the primary and secondary coils. A 30 MHz or faster two-channel oscilloscope. An AC voltmeter and a function generator.

The electromagnetic shielding effect is one of the many manifestations of the time-dependent interaction of electromagnetic fields with conductive materials.

The experimental setup for this project is illustrated in Figure 40.1. A solenoid of about 300–500 turns about 15–30 centimeters long and approximately $3 \pm 1\,\text{cm}$ internal diameter acts as primary. A second, smaller coil is placed inside, near the center of the primary. There must be enough space between the two coils to insert a metal tube, the properties of which we wish to study. The secondary coil should have 500–2,000 turns and be positioned coaxially with the primary. These coil dimensions are merely a suggestion: they may be changed with some discretion. The metal tube, which acts as an electromagnetic shield, has an inner radius R and a thickness d that must be carefully measured. It is also useful to have several tubes of the same material and of variable thicknesses, say $0.1 < d < 3\,\text{mm}$. Although it is desirable to work with several different tubes, it is not essential, and the phenomenon can be studied with a single conductive tube.

Experiments

✓ Using the circuit shown in Figures 40.1 and 40.3 (version without lock-in), study experimentally the variation of the shielding as a

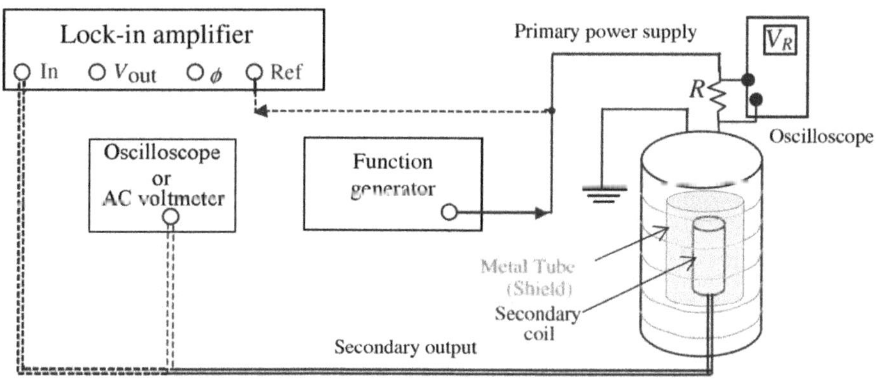

Figure 40.3 Experimental device to study electromagnetic shielding in cylindrical geometry. The function generator feeds the primary through a known resistance R. This generator can be replaced by the same lock-in reference signal (for a lock-in with a built-in function generator). The voltage V_R used to monitor the current i_1 of the primary. The voltage on the secondary can be measured by the lock-in amplifier, or with an oscilloscope, or an AC voltmeter.

function of the applied frequency. Vary the frequency from about 200 Hz to about 30 kHz.

- First, measure the voltage $V_2^{(NS)}]_{\text{peak}}$ without shielding as a function of frequency.
- Then, using the same frequencies, measure $V_2^{(WS)}]_{\text{peak}}$, with shielding.
- An oscilloscope or a wide bandwidth AC voltmeter can be used to monitor the current in the primary through the voltage drop in R. It is not crucial to keep the current constant; only it must be the same at each frequency with and without shielding.

✓ For each metal cylinder, plot the shielding factor $SF = [V_2^{(WS)}/V_2^{(NS)}]_{\text{peak}}$ as a function of the frequency f on linear and logarithmic scales. On each graph, include the predictions of the proposed model, Eq. (40.31), and mark the values of the frequencies $f_{\min}$, f_m, and $f_{\max}$, as shown in Figure 40.2. Indicate the frequency range where the proposed model may be assumed to be valid.

✓ Vary the value of the conductivity (and consequently the value of δ) used in the model, Eq. (40.31), until you find the best possible fit between the theory and the measured data.

✓ Compare the conductivity value (σ and δ) found from the fit to the tabulated values for the material used in your test. Remember that commercial materials are often not pure, so some differences are to be expected with respect to the table values.

✓ If the model reproduces your experimental data well, it means that the model used is adequate and you can therefore expect it to enable you to find the conductivity of the material without needing an electrical connection.

✓ What can you conclude regarding the goodness of the proposed model? Discuss the behavior of the model for low and high frequencies. How do you explain the behavior at high frequencies qualitatively?

Optional:

✓ If you have several tubes of the same material with different thicknesses d but the same radius R, build a plot of the product $d \cdot SF = d \cdot [V_2^{(WS)}/V_2^{(NS)}]_{\text{peak}}$ as a function of the frequency f on a logarithmic scale. Include the data of all the thicknesses d used for the same material with a similar radius R. Also include in the graph the predictions of the model Eq. (40.31).

✓ If you are able to measure the phase difference ϕ, plot the phase ϕ as a function of frequency for at least one measured thickness d. Graph $\tan(\phi)$ as a function of frequency. In the same graph, include the measured results and the prediction of the proposed model expressed by Eq. (40.30). This measurement can be made with an oscilloscope (see Annex A, Chapter 31), but it can be done more easily using a lock-in amplifier (see Appendix E).

✓ Research the literature and discuss the non-destructive testing technique for detecting transverse and longitudinal faults in metal pipes by using these eddy current principles.

✓ Explain why audio equipment and other electronic devices generally use a copper band around transformers.

40.3.2 *Project 128. Electromagnetic shielding II — Lock-in*

> **Recommended equipment:** Two concentric coils of 300–1,000 turns, arranged as in the previous project. Tubes that can be inserted between the primary and secondary coils, made of aluminum, copper, or some known conductor, with different thicknesses of approximately $0.1\,\text{mm} < d < 3\,\text{mm}$. A 30 MHz (or faster) two-channel oscilloscope. A lock-in amplifier and a function generator. If the lock-in includes a function generator, you do not need a separate function generator.

This project is nearly identical to the previous one, except that now, a lock-in amplifier [13] is used to measure the voltage in the secondary V_2 (see Appendix E) to derive the shielding factor SF and the phase shift ϕ. In addition to achieving a signal with less noise than in the previous case, the lock-in enables you to measure the components of the output signal V_2 in phase with the input (current in the primary) or reference signal, and in counter-phase with it ($\phi = 90°$). It is thereby possible to find the phase difference between the two signals.

Experiment:

- ✓ Using the circuit described in Figure 40.3 (version with lock-in), study the effect of the shielding as a function of the applied frequency. Vary the frequency from a few hundred Hz to about 35 kHz. For example, use the frequency range 200 Hz–35 kHz with a 250 or 500 Hz step. The oscilloscope is used to monitor the voltage V_R in the primary resistance. It is not crucial to keep the current constant, but it must be the same at each frequency for the cases without shielding (NS) and with shielding (WS).
- ✓ First, measure $V_2^{(NS)}]_{\text{peak}}$ and $\phi(NS)$ (phase shift between V_2 and the voltage applied to the primary) without the shield as a function of the frequency. Then, using the same frequencies, measure $V_2^{(WS)}]_{\text{peak}}$ and $\phi(WS)$ with the shield.
- ✓ Plot the shielding factor $V_2^{(WS)}(f)/\mathrm{V}_2^{(\mathrm{NS})}(\mathrm{f})]_{\text{peak}}$ as a function of the frequency f in logarithmic scale, similar to Figure 40.2, for each metal shield cylinder used. In each graph, include the prediction of the proposed model, Eq. (40.31). Mark the values of the frequencies $f_{\min}$, f_m, and $f_{\max}$. Indicate the frequency range where the proposed model is assumed to be valid. Is this prediction true?
- ✓ Vary the value of the conductivity (and consequently the value of δ) used in the model, Eq. (40.31), until you find the best possible fit between the theory and the measured data.
- ✓ Compare the conductivity value found from the previous fit to the tabulated values for the material used in your test. Remember that commercial materials are often not pure, so some differences are to be expected with respect to table values.

Optional:

- ✓ Plot the shield phase difference $\phi(f) = \phi(WS) - \phi(NS)$ as a function of frequency for at least one of the cylinders used. In the same graph, include the predictions of the proposed model.
- ✓ What can you conclude about the goodness of the model? Discuss the behavior of the model for low and high frequencies. What qualitative explanation can you provide for the behavior at high frequencies?
- ✓ Discuss the advantages of this measurement method compared to the method using an oscilloscope.

40.3.3 *Project 129. Electromagnetic shielding III — Flat plates*

Recommended equipment: Two coils of 300–2,000 turns each, forming a kind of transformer with a primary and a secondary. Plates of aluminum, copper, or some known conductor of variable thicknesses between 0.2 mm and 2 mm. The dimensions of these metal plates must be greater than the coils so that they can be interposed between the primary and secondary coils. A 20 MHz or faster two-channel oscilloscope. A function generator and a lock-in amplifier. If the lock-in includes a function generator, no other generator is required.

In this project, two cylindrical coils are used, as shown in Figure 40.4, and the shielding is provided by square-shaped flat metal plates. In order to measure the shielding factor, you can use an oscilloscope, as in Project 127, or even better, a lock-in amplifier.

Experiment:

- ✓ Using the device described in Figure 40.4 and the circuit shown in Figure 40.3, study experimentally the variation of the shielding as a function of the applied frequency. Vary the frequency from a few kHz to about 30 kHz.

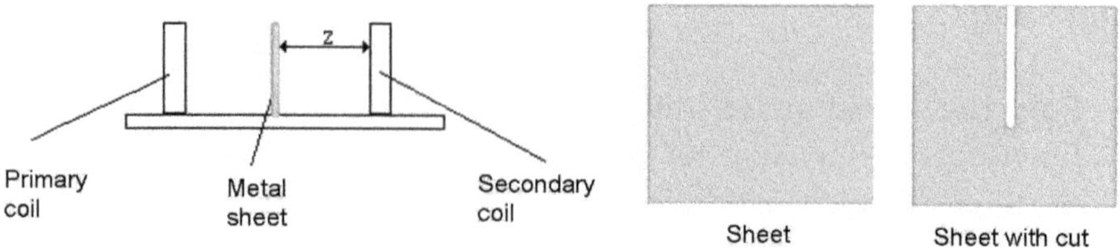

Figure 40.4 Experimental device to study electromagnetic shielding by a metal plate or sheet.

- ✓ First, measure $V_2^{(NS)}]_{\text{peak}}$ and $\phi(NS)$ without the shield as a function of the frequency. Then for the same frequency values, measure $V_2^{(cp)}]_{\text{peak}}$ and $\phi(cp)$ with the shield.
- ✓ Plot the shielding factor SF as a function of the frequency f in log-log scale for each of the metal plates of thickness d measured. In the same graphs, include the predictions of the model Eq. (40.31), using R as an adjustable parameter to optimize the fit. Note that the model in Eq. (40.31) *is not strictly adequate* for this geometry.
- ✓ For the same material, plot the product $dxSF$ as a function of the frequency f on a log-log scale. In the same graphs, include the predictions of the model Eq. (40.31), using the factor $\kappa = RxK_{\text{geom}}$ as an adjustable parameter.
- ✓ Study the shielding behavior using a smooth plate and compare it to the shielding behavior using another plate of the same thickness but with a cut, as shown in Figure 40.4. As you will see, eddy currents are partially interrupted by this cut. What can you conclude about the shape of eddy currents from this analysis?

40.4 Skin effect

Consider a homogeneous cylindrical conductor of radius a and length $l \gg a$, through which an AC current flows:

$$I(t) = I_0 e^{j\omega t}, \tag{40.34}$$

where $j = \sqrt{-1}$. Assume the cylinder is oriented in the z direction, as shown in Figures 40.5 and 40.6. Assuming that the current is increasing, the magnetic field flux inside the wire also increases

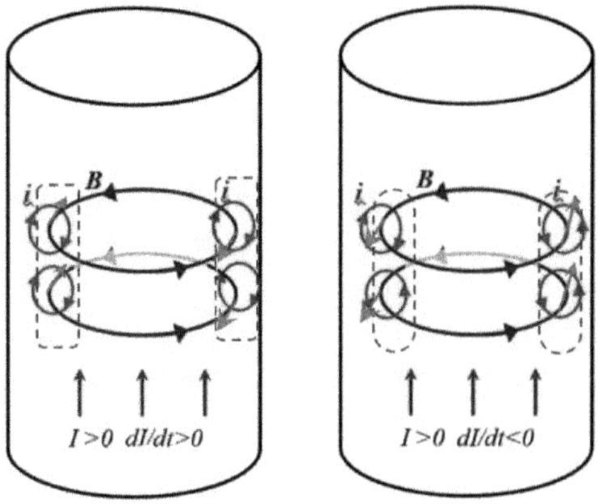

Figure 40.5 Induced currents in a wire due to a variable current I along it in the axial direction. On the left, an increasing current is moving along its axis ($dI/dt > 0$). The induced currents inside the wire (red circles) generate a magnetic field (green) that opposes the increase of the $\boldsymbol{B}$ field produced by the primary current ($dI/dt > 0$). On the right, the same situation with a decreasing current ($dI/dt < 0$).

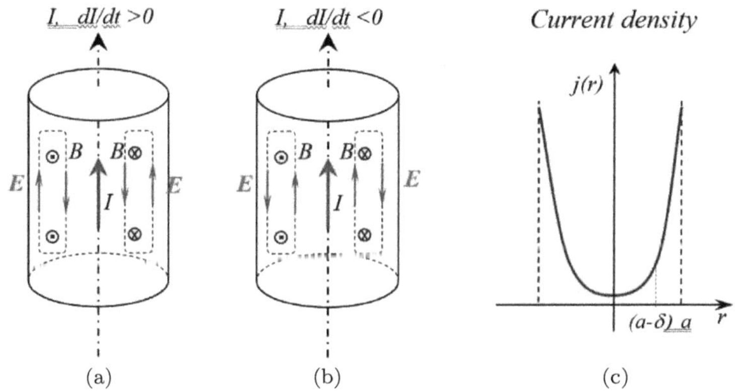

Figure 40.6 Skin effect in a cylindrical wire of radius a and length l, through which an alternating current I of frequency f circulates. In (a), the current is increasing. To counteract the decrease of the magnetic flux in the region shown by the dotted lines, an emf is induced, which generates a current that tends to decrease the current density in the central zone of the wire but favors an increase in density in the periphery. In (b), the current is decreasing, so the induced emf generates a current that tends to increase the density in the central area of the wire and favors a decrease in the periphery. The result of the calculations shows that an alternating current circulates preferentially through the periphery, in a region of thickness δ. Diagram (c) illustrates the profile of the current density.

(see Figures 40.5 and 40.6(a)). To counteract the *increase in the magnetic flux* in the dotted blue region, an emf is induced, which generates a current that tends to counteract *this increase in flux.* This induced current will tend to decrease the value of the current at the center of the cable and increase it at the periphery. Furthermore, this induced current depends on the time derivative of the changing flux. Therefore, these induced currents will increase as the frequency increases. However, this effect alone is not sufficient to understand the skin effect completely because in the case that the *current is decreasing*, the magnetic field flux inside the wire also decreases (see Figure 40.6(b)). To counteract the decrease in flux, an emf is induced in the region marked by the dotted blue line, which generates a current that tends to increase the current in the center and decrease it at the periphery [9]. This argument is not sufficient to explain this effect, but it contains two important features: (i) a time-varying current generates induced current that modifies its current distribution density and (ii) the induced current increases with frequency. To understand this effect, we need to solve Maxwell Eqs. (40.16) and (40.18), as discussed briefly in Annex A of this chapter. In this connection, it is illuminating the discussion of the skin effect in a ribbon-type conductor which is presented in Ref. [14] and analyzed in Project 132.

In the case of a cylindrical wire, when high-frequency alternating currents flow through it in axial direction, currents are generated inside the cable due to Lenz's law, favoring *circulation around the periphery.* This is called *skin effect.* The current thus tends to circulate preferably through the periphery of the cable, a region that is determined by the *penetration length*, δ, given by Eq. (40.14).

Using Maxwell's equations for quasi-static fields, Eqs. (40.16) and (40.18), it can be shown (see Annex A) that the resistance varies with the frequency for the limit of low frequencies as follows:

$$\begin{aligned} R(\omega) &\approx \left(\frac{l}{\sigma \cdot \pi a^2}\right) \cdot \left[1 + \frac{1}{3}\left(\frac{a}{2\delta}\right)^4\right] \\ &= R_o \left[1 + \frac{1}{3}\left(\frac{a}{2\delta}\right)^4\right] = R_o \left[1 + \frac{\varepsilon^2}{3}\right], \end{aligned} \tag{40.35}$$

with $\varepsilon = a/2\delta$ or also $\varepsilon^2 = (\pi/2)a^2\sigma\,\mu_0\mu_r f$ and R_0 is the *resistance under static* or DC conditions. A better approximation of $R(\omega)$ can be found by taking more terms of the full development [15]:

$$R(\omega) = R_o \times \begin{cases} \left[1 + \frac{\varepsilon^2}{3}\right] & \text{if } \varepsilon \leq 0.8 \\ [0.8813\varepsilon + 0.4315] & \text{if } 0.8 < \varepsilon < 1.2 \\ \left[\varepsilon + \frac{1}{4}\right] & \text{if } \varepsilon \geq 1.2 \end{cases} \tag{40.36}$$

Figure 40.6(c) shows the dependence of the resistance as a function of the frequency through the parameter ε. An extremely simplified or *naive* model consists of assuming that at relatively low frequencies, when $\delta > a/2$ (or $\varepsilon \leq 1$), $R(\omega) = R_0$ and at high frequencies, when $\delta < a/2$ (or $\varepsilon > 1$), the current flows only through a peripheral ring of thickness δ. That is,

$$\frac{R(\omega)}{R_0} \approx \frac{\rho l/\pi(a^2 - (a-\delta)^2)}{\rho l/\pi a^2} = \frac{a^2}{2a\delta - \delta^2} \approx \varepsilon + \frac{1}{4}, \quad \text{if } \varepsilon > 1, \tag{40.37}$$

in coincidence with the expression of Eq. (40.36), for $\varepsilon > 1$. In Figure 40.7(a), the dependence of this (naive) model is shown as a function of the frequency through the parameter ε. As can be seen, the naive model is almost as good as that of Eq. (40.36).

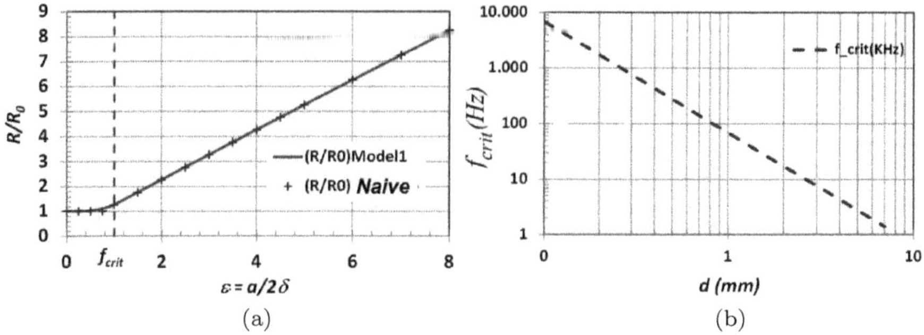

Figure 40.7 (a) Variation of the resistance of a cylindrical wire by skin effect as a function of the parameter $\varepsilon = a/2\delta$. (b) *Critical frequency* (f_{crit}) for which the skin effect begins to be noticeable, that is, $\varepsilon = 1$ or $a = 2\delta$, as a function of the wire diameter (d) in the case of copper.

Equations (40.36) and (40.37) show that $R(\omega) \approx R_0$ for frequencies lower than the critical f_{crit} (defined as those for which $\varepsilon\ (f_{\text{crit}}) = 1$). It is important to note that f_{crit} depends on the diameter and material of the wire. More specifically, we define f_{crit} as the frequency for which $\delta(f_{\text{crit}}) = a/2$ (equivalent to $\varepsilon(f_{\text{crit}}) = 1$). Figure 40.7(b) shows this dependence of f_{crit} as a function of the diameter of the wire, in the case of copper. It is clear that high frequencies are transmitted better by thin wires than by thick wires. This is the reason why cables in audio systems are usually made of many thin wires that are insulated from each other (*Litz wire*) [10]. A Litz wire enables mitigation of the skin effect at frequencies that are not too high.

On the other hand, a cylindrical wire also has an associated self-inductance L, which needs to be considered when working with high-frequency AC currents. This is because a wire carrying a current is associated with a magnetic field, which will have a certain energy $E_m = \int (B^2/2\mu_0)dv = 1/2L \cdot I^2$. For a wire of length l and radius a, the inductance will be [16]

$$L = \left(\frac{\mu_0}{2\pi}\right) \cdot l \cdot \left[\ln\left(\frac{l + \sqrt{l^2 + a^2}}{a}\right) - \sqrt{1 + \frac{a^2}{l^2}} + \frac{\mu_r}{4} + \frac{a}{l}\right], \tag{40.38}$$

where μ_r is the relative magnetic permeability of the wire. For the case where $l \gg a$ and $\mu_r \approx 1$, we can write

$$L \approx \left(\frac{\mu_0}{2\pi}\right) \cdot l \cdot \left[\ln\left(\frac{2l}{a}\right) - \frac{3}{4}\right]. \tag{40.39}$$

Although the skin effect affects the magnetic field inside the conductor, most of the wire's magnetic energy is outside the wire. Therefore, as a first approximation, it can be assumed that up to frequencies of a few tens of kHz, the self-inductance of the wire does not vary with the frequency.

40.4.1 *Project 130. Variation of the resistance of a wire with the frequency I*

Recommended equipment: A copper wire 1.7–3 mm in diameter. A 30 MHz or faster two-channel oscilloscope. A function generator and a lock-in amplifier. If you do not have a lock-in, you can use a digital multimeter (preferably 4 1/2 digits) that can measure voltages of the order of 0.5 mV with a bandwidth of about 30 kHz.

This project is intended to study the variation of the resistance r_{ab} of a conductive wire through which an AC current of variable frequency circulates [17].

In the first instance, V_R and I must be measured as a function of frequency in the range of interest. Considering that the function generator has an internal resistance of the order of 50 Ω, for a driving voltage of the order of 5 V, the maximum current is of the order of 50 mA. For such current, the voltage drop, V_{ab}, across a copper wire with a diameter of 1.7 mm and a length of the order of 35 cm (a resistance of approximately 3 mΩ), is therefore $V_{ab} \approx 0.15$ mV. To measure a small signal in the range of 100 Hz–30 kHz, a lock-in amplifier or a digital millivoltmeter sensitive to these voltages and with adequate bandwidth is required. Consider the circuit shown in Figure 40.8. To shield the cable under study from interference noise from the surroundings, you could use a metal cylinder, such as a copper or aluminum pipe of dimeter 3–5 cm, and two wooden or rubber plugs.

First, it is advisable to connect the lock-in amplifier as shown in Figure 40.8(a) to measure V_{ab} as a function of frequency. A second channel of the lock-in amplifier (or a second voltmeter) can be used to monitor the value of V_R.

Knowing the value of the limiting resistance R_0 (≈50 Ω), we can know the amplitude of the current $I_0(f)$. If $R_T = R_0 + r_{ab} + r_i \approx R_0 + r_i$, where r_i is the internal resistance of the function generator.

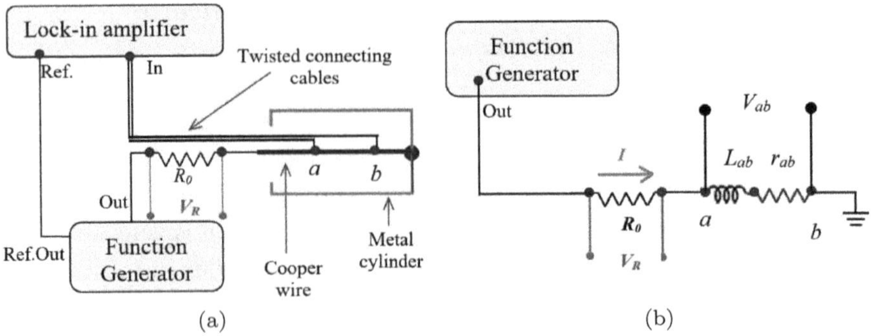

Figure 40.8 (a) Circuit to study the variation of the resistance of a conducting wire. Limiting resistance R_0 enables the current through the wire to be monitored by measuring the $V_R = R_0 I$. (b) Equivalent circuit. Voltage drop $V_{ab} = (r_{ab} + jL_{ab}\omega)I$ enables the resistance r_{ab} of the wire segment ab to be measured as a function of frequency.

The voltage V_R may be written as follows:

$$V_R = R_0 \frac{V_0}{(R_T + jL\omega)} \approx V_0 \frac{R_0}{R_T}, \tag{40.40}$$

where L is the cable inductance, and the phase shift of V_R with respect to V_0 is

$$\tan\varphi = \frac{L \cdot \omega}{R_T}. \tag{40.41}$$

Since a 1 mm copper wire has a self-inductance of the order of 1 mH/m, even at frequencies of the order of 50 kHz, we have $L\omega < 0.5\Omega$, and with $R_T \approx 100\Omega$, we get $\phi \approx 0$.

From Eq. (40.40), we see that by measuring, at low frequency, the voltages V_0 (without load) and V_R, we can find the values of R_T and r_i:

$$R_T \approx R_0 + r_i \approx R_0 \frac{V_0}{V_R} \quad \text{and} \quad r_i \approx R_0 \left(\frac{V_0}{V_R} - 1\right). \tag{40.42}$$

Next, if a lock-in is available, it can be connected as illustrated in Figure 40.8(a) to measure the voltage V_{ab} as a function of f.

The lock-in can also measure the phase difference ϕ_{ab} of the voltage V_{ab} with respect to the driving voltage V_0 or with respect to V_R, which, as we have just seen, are proportional to each other. Since segment ab has a resistance r_{ab} and a self-inductance L_{ab}, we have

$$V_{ab}^0(\omega) = (r_{ab} + jL_{ab}\omega) \cdot I_0(\omega) = (r_{ab} + jL_{ab}\omega).V_R^0(\omega)/R_0. \quad (40.43)$$

Here the super-index $^{(0)}$ indicates the amplitude of the signal. These amplitudes (or the *effective* values or rms $\frac{V_R^{\mathrm{rms}}=V_R^0}{\sqrt{2}}$) are proportional to each other. V_R is the voltage drop in R. Given that in general $r_{ab} < 1\Omega$, we cannot consider negligible the value of L_{ab},ω with respect to r_{ab}.

Therefore, we have

$$V_{ab}(\omega) = \sqrt{r_{ab}^2(\omega) + (L_{ab}\omega)^2} I_0(\omega) e^{j\phi_{ab}}, \quad (40.44)$$

and for the phase shift,

$$\phi_{ab} = \mathrm{artan}\left(\frac{L_{ab}\,\omega}{r_{ab}\,(\omega)}\right). \quad (40.45)$$

The amplitude is

$$V_{ab}^0(\omega) = |V_{ab}(\omega)| = \sqrt{r_{ab}^2(\omega) + (L_{ab}\omega)^2} I_0(\omega), \quad (40.46)$$

and at low frequencies, $f < 100\,\mathrm{Hz}$,

$$|V_{ab}(\omega)| \approx r_{ab}^0 I_0(\omega), \quad (40.47)$$

which enables us to know the value of $r_{ab}^0 \equiv r_{ab}(\omega = 0)$ at low frequencies. With this value and the expression (40.46), we can find $r_{ab}(\omega)$. Furthermore, these experimental results can be compared to the theoretical expectation of Eq. (40.36). Analogously, it is possible to compare the measurements of the phase as a function of the frequency to the theoretical model [17].

Experiment:

✓ Using the circuit described in Figure 40.8(a), study experimentally the variation of V_R^0 depending on the frequency applied.

This study enables you to find $I_0(f)$. To do so, connect the lock-in input directly to the ends of the resistor R, or if you are using a voltmeter, measure $V_R^{\mathrm{rms}} = V_R^0/\sqrt{2}$ as a function of frequency.

- ✓ Find r_i(the internal resistance of the FG) and $I_o(f)$ for the frequency range to be analyzed.
- ✓ Experimentally study the variation of V_{ab} (or $V_{ab}^0 or V_{ab}^{\mathrm{rms}}$) as a function of the frequency f.
- ✓ Using low frequency, $f < 100\,\mathrm{Hz}$, measure V_R^0 and V_{ab}^0 and find the value of r_{ab}^0. Using this value and Eq. (40.46), find the value $r_{ab}(f)$.
- ✓ Plot the measured values of V_{ab} (or V_{ab}^0 or V_{ab}^{rms}) as a function of the frequency f, Eq. (40.46). In the same plot, include the predictions of the model (40.36). Vary the value of L_{ab} using Eq. (40.46) until you find the best possible fit of the measured data.
- ✓ What can you conclude about the goodness of the proposed model? In other words, do Eqs. (40.36) and (40.46) explain the experimental results?
- ✓ Compare the value of L_{ab} obtained from the best fit to its value predicted by Eq. (40.39).
- ✓ Discuss the behavior of the model for the case ($\varepsilon \ll 1$) and high frequencies ($\varepsilon > 1$).

Optional

- ✓ Experimentally study the variation of the phase ϕ_{ab} measured as a function of f. In general, the lock-in measures the phase difference of the measured signal with respect to the voltage applied by the device's intrinsic function generator. In the same graph, compare the measured phase shift ϕ_{ab} with the values predicted by Eq. (40.45).
- ✓ What can you conclude about the goodness of the proposed model? In other words, do Eqs. (40.36), (40.45), and (40.46) explain the experimental data?

40.4.2 *Project 131. Skin effect on a wire — Expulsion of the magnetic flux*

Recommended equipment: Some 3–8 mm diameter copper, aluminum, or bronze bars. Two concentric coils of 300–1,000 turns each to form a kind of transformer with an outer primary and an inner secondary. A function generator and a lock-in amplifier. If you do not have a lock-in, you can use a digital multimeter with a frequency width of up to about 20 kHz or a good oscilloscope.

If a conducting *non-ferromagnetic* cylinder is placed in a region where there is a uniform magnetic field, B_0, which varies sinusoidally with time, with frequency f, the magnetic field is repelled from its interior and is concentrated on the surface of the cylinder [13, 17–20]. This is a consequence of the skin effect analyzed in the previous project. Consequently, if we have two concentric coils forming a transformer, the voltage measured in the secondary coil (inner coil) decreases if a conductive bar is placed inside it. The experimental setup is schematically illustrated in Figure 40.9.

Let's analyze what happens in the secondary if we introduce a *non-ferromagnetic* cylindrical conductor inside it. The induced voltage in this coil, V_2, will be reduced proportionally to the expulsion of the field and the flow will decrease. More specifically, suppose the radius of the metal sample bar is R and the inner radius of the secondary coil is R'. Assume that the magnetic field B_0, generated by the primary coil, oscillates with a frequency f with $f > f_{\text{crit}} = 4/(\pi\sigma\mu_0 R^2)$ or $\delta \ll R$. Also assume the same naive model,[d] described by Eq. (40.37), where the field only penetrates the metal cylinder a distance δ, so that the voltage in the secondary

[d]The naive model consists of assuming that the field penetrates the cylinder up to a distance equal to δ of its surface. In other words, inside the cylinder, the field on the surface down to a depth $R-\delta$ is the same as on the surface, B_0. Inside the radius $R-\delta$, the field is zero.

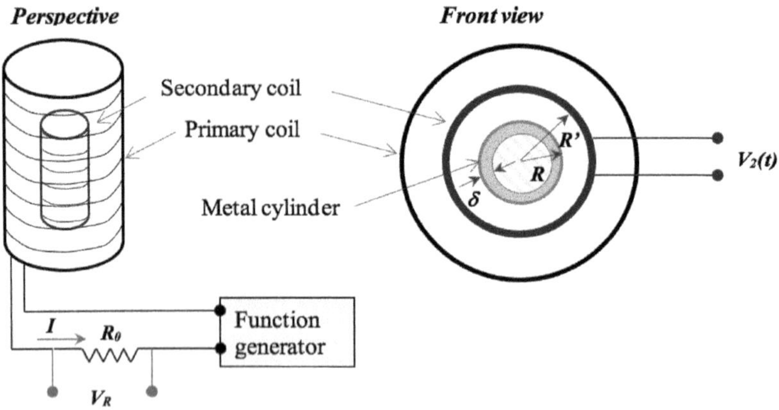

Figure 40.9 Experimental device to study the skin effect in a cylindrical metallic *non-ferromagnetic* sample. The outer (larger) coil is the primary. The inner (smaller) coil is the secondary and has a radius R'. The metal sample of radius R to be studied is placed inside the secondary coil, filling the entire length of the secondary coil.

is [21, 22]

$$V_2^0(WS) = \left|\frac{d\Phi_B}{dt}\right| \approx 2\pi N_2 f\{B_0\pi[R'^2 - (R-\delta)^2]\}$$
$$\approx 2\pi^2 N_2 f \cdot B_0\{2\cdot R\cdot\delta + (R'^2 - R^2)\}, \qquad (40.48)$$

where N_2 is the number of turns in the secondary. We call the voltage in the secondary with the metal sample (WS), $V_2^0(WS)$, from (40.48),

$$V_2^0(WS)/(N_2 B_0 f) \approx A + B/\sqrt{f}, \qquad (40.49)$$

with

$$A = 2\pi^2.(R'^2 - R^2) \quad \text{and} \quad B = 4\pi^2\cdot R/\sqrt{\pi\mu_0\sigma} \qquad (40.50)$$

When the metal sample is extracted from the interior of the secondary [we will call this case "without sample" (NS)], we have

$$V_2^0(NS)/(N_2 B_0 f) \approx A_0, \quad \text{with} \quad A_0 \approx 2\pi^2 R'^2, \qquad (40.51)$$

that is, in this case, $V_2^0(NS)/(N_2\cdot B_0)$ is proportional to f, assuming that B_0 remains relatively constant inside the secondary coil. To test this hypothesis, with constant B_0, voltage V_R can be measured in the primary as a function of the frequency f, for the system with and without a metal sample inside. V_R can be measured with a millivoltmeter with a bandwidth similar to the frequency range that you want to explore or you can use a good oscilloscope or a lock-in amplifier. If the ratio $V_R(WS)/V_R(NS) = 1$ for the measured frequencies, this means that the field $B_0(f)$ is the same in both cases.

Thus, the idea is to first measure $V_2^0(NS)$ as a function of f and then $V_2^0(WS)$ for the same frequencies. Thus, the ratio $\text{Ratio}(f) = V_2^0(WS)/V_2^0(NS)$ will be

$$\text{Ratio}(f) \equiv \frac{V_2^0(WS)}{V_2^0(NS)} = \frac{A}{A_0} + \frac{B/A_0}{\sqrt{f}} \approx \alpha_0 + \beta_0/\sqrt{f} \qquad (40.52)$$

[remember that $\delta(f) \propto 1/\sqrt{f}$] with

$$\alpha_0 = \frac{R'^2 - R^2}{R'^2} = 1 - \left(\frac{R}{R'}\right)^2 \quad \text{and} \quad \beta_0 = \frac{2R}{R'\sqrt{\pi\mu_0\sigma}} = \frac{d}{R'\sqrt{\pi\mu_0\sigma}}. \qquad (40.53)$$

Therefore, if the present model is valid, we expect that by plotting the ratio $\text{Ratio}(f)$ as a function of $1/\sqrt{f}$, we will obtain a line in which the slope is β_0 and the y-intercept is α_0. Note that β_0 depends on the diameter of the metal cylinder d, on the secondary coil radius R' and on the conductivity σ of the metal sample under study. This method thus provides a procedure for measuring the conductivity (σ) of a cylindrical sample without the need to make physical contact with it. On the other hand, it is important to bear in mind that the proposed (naive) model only applies to high frequencies, that is, when $f > f_{\text{crit}} = 4(\pi\sigma\mu R^2)$ with $\delta(f_{\text{crit}}) < R/2$. For frequencies $f < f_{\text{crit}}$, the skin effect is expected to be increasingly less effective and therefore $\text{Ratio}(f) \to 1$, for $f \to 0$.

The effective radius of the secondary coil (R') can be found by averaging the coil's internal and external diameters. If there is ambiguity in the determination of R' due to irregularities in the geometry of the secondary, a cylindrical metal sample of known conductivity can be used. Determine the slope β_0 for this sample and using Eq. (40.53), find the value of R'. Check that this value is consistent with the dimensions of the secondary coil.

It is important to consider that when finding the quotient $V_2^0(WS)/V_2^0(NS)$ in Eq. (40.52), and simplifying B_0 (the magnetic field) in both numerator and denominator, we are implicitly assuming that the field generated by the primary coil does not change when the metal sample is introduced. This assumption is not trivial because, as we can see, the metal bar modifies the self-inductance of the system under study. The following test can be performed to determine whether the field B_0 remains constant: the input of the lock-in is connected to $V_R = R_0 I_0$, or this signal is connected to a channel of the oscilloscope, and the variation of V_R with f is measured with and without the metal sample. If V_R (f) does not vary with and without the metal cylinder, this is indicative that for each frequency f, the value of $B_0(f)$ does not change much due to the presence of the metal sample.

Finally, there is a nice experiment that can be used to study the skin effect inside a solenoid. The objective of this test is to study the magnetic field distribution inside the solenoid at different frequencies when (i) it is empty (filled with air) and (ii) its interior is filled with a non-ferromagnetic metal, e.g., mercury. A small secondary pick-up coil is used to study the magnetic field distribution inside the solenoid at different frequencies and radial positions. This is done by measuring the *emf* induced in it when its interior is empty and when it is filled with Hg [22]. This interesting experiment is very illustrative of the skin effect, but it must be done with *extreme care and supervision,* since Hg is a highly toxic substance, and therefore not recommended for any student laboratory. The result of this experiment is well described in the literature [22].

Experiment:

✓ Using the circuit described in Figure 40.8, study experimentally the variation of $V_R(NS)$ (without sample) and $V_R(WS)$ (with sample) as a function of the applied frequency f. Represent both results on the same graph and $\text{Ratio}(f) = V_R(WS)/V_R(NS)$ as a function of f. Discuss whether the field $B_0(f)$ changes due to the presence of the metal sample.

✓ Using the circuit shown in Figure 40.8, experimentally study the variation of emf voltage in the secondary, $V_2^0(NS)$ as a function of the working frequency f, i.e., the system without the metal sample inside it.

✓ Perform the same study of the variation of $V_2^0(WS)$ as a function of the frequency f with the metal sample of radius R inside the secondary coil.

✓ Construct individual graphs of $V_2^0(NS)/f$ and $V_2^0(WS)/f$ as a function of $1/\sqrt{f}$ for the two cases studied. Check whether the predictions of Eqs. (40.49) and (40.51) are fulfilled.

✓ If $B_0(f)$ does not change significantly due to the presence of the metal sample, plot Ratio(f) as a function of $1/\sqrt{f}$ for the case studied. Check whether the prediction of Eq. (40.52) is fulfilled. If this relationship is linear, determine the slope β_0 and y-intercept α_0, and its reactive uncertainties.

✓ From the slope β_0 and the diameter of the cylinder d, find the value of σ and its uncertainty. Compare this value to those found by another method or with the table values for the material of which the cylinder is made.

✓ Discuss how adequate the proposed model is to explain your data.

40.4.3 *Project 132. Lateral skin effect on a ribbon-like sample*

Recommended equipment: An aluminum or copper foil ribbon, about 30 cm long, 6–10 cm wide, and 0.3–0.8 mm thick. A 30 MHz (or faster) two-channel oscilloscope. A function generator and a lock-in amplifier. If you do not have a lock-in, you can use a digital multimeter (preferably 4 ½ digits) that can measure voltages of the order of 0.5 mV with a bandwidth of about 30 kHz.

The objective of this experiment is to study the *skin effect* on a thin, wide metal ribbon that conducts an AC current of variable frequency f. We present a simple demonstration of the skin effect by observing the current distribution in a ribbon conductor with a rectangular cross-section that is much wider than its thickness. The skin effect tends to concentrate the current along the two edges of the ribbon and is often referred to as the *lateral skin effect* [14].

Here, we propose to study the current density $J(x)$ across a thin strip via the AC magnetic field that this current produces close to the surface of the conductor. The indicator of the current density $J(x)$ is the *emf* induced on a thin pickup coil that is used to scan the ribbon across its edges, for different applied frequencies. The setup is illustrated schematically in Figure 40.10. A limiting resistor R (of 30–50 Ω) is also used to monitor the current through the ribbon. The pick-up coil can be easily made using two or three plastic cards (such as credit cards) glued together and winding a thin enamel wire of about 0.1 mm–0.15 mm (AWG 33–38) with 10–25 turns, see Figure 40.10. Therefore, we assume that the pick-up coil has n turns and an effective area A.

For a flat wire (ribbon), the relation of the magnetic field and the current density on the surface of the ribbon is [2, 5]

$$B(x) \approx \frac{\mu_0}{2} J(x). \tag{40.54}$$

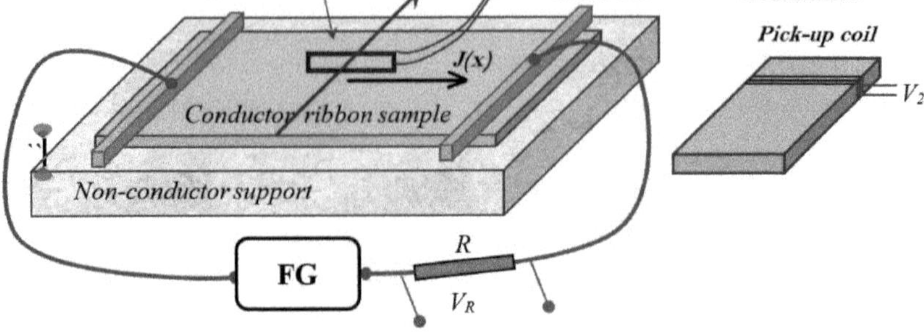

Figure 40.10 Schematic experimental setup to study the *lateral skin effect* on a ribbon-like sample. A thin, narrow pick-up coil, used as a secondary, is used to monitor the variation of the magnetic field along the ribbon, which is used as indicator of the current density $J(x)$ along the conductor. The FG generates an AC current of known frequency f. The figure on the right shows a detail of a possible configuration of the pick-up coil.

This result can be obtained using Ampere's law on an infinitesimal piece of the ribbon. If the current varies with a frequency $f = \omega/2\pi$, we expect the induced emf on the pickup coil to be

$$\varepsilon(x,t) \approx \omega.n.A.B(x,t) \approx \frac{\mu_0}{2}\omega.n.A.J(x,t). \tag{40.55}$$

In other words, the amplitude of the emf induced on the pick-coil will be

$$\varepsilon^0(x) \approx \omega.\beta.J^0(x). \tag{40.56}$$

Thus, measuring the emf induced on the pick-coil provides a signal that can be used as a proxy of the current density $J^0(x)$ on the ribbon, enabling us to study the variation of the current density with the frequency. Figure 40.11 shows schematically the theoretical expectation of the effect.

Experiment:

✓ Using the circuit like the one shown in Figure 40.8, study experimentally the variation of the emf $V_2^0(x)$ induced on the pick-coil, as a function of x, from one edge to the other of the ribbon. It may be useful to employ a voltmeter connected to the limiting

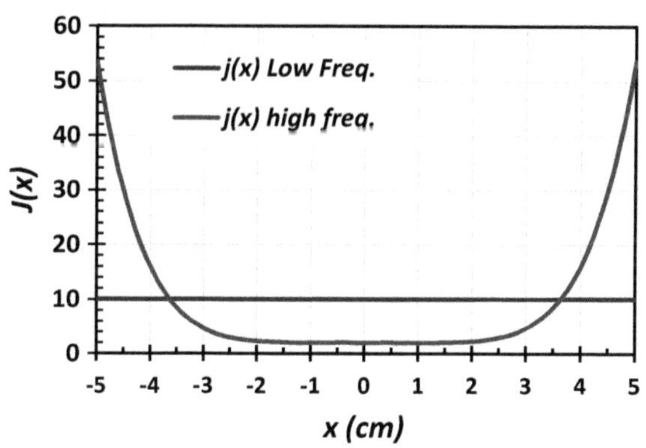

Figure 40.11 Schematic plot of the *lateral skin effect* we may expect to find [14]. For low frequencies, we expect a flat o uniform current density (blue line). As the frequency increases, the skin effect favors an increase in current density as we move closer to the edge of the ribbon (red curve).

resistor R,V_R, to monitor the possible variation of primary current $I(f)$ with the frequency.

✓ First, use a relatively low frequency, such as 70–200 Hz, and measure the amplitude of the emf $V_2^0(x)$ from one edge to the other.

✓ Next, repeat the same measurement for other frequencies, such as $f = 1\,\text{kHz}$, 2 kHz, 3 kHz, 4 kHz, 6 kHz, and 8 kHz. In all cases, try to maintain V_R constant or at least record its value for each frequency.

✓ Plot the measured values of $V_2^0(x)$ as a function of x for each frequency on the same graph.

✓ Using the values of V_R recorded for each frequency, find $J(x)\left(\propto\left(\frac{V_2^0(x)}{\omega\cdot V_R}\right)\right)$.

✓ Plot the values of $J(x)$ as a function of x for each frequency on the same graph.

✓ Discuss whether the *lateral skin effect* is present in the system studied.

40.5 Annex A: Theory of skin or skin effect

Consider a homogeneous cylindrical conductor of radius a and length $l \gg a$, through which a current flows:

$$i(t) = I_0 e^{j\omega t}, \tag{40.57}$$

where $j = \sqrt{-1}$. Assume that the cylinder is oriented in the z direction, as shown in Figure 40.12.

In this way, by symmetry, the current density will have a single component, z, that is, $\vec{J} = J(r)\hat{k}$, where we assume that the z component of the current J can only have a dependence on r, due to the cylindrical symmetry of the problem. According to (40.18), using cylindrical coordinates, $J(r)$ satisfies the following equation:

$$\nabla^2 J(r) + j\frac{2}{\delta^2}J(r) = \frac{1}{r}\frac{d}{dr}(rJ(r)) + j\frac{2}{\delta^2}J(r) = 0 \tag{40.58}$$

or

$$\left(\frac{d^2}{dr^2} + \frac{1}{r}\frac{d}{dr} + k^2\right)J(r) = 0 \quad \text{with } k^2 = j\frac{2}{\delta^2} = j\,m^2, \tag{40.59}$$

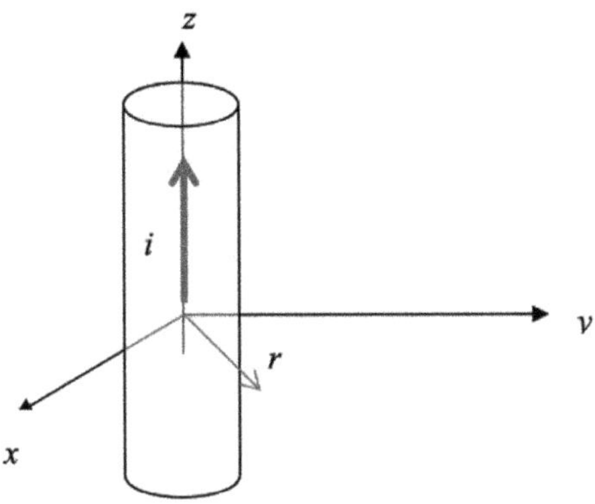

Figure 40.12 Cylindrical wire of radius a and length l, through which an alternating current $i(f)$ of frequency f circulates in axial direction. The direction of the z axis coincides with the axis of the wire.

where

$$m^2 = \mu_0 . \sigma \omega \quad \text{and} \quad k = j^{1/2} m = (1+j)m/2. \tag{40.60}$$

The solution $J(r)$ must satisfy the following condition:

$$I_0 = 2\pi \int_{r=0}^{r=a} J(r) r dr, \tag{40.61}$$

where I_0 is the total current flowing through the cable. Equation (40.59) is a Bessel differential equation [24] whose solution, with the condition of being regular at the origin ($r = 0$), is

$$J(r) = A\, J_0(kr). \tag{40.62}$$

By applying the condition (40.61), we can determine A. The solution is [18, 25]

$$I_0 = \left| \frac{2\pi a J_1(j^{3/2} ma)}{j^{3/2} m} \right| \cdot A \tag{40.63}$$

so that

$$J(r) = \frac{I_0}{2\pi a \cdot |J_1(j^{3/2} ma)|} \cdot J_0(j^{1/2} mr). \tag{40.64}$$

In the extreme case of low frequencies, that is, when $a/2\delta \ll 1$, it can be proved that

$$J(r) \propto \text{Constant}\cdot \left[1 + \left(\frac{r}{2\delta}\right)^4 + \cdots\right], \tag{40.65}$$

and at the limit of high frequencies, we have ($a/\delta \gg 1$)

$$J(r) = J_0 \cdot \left[\text{Exp}\left(\frac{r-a}{\delta}\right)\right]. \tag{40.66}$$

That is, in both cases, the current increases as the surface is approached (skin effect). The penetration distance is given precisely by the parameter δ. From these relationships, it is possible to obtain the resistance of the wire as a function of frequency. For the low frequency limit $a/2\delta \ll 1$, the result is

$$R(\omega) = \left(\frac{L}{\sigma \cdot \pi a^2}\right) \cdot \left[1 + \frac{\omega^2\tau^2}{192}\right] = R_o\left[1 + \frac{1}{3}\varepsilon^4\right], \tag{40.67}$$

where $\varepsilon = a/2\delta$, in accordance with Eq. (40.36).

On the other hand, at the high-frequency limit ($a \gg \delta$), from (40.66), we have

$$I_0 = \int_0^a J(r) 2\pi r dr \approx 2\pi J_0 \cdot [\delta(a-\delta)] \approx J_0 \cdot \pi\left[a^2 - (a-\delta)^2\right], \tag{40.68}$$

which justifies the *naive model*, which consists of assuming that at high frequencies, the current in cylindrical cables circulates through a peripheral ring of thickness δ.

In the case of a metal cylinder with its axis oriented in the field direction $\vec{B} = \boldsymbol{B}(\boldsymbol{r})\widehat{\boldsymbol{k}}$, according to Eq. (40.13), the differential equation of $B(r)$ is of the form (40.59). Therefore, the solution of $B(r)$ is similar to the solution of $J(r)$, Eq. (40.64) [18, 21, 22]. This justifies the introduction of the naive model for this case.

40.6 Annex B: Bessel functions

The differential equation

$$z^2 \frac{d^2y}{dz^2} + z\frac{dy}{dz} + (z^2 - \upsilon^2)y = 0 \tag{40.69}$$

with real z and υ, is known as the Bessel differential equation [18, 24, 25]. Its solution defines the Bessel function of the first kind of order υ and argument z.

References

[1] R. Feynman, *et al.*, *The Feynman Lectures on Physics, Boxed Set: The New Millennium Edition*, vol. 1, Basic Books, 2011.

[2] E. M. Purcell and D. J. Morin, *Electricity and Magnetism, Berkeley Physics Course*, 3rd Ed., NY: Cambridge University Press, 2014, p. 839.

[3] L. Landau and E. Lifshitz, *Electrodynamics of Continuous Media* (vol. 8), 2nd Ed., Oxford: Pergamon Press, 1985.

[4] W. Smythe, *Static and Dynamic Electricity*, NY: McGraw-Hill, 1950.

[5] J. Jackson, *Classical Electrodynamics*, New York: Wiley and Sons, 1991.

[6] Wikipedia, "Induction cooking," 2022. Available: https://en.wikipedia.org/wiki/Induction_cooking.

[7] G. Wouch and . A. Lord Jr., "Eddy Currents: Levitation, metal detectors, and induction heating," *Am. J. Phys.*, vol. 46, p. 464, 1978.

[8] W. Saslow, "Maxwell's theory of eddy currents in thin conducting sheets, and application to electromagnetic shielding and MAGLEV," *Am. J. Phys.*, vol. 60, p. 693, 1992.

[9] Wikipedia, "Skin effect," 2021. Available: https://en.wikipedia.org/wiki/Skin_effect.

[10] Wikipedia, "Litz wire," 2021. Available: https://en.wikipedia.org/wiki/Litz_wire.

[11] S. Fahy, C. Kittel and S. Louie, "Electromagnetic screening by metals," *Am. J. Phys.*, vol. 56, p. 989, 1988.

[12] J. Gray, *American Institute of Physics Handbook*, 2nd Ed., NY: Ed. McGraw Hill, 1975.

[13] J. Scofield, "Frequency-domain description of a lock-in amplifier," *Am. J. Phys.*, vol. 62, p. 129, 1994.

[14] A. Blackwell, A. Rotunno and S. Aubin, "Demonstration of the lateral AC skin effect using a pickup coil," *Am. J. Phys.*, vol. 88, p. 676, 2020.

[15] N. Koshkin and M. Shirkevich, *Physics Handbook*, Moscu: MIR, 1975.

[16] E. Rosa, "The Self and mutual inductances of linear conductors," *Bull. Bur. Stand.*, vol. 4, no. 2, pp. 301–344, 1908. Available: https://nvlpubs.nist.gov/.

[17] J. Gosselin, P. Rochon and N. Gauthier, "Study of eddy current in cylindrical wire: An undergraduate laboratory experiment," *Am. J. Phys.*, vol. 50, p. 440, 1982.

[18] L. A. Pipes, *Higher Mathematics for Engineers and Scientists*, NY: Mc Graw-Hill, 1970.

[19] Y. Kraftmakher, "Eddy currents: Contactless measurement of electrical resistivity," *Am. J. Phys.*, vol. 68, p. 375, 2000.

[20] L. Juri, V. Bekeris and R. Steinmann, "Skin depth and complex magnetic susceptibility: An experimental alternative approach," *Am. J. Phys.*, vol. 54, p. 836, 1986.

[21] H. Wiederick and N. Gauthier, "Frequency dependence of the skin depth in a metal cylinder," *Am. J. Phys.*, vol. 51, p. 175, 1983.

[22] J. Wagner and M. Syed, "Investigation of skin effect in mercury using a simple solenoid setup," *Am. J. Phys.*, vol. 79, p. 850, 2011.

[23] J. Wagner and M. Syed, "Investigation of skin effect in mercury using a simple solenoid setup," *Am. J. Phys.*, vol. 79, no. 8, pp. 850– 855, 2011.

[24] Wikipedia, "Bessel functions," 2022. Available: https://en.wikipedia.org/wiki/Bessel_function.

[25] G. Arfken, *Mathematical Methods for Physicist*, NY: Academic Press, 1995.

[26] Wikipedia, "Dielectric spectroscopy," 2021. Available: https://en.wikipedia.org/wiki/Dielectric_spectroscopy.

Chapter 41

STANDING WAVES IN ONE DIMENSION

This chapter explores some experiments to study standing waves in one dimension. First, we consider the case of a string and analyze the normal modes of vibration and the characteristic frequencies of the system. These studies explore the relation between the speed of the waves and the tension and density of the string. Then, we suggest some experimental studies of standing waves in open, semi-closed, and closed tubes. We also discuss the mode of determining the resonance frequencies or characteristic frequencies of these systems. Based on these studies, we explore several techniques to determine the speed of sound.

Goals

- ✓ Standing waves in a string
- ✓ Speed of a wave on a string
- ✓ Sound waves in a tube
- ✓ Kundt tube
- ✓ Speed of sound

41.1 Standing waves on a string

When a taut string with fixed ends vibrates, it assumes well-defined oscillation patterns, each with a characteristic frequency. These types of vibration modes are called *standing waves* [1]. For each frequency of vibration, there is a characteristic *pattern* or *normal mode* and vice versa. For a vibrating string, these modes can be characterized either by the number n of nodes between the fixed ends (the number of nodes excluding the ends) or the vibration frequency f_n.

The wavelength λ_n in each mode is determined by the length of the string L, which is the distance between the fixed ends. The modes of vibration are characterized by the fact that the length L must always contain an integer number of half wavelengths, that is,

$$L = n \cdot \frac{\lambda_n}{2}. \tag{41.1}$$

Since the product of the wavelength multiplied by the frequency is equal to the speed of propagation of the wave c, we have the following relationship [2–4]:

$$c = \lambda_n \cdot f_n = \frac{2 \cdot L}{n} \cdot f_n. \tag{41.2}$$

Since the speed of propagation c of the wave depends on the physical characteristic of the string, this expression enables us to find the normal frequencies f_n of oscillation of the system. Alternatively, it is possible to determine the speed of propagation of the wave c on the string by measuring the wavelength or, more simply, by knowing its length L and the number of nodes n of each mode. Measuring the corresponding vibration frequency f_n, Eq. (41.2) enables us to find c.

Moreover, by dynamical analysis of the movement, i.e., applying Newton's laws to the string, it can be shown that the speed of propagation of the wave of a taut string is given by [2, 3]

$$c = \sqrt{\frac{T}{\mu}}, \tag{41.3}$$

where T is the tension of the string and μ is its linear mass density.

The goal of this activity is to determine the speed of propagation of waves in a taut string and to test the relationship experimentally (41.3).

41.1.1 *Project 133. Standing waves in strings*

Recommended basic equipment: A cotton string approximately 2 m long and 5 mm in diameter. A mechanical actuator capable of producing pulses at variable frequencies. A function generator. A set of masses between 50 g and 500 g.

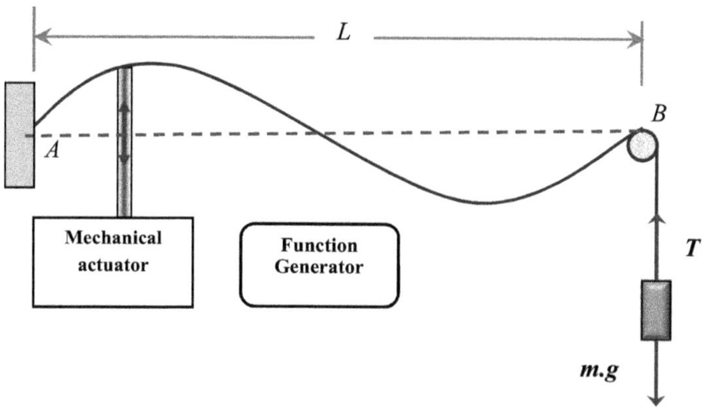

Figure 41.1 Schematic diagram of the device for studying standing waves in a string. The string has two fixed points, A and B. The tension T is determined by the weight mg hung from one of its ends. A mechanical actuator excites the string at a frequency defined by the function generator.

This study uses a string fixed at both ends as shown schematically in Figure 41.1. One end of the string is tied to a fixed point (point A), while the other end passes over a horizontal cylindrical bar or pulley of approximately 1 or 2 cm in diameter (point B). A mass m is hung from the free end of the string. Point B is at the same height as point A. The string has a mass per unit length that we call μ. This value is determined by weighing the string and measuring its length. The string tension, T, is determined by the weight mg of the mass that hangs from one of its ends. A function generator (FG) drives a mechanical actuator [5] at the frequency f determined by FG. In turn, this actuator excites the string at the same frequency f. More details on how to build one of these actuators are provided in Annex A. When the applied frequency f is close to the resonance frequency of the string, f_n, the characteristic pattern of the mode becomes clearly visible. Adjust the applied frequency f so as to maximize the amplitude of each mode — this is the resonance frequency f_n, that we want to find.

Suggested tasks:

✓ Vary the excitation frequency with the FG and observe well-defined *resonant frequencies*, f_n. This resonance situation is characterized by a well-defined pattern of nodes and antinodes (bellies). Under resonance conditions, a stable vibration pattern is generated in the string. Note that by finely varying the excitation frequency, the amplitude of the antinodes passes through a maximum at the resonant frequency f_n, which has a well-defined number n of nodes.

✓ For a given value of T and μ, determine the frequencies f_n for at least five normal modes. Plot f_n as a function of n. From Eq. (41.2), determine c.

✓ For the same value of T and μ, determine λ_n $(=2L/n)$ and plot λ_n as a function of $(1/f_n)$. According to Eq. (41.2), it is expected that the relationship between these variables will be linear and that their slope will be the speed c of the wave. Check whether your data agree with this expectation. If so, determine the best value of c and its uncertainty.

✓ For a given string, perform the same analysis as above, and determine the velocity of the wave c for different values of the tension T. To do so, vary the mass m hanging at one of its ends. Does c vary with T? Explain your results.

✓ Plot c as a function of $\sqrt{T/\mu}$. What can you conclude about the validity of Eq. (41.3)?

41.2 Standing waves in tubes (Kundt tube)

Sound waves are an important example of pressure waves in a compressible fluid. Pressure waves with frequencies in the range of about 20 Hz–20 kHz can excite our hearing and give rise to *sound sensation.* Pressure waves at higher frequencies are called *ultrasonic*, and lower frequency pressure waves are called *subsonic* [2, 3].

A tube with a length much greater than the diameter can host *standing sound waves* in one dimension. A tube like this is the acoustic analog of a taut string. In a tube with open ends, the pressure waves have a node at the ends [2, 4, 6].

When a traveling wave in a tube meets a boundary where the properties of the medium (compressibility or density) change or the geometric restrictions imposed on the wave vary, the waves are partially transmitted and partially reflected, according to the

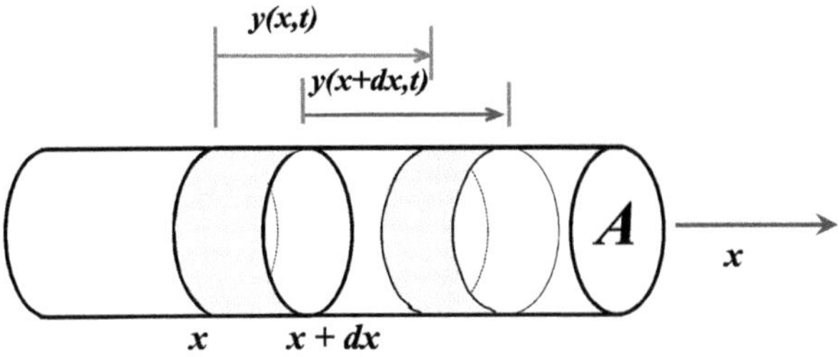

Figure 41.2 Cylindrical volume of a gas through which a sound wave passes.

characteristics of the interface. The boundary can be the interface between two media, such as air and a solid or a liquid. A boundary can also be a change in the geometry of the tube such as a change in tube diameter or an open or closed end of the tube [4, 7].

Consider a wave that propagates along a uniform tube in the positive direction of the x-axis, coinciding with the axis of the tube, as illustrated in Figure 41.2.

Let $y(x,t)$ be the function that describes the *displacement* (along the x axis) of the particles that are originally at the point of coordinates x at the instant t. Note that the particles move in the same direction as the propagation of the wave, along the x axis. Analogously, $p(x,t)$ denotes the function that describes the pressure variation at the point of coordinates x at time t. Thus, acoustic waves can be described by means of displacement waves, $y(x,t)$, or by pressure waves, $p(x,t)$. Of course, these two descriptions are related and complementary. This relationship is discussed in Annex B. This analysis shows that the displacement and pressure sinusoidal waves have a phase shift of $\pi/2$ with each other.

A sound wave that propagates through a tube has a pressure node at the *open end*, since the air pressure at this point is the atmospheric or equilibrium pressure [4] P_0. Therefore, the reflection of an acoustic pressure wave at the open end of a cylindrical tube is similar to the reflections that occur at the fixed end of a string. Similarly, the *closed end* of a tube has reflection properties equivalent to the wave in a string with a free end, as shown in Figure 41.3 (see Annex B).

However, when pressure waves are generated in an open-ended tube, the pressure in the tube does not equal atmospheric pressure

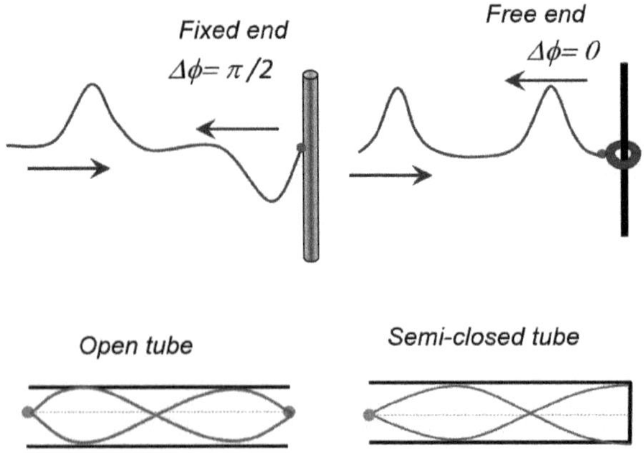

Figure 41.3 When a pulse propagating down a string is reflected at a fixed end (node), the pulse changes phase. At a free end, the reflected pulse has the same phase as the incident pulse. Pressure waves at the open end of a tube behave in the same way as at the fixed end of a string (node), while the closed end of the tube is analogous to the free end of the string (anti-node).

immediately at the end of the tube. The pressure waves emerge from the tube, so these ends are radiation zones through which we can hear the sounds that travel through the tube. As soon as a region of compression/expansion reaches the open end, the air can expand outwards, while inside the tube, the air is confined in a single direction. Consequently, compression/expansion is rapidly "damped" as the distance from the end of the tube increases. This distance is of the order of the radius of the tube. As the distance from the end of the tube increases, the pressure tends to atmospheric pressure [4, 7]. Thus, the effective length L_{ef} of a tube with an open end is greater than the nominal length L_0 of the tube. Empirically, the *effective length* of a tube with an open end is found to be

$$L_{\text{ef}} = L_0 + f \cdot d, \tag{41.4}$$

where d is the diameter of the tube and the coefficient $f \approx 0.6 \pm 0.1$ [4].

The boundary condition for an open end is [3, 4, 7]

$$(p)_{\text{open}} = 0 \tag{41.5}$$

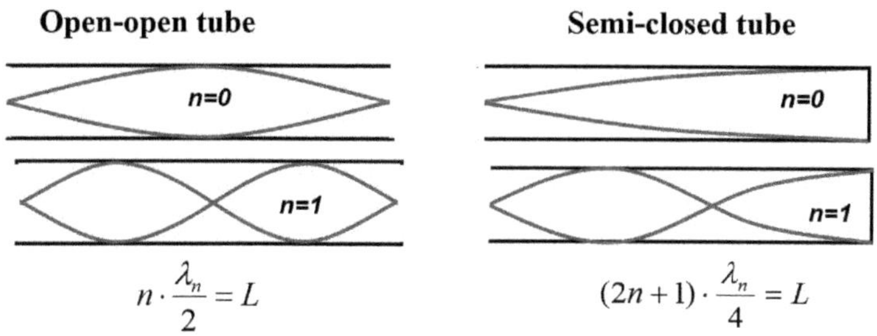

Figure 41.4 First oscillation modes of the pressure wave for an open–open tube (left) and for open–closed tube (right).

and the edge condition for a closed end is

$$\left(\frac{\partial p}{\partial x}\right)_{\text{closed}} = 0, \qquad (41.6)$$

which means that there is a pressure peak at the closed ends of a tube. The pressure to which we are referring is the *gauge pressure* (i.e., the pressure variation with respect to atmospheric pressure), not the *absolute pressure.*

The first oscillation modes of a standing pressure wave in closed and semi-closed tubes are illustrated in Figure 41.4.

From the boundary conditions at the ends, it is easy to see that for a closed–closed tube (pressure peaks at both ends), or open–open tube (pressure nodes at both ends), the resonance frequencies are given by [2, 4]

$$f_n = \left(\frac{c}{2 \cdot L_{ef}}\right) \cdot n. \qquad (41.7)$$

For semi-closed tubes [3, 4],

$$f_n = \left(\frac{c}{4 \cdot L_{ef}}\right) \cdot (2n+1). \qquad (41.8)$$

Applying Newton's laws to a column of gas, and assuming that the compressions are fast, so that the process may be assumed to be adiabatic, the speed of sound c in the gas is given by [2, 4, 7]

$$c = \sqrt{\frac{\gamma \cdot R \cdot T}{M}}, \qquad (41.9)$$

where $\gamma = c_p/c_v$ (ratio of heat capacity at constant pressure and that corresponding to constant volume) is called adiabatic compressibility coefficient, T is the absolute temperature, R is the universal gas constant, and M is the molecular mass of the gas (see Annex B).

41.2.1 *Project 134. Standing waves in a semi-closed tube — Kundt tube*

Recommended equipment: A tube approximately 0.6 m in length and 2–3 cm in diameter, made of acrylic, PVC, or glass or a glass measuring cylinder. A small speaker or earphone and a microphone. A function generator and a two-channel oscilloscope, or a data acquisition system connected to a PC. You can also use a PC with a sound card or a smartphone, as indicated in Chapter 11.

This experiment requires an acoustic emitter (speaker or earphone excited by a function generator or a smartphone with a sound generator [8]) that can emit *pure* sounds, i.e., sounds of well-defined frequencies. The frequency must be variable within the range of audio frequencies (20 Hz–10 kHz). The experiment also requires sound detectors (microphones) connected to an oscilloscope (or data acquisition system connected to a PC or to a PC sound card or a smartphone with audio recording).

For this experiment, you can use a glass or plastic cylindrical tube 20–40 cm in length with an internal diameter of less than about 3 cm. An acrylic or PVC tube approximately 0.6 m long and 1–3 cm in diameter can also be used. Carefully measure the dimensions of the tube: length l and internal diameter d. To determine the tube resonance frequencies unambiguously, place the sound emitter and receiver facing each other, exactly at the open end of the tube, as shown in Figure 41.5.

Perform a frequency sweep and try to locate the resonance frequencies by controlling the frequency of the function generator that drives the emitter. Make sure that the amplitude of the generator output voltage is constant. Resonances appear as a strong increase in the amplitude of the receiver's output signal. We can

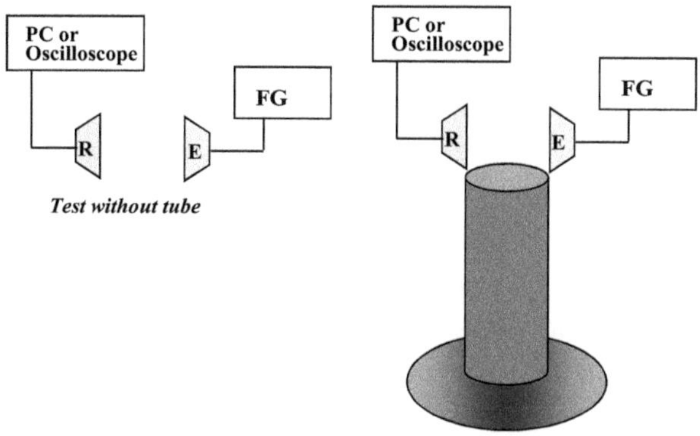

Figure 41.5 Experimental setup to study the modes of resonance in a tube.

generally hear resonance clearly. An increase in the intensity can be heard as well as seen on the instrument that registers the sound amplitude, oscilloscope, or similar sound detection device. In other words, at resonant frequencies, for a given amplitude of the emitter drive, the response of the receiver has a relative maximum. If you are using a two-channel oscilloscope and connect the receiver to one channel and the emitter input to the other channel, when you operate the oscilloscope in the X–Y mode, the phase relationship between these signals at resonance shows a characteristic shape (see Chapter 37).

As discussed above, Eq. (41.4), due to the finite diameter of the tubes, their *effective length* L_{ef} is greater than their *geometric length* L_0.

Suggested tasks:

✓ Determine at least the first five resonances in the tube you use. Plot the amplitude of the receiver as a function of the applied frequency. For this study, try to keep the geometry of the system (relative positions of the tube, emitter, and receiver) unchanged as the frequency varies (Figure 41.5).

✓ To verify that the resonances found are actually associated with the tube and not to particular characteristics of the emitter–receiver system, remove the tube and repeat the previous study, taking care to pass through the same frequencies at which the resonances were detected. Make sure that the geometry of the emitter–receiver pair is the same as when the tube was present.

✓ Plot the resonance frequencies of the tube as a function of order n of each resonance, i.e., the index that identifies its appearance as the frequency increases. Assign the order $n = 0$ to the fundamental frequency (the lowest resonant frequency).

✓ Plot the parameter $z_n = 4L_{\mathrm{ef}} \cdot f_n$ as a function of $(2n + 1)$ and verify whether the expectation of Eq. (41.8) is fulfilled. If so, determine the speed of sound c and its uncertainty.

41.2.2 *Project 135. Effect of the variation of tube length*

> **Recommended equipment:** A tube approximately 0.6 m in length and 2–3 cm in diameter, made of acrylic, PVC, or glass or a glass measuring cylinder. A small speaker or earphone and a microphone. A function generator and a two-channel oscilloscope or a data acquisition system connected to a PC. You can also use a PC with a sound card or a smartphone, as indicated in Chapter 11.

If you use a graduated cylinder, you can vary the length of the column of air above the water by adding water. If you are using an open tube, you can vary the length by inserting a rigid wooden or rubber sliding plug, to act as a piston. Repeat the study performed in the previous project for at least two water heights in the measuring cylinder or two tube lengths.

Suggested tasks:

✓ For each tube length (or water height in a graduated cylinder), determine the resonance frequencies f_n as a function of the resonance order n.

✓ Plot the parameter $z_n = 4 \cdot L_{\mathrm{ef}} \cdot f_n$ as a function of $(2n + 1)$ for each length L_{ef} used. Determine the speed of sound c and its uncertainty.

✓ How do your results compare to accepted values of the speed of sound for the working temperature?

41.2.3 *Project 136.* ♣♣ *Study of resonances in a tube using a lock-in amplifier*

Recommended equipment: A tube approximately 0.6 m long and 2–3 cm in diameter, made of acrylic, PVC, or glass or a graduated cylinder 40 or 50 cm long. A small speaker or earphone and a microphone. A function generator and a lock-in amplifier.

The aim of this project is to analyze the resonances in a tube, in a manner similar to the two previous projects, using a *lock-in amplifier* (see Appendix E). The advantage of using a lock-in is the possibility of making a continuous sweep of frequencies and simultaneously analyzing the amplitude and the phase relation between the excitation and the response. A lock-in also automatically suppresses any noises that may be present in the laboratory during the test. A possible experimental arrangement is illustrated in Figure 41.6. Choose a frequency sweep range that covers at least the first five resonances of the tube. Choose a sweep step that allows you to determine the shape of the resonances.

Suggested tasks:

✓ For a tube of a given length, determine the resonance frequencies f_n as a function of the resonance order n.

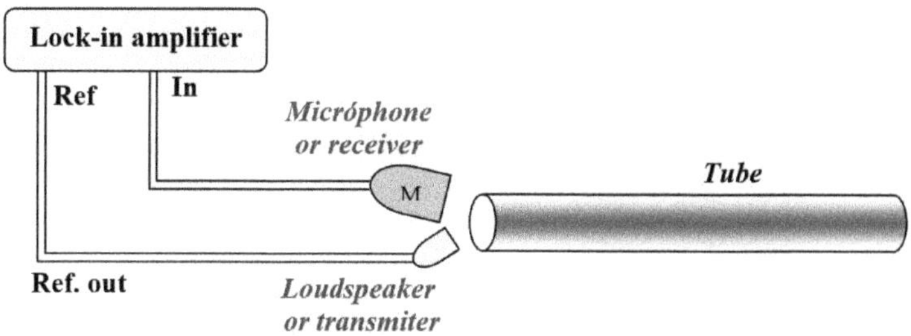

Figure 41.6 Experimental setup to study the resonances in a tube using a lock-in amplifier. The Ref. out signal drives the speaker. The signal from the microphone is connected to the input of the lock-in. A frequency sweep range is chosen to study the response of the tube.

✓ Analyze the shape of the first two resonances, and determine their corresponding half-widths.

✓ Graph the parameter $z_n = 4 \cdot L_{ef} \cdot f_n$ as a function of $(2n+1)$. Determine the speed of sound c and its uncertainty.

Optional: If possible, insert a cloth into the tube and study how the resonance frequencies and corresponding half-widths vary.

41.3 Annex A: Variable frequency mechanical actuator

There are many ways to build a mechanical actuator. There are also commercial versions that can be purchased from laboratory product distributors [5]. A simple way to build an actuator is to use an audio speaker of about 30–100 W and 15–40 cm in diameter (subwoofer). Take a light, rigid cylindrical plastic cap (such as the lid of a coffee jar) center it carefully on the speaker, and glue it to the speaker. Glue this cap to a rigid plastic bar along the common axis of the cylinder and cap, as illustrated in Figure 41.7 (right). In essence, this is a mechanical actuator. Then, glue the rim of the speaker to the inverted plastic cup, with a hole for the first plastic cylinder that is the actual actuator. Remember that audio speakers generally have very low impedance and that function generators cannot deliver much power,

Figure 41.7 Schematic diagram of a mechanical actuator. On the left, a commercial version of a Pasco device [5]. On the right, a homemade actuator built with a speaker.

so the function generator should not be connected directly to the speaker. One suggestion is to connect an audio amplifier between the function generator and the speaker. Before connecting, it is advisable to check that the amplifier power does not exceed the maximum power that the speaker can support.

41.4 Annex B: One-dimensional pressure waves

Consider a one-dimensional wave that propagates along a uniform tube of section A in the positive direction of the axis x, as illustrated in Figure 41.2. Consider a small test volume $V = Adx$. Here, $y(x,t)$ designates the function that describes the displacement along the x-axis of the particles at the x-coordinate point at the instant t. The test volume variation when the wave passes will be given by

$$\delta V(x,t) = A \cdot (y(x+dx,t) - y(x,t)) \tag{41.10}$$

and

$$\delta V\,(x,t) = A \cdot \delta x \cdot \partial y/\partial x \sim V \cdot \partial y/\partial x, \tag{41.11}$$

where P is the absolute pressure of the gas. If P_0, is the atmospheric pressure, then the total (or absolute) pressure of the gas is $P = P_0 + p$, where p is the pressure in excess of the atmospheric pressure (gauge pressure), due to the wave.

For almost any wave, its speed c can be written as follows [4, 7]:

$$c = \sqrt{\frac{B}{\rho}}, \tag{41.12}$$

where B is the modulus of compressibility ($B = -\Delta P/(\Delta V/V)$) and ρ is the density of the medium. In adiabatic compression [2, 4], $PV^{\gamma} =$ constant, where γ is the adiabatic coefficient of the gas, equal to the quotient of the molar heat capacities of the gas at constant pressure and volume [4, 7]. Thus, $\delta V/V = -(1/\gamma)(\delta P/P)$, and therefore,

$$B = \gamma P. \tag{41.13}$$

From the equation of state of an ideal gas,

$$\rho = \frac{P}{RT}M, \tag{41.14}$$

where M is the molar mass of the medium, T is its absolute temperature, and R is the universal gas constant. Thus from Eq. (41.12),

$$c = \sqrt{\frac{\gamma P}{\rho}} = \sqrt{\frac{\gamma \cdot RT}{M}}. \tag{41.15}$$

References

[1] Wikipedia, "Standing waves," 2021. Available: https://en.wikipedia.org/wiki/Standing_wave.

[2] R. Halliday, D. Resnick and M. Krane, *Fundamentals of Physics*, NY, John Wiley & Sons, Inc., 2002.

[3] F. Sears, M. Zemansky, H. Young and R. Freedman, *College Physics*, NY: Pearson Ed., 2009.

[4] F. S. Crawford Jr., *Waves (Berkeley Physics Course, Vol. 3)*, Cambridge University Press, 2013.

[5] PASCO, "Mechanical wave driver," 2022. Available: https://www.pasco.com/products/lab-apparatus/waves-and-sound/ripple-tank-and-standing-waves/sf-9324.

[6] R. Peterson, "Coupling a speaker to a closed–tube resonator," *Am. J. Phys.*, vol. 63, p. 486, 1995.

[7] R. Feynman, *et al.*, *The Feynman Lectures on Physics, Boxed Set: The New Millennium Edition*, vol. 1, Basic Books, 2011.

[8] Physphox — RWTH Aachen University, "The smartphone as a mobile lab," RWTH Aachen University, 2022. Available: https://phyphox.org/.

[9] Wikipedia, "Compressibility," 2022. Available: https://en.wikipedia.org/wiki/Compressibility.

[10] Wikipedia, "Bulk modulus," 2022. Available: https://en.wikipedia.org/wiki/Bulk_modulus.

Chapter 42

ACOUSTIC WAVE INTERFERENCE — BEATS

The goal of this chapter is to study the principle of superposition and the phenomenon of wave interference. We begin by studying the superposition of sound waves generated with tuning forks that vibrate at audible frequencies. When the vibration frequencies are close, the beat phenomenon is observed (and heard).

Goals

- ✓ Principle of superposition
- ✓ Interference
- ✓ Beats

42.1 Principle of superposition

A fundamental characteristic of waves is that they satisfy the principle of superposition[1]. This means that if a point is simultaneously disturbed by two or more waves of the same type, the resulting amplitude will be equal to the sum of the individual amplitudes, while the energy transported by the wave per unit area and time (its intensity I) is proportional to the square of the resulting amplitude. This property of wave motions is known as the *principle of superposition* and explains some important and sometimes curious phenomena that occur with waves, one of which is *interference.* Interference can sometimes make the resulting intensity at a given point exceed the intensity of the individual waves, while at other times, the resulting intensity may be null.

This chapter studies *acoustic wave interference* using pure sounds of two monochromatic sources with frequencies that are close to each

Figure 42.1 On the left, a simple tuning fork. On the right, tuning forks with soundboards or resonance boxes. In the image on the right, an extra weight has been added to one of the tuning forks to alter its vibration frequency slightly.

other, which gives rise to the phenomenon of *beats*. Beats are audible and can be recorded by a data acquisition system.

42.2 Beat

A plane wave traveling in the positive direction of the z axis with speed c and frequency f can be represented by

$$X(z,t) = A\sin\left[2\pi f\left(t - \frac{z}{c}\right)\right] = -A\sin[k(z - ct)]. \qquad (42.1)$$

In this expression, A is the amplitude and $k = 2\pi f/c = 2\pi/\lambda$ is the wave number, $\lambda = f/c$ is the wavelength, and t is the time. The intensity of this wave at a given point z_0 is

$$I(z_0,t) \propto |X(z_0,t)|^2, \qquad (42.2)$$

and the average *intensity*, which is what the ear perceives as volume, is the temporal average of the previous expression, that is,

$$\bar{I}(z_0) = I_0 = \langle I(z_0)\rangle = \frac{|A|^2}{2}. \qquad (42.3)$$

Consider the case of two waves of equal intensity and slightly different frequency f_1 and f_2 that overlap. For simplicity, we will consider the special case where both waves have the same amplitude A. According to the superposition principle, the resulting amplitude will be [1]

$$X_{1,2}(z,t) = X_1 + X_2 = A\sin\left[2\pi f_1\left(t - \frac{z}{c}\right)\right] + A\sin\left[2\pi f_2\left(t - \frac{z}{c}\right)\right]. \tag{42.4}$$

Or also

$$X_{1,2}(z,t) = 2A\sin\left[2\pi\left(\frac{f_1 + f_2}{2}\right)\left(t - \frac{z}{c}\right)\right] \times \cos\left[2\pi\left(\frac{f_1 - f_2}{2}\right)\left(t - \frac{z}{c}\right)\right] \tag{42.5}$$

or

$$X_{1,2}(z,t) = 2A\sin\left[2\pi\bar{f}\left(t - \frac{z}{c}\right)\right]\cos\left[2\pi\left(\frac{\Delta f}{2}\right)\left(t - \frac{z}{c}\right)\right], \tag{42.6}$$

with

$$\bar{f} = (f_1 + f_2)/2 \quad \text{and} \quad \Delta f = (f_1 - f_2) = c/\lambda_{12}. \tag{42.7}$$

The average intensity, at a given point $z = z_0$, will be

$$\bar{I}(z_0,t) = \overline{|X_{1,2}(z_0,t)|^2} = A\cos^2\left[2\pi\frac{\Delta f}{2}\left(t - \frac{z_0}{c}\right)\right] = 2I_0\cos^2\left[2\pi\left(\frac{\Delta f}{2}t - \phi_0\right)\right]. \tag{42.8}$$

This result indicates that the resulting intensity is no longer constant but will vary over time. The frequency of pulsation will be $\Delta f/2 = (f_1 - f_2)/2 = c/\lambda_{12}$, where λ_{12} is the wavelength associated with the beat (see Figure 42.2). This effect will produce a modulation of the intensity of the sound called a *beat* and is easy to perceive. This phenomenon is partly the cause of the unpleasant sound heard on poorly tuned instruments.

42.2.1 *Project 137. Listening to the wave superposition — Beating*

Recommended equipment: Two tuning forks vibrating at the same frequency. A sound synthesizer on a PC or two smartphones with sound synthesizers.

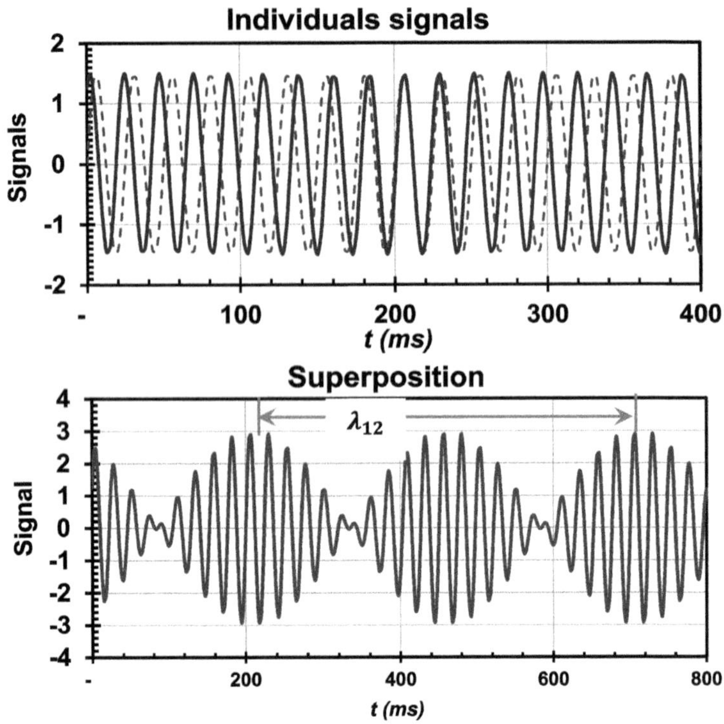

Figure 42.2 Above, two individual signals of equal amplitude and slightly different frequencies. Below, signal resulting from the sum of the two signals shown above. The beat is the amplitude variation of the resulting signal.

One way to produce pure sounds is with tuning forks. Tuning forks are devices that emit pure sounds corresponding to musical notes and are used to tune musical instruments. When struck, they produce sounds of very well-defined frequencies. Sometimes they are mounted on a soundboard or resonant box to increase their loudness (see Figure 42.1). For this experiment, you will need two similar tuning forks (i.e., which produce sounds of very close frequencies). Another way to generate pure sounds is with a sound synthesizer. There are many programs, several of which are free to use, e.g., Audacity[a] and

[a]See also http://en.wikipedia.org/wiki/Audacity and www.audacity.com

Sound Forge,[b] which generate sounds of pure frequencies in each output channel (speaker) of a PC[1]. In particular, the University of Colorado at Boulder has an excellent set of interactive simulations PhET,[c] which are free to use. Likewise, there are applications to generate sound interference for cell phones.[d] If you want to perform this experiment using two smartphones, a very convenient application is Phyphox [3], which enables you to generate a pure sound very easily. Two smartphones with Phyphox programs can be used instead of tuning forks to study the effects described here.

Suggested tasks:

✓ With a tuning fork or whichever sound source you are using, first generate a ***single pure*** sound and listen to it. *Train* your ear to recognize how the intensity of sound changes over time as the vibration of the tuning fork dampens and fades or if using a smartphone, if you change the intensity (amplitude) of the signal. Then make the two identical tuning forks (or two pure signals of the same frequency on the synthesizer or smartphone that you are using) vibrate together, and listen to the resulting sound. Describe whether you hear (observer) any noticeable changes in the resulting sound compared to when only one source vibrates.

✓ Then add a small weight to one arm of one of the tuning forks, e.g., a metal clip weighing a few grams attached firmly to the arm. Adding this weight will produce a slight change in the oscillation frequency of the tuning fork, and the sound it generates will be of a different (smaller) frequency with respect to the original one. If you use a synthesizer of sound or two smartphones, change slightly the output frequency of one of them. To further change the frequency of this tuning fork, change the location of the weight on the arm. If you are using a synthesizer, slightly vary (by a few Hz) the frequency on one of the output channels. In this case, you will

[b] http://en.wikipedia.org/wiki/Sound_Forge

[c] Interactive simulations of physical phenomena, of the project PhET™ at Colorado university. https://phet.colorado.edu/es/

[d] Physics Beat Phenomena Simulator By Fadi Shaghlil, Apple, iTunes. https://itunes.apple.com

hear a resulting beat (intensity modulation of the sound). This effect is a consequence of the superposition of two sound waves of close frequencies.

42.2.2 *Project 138. Quantitative Experiments — Beating*

Recommended equipment: Two tuning forks vibrating at the same frequency. A sound synthesizer on a PC or two smartphones with sound synthesizers can also be used. A microphone connected to a computer data acquisition system, or a second PC with a sound card, or a third smartphone to record the sound.

This project proposes to *measure* what you observed in the previous project.

Suggested tasks:

✓ With the tuning forks operating at slightly different frequencies, or with a synthesizer operating at two different frequencies, record the resulting sound with a data collection system or with a second PC or a third smartphone in recording mode. You could use Phyphox (measure the amplitude of sound) for this operation [3].

✓ A microphone can be used to convert sound into electrical signal (voltage). The microphone can be connected to a computer data collection system. You can also connect the microphone to the sound card of another PC and use a program for digitization and analysis of the resulting sound.

  ✓ Once the beat effect is achieved, alternately turn off one of the sources and record each signal operating alone. Record the sound emitted by each tuning fork or source and measure their corresponding frequencies. If you use a data acquisition system, you may use the frequency analyzer (FFT = Fast Fourier Transform), which enables you to easily determine the frequency spectrum of a signal. There are also many programs that can do this mathematical operation and return the Fourier spectrum of the recorded signal. The discrete frequencies in the Fourier spectrum describe the frequencies present in the analyzed signal. Programs such as Microsoft Excel®, MathWorks Matlab®, Wolfram Research Mathematica®, and OriginLab Origin®, among others can perform this type of operation.

✓ If you are using a data acquisition card, make sure that its sampling rate f_m is high enough to be able to measure the wave in great detail (for a wave of frequency f it should be $f_m > 5f$). The experiment should be conducted in a room with low background noise to minimize the effects of spurious signals. You may use a second PC to record sound because the acquisition rates of sound cards are generally higher than 40 KHz. Signals recorded in *.wav* format can easily be transformed into a table of time and voltage numbers (programs like Matlab or Origin, among others, can do this transformation):

- Fit a sinusoidal function in the form of Eq. (42.1) to your experimentally obtained voltage versus time data. Manually vary the parameters (A, f, and ϕ) until the best possible fit is achieved. In general, it is easy to obtain a very reliable value of f if there are several cycles of the oscillation and care is taken that in all of them the maximums, minimums, and zero crossings coincide.
- Determine the frequencies f_1 and f_2 separately.

✓ Once the frequencies of each tuning fork or channel have been characterized, set them to oscillate simultaneously so that you can hear the beats. Record this signal with the microphone and determine the beat frequency:

- Determine the frequencies $\bar{f}$ and Δf.
- Compare your experimental results to those predicted by Eq. (42.7).

References

[1] R. Feynman, *et al.*, *The Feynman Lectures on Physics, Boxed Set: The New Millennium Edition*, Vol. 1, Basic Books, 2011.
[2] F. Sears, M. Zemansky, H. Young and R. Freedman, *College Physics*, NY: Pearson Ed., 2009.
[3] Physphox — RWTH Aachen University, "The Smartphone as a mobile lab," RWTH Aachen University, 2022. Available: https://phyphox.org/.
[4] Wikipedia, "Software synthesizer," 2021. Available: http://en.wikipedia.org/wiki/Software_synthesizer.

Chapter 43

SQUARE BOX — HELMHOLTZ RESONATORS

Resonant cavities are especially interesting for analyzing the behavior of sound waves and investigating the properties of the standing waves or stationary waves and resonant frequencies produced in some systems. Simple experiments can be performed to observe the resonance phenomenon in different cavities and determine their characteristic frequencies. Chapter 41 discussed one-dimensional waves; this chapter proposes to extend this study to three-dimensional systems. These experiments also enable indirect determination of the speed of sound. First, we will study the characteristics of standing waves in a rectangular box. This model is particularly interesting because it is the classical analog of a quantum particle in a square box. Finally, we will analyze the normal modes of oscillation in a bottle, which is an example of Helmholtz resonator.

Goals

- ✓ Resonant cavities
- ✓ Standing waves in a rectangular box
- ✓ Resonance in a bottle
- ✓ Helmholtz resonators

43.1 Standing waves in a square box

Previous chapters discussed standing waves in one dimension (open and closed tubes). This chapter addresses the different modes of oscillation of a rectangular box [1–4], a topic of great theoretical and practical interest. It is one of the few systems that can be solved analytically, and its results are compared directly with measurements. Moreover, this system is a paradigm that provides an understanding of many other important problems, such as the discrete levels of energies of a particle confined in a box, which is frequently found in many quantum systems. In fact, it is the classical analog of the quantum system of a particle confined to a finite volume, e.g., nucleons in a nucleus, electrons in a crystal, electrons in an atom, and photons in a cavity.

Assume that we have a rectangular box of dimensions a, b, and c. As discussed in Chapter 41, the pressure of the gas in the box must satisfy the wave equation [5]:

$$\nabla^2 p + \frac{1}{c^2}\frac{\partial^2 p}{\partial t^2} = 0, \tag{43.1}$$

where c is the speed of sound. The pressure must meet the following boundary conditions on each wall:

$$\left(\frac{\partial p}{\partial x}\right)_{\text{wall}(y,z)} = 0, \quad \left(\frac{\partial p}{\partial y}\right)_{\text{wall}(x,z)} = 0, \quad \left(\frac{\partial p}{\partial z}\right)_{\text{wall}(x,y)} = 0. \tag{43.2}$$

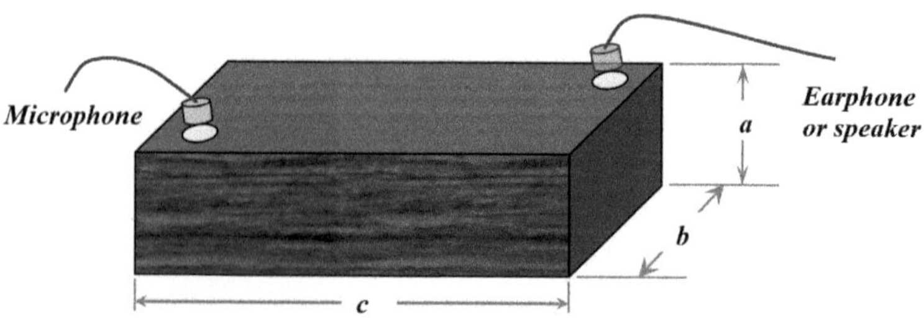

Figure 43.1 Square box with two openings in which to place a speaker and a microphone to study the normal modes of oscillation (natural frequencies of the box).

By using the method of separation of variables, it is easy to see that a solution of the wave Eq. (43.1) is

$$p = Ae^{i\omega \cdot t} \cos\left(\frac{\pi}{a} n_1 x\right) \cos\left(\frac{\pi}{b} n_2 y\right) \cos\left(\frac{\pi}{c} n_3 z\right), \quad (43.3)$$

where n_1, n_2, and n_3 must be natural numbers, 0, 1, 2, 3, etc. By replacing Eq. (43.3) in Eq. (43.1), it can be verified that this equation is indeed the solution sought. Furthermore, it satisfies the boundary condition of Eq. (43.2) on all walls of the enclosure. By replacing Eq. (43.3) in Eq. (43.1), we obtain a condition for ω, namely

$$\omega_{1,2,3}^2 = \pi^2 c^2 \left[\left(\frac{n_1}{a}\right)^2 + \left(\frac{n_2}{b}\right)^2 + \left(\frac{n_3}{c}\right)^2\right], \quad (43.4)$$

or, in terms of the frequency,

$$f_{1,2,3} = \frac{c}{2}\sqrt{\left[\left(\frac{n_1}{a}\right)^2 + \left(\frac{n_2}{b}\right)^2 + \left(\frac{n_3}{c}\right)^2\right]}. \quad (43.5)$$

This relationship tells us that the *natural frequencies* of oscillation of the gas in the box are *quantized*, that is, only certain discrete frequencies are allowed, as determined by Eq. (43.5). The triad of numbers (n_1, n_2, n_3) characterizes the allowed frequency $f_{1,2,3}$ and the oscillation modes or *normal modes* associated with these frequencies. One way to falsify these predictions is to build a square box resonator and study its natural frequencies.

43.1.1 *Project 130. Standing waves in a box*

Recommended equipment: A rectangular box made of wood, glass, or acrylic, measuring approximately 17 cm × 25 cm × 32 cm, with two holes of about 2 cm in diameter in two of its opposite corners. A small speaker or earphone connected to a function generator and a microphone. A function generator (FG) and a two-channel oscilloscope or a data acquisition system connected to a PC. Alternatively, two smartphones with Phyphox [6].

Ensure that the dimensions a, b, and c of the rectangular box are not multiples of each other, in order to prevent the natural frequencies from overlapping. It is advisable for the top of the box to have two holes at opposite ends, through which to insert the emitter and the receiver. For this experiment, you need an acoustic emitter (loudspeaker or earphones connected to a function generator or a smartphone with Phyphox that generates a monochromatic tone). To determine the resonance frequencies or normal modes of the oscillating gas inside the box, place the emitter near one of the holes (or the speaker of the smartphone close to this hole) and the microphone in the other opening and perform a frequency sweep with the function generator (FG). Try to locate the resonance frequencies, controlling the frequency with the function generator. Take care to keep the amplitude of the generator output constant. Resonances appear as pronounced increases in the amplitude of the receiver's output signal and are usually loud enough to be heard. If you use a two-channel oscilloscope operated in $X - Y$ mode, with one channel connected to the receiver and the other to the emitter, when the system reaches a resonance, the $Y(X)$ graph will be linear (zero phase shift, see Chapter 41).

Suggested tasks:

✓ Determine the first 10 or 12 resonance frequencies of the box.

- Record the following in a four-column table:
  - In the first column, the measured frequencies, in ascending order.
  - In the second column, the corresponding frequencies obtained using Eq. (43.5), trying to ensure that in each row, the frequencies have the best agreement.
  - In the third column, write down the value of the triad (n_1, n_2, n_3) used to calculate the values in the second column.
  - In the fourth column, record the new variable (or pseudovariable) N, defined as follows:

$$N = \sqrt{\left[\left(\frac{n_1}{a}\right)^2 + \left(\frac{n_2}{b}\right)^2 + \left(\frac{n_3}{c}\right)^2\right]}. \tag{43.6}$$

Plot the frequencies found experimentally and those calculated with expression (43.5) (first and second columns of the table) with different symbols and color as a function of N, Eq. (43.6). Discuss the degree of agreement or discrepancy between the frequencies determined experimentally and those found using expression (43.5). What can you conclude about the validity of the proposed model?

✓ Using the graph, vary the value of the speed of sound c until you find the best agreement between the experimental and theoretical data, see Eq. (43.5). What value do you find for c? How does this value compare to the expected values of c or the tabulated values?

43.1.2 *Project 140.* ♣♣ *Standing waves in a box using a lock-in amplifier*

Recommended equipment: A prismatic box made of wood, glass or acrylic, measuring approximately 17 cm × 25 cm × 32 cm, with two holes of about 2 cm in diameter in two of its opposite corners. A small speaker or earphone and a microphone. A function generator and a lock-in amplifier.

The aim of this project is to conduct an analysis of the resonances in a box, similar to the one described in the previous project, using a lock-in amplifier, as suggested in Chapter 41 (see also Appendix E). The advantage of using the lock-in amplifier is that it enables you to perform a continuous frequency sweep while simultaneously analyzing the amplitude and the phase relationship between the excitation and the response. The lock-in amplifier also automatically suppresses any noises that may be present in the laboratory during the test. A possible experimental arrangement is illustrated in Figure 41.6. Choose a frequency sweep range that covers at least the first 10 or 12 resonances. Choose a sweep step that allows you to determine the shape of the resonances.

Suggested tasks:

✓ Determine the first 10 or 12 resonance frequencies of the box.

✓ Following the suggestions made in the previous project, analyze the degree of agreement between the measured natural frequencies of the box and those predicted by the model of Eq. (43.5).

✓ **Optional:** Study the half-widths of the measured resonances. Half-widths are defined as the separation in frequency at which the amplitude falls to half of its value at the peak.

✓ **Optional:** If possible, insert a cloth into the box and study how the resonant frequencies and half-widths change. Try to explain your results using the friction in a spring-mass system as an analogy.

43.2 Resonance of a bottle — Helmholtz resonator

It is well known that blowing on the edge of a closed tube or bottle will make it vibrate and emit a characteristic sound [7]. This is a general property of many closed systems, which have well-defined frequencies of vibration. Some wind instruments, such as the panpipe [8], are made of hollow tubes (reeds), and their sound is based on this characteristic.

While it is relatively easy to determine the normal frequencies of oscillation in tubes, it is less simple to do so in bottles, which have non-uniform cross-sections. For bottle-shaped oscillators (i.e., consisting of a large volume [the body of the bottle] and a small neck), the fundamental frequency of oscillation can be calculated using an ingenious procedure devised by Helmholtz. Hence, a resonator of this type is known as a *Helmholtz resonator*.

The model that we will use consists of assuming that the bottle is composed of two volumes, as shown in Figure 43.2. More specifically, we will use the bottle model shown on the right side of Figure 43.2. Let V be the volume of the body of the bottle (the larger cylinder, with height L and cross-section A), and v the volume of the neck of the bottle (the smaller cylinder, with length l and cross-section a).

By adding water to the bottle, we can easily vary the dimensions of its larger volume, while keeping the geometry of the neck constant. To calculate the fundamental frequency of oscillation of this system, we will assimilate it to a mass-spring system, in which the vibrating mass is the volume of the neck (v) and the larger volume of air acts as a spring. This system has an oscillation frequency given by

$$\omega_0^2 = (2\pi f_0)^2 = \frac{k}{m}. \tag{43.7}$$

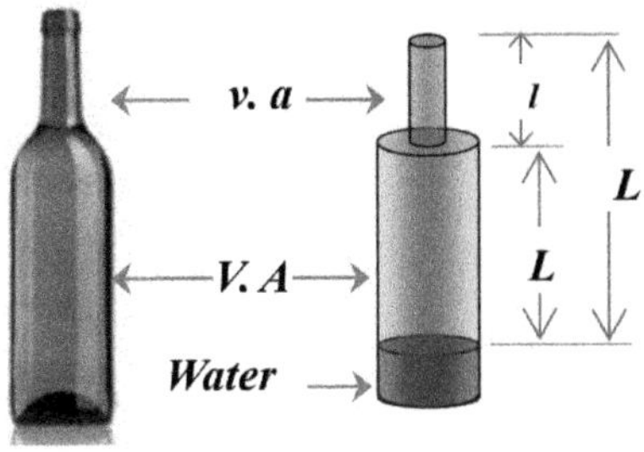

Figure 43.2 Schematic model (right) of a real bottle (left). Assume that the bottle contains water, so the model refers to the region of the bottle occupied by air. A, V, and L represent the cross-sectional area, volume, and height of the largest volume; a, v, and l are the analogous values for the neck.

In Eq. (43.7), f_0 is the fundamental frequency, k is the spring constant, and m is the mass of the volume of air in the neck of the bottle. A more detailed explanation of this model can be seen in the bibliography [9–11]. In our case, we will evaluate the validity of the model by its degree of agreement with the experimental results that we obtain.

Remember that the compressibility coefficient B of a gas is defined as specified in Annex B of Chapter 41 [1, 5, 12]:

$$B = -V\frac{\Delta P}{\Delta V} \approx \gamma \cdot P, \tag{43.8}$$

where P is the pressure and γ the adiabatic coefficient of air. The speed of sound c can be expressed in terms of the coefficient of compressibility B and the density ρ as follows [1, 5]:

$$c^2 = \frac{B}{\rho} = \frac{\gamma \cdot P}{\rho} = \frac{\gamma \cdot RT}{M}. \tag{43.9}$$

Returning to the bottle, if the volume v is displaced a distance Δx, the air volume of the lower volume (larger volume) V will vary by $\Delta V = a\Delta x$, and the force F that will act on the neck volume will be, using Eq. (43.8),

$$F = -a\Delta P = a \cdot B \cdot \frac{\Delta V}{V} = -\left(B\frac{a^2}{V}\right)\Delta x = -k\Delta x. \tag{43.10}$$

Therefore, from Eq. (43.7), we have

$$\omega_0^2 = \frac{f_0^2}{4\pi^2} = \frac{B \cdot a^2}{\rho \cdot v \cdot V} = \frac{a^2c^2}{vV} = \left(\frac{a}{A}\right)\left(\frac{c^2}{l \cdot L}\right). \tag{43.11}$$

This expression can also be written in terms of the height L', measured from the water level inside the bottle to the edge of the neck, as shown in Figure 43.2. By finding L from Eq. (43.11), we have

$$L' = L + l = 4\pi^2\left(\frac{ac^2}{A.l}\right)\left(\frac{1}{f_0^2}\right) + l. \tag{43.12}$$

Therefore, to test this model experimentally, we can vary the volume of air in the bottle by adding water to it. For each configuration of the bottle, measure the height L' and the corresponding fundamental frequency of oscillation f_0. If the model is adequate, a plot of L' as a function of the experimental data of $(1/f_0^2)$, should be aligned. Furthermore, the slope of the line will be related to the geometry of the bottle and the speed of sound, which can therefore be determined.

43.2.1 *Project 141. Resonances in a bottle — Helmholtz resonators I*

Recommended equipment: A glass bottle. A small speaker or earphone connected to a function generator and a microphone. A two-channel oscilloscope or a data acquisition system connected to a PC or two smartphones, one with a white noise generator and the other with a spectrum analyzer.

A bottle with a volume of 350 cm^3 – 1,000 cm^3 may be adequate. To study the fundamental frequencies of cylinder air oscillation, use the device schematically illustrated in Figure 41.5. Preferably choose a bottle with a shape as close as possible to that of the model in Figure 43.2, i.e., a body and a neck in the shape of regular cylinders.

The values of the bottle geometry, a, A, L, and l can be measured directly, although a more suitable method is as follows: Starting with

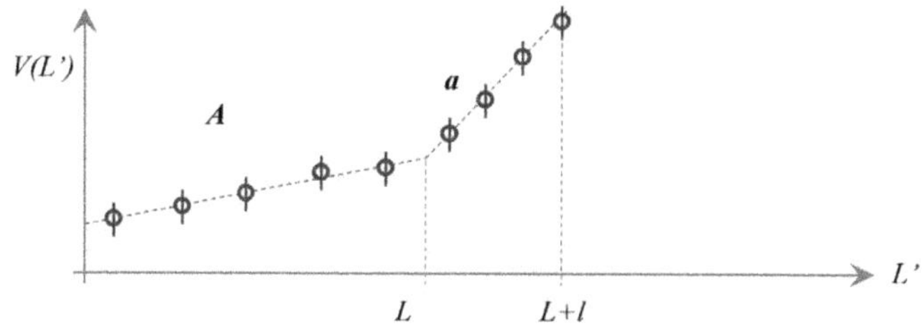

Figure 43.3 Schematic model to determine the parameters of the bottle. The parameters a and A can be determined from the slopes of the lines.

the empty bottle, repeatedly add constant volumes of water, ΔV_0, and measure the height L' simultaneously. Plot the accumulated volume of water $V(L') = n\Delta V_0$ as a function of L'. For a bottle with a simple geometry (i.e., made up of two regular cylinders), this graphic representation, as shown in Figure 43.3, can be well characterized by two lines with different slopes. Each of those slopes is directly A and a (the cross-sectional area of the body and the neck, respectively). The value of L', where these two lines intersect, provides a good estimate of L, while the value of the fully filled volume provides the value of l. Of course, the value of $L + l$ can be measured directly. Similarly, the values found for a and A can be contrasted with direct measurements of the corresponding diameters. The volumes of added water may be measured by placing the bottle on a balance during the filling process.

To determine the fundamental resonant frequency of the bottle, we can proceed as follows (see also Project 134 of Chapter 41). With the emitter and receiver located near the mouth of the bottle (Figure 43.5), perform a frequency sweep and try to locate the resonances, controlling the frequency of the function generator output that drives the emitter or microphone (E). Ensure that the amplitude of the function generator is constant, by monitoring the amplitude of the signal fed to the microphone, using the second channel of the oscilloscope as a sound detector or acquisition system, that is connected to an earphone (R). Resonances appear as pronounced increases in the amplitude of the receiver's output signal. Resonance

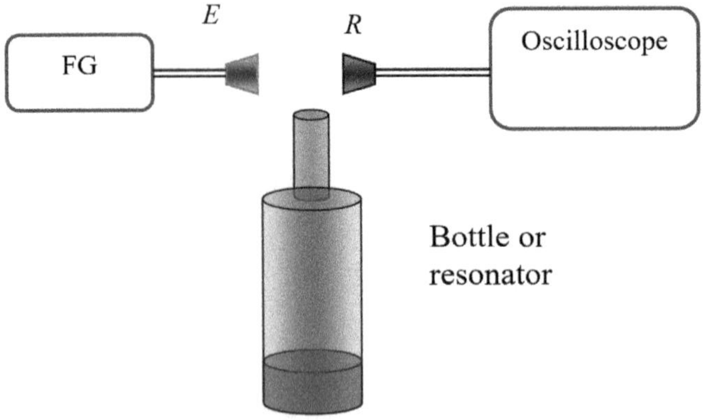

Figure 43.4 Experimental device for studying the resonance modes of a bottle, or Helmholtz resonator. The function generator (FG) excites the emitter or speaker (E) and a receiver or microphone (R) connected to an oscilloscope records the output signal.

frequencies are usually less than 2 kHz, and in general, at resonance, the increased sound intensity can be clearly heard.

With the empty bottle, study the first two or three resonances. Then proceed to add water gradually and, for each water level, measure the frequencies and the heights L'.

This experiment can also be carried out using two smartphones: one has a *white noise generator* [13] and the second smartphone has a *Spectrum Analyzer* [14]. In these cases, the resonance frequencies will show up as very conspicuous peaks in the spectrum analyzer. Further detail on this technique can be found in Refs. [15] and [16].

Suggested tasks:

✓ As you fill the bottle with water, measure the height value L' and the lowest resonant frequency (fundamental frequency f_0). Plot L' as a function of $1/f_0^2$. What relationship do you find? Does this graphical representation agree with the predictions of Eq. (43.12)?

✓ From the graph of L' as a function of $1/f_0^2$, estimate the value of the speed of sound and its uncertainty. Compare the value obtained to the sound velocity values obtained from tables, for the same temperature used in the experiment.

43.2.2 *Project 142. Helmholtz Resonators II*

Recommended equipment: A bottle model made up of two cylinders of different diameters. A small speaker or earphone connected to a function generator and a microphone. A function generator and a two-channel oscilloscope or a data acquisition system connected to a PC or two smartphones: one with a white noise generator and the other with a spectrum analyzer.

One way to conduct a more detailed test of the previous model is to make a bottle model using two plastic, PVC, or acrylic tubes, with a bottom cap (see Figure 43.2, right). In addition, it is advisable to have a system with different neck lengths to control the effect of the variation of this parameter. Another way to conduct this experiment is to have several glass bottles with a long, even neck, which can be gradually cut to different lengths. To do this, you need to be very careful and have experience in cutting glass. Preferably, it should be done by a specialist in a glass workshop.

Suggested tasks:

- ✓ For each neck length, measure the lowest resonance frequency f_0. Graph L' as a function of $1/f_0^2$.
- ✓ For each neck length, make a graph like the previous one and discuss your results.
- ✓ Discuss the validity of the proposed model.

References

[1] M. Alonso and E. J. Finn, *Physics, Vol. II, Fields and Waves, and Vol. III, Quantum and Statistical Foundations*, NY: Addison-Wesley, 1967.

[2] F. S. Crawford Jr., *Waves (Berkeley Physics Course, Vol. 3)*, Cambridge University Press, 2013.

[3] H. Armstrong, "An experiment on sound in an enclosure," *Am. J. Phys.*, vol. 51, p. 1052, 1983.

[4] U. Ingard and D. C. Galehouse, "Second-order pressure distribution in an acoustic normal mode in a rectangular cavity," *Am. J. Phys.*, vol. 39, p. 811, 1971.

[5] F. Sears, M. Zemansky, H. Young and R. Freedman, *College Physics*, NY: Pearson Ed., 2009.
[6] Physphox — RWTH Aachen University, "The smartphone as a mobile lab," RWTH Aachen University, 2022. Avaialble: https://phyphox.org/.
[7] Youtube, "Berlin's bottles only quintet," 2022. Available: http://www.youtube.com/watch?v=IAHGtzZY5Yg.
[8] Youtube, "How to make a panpipe," 2022. Available: https://www.youtube.com/watch?v=NYcl7f7RQlk.
[9] G. Smith and P. D. Loly, "The great bear bottle experiment," *Am. J. Phys.*, vol. 47, p. 515, 1979.
[10] M. P. Silverman and E. R. Worrthy, "Musical mastery of a Coke bottle: Physical modeling by analogy," *Phys. Teach.*, vol. 36, p. 70, 1998.
[11] P. M. Morse, *Vibration and Sound*, 2nd Ed., NY: Mc Graw Hill, 1948.
[12] Wikipedia, Compressibility, 2022. Available: https://en.wikipedia.org/wiki/Compressibility.
[13] BestApp.com, "The best white noise apps of 2022," 2022. Available: https://www.bestapp.com/best-white-noise-apps/#mynoise____top_pick.
[14] W. Czaplejewicz, "Advanced spectrum analyzer PRO," GooglePlay, 2022. Available: https://play.google.com/store/apps/details?id=com.vuche.asap&gl=US.
[15] M. Monteiro, A. C. Marti, P. Vogt, L. Kasper and D. Quarthal, "Measuring the acoustic response of Helmholtz resonators," *Phys. Teach.*, vol. 53, p. 247, 2015.
[16] M. Monteiro, C. Stari, C. Cabeza and C. Martí, "A bottle of tea as a universal Helmholtz resonator," *Phys. Teach.*, vol. 56, p. 644, 2018.

Chapter 44

ULTRASOUND WAVES

Ultrasonic transducers are low-cost devices which are simple to use and very suitable for studying the properties of ultrasound waves. The aim of the experiments in this chapter is to study the frequency response of an ultrasound emitter–receiver system. This information enables ultrasound to be used to analyze different wave phenomena, such as reflection, interference, and wave diffraction.

Goals

✓ Frequency response of an ultrasound emitter–receiver system

✓ Speed of ultrasonic waves

✓ Reflection and interference of ultrasound waves

44.1 Ultrasound emitting–receiving pairs

Ultrasound waves are a special case of pressure waves in a compressible medium. They are generated by ultrasonic transducers. The nature of ultrasound waves is identical to that of normal sound, except that their frequencies are greater than about 20 kHz, and we cannot hear them. An ultrasound emitting–receiving pair is essentially a vibrating metallic membrane whose oscillations can be forced at frequencies above 20 kHz.

Nowadays, ultrasound emission-detection systems have a wide range of applications in many fields of science, technology, and everyday life. Some common applications are ultrasound imaging in medical diagnostics [1], alarms that detect movements, several non-destructive tests, motion detectors, sonar, etc. Ultrasound waves can be used to study the propagation of pressure waves in gas and to analyze common wave propagation phenomena usually studied with light, such as reflection and diffraction, among others [1–4].

Just like any other wave, sound waves are reflected and refracted and can therefore be used to reconstruct images. The problem is that the resolution of an image is limited by wavelength. At 1 kHz, a sound wave in air ($c \approx 330$ m/s) has a wavelength $\lambda = c/f \approx 33\,\text{cm}$. On the other hand, in soft human tissues, the speed of sound is much higher ($c \approx 1,540$ m/s), so at this frequency, the wavelength in this type of tissue is $\lambda \approx 1.5$ m. In medical applications, it is therefore necessary to use frequencies of the order of 1 MHz or greater, for which $\lambda < 1.5$ mm. The reflection of sound waves is also used in locating systems, such as sonars. Bats, dolphins, and other animals emit and perceive ultrasound, which they use as an echolocation system.

The experiments proposed here require an ultrasonic emitter that can emit pure high-frequency waves ($f \geq 20\,\text{kHz}$, often $f \sim 44\,\text{kHz}$) and a receiver capable of detecting them. These two elements are used in an emitter–receiving pair in a circuit like the one shown in Figure 44.1.

The emitter radiates ultrasonic waves when excited or driven by an alternating signal provided by a function generator (FG). The driving signal is routed to one of the channels of an oscilloscope and the output of the receiver is measured on another channel of the

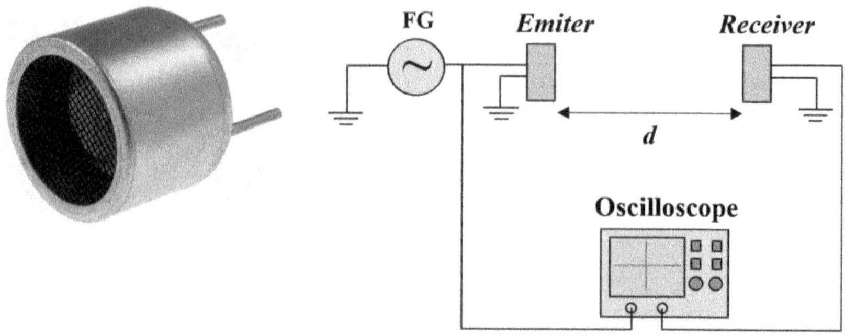

Figure 44.1 On the left, an ultrasonic transducer with its connectors. On the right, an experimental setup to study the propagation of ultrasound waves.

oscilloscope. In this way, the waveforms of the two signals (emitted and received) can be compared, and their amplitudes and relative phases measured, which enables the system to be characterized and provides a complete picture of the propagation of these waves.

44.2 Ultrasonic pair frequency response

In general, the emitter and receiver must be mutually compatible, i.e., the maximum response in frequency of the emitter must coincide with the receiver's maximum sensitivity. In the following experiment, we will study the frequency response curve of the receiver-emitter system. When purchasing an emitter–receiver pair, check the data sheets provided by the manufacturer to make sure that the emitter–receiver pair is indeed compatible. These data sheets also specify the value of the optimal working frequency, angular aperture of the emitter, etc. For the experiments described here, it is advisable to use a pair with an optimal operating frequency in the range of about 36–45 kHz. In air, a wave of about 40 kHz has a wavelength $\lambda \approx 0.85\,\text{cm}$.

Since at each point in space, the wave generated by the emitter has a well-defined phase relation with respect to the position x of the receiver, the relative phase between the emitter and the receiver is expected to vary by 2π each time x increases by one wavelength λ. This can be used to determine the wavelength and, knowing the frequency f of the signal, the speed of sound can be calculated as $c = \lambda \cdot f$.

44.2.1 *Project 143. Frequency response of an ultrasonic pair*

Recommended basic equipment: An ultrasound emitter–receiver pair. A function generator with a range greater than 45 kHz. A two-channel oscilloscope.

Using the arrangement shown in Figure 44.1, place the receiver facing the emitter at a distance of a few centimeters. Excite the emitter with a sinusoidal voltage. Keeping the amplitude of the voltage constant, vary the frequency and obtain the response curve of the emitter–receiver pair as a function of frequency [1, 2].

Suggested tasks:

✓ Using the oscilloscope, determine the amplitude of the signal received by the receiver, V_r, and the amplitude of the excitation on the emitter, V_e. Plot V_r/V_e as a function of the driving frequency f. Determine the optimal working frequency f, which is the frequency f_0 for which the response V_r/V_e is maximum. Also determine the half-width of the curves, defined as the difference $\Delta f = f_1 - f_2$ between the frequencies for which the response reaches half its maximum value at f_0. Since the optimal working frequencies of ultrasound pairs are between 35 kHz and 45 kHz, cycle through this frequency range in steps of about 250 Hz.

✓ Determine the quality factor of the system, defined as $Q = f_0/\Delta f$. Study the response in a frequency around f_0 in steps of about 25 Hz. Estimate the value of Q. Do you find any other resonance in the range of $f_0 \pm 15$ kHz? Is the quality factor a property of the emitter, the receiver, or both?

✓ Study the variation of the relative phase of the signals V_r and V_e around resonant frequencies, at fixed geometry of the emitter–receiver pair, but varying the frequency. Using the oscilloscope in XY mode, observe the V_r signal as a function of V_e and discuss the shape of the figures observed at different frequencies close to f_0. Describe these figures when signals are in phase ($\Delta\varphi = 0$) and when the phase shift is $\pi/2$ and π. Plot the phase difference of the signals $\Delta\varphi$ as a function of frequency (see Annex A of Chapter 31).

✓ What is the value of the phase difference for the frequency f_0?

44.2.2 *Project 144. Determining the speed of sound*

Recommended basic equipment: An ultrasound emitter–receiver pair. A function generator with a range greater than 45 kHz. A two-channel oscilloscope.

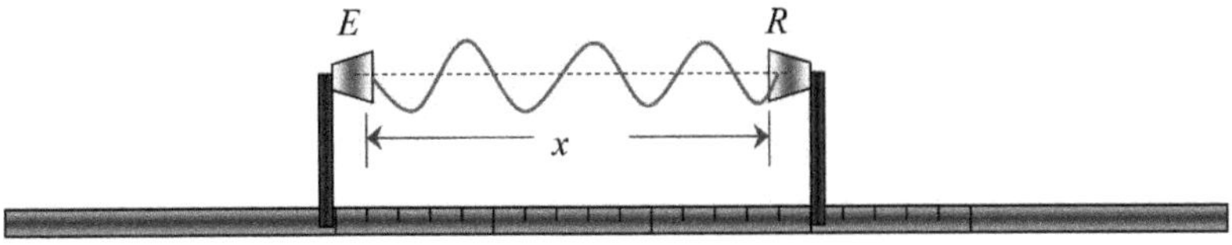

Figure 44.2 Diagram of the operation of an emitter–receiver pair separated by a given distance x. The emitter radiates a wave which, at each point in space, has a well-defined phase relationship with respect to the position x of the receiver.

For this experiment, it is advisable to mount an experimental setup like the one shown in Figure 44.2 on a wooden ruler, forming a sort of optical bench [6], so that the position or distance of the emitter with respect to the receiver can change along a line, but making it possible to measure the distance, x, between emitter and receiver with a precision of the order of a millimeter. Make sure that when the emitter is moved away from the receiver, the relative orientations remain constant. Drive the emitter at the optimal frequency f_0 determined in the previous project. Connect the oscilloscope so that you can determine the phase difference $\Delta\varphi$ between the driving signal of the FG and the signal of the receiver, as indicated in the previous project.

Suggested tasks:

✓ Study the variation of the relative phase $\Delta\varphi$ by varying the distance x between the emitter and the receiver. Qualitatively describe what you observe based on the spatial periodicity of the results.

✓ Plot the emitter–receiver distance, x, for which the signals of the emitter and receiver are in phase ($\Delta\varphi = 0$), as a function of the order n ($n = 1, 2, 3, \ldots$) at which such condition is fulfilled. Remember that every time $\Delta\varphi = 0$ or $n \cdot 2\pi$, the figure of the two signals in the XY mode of the oscilloscope becomes a straight line, with the same orientation. How can you determine the wavelength λ of the ultrasound wave from this graph?

✓ From the graph of x versus n, determine the wavelength λ of the ultrasound waves. Using the relationship $c = \lambda f$, where f is the frequency of the FG, determine the speed of sound and its uncertainty.

44.3 Physical properties of ultrasound waves

As mentioned above, the emitter–receiver in the pair to be used must be mutually compatible. In general, both the emitter and the receiver are very directional in their operation. This property should be analyzed using their corresponding datasheets. The ultrasound reflection properties can be easily studied using a reflector (mirror), which can be a cardboard, wooden, or any other solid plate. Furthermore, as in all waves in general, the law of reflection is fulfilled, which establishes that the angle of reflection is equal to the angle of incidence [5, 6]. The following projects are intended to study some physical properties of ultrasound waves. First, we analyze the reflection and then the interference [1, 2, 4].

The goal of the following activities is to study the reflection and interference of ultrasonic waves in the ultrasonic emitter–receiver pair.

44.3.1 *Project 145. Ultrasonic wave reflection*

Recommended basic equipment: An ultrasound emitter–receiver pair. A function generator with a range greater than 45 kHz. A two-channel oscilloscope.

This experiment requires a flat screen, consisting of a hard plate measuring at least 40 cm × 40 cm. Mark the center of the plate. Place the emitter facing the flat screen and on the same side and place the receiver at a distance between 30 and 50 cm from the screen. See Figure 44.3.

Point both the emitter and the receiver at the mark in the center of the plate. Monitor the reflected wave with the receiver. One way to determine the angles of incidence and reflection is to place the emitter and receiver pointing to the center of the screen at approximately the same distance from this point. Rotate the screen around a vertical axis that passes through the center and find the angle at which the

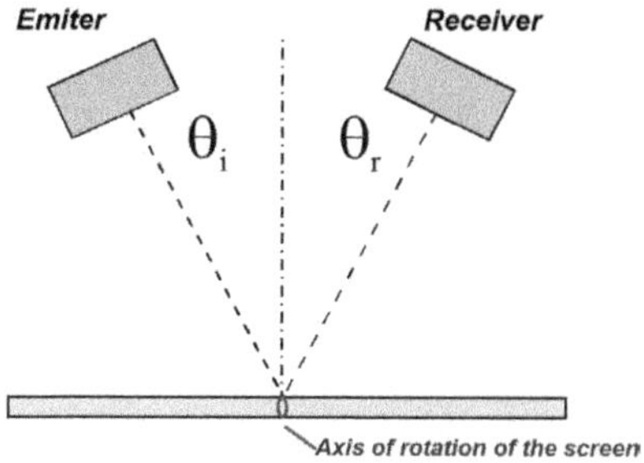

Figure 44.3 Location of the emitter and receiver with respect to a screen and definition of angles to study the reflection of the waves.

receiver signal is maximized. This means that the signal from the transmitter is reaching the receiver.

Suggested tasks:

✓ Study how the angle of reflection θ_r varies depending on the angle of incidence θ_i (see Figure 44.3).
✓ Does the reflection of ultrasound waves comply with the law of reflection $\theta_r = \theta_i$?

44.3.2 *Project 146. Ultrasound interference measurement*

Recommended basic equipment: Two similar ultrasonic emitter–receiver pairs. A function generator. A two-channel oscilloscope.

Suggested tasks: This experiment requires *two similar emitters* and a compatible receiver. Connect the FG to the two emitters in parallel. Take care with the connections so that they both oscillate in phase. For this, the two grounds of the emitters must be connected to the ground of the FG. If the connection is reversed in one of them, they can oscillate in counter-phase. The experimental setup is illustrated in Figure 44.4.

The distance d between the emitters must be of the order 2λ or 3λ approximately ($d \approx 3\,\text{cm}$–$5\,\text{cm}$). The receiver should be placed at

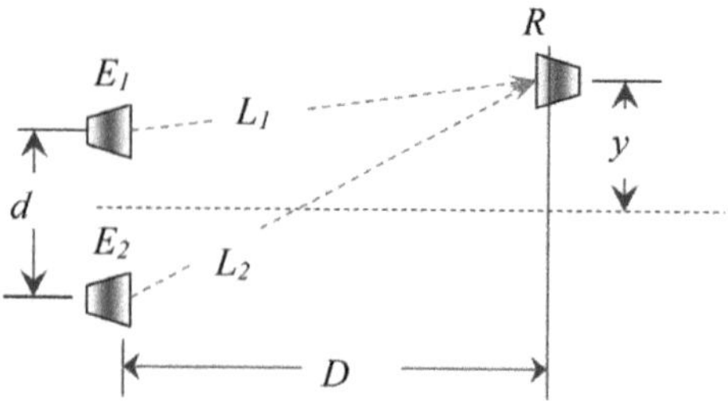

Figure 44.4 Diagram of the system for studying the interference of ultrasonic waves.

a distance $D \approx 25\,\text{cm}$–$45\,\text{cm}$. The idea is to move the receiver along the y-axis to detect the minima and maxima. These minima occur each time the difference in $|L_1 - L_2| = n\lambda/2$, that is,

$$|L_1 - L_2| = |\sqrt{D^2 + (y - d/2)^2} - \sqrt{D^2 + (y + d/2)^2}| = n \cdot \frac{\lambda}{2}, \quad (44.1)$$

where n is an integer that designates the order of the minima. The technique consists of carefully registering the first 4 or 6 minima in the detected signal on each side of the bisector of the two emitters. In each case, record the order n of the minima, and the corresponding value of y_n. Plot $|L_1 - L_2|$as a function of n. If the experimental results show a linear tendency, fit a line through them. According to Eq. (44.1), its slope provides the value of $\lambda/2$ [1, 2, 4].

Suggested tasks:

- ✓ Using the experimental setup in Figure 44.4, identify 4 or 5 minima at the receiver. Record the nth orders of the minima and the corresponding values of y_n.
- ✓ Using Eq. (44.1), plot $|L_1 - L_2|$as a function of n, check whether the experimental results line up. If so, find the slope and its uncertainty.
- ✓ Find the best value of λ and its uncertainty.
- ✓ Compare this value to that obtained using the speed of sound at the working temperature and frequency.
- ✓ Repeat the experiment using another distance d. How do the results for the wavelength compare to those obtained at the previous distance?

Low-cost ultrasonic sensors that work with Arduino are available on the market, such as the HC-SR04 sensor, which can be used to measure distance, speed of sound, etc. [7, 8]. They emit an ultrasound wave of about 40 kHz, which is reflected on solid surfaces, as we have studied in our projects.

References

[1] Wikipedia, "Medical ultrasound," 2022. Available: https://en.wikipedia.org/wiki/Medical_ultrasound.

[2] T. B. Greenslade Jr., "Experiment with ultrasonic transducers," *Phys. Teach.*, vol. 32, pp. 392–298, 1994.

[3] T. B. Greenslade Jr, "Ultrasonic interferometers," *Phys. Teach.*, vol. 36, pp. 249–252, 1998.

[4] E. Hecht, *Optics*, NY: Addison Wesley, 1990.

[5] T. B. Greenslade Jr., "Ultrasonic interferometers revisited," *Phys. Teach.*, vol. 45, p. 145, 2007.

[6] "Introduction to optical bench," Ann Harbor scientific, 2022. Available: https://www.youtube.com/watch?v=AfLqWR95IcE.

[7] F. Sears, M. Zemansky, H. Young and R. Freedman, *College Physics*, NY: Pearson Ed., 2009.

[8] F. S. Crawford Jr., *Waves (Berkeley Physics Course, Vol. 3)*, Cambridge University Press, 2013.

[9] D. Nedelkovski, "Ultrasonic sensor HC-SR04 and Arduino," 2022. Available: https://howtomechatronics.com/tutorials/arduino/ultrasonic-sensor-hc-sr04/.

[10] RandomNerdTutorials.com, "Complete guide for ultrasonic sensor HC-SR04 with Arduino," 2022. Available: https://randomnerdtutorials.com/complete-guide-for-ultrasonic-sensor-hc-sr04/.

Chapter 45

DOPPLER EFFECT

This chapter presents an experimental study of the Doppler effect — a phenomenon that is very easy to perceive in everyday life and has many practical applications in science, technology, and medicine. The experiment proposed is easy to implement, is inexpensive, and enables the effect to be studied quantitatively.

Goals

✓ Doppler effect
✓ Analysis of the Doppler effect with a rotating sound source

45.1 Doppler effect — Introduction

When a vehicle sounding a horn or siren passes us on the road, we perceive a change in the pitch of the sound as it overtakes us. At the exact moment when the vehicle passes us, we can clearly hear the pitch changing to a lower one. However, if we were traveling in the car or ambulance sounding the horn or siren, we would not notice any change. This phenomenon is known as the *Doppler effect* and occurs with all types of waves.

The Doppler effect is the change in frequency that occurs when there is relative motion between an observer and a source emitting the wave [1, 2]. As mentioned above, it is easily detectable. This phenomenon was discovered by the Austrian mathematician and physicist Christian Doppler (1803–1853) [3].

The Doppler effect also occurs in electromagnetic waves, such as light, radio waves, and TV, among others. It has many practical applications in science, technology, and medicine and may be the

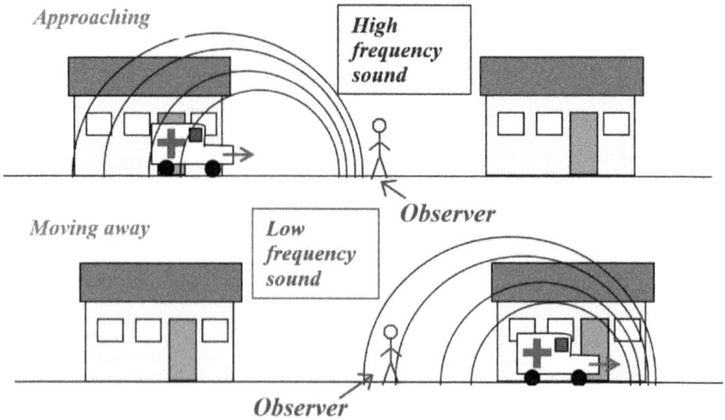

Figure 45.1 The pitch of the siren of a passing ambulance changes from high, as it approaches, to low, as it moves away. In contrast, the ambulance driver does not notice any change in pitch. In this Figure, the circles centered on the instantaneous positions of the source (ambulance) indicate the position of the different wavefronts perceived by an observer at rest.

most common physical phenomenon to come up in everyday conversation [3]. If a light source approaches us at high speed (compared to the speed of light), we would see a change in color toward blue (blue shifted), while if the source is receding from us, the perceived color changes toward the red (red shifted). Depending on the relative velocity between the source and the observer, these shifts in frequency in either direction are more noticeable, as the relative velocity is similar to the speed of the wave in the medium of propagation.

Making use of this effect, Edwin Hubble made one of the most remarkable discoveries while using the Mount Wilson telescope. Hubble was studying the properties of the Cepheids [4], which are stars of varying intensity located at known distances from Earth. Cepheids are regular radial-pulsating stars, with well-defined period-luminosity curves, which makes them ideal for use as primary distance indicators. They can be thought of as reference lights or candles. Hubble used these relationships to deduce the distance to Cepheids in other galaxies [4]. In 1929, Hubble discovered that the spectra of the elements coming from distant extragalactic nebulae showed a red shift which increased with the distance of the nebula (galaxy) from Earth. This led him to infer that the speed of recession of distant galaxies not only increased, but it turned out to be proportional to their distance from us. By plotting the speeds of some 24 nebulae (obtained from the red-shift of their spectra) as a function of their distances, he found a suggestive linear relationship $v = Hd$ between the speed v of the galaxies and their distance d from us. This law constitutes one of the cornerstones of the Big Bang theory [5]. The constant $H \approx 1/1.4 \times 10^{10}$ years is known as the *Hubble constant* and is related to the age of the universe.

Many of the stars that we see on a clear night are binary stars that rotate around a common center of mass [5]. It is possible to infer important information about the star masses and speeds by observing the Doppler shift in their spectra. This approach, often known as radial-velocity method or Doppler spectroscopy, has been used to estimate masses and other interesting properties of binary stars [3, 6]. More recently, the same technique with improved resolution has been used to discover thousands of exoplanets by observing the wobble of some stars due to the presence of Jupiter-like planets orbiting around them [7].

The Doppler effect is widely used in technology and everyday applications ranging from speed detection cameras (for monitoring vehicle speeds on roads) to the Mossbauer effect to astrophysics, or from acoustic phenomena such as the pitch change of a passing ambulance siren to atomic and nuclear physics. This effect has many practical applications in science, technology, and medicine.

The change in frequency f or wavelength λ due to the Doppler effect is simple to understand. Consider the case of a sound wave that has a velocity of propagation in air denoted by c. This speed is constant and does not vary, whether or not the emitter or observer moves [1, 2, 9].

Specifically, consider consecutive wavefronts that have the same phase, that is, they are separated by a distance equal to the wavelength λ. If the source is at rest at point P (see Figure 45.2(a)), these wavefronts are concentric spheres (centered on P) with a separation equal to λ. If the sound source moves toward the observer at velocity u_s, and this velocity is measured with respect to a fixed

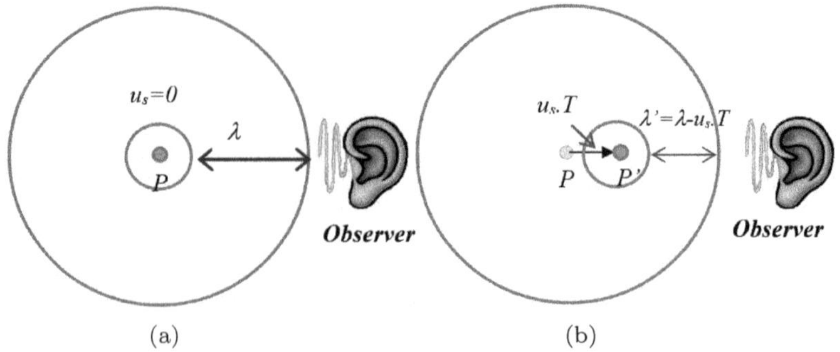

Figure 45.2 Change in the perceived wavelength due to the movement of the emitting source, with speed u_s. (a) Two consecutive wavefronts with the same phase, that is, spaced by a wavelength λ. (b) The source moves with speed u_s. The first wavefront, which was emitted when the source was at P, is the same as before, but when the next one is emitted, the source moves a distance u_sT. The distance between two wavefronts of the same phase will therefore be $\lambda' = \lambda - u_sT$, where T is the period of the wave.

system (see Figure 45.2(b)), the wavefront that was emitted when the source was at P propagates as before, forming a sphere centered on P. The next wavefront is emitted when the source is at P' and forms a sphere centered on P'. The distance between P and P' is equal to u_sT, where T $(= 1/f)$ is the period of the wave. The distance between the two wavefronts for the observer will be the wavelength λ' given by

$$\lambda' = \lambda \mp u_sT. \tag{45.1}$$

The sign depends on whether the source approaches (−) or moves away (+) from the observer. Regarding the frequency f' that the observer perceives,

$$f' = c/\lambda' = \frac{c}{\lambda \mp u_sT} = \frac{f}{1 \mp u_s/c}. \tag{45.2}$$

In general, if both the source and the observer move along a line, it can be shown that the frequency that the observer perceives is [10–12]

$$f' = f\frac{1 \mp u_o/c}{1 \mp u_s/c}, \tag{45.3}$$

where u_s and u_o are the velocities of the source and the observer, respectively, with respect to the medium (air), considering the sign $(-)$ if the observer moves toward the source and $(+)$ if the observer moves away from it.

If the speed u_s is small compared to c, expression (45.3) can be approximated as follows:

$$f' = f(1 \mp u_o/c)(1 \mp u_s/c). \tag{45.4}$$

45.2 Source in circular motion

A practical arrangement for studying the Doppler effect in the laboratory is to attach a "monochromatic" sound source (emitting a frequency f_0) to the end of a rotating rod, as illustrated in Figure 45.3.

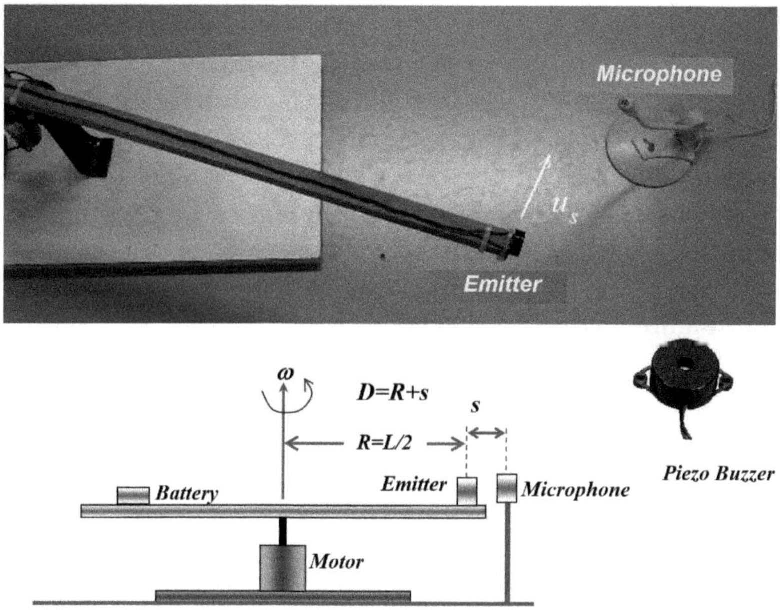

Figure 45.3 Top: Photo of a possible experimental arrangement. Bottom: Schematic diagram of the device to achieve a source in circular motion. The motor determines the angular frequency ω of the spinning bar. As mentioned in the text, the emitter can be a buzzer or a smartphone.

This monochromatic sound source can be a piezoelectric buzzer.[a] Buzzers are used in many electronic devices and are very cheap. When connected to a 9 V battery, they generate a sound with a well-defined frequency, so they are almost monochromatic sound sources. A Smartphone can also be used (see Chapter 42, Project 136) as a monochromatic sound source [13, 14]. Almost any smartphone can produce a monochromatic sound: all you need to do is install a frequency sound generator [15] and/or Phyphox application [16]. Thus, any smartphone can be converted into a basic Physics Lab. A smartphone can also be used to measure the speed of rotation of the device. A laptop or PC with a microphone records the sound of the moving source (see Chapter 11) as a *.wav file using any utility: Windows Sound Recorder [17], Audacity [15], etc. A public domain time-frequency analysis program, available on the Web, [14] can be used to analyze the sound recorded. Of course, any time-frequency analysis program could be used. In principle, a second smartphone could also be used to record the sound, but a laptop with a larger screen makes it easier to monitor and visualize the results online.

The path of the sound source motion is illustrated in Figure 45.4. The source tangential velocity is $v_s = R\omega$. To calculate the variation of the detected frequency in the lab system, what matters is the component u_s of the source velocity v_s in the source-observer direction.

From Figure 45.4,

$$u_s = -v_s \cdot \sin(\omega t + \delta) = -R\omega \cdot (\cos(\delta)\sin(\omega t) + \sin(\delta)\cos(\omega t)). \tag{45.5}$$

The sign $(-)$ in this equation indicates that the source is moving away from the observer. If the direction of rotation were opposite, the sign would change. From Figure 45.4, we have $\rho \sin\delta = R\sin(\omega t)$.

[a]A buzzer emits sounds at a specified frequency, f_0, with a half-width Δf on the order of $\Delta f/f_0 \approx 10\%$. Sometimes, if the voltage is high, two or more frequencies are observed, generally multiples of the main one and therefore very easy to distinguish from f_0. The fundamental frequency of these devices is a few kHz and their cost is of the order of 1 US$.

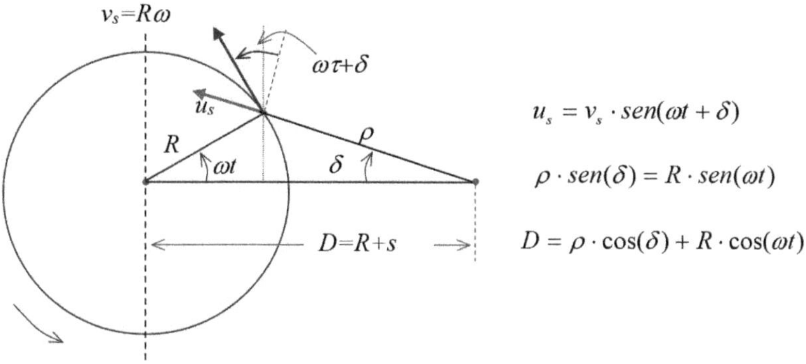

Figure 45.4 Diagram of the top view of the experimental device.

Therefore,

$$u_s = -R\omega \cdot \sin(\omega t)\left(\cos(\delta) + \frac{\mathrm{R}}{\rho}\cos(\omega t)\right). \tag{45.6}$$

As $D = \rho \cdot \cos\delta + R \cdot \cos(\omega t)$,

$$u_s = -R\omega \cdot \frac{D}{\rho}\sin(\omega t). \tag{45.7}$$

From Figure 45.4, $\rho^2 = R^2 + D^2 - 2RD\cos(\omega t)$. From where

$$u_s = -R\omega \cdot \sin(\omega t)\frac{1}{\sqrt{1+\varepsilon^2 - 2\varepsilon\cos(\omega t)}}, \tag{45.8}$$

with $\varepsilon \equiv R/D = R/(R+s)$. If at the instant $t = 0$ the source is not in front of the detector, the phase should be changed from ωt in Eq. (45.8) to $\omega t + \phi$, where ϕ is a phase factor that depends on the position of the source at time $t = 0$. Combining this result with the fact that $u_0 = 0$ and Eq. (45.4), we have

$$f'(t) \approx f_0\left(1 - \frac{R\omega}{C} \cdot \sin(\omega t + \varphi)\frac{1}{\sqrt{1+\varepsilon^2 - 2\varepsilon\cos(\omega t + \varphi)}}\right). \tag{45.9}$$

Figure 45.5 is a schematic illustration of the expected frequency change according to the theoretical model described by Eq. (45.9). When the source passes in front of the microphone, $\omega t = n\pi$ ($\phi = 0$)

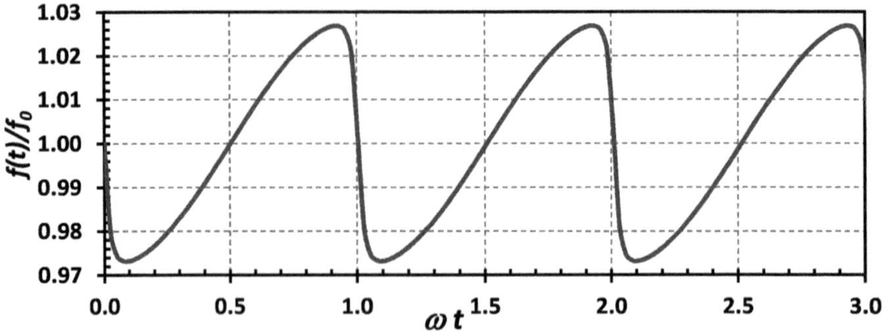

Figure 45.5 Frequency of the sound signal detected by the observer, according to the theoretical prediction of Eq. (45.9). In this example, $L \approx 1.8$ m, $\varepsilon \approx 0.99$, $\phi = 0$, and at a rotation frequency of 1.2Hz, the frequency modulation is of the order of 2%.

with $n = 0, 1, 2, \ldots$ the frequency plot presents abrupt jumps that can easily be observed in the graph, which correspond to the frequency variations when the speed of the source goes from maximum *approaching velocity* to maximum *velocity of distancing* from the microphone (observer).

For this experiment, we will need a Fourier analysis program, which can be used to determine the frequency of the signal detected by the microphone. Fourier analysis enables us to find the frequencies present in the recorded signal [9]. More specifically, we need an application that can perform a Fourier-time analysis, that is, a *sonogram* [17] of the recorded .wav file. There are many programs that can do this, such as Wolfram Research Mathematica (R), MathWorks Matlab (R), and Audacity [16], several of which can be downloaded from the Internet.

Figure 45.6 shows a sonogram obtained from an experiment like the one described here using the program SpectroGram [17]. The vertical axis gives the frequency of the signal detected by the microphone and the horizontal axis is the time. In general, these types of programs only provide a graph similar to the one illustrated in Figure 45.6. However, it is possible to extract from them the values

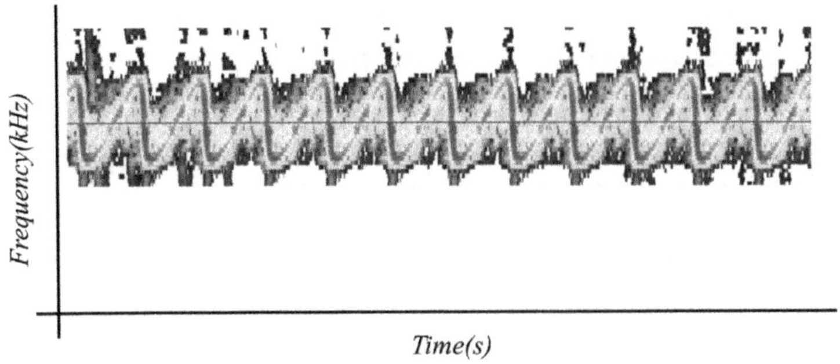

Figure 45.6 Sonogram of a signal detected by the microphone in an experiment similar to the one discussed here. The vertical axis gives the frequency of the signal detected by the microphone and the horizontal axis is the time. Frequency changes of the sound signal detected by the observer may be compared to the theoretical prediction of Eq. (45.9).

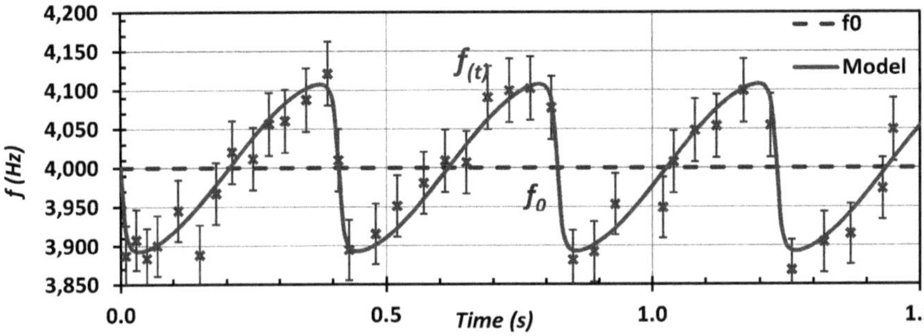

Figure 45.7 Sonogram of the signal illustrated in Figure 45.6, entered as the background of a spreadsheet-generated graph. The solid line is the prediction of the theoretical model.

of the most representative frequency at each time, using graphic programs (see Chapter 10 and Ref. [18, 19]).

We can also cut out the spectrogram and insert it as the background of a graph generated with a spreadsheet, as discussed in Chapter 10. The experimental results can thus be compared to the theoretical model, as illustrated in Figure 45.7.

45.2.1 Project 147. Study of the Doppler effect of a sound source in circular motion

Recommended equipment: A bar of length L ≈ 1.5 m rotating about its center on a horizontal plane. We suggest using a motor with a rotation frequency that can be varied from 1 to 5 Hz. A piezoelectric buzzer with frequency f_0 of a few kHz or a smartphone with an application that produces monochromatic signals. A microphone connected to the sound card of a PC with the capacity to record .wav files. A Fourier analysis program.

The suggested experimental setup is illustrated in Figure 45.3. The microphone records the sound signal in a file for subsequent analysis. The distance s can be varied, although it is recommended that it be as short as possible, to achieve detectable frequency jumps, but without touching the rotating bar. It is also advisable that the system should be located a few meters away from the nearest wall to prevent the sound emitted by the source from bouncing off a wall and being detected, which would generate spurious signals in the experiment.

Suggested tasks

✓ Measure the characteristic parameters of the described experimental arrangement, that is, L and s.

✓ Position the microphone close to the sound source (buzzer) and record the sound emitted when the source is stationary. Analyze the spectrum of the measured signal and determine the characteristic frequency f_0 of the source and its half-width Δf. If the voltage applied to the buzzer is large, you may see some harmonics in the spectrum ($2\,f_0$, $3\,f_0$, etc.). If so, select the lowest frequency signal to work with. You can also use one of the harmonics; it is only important to select the frequency on which you work. If you use a smartphone as a sound source, you can select the desired frequency digitally.

✓ Set the bar in rotation and make sure the speed reaches steady-state conditions. Under these conditions, record the sound emitted by the source after making about 5 or 6 turns. Analyze the sonogram, concentrating on the frequency selected for the study.

✓ From the sonogram, locating the times at which the frequency passes through the value f_0, call these values t_i. See Figure 45.6. Plot these time values, t_i, based on their order of occurrence i, and determine the rotation period of the source, T_{rot}, and its uncertainty. Also estimate the angular frequency of rotation ω and its uncertainty.

✓ From the sonogram, determine the variation of the frequency detected by the microphone as a function of time, $f_{\text{exp}}(t)$, as shown in Figure 45.6. On the same graph, include the experimental values $f_{\text{exp}}(t)$ as a function of t and the theoretical prediction of Eq. (45.9), similar to Figure 45.7. Vary the phase ϕ of Eq. (45.9) until the best fit is achieved.

✓ Repeat the experiment and analysis for at least 3 or 4 speeds of rotation of the bar.

✓ Discuss the degree of agreement between the results obtained and the theoretical model. How adequate is the model to describe your measurements?

✓ What can you conclude about the expression (45.3) to describe the Doppler effect?

45.2.2 *Project 148. Study of the Doppler effect of an observer in circular motion*

Recommended equipment: A bar of length $L \approx 1.5$ m that rotates around its center on a horizontal plane. We suggest using a motor with a rotation frequency that can be varied from 1 to 5 Hz. A piezoelectric buzzer with frequency f_0 of some kHz. A wireless microphone connected to the sound card of a PC capable of recording *.wav files or a smartphone that records sound. A time Fourier analysis program.

The proposed experimental arrangement is illustrated in Figure 45.3, except that this time it is the wireless microphone that rotates, while the sound source remains at rest. The microphone or cell phone records the sound signal in a file for subsequent analysis. In this case, too, the system should be located away from walls.

Suggested tasks

- ✓ Using expression (45.4) as a starting point, develop a theoretical model to describe the variation in the frequency detected by the observer as a function of t, when the observer is moving.
- ✓ For the constructed arrangement, measure the characteristic parameters of the experimental arrangement, that is, L and s.
- ✓ Analyze the spectrum of the signal obtained for the static system and determine the characteristic frequency f_0 from the source.
- ✓ Set the bar into rotation and once it has reached a stable speed, record the sound emitted by the source after 5 or 6 turns. Analyze the sonogram, concentrating on the frequency selected for the study.
- ✓ From the sonogram, determine the variation of the frequency detected by the microphone as a function of time, $f_{\mathrm{exp}}(t)$. On the same graph, plot the theoretical prediction, Eq. (45.9). Vary the phase ϕ of Eq. (45.9) until the best fit is achieved.

45.3 Binary star simulation

45.3.1 *Project 149. Binary star simulation using two smartphones and a PC*

Recommended equipment: A bar of length $L \approx 1.5$ m that rotates around its center on a horizontal plane. We suggest using a motor with a rotation frequency that can be varied from 1 to 5 Hz. Two smartphones as monochromatic sound sources. A microphone connected to the sound card of a PC capable of recording *.wav files or a third cell phone that records sound. A time Fourier analysis program.

The purpose of this experiment is to simulate a binary star system and study the Doppler effect using two smartphones and a PC [18]. The smartphones mounted on a rotating bar are used as monochromatic sound sources. A microphone connected to a PC records these signals in the lab frame of reference. The Doppler effect is readily observed, and the experimental results can be compared quantitatively to the expectation of a theoretical model. This experiment can simulate the Doppler spectroscopy technique used to measure radial velocities and masses in rotating binary stars and the exploration of exoplanets. The arrangement is illustrated in Figure 45.8.

Figure 45.8 shows the experimental setup. Two smartphones are placed on opposite sides of a rotating bar or turntable. The smartphones can emit monochromatic sound. Optionally, they could also measure the speed of rotation of the device. A laptop or PC with a microphone records the sound of the sources. A public domain time-frequency analysis program, available on the Web [14], is used to analyze the sound recorded. Of course, any time-frequency analysis program could be used. In principle, a third smartphone could also be used to record the sound, but a laptop with a larger screen makes it easier to monitor and visualize the results online. The smartphones used here are inexpensive and robust. Each device was installed with

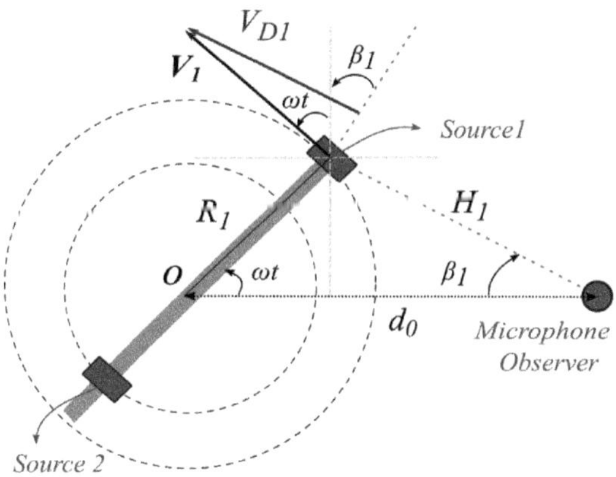

Figure 45.8 Diagram of the experimental setup. One smartphone (violet rectangle in the diagram) is attached to each end of a rotating bar. Both these smartphones can emit sound at a well-defined frequency. Alternatively, the second smartphone can be set to record the sound of the other phone in the same system of reference. A different microphone, in the lab system of reference (Observer) records the sound of the emitting sources.

a frequency sound generator [15] and Phyphox [16], which converts almost any smartphone into a basic Physics Lab. The program Spectrogram is used to analyze the sound signal recorded by the fixed microphone [14]. Furthermore, the program xyExtract [17] can be used to digitalize the data of the time-frequency analysis.

Using the same logic that leads to Eq. (45.9), assuming that $R\omega < c$ (and $R_2\omega < c$), we have

$$f_1(t) \approx f_{10}\left(1 - \frac{R_1\omega}{c} sen(\omega t + \phi)\frac{1}{\sqrt{1 + \alpha_1^2 - 2\alpha_1 \cos(\omega t + \phi)}}\right). \tag{45.10}$$

and

$$f_2(t) \approx f_{20}\left(1 + \frac{R_2\omega}{c} sen(\omega t + \phi)\frac{1}{\sqrt{1 + \alpha_2^2 + 2\alpha_2 \cos(\omega t + \phi)}}\right). \tag{45.11}$$

Here, $\alpha_1 = \frac{R_1}{d_0}$ and $\alpha_2 = \frac{R_2}{d_0}$, analogous to the parameter ε of Eq. (45.8).

Suggested tasks

✓ Using an experimental setup like the one illustrated in Figure 45.9, study the Doppler variation of the frequencies f_1 and f_2 of the two rotating sound sources. Their frequencies at rest can be set, for example, at $f_{10} = 3.5$ kHz and $f_{20} = 4$ kHz.

✓ Record the resulting sound with the microphone of a PC at rest (or a third smartphone at rest), and using a sonogram like the one obtained in Project 147, extract the variation of the frequencies f_1 and f_2 as a function of time.

✓ Plot the experimental values of the frequencies f_1 and f_2 as a function of $\omega t/2\pi$. On the same figure, plot the theoretical expectations of Eqs. (45.9) and (45.11).

✓ Discuss the adequacy of the theoretical model to describe the variation in the frequency detected by the observer as a function of t.

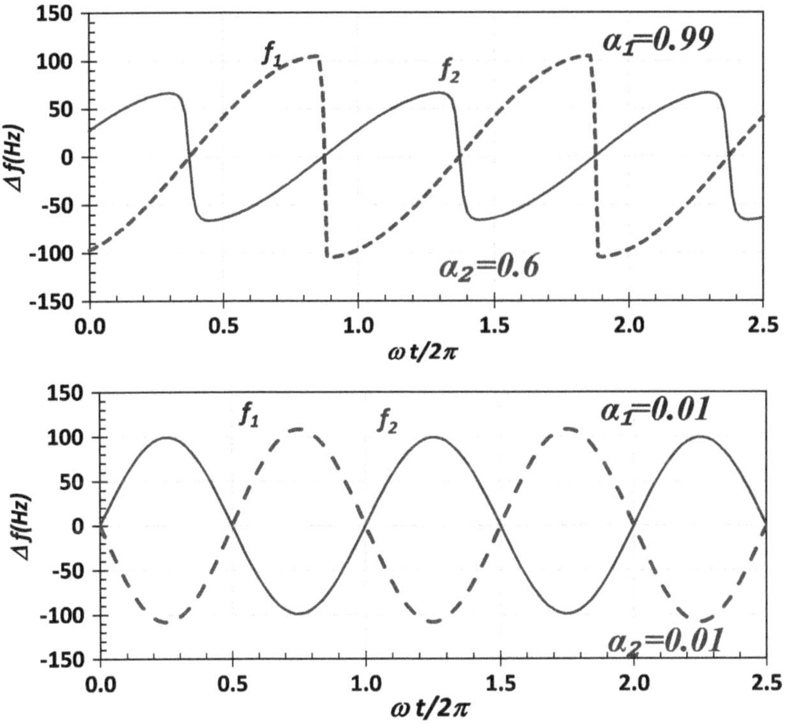

Figure 45.9 Theoretical prediction of a model, Eqs. (45.9) and (45.11), for the difference in frequency $\Delta f_i = f_i - f_{i0}$, observed in the lab-frame for two sources located at opposite ends of a rotating bar. The dashed blue curve is the difference in frequency seen from source 1 and the solid red curve is the expectation for source 2, both measured by an observer at rest in the laboratory frame of reference. The top panel corresponds to the case of an observer close to the sources, $\alpha_1 = 0.99$, while the lower panel applies to a case of a distant source $\alpha_1 = 0.01$. Here we have used $f_{10} = 4$ kHz and $f_{20} = 4.4$ kHz.

Compare your experimental results to those obtained from the measurement of binary stars orbiting around their common center of mass. The study of binary systems provides a wealth of information on the masses of the stars, the distance between them, the period of rotation, and many other useful data. Some binary stars can be studied by direct observation with telescopes or even space telescopes such as Hubble. If their plane of rotation is quasi-perpendicular to the line of sight, and the system is close by and within our Galaxy,

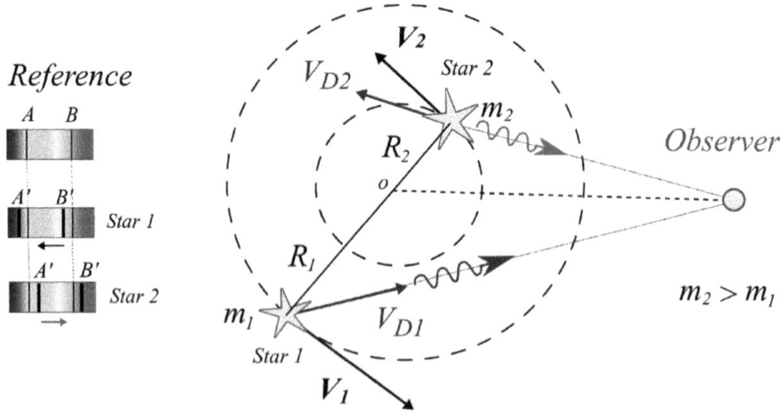

Figure 45.10 The spectroscopic method employs the Doppler shift of the absorption spectra to obtain the relative velocity of the stars, using the same conceptual scheme as the one used in the experiment of the smartphones. From their relative velocities, the masses of the stars m_1 and m_2 can be deduced. The component of the velocity in blue is the *observational speed* or radial velocity.

the rotation may be observed and studied over time. Using this type of method, it has been possible to study and analyze binary systems, such as the Sirius star, one of the brightest in the sky, and α Centauri, the closest star, among many others [5, 8, 18]. See Figure 45.10 [18].

References

[1] D. Halliday, R. Resnick and J. Walker, *Fundamentals of Physics: Extended*, 11th Ed., John Willey and Sons Inc., 2018.
[2] H. Young, R. Freedman and F. Sears, *Sears and Zemansky's University Physics*, NY: Pearson Pub., 2004.
[3] D. Nolte, "The fall and rise of the Doppler effect," *Phys. Today*, vol. 73, no. 3, pp. 30–35, March 2020.
[4] J. Pasachoff and J. Mativi, "Demonstrating the cosmic distance ladder with cepheids," *Phys. Teach.*, vol. 58, pp. 6–7, 2020.
[5] Wikipedia, "Big Bang theory," 2022. Available: https://en.wikipedia.org/wiki/Big_Bang.
[6] R. Hilditch, *An Introduction to Close Binary Stars* (Cambridge Astrophysics) 1st Ed., NY: Cambridge University Press, 2001, p. 381.
[7] B. B. Youtube, "Binary star masses — Part 1, 2020. Available: https://www.youtube.com/watch?v=EWSobgpcZl0.

[8] A. Grant, "Half of Nobel Prize in Physics honors exoplanet trailblazers," *Phys. Today*, vol. 72, no. 12, pp. 17–19, 2019.
[9] F. S. Crawford Jr., *Waves (Berkeley Physics Course, Vol. 3)*, Cambridge University Press, 2013.
[10] R. Halliday, D. Resnick and M. Krane, *Fundamentals of Physics*, NY: John Wiley & Sons, Inc., 2002.
[11] F. Sears, M. Zemansky, H. Young and R. Freedman, *College Physics*, NY: Pearson Ed., 2009.
[12] R. Feynman, *et al.*, *The Feynman Lectures on Physics, Boxed Set: The New Millennium Edition*, Vol. 1, Basic Books, 2011.
[13] C. Wörner and R. Rojas, "Classical Doppler effect in some accelerated systems," *Phys. Teach.*, vol. 59, p. 239, 2021.
[14] J. L. Di Laccio, L. Vitale, R. Alonso-Suárez, N. Perez and S. Gil, "Estudio del efecto Doppler utilizando teléfonos inteligentes," *Eureka*, vol. 14, no. 3, p. 637–646.
[15] Google Play-LuxDeLux, "Google play: Frequency sound generator," 2020. Available: https://play.google.com/store/apps/details?id=com.luxdelux.frequencygenerator&hl=es_UY&gl=US.
[16] RWTH Aachen University, "Phyphox," 2020. Available: https://phyphox.org/.
[17] Phyphox, "Smartphone-experiment: Pendulum," Phyphox App., 2021. Available: https://www.youtube.com/watch?v=xY3NFcDG3ZU.
[18] Audacity, "Free audacity audio editor for windows, Mac or Linux," 2022. Available: https://sourceforge.net/projects/audacity/.
[19] W5Big, "Spectrogram," 2020. Available: http://w5big.com/spectrogram.htm.
[20] J. DiLaccio, P. Nuñez and S. Gil, "Binary stars simulation using smartphones — A Doppler effect experiment," *Brasileira de Ensino de Física*, vol. 43, p. e20210241, 2021.
[21] M. Saba and R. Rosa, "The Doppler effect of a sound source moving in a circle," *Phys. Teach.*, vol. 41, pp. 89–91, 2003.
[22] Wilton Pereira da Silva, Cleide Pereira da Silva, "xy-extract graph digitizer," 2011. Available: http://zeus.df.ufcg.edu.br/labfit/index_xyExtract.htm.
[23] V. Dhillon, "The lives of binary stars," University of Sheffield, 2012. Available: http://www.vikdhillon.staff.shef.ac.uk/seminars/lives_of_binary_stars/title.html. [Accessed 2020].
[24] H. Bond, *et al.*, "The Sirius system and its astrophysical puzzles: Hubble space telescope and ground-based astrometry," *Astrophys. J.*, vol. 840, no. 2, pp. 1–17, 2017.

Chapter 46

GEOMETRIC OPTICS EXPERIMENTS

This chapter proposes an experimental study of the basic laws of geometric optics and the laws of reflection and refraction of light. It also analyzes the phenomenon of total internal reflection. Next, it discusses the properties of thin lenses and various methods to determine the focal lengths of different lenses. It analyzes the properties of a converging and a diverging lens and ways to determine their focal lengths.

Goals

✓ Geometric optics
✓ Reflection and refraction of light — Snell's law
✓ Thin lenses
✓ Determining the focal length of a lens

46.1 Geometric optics — Laws of reflection and refraction

When a light beam falls on the surface that separates two different media in which the light propagates at different speeds, it is partially transmitted and partially reflected [1], as shown schematically in Figure 46.1. For any medium, the refractive index n is defined as follows:

$$n = \frac{c}{v}, \tag{46.1}$$

where c is the speed of light in vacuum ($c = 2.99792458 \times 10^8$ m/s) and v the speed of light in that medium.

In a homogeneous, isotropic medium, light travels in a straight line in the form of light rays [2]. This statement is the basic assumption of the ray model of light. This model assumes that light travels in straight-line paths called light rays [3, 4] and is the central concept of geometric optics. There is a great variety of phenomena that can be explained by means of rays, such as the formation of shadows and images. The basic laws of geometric optics can be summarized as follows:

✓ **Law of reflection**: When a ray of light is reflected from the flat surface that separates two media of different refractive indices, the *incident ray*, the *normal to the surface*, and the *reflected ray* are in the same plane, as indicated in Figure 46.1. In other words, the angle of reflection (θ_2) equals the angle of incidence (θ_1):

$$\theta_1 = \theta_2. \tag{46.2}$$

✓ **Snell's law:** When light passes from one medium to another separated by a flat surface, the incident ray, the normal to the surface, and the refracted (transmitted) ray are in the same plane and the angle of incidence (θ_1) and the angle of refraction (θ_3) are related by

$$n_1 \sin(\theta_1) = n_3 \sin(\theta_3), \tag{46.3}$$

where n_1 is the refractive index of the medium of the incident ray and n_3 is the refractive index of the transmitted ray.

As a consequence of these laws, there is the phenomenon of *total reflection*, when the light ray passes from a more refractive medium

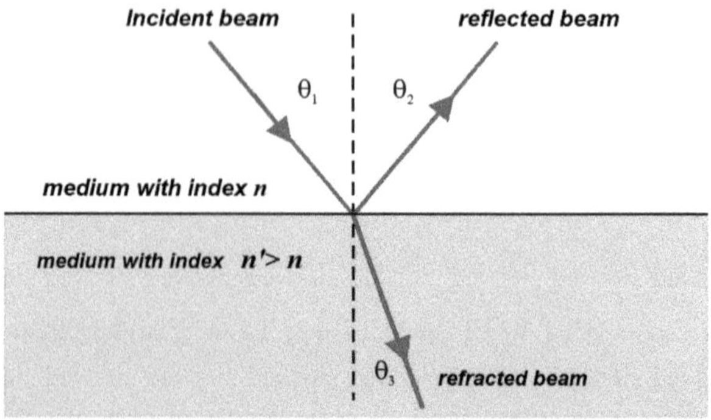

Figure 46.1 Reflected and refracted light beam.

(greater n) to a less refractive medium (smaller n, that is, $n_1 > n_2$); here the refracted ray moves further away from the normal to the surface. There is an incident angle θ_c for which the refracted ray reaches the angle $\theta_3 = 90°$. This incident angle is called the *limit angle* or *critical angle*; if $\theta_1 > \theta_c$, there is no transmission but total reflection of the light. This phenomenon is used in optical fibers to confine a beam of light inside them and is very useful for transmitting information through them.

The goal of the following experiments is to falsify these laws and study the general properties of the reflection and refraction of light.

Note: Be very careful when you use a laser. Do not allow any rays or their reflections to touch your face or anyone else's. A laser beam can cause irreversible damage to the retina. Read the safety recommendations in Appendix B and the safety regulations of your institution or country regarding the use of lasers [5].

46.1.1 *Project 150. Study of reflection and refraction*

Recommended equipment: A glass or acrylic half cylinder approximately 15 cm in diameter and 3 cm in height. It can be a solid cylinder or a hollow one that can be filled with water or another liquid. Pins and a soft wooden or cardboard plate to stick pins into. Optional: A laser pointer.

The objective of this experiment is to investigate the relationship between the angle of reflection, θ_2, and the angle of refraction, θ_3, according to the angle of incidence θ_1. All the angles are defined with respect to the normal line of the separation surface between the two media. For this experiment, it is suggested to use the device shown in Figure 46.2 [6].

Refraction: Take the point O, coinciding with the center of the half cylinder, as the origin of the coordinates. Choose the x-axis so that it coincides with the flat face of the half cylinder and take the normal to the flat face as the y-axis, with respect to which you will measure the angles. Set the half cylinder on a flat surface on which graph paper is placed [7] to determine the coordinates of each point

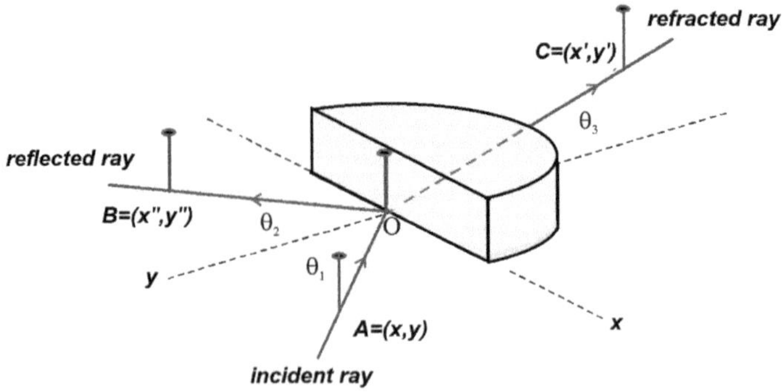

Figure 46.2 Suggested device for studying the laws of reflection and refraction.

on the plane. A simple way to determine the direction of light rays is to use three pins, one of which remains fixed at the center O. To determine the direction of the incident and refracted rays, place a pin at position A [whose coordinates are (x, y)], another at O (center of the half cylinder, just outside it), and the third in position C [whose coordinates are (x', $y\prime$)]. The position of the third pin is chosen such that when viewed horizontally, all three pins appear aligned. In other words, the light that comes from the pin in C passes through O and A and finally reaches our eyes. Note that the light that reaches our eyes does not always follow a straight line.

Suggested tasks:

✓ Place a sheet of graph paper on the table, clearly marking the x and y axes (normal). Keep the pin in the center of the half cylinder in a fixed position, point O. The pin at position A defines the angle of incidence.

✓ Looking from the line formed by pins A and O, place the third pin at point C so that all three are aligned. When such alignment has been achieved, record the coordinate values (x, y) and (x', y') and determine the angles θ_1 and θ_3. It is advisable to vary the angle of incidence θ_1 from 0° to 80° in steps of approximately 10°. If you have a laser pointer (or an ordinary laser), it is possible to visualize the progress of the rays. To do this, point the laser so that the beam strikes in the direction defined by AO through the pins. Verify that the refracted beam actually passes through point C. To see the laser light more clearly, you could use a fog machine that generates vapor or smoke [8].

✓ Plot θ_3 as a function of θ_1 and $\sin(\theta_1)$ as a function of $\sin(\theta_3)$. Analyze these dependencies. Check whether the latter graphical representation can be described by a line. If so, determine its slope and its uncertainty.

✓ Do your experimental results agree with the prediction of Snell's law? If so, determine the refractive index of the half-cylinder material (n_3). Compare the value obtained in your experiment to the values in the tables for the material under study. Remember that the refractive index of air under normal conditions of pressure and temperature is $n_1 = 1.00029$. Using Eq. (46.1), determine the speed of propagation of the light used in the medium of the index n_3.

✓ ♣♣ Reverse the direction of incidence of the laser beam, that is, make it go in the direction CO and verify that it passes through A. Likewise, if the incident beam follows the BO direction, verify that the reflected beam follows the direction OA. This remarkable property of light is known as the *principle of optical reversibility*, which states that in the absence of absorption, if the directions of the light rays are reversed, the trajectories are not altered. More generally, this principle is a consequence of the *principle of temporal reversibility* of the equations of physics [4].

Reflection: Perform the same study described above but with the ray reflected. To do this, you can use the three pins technique described above or a laser pointer that points from the AO direction and determine the coordinates (x'', y'') from point B. For this approach (with a laser), you may want to remove the pin in front of O. A small mark on the acrylic will help you locate this point O at which to point the laser beam. To determine the coordinates of B, you can use a sheet of paper as a screen, to see where the reflected light falls, thereby locating point B more easily.

Suggested tasks:

✓ Graph θ_2 as a function of θ_1.

✓ Are your data in accordance with the law of reflection, Eq. (46.1)?

46.1.2 *Project 151. Total internal reflection*

Recommended equipment: A glass or acrylic half cylinder (half round) approximately 15 cm in diameter and 3 cm in height. A laser pointer.

As stated above, the phenomenon of total internal reflection occurs when light passes from a medium with a higher refractive index to another with a lower index.

For this experiment, point a laser along the curved surface side of the setup in Figure 46.2 so that it passes through point O (radial direction). The light will hit the flat face of the half cylinder radially from the acrylic, which has a refractive index (n_3) greater than air (n_1). Observe the transmission of the light from the half cylinder to the air as you change the angle of incidence (θ_3) of the light on the acrylic–air interface. When the refracted ray in the air is close to 90°, you will notice that the reflected ray intensifies. Determine the critical angle, θ_c, as precisely as possible and compare it to the expected value of Eq. (46.3), that is,

$$\sin(\theta_c) = n_1/n_3. \tag{46.4}$$

Figure 46.3 illustrates these situations.

Suggested tasks:

- ✓ Using the laser pointer, point radially into the half cylinder, as illustrated in Figure 46.3, estimate the value of θ_c *critical angle* (or *limit angle*), for which the transmitted ray ceases to exist.
- ✓ Compare your result to the expectation of Eq. (46.3), with the value of n_3 found in the previous project.
- ✓ For $\theta_3 > \theta_c$, there is no transmitted beam. Using the laser pointer, check that this condition is fulfilled in the case under study and check that all incident light is reflected on the inside of the half cylinder.

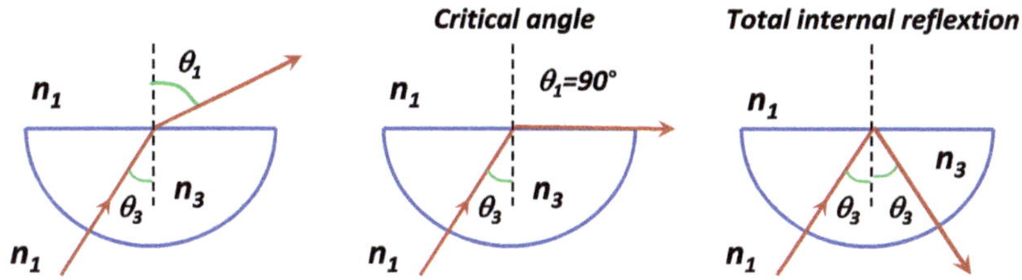

Figure 46.3 Light reflection from a medium of higher refractive index (n_3) onto a flat surface that connects to a medium of lower refractive index (n_1). If the incident angle $\theta_3 > \theta_c$, there is a total internal reflection, as shown in the figure on the right.

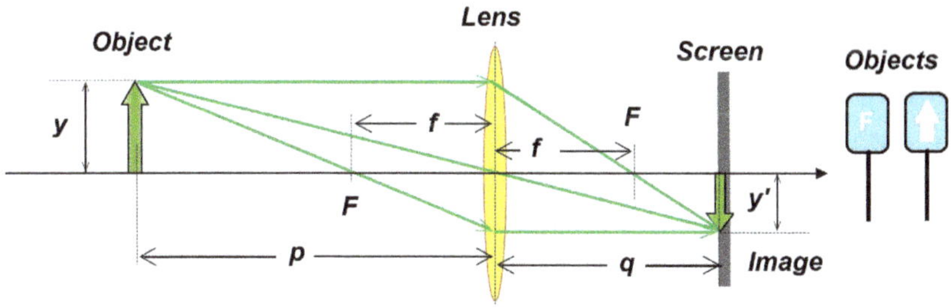

Figure 46.4 Optical bench for studying the properties of a thin converging lens and an object, and definition of important distances. The figure on the right illustrates some examples of possible objects.

46.2 Thin lenses

A lens is an optical system with two or more refracting surfaces. Lenses are essential elements of many optical instruments: microscopes, telescopes, cameras, etc. [1, 4, 6]. Their operation can be understood using the laws of geometric optics. If a parallel beam of light, after passing through a lens, converges to a point, the lens is said to be converging. The distance from the lens to the convergence point defines its *focal length*. If the parallel rays diverge after passing through the lens, the lens is diverging [1, 4, 6]. One way to study the properties of a lens is to use an *optical bench* [9, 10], with the elements schematically shown in Figure 46.4.

An optical bench consists of a horizontal rail, with a graduated scale attached to record the positions of the different elements. These elements can slide along the axis of the bench. Optical benches usually have supports that hold the elements to be used: lenses, screens, light sources, objects, diaphragms, etc. Almost any thin object or a screen can be used as an *object*. A good object can be made with a screen made of hard cardboard or metal sheet with a cross-shaped opening (for example, the orientation of the object can be clearly indicated with arrows), behind which a light source is placed, as shown on the right of Figure 46.4. An electric lightbulb or a lamp can also be used (for example, a 12 V straight filament bulb for car headlights is suitable). A simple way to build an optical bench is to use a 1-meter ruler graduated in millimeters on which the optical elements can move [9, 10]. If there are no suitable mounts to hold the optical elements, they can be supported with plasticine. The image is formed on a sliding screen. Care must be taken that the axes of the optics are properly aligned.

46.2.1 *Project 152. Converging lenses — Qualitative Observations I*

> **Recommended equipment:** A converging lens or ordinary magnifying glass.

The goal of this project is to make some qualitative observations using a converging lens. There are many commercial providers of good, inexpensive lenses [11, 12].

Observation of the image formed by a converging lens: Observe an object through a converging lens. Observe nearby objects (printed text, for example) and distant objects (a landscape, a building, or a tree). Describe your observations qualitatively.

✓ Describe how the characteristics of what you observe vary as the observer-object distance varies. Is the image larger, smaller, or the same as the object itself? Is the image upright or inverted? Do these images vary by varying the observer-lens distance?

✓ Another interesting property of converging lenses is that they form *real images* [1, 4], that is, images that can be projected onto a screen. For this observation, it is advisable to have a well-lit object, for example, a tree or a landscape, as shown in Figure 46.4. Place the lens between the object and the screen and vary the lens-to-screen distance until a clear image of the object appears. The image on the screen will appear inverted. This is the principle of operation of a camera, in which the photographic film constitutes the "screen" where the image is formed.

Determining the focal length of a converging lens: The *focus* of a converging lens is the point on the optical axis at a distance f from the lens, where all the incident paraxial rays (parallel and close to the axis) converge after refraction [1, 4, 6]. A simple way to determine f, called the *focal length* of the lens, consists of forming on a screen an image of a light source that is very far from the lens. With this condition, the incident rays can be considered as parallel, and the point image of this distant source will be at a distance from the lens equal to f.

✓ Choose a distant light source, for example, the Sun or a distant lamp, and place your converging lens between this source and a screen. Vary the lens-to-screen distance until a spot image appears sharply on the screen (or as small as possible). Under these conditions, the lens-screen distance is the focal length of the lens. To achieve clear images, it is important that the lens plane is parallel to the screen and both planes are perpendicular to the direction determined by the screen and the light source.

✓ Estimate the focal length of your lens using this procedure.

Note: If you use the Sun as a distant light source, note that the lens concentrates all the sunlight falling onto the lens in its focus. This can be demonstrated by burning paper (which burns at approximately 240° C) using the heat produced by this high concentration of light at a single point. To burn the paper, you need a certain exposure time of about 1 min.

46.2.2 *Project 153. Properties of lenses — Qualitative observations II*

Recommended equipment: A converging lens with focal length between 5 cm and 10 cm. An incandescent lamp or another luminous object. Transparent colored sheets to use as filters. A 1 m ruler or optical bench.

An interesting property of lenses is the following. Imagine that you have an object, which, by means of a converging lens, forms a real image on a screen, as shown in Figure 46.4. Without experimenting, predict how the image will change if the top half of the lens is covered with an opaque mask (which does not allow light to pass through) and what the image will look like if the left half is covered. Make a diagram illustrating the shape of the object and its image in each case. What would the image look like if three-quarters of the lens were covered?

Then do the experiment and compare your predictions with your observations. How can these results be explained? Try to understand your observations using Fermat's principle [1, 3, 4, 6]. To do so, remember that, according to this principle, the path (or paths) that light follows to go from one point A to another point B are those that make the transit time between A and B minimum, maximum, or stationary. In an ideal lens, the transit time of light from one point on the object to the corresponding point in its image is the same for all (paraxial) rays passing through the lens. In other words, the rays that reach any point in the image, come from the corresponding point on the object and pass through *all points* on the lens. This

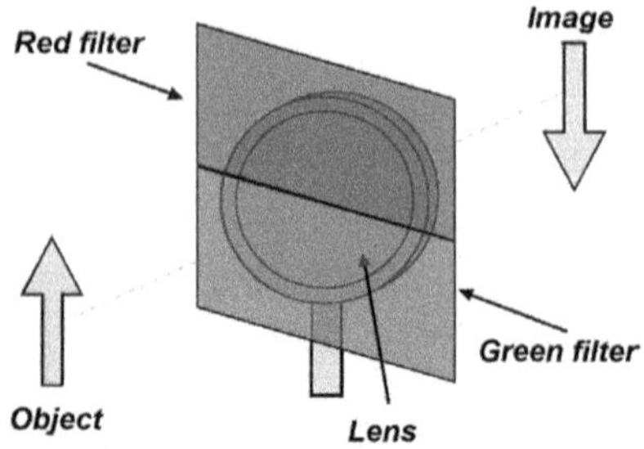

Figure 46.5 Exploring lens properties.

property of lenses is similar to the relationship between image and object in the case of elliptical mirrors. If we place an object in one of the foci of the ellipse, the image will be produced in the other focus. As the focus-mirror-focus distance is the same for all points on the ellipse, so is the time it takes for light to go from one focus (object) to another (image). Therefore, the rays that reach the image come from all the points of the ellipse; if one part of the elliptical mirror is obstructed, the image is still formed equally completely in the other focus, though the amount of light that will arrive will be less.

Suggested tasks:

- ✓ Block different parts of the lens and see how it affects the image. State your conclusions.
- ✓ Another interesting lens property can be seen by covering the upper half of the lens with a red filter and the lower half with a green one [13]. You can use two pieces of transparent colored paper or two pieces of colored acrylic as filters. Before doing the experiment, guess what you expect to see (in terms of the color of the image that will be formed). Then do the experiment and discuss your results.
- ✓ What is the difference between a real image and a virtual image? To make a comparison, what kind of image is seen in a flat mirror? And in a concave mirror?
- ✓ What kind of image can be projected on a screen: a real image or a virtual one? Where must the object be located relative to the lens to get an image that can be seen on a screen? Investigate these points by working on the optical bench.

46.2.3 *Project 154. Converging lenses — Quantitative study*

> **Recommended equipment:** A converging lens with focal length between 5 cm and 10 cm. An incandescent lamp or another luminous object. Transparent colored sheets to use as filters. A ruler of approximately 1 m or optical bench.

The aim of the following experiment is to study quantitatively what was discussed qualitatively in the previous projects. Use an optical bench to align the elements (object, lens, and screen), as shown in Figure 46.4.

Suggested tasks:

✓ For various object-screen distances, find as many images as you can by varying the position of the lens. For how many lens positions do you see sharp images on the screen? Every time you see sharp images, measure

- the object-lens distance (p),
- the screen-lens distance (q),
- the object and image sizes and their respective orientations (upright or inverted).

✓ Represent q^{-1} as a function of p^{-1}. What relationship do you find between these variables? Discuss whether your data can be described by the *thin lens equation* [1, 3, 4]:

$$\frac{1}{p} + \frac{1}{q} = \frac{1}{f}, \tag{46.5}$$

where f is the focal length of the lens.

✓ From the graph q^{-1} versus p^{-1}, find the focal length f of the lens. Determine the uncertainty of the determination of f. One way to estimate the uncertainties of the distances p and q is to move the lens while keeping the object-screen distance fixed ($D = p+q$). Varying the lens position determines the range of positions for which the image sharpness does not vary. This range enables estimation of the uncertainties of the distances p and q. If there are more factors that influence the determination of distances, specify them and discuss their weight in determining the final uncertainties.

✓ Lateral magnification m is defined as the ratio of image size to object size. Determine experimentally the magnification of the image that results from different relative positions between object and lens. Compare the result of your measurements with the predictions of geometric optics. Graph m and the quotient q/pas a function of p on the same graph and discuss your results.

✓ Another interesting property of converging lenses is that they only form an image of an object on a screen when the object-screen distance, $D = p + q$, meets the condition $D > 4f$ (Figure 46.4), where f is the focal length.

✓ Experimentally investigate the validity of this statement using an optical bench and a lens of known focal length f.

✓ Using Eq. (46.5), it is possible to show that if $D < 4f$, the lens does not form image. To do this, write Eq. (46.5) in terms of D and q and find q as a function of the other parameters (this results into a quadratic equation in q). Analyze the solution and show that imaging is not possible if $D < 4f$. Consider the case of $D = 4f$. Where is the image formed, and with what magnification, in this case? Verify your preliminary expectations experimentally.

46.2.4 *Project 155. Simple method to estimate f for a diverging lens*

Recommended equipment: A diverging lens.

Diverging lenses do not form a real image of an object, but a *virtual image* can be observed. A simple method to estimate the value of the focal length of a diverging lens is to use a set of parallel, equally spaced lines [14], as shown in Figure 46.7, and view them through the lens. Place the diverging lens between your eyes and the paper and try to simultaneously see part of the object directly and part through the lens; with a little practice, you will soon be able to do this. Vary the object-lens distance until the magnification (of the virtual image) is $1/2$, which is characterized by the fact that in this condition ($1/2$ magnification), three parallel lines of the virtual image coincide with two of the objects. In this situation, the object-lens distance is the focal length of the diverging lens. There are other techniques to find the focal length of diverging lenses [15–17].

Suggested tasks:

✓ Using the lens formula, Eq. (46.5), justify the procedure described for determining the focal length of a diverging lens.

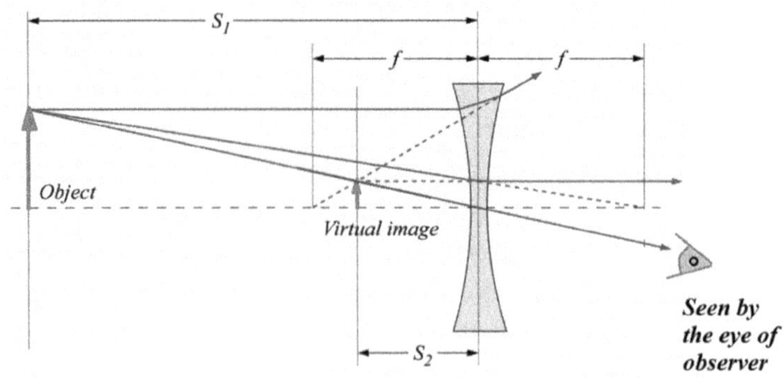

Figure 46.6 Diverging lens properties. The image (virtual) can be seen by the eye of the observer, but it cannot be projected onto a screen.

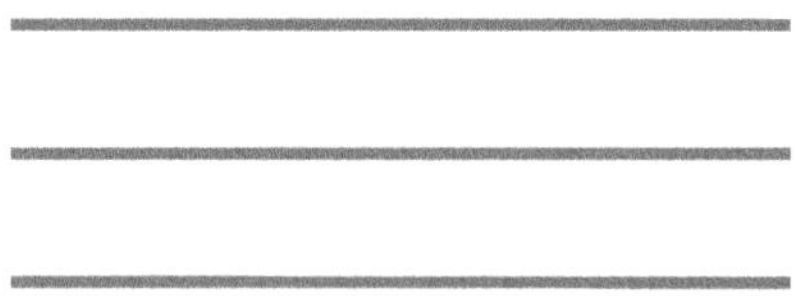

Figure 46.7 Three evenly spaced parallel lines used to estimate the focal length of a converging lens.

✓ Using a diverging lens (you can use spectacles belonging to someone with myopia), determine the focal length of the lens. Compare it to the nominal value of the diopter prescribed by the ophthalmologist to the owner of the glasses. Remember that diopter is the inverse of the focal length measured in meters.

46.2.5 *Project 156. Quantitative method to estimate f for a diverging lens*

Recommended basic equipment: A diverging lens and a converging lens. An optical bench.

A more elaborate method of determining the focal length of a diverging lens is to measure the object-image distances, as in the case of converging lenses. To determine the lens-image distance, q,

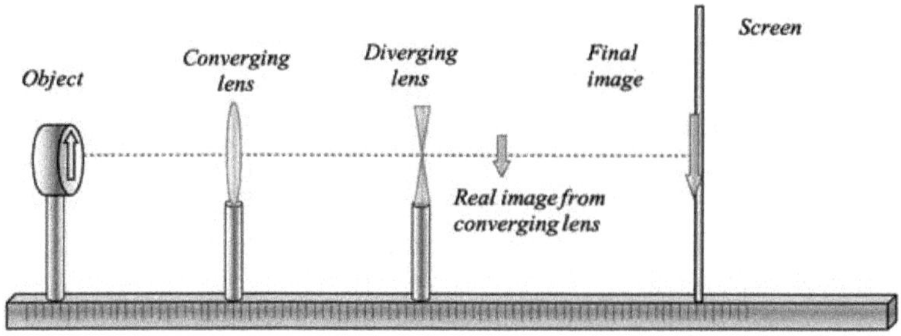

Figure 46.8 A way to measure the focal length of diverging lenses.

it is necessary for the image to be real in order to collect it on a screen. Therefore, a *virtual object* is used, produced by the real image formed by a converging lens. The experimental setup is shown in Figure 46.8 [15–17].

Suggested tasks:

- ✓ Collect the actual image of the object formed by the converging lens alone on the screen. Read the position of this image on the optical bench ruler and estimate the uncertainty of that position.
- ✓ Sandwich the diverging lens between the converging lens and the actual image produced by it, as shown in Figure 46.8. Determine the position of the virtual object, relative to the diverging lens and the final real image resulting from the two combined lenses.
- ✓ Adjust the position of the screen until the final image is clear. Read the position of the final image on the ruler of the optical bench. With the readings, determine p and q corresponding to the diverging lens, and use these values to estimate the focal length f of the diverging lens, using Eq. (46.5). Estimate the uncertainty of this magnitude.
- ✓ How does this measurement of f compare to the measurement taken using the method from the previous project?

References

[1] F. Sears, M. Zemansky, H. Young and R. Freedman, *College Physics*, NY: Pearson Ed., 2009.

[2] Wikipedia, "Ray - optics," 2022. Available: https://en.wikipedia.org/wiki/Ray_(optics).
[3] R. Halliday, D. Resnick and M. Krane, *Fundamentals of Physics*, NY: John Wiley & Sons, Inc., 2002.
[4] E. Hecht, *Optics*, NY: Addison Wesley, 1990.
[5] University of Princeton, NJ, "Safety recommendations for laser pointers, 2021. Available: http://web.princeton.edu/sites/ehs/labpage/laserpointersafety.htm.
[6] U. Haber-Schaim, J. H. Dodge, E. Shore and R. Gardner, *PSSC Physics — Lab Guide*, 7th Ed., NY: Kendall Hunt Pub Co., 1997.
[7] "Graph paper generator," MathPolate, 2022. Available: https://mathpolate.com/graph.
[8] "10 ways to make a laser beam more visible," 2022. Available: https://www.youtube.com/watch?v=7x0SV0oidMk.
[9] L. Dvořák, "A do-it-yourself optical bench," *Phys. Tech.*, vol. 49, no. 7, pp. 452–455, 2011.
[10] "Introduction to optical bench," Ann Harbor Scientific, 2022. Available: https://www.youtube.com/watch?v=AfLqWR95IcE.
[11] Edmund Optics Co., "Edmund optics," 2022. Available: https://www.edmundoptics.com/.
[12] Amazon, "Best sellers in science classroom optics kits," 2022. Available: https://www.amazon.com/Best-Sellers-Science-Classroom-Optics-Kits/zgbs/industrial/6651007011.
[13] H. Drewit and D. Heckathorn, "Figuring physics," *Phys. Teach.*, vol. 37, p. 104, 1999.
[14] S. C. Dudley, "Tick of the trade — How to quickly estimate the focal length of a diverging lens," *Phys. Teach.*, vol. 37, p. 94, 1999.
[15] D. Crandles and R. Kauffman, "Locating images formed by diverging lenses," *Phys. Teach.*, vol. 35, p. 365, 1997.
[16] H. Roy, "A new method for measuring the focal length of a diverging lens," *Am. J. Phys.*, vol. 40, p. 1869, 1972.
[17] J. Wang and W. Sun, "Measuring the focal length of a camera lens in a smart-phone with a ruler," *Phys. Teach.*, vol. 57, p. 54, 2019.

Chapter 47

PHYSICAL OPTICS EXPERIMENTS

This chapter analyzes the phenomena of interference and diffraction of light using a laser as a coherent monochromatic light source. Two experiments — diffraction through a slit and from an opaque solid object — provide an opportunity to investigate Babinet's principle. This chapter also analyzes interference and diffraction through a pair of slits and through a diffraction grating. The experiments can be used to determine the wavelength of a laser or to measure small lengths by using optical techniques. This chapter complements the phenomena discussed in Chapter 42. It also investigates some properties of polarized light, in particular Malus's law.

Goals

- ✓ Physical optics
- ✓ Interference and diffraction through slits and through a grating
- ✓ Young's experiment
- ✓ Determining wavelength
- ✓ Malus's law

47.1 Diffraction and interference of light — Nature of light

In classical physics, there are two ways in which energy can be transmitted from one point to another. One is associated with the transport of matter, e.g., the kinetic energy associated with a moving particle or body. The other is associated with the propagation of waves [1], e.g., the energy of a sound that reaches our ears as vibrations that propagate through the air in the form of acoustic waves.

These two modes of energy transmission differ in what happens when the intensities associated with the energy that they propagate are added. In the case of particles, the intensities always add up. For example, if we throw N darts at a target, we will accumulate the N darts that hit it. If we identify the intensity of impacts on the target with the number of darts received, the intensity will be greater when more darts are thrown. On the other hand, when two or more waves overlap, the situation is completely different. According to the *principle of superposition*, as discussed in Chapter 42, what is added in the case of waves are their amplitudes, and the amplitude of the resulting wave is the sum of the individual components. To obtain the *intensity* of the resulting wave (energy that arrives per unit area and per unit time), the resulting amplitude must be squared. Thus, when two or more waves are superimposed, the resulting intensity may be reinforced and increase in some zones and decrease or even be canceled out in others.

This *interference* phenomenon is a characteristic property of waves and a consequence of the principle of superposition. In classical theory, interference is used to distinguish waves from particles operationally. In a situation in which it is unknown whether the nature of a phenomenon is "corpuscular" or "wave", a superposition experiment can be performed: interference indicates the presence of waves, while the absence of interference indicates a corpuscular or particle phenomenon [1].

In about 1800, Thomas Young performed an interference experiment that produced results which were inexplicable in terms of the "corpuscular theory" of light prevalent at the time. From these experiments, it was deduced that light is a wave phenomenon [1–4]. In his experiment, Young observed the image produced by light as it passed first through one slit and then through two other very close slits. He used filtered light from a mercury arc to make sure he was working with monochromatic light.

As a result of the experiment, Young observed a series of light and dark areas, and furthermore, that a certain point on the screen lit up when one of the slits was covered while it became a dark point when both slits were uncovered. In other words, he observed that "light + light" sometimes produces a "brighter area" (a more illuminated area) and other times a "less bright area" (a less illuminated or dark area). It would be impossible to explain this phenomenon if light had a corpuscular nature because the individual intensities would add up and there would always be light.

A fundamental condition for observing the phenomena of diffraction [5] and interference is that the incident light beam on the slits must be coherent, i.e., the phase relationship between any two points on the wavefront must maintain a constant phase relationship. This can be easily accomplished by using a laser as the source of the incident beam.

If a coherent beam of light with wavelength λ passes through a slit of width a, of size comparable to λ, a diffraction pattern will be produced on the screen, and its intensity can be described by [1–4]

$$I_d(\theta) = I_0 \left(\frac{\sin(z)}{z} \right)^2, \tag{47.1}$$

where

$$z(\theta) = \pi \frac{a}{\lambda} \sin(\theta). \tag{47.2}$$

The angle θ determines the position of the point on the screen where the intensity is calculated (point P in Figure 47.1). If there are two

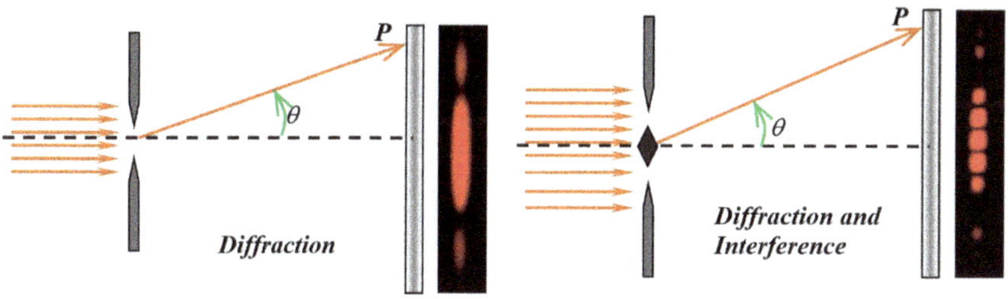

Figure 47.1 On the left, diffraction through a slit of width a. On the right, the combined diffraction and interference effect, generated by two slits of width a, separated by a distance d. In all cases, assume that the distance from the slot to the screen is much greater than the other dimensions, a or d. The angle θ determines the position of point P on the screen.

slits of width a separated by a distance d, the intensity pattern is described by [1–4]

$$I_I(\theta) = I_d(\theta) \cdot \cos^2 u_\theta, \tag{47.3}$$

with

$$u_\theta = \pi \frac{d}{\lambda} \sin(\theta). \tag{47.4}$$

On the other hand, the value of $I_d(\theta)$ in Eq. (47.3) is given by Eq. (47.1). The resulting patterns are shown in Figure 47.2. These equations are valid as long as the distance to the screen is much greater than λ, a, and d. This condition characterizes *Fraunhofer diffraction* [6].

The double-slit interference experiment is known as Young's experiment. In 1801, Thomas Young found that light from a distant source passing through a double slit produced an interference pattern

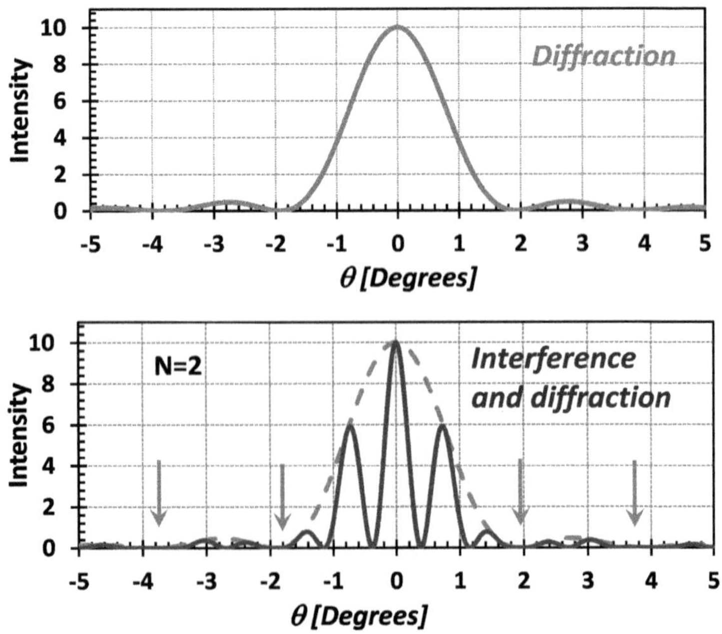

Figure 47.2 Above: Intensity distribution produced by a slit of width a. Below: Interference pattern generated by two slits of width a, separated by a distance d. Superimposed on the intensity pattern generated by diffraction (dashed line), the interference pattern is observed.

similar to the one illustrated in Figure 47.1, thereby showing that the behavior of light is consistent with the behavior of a wave, i.e., light seems to have a wave nature. Indeed, *interference and diffraction* are considered the fundamental signatures that reveal the wave nature of a given phenomenon and are often used to distinguish between wave or particle. It is interesting to note that when these experiments were conducted with electrons, protons, and neutrons, it was found that they produced interference patterns similar to those of light [1].

The purpose of the following projects is to study phenomena in which light reveals its wave nature. We will analyze the interference and diffraction of light using a laser as a coherent monochromatic source. We will also discuss some applications of the interference and diffraction effects [3, 4].

47.1.1 *Project 157. Diffraction through a slit or a thin wire*

Recommended equipment: A laser pointer and a set of slits of various widths, between 150 μm and 20 μm, or a microscope slide with copper wires of various diameters.

A device like the one illustrated in Figure 47.3 can be constructed using a laser and a narrow slit (150–20 μm). The goal is to study the characteristics of the intensity distribution observed on a screen placed at a distance D (50–200 cm) from the slit. The slit width

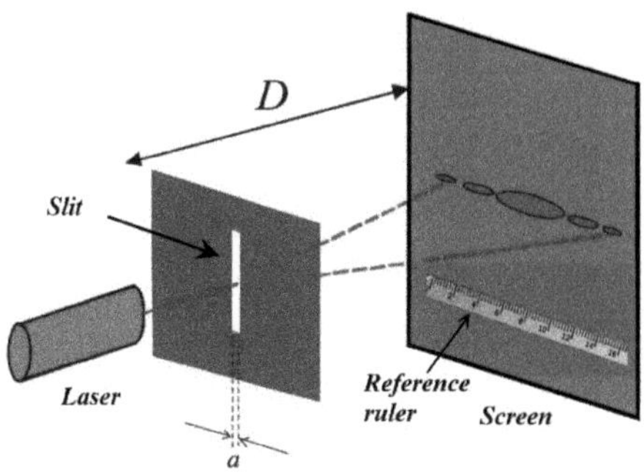

Figure 47.3 Experimental setup consisting of an optical bench, a laser, a slit, and a projection screen with a reference ruler.

should be a few tens or hundreds of microns. These types of slits can be purchased from several commercial suppliers [7–9] or made as printed patterns with a good quality printer, on a scale 10 times larger. To make your own slits, take a sheet of white paper and draw on it a black square of about $5\,\mathrm{cm} \times 5\,\mathrm{cm}$ with a white slit about 1mm wide and 2 cm high, as shown in Figure 47.3. Take a photograph of the drawing with a traditional camera using high-contrast lithographic film. The photo is taken to reduce the actual dimension of the object (drawing with the slit) by a factor of about 10. It is thus possible to obtain a photographic slit with well-defined edges that are very suitable for these experiments. Another alternative is to use thin copper or other metal wires of comparable thickness attached to a microscope slide or stretched tautly across a plastic picture slide frame [10]. Several wires of well-known thicknesses can be purchased at very affordable prices.

A remarkable feature of the phenomenon of diffraction is that the patterns observed on a screen are the same whether a slit or a wire of equal thickness is illuminated [3, 11], i.e., when complementary diffracting objects are used. Two figures are said to be complementary if they form an opaque plate when superimposed. The fact that complementary diffracting objects form the same diffraction pattern is known as *Babinet's principle* [3]. This phenomenon suggests that the diffraction pattern depends fundamentally on the locations of the *edges* of the obstacles that produce diffraction [8, 9, 12].

To perform this experiment, position the laser a few centimeters from the slit and keep the distance between the slit and the wall, D(>50 cm), constant throughout the test. Making the distance between the slit and the screen much larger than the slit ($D \gg a$) ensures that the wave fronts that reach the screen are flat, i.e., they meet the *Fraunhofer condition* [6].

Suggested tasks

✓ Observe how the light intensity is distributed on the screen. Vary the width of the slit (or thickness of the wire used) and study how the diffraction image changes. Observe how the width of the illuminated central zone of maximum intensity increases as the width of the slit (or the thickness of the wire) decreases. Describe what would happen if a large slit (or a thick wire) was used instead. Explain how to go from a diffraction pattern to a simple shadow of a large slit or wire.

✓ Place a reference ruler on the screen to enable you to measure real distances on the projection screen. Determine the position of the maxima ($y_n^{(\mathrm{max})}$) and minima ($y_n^{(\mathrm{min})}$) of intensity on the screen as a function of the order n in which they appear from the central maximum ($y = 0$). That is, map the first minimum to one side with the order $n = 1$, the second with $n = 2$, etc.

✓ Plot the position of the minima of intensities ($y_n^{(\mathrm{min})}$) on the screen as a function of order n and the maxima of intensities ($y_n^{(\mathrm{max})}$) as a function of n.

✓ Show theoretically that the positions of the maxima and minima are given approximately by [2, 3]

$$y_n^{(\mathrm{max})} = \left(n + \frac{1}{2}\right)\frac{D\lambda}{a}, \quad n = \pm 1, \pm 2, \tag{47.5}$$

and

$$y_n^{(\mathrm{min})} = n\frac{D\lambda}{a}, \quad n = \pm 1, \pm 2, \ldots, \tag{47.6}$$

where n is the order of the nth minimum or maximum, D is the slit-screen distance, a is the width of the slit, and λ is the wavelength of the laser used.

✓ Using this theoretical framework and the experimental plots of $y_n^{(\mathrm{max})}$ and $y_n^{(\mathrm{min})}$ as functions of n, try to fit a straight line through the data. According to Eqs. (47.5) and (47.6), the slope of y_n as function of n, knowing D and a, enables you to determine λ. Determine λ and estimate its uncertainty.

- This procedure is suitable for determining the wavelength if the thickness a of the slit is known.
- If you know the wavelength λ, you can use the diffraction pattern to calculate the value of slit width a. This procedure is suitable when using, for example, a He-Ne laser for which the wavelength is known precisely ($\lambda = 632.8$ nm).

✓ Discuss the validity of the proposed model to explain your observations. What can you conclude about the validity of the wave theory of light?

✓ ♣♣ **Babinet's principle:** If you have a slit and a wire with a diameter equal to or comparable to the width of the slit, study the diffraction figures produced by both systems. Systems such as this, which complete an opaque screen when they are overlapped, are called *complementary systems*. One of their notable features is that they form similar diffraction figures.

✓ Do the diffraction figures produced by the complementary systems used in your experiments (slit and wire) follow *Babinet's principle*?

47.2 Intensity variation of a diffraction pattern

It is often useful to determine the *intensity variation* of a diffraction or *interference pattern* that is generated on a screen. This type of study allows a more quantitative analysis of the phenomena. There are several ways to achieve a spectrum of intensities of a given pattern. One way to carry out this type of measurement is to use a photometer and measure the distribution of intensities as a function of position. Photometers can be purchased from commercial vendors [8, 9] or may be home-made [13, 14].

Another method to determine the intensity distribution of a certain pattern is to use a digital camera or a smartphone [14]. Use the camera to capture the complete image of the diffraction or interference pattern on a screen. As discussed in Chapter 10, it is advisable to place on the screen a reference ruler graduated in metric scale (mm), see Figure 47.3. The digitized image can be analyzed using a program that can measure pixel intensity along a line (intensity profile) [15]. This type of analysis can be performed with MathWorks Matlab® [14] or with specific programs, such as ImageMeasurements [16], [17], among others. The ruler on the screen enables you to convert the pixel scale into real distances. You can use programs such as Matlab or ImageMeasurements to determine the intensity of the pixels along a line in the image. It is thus possible to transform the image into an intensity graph as a function of position. To achieve a suitable image, it is advisable to take several shots of the interference or diffraction pattern, varying the exposure time. With a little practice, you can achieve adequate images to obtain good intensity spectra [14].

There are also several commercial devices that can be used to sweep or "scan" a pattern and determine the intensity value directly as a function of the position [9].

47.2.1 *Project 158. Intensity distribution of the diffraction patterns*

Recommended equipment: A laser pointer and a set of slits of various widths, from 150 μm to 20 μm. A calibrated photometer with small sensitive area ($\phi_{\text{sens}} \approx 1$ mm) mounted on a linear scale or a smartphone or digital camera.

With the intensity analyzer you have chosen (photometer or digital camera), determine the intensity distribution that is formed on the screen by diffraction through a slit. The objective is to obtain a graph of intensity as a function of the position along a line of the pattern under study.

Suggested tasks

✓ Following the instructions discussed above, obtain a very sharp diffraction pattern on the screen. Place a graduated scale on the screen.

✓ If you use a photometer, it is advisable to mount it on a sliding device along a line to be explored. Another possibility is to place the detector on a ruler and at each detector position, determine the intensity at each position x. You can thus determine the experimental variation of the intensity as a function of the position $I_{\text{ex}}(x)$. If you are using a digital camera or smartphone, take the image of the target under study under different exposure conditions until you achieve the most appropriate conditions for determining $I_{\text{ex}}(x)$. Remember to include a ruler or some kind of scale on the screen where the diffraction pattern is projected.

✓ On the same graph, plot both the measured intensity data $I_{\text{ex}}(x)$ and the prediction from Eq. (47.1) of the theoretical model for the diffraction from a slit $I(x)$, using the values of a, λ, and D that characterize this experiment. The angle θ is given by $\tan(\theta) = x/D$, where x is the coordinate on the screen.

✓ What can you conclude from this comparison of experimental results versus theoretical model? Is the proposed model adequate to explain your experimental results?

✓ ♣♣ **Babinet's principle:** If you have an object complementary to the slit, such as a fine wire with a diameter equal to the slit width, measure the light intensity distribution on the screen and compare the diffraction patterns obtained with the slit and the wire.

- Make a graph showing intensity versus position for the two systems.
- Analyze the results and discuss the validity of Babinet's principle in this case.

47.2.2 *Project 159. Interference of two or more slits — Young's experiment*

> **Recommended equipment:** A laser pointer and a screen with one, two, or more parallel slits of equal dimension. A set of copper wires arranged in parallel on a microscope slide can also be used.

Young's interference experiment can be replicated using a laser as a light source and a screen with two parallel slits. The opening a of the slits should be between 20 μm and 150 μm, and the spacing d between the slits should be 2–6 times greater than the thickness a. The twin-slit set can be purchased from commercial laboratory instrument suppliers or handcrafted [13, 14, 18].

Suggested tasks

✓ Study the diffraction pattern of only one of these slits. Determine the position of the maxima and minima of the diffraction pattern by making marks on a sheet of graph paper used as a screen.

- ✓ Without modifying the geometry, replace the single slit with the double slit. Using the same technique as described above, determine the position of the maxima and minima from the interference pattern.
- ✓ Study the interference from more than two slits. Determine the position of maxima and minima from the observed interference pattern.
- ✓ Compare the diagrams of the interference patterns obtained for one, two, and more slits.
- ✓ In the case of the two-slit experiment, from the position of the maxima and minima, estimate the value of the wavelength λ of the laser used and its uncertainty. If you know the λ value from the laser specifications, compare it to the estimated value.
- ✓ Using Eqs. (47.1)–(47.4), try to explain your experimental results.

47.2.3 *Project 160. Measuring λ using diffraction grating*

Recommended equipment: A laser pointer and a diffraction grating with known line spacing.

Using a laser and a diffraction grating with known line spacing d, determine the wavelength λ of a laser. If you are using a *transmission grating*, use the arrangement shown in Figure 47.3.

If you use a *reflection grating*, strike the laser on the grating at a certain angle of incidence so that after reflection in the grating, the diffraction image can be projected onto a wall or screen. A CD or DVD can be used as a *reflection grating.*

Suggested tasks

- ✓ Study the diffraction pattern of the laser as it passes through or is reflected by the diffraction grating.
- ✓ From the measurements of the angular position of the central maximum and the first maxima (first order, second order, etc.) and using the expression [2–4],

$$\sin\theta_{\max} = m\frac{\lambda}{d}, \quad \text{with} \quad m = 0, 1, 2, \ldots \text{(order)}, \qquad (47.7)$$

which relates the location of the maxima with the different orders m of diffraction, determine the value of the laser wavelength and estimate its uncertainty.

✓ If you know the wavelength of your laser, determine the d spacing of your *diffraction grating.*

47.3 Polarization — Malus's Law

When the vibrating property of a wave has a *vector characteristic* and therefore has a direction of oscillation, there is an important difference depending on whether these vibrations are along the direction of propagation of the wave or perpendicular to it. When the vibrating property is perpendicular to the direction of propagation, we say that the wave is *transverse.* A conspicuous example is the case of electromagnetic waves (light), where the vibrating properties are the electric and magnetic fields, both of which are perpendicular to the propagation of the wave [1–3]. In a transverse wave, if the vibration direction is predictable, we say that the wave is polarized. For example, if the vibration is always parallel to a fixed direction, then we have *linear polarization.* If the vector that describes the vibration rotates at a given frequency perpendicular to the direction of propagation, we have a wave with *circular polarization.* Another simple case is elliptical polarization [3]. It is clear that only transverse waves can be polarized. If a wave has a scalar character, like temperature waves or pressure waves, they cannot be classified as transverse or longitudinal. When the vibration is in the direction of propagation, we say that the wave is *longitudinal.* For example, sound is a longitudinal wave [1, 2].

Light, like all electromagnetic waves, is transversal, with the electric and magnetic fields perpendicular to each other and perpendicular to the direction of propagation. In general, in ordinary light sources, such as the Sun and a lamp, light is emitted by many atoms and molecules that act independently of each other, that is, *incoherently.* Thus, the planes of vibration of these sources vary randomly. In these cases, we say that the light is not polarized. There are devices called *linear polarizers* [19] (dichroic crystals, polaroid sheets, etc.) that only allow light vibrations to pass in one direction.

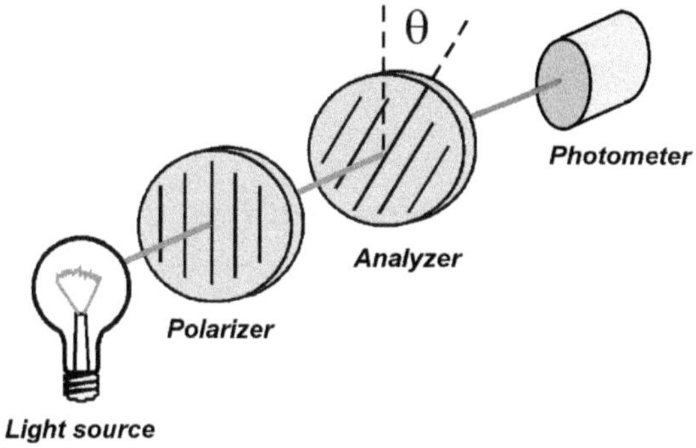

Figure 47.4 Diagram of the experimental device with a light source, a polarizer, an analyzer, and a photometer.

If a non-polarized beam of light passes through a linear polarizer, only the component of light that has an electric vector parallel to the axis of polarization is transmitted. In this way, it is possible to polarize the light coming from any source. If the original source was not polarized, the intensity of the transmitted light does not change if we rotate the polarizer. However, if we use two polarizers placed sequentially along the beam of light, when one is rotated with respect to the other, a variation in intensity is observed.

A key experiment to test the transverse character of a wave consists of using two polarizers consecutively, forming an angle θ between their polarization axes (Figure 47.4), and measuring the intensity of the wave that is transmitted as a function of θ.

The first polarizer polarizes the incident wave linearly; the second polarizer, called the *analyzer*, allows the field component in line with its polarization axis to pass through. If E_0 is the amplitude of the polarized wave at the output of the first polarizer, the amplitude transmitted by the analyzer will be $E_0 \cos(\theta)$, since only the component of the electric field in the direction of the polarization axis of the analyzer will be transmitted. As the intensity of the wave (energy per unit area and time) is proportional to the square of the amplitude, we expect the transmitted intensity to vary with the square of the $\cos(\theta)$, that is,

$$I(\theta) = I_0 \cos^2 \theta. \tag{47.8}$$

This relationship is known as *Malus's law*.

The goal of the following experiments is to study the basic properties and characteristics of polarized light and to falsify Malus's law [20, 21].

47.3.1 *Project 161. Experimental test of Malus's Law*

Recommended equipment: Two polarizing sheets (polaroid). An incandescent lamp. A calibrated photometer. A smartphone with ambient light sensor (ALS), which is a common device that can be used as an effective photometer.

A key experiment to detect the transverse character of a wave is to check whether the light obeys Malus's law. In fact, this is an operational way of finding out whether or not a wave is transverse.

To study this phenomenon, use two *linear polarizers*. Although there are many types of polarizers, polaroid sheets are very suitable for this experiment and very low cost [7, 8]. You will also need a calibrated photometer to measure light intensity.

The experimental device is shown schematically in Figure 47.4. The light source can be an incandescent lamp, and two polarizers (the second polarizer is generally called an *analyzer*) are placed between the source and the photometer. The first polarizer defines the state of polarization of light. One of the two polarizers must have a goniometer to measure the direction of its polarization axis with respect to the polarization axis of the other. Care must be taken to ensure good alignment of all these elements.

These experiments can also be performed using a solid-state laser as a light source. These lasers usually emit *linearly polarized light* so that one of the polarizers can be omitted. Nonetheless, He-Ne lasers do not always generate linearly polarized light. Even those that have polarization devices (Brewster windows [22]) can have random variations of the polarization plane during the laser heating period, which can be several minutes, so even in these cases, it is advisable to use both a polarizer and an analyzer.

If a smartphone is used as a photometer [23], and if the analyzer is taped onto the *light sensor* of the phone, the orientation (angle θ) can also be measured using the *tilt detector* of the same phone [21, 24].

An application such as Phyphox [25, 26] does this and converts almost any smartphone into a basic Physics Lab. Reference [21] provides further details on this technique.

Suggested tasks

✓ Using the device shown in Figure 47.4 and a calibrated photometer, study how the transmitted light intensity varies as a function of the angle between the two polarizers. To do this, keep the source-detector distance constant and rotate the polarizer (or the analyzer) until you observe that the transmitted intensity is maximum (maximum photometer response). Take this angle as the origin ($\theta = 0°$) to measure the angle between the two polarizers. Verify that when you rotate the polarizer 180°, the intensity is the same as when $\theta = 0°$. If you note significant asymmetry between these intensities, it is advisable to check the device, for example, check the alignment of the elements and check for spurious light sources.

✓ Using a photometer, measure the light intensity, I, as a function of θ. If possible, vary θ between 0° and 360° in steps of ≈5°.

✓ Plot the intensity of transmitted light I:

- as a function of $\cos(\theta)$,
- as a function of $\cos^2(\theta)$.

✓ Discuss, from these graphs, whether the light obeys Malus's law.

✓ What can you say about the transversal character of light?

References

[1] R. Feynman, *et al.*, *The Feynman Lectures on Physics, Boxed Set: The New Millennium Edition*, Vol. 1, Basic Books, 2011.

[2] R. Halliday, D. Resnick and M. Krane, *Fundamentals of Physics*, NY: John Wiley & Sons, Inc., 2002.

[3] E. Hecht, *Optics*, NY: Addison Wesley, 1990.

[4] F. Sears, M. Zemansky, H. Young and R. Freedman, *College Physics*, NY: Pearson Ed., 2009.

[5] Wikipedia, "Diffraction," 2022. Available: https://en.wikipedia.org/wiki/Diffraction.

[6] Wikipedia, "Fraunhofer diffraction," 2022. Available: https://en.wikipedia.org/wiki/Fraunhofer_diffraction.

[7] AIP, "Physics today Buyer's guide," 2022. Available: https://buyersguide.physicstoday.org/.

[8] Edmund Optics Co., "Edmund optics," 2022. Available: https://www.edmundoptics.com/.

[9] Pasco Co., "Light & optics," 2022. Available: https://www.pasco.com/products/lab-apparatus/light-and-optics.

[10] Physics and the National STEM Learning Centre and Network, "Demonstrating diffraction using laser light," 2022. Available: https://www.youtube.com/watch?v=71Rp-jG6Eek.

[11] Wikipedia, "Babinet principle," 2022. Available: https://en.wikipedia.org/wiki/Babinet%27s_principle.

[12] A. Hitachi and M. Takata, "Babinet's principle in the Fresnel regime studied using ultrasound, "*Am. J. Phys.*, vol. 78, p. 678, 2010.

[13] M. Peiris and I. Perera, "A simple and inexpensive photometer using a light- dependent resistor," *Am. J. Phys.*, vol. 55, p. 1147, 1987.

[14] A. Ramil, A. J. López and F. Vincitorio, "Improvements in the analysis of diffraction phenomena by means of digital images," *Am. J. Phys.*, vol. 75, p. 999, 2007.

[15] The MathWorks, Inc., "Intensity profile of images," 2022. Available: https://www.mathworks.com.

[16] Vernier Software, "Logger pro and Vernier video analysis," 2021. Available: https://www.vernier.com.

[17] "xyExtract graph digitizer," 2021 Shareup LLC., 2021. Available: http://www.shareup.com/xyExtract-download-19171.html.

[18] M. D. Levenson, "Sensitive small area photometer," *Am. J. Phys.*, vol. 38, p. 1987, 1970.

[19] Wikipedia, "Polarizer," 2022. Available: https://en.wikipedia.org/wiki/Polarizer.

[20] H. Gottlieb, "The law of Malus using polaroid polarizers," *Phys. Teach.*, vol. 18, no. 11, pp. 612–614, 1980.

[21] M. Monteiro, C. Stari, C. Cabeza and A. Martí, "The polarization of light and Malus' Law using smartphones," *Phys. Teach.*, vol. 55, p. 264, 2017.

[22] Wikipedia, "Brewster windows," 2022. Available: https://en.wikipedia.org/wiki/Brewster%27s_angle#Brewster_windows.

[23] Google Play- LuxDeLux, "Google Play: Frequency sound generator," 2020. Available: https://play.google.com/.

[24] C.-M. Chiang and H.-Y.Cheng, "Use smartphones to measure Brewster's angle," *Phys. Teach.*, vol. 57, p. 118, 2019.

[25] RWTH Aachen University, "Phyphox," 2020. Available: https://phyphox.org/.
[26] R. Carroll and J. Lincoln, "Phyphox app in the physics classroom," *Phys. Teach.*, vol. 58, no. 11, p. 606, 2020.

MODULE IV

EXPERIMENTS WITH FLUIDS AND THERMAL PHYSICS

Chapter 48

SURFACE TENSION

This chapter discusses surface tension of a liquid, some basic ways to measure it, and how it is modified by detergents. This chapter also presents a simple model for determining Avogadro's number and the molecular size of the surface tension value and the latent heat of vaporization.

Goals

- ✓ Surface tension of a liquid
- ✓ Capillary rise
- ✓ Determination of surface tension
- ✓ Estimation of Avogadro's number

48.1 Forces of cohesion and adhesion

Surface tension is a property of the separation surface of two media caused by the attraction among molecules of the same type that interact through the various intermolecular forces [1]. There are short range interactions (forces) among molecules and atoms, known as *van der Waals forces*, caused by fluctuating correlated polarizations among nearby particles [2–4]. The forces among molecules or atoms of the same species are generally different from those among atoms and molecules of different species. The force of attraction exerted by molecules of the same type is called *cohesion*, and the force between different molecules or atoms is called *adhesion*.

The phenomenon of surface tension is most visible at a liquid–gas interface such as the interface between water and air. At the surface of the liquid, the molecules are drawn into the liquid by analogous

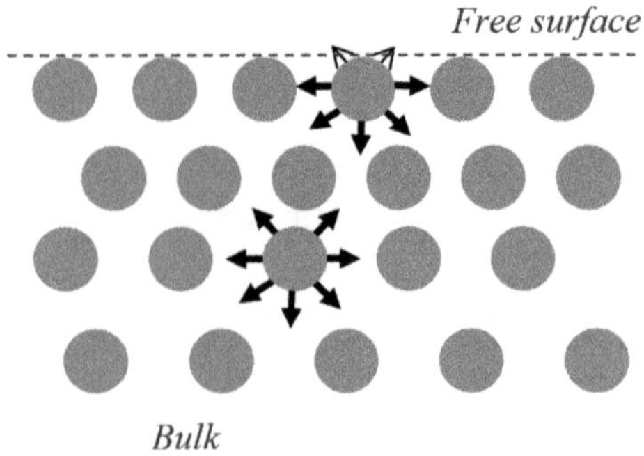

Figure 48.1 Diagram of the intermolecular forces on molecules inside and on the surface of a liquid.

molecules and outwards by the molecules present in the gas phase (see Figure 48.1).

As the cohesion among the molecules of liquid is stronger, the molecules of the liquid surface are more strongly attracted by their neighbors in the liquid phase than by the molecules of the gas phase, which are fewer per unit volume and further away from each other. As a result of these interactions, the surface resembles a *tight membrane* and acquires membrane-like properties. In particular, it acquires a form such that it minimizes its area, and thus, the surface energy is minimized. Work is required to increase that surface, and the work per unit area is called the *surface tension* γ. Therefore, γ can be also defined as the *resultant force that per unit length acts on the molecules on the surface* [1]:

$$\gamma = \frac{F}{l} = \frac{\Delta W}{\Delta A}, \tag{48.1}$$

where F is the force on a length l of its surface. ΔW is the work required to increase the area of a membrane by ΔA. This expression can be used as an operational definition of surface tension.

The adhesion of the molecules of a liquid to the solid walls of a container that holds it determines how easy it is for the liquid to wet the walls of this solid. A liquid's *wetting* properties can be modified by adding *surfactants* [5] such as soaps or detergents, which essentially decrease the surface cohesion and therefore the value of γ. This property of detergents or soaps improves the wetting and cleansing properties of pure water.

There are several methods for measuring the surface tension of a liquid [5–9]. Some use the rise (or fall) of a liquid through a capillary tube, others measure the strength of surface tension. References [7] and [10] describe various experimental techniques. There is also commercial equipment that can measure surface tension quickly and accurately, such as the Du Noüy blood pressure monitor [11].

48.2 Capillary rise

When a liquid is in contact with a solid, if cohesion prevails, we say that the liquid does not wet the surface (an example is mercury in glass). If adhesion prevails, then the liquid wets the surface (as in the case of water in a glass container). If a capillary tube (submillimeter diameter) is placed in water, the water wets the tube and rises through it to a certain height h. Conversely, if the liquid does not wet the tube material, the liquid level inside the tube will drop. When a glass capillary tube is placed in water, the first effect (rise) is observed, and when the same tube is placed in mercury, the second (fall) is observed (see Figure 48.2).

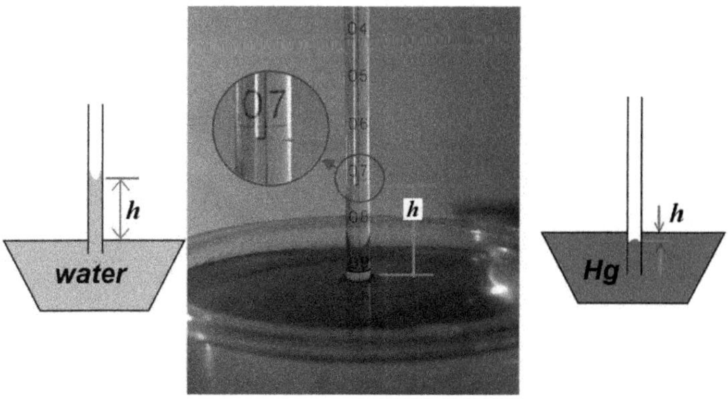

Figure 48.2 Left and center: Example of rise (water). On the right: Capillary fall (mercury).

In the case of capillary rise, if we assume that, in equilibrium, the weight of the ascending column of water is balanced by the force exerted by the surface tension on the contour of the tube, we have

$$\pi d\gamma = \text{weight} = \rho g \frac{\pi}{4} d^2 h \tag{48.2}$$

or also

$$\gamma = \frac{\rho g}{4} d \cdot h, \tag{48.3}$$

where g is the acceleration of gravity, ρ is the density of the liquid, d is the internal diameter of the capillary tube, and h is the height to which the liquid rises. Implicitly, we assume that the contact angle of the liquid [12, 13] with the tube is zero [7, 9].

A frequent difficulty in this method is the determination of d and h. It is not easy to obtain small-diameter tubes or to determine their internal diameter, d, accurately. A solution to both problems is to use glass pipettes of small volumes, e.g., 0.1 or 1 ml, which are easy to obtain and very inexpensive (see Figure 48.3). If you use a ruler to measure the distance l corresponding to a known volume of the same V, you can calculate the interior diameter as follows:

$$d = \sqrt{\frac{4}{\pi} \frac{V}{\ell}}. \tag{48.4}$$

Furthermore, this measurement enables "calibration" of the divisions of the pipette in units of "length" or "height". If a *digital photo* is taken of the pipette in the liquid (see Figure 48.3), the techniques

Figure 48.3 Example of 0.1 and 1 ml glass pipettes. By using a ruler to measure the lengths l_1 and l_2 of volumes V_1 and V_2, you can calculate the corresponding internal diameters and at the same time obtain a calibration in centimeters of the division marks on the pipettes.

discussed in Chapter 10 can be used to determine the value of h in each case. In all these methods for measuring the surface tension of a liquid, it is important to take the precaution of cleaning everything carefully (containers, pipettes, etc.) and rinsing several times with distilled water. Likewise, the liquids must have the least possible contamination. In the case of water, use good quality distilled water.

48.2.1 *Project 162. Determining surface tension by capillary rise*

> **Recommended basic equipment:** A 1 ml pipette and a 0.1 ml pipette. A tray of water. A digital camera or a smartphone.

After applying the cleaning precautions described above, use two pipettes (it is not essential to have two, but it helps to compare results) to build a device like the one illustrated in Figure 48.2, which allows you to insert the two pipettes in the same container. You can use a lab stand with an ordinary grip.

Suggested tasks

- ✓ Measure the water temperature.
- ✓ Measure the internal diameters of each pipette to be used and their corresponding uncertainties.
- ✓ Place the pipettes in the water. Make sure that the water level coincides with a division of the scale and record this value.
- ✓ If you have a digital camera or a smartphone, take several pictures of the system, similar to the one shown in Figure 48.2. Vary the lighting conditions until your pictures clearly show the water meniscus inside the pipette.
- ✓ Measure the heights h in each tube and their corresponding uncertainties.
- ✓ Calculate the value of the surface tension coefficient using Eq. (48.3) for each pipette and its corresponding uncertainties. Find the best value of γ and its uncertainty. Note that the errors of the determinations for the two pipettes are not the same, therefore the values are "weighted" differently. See Chapter 5.

✓ Compare your results with the corresponding published values [15]. What can you conclude from the method used?
✓ Prepare various solutions of water and detergent, for example, 0.5%, 1%, 2%, 3%, 5%, 10%, and 20%. Measure the surface tension of each solution at room temperature.
✓ Plot the measured γ as a function of the concentration of the detergent. Describe how the value of γ changes with the addition of detergent. Research the literature and discuss why this property improves the cleansing properties of water.

48.2.2 *Project 163. Capillary rise through a wedge-shaped wall*

Recommended basic equipment: A flat dish filled with water. Two glass plates 3 or 4 mm thick and 20 cm × 15 cm, approximately. A digital camera or smartphone.

A very easily built device that has been used for a long time [14] illustrates an interesting aspect of capillary ascension. It is made with two glass plates measuring approximately 10 cm × 15 cm, arranged in the shape of a wedge, as shown in Figure 48.4. The plates are positioned to form a wedge angle β by placing a spacer of known thickness d at a distance l from the joint line. Thus, $\tan\beta = d/l$. In order to achieve a wedge that is suitable for detectable "capillary rise", the thickness d should be of the order of 1 mm if l is of the order of 10 cm.

To have a reference scale, a grid with known spacing can be drawn on the plate closest to the observer (camera). The water should be colored with blue ink or another dye to make the meniscus more visible. Define the axis of the vertical edge as the z axis and the horizontal as x. Taking an element of infinitesimal prismatic volume, with base area $x\ \tan(\beta)dx$ and height z, its weight will be $(\rho\ \tan(\beta)\ z\ x\ dx)$g. The force due to surface tension that supports this column is $\gamma 2\ dx$, and g is the gravity acceleration. From the balance of these forces, we have

$$z = \left(\frac{2\gamma}{\rho g \cdot \tan\beta}\right)\frac{1}{x}. \tag{48.5}$$

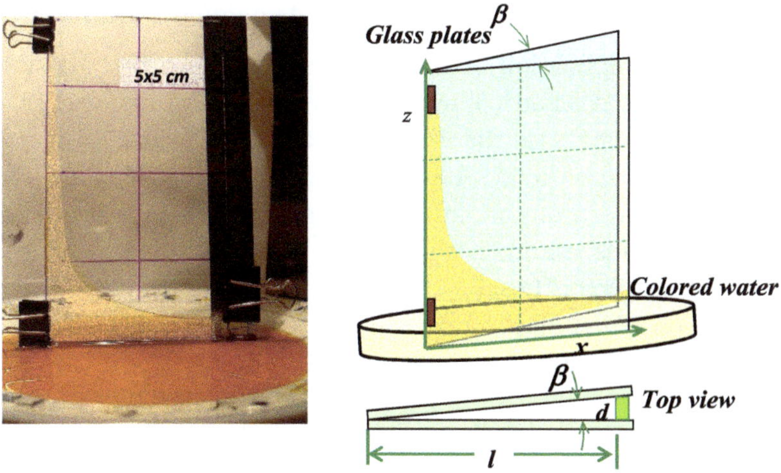

Figure 48.4 Two glass plates positioned to form a wedge shape. On the right, a diagram of the arrangement, and on the left, a photo of the actual setup. The glass plates can be glued together with glass cement or simply held with paper clamps.

In this experiment, we wish to falsify this expression. As can be seen, Eq. (48.5) predicts a hyperbola-shaped filling of the wedge.

Build the wedge-shaped device described and place it in a dish filled with colored water.

Suggested tasks

✓ Measure the water temperature.
✓ Determine the angle β of the wedge.
✓ Using the capillary ascension method, determine the surface tension of the colored water.
✓ Take a picture of the wedge in water. Vary the lighting conditions until you get a photo that shows the shape of the liquid between the plates as clearly as possible.
✓ From this image, using the image analysis techniques discussed in Chapter 10, check to what extent your data agree or disagree with the prediction of Eq. (48.5). Discuss your results.

48.2.3 *Project 164. Estimation of Avogadro's number*

An interesting application of the phenomenon of surface tension is the estimation of Avogadro's number [15–18]. As we saw earlier, the work required to increase the area of a liquid surface by dA is $\gamma\ dA$. Consider one mole of liquid. Its volume will be $V_0 = M_{\rm mol}/\rho$, where $M_{\rm mol}$ is the mass of one mole of that substance and ρ its density. Assuming that this volume is shaped like a regular cube, its side will be $L_0 = (M_{\rm mol}/\rho)^{1/3}$. If each side of the cube is divided into n equal parts, as shown in Figure 48.5, they generate n^3 partitions of the original cube, and the area increment will be

$$\Delta A = n^3 6(L_0/n)^2 - 6(L_0)^2 = 6L_0^2 \cdot (n-1). \tag{48.6}$$

If the total number of partitions coincides with the number of molecules, that is, if $n^3 = N_A$ (Avogadro's number), the energy required to generate this partition will be

$$\Delta E = \gamma \Delta A = \gamma 6 L_0^2 (n-1) \approx 6\gamma N_A^{1/3} (M_{\rm mol}/\rho)^{2/3}. \tag{48.7}$$

On the other hand, the energy required to bring the system from its original liquid form to individual molecules is precisely what is called the *latent heat of vaporization per mole* L_v.[a] Therefore,

$$L_v \approx N_A^{1/3} [6\gamma (M_{\rm mol}/\rho)^{2/3}] \tag{48.8}$$

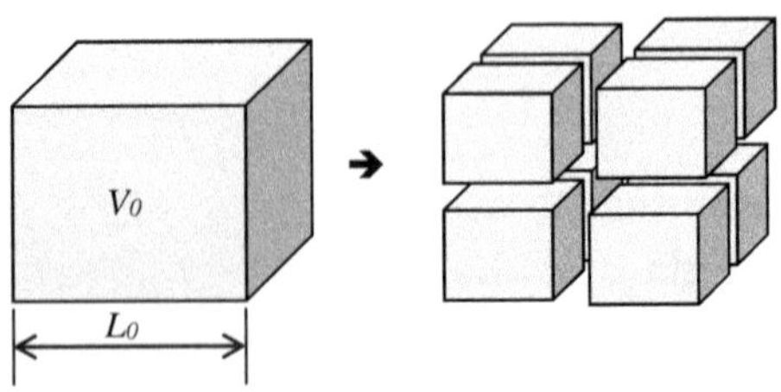

Figure 48.5 Partition of a cube into $2^3 = 8$ equal parts.

---

[a]At this point, it is important to distinguish the latent heat of vaporization L_v from the enthalpy change ΔH_v that is generally provided in tables. Enthalpy change includes both the latent heat and the work that steam does when expanding in the air, that is, $\Delta H_v = L_v + P\Delta V$, for one mole ($P\Delta V \approx RT_v$). Therefore, $L_v \approx \Delta H_v - RT_v$, where T_v is the boiling temperature in Kelvin. The surface tension value that should be used is close to T_v.

or also

$$N_A \approx \left(\frac{L_v}{6\gamma}\right)^3 \left(\frac{\rho}{M_{\text{mol}}}\right)^2. \tag{48.9}$$

An estimate of the size, or intermolecular distance, d, can be expressed as follows:

$$d \approx \sqrt[3]{V_0/N_A}. \tag{48.10}$$

Combining these two expressions, we get

$$d \approx \left(\frac{6\gamma}{L_v} \cdot \frac{M_{\text{Mol}}}{\rho}\right). \tag{48.11}$$

It is interesting to note that although ancient Greek philosophers were already speculating on the atomic nature of matter, it was Thomas Young (1773–1829) who, using this type of argument, was able to make one of the first estimates of molecular dimensions [17].

Suggested tasks

For a set of simple substances which are liquid at room temperature, for example, water, carbon tetrachloride, acetone, chloroform, methanol, ethanol, hexane, neon, and argon, find the values of γ, L_v, ρ, and M_{mol}. Remember to correct the value of ΔH_v to find L_v.

- ✓ Using expression (48.9), estimate the value of N_A in each of these cases.
- ✓ How do the values and orders of magnitude of N_A compare to its accepted value $N_A = 6.022 \times 10^{23}$? What can you conclude about the model proposed to connect the surface tension with the latent heat of a liquid?
- ✓ Using expression (48.11), estimate the value of d for each of the substances studied.
- ✓ How do the values and orders of magnitude of d compare to their accepted values?

References

[1] D. Giancoli, *Physics Principles with Applications*, 7th Ed., NY: Pearson, 2014.

[2] D. Kleppner, "With apologies to Casimir," *Phys. Today*, vol. 43, no. 10, p. 9, 1990.
[3] "Surface tension — What is it, how does it form," Crash Chemestry Academy, 2022. Available: https://www.youtube.com/watch?v=zMzqiAuOSz0.
[4] W. Thoburn, "Role of intermolecular attractions in liquids and gases," *Am. J. Phys.*, vol. 34, p. 132, 1966.
[5] Wikipedia, "Surfactant," 2022. Available: https://en.wikipedia.org/wiki/Surfactant#:~:text=Surfactants%20are%20compounds%20that%20lower,%2C%20foaming%20agents%2C%20or%20dispersants..
[6] S. Mak and K. Y. Wong, "The measurement of surface tension by the method of direct pull," *Am. J. Phys.*, vol. 58, p. 791, 1990.
[7] F. Román, J. Faro and S. Velasco, "A simple experiment for measuring the surface tension of soap solutions," *Am. J. Phys.*, vol. 69, p. 920, 2001.
[8] The Kibron Inc., "Measurement of surface tension," 2022. Available: https://www.kibron.com/surface-tension-measurement.
[9] J. Grigera, *Elements of Biophysics*, Buenos Aires: Editorial Hemisferio Sur S.A, 1980.
[10] B. Worsnop and H. Flint, *Advanced Practical Physics for Students*, London: Methuen Educ., 1957.
[11] F. Behroozi and P. Behroozi, "Determination of surface tension from the measurement of internal pressure of mini soap bubbles," *Am. J. Phys.*, vol. 79, p. 1089, 2011.
[12] Wikipedia, "Surface tension," 2022. Available: https://en.wikipedia.org/wiki/Surface_tension#Surface_curvature_and_pressure.
[13] Wikipedia, "Contact Angle," 2022. Available: http://en.wikipedia.org/wiki/Contact_angle.
[14] M. Durand, "Mechanical approach to surface tension and capillary phenomena," *Am. J. Phys.*, vol. 89, p. 261, 2021.
[15] Engineering ToolBox, "Surface tension of water in contact with air," 2022. Avalable: https://www.engineeringtoolbox.com/water-surface-tension-d_597.html?msclkid=1bc45f0ec01711ec882261db06c7f618.
[16] M. Piva, "Capillary rise in a wedge," *Phys. Teach.*, vol. 47, p. 528, 2009.
[17] F. R. Stauffer, "An estimate of Avogadro's Number," *Phys. Teach.*, vol. 29, p. 252, 1991.
[18] V. Weisskopf, "Search for simplicity," *Am. J. Phys.*, vol. 53, pp. 19–20, 1985.
[19] V. Weisskopf, "Search for simplicity," *Am. J. Phys.*, vol. 53, p. 618, 1985.

[20] R. Crandall and T. Dunne, "Heuristic methods for counting and sizing of molecules in the liquid state," *Am. J. Phys.*, vol. 57, p. 346, 1989.
[21] N. Goy, *et al.*, "Surface tension measurements with a smartphone," *Phys. Teach.*, vol. 55, p. 498, 2017.

Chapter 49

FLUID EXPERIMENTS — TORRICELLI EXPERIMENT

This chapter studies some properties of liquids in motion. It begins with an application of Bernoulli's theorem to the case of a liquid draining through an orifice, an updated version of the classical Torricelli experiment. To explain the phenomenon quantitatively, we will see that it is necessary to include the loss effects in Bernoulli's equation. Finally, the characteristics of the movement of fluids in laminar and turbulent regimes are explored.

Goals

- ✓ Shape of a jet of water
- ✓ Torricelli experiment
- ✓ Bernoulli's theorem
- ✓ Time taken to empty a container through a hole
- ✓ Laminar and turbulent regimes

49.1 Ideal fluids and Bernoulli's theorem

The movement of fluids is one of the most complex physical phenomena in nature. However, it is possible to understand many of its characteristics using Newton's laws and conservation laws: conservation of mass, momentum, energy, etc. [1–4].

Flow, Q, is defined as the volume of fluid that passes through a cross-sectional area of the tube per unit of time. If the fluid is a liquid, it can be treated as incompressible because its density at a given temperature is generally constant. The principle of conservation of mass, applied to a fluid moving through the tube of variable cross-section, implies that

$$Q_1 = v_1 \cdot A_1 = Q_2 = v_2 \cdot A_2, \tag{49.1}$$

where v_1, A_1, v_2, and A_2 are the velocities and the cross-sections at points 1 and 2, respectively, along the tube. Here, it is assumed that there are no fluid losses or leaks between points 1 and 2. Eq. (49.1) is a way of writing the *continuity equation* for incompressible fluid in stationary state.

Another very useful expression in various areas of science and engineering is the *Bernoulli equation.* In its simplest form, the Bernoulli equation is basically a formulation of the conservation of mechanical energy per unit volume along a fluid line, which is expressed as follows [1, 3, 4]:

$$P_1 + g\rho \cdot h_1 + \frac{1}{2}\rho \cdot v_1^2 = P_2 + g\rho \cdot h_2 + \frac{1}{2}\rho \cdot v_2^2. \tag{49.2}$$

Here, ρ is the density of the fluid, P_1 and P_2 are the pressures at two points 1 and 2, respectively, along a streamline, g is the gravity acceleration, v_1 and v_2 are the mean fluid velocities at points 1 and 2, respectively, and h_1 and h_2 are the heights of each point. This expression is applicable to ideal fluids (incompressible and non-viscous).

49.2 Viscous fluids

Viscosity is a property of fluids associated with resistance to flow. We say that water is not very viscous, while honey has high viscosity. More specifically, if a fluid is confined between two parallel solid plates separated by a distance y_0, as seen in Figure 49.1, to move the upper plate with velocity v_0 with respect to the lower one, a (shear) force F must be exerted, which is proportional to v_0 and to the area A of the plate, and inversely proportional to y_0. The constant of proportionality, η, is the dynamic viscosity of the fluid, that is,

$$F = \eta \cdot A \cdot \frac{v_0}{y_0}, \quad \text{or} \quad F = \eta \cdot A \cdot \frac{dv}{dy}. \tag{49.3}$$

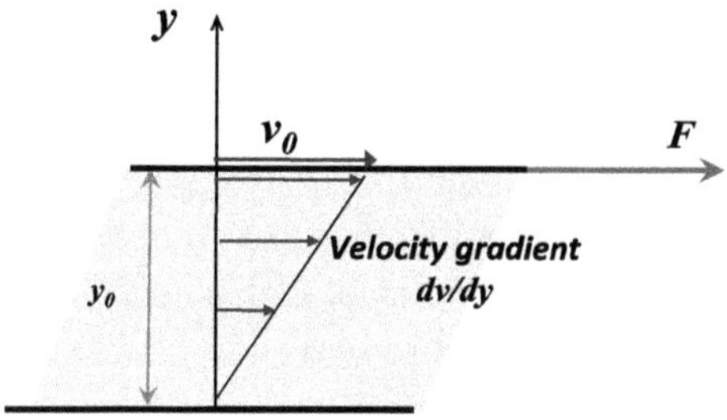

Figure 49.1 To keep the top plate moving with speed v_0, a shear force F is required, which is proportional to the area of the plate and to the velocity gradient.

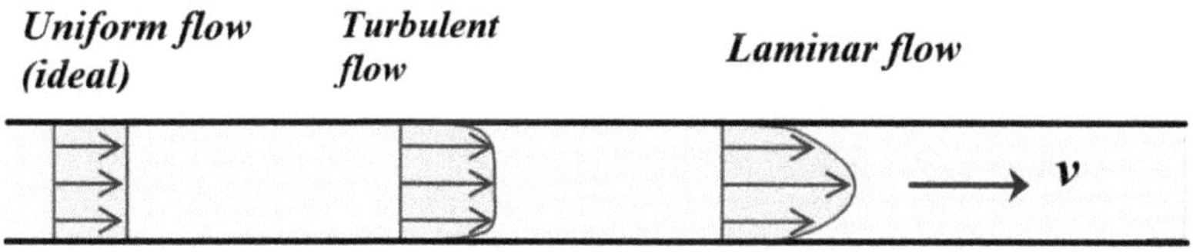

Figure 49.2 Velocity profiles in a tube through which a liquid flows. On the left, a uniform profile corresponding to an ideal fluid. On the right, a characteristic profile of a viscous fluid in a laminar regime. In the center, the shape of the profile of the average velocity of a turbulent flow.

In other words, the shear stress F/A necessary to move one fluid layer relative to another is proportional to the viscosity and to the velocity gradient dv/dy in the direction perpendicular to the layers [4, 5].

In the case of real liquid flowing through the interior of a tube or in a channel, there are two important effects that arise as a consequence of viscosity: the speed of the fluid at the surface of the solid is zero and the mechanical energy is not conserved. Consequently, the velocity profile has a maximum in the center of the tube and is zero on the tube walls, as illustrated on the right of Figure 49.2. Otherwise, the velocity gradient and shear stress would be infinite.

As discussed in Annex A of Chapter 10, in the motion of fluid, there are two important regimes to be distinguished: *laminar* flow and *turbulent* flow. When the flow is laminar, the fluid particles move in an orderly manner following stable trajectories, approximately parallel to each other. This is the regime observed in the flow of a channel in which the fluid moves slowly (see Chapter 10, Figure 10.10). In the turbulent regime, the fluid particles move in a disorderly manner, changing their trajectory in an irregular way or in vortices; the fluid motion at a given point in space appears to vary in time. Both these regimes can be easily observed in the smoke issuing from burning incense or a lit cigarette. At first, the flow is laminar, but as the smoke rises, it gradually becomes turbulent. The value of the Reynolds number (Re), which represents the quotient between the inertial and viscous forces, makes it possible to differentiate these two regimes.

49.2.1 *Project 165. Shape of a water jet*

> **Recommended equipment:** A digital camera or a smartphone, access to a water faucet.

When a faucet or tap is slowly turned on, a cone-shaped stream of water forms, and its diameter decreases as the distance to the mouth of the faucet increases. (See Figure 49.3.)

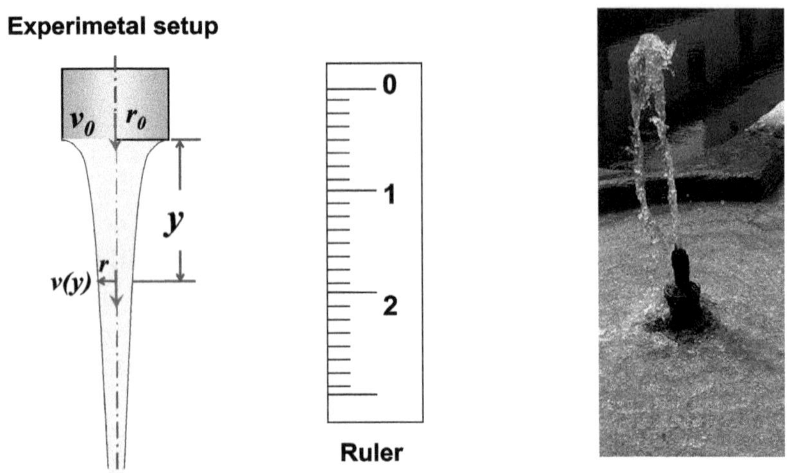

Figure 49.3 Left: Diagram of the proposed experimental device. Right: Image of a vertical water jet.

Finally, the jet begins to break into droplets. Assume that the exit velocity of the water jet at the mouth of the faucet is v_0. If the faucet outlet area is $A_0 = \pi \cdot r_0^2$, the initial flow is $Q_0 = A_0 v_0$, which is conserved throughout the jet. Due to the acceleration of gravity, the speed of the jet increases, and by conservation of energy,

$$v^2(y) = v_0^2 + 2g \cdot y, \tag{49.4}$$

where y represents the distance from the faucet mouth. By the conservation of mass, Eq. (49.1) gives

$$v(y) \cdot \pi \cdot r^2 = v_0 \cdot \pi \cdot r_0^2, \tag{49.5}$$

Therefore,

$$r^2 = \frac{v_0 r_0^2}{v(y)} = \frac{v_0 r_0^2}{\sqrt{v_0^2 + 2gy}} \quad \text{or} \quad r(y) = r_0 \cdot \left[\frac{v_0^2}{v_0^2 + 2gy}\right]^{1/4}. \tag{49.6}$$

This expression enables the formulation of a model of the variation of the radius of the jet with the fall distance y. To calculate v_0, the flow Q_0 is measured as the volume ΔV poured in a given time interval Δt, that is,

$$Q_0 = v_0 \cdot \pi \cdot r_0^2 = \frac{\Delta V}{\Delta t}. \tag{49.7}$$

On the other hand, the radius output r_0 can be measured directly from the faucet nozzle or from a digital photo of the jet. For the second option, a millimeter scale ruler located near the faucet should be included in the photogram to calibrate the distances in the digital photo, following the recommendations discussed in Chapter 10. Reference [6] provides further details on this experiment.

Suggested tasks:

✓ Select a faucet that provides even streams of water (without turbulence), and select lighting and background conditions that let you take sharp photos of the stream of water.

✓ For the chosen flow rate, determine the flow of water, Eq. (49.7), and the initial diameter of the jet ($2r_0$).

✓ Take at least a couple of digital photos of the jets. Try to take photos that include both the jet of water and the calibration scale, as shown in Figure 49.3, with suitable illumination to provide sharp images.

✓ For each flow, try to describe the shape of the water jet using the proposed model, Eq. (49.6). Plot your observations and theoretical predictions on one graph.

✓ Vary the flow rate and repeat this analysis for each case, using low flow rates for which the jet can be clearly seen without breaking and the highest values you can achieve under laminar flow conditions.

✓ What can you conclude about the goodness of the proposed model to describe your data? Discuss your results.

✓ List the assumptions and physical laws implicit in the model, Eq. (49.6).

Additional reflection: Consider a jet of water that spurts vertically upward. Its velocity decreases as it rises, and at a certain height, the vertical component of velocity will be zero. According to the continuity equation, Eqs. (49.1) and (49.7), it is concluded that the jet should have an infinite radius when it reaches the highest point. If you look at the jet of water from a fountain, like the one illustrated in Figure 49.3, you quickly realize that this is not the case.

✓ How can you solve this paradox? Discuss different physical arguments that allow you to explain this phenomenon.

49.3 Torricelli experiment

Consider the case of a cylindrical container of diameter d_1, whose cross-sectional area is S_1, that contains a fluid up to a certain level h, as shown schematically in Figure 49.4. The container drains through a small hole in the bottom, of diameter d_2 and section $S_2 (S_2 \ll S_1)$. Call the evacuation speed of the fluid at the exit of this hole u_2. This section discusses the characteristics of an idealized model (ideal liquid) and therefore uses the standard Bernoulli equation given by Eq. (49.2). In Annex A, the model is improved to introduce the effects that arise from the non-conservation of mechanical energy in the fluid.

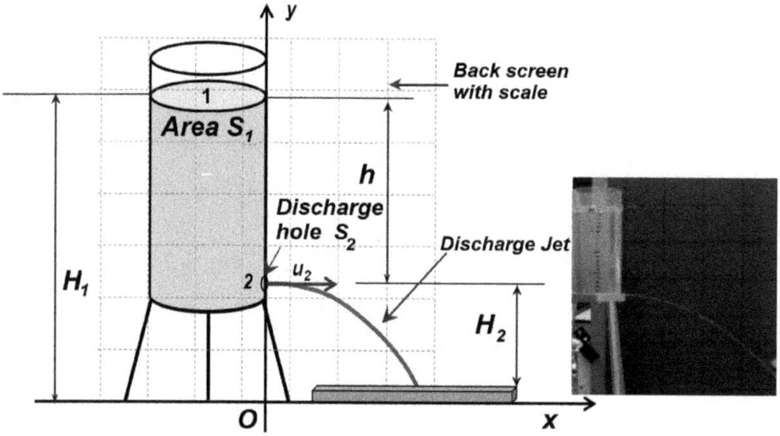

Figure 49.4 Torricelli experiment. On the left, a diagram of the proposed experimental device. On the right, an image of a real device, with the scale in the background, which serves as a reference to determine distances.

Applying Bernoulli's theorem at points 1 and 2, we can write

$$P_1 + g\rho \cdot H_1 + \frac{1}{2}\rho \cdot u_1^2 = P_2 + g\rho \cdot H_2 + \frac{1}{2}\rho \cdot u_2^2. \tag{49.8}$$

The pressure at the upper air–water interface (point 1) is atmospheric ($P_1 = P_{atm}$). Similarly, it is possible to identify P_2 with P_{atm}, that is,

$$P_1 = P_2 = P_{atm}. \tag{49.9}$$

Therefore, Eq. (49.8) can be written as follows:

$$gH_1 + \frac{1}{2}u_1^2 = +gH_2 + \frac{1}{2}u_2^2. \tag{49.10}$$

From the continuity equation, we have that

$$u_1 S_1 = u_2 S_2, \tag{49.11}$$

or in terms of the diameters, we have

$$u_2 = u_1 (d_1/d_2)^2. \tag{49.12}$$

Combining this result with Eq. (49.10), we see that the evacuation speed is given by

$$u_2 = \sqrt{\frac{2g(H_2 - H_1)}{(1 - (d_2/d_1)^4)}} = \gamma \cdot \sqrt{2gh}, \tag{49.13}$$

with

$$\gamma = \frac{1}{\sqrt{1-(d_2/d_1)^4}}. \tag{49.14}$$

The Torricelli model consists of applying the approximation $d_2 \ll d_1$, by which $\gamma \approx 1$. This approximate result is known as Torricelli's theorem [1, 3, 7].

Annex B generalizes this result to include the effects of viscosity and other energy losses.

49.3.1 *Project 166. Path of a water jet — Output speed*

Recommended equipment: A smartphone or digital camera. A transparent container of water with a hole in its side.

Consider a water jet issuing horizontally, like the one shown in the device in Figure 49.4. The fluid evacuation speed, u_2, can be determined by measuring the discharge flow rate and the outlet diameter of the container, Eq. (49.7).

Considering that the water particles behave like solid particles, which follow the equations of motion in the x (horizontal) and y (vertical) directions given by

$$x(t) = u_2 t \quad \text{and} \quad y(t) = y_0 + \frac{1}{2}at^2, \tag{49.15}$$

with $a = -g$ and $y_0 = H_2$, we get

$$y(x) = H_2 - \frac{g}{2 \cdot u_2^2} \cdot x^2. \tag{49.16}$$

Therefore, the distance x_f where fluid flow descends a height H_2 from the hole (Figure 49.4) is given by

$$x_f^2 = \left(\frac{2 \cdot u_2^2}{g} H_2\right). \tag{49.17}$$

One way to test the hypotheses of our study of the trajectory of the water jet is to compare the digital image of a jet with the theoretical prediction, Eq. (49.16). More specifically, try to find out whether the trajectory of the liquid issuing through a lateral opening is indeed parabolic, as indicated by Eq. (49.16). If it is, then from the fitting of this curve, the value of u_2 can be found, which can be compared to the value obtained by measuring the flow combined with Eq. (49.7). Incidentally, the validity of Eqs. (49.15) suggests that the kinematics of a fluid is equivalent to that of solids, that is, liquids and solids follow the same physical laws.

Suggested tasks:

✓ Using a transparent container with a hole of precisely known diameter (ranging from about 2–5 mm) in one side, construct an experimental device as illustrated in Figure 49.4. The diameter of the cylindrical vessel should be 10 cm or more so that the level of the liquid will not change too rapidly.

✓ In the same plane as the water jet, place a ruler or a clearly visible background scale, to serve as a reference to transform the pixels of the image into real distances; see Chapter 10. It is also advisable to mark the heights on the container so that the height h and the trajectory of the liquid are recorded in each photogram. Preferably use a tripod to position the camera or smartphone.

✓ Fill the container with water and take pictures of the jet for different heights h of the water level in the container.

✓ Superimpose the pictures of the jet and the theoretical shape of its trajectory, Eq. (49.16), on the same graph, as discussed in Chapter 10. Vary the value of parameter u_2 until you achieve the best fit or match between the actual and the theoretical path.

✓ For some heights h of water in the container, estimate the outlet flow by collecting the poured liquid in a graduated cylinder. Because the flow rate varies with height, select a volume of water to collect such that during discharge, h does not vary by more than about 1 cm. Using Eq. (49.7), determine the flow rate and estimate the value of u_2 and its uncertainty. How do the values of u_2 obtained with the two procedures used compare? Discuss the goodness of the proposed model represented by Eq. (49.16) to determine the value of u_2.

49.3.2 *Project 167. Container emptying time*

Recommended equipment: A transparent container of water with a hole in the bottom. A stopwatch and/or a digital camera or a smartphone. Alternatively, this experiment can be performed with a cylindrical container with an outlet hole at the bottom hanging from a force sensor.

Consider a cylindrical container with a hole of known diameter at the bottom. The container is filled to a certain height with water or another liquid. As fluid is lost, the height h of the fluid will vary over time. The aim of this activity is to study experimentally the variation of h with time and to try to build a theoretical model that explains the experimental results.

One possible experimental setup is a cylindrical container like the one illustrated in Figure 49.4, on which heights have been marked every centimeter or half centimeter. An observer with a stopwatch records the heights h_i and the times t_i as the liquid level drops. This table of values provides h as a function of t.

Another way to perform this experiment is with a force sensor connected to a PC. The container of water hangs from the force sensor. If the outlet hole is covered, and the container is filled with different volumes of water, the height h can be calibrated as a function of the mass or weight of the container, since

$$W = m_0 \cdot g + \rho g A_2 \cdot h, \tag{49.18}$$

where ρ is the density of the liquid, g is the acceleration of gravity, A_2 is the area of the base of the cylindrical container, h is the height of the liquid, W is the force or weight registered by the force sensor, and m_0 is the mass of the container. Once the force sensor has been calibrated to measure h, the computer records the data of $h(t)$. This experimental arrangement is schematically illustrated in Figure 49.5.

Annex A presents a possible model to determine the function $h(t)$ theoretically and use it to determine the evacuation time of a container [8].

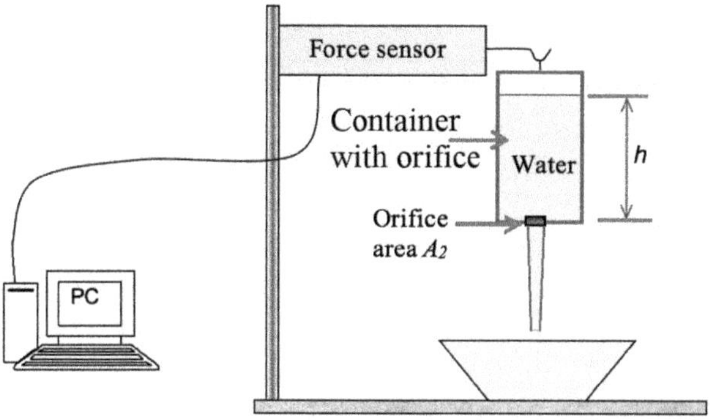

Figure 49.5 A possible setup of the proposed experiment to study the emptying of the container. In this arrangement, the force sensor is used to measure the mass of water as a function of time. These data can be used to find the value of the height $h(t)$.

Suggested tasks:

✓ Carefully measure the inside diameter of the cylindrical container and the diameter of the outlet hole. For each of these quantities, estimate the uncertainties and calculate the areas of the base of the container, A_1, and the exit hole area, A_2. Estimate the uncertainties in these areas.

✓ With the selected experimental setup, measure the experimental variation of h as a function of t.

✓ In the same graph, plot the measured values of $h(t)$ and the corresponding prediction, Annex A. Vary the size of the drain hole and repeat the comparison of the experimental data with the theory.

✓ Plot the measured values $\sqrt{h(t)/h_0}$ as a function of t. Do these data show a linear trend? If so, fit a line to the data and extract the drain time t_e from the fit; see Annexes A and C.

✓ How do the measured values of $h(t)$ and $\sqrt{h(t)/h_0}$. as a function of t compare to those predicted by your model?

49.3.3 *Project 168. Torricelli experiment*

Recommended equipment: A smartphone or a digital camera. A chronometer. A transparent container of water with a hole in its side.

Build a device like the one in Figure 49.6, and use the smartphone or digital camera to take pictures of it. The chronometer in each frame will indicate the time t_i, and the heights h_i of the liquid will be registered in each photogram.

Suggested tasks:

✓ Take photos when the level of liquid is at different heights, simultaneously record the time t_i and the height h_i from the free surface of the liquid, as well as the trajectory of the output jet. One way to get all this information in each photogram is to attach a chronometer to the side of the experimental device so that each photo i contains the height h_i and the time t_i. The height h_i and the coordinates of the output jet can also be obtained from the photo.

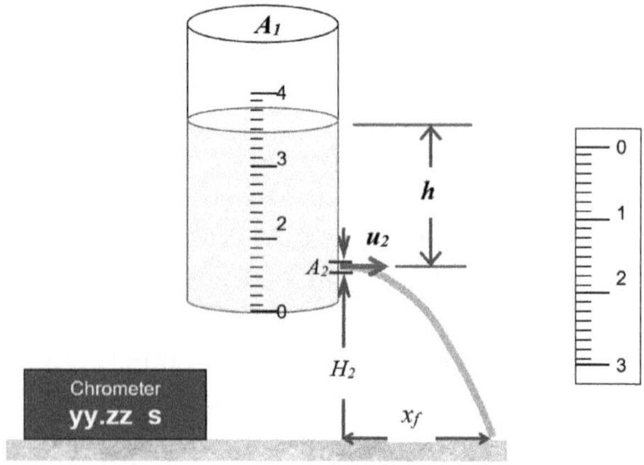

Figure 49.6 Diagram of the proposed experimental device. The scale on the right is used to calibrate the jet dimensions. The chronometer indicates the time when each photo is taken.

- ✓ From the fit of the trajectory of the output jet using Eq. (49.16), the velocity u_2 is calculated for each height h. To find this fit, it is advisable to plot $[H_2 - y(x)]$ as a function of x^2.
- ✓ Then, use the data obtained to graph u_2^2 as a function of h and study whether the approximation made by Torricelli, Eq. (49.13), is valid.
- ✓ A more elaborate model of drainage is discussed in Annex C or Eq. (49.37). Compare the two models and discuss which one best describes the data.
- ✓ From the fit of u_2^2 as a function of h, determine the value of μ. From the previous point, u_1 is already known, therefore Re and can be calculated, and this value of μ can be calculated from Eqs. (49.23) and (49.24). Determine whether the values of μ found using the two methods coincide.
- ✓ What can you conclude from this study regarding Torricelli's theorem and Bernoulli's theorem?
- ✓ Graph $\sqrt{h}$ as a function of t, taking as the origin of time the initial measurement, that is, for $t = 0, h = h_0$. Is the relationship linear? If so, find the value of t_e, μ, and g_{ef}, Eq. (49.29), for each hole used in the experiment, and compare it to the one actually observed, that is, the time it took for the system to empty.
- ✓ Plot u_1^2 as a function of h. If the dependence between these variables is linear, using Eqs. (49.38) and (49.39), find the values of $\Delta Z, \mu$, and k. What can you say about the model presented in Annex B to interpret your experimental results?
- ✓ Graph $\sqrt{h - \Delta Z}/\sqrt{h_0 - \Delta Z}$ as a function of t, taking the initial measurement as the origin of time ($t = 0, h = h_0$). Is the relationship linear? If so, find the value of t_e, μ, and k, Eq. (49.43). On the same graph, indicate what the dependence of $\sqrt{h - \Delta Z}/\sqrt{h_0 - \Delta Z}$ with t would be if $C_v = 1$.

From the above analysis, estimate the value of the vena contracta coefficient Cv, Eq. (49.44), and compare it to the prediction $C_v \approx 0.5 - 0.6$.

49.4 Annex A: Container evacuation time

To estimate the emptying time, t_e, of a container with an opening of area A_2, as illustrated in Figures 49.5 or 49.6, start from Eq. (49.13), which provides the value of the output velocity u_2. The outgoing flow, Q_2, will be

$$Q_1 = A_1 \cdot u_1 = \mu \cdot A_2 \cdot u_2. \tag{49.19}$$

In this equation, a coefficient μ has been introduced to take into account the possible narrowing of the water jet at the outlet of the orifice as well as the energy losses, which will be analyzed in Annex C. In the ideal case, $\mu = 1$, but as we shall see in a more elaborate model, its value may be different. Referring to Figures 49.5 or 49.6, $u_1 = dh/dt$, therefore, making use of Eq. (49.13), we have

$$\frac{dh}{dt} = \frac{A_2}{A_1} \cdot \mu \cdot \gamma \cdot \sqrt{2gh}. \tag{49.20}$$

Integrating this expression we obtain

$$\sqrt{h} - \sqrt{h_0} = -\frac{1}{2} \cdot \frac{A_2}{A_1} \cdot \gamma \cdot \mu \cdot \sqrt{2g} \cdot t. \tag{49.21}$$

Here, h_0 is the value of h for $t = 0$. Thus, emptying time t_e is determined by calculating the value of t for $h = 0$, which is given by

$$t_e = \sqrt{\frac{2 \cdot h_0}{g}} \cdot \frac{A_1}{A_2} \cdot \frac{1}{\gamma \cdot \mu} = \sqrt{\frac{2 \cdot h_0}{g_{ef}}}, \tag{49.22}$$

where we have defined

$$g_{ef} = \frac{g \cdot \mu^2}{[1 - (A_2/A_1)^2]} \left(\frac{A_2}{A_1}\right)^2 = \frac{g \cdot \mu^2}{[(d_1/d_2)^4 - 1]} = \frac{d^2 h}{dt^2}. \tag{49.23}$$

Thus, g_{ef} represents the effective acceleration with which the free surface of the liquid in the container descends. With the introduction of this parameter g_{ef}, Eq. (49.22) is similar to the expression that describes the free fall of a body from a height h_0. However, there is a notable difference: while a body in free fall accelerates as it descends, the free surface of a liquid slows down as it descends [8]. Physically, this corresponds to the decrease in hydrostatic pressure that "pushes" the liquid through the orifice.

Using this definition of t_e, the expression (49.22) can be written as follows:

$$h(t) = h_0 \cdot \left(1 - \frac{t}{t_e}\right)^2 \quad \text{or} \quad \sqrt{h(t)/h_0} = \left(1 - \frac{t}{t_e}\right). \tag{49.24}$$

Therefore, if the hypotheses formulated here are correct, by graphing the square root of the height h/h_0 as a function of time, we should obtain a line whose slope would give us the value of t_e, which in turn enables us to estimate g_{ef} and μ.

49.5 Annex B: Vena contracta

When a fluid exits an orifice, its cross-section generally contracts, as illustrated schematically in Figure 49.7. This phenomenon is known as *vena contracta* [4, 5, 7–9]. The relationship between the area of the narrowing of the jet and the area of the hole is called the coefficient of *vena contracta* $C_v (= A_{vc}/A_2)$. Its value depends on the Reynolds number Re [4, 5, 7]. Usually $C_v \leq 1$, and for many liquids, $C_v \approx 0.5 - 0.6$.

A heuristic justification to understand why this contraction occurs can be achieved by referring to Figure 49.7. The variation of the linear momentum of the jet at the discharge along the horizontal axis is $dp/dt = Q_2 u_2 \rho$, where $Q_2 = C_v A_2 u_2$ is the outflow and A_2 is the hole area. The force responsible for this change in momentum can be associated with the hydrostatic pressure, $F = PA_2 = \rho\, g\, h\, A_2$.

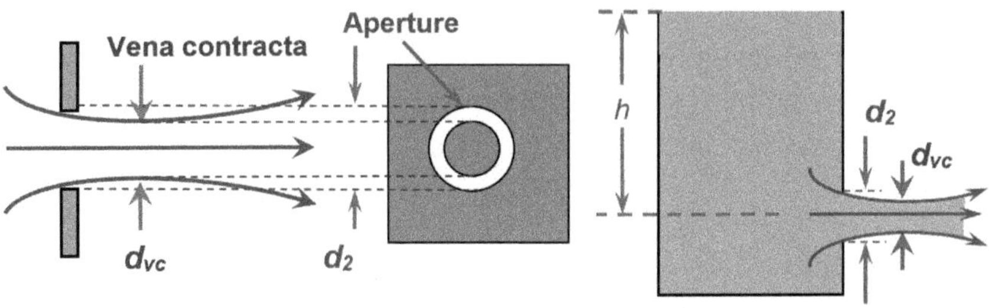

Figure 49.7 Vena contracta. Diagram of the characteristics of a flow of a fluid when exiting through an orifice. The *coefficient of vena contracta* $C_v = A_{vc}/A_2 = (d_2/d_{vc})^2$.

Therefore,

$$\rho\, g\, h = C_v \rho u_2^2. \tag{49.25}$$

Combining this expression with Eq. (49.13), we have

$$C_v = \gamma^2 \frac{1}{2} \approx 0.5. \tag{49.26}$$

49.6 Annex C: ♣♣ Torricelli's theorem — Theoretical model

Bernoulli's equation, in its simplest form, is a way of writing the conservation of mechanical energy in a fluid. In real fluids, mechanical energy is not conserved due to viscous effects, small losses in constrictions, etc. In the presence of viscosity, mechanical energy is not conserved and Bernoulli's equation becomes an expression of the energy balance. In these cases, the energy losses can be written as the sum of two terms: a speed-dependent term, usually called "small losses" [4, 5], and another one independent of speed, that is,

$$\Delta w_{\text{loss}} = \frac{\rho}{2} k \cdot u_1^2 + \rho\, g\, \Delta Z, \tag{49.27}$$

where g is gravity acceleration, k is the loss coefficient, and ρ is the density of the fluid.

Taking these losses into account, the *Bernoulli equation* for points 1 and 2 of a fluid flow can be written as follows [4, 5, 10]:

$$\frac{1}{2}\alpha_1 u_1^2 + \frac{P_1}{\rho} + g\, z_1 = \frac{1}{2}\alpha_2 u_2^2 + \frac{P_2}{\rho} + g\, z_2 + \Delta w_{loss}. \tag{49.28}$$

Here, u_i and z_i designate the average velocities of the fluid and the heights at the point $i(=1,2)$. The mean velocity is defined in terms of the flow Q as follows:

$$Q = \iint_S \vec{v} \cdot d\vec{S} = u \cdot A, \tag{49.29}$$

where A is the cross-sectional area of the flow (normal to the flow) and $\vec{v}$ is the local speed that can vary from point to point. Equation (49.29) can be considered as the definition of the average velocity u, which represents the average velocity that the fluid (with a uniform profile) should have for the flow to be the same as the real one. The coefficients α_i in Eq. (49.28) represent the relationship between the kinetic energy of the fluid that passes through the control surface of the tube and that of the same fluid if it moves with a uniform velocity profile (u) that carries the same flow. More concretely,

$$\alpha = \iint_S v^3 \cdot dS/u^3 \cdot A. \tag{49.30}$$

So, the term $\alpha u^2/2$ in Eq. (49.28) represents the kinetic energy per unit mass through a surface normal to the pipe. For a parabolic velocity profile inside the tube (in laminar regime), $\alpha = 2$. For a uniform profile (turbulent flow), $\alpha = 1$ and in general, $\alpha > 1$. If we apply these considerations to the container in Figure 49.4, we have

$$\frac{1}{2g}\alpha_1 u_1^2 + \frac{1}{2g}ku_1^2 + \Delta Z = h + \frac{1}{2g}\alpha_2 u_2^2, \tag{49.31}$$

where h is the variation in the height of the fluid between points 1 and 2 and u_1 and u_2 are the average velocities at the same points. The coefficients of α are given by Eq. (49.30).

When a fluid issues from a tube in a laminar regime [4, 5, 8], it acquires the parabolic velocity profile gradually, as shown schematically in Figure 49.8. The characteristic length at which the flow changes its velocity profile from plane to laminar is denoted as L_{Sch} (*Schiller's length*). This length can be calculated by the following empirical relationship [4, 5, 7, 8]:

$$L_{\text{Sch}} = 0.029 \cdot \text{Re} \cdot d, \tag{49.32}$$

where Re is the Reynolds number and d is the diameter of the hole. In general, the value of L_{Sch} is much greater than the thickness of

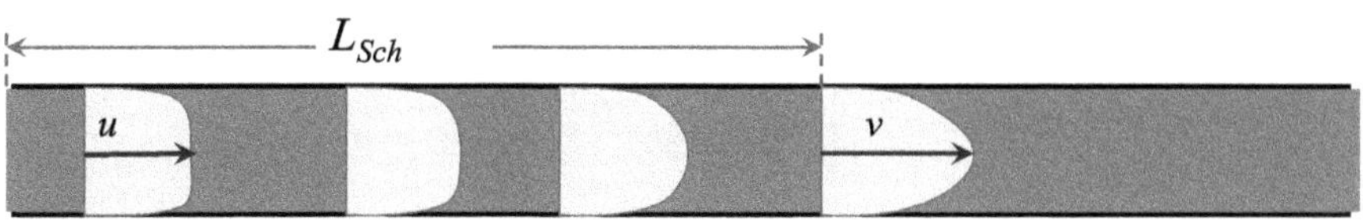

Figure 49.8 Diagram of how turbulent flow is transformed into laminar flow. L_{Sch} (Schiller's length) is the characteristic length where this transition occurs.

the container. This indicates that the velocity profile at the exit of the opening is generally flat, that is, the velocity is constant through the hole. The fact that the profile is flat allows us to consider the coefficient α_2 of the order of one ($\alpha_2 \approx 1$). Furthermore, since the mouth of the container is generally very wide and the air–water interface has very low speed, we can assume that the other coefficient, ≈1, is also of the order of the unit ($\alpha_1 \approx 1$).

On the other hand, considering the conservation of mass or conservation of flow, for surfaces of circular section, we have the following relationship:

$$d_1^2 \cdot u_1 = d_2^2 \cdot u_2 \cdot C_v, \tag{49.33}$$

where d_1 and d_2 are the upper and lower diameter, respectively, and the coefficient C_v is the so-called *vena contracta* coefficient discussed in Annex B [4, 5].

From Eq. (49.33), we can solve for the value of the speed with which the height of the liquid changes as a function of the geometry of the container, the evacuation speed u_1, and the coefficient C_v:

$$u_1 = \left(\frac{d_2}{d_1}\right)^2 \cdot u_2 \cdot C_v. \tag{49.34}$$

If we replace the value found for u_1 in the Bernoulli equation with loss, Eq. (49.31), we can write the following expression:

$$\frac{1}{2g}u_2^2\left[1 - \left(\frac{d_2}{d_1}\right)^4 C_v^2 + k\right] = h - \Delta Z, \tag{49.35}$$

from which we can determine the value of the evacuation speed as follows:

$$u_2^2 = \frac{2g \cdot (h - \Delta Z)}{\left[1 - \left(\frac{d_2}{d_1}\right)^4 C_v^2 + k\right]}. \tag{49.36}$$

Since d_1 is generally significantly greater than d_2, then $d_2/d_1 \approx 0$, and if these are raised to the fourth power, then $(d_1/d_2)^4 \ll 1$. Also, as $C_v < 1$, the term $C_v^2(d_1/d_2)^4 \approx 0$ and is negligible compared to the other terms. This makes it possible to write, in the first order, the evacuation speed as follows:

$$u_2^2 = \frac{2g}{[1+k]}(h - \Delta Z). \tag{49.37}$$

Based on this equation, we can find the value of the loss coefficient k and the energy loss ΔZ.

This problem can also be addressed from the standpoint of the evacuation time of the fluid in the container. This analysis can be based on the previous results. If in Eq. (49.20) we replace $\mu = \frac{1}{1+k}$, we can write the evacuation speed as follows:

$$u_2^2 = 2g\mu \cdot (h - \Delta Z). \tag{49.38}$$

Therefore, by plotting u_2^2 as a function of h, we can find the coefficients $\mu, \Delta Z$, and k. On the other hand, the speed of the interface, u_1, can be written as follows:

$$u_1 = \frac{dh}{dt} = -C_v \left(\frac{d_2}{d_1}\right)^2 \sqrt{2g\mu}\sqrt{h - \Delta Z}, \tag{49.39}$$

using the continuity equation and replacing the value of u_1 with the one found in Eq. (49.38). The minus sign in this equation is explained by the fact that u_1 and h are oriented in opposite directions.

To simplify the analysis, we introduce the parameter Λ, which is defined as follows:

$$\Lambda = C_v \left(\frac{d_2}{d_1}\right)^2 \sqrt{2g\mu}, \tag{49.40}$$

with which we can write the following differential equation:

$$\frac{dh}{\sqrt{h - \Delta Z}} = -\Lambda \cdot \mathrm{dt}. \tag{49.41}$$

By integrating and defining t_e (emptying time) as the time elapsed from $h = h_0$ (container full) to $h = 0$ (empty), we can write the following equation that allows us to calculate the height of the fluid as a function of time:

$$\frac{\sqrt{h - \Delta Z}}{\sqrt{h_0 - \Delta Z}} = \left(1 - \frac{t}{t_e}\right). \tag{49.42}$$

From this analysis, we can write the value of the emptying time as follows:

$$t_e = \frac{\sqrt{h_0 - \Delta Z}}{C_v \left(\frac{d_2}{d_1}\right)^2 \sqrt{2g\mu}}, \tag{49.43}$$

which, based on the previous equations and the values of ΔZ and μ found by the previous analysis, Eq. (49.39), allows us to calculate the value of the coefficient of *vena contracta*:

$$C_v = \frac{\sqrt{h_0 - \Delta Z}}{t_e \left(\frac{d_1}{d_2}\right)^2 \sqrt{2g\mu}}. \tag{49.44}$$

References

[1] D. Giancoli, *Physics Principles with Applications*, 7th Ed., NY: Pearson, 2014.
[2] R. Halliday, D. Resnick and M. Krane, *Fundamentals of Physics*, NY: John Wiley & Sons, Inc., 2002.
[3] F. Sears, M. Zemansky, H. Young and R. Freedman, *College Physics*, NY: Pearson Ed., 2009.
[4] B. Munson, D. Young and T. Okiishi, *Fundamentals of Fluid Mechanics*, New York: Wiley, 1994.
[5] A. A. Çengel and J. M. Cimbala, *Fluid Mechanics: Fundamentals and Applications*, 3rd Ed, NY: McGraw-Hill Co., 2014.
[6] J. Libii, "Mechanics of the slow draining of a large tank under gravity," *Am. J. Phys.*, vol. 71, p. 1204, 2003.
[7] T. Massalha and R. Digilov, "The shape function of a free-falling laminar jet: Making use of Bernoulli's equation," *Am. J. Phys.*, vol. 81, p. 733, 2013.
[8] M. E. Saleta, D. Tobia and S. Gil, "Experimental study of Bernoulli's equation with losses," *Am. J. Phys.*, vol. 73, p. 598, 2005.
[9] Wikipedia, "Vena Contracta," 2022. Available: https://en.wikipedia.org/wiki/Vena_contracta.
[10] P. L. L. Escamilla, "Surface tension influence in vessel discharge," comment on M. E. Saleta, D. Tobia, and S. Gil, "Experimental study of Bernoulli's equation with losses," *Am. J. Phys.*, vol. 77, no. 5, p. 477, 2009.

Chapter 50

THERMOMETRY — TEMPERATURE SENSORS

This chapter studies the basic characteristics of different thermometers and temperature sensors commonly used in the laboratory, such as thermistors, thermocouples, thermo-resistances (RTD), IC temperature sensors, and diodes, used for this purpose. It provides calibration techniques for some of them. It also describes the design of a gas thermometer, in which the dependence of pressure on temperature can be studied.

Goals

✓ Thermometers
✓ Temperature sensors
✓ RTD
✓ Thermocouples
✓ Integrated circuits to measure temperatures

50.1 Thermometers — Temperature sensors

Qualitatively, temperature is associated with the sensations of heat and cold. However, defining this concept operationally requires a thermometric property, i.e., a property that varies with temperature and allows a temperature scale to be defined [1]. Tentatively, we could say that temperature "is a property that is measured by thermometers". Next, we must define the physical property that will be adopted as a thermometer. Examples of thermometric properties are, among others, the volume of a liquid, the resistance of a conductor, the electrical voltage generated by a thermocouple, etc. There are numerous types of thermometers or temperature sensors that can be used for different applications. Table 50.1 lists some

Table 50.1 Types of thermometers and their basic characteristics.

Thermometer type	Range [°C]	Cost	Linearity	Notable features
Mercury in glass	−10 to 300	Low	Good	Slow, manual reading
RTD	−200 to 500	Low	High	Accurate
Thermocouple	−200 to 2000	Low	Good	Fast response, requires reference T
IC Thermometer	−15 to 120	Low	Good	Easy connection
Thermistor	−15 to 120	Low	Nonlinear	Fast response, very sensitive
Gas thermometer	−20 to 100	Medium	Good	Not very versatile, large size
Diodes	−200 to 50	Low	Good	Low cost

common types of thermometers and some of their more notable features. A more complete list can be found in Refs. [2, 3].

The aim of the experiments proposed in this section is to study the basic characteristics and to calibrate different types of the most common thermometers used in laboratories with respect to a *reference or standard thermometer.* A well-calibrated thermometer, such as a good mercury thermometer, will probably be available in the laboratory and can be used as a secondary standard for these experiments. Some pure substances with well-known fusion or boiling temperatures can also be used to calibrate thermometers.

Resistive Temperature Detectors (RTD) may be metal, semiconductor, or carbon resistors. Metal RTDs are usually made of copper, nickel, or platinum. Platinum RDTs are the most reliable because Pt wire can be produced with very small impurity content, which makes the temperature coefficient of the sample highly reproducible (though they are often very expensive). These devices are constructed of fine metal wire, protected by an alumina (Al_2O_3) capsule, from which two thicker wires protrude for connection. A recent RDT construction technique consists of depositing thin films of any of the metals used. By using laser techniques to give the film exactly the right geometry, the resistance can be adjusted to the desired value.

At temperatures that are not too low, the resistivity of a pure metal approximately follows the linear law $\rho(T) = \rho_0(1+\alpha(T-T_0))$,

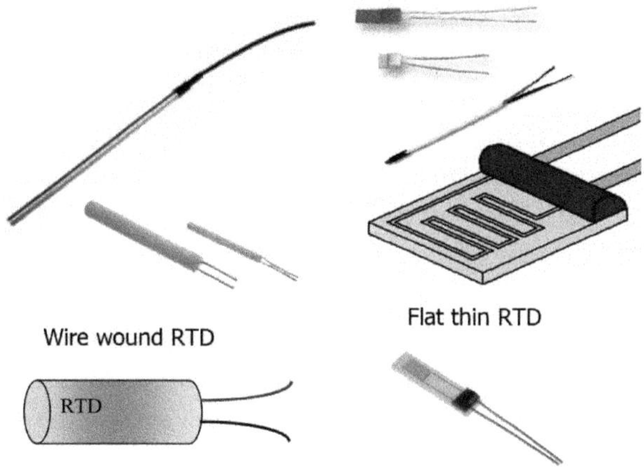

Figure 50.1 Different types of RTDs. Left: Cylindrical wire wound RTD. Right: Flat thin film RTD.

where ρ_0 is the resistivity at $T = T_0$ that usually depends on the type on material, and $\alpha = (\partial\rho/\partial T)/\rho_0$ is the temperature coefficient of resistance [4]. For platinum, $\alpha \approx 3.85 \cdot 10^{-3}\mathrm{K}^{-1}$, for copper, $\alpha \approx 3.9 \cdot 10^{-3}\mathrm{K}^{-1}$, and for nickel, $\alpha \approx 5 - 7 \cdot 10^{-3}\mathrm{K}^{-1}$.

RDTs come in different sizes and shapes (see Figure 50.1). They usually have small mass (and therefore fast response) and good linearity over a large temperature range. They must be biased by a DC or an AC constant current [5]. Sensors with small dimensions have low electric resistance, which requires care in the interfacing technique in order to minimize the error due to the resistance of the cables. Their sensitivity is limited by Joule self-heating, which requires reducing the bias current and therefore the signal amplitude. Typical useful ranges are platinum 10 K–800 K, nickel −60°C to +300°C, and copper −70°C to +150°C.

RTDs are rated by their nominal resistance at 0°C. Typical values are 10 Ω, 100 Ω, and 1,000 Ω. The relationship between resistance and temperature is almost linear. The most common RDTs are called Pt–100, which means they are made of platinum and have an electrical resistance of 100 Ω at 0°C. The variation of resistance with temperature is expressed by a polynomial whose coefficients are provided by the manufacturer.

RTDs provide highly repeatable readings. In the temperature range between −20°C and 200°C, the dependence of Pt resistance with temperature is highly linear. To calibrate an RTD, it usually suffices to use two temperatures: the melting temperature, $T_f \approx 0°\mathrm{C}$, and the boiling temperature, $T_e \approx 100°\mathrm{C}$, of water. To avoid interference by the resistance of the connection cables, it is advisable to measure the resistance of the RTD using the *four-wire method.* This technique to measure resistances, also known as the "four-point resistance measurement or Kelvin method," was discussed in Chapter 27.

A *thermistor* is a type of semiconductor device for measuring temperatures. Thermistors are very useful, sensitive, and inexpensive, see Chapter 28. They may have negative (NTC) or positive (PTC) temperature coefficients, i.e., their resistance may decrease (NTC) or increase (PTC) with temperature. NTC *thermistors* normally have exponential temperature dependence, which implies high sensitivity and nonlinearity. The advantage of these sensors is their small size and wide range of resistance value. As we have seen in Chapter 28, a thermistor characteristic equation is frequently written as $R(T) = R(T_0)e^{\beta(1/T-1/T_0)}$, where $R(T_0)$ is the value of the resistance at a given reference temperature T_0 and the constant β (typically from 2,000 to 5,000 K) indicates the sensitivity of the device, see Figure 50.2.

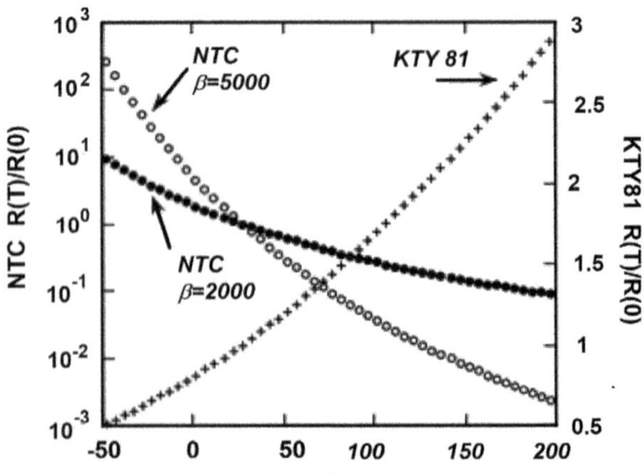

Figure 50.2 Reduced resistance versus temperature for two NTCs with different values of β and for one PTC (model KTY81)

A *thermocouple* consists of two different metal wires welded at a point called a "junction". The other ends of the pair are kept at a common *reference temperature* (T_{Ref}) but electrically isolated from each other, see Figures 50.3(b) and (c).

Note that when two wires of a thermocouple are connected to the voltmeter (usually made of copper), two more junctions (usually at T_a and T_b, respectively) are made (see Figure 50.3(a)); therefore, any measurement of V_{AB} is the sum of three voltages.

Consider three materials 1, 2, 3. 1 and 2 are thermocouple wires, and 3 is the common material of the voltmeter terminals (copper). The *measured* $V_{AB}(T)$ emf is the result of the voltages generated in the three junctions: $V_{AB}(T) = V^{31}(T_a) + V^{12}(T) + V^{23}(T_b)$. To simplify the measurements, it is advisable to choose

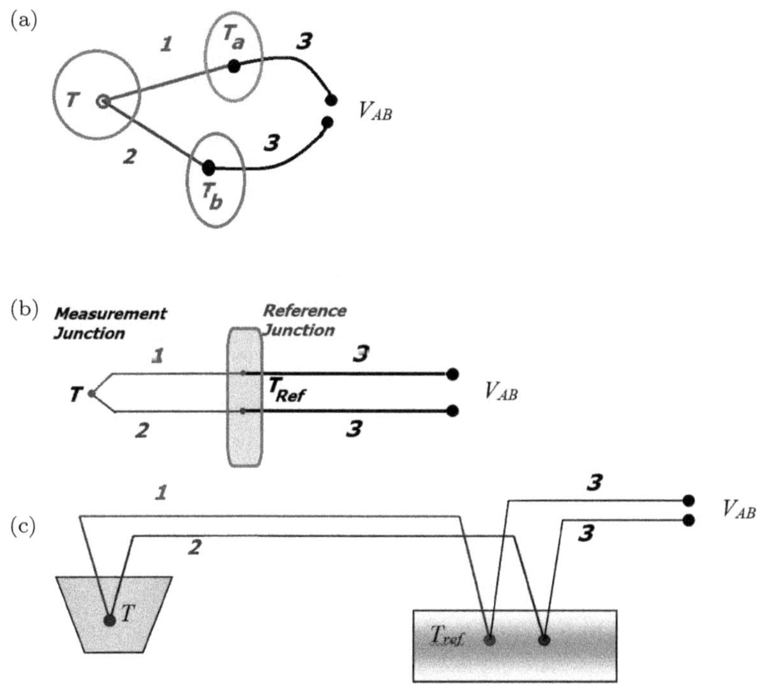

Figure 50.3 (a) Thermocouple configuration; T is the temperature to be measured, connection 1–3 is at a temperature T_a, and connection 2–3 is at a temperature T_b. (b) and (c) are alternative configurations with a single (common) reference temperature T_{Ref}.

a single *reference junction*, $T_{\rm ref}$, for the junctions 2–3 and 1–3, as shown in Figure 50.3(b). In this case, $V^{31}(T_a) = -V^{13}(T_a)$, therefore their contributions cancel out in the final measurement of $V_{AB}(T) = V^{12}(T)$. In Figure 50.3(c), $T_{\rm ref}$ is fixed at a well-known temperature (e.g., 0°C ice-bath), where the values of $V^{12}(T)$ can be obtained from the literature [6]. When the measurement junction (2–3) is at a temperature T, and the others (2–3 and 1–3) are set at the same reference temperature $T_{\rm Ref}$, an electric potential difference, V_{ab}, depends on the temperature difference $(T - T_{\rm Ref})$. This potential difference can be measured with a high-impedance voltmeter connected to the free ends, Figures 50.3(b) and (c).

The value of the emf $V^{12}(T, T_{\rm Ref})$ generated in a thermoelectric couple 1–2, with the measurement junction at temperature T and a reference point at $T_{\rm Ref}$ (as in Figure 50.3(b)), is

$$V^m(T, T_{\rm Ref}) = a_0 + a_1(T - T_{\rm Ref}) + a_2(T - T_{\rm Ref})^2 + \cdots, \qquad (50.1)$$

where the values of $a_0, a_1, a_2, \ldots$ of expansion in powers of $T - T_{\rm Ref}$ depend on the characteristics of the materials. These values are generally tabulated for different pairs of metals [5, 6]. Thermocouples are often made of pure metals or alloys, and their sensitivity depends on the junction materials. Some of the most frequently used types are ***J*** (iron+, constantan–) and ***K*** (chromel+, alumel–), where constantan is an alloy consisting of 60% Cu-40%Ni (also 55%Cu-45%Ni), chromel is 90% Ni-10% Cr, and alumel is 95% Ni-2% Mn-2% Al-1% Co.

An important characteristic of thermocouples is that in addition to being very inexpensive, they can be used to measure temperatures in a very wide range compared to other thermometers. The typical values of the temperature range are between 70 K and 1,700 K, with special alloys having ranges of up to 2,000 K.

Another simple, inexpensive way to measure temperatures is to use an ordinary diode [7]. The basic characteristic of a diode is that it allows the passage of electrical current in only one direction. The potential difference, V, between the diode terminals is related to the current, I, that passes through it and to the (absolute) temperature

T, which is given by the Shockley equation [8]:

$$I = I_0 \exp\left(-\frac{e\,V_0}{k\,T}\right)\left[\exp\left(\frac{e\,V}{k\,T}\right) - 1\right]. \tag{50.2}$$

In this expression, I_0 and V_0 are constants, e is the charge of the electron, k is the Boltzmann constant, and T is the absolute temperature [8]. When $e\,V \gg k\,T$(usual case at temperatures below 500°C; 1 eV/k $\sim$ 1,1600 K), we have

$$I \approx I_0 \left[\exp\left(\frac{e\,(V - V_0)}{k\,T}\right)\right]. \tag{50.3}$$

If a *constant I circulates through the diode*, V must be proportional to T, that is,

$$V = V_0 - b \cdot T. \tag{50.4}$$

One way to make the current constant is to connect the diode to a stable current source, or simply connect it to a voltage source in series, with a resistor with resistance several orders of magnitude greater than that of the diode. Figure 50.4 shows a diagram of the circuit that can be used in the latter case. A current between 0.01 mA and 1 mA is adequate.

There are several integrated circuit thermometers (*IC thermometers*) on the market for measuring temperatures in the laboratory and industrial applications, for example, LM35 or AD590 [9]. These ICs are very suitablo for using with a data collection system connected

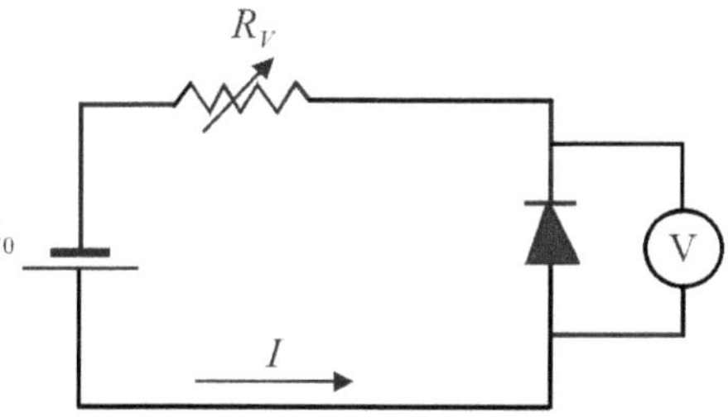

Figure 50.4 Schematic diagram of the circuit to measure the variation of the voltage of a diode with temperature.

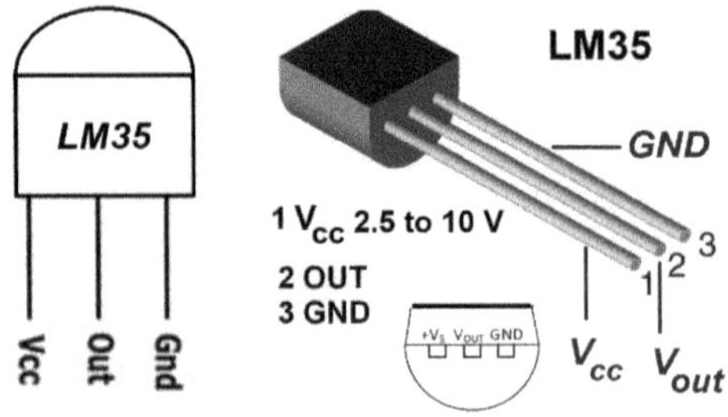

Figure 50.5 The LM35 is a precision integrated-circuit temperature sensor, whose output voltage is linearly proportional to the Celsius (Centigrade) temperature.

to a computer or to Arduino [10, 11]. They generally have three terminals: one is the ground connection, another is the voltage supply (V_c between 2.7 V and 5 V), and the third is the voltage output $V_{\text{out}}(T)$ that changes with the temperature, T (see Figure 50.5). These circuits generate a very linear signal which is proportional to temperature. They generally produce between 1 mV/K and 10 mV/K in a typical range of −10°C–120°C. There are also integrated circuits with calibrations expressed in the various systems of units (°F, K, °C).

50.1.1 *Project 169. Calibration of a gas thermometer*

> **Recommended basic equipment:** A metal or plastic container of about 250 cm^3. A calibrated thermometer. A pressure gauge capable of measuring manometric pressures in the range of 0–0.3 atm or a pressure sensor connected to a PC.

This experiment uses an airtight metal or rigid plastic container of approximately 250 cm^3 containing air or another gas, connected either to a pressure gauge or through a flexible tube to a pressure sensor, as shown in Figure 50.6. The pressure signal can be read with a PC connected to the pressure sensor or directly by reading the scale on the pressure gauge. The experiment also requires a thermal bath, e.g., a vessel with water that covers the container completely, to ensure that in equilibrium, the container is at the same temperature

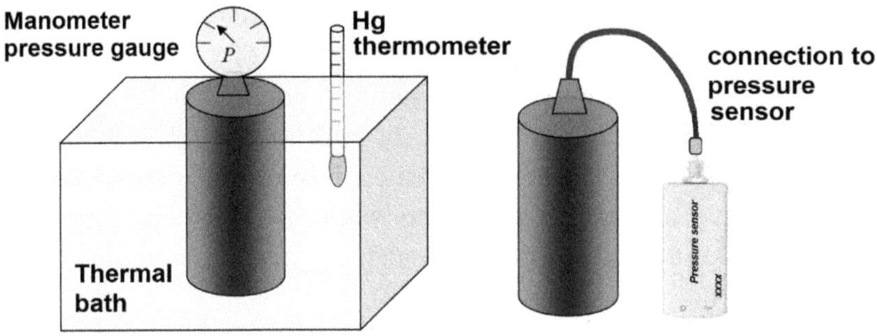

Figure 50.6 A gas thermometer. The thermal bath is a container with water, the temperature of which can be varied. Left: A diagram of a pressure gauge and Hg thermometer. Right: Setup with pressure sensor and temperature sensor.

Figure 50.7 Example of a homemade gas thermometer.

as the water. A mercury (Hg) thermometer is used to monitor this temperature.

How to make a simple gas thermometer at home [12]?
One way to make a simple homemade gas thermometer is illustrated in Figure 5.7. It can be connected directly to a pressure gauge, as in the figure, or to a pressure sensor connected to a data logger. If you use a small plastic soda bottle (preferably 250 ml or less), modify the plastic cap so that it can be connected to a pressure gauge or to the input of the pressure sensor. Perforate the plastic cap (hole of small diameter, e.g., 3 mm) and thread it so that you can insert a 1/4″

(4MA) axially perforated brass screw (hole diameter approximately 1.5 mm). Connect a pressure gauge to the end of the screw using a regular coupling. Otherwise, if you want to connect it to a pressure sensor, you can use a coupling to a hose or remove the last cm of the thread of the screw (using a lathe or drill and sandpaper), in order to achieve air-tight connection with the tube. The pressure sensor can be connected to the bottle using a plastic tube of the right diameter.

A standard temperature sensor can be attached with a little adhesive tape to the side of the bottle. The gas thermometer can be calibrated by immersing the whole bottle and the standard thermometer in an isothermal bath. You can use a Pyrex container of about 1 liter filled with water as the isothermal bath. Archimedes' thrust would tend to make the device float, so you may need to hold it in place by means of a clamp attached to a vertical rod.

To make sure there is no vapor in the gas, it is advisable to dislodge any moisture from the container by heating it. Also, an automobile or bicycle tire valve can be connected to the container so that it can be pressurized with a pump, as shown in Figure 50.7. Initially, the pressure inside the container should be somewhat higher than the barometric pressure of the place. The experiments can be carried out using air, argon, nitrogen, helium, etc. as a gas. Remember that gauges generally measure manometric pressure, P_m, and that the absolute pressure P is

$$P = P_m + P_{\text{bar}}, \tag{50.5}$$

where P_{bar} is the local atmospheric pressure or barometric pressure of the place.

Suggested tasks:

✓ Measure the pressure in the container, P_m, as a function of the gas temperature, T.

✓ Plot the absolute pressure P as a function of the absolute temperature T and analyze the dependence found. Is the dependence of P on T linear? If so, determine the slope and the y-intercept, and their uncertainties. Discuss the physical meaning of the y-intercept.

✓ Compare your results to the prediction given by the ideal gas law, $PV = nRT$, when the number of moles of gas n and the volume of the container V are constant.

50.1.2 *Project 170. Calibrating a thermocouple*

Recommended basic equipment: A thermocouple. A voltmeter that measures mV. A calibrated thermometer. A thermal bath.

This experiment uses a setup like the one shown in Figure 50.3(c). To avoid electrical contact in the thermal bath, a container of oil inside a box of ice can be used as T_{0f}. Study how the voltage V^m varies as the temperature T of the heat source varies.

Suggested tasks:

✓ Place a thermocouple junction in contact with a heat source and place a calibrated thermometer in the source. Plot the emf V^m generated by the thermocouple as a function of the temperature T measured by the thermometer. Vary T between the lowest possible temperature and the highest that you can achieve with your heat source. To extend the temperature range, you can use a vessel with automobile engine oil as a heat source. Take care that the oil does not boil or burn. You can also use tap water, but in this case, the temperature range will be 0°C to 100°C.

✓ Plot V^m as a function of T. For this experiment, the temperature can be in degrees centigrade. Fit with a polynomial of degrees 1 and 2. Discuss the significance of the parameters of these polynomials, see Chapter 7 and Appendix C of this book.

✓ Find the first coefficients of variation of voltage with temperature, represented by $a_0, a_1, a_2, \ldots$, and compare them to the table values for the thermocouple used [6].

Note: For more precise calibration, fixed temperature points are used which are well defined by the melting points of pure metals or the melting points of some pure substances.

50.1.3 *Project 171. Calibration of an RTD*

> **Recommended basic equipment:** An RTD temperature sensor. A circuit for measuring resistance by the four-lead method.

Build a circuit to measure resistance using the Kelvin or four-lead method discussed in Chapter 27. Place the RTD in a container of water with a well-calibrated mercury thermometer.

Suggested tasks:

✓ Measure the RTD resistance value as a function of temperature for the widest possible range.
✓ Determine the RTD calibration line, i.e., make a plot of resistance of the RTD as a function of the temperature, valid in the approximate range $0°C < T < 100°C$.
✓ Is the relationship of R versus T linear? Determine the temperature coefficient of RTD and its uncertainty and compare them to their value from the literature.
✓ Compare the values of the parameters of your line to the RTD specifications provided by the manufacturer. What can you conclude from this comparison?

50.1.4 *Project 172. Thermometer based on a diode*

> **Recommended basic equipment:** A diode. A source of current up to 1 mA or voltage of 5 V. A variable resistance of $1 - 10$ kΩ. A voltmeter that measures mV. A calibrated thermometer. A thermal bath.

For this experiment, you can use a resistor of about 5 kΩ connected in series with the diode, polarized so as to allow the passage of current, as indicated in Figure 50.4. Connect a millivoltmeter to the diode terminals to measure its voltage drop as a function of temperature.

Suggested tasks:

✓ Study the temperature response of a diode by placing the diode beside a calibrated thermometer in the same thermal bath. If the thermal bath consists of water, take care to electrically insulate the wires connected to the diode. Alternatively, you can use distilled water (or oil) in the thermal bath. By measuring the voltage drop across the resistor, determine the current through the circuit.

✓ Plot V as a function of T. Do your data agree with the theoretical expectation given by Eq. (50.4)?

✓ Vary the value of R_V (the limiting resistance of Figure 50.4) by at least 20% and repeat the analysis.

✓ Does the value of the slope of the line of V versus T vary with the value of the current that passes through the diode (or the value of R_V)? Explain your results.

50.1.5 *Project 173. Thermometer based on an integrated circuit*

Recommended basic equipment: An IC temperature sensor (LM35, AD590, or other). A calibrated thermometer. A thermal bath.

Connect the IC temperature sensor according to the manufacturer's specifications, which will be similar to those shown in Figure 50.5. Precaution: Note that this type of device cannot be placed in tap water, which would short-circuit its terminals. You therefore need to use very good electrical insulation but without it generating too much thermal insulation. One possibility is to use oil or *distilled* water in the thermal bath. You can use a well-calibrated thermometer to measure the temperature of the thermal bath. Measure the voltage output with a good millivoltmeter.

Suggested tasks:

✓ Measure the output voltage, V_{out}, of the sensor as a function of T for a range of approximately 5°C–95°C. Check the specifications of the sensor and do not exceed the allowed limits.

✓ Plot V_{out} as a function of the temperature measured with the calibrated thermometer.

✓ Using two containers (one with cold liquid and one with hot liquid), study the "time constant" of the sensor. To do so, immerse the sensor in the cold bath (temperature T_1) and wait for the V_{out} reading to stabilize. Then remove it and transfer it to the hot bath (temperature T_2). Measure V_{out} as a function of time until the thermometer has reached a final equilibrium temperature (hot bath temperature). The time constant, τ, can be defined, for example, as the time it takes for the thermometer reading to change from the initial value $T(t = 0) = T_1$ to 70% of the final temperature T_2. Compare the response time of this thermometer with the response time of a thermocouple.

References

[1] M. Zemansky, *Calor y Termodinámica* (Trad. de: Heat and Thermodynamics — 5th Ed. NY: McGraw Hill, 1968), Madrid: Aguilar, 1975.
[2] P. Childs, J. Greenwoods and C. Long, "Review of temperature measurement," *Rev. Sci. Instrum.*, vol. 71, p. 2959, 2000.
[3] OMEGA Engineering inc., "Introduction to temperature measurement," 2022. Available: https://www.omega.com/en-us/resources/measure/temperature.
[4] Wikipedia, "Temperature coefficient," 2022. Available: https://en.wikipedia.org/wiki/Temperature_coefficient.
[5] PR Electronics, "The fundamentals of RTD temperature sensors," 2022. Available: https://www.prelectronics.com/the-fundamentals-of-rtd-temperature-sensors/.
[6] D. R. Lide, *Handbook of Chemistry and Physics*, 81st Ed., CRC Press, 2000.
[7] F. W. Inman and D. Woodruff, "The thermometric properties of a diode," *Phys. Teach.*, vol. 33, p. 120, 1995.
[8] S. Sze and K. K. Ng, *Physics of Semiconductor Devices*, NY: John Wiley & Sons, 2006.
[9] Digi-Key Electronics, "Temperature sensors — Analog and digital output," 2022. Available: https://www.digikey.com/.
[10] "Making LCD thermometer with Arduino and LM35/36," 2022. Available: https://www.arduino.cc/.

[11] Autodesk, Inc., "Arduino based digital thermometer," 2022. Available: https://www.instructables.com/Arduino-Based-Digital-Thermometer/.
[12] Labtrek, "Complete kit, including datalogger to use with smartphone or PC," 2022. Available: www.labtrek.it/isochore.

Chapter 51

THERMAL EXPANSION OF SOLIDS

The goal of this activity is to study how the dimensions of a solid change when its temperature changes. Of the various possibilities that exist to study the thermal expansion of a solid, we have chosen two simple variants. In the first, the effect of the small thermal expansion of a metal tube is visualized through a mechanical amplifying device. In the second version, changes in length are measured directly with a micrometer dial. The coefficient of linear expansion α of the metal used can be found in both experiments.

Goals

✓ Thermal expansion
✓ Measurement of the coefficient of thermal expansion

51.1 Thermal expansion of a bar

In general, solids expand when heated and contract when cooled. The linear expansion of a solid is given by [1]

$$\Delta L = L_0 \alpha (T_f - T_0), \tag{51.1}$$

where ΔL is the change in length of a body as the temperature changes from T_0 to T, the temperature variation is $\Delta T = T - T_0$, L_0 is the length at T_0, and α is the coefficient of linear expansion, which depends on the type of material analyzed. The length $L(T)$ can be written as follows:

$$L(T) = L_0(1 + \alpha \Delta T). \tag{51.2}$$

In general, the coefficient α depends on the temperature, although in a limited temperature interval, it can be taken as constant. This coefficient can be determined very precisely using devices known as dilatometers [2]. This activity proposes to determine the coefficient of linear expansion of some metals using a simple device. Of course, there are many other experimental techniques for determining the coefficient of linear expansion that can be found in the literature [3]. A traditional method for measuring thermal expansion is mechanical dilatometry [4]. The sample is placed in an oven with temperature regulation, and the expansion of the sample with temperature is transmitted mechanically to a sensor located away from the oven. This type of push rod dilatometer is used in several commercial devices [5, 6].

51.1.1 *Project 174. Determining the coefficient of thermal expansion I*

> **Recommended basic equipment:** A metal tube (Cu, Al, or other material) of about 100 cm in length, with walls less than 2 mm thick. A thermocouple. A hot air gun or steam generator.

To study the thermal expansion of a solid, we propose a device like the one shown in Figure 51.1. The solid to be studied is a tube made of aluminum, copper, or other metal, approximately 100 cm long, with a diameter of approximately 10 mm–30 mm and thin walls (not more than 2 mm thick). Place a thin nail or pin (diameter

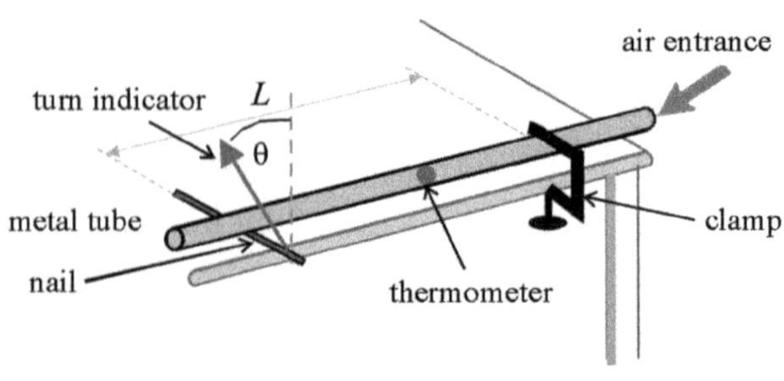

Figure 51.1 Device for studying the thermal expansion of a metal. The tube is fixed to a table by means of a clamp and heated with hot air or steam.

$d \approx$ 1–2 mm) under the tube near one end and fix the other end of the tube to the table with a clamp. Hot air or steam is blown through this end. You can place a metal funnel at the entrance of the tube so that a significant fraction of the air produced by a hot air gun penetrates through the tube.

When the length of the tube increases due to an increase in temperature, it will cause the nail to rotate (for which you need to make sure that the nail is firmly held by the tube). Attach the nail to a pointer (turn indicator) made of a thin plastic or wooden rod (e.g., a toothpick) which will account for the change in length with temperature. Place a goniometer behind the pointer to measure the angle θ between the pointer and the vertical. The nail acts as a "length variation amplifier". If ΔL is the elongation of the tube and d is the diameter of the nail, the angle θ that the nail rotates is

$$\theta = \frac{\Delta L}{d}. \tag{51.3}$$

This means that if $d \approx 1$ mm, for a variation of $\Delta L \approx 0.1$ mm, the angle θ is approximately 5°.

To measure the mean temperature, T, of the tube, attach a temperature sensor (calibrated thermocouple or any other small thermometer) halfway between the clamp and the nail. Make sure that the thermometer remains in good thermal contact with the tube throughout the experiment. You can do this by covering or wrapping the temperature sensor and tube with copper tape, thermal silicone, or thermal putty to improve the contact of the tube with the sensor.

Suggested tasks:

✓ Show that in the device shown in Figure 51.1, if d is the diameter of the nail, by combining Eqs. (51.1) and (51.3), we have

$$\theta = \frac{L_0}{d}\alpha\Delta T = \frac{L_0}{d}\alpha(T_f - T_0), \tag{51.4}$$

where L_0 is the length of the tube.

✓ Heat the tube by blowing through it a continuous jet of hot air (using an air gun or a hair dryer) or steam (produced by a steam generator).

- Measure θ as a function of the measured temperature, T.
- Plot θ as a function of T.
- What relationship do you find between these variables?

✓ If the relationship between θ and T is linear, Eq. (51.4) can be used to determine the coefficient of thermal expansion α of the metal. Determine the slope and its uncertainty in the plot of θ as a function of T.

✓ Measure L_0 (clamp-nail distance) and d and find their respective uncertainties. From these values, determine α from Eq. (51.4) and its uncertainty. Compare your result to the tabulated value for the material you used.

✓ Another way to proceed is to heat the tube evenly by blowing hot air on the outside with the air gun or hair dryer and raising the tube temperature to a value of about 40°C above room temperature. Then, as the tube slowly cools, the angle θ can be read for various decreasing temperatures. For example, the tube temperature can be measured each time the angle θ varies by 1° or 2° to take a large number of data.

Microscopic interpretation of the thermal expansion: An interesting aspect of the study of thermal expansion is its connection with the atomic and molecular properties of the solid. It is possible to connect the coefficient of linear expansion with the parameter that measures the asymmetry of the interatomic interaction potential. References [4] and [5] discuss this topic in detail, considering whether or not the interatomic potential in the lattice has to be asymmetric in order to observe a positive thermal expansion of the material.

51.1.2 *Project 175. Determining the coefficient of thermal expansion II*

Recommended basic equipment: A metal tube (Cu, Al, or other material) of at least 50 cm in length, with walls 1–2 mm thick. A thermistor or thermocouple. A hot air gun or a system that produces steam at an approximately constant rate (steam generator). A micrometer comparator.

This experiment is essentially the same as the previous one, except this time it is proposed to use a micrometer comparator to measure elongation ΔL directly. For this experiment, place a stainless steel nail or bolt through the tube near its free end. This nail or bolt rests on the dial gauge [9] (for example, a digital displacement gauge, with a resolution of about 0.01 mm). Fix the other end of the tube to the table by means of a clamp (see Figure 51.1). The nail or bolt enables you to measure the elongation of the tube without the hot air or the steam jet striking the comparator directly. This is important to avoid damaging the instrument and to allow more accurate measurements. Measure temperatures in the same way as in the previous experiment. There are some commercial devices available on the market that are ready to use for performing this type of measurement [6]. Some models enable you to connect the digital gauge displacement output to a data acquisition system. If the temperature signal is also connected to the same data acquisition system, it is possible to make an online plot of the elongation ΔL as a function of T.

✓ Using whichever device is available to you, plot ΔL as a function of T.
✓ If the relationship of ΔL versus T is linear, determine α and estimate its uncertainty.
✓ Discuss how your experimental results of α compare to tabulated values for the material of which the tube is made.

Lastly, Ref. [3] presents an interesting method for measuring the thermal expansion of a solid. Here, the thermal expansion of a wire is analyzed by passing a current through the wire to heat it. A load is suspended from the midpoint of a taut wire supported by two clamps. A weight is hung on the wire to produce a V shape in it. By measuring the variation of the vertex of the V as a function of the current, it is possible to infer the thermal expansion coefficient of the wire.

References

[1] F. Sears, M. Zemansky, H. Young and R. Freedman, *College Physics*, NY: Pearson Ed., 2009.

[2] M. Zemansky, *Calor y Termodinámica* (Trad. de: Heat and Thermodynamics - 5th Ed. McGraw Hill, N.Y. 1968), Madrid: Aguilar, 1975.
[3] D. Dounas-Frazer, P. Gandhi and G. Iwata, "Uncertainty analysis for a simple thermal expansion experiment," *Am. J. Phys.*, vol. 81, no. 5, p. 338, 2013.
[4] J. James, *et al.*, "A review of measurement techniques for the thermal expansion coefficient of metals and alloys at elevated temperatures," *Meas. Sci. Technol.*, vol. 12, no. 3, pp. R1–R15, 2001.
[5] TA Instruments, "Differential dilatometer," 2022. Available: https://www.tainstruments.com/products/dilatometers-2/.
[6] Pasco Scientific, "Thermal expansion apparatus — TD-8856," 2022. Available: https://www.pasco.com/.
[7] D. Thompson, "Thermal espansion in a symmetrical environment," *Am. J. Phys.*, vol. 62, p. 728, 1994.
[8] V. Weisskopf and H. Bernstein, "Search for simplicity: Thermal expansion," *Am. J. Phys.*, vol. 53, p. 1140, 1985.
[9] Wikipedia, "Indicator (distance amplifying instrument)," 2022. Available: https://en.wikipedia.org/wiki/Indicator_(distance_amplifying_instrument).

Chapter 52

BODY COOLING EXPERIMENTS — EXPONENTIAL DECAY

This chapter studies the cooling and heating processes of a body submerged in a fluid. These physical processes illustrate some characteristics of the natural convection process. In particular, this chapter analyzes Newton's law of cooling.

Goals

✓ Natural convection processes

✓ Newton's law of cooling

52.1 Heat transfer

Heat is basically transferred by three different processes: *conduction*, *convection*, and *radiation*. In nature, all these transmission mechanisms intervene simultaneously with varying degrees of importance. However, properly designed experiments can make only one of these mechanisms become dominant.

In *conduction*, heat is propagated through a material medium without transport of matter [1, 2]. Liquids and gases are not usually very good conductors of heat, however, they can transmit it efficiently by *convection*, which involves a fluid medium through which heat propagates with transport of matter. The density of most fluids varies with temperature, and gravity plays an important role: without it there would be no convection. One example of convection occurs in a room with a stove or heat radiator. When the air surrounding the heat source warms, it expands and its density decreases, generating a macroscopic flow of air currents in the room. Ocean currents, hot or cold, such as the Gulf Stream, are an example of natural convection on a large scale. Wind is another example of convection [1].

In the case of natural convection, a thin layer of air (boundary layer) is formed around the hot object, which by conduction transfers energy from the hot surface to the environment [2]. The power dissipated by natural convection P_{conv} may be described by the following relation:

$$P_{\text{conv}} = A \cdot h \cdot (T_{\text{surf}} - T_{\text{fluid}}), \tag{52.1}$$

where A is the surface of the body exposed to the medium, h is a characteristic constant, *convection coefficient*, which depends on the flow regime (laminar or turbulent), the thermal conductivity of the fluid medium, and the state of the surfaces. T_{surf} and T_{fluid} designate the temperatures of the surface of the body and the fluid medium, respectively.

In conduction and convection, the presence of a material medium is necessary for the propagation of heat. However, life on Earth depends on the transfer of energy from the Sun, which passes through essentially empty space. This form of energy transfer is called *radiation*. Heat propagates through radiation: (a) *without the need for a material medium* and (b) *without the transport of matter*. In radiation, energy is transmitted by means of electromagnetic waves [1, 3].

52.2 Cooling a body

If we have a body of mass m and area A, at a certain initial temperature T_i, in a fluid (air or water) at a temperature T_f (which we assume is lower than T_i), the body will cool. If $T(t)$ represents the body temperature as a function of time t, according to Eq. (52.1), we have

$$P_{\text{conv}} = mc\frac{dT}{dt} = A \cdot h \cdot (T(t) - T_f), \tag{52.2}$$

where c is the specific heat of the body. Therefore, we can write

$$\frac{dT(t)}{dt} = -k \cdot (T(t) - T_f), \quad \text{with} \quad k = \frac{A \cdot h}{m.c}. \tag{52.3}$$

This relationship is known as *Newton's law of cooling* and roughly describes how a body at a temperature T is cooled (or heated) when the surrounding medium is at a temperature T_f. Equations (52.1) and (52.3) are approximate empirical expressions and hold that the speed with which a body changes its temperature over time, that is, dT/dt, is proportional to the difference in temperature between the body and its environment, that is, $(T - T_f)$. Equation (52.3) assumes that the effects of heat loss by radiation are negligible compared to those of convection [4, 5]. This will be true as long as the temperature difference between the body and the environment is not very large, typically $\Delta T < 100$ K.

It is easy to verify that

$$T(t) = T_f + B \cdot e^{-\frac{t}{\tau}} \tag{52.4}$$

satisfies Eq. (52.3), where B is a constant that depends on the initial conditions. The parameter $\tau \equiv k^{-1} = mc/Ah$ is measured in units of time and represents the *characteristic time* of cooling. If we assume that initially the body is at a temperature T_i, that is to say for $t = 0$, $T(t) = T_i$, the expression (52.4) reduces to

$$T(t) - T_f = (T_i - T_f) \cdot e^{-\frac{t}{\tau}}. \tag{52.5}$$

Taking log of this expression,

$$\ln(T(t) - T_f) = \ln(T_i - T_f) - \frac{1}{\tau}t. \tag{52.6}$$

Equation (52.6) indicates that a *semi-logarithmic plot* of $(T(t) - T_f)$ as a function of time t gives a line of slope $-1/\tau$.

In particular, if we consider a sphere of a certain material, of radius a, density ρ, and specific heat c_s in a given medium with a given convection coefficient, h, according to Eqs. (52.3) and (52.5), we would expect that

$$\tau(a) = \frac{\rho \frac{4}{3}\pi a^3 c_s}{4\pi a^2 h} = \frac{\rho \cdot c_s}{3 \cdot h} a, \tag{52.7}$$

therefore, a smaller sphere will cool faster than a larger sphere of the same material in a given medium [6]. The purpose of this section is to study experimentally the cooling process of a body in air and water in order to test Eqs. (52.1)–(52.7).

52.2.1 *Project 176. Cooling a glass thermometer in air*

Recommended equipment: A glass thermometer with a measuring range between 0°C and 100°C and a watch. A container of water of about 250 ml.

This activity proposes to use a thermometer and observe how it cools once it is removed from a container of hot water ($T \approx 90°\mathrm{C}$). The thermometer will cool until, after a certain time, it reaches room temperature.

Place approximately 200 ml of water in a container and bring it to boil.

Suggested tasks:

✓ Immerse the thermometer in hot water until its temperature is in equilibrium with the water temperature. Record this value T_i, which is the initial thermometer temperature. Remove the thermometer from the hot water, moving it as little as possible so that it does not agitate the surrounding air, and without letting it touch anything else. You could do this by hanging it from a thread. As it cools, read the temperature every two to three seconds for the first minute and then every 10 s, until it reaches a final stable value close to room temperature T_f.

✓ Plot the data for temperature, $T(t)$, as a function of time, t, on a graph with linear scales.

✓ Using the measured values T_i and T_f, plot on a semi-logarithmic graph $(T(t) — T_f)$ as a function of time t and determine whether the graph is linearized. If so, construct a graph of $\ln(T(t) — T_f)$ as a function of t. If the trend of the data is linear, fit them with a line and find the constant k, Eq. (52.3), and its error, see Chapter 7. Also determine the characteristic time τ and its uncertainty. Verify that the y intercept corresponds to ln $(T_i - T_f)$; see Eq. (52.6).

✓ Is Newton's law of cooling a good representation of the cooling you have studied?

Questions

✓ Could you use the thermometer from the experiment to measure the temperature of an animal such as a pigeon? And the temperature of a mosquito? Explain your answers.

52.2.2 *Project 177. Cooling a body in air and in water*

> **Recommended equipment:** A glass thermometer with a measuring range between 0°C and 100°C. A thermometer connected to a data acquisition system connected to a PC with a range from 0°C to 100°C. About 300 g of putty or modeling clay. Two containers of water of approximately 250 ml.

The goal of this experiment is to study how a body cools differently in air and in water. Use the putty or modeling clay to form a ball of approximately 50 g, and insert the thermometer as illustrated in Fig. 52.1. Make sure there is a tight seal between the thermometer and the ball of putty or modeling clay so that there is close contact between them and no air or water penetrates.

In this activity, we propose to study how a body cools in air and in water. Fill one of the containers with hot water ($T \approx 40 - 50$°C) and the other with water at room temperature. Choose a measurement rate in the acquisition system of approximately 0.1 Hz, that is, a temperature measurement every 10 s and a measurement range of about 30 min. Place the body (putty with the thermometer) in hot water until thermal equilibrium is reached. With the glass

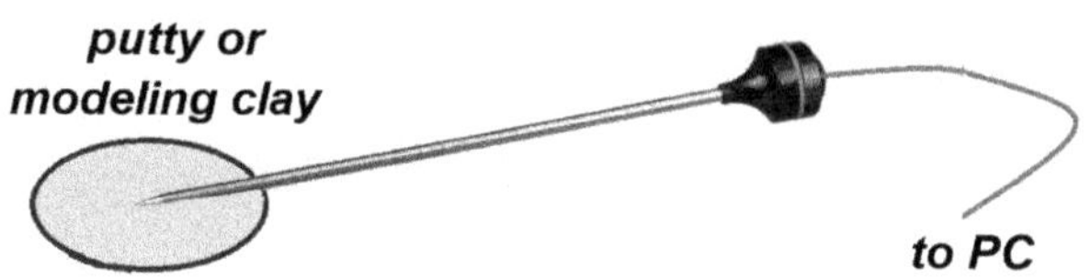

Figure 52.1 Setup for studying the law of cooling. Adding putty or modeling clay to the thermometer makes the process slower and simpler to study.

thermometer, measure the temperature of the hot water. Start data collection.

Suggested tasks:

- ✓ Remove the body with the thermometer, taking care that the clay ball does not become detached. Allow the system to cool in the air. While you wait for the body to reach room temperature, measure the room temperature with the glass thermometer T_f.
- ✓ Plot $T(t) - T_f$ as a function of t on a linear and on semi-logarithmic scales. Do either of these graphs show a linear trend?
- ✓ Plot $\ln(T(t)–T_f)$ as a function of t. If the data show a linear trend, fit a line through them. Calculate the characteristic time τ and its uncertainty.
- ✓ Repeat the above experiment, but this time, instead of cooling the body in air after removing it from the hot water, immerse it in cold water and allow it to cool there. To ensure that the temperature of the cold bath does not vary significantly, the cold water mass should be 10 times greater than the mass of the body under study (putty or modeling clay). While you wait for the body to reach the cold water temperature, use the glass thermometer to measure the temperature of the cold water in the container, T_f.
- ✓ Plot $T(t) - T_f$ as a function of time, t, on a semi-logarithmic scale.
- ✓ Plot $\ln(T(t) - T_f)$ as a function of t and fit a line to the data. Calculate the characteristic time τ and its uncertainty. Compare the τ values found for cooling in air and in water.
- ✓ Based on your observations, explain why a person can survive without major problems at a temperature of 5°C in air, but if they fall into water at the same temperature, they will only survive for a few minutes.

52.2.3 *Project 178. Variation of cooling with mass*

Recommended equipment: A glass thermometer with a measuring range between 0° and 100°C. A thermometer connected to a data acquisition system connected to a PC, with a range of 0°C–100°C. About 500 g of putty or modeling clay. A scale ranging from 0 to 300 g @ 1g or better. Two containers of water of approximately 250 ml.

The goal of this activity is to study how the cooling of a body varies with its mass. Use the putty to make 4 or 5 approximately spherical balls weighing about 50–300 g. Measure the mass m and estimate the *radius* a of each ball. Insert the thermometer into each ball, as illustrated in Figure 52.1. Make sure there is a tight seal between the thermometer and the ball.

If we assume Newton's law of cooling, Eq. (52.3), we have

$$k = \frac{Ah}{mc} = \frac{h}{c}\frac{4\pi \cdot a^2}{4\pi\rho \cdot a^3/3} = \frac{h}{c\rho}\frac{3}{a} \propto \frac{1}{a} \propto \frac{1}{m^{1/3}}, \tag{52.8}$$

where ρ is the putty density and a is the mean radius of the ball. The cooling characteristic time will therefore be

$$\tau = \frac{c_s\rho}{3.h}a \propto a \propto m^{1/3}. \tag{52.9}$$

Fill one of the containers with hot water (T $\approx$ 50°C) and the other with water at room temperature. Choose a measurement rate in the acquisition system of approximately 0.1 Hz and a measurement range of approximately 30 min. Place each of the bodies with the thermometer in hot water until thermal equilibrium is reached. With the mercury thermometer, measure the temperature of the hot water. Start data collection.

Suggested tasks:

- ✓ For each ball, measure the cooling temperature as a function of time. Use the mercury thermometer to measure the temperature of the cold water T_f.
- ✓ Plot $\ln(T(t) - T_f)$ as a function of t and fit the data with a line. Find the characteristic time τ and its uncertainty for each ball.
- ✓ Next, plot the values of τ as a function of m and as a function of $m^{1/3}$.
- ✓ Plot the values of τ as a function of a (the radii of the balls).
- ✓ Discuss whether your data can be explained by the model described by Eqs. (52.7) and (52.9).

References

[1] D. Giancoli, *Physics Principles with Applications*, 7th Ed., NY: Pearson, 2014.

[2] Y. Cengel, *Heat Transfer: A Practical Approach*, 2nd Ed., New York: McGraw-Hill, 2002.

[3] R. Halliday, D. Resnick and M. Krane, *Fundamentals of Physics*, NY: John Wiley & Sons, Inc., 2002.

[4] M. P. Silverman and C. R. Silverman, "Cool in the kitchen: Radiation, conduction, and the Newton "hot block" experiment," *Phys. Teach.*, vol. 38, no. 1, p. 82, 2000.

[5] C. T. O'Sullivan, "Newton's law of cooling — A critical assessment," *J. Phys.*, vol. 58, no. 10, p. 956–960, 1990.

[6] K. Vandervoort, "Animal size and heat transfer," *Phys. Teach.*, vol. 58, p. 104, 2020.

Chapter 53

CONSERVATION OF ENERGY AND CALORIMETRY

This chapter proposes some experiments on heat transfer processes and the application of the first principle of thermodynamics, which is used to estimate the equilibrium temperature reached by a mixture of two liquids originally at different temperatures. Next, conservation of energy is used to measure the specific heat of some materials. The chapter also proposes two experiments to determine the latent heat of vaporization and fusion of water.

Goals

✓ Calorimeters
✓ Specific heat of solids and liquids
✓ Phase transitions
✓ Latent heat of fusion and vaporization

53.1 First principle of thermodynamics

The first principle of thermodynamics states that energy is neither created nor destroyed, it only transforms [1]. When two systems at different temperatures are brought into thermal contact, energy is transferred in the form of heat from the warmer to the cooler system. The transfer of thermal energy ceases once the two systems reach a common equilibrium temperature, which depends on the masses and the specific heat of the systems involved.

The specific heat c_{sub} of a substance is the amount of heat required to increase the temperature of a unit mass of the substance by one degree Celsius.

As the first example, consider two masses of water: one hot, at an initial temperature T_{hot}, and another cold, at an initial temperature T_{cool}. These two masses of water are mixed in a calorimeter, and the mixture reaches a final temperature T_{final}. Applying the first principle, these bodies of water undergo the following heat exchanges [2, 3]:

$$\begin{aligned}\Delta Q_{\text{hot}} &= M_{\text{water\_hot}} c_{\text{water}} \Delta T_{\text{hot}} \\ &= M_{\text{water\_cool}} c_{\text{water}} (T_{\text{hot}} - T_{\text{final}}), \qquad (53.1)\end{aligned}$$

$$\begin{aligned}\Delta Q_{\text{cool}} &= M_{\text{water\_cool}} c_{\text{water}} \Delta T_{\text{cool}} \\ &= M_{\text{water\_cool}} c_{\text{water}} (T_{\text{final}} - T_{\text{cool}}), \qquad (53.2)\end{aligned}$$

where c_{water} is the specific heat of water, which we will assume independent of temperature and equal to 1 cal/(g°C). $\mathrm{M}_{\text{water\_hot}}$ and $\mathrm{M}_{\text{water\_cool}}$ are the masses of hot and cold water, respectively. ΔQ_{hot} and ΔQ_{cool} are the heat variations of each body of water separately. If energy is conserved, we have

$$\Delta Q_{\text{hot}} = \Delta Q_{\text{cool}}. \qquad (53.3)$$

Since all the parameters involved in these equations can be measured directly, it is possible to verify experimentally whether Eq. (53.3) is fulfilled.

Several of the experiments discussed in this chapter make use of a calorimeter. A calorimeter (a vessel with good thermal insulation) can be made using an expanded polyurethane container with a volume of about $250\,\text{cm}^3$ and a wall thickness of about 1 cm (e.g., an ice-cream container). The low thermal conductivity of expanded polyurethane inhibits heat exchange with the outside environment.

Many experiments require knowledge of the temperature jump of a process. If we can follow the evolution of temperature as a function of time, we can use the following graphical method to determine temperature changes when there are dissipation effects or heat losses. This procedure is particularly simple if we use a data acquisition system (DAS) to measure the temperatures. Imagine, for example, that we are measuring the temperature of the cold-water mass in our system, to which we have added the hot-water mass. The

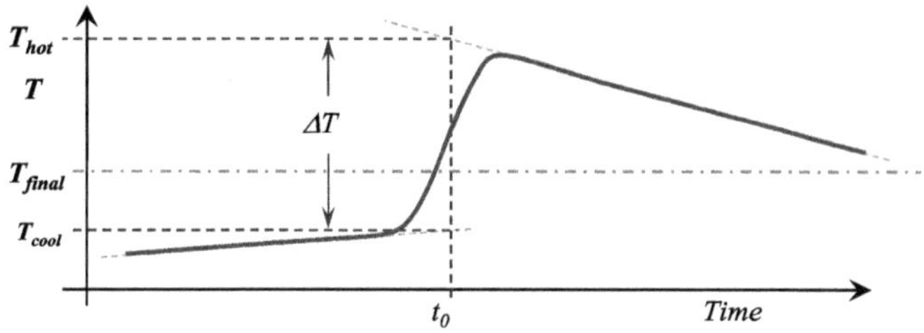

Figure 53.1 Graphical procedure to determine the temperature change ΔT in an experiment with heat dissipation effects. The solid thick red line shows the time evolution of the temperature. The dotted orange lines are linear extrapolations of the trends before and after the mixing, performed at the time t_0. ΔT is the temperature jump inferred at the time of mixing t_0.

expected temperature will evolve as illustrated in Figure 53.1. Due to possible heat losses, the temperature of the mixture will not remain constant.

The procedure for determining the temperature jump ΔT and the final temperature of the mixture T_{final} is described in Figure 53.1. The *nominal* time at which the mixture is performed is t_0. The temperature jump inferred at this instant is taken as ΔT. We will also use this procedure in the study of the heat exchange between solid metal bodies and water to determine the specific heat of some metals.

53.2 Water equivalent of the calorimeter

Several of the experiments in this chapter use calorimeters, which are containers with good thermal insulation. However, calorimeters also contain thermometers or other devices whose heat capacity cannot usually be neglected in the energy balance. In this case, it is useful to introduce the concept of the *water equivalent of the calorimeter* M_{EW}, which is defined as a mass of water whose heat capacity is equivalent to the total heat capacity of the calorimeter, including the elements it contains (thermometers, stirrers, etc.). Thus, consider, as before, two masses of water: one hot ($M_{\text{water\_hot}}$) at an initial temperature T_{hot}, in its own container and another cold ($M_{\text{water\_cool}}$) at an initial temperature T_{cool}, inside the calorimeter, including all its elements (thermometer, stirrer, etc.), whose *equivalent in water* (M_{WE}) we wish to determine. The hot water is poured into the calorimeter and the mixture reaches a final temperature T_{final}. The thermal balance this time is

$$\begin{aligned}(M_{\mathrm{water\_cool}} + M_{\mathrm{WE}})c_{\mathrm{water}}(T_{\mathrm{final}} - T_{\mathrm{cool}}) \\ = M_{\mathrm{water\_hot}}c_{\mathrm{water}}(T_{\mathrm{hot}} - T_{\mathrm{final}}).\end{aligned} \tag{53.4}$$

Therefore,

$$M_{\mathrm{WE}} = M_{\mathrm{water\_hot}}\left(\frac{T_{\mathrm{hot}} - T_{\mathrm{final}}}{T_{\mathrm{final}} - T_{\mathrm{cool}}}\right) - M_{\mathrm{water\_cool}}. \tag{53.5}$$

53.2.1 *Project 179. Conservation of energy in a mixture of two water masses*

Recommended basic equipment: A calorimeter. Two standard thermometers or temperature sensors connected to a DAS, measuring in the range 0–100°C.

The goal of this activity is to falsify the conservation of energy in a mixture of two volumes of water at different initial temperatures, i.e., to test the result of Eqs. (53.3) and (53.4) experimentally.

To do so, determine the masses of the calorimeters when empty and dry. Fill one calorimeter with hot water and the other with cold water to about a third of their capacity, and determine their masses again. By difference, it is possible to determine the values of $M_{\mathrm{water\_hot}}$ and $M_{\mathrm{water\_cool}}$. Hot water should be around 20°C above room temperature and cold water around 10°C below room temperature but above 0°C. Measure the temperatures T_{cool} and T_{hot}. Immediately *add the hot water to the cold water* and stir until the thermometer temperature stabilizes. Measure the final temperature of the mixture, T_{final}.

If you have a temperature sensor connected to a DAS, continuously measure the temperature of the water in the calorimeter and obtain a plot of temperature as a function of time. From this curve, evaluate the heat losses from the calorimeter qualitatively by observing the variation in temperature over time. If you can follow the evolution of temperature as a function of time, you can determine the value of the final equilibrium temperature T_{final}, using the procedure described in Figure 53.1.

Suggested tasks:

✓ Determine the value of T_{final} for the water mixture. Using Eqs. (53.1) and (53.2), calculate ΔQ_{hot} and ΔQ_{cool} from your experimental data.

✓ Repeat the experiment for different masses of water at different starting temperatures, and plot ΔQ_{hot} as a function of ΔQ_{cool}. If energy is conserved, that is, if Eq. (53.4) is verified, we would expect the data to be distributed around a 45° line.

✓ Can you conclude from your experiment that "energy is conserved" in the process studied?

✓ Determine the water equivalent of your calorimeter. Repeat the analysis using the same data, but this time graph ΔQ_{hot} as a function of ΔQ_{cool}, using the definition

$$\Delta Q_{\text{cool}} = (M_{\text{water\_cool}} + M_{\text{WE}}) \cdot c_{\text{water}} \cdot (T_{\text{final}} - T_{\text{cool}}). \quad (53.6)$$

✓ How are your conclusions modified by the incorporation of the equivalent in water?

53.2.2 *Project 180. Measuring the specific heat of a solid I*

> **Recommended basic equipment:** A calorimeter. Two standard thermometers or temperature sensors connected to a DAS, measuring in the range 0°–100°C. Blocks of Cu, Fe, Al, or other metals, with masses between 400 and 500 g.

This experiment measures the specific heat of some metals (Al, Cu, Fe, etc.). To do so, you need to know the mass M_{sample} of each metal sample and water equivalent mass of the calorimeter, M_{WE}. Place hot water in a container and immerse the sample body completely. Use a thermometer to measure the temperature of the hot water T_{hot} when the equilibrium temperature is reached. Measure the mass M_C of the empty calorimeter. Fill the calorimeter halfway with cold water, a few degrees lower than room temperature. To have greater sensitivity, the masses of the cold water and the body should be comparable, i.e., neither of them should be much greater than the other, but they should be sufficient for the water to cover the body completely when it is placed inside the calorimeter, whose M_{WE} you already have determined. Measure the total mass of the calorimeter and the cool water $M_T = M_C + M_{\mathrm{water\_cool}}$ and use the difference to calculate the mass of cold water, $M_{\mathrm{water\_cool}} = M_T - M_C$. Measure the water temperature, T_{cool}, before transferring the body from the hot water container to the calorimeter and, if possible, use a temperature sensor connected to DAS that allows you to record time evolution of the temperature, as shown in Figure 53.1. If possible, position the sample without it touching the walls or bottom of the calorimeter, e.g., by hanging it from a thin thread before immersing it in hot water, which will facilitate subsequent operations. Mix the water slowly and measure the time evolution of the temperature. Determine the highest temperature T_{final} reached by cold water with the immersed body (see Figure 53.1).

By energy conservation, we have

$$M_{\mathrm{sample}} c_{\mathrm{sample}} (T_{cal} - T_{\mathrm{final}}) \\ = (M_{\mathrm{water\_cool}} + M_{\mathrm{WE}}) c_{\mathrm{water}} (T_{\mathrm{final}} - T_{\mathrm{cool}}), \qquad (53.7)$$

where c_{sample} is the specific heat of the sample and c_{water} the specific heat of water. From here, the specific heat of the sample can be deduced for each sample used as follows:

$$c_{\mathrm{sample}} = c_{\mathrm{water}} (M_{\mathrm{water\_cool}} + M_{\mathrm{WE}}) \left(\frac{\Delta T}{T_{\mathrm{hot}} - T_{\mathrm{final}}} \right). \qquad (53.8)$$

Suggested tasks:

✓ Using Eq. (53.8), determine the specific heat of each sample and its corresponding error, applying the technique discussed in Chapter 6.

✓ Compare the values found to the tabulated values for each sample.

53.2.3 *Project 181. Measuring the specific heat of a solid II*

> **Recommended basic equipment:** Two beakers of about 250 cm^3. A thermometer and temperature sensors connected to a PC, which measure in the range 0 – 100°C. Blocks of Cu, Fe, Al, or other metals, with masses between 400 g and 500 g.

This experiment measures the specific heat of some metals using a slightly different method, with a thermometer connected to a data acquisition system (DAS). While in the previous experiment we have introduced the heated sample into a cold calorimeter, here we transfer a cold sample into a hot calorimeter. This method makes use of the cooling that occurs when a container of hot water is left at room temperature. Assume that you know the mass M_{sample} of each metal sample you will use and the mass $M_{\text{water\_hot}}$ that you place in a beaker. The mass of hot water in the beaker must completely cover the body placed in it. Place a thermometer connected to a computer data collection system in the hot water and study the thermal evolution of the system. Fill a second beaker with cold water a few degrees lower than room temperature. Make sure the cold water completely covers the body when it is placed in the beaker. When the cool water and the body immersed in it have reached thermal equilibrium, measure this equilibrium temperature and call it T_{cool}. You already have measured the mass of the empty calorimeter M_C, so by weighing it after filling it with hot water, you can calculate the hot water mass, $M_{\text{water\_hot}}$, by difference. Monitor the temperature of this calorimeter with the hot water in it and the thermometer connected to the DAS. It is enough to collect the temperature data at a rate of about once per minute. Once you have recorded the variation of the hot water temperature for about 15 minutes, remove the body from the cold water, dry it quickly with the least possible thermal interference, place it in the hot water inside the calorimeter (whose M_{EW} you already know), and continue to measure

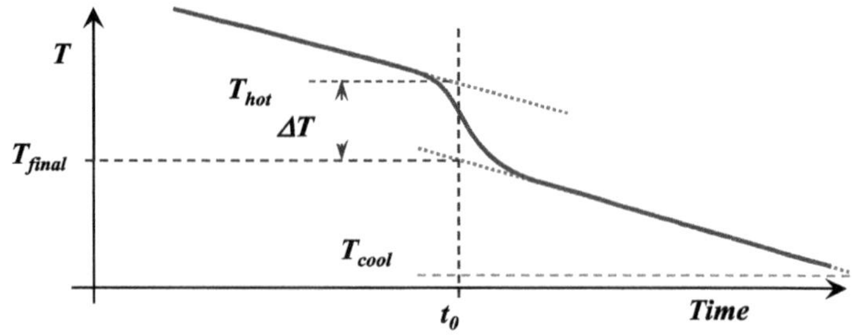

Figure 53.2 Evolution of the temperature of a volume of hot water when a cold body is placed in it. The dotted lines are linear extrapolations of the trends before and after the change at time t_0. The temperature jump inferred at this time is taken as ΔT. T_{cool} is the equilibrium temperature of the cool water with the immersed metal body.

this temperature for another 20 min. In the resulting plot, you will see a drop in temperature ΔT due to the introduction of the body (see Figure 53.2). From the energy balance,

$$M_{\text{sample}} c_{\text{sample}} (T_{\text{final}} - T_{\text{cool}}) = (M_{\text{water\_hot}} + M_{\text{WE}}) c_{\text{water}} \Delta T, \tag{53.9}$$

where c_{sample} is the specific heat of the sample and M_{EW} is the water equivalent of the calorimeter.

Suggested tasks:

✓ Using Eq. (53.9), determine the specific heat of each metal sample and its corresponding error, applying the technique discussed in Chapter 6.

✓ Compare the values obtained with the tabulated values for each sample.

53.3 Phase transitions

When a substance changes its state of aggregation, for example, from liquid to vapor or from solid to liquid, we say that it undergoes a *phase transition*. The heat released or absorbed during this transition, at constant temperature, is called *latent heat*. The concept of phase transition is broader than the concept of a change in the state of aggregation, and includes other types of transformations

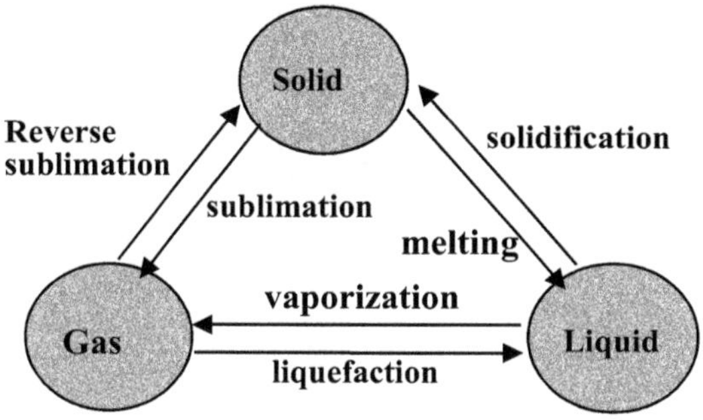

Figure 53.3 Phase transitions of a substance associated with changes in the state of aggregation.

of matter, for example, between different crystalline states, from paramagnetic to ferromagnetic, or when a metal passes from a normal state to a superconductor. Figure 53.3 illustrates the changes in the aggregation state of a substance. For simple substances, these transitions occur at well-defined temperatures, which are generally pressure-dependent [4, 5].

When a solid substance is heated, it goes from the solid state to the liquid state in a process called *fusion*, which occurs, for a given pressure, at a melting temperature T_m. Similarly, if the liquid is cooled below T_m, it will become solid in a process called *solidification*. When a liquid is heated to its *boiling point* T_b, the liquid undergoes a transition to the vapor state called *vaporization*. Conversely, if a vapor that is at a temperature higher than T_b cools to a liquid state, the process is called *liquefaction*. Another type of transition is *sublimation*, which is the change from a solid to a vapor state, without going through an intermediate liquid state. In all these processes, a quantity of heat is involved, which is called *latent heat of transformation*. Latent heat is associated with the energy required for the substance to make the phase change from one state to another. In this chapter, we will study some methods to determine the latent heat of vaporization and fusion of a substance.

Table 53.1 Data for some simple substances.

Substance	Latent heat of fusion kJ/kg	Melting point °C	Latent heat of vaporization kJ/kg	Boiling point °C
Ethyl alcohol	108	−114	855	78.3
Water	334	0	2260	100
Nitrogen	12.87	−210	198.9	−197

53.3.1 *Project 182. Liquid-vapor transition — Latent heat of evaporation I*

Recommended basic equipment: A steam generator. A calorimeter. A glass thermometer and a temperature sensor connected to a PC, which measure in the range 0–105°C.

The purpose of this project is to measure the latent heat of evaporation L_v of water, associated with the liquid-vapor transition. The latent heat L_v is alternatively defined as the quantity of heat Q that must be supplied to a unit mass of liquid at boiling temperature, T_b, to pass to the vapor state, that is,

$$L_v = \frac{\Delta Q}{\Delta m} \text{ at } T = T_b. \tag{53.10}$$

There are several ways to make this determination. A classic design is shown in Figure 53.4. This method uses a device that generates water vapor, such as a boiling kettle attached to a heat-resistant hose to conduct the steam produced to a nearby calorimeter. You could also use an air humidifier or a commercial laboratory steam generator [5].

The steam generated is conducted to the calorimeter containing a cold water mass at the initial temperature T_{cool}. When steam condenses in the water in the calorimeter, the heat energy from the steam is released into the water in two ways. One part of the energy is contributed by the condensation of steam at the temperature T_{vapor}. If the mass of condensed vapor is M_{vapor}, according to Eq. (53.10),

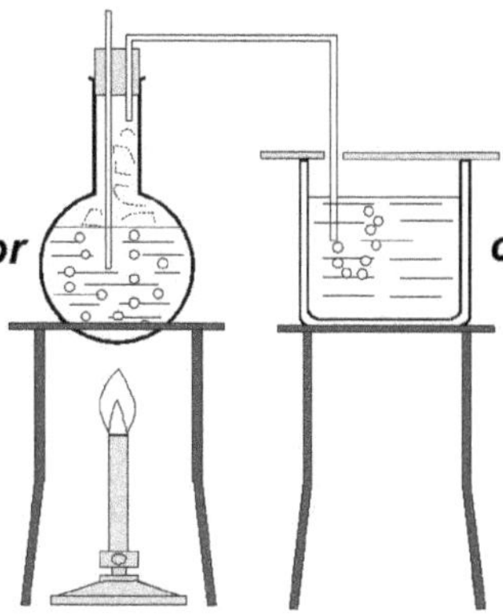

Figure 53.4 Schematic diagram of the experiment to determine the heat of evaporation of water.

this heat is

$$Q_1 = M_{\text{vapor}} L_v. \tag{53.11}$$

The other part of the energy is provided by condensed water, which lowers its temperature from the temperature T_{vapor} to the final temperature T_f, which will be the final temperature of all the water in the calorimeter:

$$Q_2 = M_{\text{vapor}} c_{\text{water}} (T_{\text{vapor}} - T_f), \tag{53.12}$$

where c_{water} is the specific heat of water. From the conservation of energy, we have

$$\begin{aligned} M_{\text{vapor}} L_v &+ M_{\text{vapor}} c_{\text{water}} (T_{\text{vapor}} - T_f) \\ &= (M_{\text{water\_cool}} + M_{\text{WE}}) c_{\text{water}} (T_f - T_{\text{cool}}), \end{aligned} \tag{53.13}$$

where M_{WE} is the water equivalent of the calorimeter. From this expression, we can find L_v by measuring M_{vapor}, $M_{\text{water\_cool}}$, T_{cool}, and T_f.

Suggested tasks:

✓ Place in the calorimeter approximately 250 cm^3 of cold water at about 10°C below room temperature. Measure the mass of this cold water and its initial temperature, $M_{\text{water\_cool}}$ and T_{cool}. To find the mass of cold water, weigh the calorimeter without and with water and determine $M_{\text{water\_cool}}$ by difference.

✓ Activate the steam generator and wait for it to produce steam. Measure the temperature of the boiling water, T_b. When steam production stabilizes and steam has been flowing continuously for about 5 min, insert the hose into the calorimeter containing the cold water.

✓ Using a thermometer in the calorimeter, preferably connected to a computer (though this is not essential), follow the evolution of the calorimeter's temperature. When the temperature of the water, T, rises above the ambient temperature by a magnitude such that

$$T - T_{\text{cool}} \approx T_{\text{amb}} - T_{\text{cool}}, \tag{53.14}$$

remove the hose and stop supplying steam. Continue stirring the water and record the final temperature, T_f, reached by the water.

✓ Finally, weigh the calorimeter again with the water. Find, by difference, the mass of the water M_{vapor} added by the condensation of the steam.

✓ Using Eq. (53.13) and the measured values of mass and temperature, estimate the value of L_v for water and estimate the uncertainty of the determination.

✓ Compare the value obtained to the table value of L_v for the same temperature T_b which was measured in the steam generator.

53.3.2 *Project 183. Liquid-vapor transition — Latent heat of evaporation II*

Recommended basic equipment: A calorimeter. A temperature sensor that measures in the range 0–105°C. An electronic scale or balance with a sensitivity of 1 g or better and a range of approximately 600 g. A stopwatch. An electric immersion heater or electric power resistance. An electric power meter.

The purpose of this project is to measure the latent heat of evaporation L_v of water, liquid nitrogen, or other non-flammable liquid. The experimental setup is illustrated in Figure 53.5.

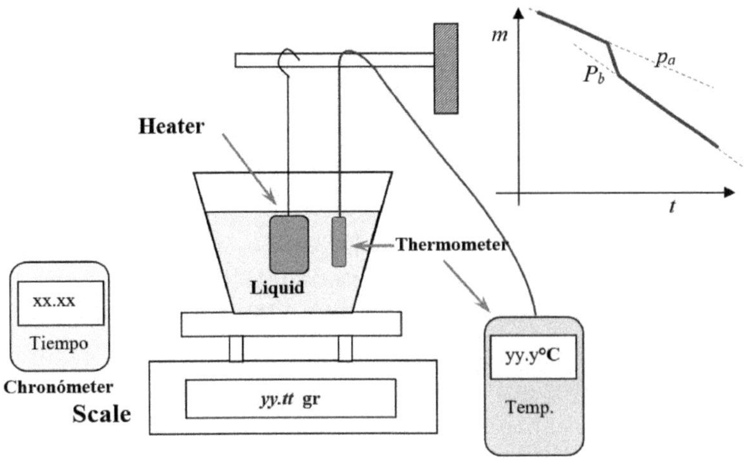

Figure 53.5 Experimental arrangement for measuring the latent heat of evaporation of liquid nitrogen. The insert shows a schematic diagram of the variation of the mass as a function of time, before and after activating the heater.

Basically, the experiment consists of measuring the *change in mass* of the liquid sample when the electric heater is turned on and the liquid boils. First, this arrangement is illustrated for liquid nitrogen (LN). The idea is to characterize the power of the heating resistance that is placed in the liquid. Assume that there are initially about 250 g of LN, and a resistance of approximately 10 Ω and 10 W is used. By measuring the voltage drop on this resistor and the current that flows through it, you can know the power P_R supplied to the resistance. For LN, there is an evaporation rate due to heat flux from the environment, characterized by the slope p_a of the mass as a function of time, before turning on the heating. Once the heating is turned on, the evaporation rate will be given by the difference in slopes $(p_a - p_b)$.

Thus, the evaporation rate $dm/dt = (p_a - p_b)$ can be associated with the energy supplied by the resistance, i.e.,

$$\frac{\mathrm{dm}}{dt} L_v = (p_a - p_b) L_v = P_R, \tag{53.15}$$

from which, by measuring the slopes before (p_a) and after (p_b), the heater is activated, and from the power of the heater (P_R), it is possible to estimate L_v. To determine these slopes, measure the change in mass (according to the scale) as a function of time (measured by the stopwatch). If you use a scale connected to a DAS, you can obtain this type of plot directly. Use the thermometer to determine when the liquid begins to boil. This technique can also be used for water. First, bring the water to boiling, and immediately pour it into a beaker containing a thermometer and the heater. In this case, it is advisable to use a heater that can deliver more power than the one used for LN, say about 0.5 kW. In this case, the slope $p_a = 0(dm/dt)$ before boiling is assumed. With a 0.5 kW heater, you would expect an evaporation rate of the order of 10 g per minute. In this case, to measure the power of the immersion heater, it is advisable to use a power meter, a device usually used to measure the power of household appliances. However, the power may also be calculated as $P_R = VI$, by measuring the voltage drop V across the heater resistance, and the current I fed to the resistance.

Suggested tasks:

- ✓ **For *LN*:** Place about 250 cm^3 of liquid nitrogen in the calorimeter.
- ✓ Using a resistance of 10 Ω and 10 W, regulate the current in the resistor so that it delivers about 10 W. Determine this power P_R as accurately as possible and its uncertainty.
- ✓ Measure the mass of the *LN* as a function of time, at a rate of approximately one reading per minute. Measure the change in mass as a function of time for about 15 min without passing current through the resistor. Once the slope p_a is well defined, turn on the current in the resistor and measure the mass as a function of t for another 15 or 20 min. Draw a graph and determine p_a and p_b and their uncertainties.
- ✓ Use Eq. (53.15) to find the value of L_v for *LN* and estimate the uncertainty of the determination.
- ✓ Compare the value found to the table value of L_v for *LN*.
- ✓ **For water:** Place in the calorimeter approximately 300 cm^3 of hot water, at a temperature close to boiling point.

- ✓ Use a resistor capable of delivering about 500 W. If necessary, you can regulate the input voltage using a Variac. Determine this power P_R with a wattmeter or power meter or by measuring current and voltage. Estimate its uncertainty. Be careful with this type of resistance, as in general it works at voltages close to line voltages and is submerged in water. If possible, use a water heater that operates at low voltage and avoid electrical contact with the water.
- ✓ Measure the mass of water as a function of time at a rate of approximately one record per minute. After about 5 min without passing current through the resistor, turn on the power supply to circulate current through the resistor, at the same value (P_R) previously measured. When the water begins to boil, record the time and temperature of the water. Continue measuring the mass of the water as a function of time for about 20 min longer or until the remaining water no longer covers the resistor completely. From this graph, determine p_b and its uncertainty.
- ✓ Use Eq. (53.15) to find the value of L_v for water and estimate the uncertainty of the determination.
- ✓ Compare the value found to the corresponding table value.

53.3.3 *Project 184. Solid-liquid transition — Latent heat of fusion*

Recommended equipment: A glass thermometer with a measuring range between −5°C and 100°C. A timer. A container of water of about 250 ml. A similar amount of ice. A scale with a sensitivity of 0.1 g.

The purpose of this experiment is to measure the latent heat of fusion L_{fus} of water, that is, the amount of heat Q necessary for one unit of mass of ice, at the melting temperature T_{fus}, to pass to liquid state at the same temperature. This experiment is similar to Project 183, except that it uses ice, which is mixed with the liquid water contained in a calorimeter. Mix a mass M_{ice} of solid water with a mass M_{water}

of liquid water that is at a temperature T_{hot}, about 20°C higher than room temperature. The water will deliver an amount of heat to the ice that will melt it and, once liquid, will heat to a final equilibrium temperature T_f. By applying the principle of conservation of energy to this process, we have

$$M_{\text{ice}}L_{\text{fus}} + M_{\text{ice}}c_{\text{water}}(T_f - T_{\text{ice}}) = (M_{\text{water}} + M_{\text{WE}})c_{\text{water}}(T_{\text{hot}} - T_f), \tag{53.16}$$

where the first term is the heat that the ice receives to melt it at $T_{\text{ice}} = T_{\text{fus}}$. The second term is the heat delivered to warm it to the final temperature T_f. As usual, M_{WE} is the water equivalent of the calorimeter. The value of L_{fus} for water can be found from this expression.

Suggested tasks:

- ✓ Place in the calorimeter approximately 250 cm^3 of water at a temperature about 20°C above room temperature. Put the ice in another container. To measure T_{fus}, you can mix the ice with cold water and wait for a thermal equilibrium in this mixture, *without the ice melting completely.* This equilibrium mixture will be at the temperature T_{fus} which can be measured with a thermometer immersed in water.
- ✓ Measure the mass of hot water M_{water} placed in the calorimeter and its temperature T_{hot}.
- ✓ Measure the temperature of the ice, $T_{\text{ice}} = T_{\text{fus}}$, in its own container, which may have some water in thermal equilibrium.
- ✓ Use tweezers to pick up small pieces of ice and add them slowly to the water, while continuously stirring the mixture in the thermostat. Try to make sure you are adding ice only to the calorimeter with the hot water, by holding each piece with the tweezers and shaking off any excess liquid before transferring it. Measure the evolution of the temperature of the mixture. It may be helpful to use a thermometer connected to a DAS to get a continuous reading. When the temperature of the mixture T is such that

$$T_{\text{amb}} - T \approx T_{\text{hot}} - T_{\text{amb}} \tag{53.17}$$

and all the ice is melted, measure the final temperature of the mixture (which is now liquid water). This is T_f.

✓ Weigh the calorimeter to determine by difference the mass of ice added, M_{ice}.

✓ Use Eq. (53.16) to find the value of L_{fus} for water and estimate its uncertainty. Compare the value found to the corresponding table value.

References

[1] D. Giancoli, *Physics Principles with Applications*, 7th Ed., NY: Pearson, 2014.
[2] M. W. Zemansky and R. H. Dittman, *Heat and Thermodynamics Subsequent Edition*, NY: McGraw-Hill, 1996.
[3] Y. Cengel, *Heat Transfer: A Practical Approach*, 2nd Ed., New York: McGraw-Hill, 2002.
[4] S. Glasstone, *Thermodynamics For Chemists*, NY: Van Nostrand Company, 1947.
[5] Pasco Scientific, "Thermal expansion apparatus — TD-8856," 2022. Available: https://www.pasco.com/.

Chapter 54

IDEAL GASES — DETERMINATION OF MOLECULAR WEIGHT

The goal of this chapter is to review the properties of ideal gases and discuss some techniques for determining the molecular weight and density of a gas. To do so, it proposes to study how the pressure of a gas varies with the number of molecules (mass) at constant volume and temperature. These experiments will enable us to test how well the ideal gas equation of state describes actual observations.

Goals

✓ Ideal gases
✓ Density of gases
✓ Molecular weight of air
✓ Molecular weight of butane

54.1 Ideal gases

Most real gases at moderate pressures ($P \leq 5\,\text{bar}$) and not very high temperatures ($T \leq 1{,}000\,\text{K}$) have similar behavior, which can be described by the *ideal gas equation of state* [1, 2],

$$PV = nRT, \tag{54.1}$$

where n is the number of moles of the gas, V is the volume, P is the absolute pressure, T is the absolute temperature (in Kelvin), and R is the universal gas constant.[a] In turn, the number of molecules in

[a]The universal gas constant can be expressed in terms of Boltzmann's constant, k_B, and Avogadro's number N_A: $R = N_A \cdot k_B = 0.082057\,\text{atm} \cdot \text{l/mol} \cdot \text{K} = 8.3145\,\text{J/mol} \cdot \text{K}$.

the gas, N, is related to n by

$$N = nN_A, \tag{54.2}$$

where $N_A(= 6.022 \times 10^{23}$ 1/mol) is Avogadro's number. According to Eq. (54.1), the density ρ of a gas (the mass per unit volume) can be written as follows:

$$\rho = nM_R/V = M_R\frac{P}{\mathrm{RT}}, \tag{54.3}$$

where M_R is the molecular mass of the gas. It is important to bear in mind that the pressure measured by conventional gauges is the *manometric pressure*, P_m, that is relative to atmospheric or barometric pressure, P_{bar}. The absolute pressure P involved in Eqs. (54.1) and (54.3) is

$$P = P_m + P_{\mathrm{bar}}. \tag{54.4}$$

We often work with gases that are mixtures of components. For example, air is composed of N_2 (78.08%), O_2 (20.946%), Ar (0.946%), CO_2 (0.35%), and other trace gases. In these cases, the total pressure of the gas is the sum of the partial pressures, and the molecular weight and density are obtained from a weighted average of its components, that is,

$$M_R(\mathrm{air}) = q_{\mathrm{N_2}}M_R(N_2) + q_{\mathrm{O_2}}M_R(O_2) + q_{\mathrm{Ar}}M_R(Ar) + \cdots, \tag{54.5}$$

where q_{N2} is the mole fraction of $N_2(q_{\mathrm{N2}} = 78.08\%)$, q_{O2} is the mole fraction of O_2 $(q_{\mathrm{O2}} = 20.946\%)$, and so on. Thus, the molecular weight of dry air is approximately 28.84 g/mol.

54.1.1 *Project 185. Molecular weight of air*

Recommended equipment: Precision scale or balance with a range of 300 g or more and sensitivity 0.1 g or better. A container of approximately 2 liters with pressure gauge and valve to pressurize. A pump to pressurize air or a pressurized cylinder containing a gas, such as air, N_2, He, or CO_2.

This experiment proposes to measure the density of air by an almost direct method, that is, to measure its mass and volume. Figure 54.1

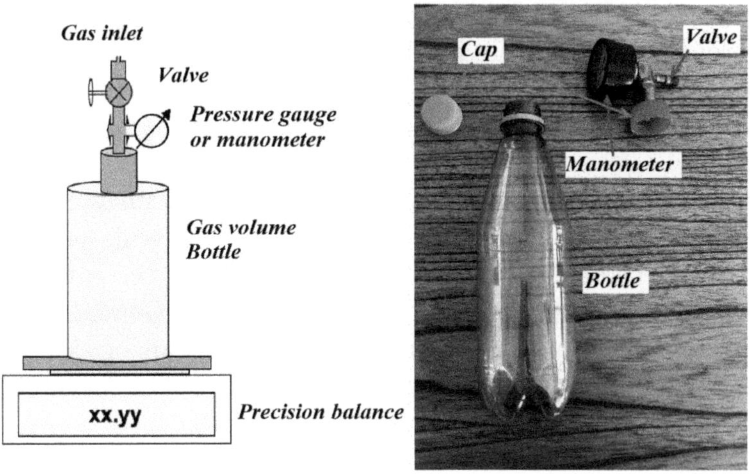

Figure 54.1 On the left, a schematic diagram of the experimental device. On the right, a possible setup: bottle with valve and pressure gauge.

shows a schematic diagram of the proposed device for determining the molecular weight and density of air. It consists of a plastic (soda) bottle with a volume of about 2 liters, preferably with rigid walls so that the volume does not vary appreciably when the bottle is pressurized. A pressure gauge is used to measure the internal pressure and can be the Bourdon type or simply a car tire pressure gauge. The setup also needs a mechanism to pressurize the bottle to approximately 2–3 atm[b] (less than the bottle's burst pressure, which is usually more than 5 atm). If you want to experiment with air, you can use compressed air from a compressor or a hand pump. If you use a gas that is not air, such as N_2, He, or CO_2, use a cylinder of the compressed gas and take care with the use of the valves of these pressurized systems during the bottle filling operation.

Note: Before starting the experiment, pressurize the bottle to about 4 atm while using a protective barrier and goggles. Once you have verified that the bottle withstands this pressure, continue with the

[b]Pressure may be measured in pascals (Pa), bar, atm, PSI: $1\text{bar} = 0.987\,\text{atm} = 10^5\,\text{Pa} = 1.0197\ \text{kg/cm}^2 = 14.50377$ PSI.

experiment. If you want more protection, you can use a protective mesh around the bottle. To retain the gas inside the bottle, you need a valve (one-way) such as those used for car or bicycle tires, see Figure 54.1. To depressurize, use a needle valve or simply loosen the cap slightly.

The goal of this experiment is to measure the mass of the gas inside the bottle as a function of its pressure. A scale or balance, preferably electronic, with sensitivity 0.1 g or better can be used to measure the mass. You also need to know the volume V_0 of the bottle (including connections and valve), which can be determined by weighing it before and after filling it with water. Knowing the density of water, you can calculate the volume V_0.

Let T_0 be the ambient (absolute) temperature where the experiment is carried out and which will be assumed to be the gas temperature. Let $m_0 = m_{\text{gas}}(P_{\text{bar}})$ be the mass of the gas at atmospheric pressure (P_{bar}). In this experiment, the mass of the gas, $m_{\text{gas}}(P_m)$, is measured as a function of gauge pressure P_m. When working with air masses, it is important to take air pressure into account. As we have seen in Chapter 8, the air thrust can be written as $E = m_E g$, where m_E represents the equivalent mass of the thrust,

$$m_E = E/g = \rho^0_{\text{air}} V_0. \tag{54.6}$$

Here, ρ^0_{air} is the density of air at barometric pressure. If M_{Bottle} represents the mass of the system, the bottle with pressure gauge, valve, gas, etc., then at the pressure P_m, we have for the weight

$$W = g \cdot m_{\text{System}}(P_m) = g \cdot (M_{\text{Bottle}} + m_{\text{gas}}(P_m) - m_E). \tag{54.7}$$

Since most scales are calibrated in units of mass, we have

$$\Delta m(P_m) = m_{\text{System}}(P_m) - m_{\text{System}}(P_m = 0) = m_{\text{gas}}(P_m). \tag{54.8}$$

Therefore, $\Delta m(P_m)$ represents the change in mass of the bottle pressurized at P_m with respect to the same system at $P_m = 0$. Note that by making this difference, the air thrust is canceled out, in other words,

$$\Delta m(P_m) = m_{\text{gas}}(P) - m_{\text{gas}}(P_{\text{bar}}) = M_R(V_0/RT_0)(P - P_{\text{bar}})$$
$$= M_R(V_0/RT_0)P_m. \quad (54.9)$$

Using Eq. (54.1), we can write

$$n = m_{\text{gas}}(P)/M_R = (PV_0/RT_0) \quad (54.10)$$

or also

$$m_{\text{gas}}(P_m) = \Delta m(P_m) = M_R\left(\frac{V_0}{RT_0}\right)P_m. \quad (54.11)$$

Therefore, when $\Delta m(P_m)$ is plotted as a function of P_m, if Eq. (54.1) is valid, a linear relationship is expected. The slope of this line enables you to calculate the molecular mass M_R of the gas. On the other hand, from Eq. (54.10), we have

$$\rho^0_{\text{air}} = \frac{m_0}{V_0} = \frac{M_R}{RT_0}P_{\text{bar}}. \quad (54.12)$$

Thus, knowing the barometric pressure, P_{bar}, you can estimate the density of the air in the laboratory, at this barometric pressure. If a barometer is not available locally, you can use the data from the nearest weather service, usually available through the local news website or any nearby meteorological station. Local atmospheric pressure and temperature data are available on the Internet.

Suggested tasks:

- ✓ Using an experimental setup like the one in Figure 54.1, measure $\Delta m(P_m)$ as a function of P_m, for $0 \leq P_m < 3\,\text{atm}$. Plot $\Delta m(P_m)$ as a function of P_m. Evaluate and discuss its linearity.
- ✓ If this plot is consistent with a linear relationship, determine the slope and its corresponding uncertainty, see Chapter 7 and Appendix C.
- ✓ From your results and using Eq. (54.11), determine the molecular mass of air M_R and its density ρ^0_{air}, with their uncertainties.
- ✓ Compare the experimental results to the values known from literature, at the working temperature T_0.

✓ Calculate the (average) molecular mass of air and its density at ambient temperature and pressure, Eq. (54.5). Compare the values obtained experimentally with those known from literature.

✓ ♣♣ If you have another gas, such as He, N_2, or Ar, use it to perform the same experiment. Take care to purge your system carefully to minimize any contamination with air.

✓ ♣♣ Compare your experimental results with the corresponding values from the literature.

54.1.2 *Project 186. Measuring the molecular weight of butane*

Recommended equipment: Precision scale or balance with a range of 300 g and sensitivity 0.1 g or better. A water container of approximately 5 liters. A transparent plastic or glass graduated cylinder of approximately 1 liter. A disposable plastic butane gas lighter.

The objective of this experiment is to determine the molecular mass M_R and density of butane gas [3] contained in a commercial disposable gas lighter. Butane at room temperature is a vapor and can be liquefied by simple compression [1, 4, 5]. Its boiling point at atmospheric pressure is $\sim -6°C$ and its critical temperature is 152°C. When butane is at pressures close to atmospheric pressure and at room temperature, $T > 10°C$, its properties can be described by the ideal gas equation of state, Eq. (54.1).

At a given pressure and temperature, Eq. (54.10), the volume of the gas is proportional to its mass. If butane obeys Eq. (54.1), when we plot $m_{gas}(P_{bar})$ as a function of volume V, at barometric pressure P_{bar} and room temperature T_0, we expect a linear relationship:

$$m_{gas}(P_{bar}) = M_R(P_{bar}/RT_0) \cdot V, \tag{54.13}$$

whose slope will be given by $k = M_R(P_{bar}/RT_0)$, from which we can find M_R.

For this experiment (see Figure 54.2), you can use a disposable gas lighter as a source of butane, a plastic or glass graduated cylinder

Figure 54.2 Items needed for this experiment: graduated cylinder, disposable gas lighter and scale.

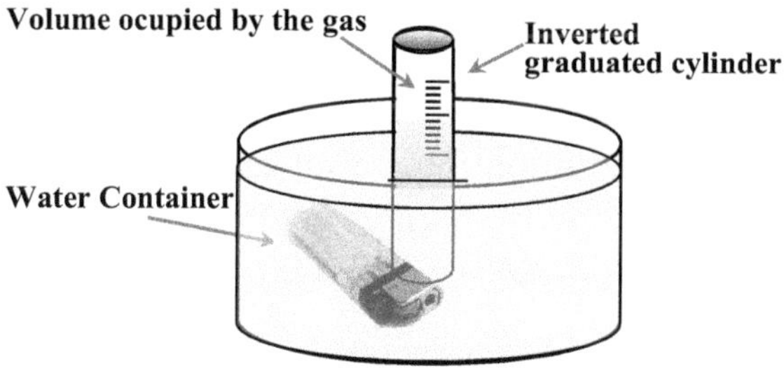

Figure 54.3 Device for measuring the volume of water displaced by gas.

of approximately $^1/_2$ or 1 liter with a graduation of 10 ml or better, and a digital scale with a sensitivity of 0.1 g or better.

Measure the atmospheric pressure with a barometer or obtain it from the local meteorological service, and measure the ambient temperature T_0 with a common thermometer, but express it in Kelvin $T(\mathrm{K}) = T(^\circ\mathrm{C}) + 273.15$.

Fill the graduated cylinder completely with water and place it, carefully inverted, in a larger container half filled with water, as shown schematically in Figure 54.3. Next, place the previously weighed lighter in the water and press the stopcock to release the butane.

All the gas should bubble into the inverted graduated cylinder. When a volume of the order of $^1/_2$ liter has been displaced, which can be easily measured, remove the lighter from the water. Carefully move the cylinder vertically until the water inside and outside the graduated cylinder are at the same level. Under these conditions, you know that the pressure inside the cylinder is equal to the outside pressure, that is; inside the cylinder the gas is at the pressure P_{bar}. In this situation, determine the volume, V, of the gas in the graduated cylinder.

Then dry the lighter as well as you can with a hair dryer on cold setting or use a paper towel, and weigh it again. By weight difference, you can determine the missing gas mass, m_{gas}, which is what remained inside the graduated cylinder. To minimize the measurement uncertainties of the masses, it is advisable to weigh the lighter 3–5 times, both before placing it in the water and after drying it. Do not release gas from the lighter while performing these measurements. Averaging these mass measurements reduces statistical errors in determination. From Eq. (54.13), we have

$$M_R = \frac{m_{\text{gas}}}{P_0 V} R T_0, \tag{54.14}$$

where m_{gas} corresponds to the decrease in the weight of the lighter, V designates the volume of water displaced by the gas, and T_0 is the ambient temperature in Kelvin. The value of P_0 is the barometric pressure of the gas corrected for the pressure of the water vapor:

$$P_0 = P_{\text{bar}} - P_{\text{vapor}}(T_0). \tag{54.15}$$

Here $P_{\text{vapor}}(T_0)$ is the pressure of water vapor at the working temperature, T_0. The value of the vapor pressure can be obtained from tables or calculated using the *Clausius–Clapeyron* equation [4, 5]:

$$P_{\text{vapor}}(T) = P_{00} \exp\left(\frac{L_v}{R}\left[\frac{1}{T} - \frac{1}{T_{00}}\right]\right) = P_{00} \exp\left(\beta \cdot \left[\frac{1}{T} - \frac{1}{T_{00}}\right]\right), \tag{54.16}$$

where P_{00} is the vapor pressure at temperature $T = T_{00}$. L_v represents the latent heat of vaporization of water per mole.

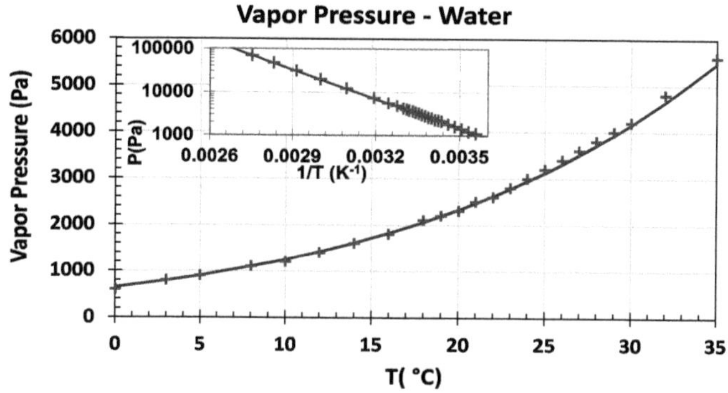

Figure 54.4 Vapor pressure of water in Pa as a function of the temperature in °C. The insert shows vapor pressure of water in Pa as a function of inverse of absolute temperature 1/T(K), for temperatures between 0°C and 300°C. The symbols represent the measured vapor pressure data and the trend line the prediction of the Clausius–Clapeyron equation, with the parameters $P_{00} = 101.3$ kPa, $T_{00} = 273.15$K, $\beta = 5160$ K, and $L_v = 42.9$ kJ/mol. The experimental data were taken from Ref. [6].

Figure 54.4 shows in a semi-logarithmic graph the quality with which this expression describes the water vapor pressure data.

Suggested tasks:

✓ After measuring the gas volumes for various masses of gas used, plot the gas mass m_{gas} as a function of volume V. Is the relationship between the variables m_{gas} and V linear? If so, determine the slope s of this graph and its uncertainty.

✓ Measure the water temperature and use it as the gas temperature T_0. From your measurements and from Eq. (54.14), determine the molecular mass of butane:

$$M_R = s\frac{RT_0}{P_0}. \tag{54.17}$$

✓ From your measurements and the determination of the slope s obtained with the least squares method, estimate the best value of M_R and its uncertainty.

✓ Compare the value obtained for M to the expected value from the chemical formula of butane: C_4H_{10}. What can you conclude from this method of estimating the molecular mass of this gas or vapor?

✓ Calculate the density of butane and estimate its error. Compare it to the value from the literature for butane under the same measurement conditions.

54.2 Isothermal transformation

An *isochoric process* [1, 7, 8] is a process which takes place at constant volume V. This type of process occurs when the thermodynamic system (in this case an ideal gas) is enclosed in a container with rigid walls and constant volume, V.

Alessandro Volta (1745–1827), known above all for the invention of the batteries, also measured the expansion coefficient of the air, which he estimated to be $\alpha = 0.0037°\text{C}^{-1}$. He was the first to show the validity for air of what is now called *Charles's law* or *Gay-Lussac's law*, after the two French physicists [Charles (1746–1823) and Gay-Lussac (1778–1850)] who showed that for all gases, volume (at constant pressure) increases linearly with temperature.

Today, this law is written as follows:

$$V_t = V_0(1 + \alpha t), \tag{54.18}$$

where t is the temperature expressed in Celsius and V_0 is the volume at zero degrees. About a century earlier, the Irishman Robert Boyle (1627–1691) had discovered the relationship (*Boyle's law*[1]) between pressure p and volume V (at constant temperature t)[c]:

$$P \cdot V = \text{constant}. \tag{54.19}$$

The set of these three laws now goes under the name of the *equation of state of ideal gases*, in the form of Eq. (54.1). In this context, mention is rarely made of the various technological applications of this law, including the *gas thermometer*, as we have seen in

[c] Also known as Boyle–Mariotte law, from the name of the French physicist Edme Mariotte (1620–1684), perhaps better known for his work on surface tension than on fluid mechanics.

Chapter 50 and Project 170. If gas in a cylinder with a piston is compressed slowly so that it always has time to reach equilibrium with the surroundings, an isothermal process can be carried out [5, 8].

54.2.1 *Project 187. An isothermal process using a syringe*

Recommended equipment: a plastic syringe (10–25 cc) and a pressure sensor connected to a data acquisition system or to a pressure gauge.

Suggested tasks:

✓ Connect the syringe to a manometer (pressure sensor) as shown in Figure 54.5. The manometer could be connected to a DAS [9–11]. First, practice compressing the gas in the syringe slowly in small steps (1 or 2 ml). Once you are able to do this, while you compress the syringe, have someone else record the pressure, P_m(manometric pressure), for each volume of the syringe. Also register the volume, V_0, when the piston is not compressed, that is, when the gas inside the syringe is at atmospheric or barometric pressure, $P_0 = P_{\text{bar}}$.

✓ From your measurements, make a plot of $P = P_m + P_{\text{bar}}$ versus V and another of $P \cdot V$ versus V.

✓ What can you conclude about the shape of these plots? How do your experimental results compare with the expectation from Eq. (54.1) if T is constant?

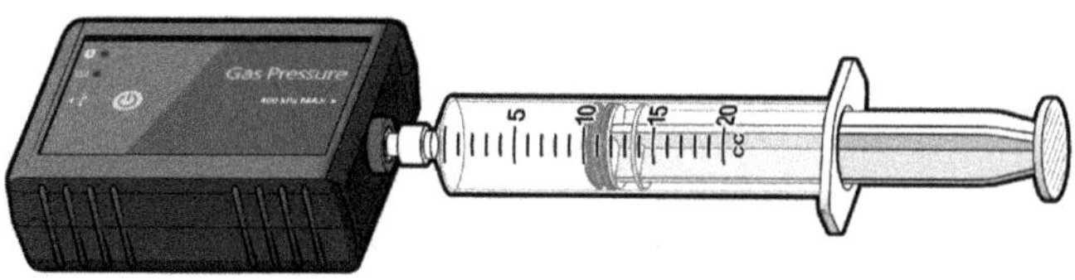

Figure 54.5 Syringe connected to a pressure sensor, connected to a DAS.

References

[1] F. Sears, M. Zemansky, H. Young and R. Freedman, *College Physics*, NY: Pearson Ed., 2009.

[2] Wikipedia, "Ideal gas law," 2022. Available: https://en.wikipedia.org/wiki/Ideal_gas_law.

[3] Wikipedia, "Butane," 2022. Available: https://en.wikipedia.org/wiki/Butane.

[4] S. Glasstone, *Thermodynamics for Chemists*, NY: Van Nostrand Company, 1947.

[5] R. Chang and J. Overby, *ISE Chemistry*, NY: Mc Graw-Hill, 2021.

[6] "Wired Chemist," 2022. Available: https://www.wiredchemist.com/chemistry/data/vapor-pressure.

[7] M. W. Zemansky and R. H. Dittman, *Heat and Thermodynamics Subsequent Edition*, NY, McGraw-Hill, 1996.

[8] Y. Cengel and M. Boles, *Thermodynamics: An Engineering Approach*, 8th Ed., NY: McGraw-Hill Education, 2014.

[9] LabTrek, "LabTrek designs and produces experimental apparatuses for advanced Physics Lab- Padova It," 2022. Available: https://www.labtrek.it/.

[10] MakerPortal, "Pressure sensor calibration with Arduino," 2022. Available: https://makersportal.com/blog/2020/6/4/mps20n0040d-pressure-sensor-calibration-with-arduino.

[11] Vernier Software, "Gas pressure sensor," 2022. Available: https://www.vernier.com/product/gas-pressure-sensor/.

Chapter 55

KINETIC THEORY OF GASES — RATIO OF SPECIFIC HEAT OF IDEAL GASES

The goal of this chapter is to experimentally study some consequences of the kinetic theory of gases. In particular, the chapter discusses how to determine the relationship between the specific heats at constant pressure (C_p) and at constant volume (C_v) for air and other gases. For ideal gases, this ratio can be calculated using the kinetic theory of gases. This experiment therefore tests an important result of this theory and shows how molecular properties have direct influence on the macroscopic properties of a gas, such as its compressibility. Furthermore, this experiment illustrates how the ratio $\gamma = C_p/C_v$ determines the compressibility properties of a gas.

Goals

✓ Specific heats of ideal gases
✓ C_p/C_v ratio
✓ Clement–Desormes experiment
✓ Rüchardt's experiment

55.1 Kinetic theory and heat capacity of ideal gases

As we have seen in the previous chapter, most real gases at moderate pressures ($P \leq 5\,\text{bar}$) and not very high temperatures ($T \leq 1,000\,\text{K}$) have similar behavior, which can be described by the *ideal gas equation* [1–3].

$$PV = k_B NT, \tag{55.1}$$

where k_B is a universal constant known as *Boltzmann's constant*, $k_B = 1.38 \times 10^{-23}$ J/K, N is the number of molecules in volume V, P is their total pressure, and T is the absolute temperature (in Kelvin). The kinetic theory of gases was originally developed to provide a microscopic explanation of the equation of state for ideal gases and was one of the first molecular theories to quantitatively explain the experimental results [1]. The basic assumptions of this theory are as follows:

- Gases are made up of molecules in constant motion that collide elastically with each other and with the walls of the container that confines them.
- The pressure on the walls is due to these collisions.

To establish this connection between the mechanical properties of molecules, the kinetic theory introduces the *principle of equipartition of energy*. This principle associates with each degree of freedom of the molecules an average energy equal to $^1/_2\, k_B T$[1]. Thus, for a molecule with f degrees of freedom, internal energy per mole, U, will be

$$U = \frac{f}{2} \cdot N_A \cdot k_B \cdot T = \frac{f}{2} \cdot R \cdot T. \tag{55.2}$$

In particular, for diatomic molecules, there will be 5 *degrees of freedom* (3 of translation and 2 of rotation). The internal energy per mole of this type of molecule will be [1]:

$$U = \frac{5}{2} \cdot R \cdot T. \tag{55.3}$$

Molar-specific heat at constant volume is defined as $C_V = (\partial U/\partial T)_{V,\mathrm{mol}}$ [2]. Therefore,

$$C_V = \frac{f}{2} \cdot R. \tag{55.4}$$

Also, molar-specific heat at constant pressure, C_p, is the heat needed to raise the temperature of a molar gas at constant pressure by one degree Kelvin. It can be shown that for ideal gases [1, 2], the following relation is satisfied:

$$C_p = C_V + R = (1 + \frac{f}{2})R. \tag{55.5}$$

According to the kinetic theory of ideal gases, the ratio $\gamma = C_p/C_v$ is given by

$$\gamma = \frac{C_p}{C_V} = 1 + \frac{2}{f} \quad \text{and} \quad f = \frac{2}{\gamma - 1}. \tag{55.6}$$

The coefficient γ is called *adiabatic gas coefficient.* If we can measure $C_v, C_p,$ or γ, we have a direct way to compare the predictions of the kinetic theory with the experimental results.

When a gas undergoes a process in which there is no heat exchange between the gas and the environment around it, we say it performs an *adiabatic process* [1, 3]. For an ideal gas which undergoes a reversible adiabatic process, its path in the PV plane can be described by the Poisson equation [1–3]:

$$PV^{\gamma} = \text{Constant}, \tag{55.7}$$

where the exponent γ is the *adiabatic coefficient* of the gas. For a monatomic gas whose molecule has only three translational degrees of freedom, $f = 3$ and $\gamma = 5/3 = 1.66$. For a diatomic gas, $f = 5$, since diatomic molecules have three translational degrees of freedom and two rotational, so $\gamma = 7/5 = 1.4$.

There are many experiments in which the adiabatic coefficient of a gas can be measured [1, 4, 5]. In fact, in Chapters 11 and 41, when we measured the speed of sound, we were indirectly determining γ. This chapter discusses two experimental techniques: the *Clement–Desormes method* and the *Rüchardt method* for determining the quotient between the specific heats of a gas [1, 6]. The experiments in this chapter are updated versions of these classic experiments [2, 4, 5], using a pressure sensor connected to a data acquisition system. Rüchardt's experiment will enable us to integrate previously studied mechanical concepts (free and damped oscillations) with thermodynamic concepts (specific heat, thermodynamic processes) through a simple experiment that is both attractive and elegant.

Furthermore, this chapter provides two examples illustrating the many interesting thermodynamic experiments that can be performed using a syringe connected to a pressure sensor [7].

55.2 Clement–Desormes experiment

This is a very suitable classical experiment to determine γ using a pressure gauge, a plastic syringe, and a glass bottle [1, 7]. The bottle has an airtight rubber stopper with two outlets: one connected by means of a flexible tube to the syringe, which is expanded to its maximum volume, and the other connected to the pressure gauge, see Figure 55.2. In conditions of equilibrium with the surrounding medium at temperature $T = T_1$, the pressure P_1 is measured. At a certain instant, $t = 0$, the piston of the syringe is rapidly compressed, in an adiabatic process.

The pressure instantly rises to the P_2 value, and the gas temperature also rises to $T = T_2$. Therefore, when thermal equilibrium with the surrounding environment is reached, the pressure decreases to the equilibrium value P_3. Hence the pressure P_1 and P_3 are on the same isotherm in a PV diagram corresponding to the temperature $T = T_1$, such as the one shown schematically in Figure 55.2.

With reference to the PV diagram of the physical process, illustrated at the lower right of Figure 55.2, if we call the original

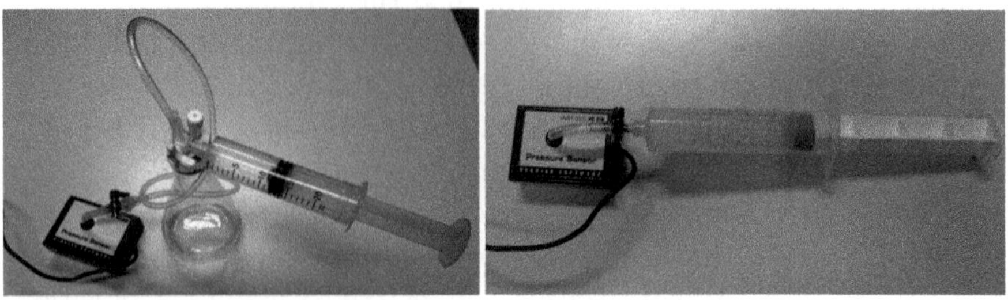

Figure 55.1 Two possible configurations to perform the Clement–Desormes experiment. On the left, a syringe connected to a glass bottle and a pressure sensor. On the right, the device consists only of a syringe connected to a pressure sensor.

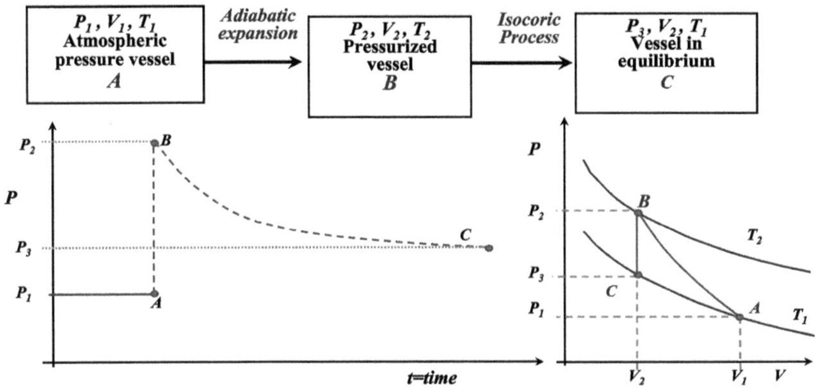

Figure 55.2 The Clement–Desormes experiment. Top: Conceptual scheme of the experiment; Bottom left: Pressure variation over time. Bottom right: PV diagram of the physical process. A represents the initial state, B is the point immediately after adiabatic compression, and C is the final equilibrium state.

volume V_1 and the final volume V_2, for the adiabatic process AB we have

$$P_1 V_1^{\gamma} = P_2 V_2^{\gamma}. \tag{55.8}$$

Since A and C are on the same isotherm ($T = T_1$),

$$P_1 V_1 = P_3 V_2. \tag{55.9}$$

Combining these last two expressions, we have

$$\gamma = \frac{\ln(P_2/P_1)}{\ln(P_3/P_1)}. \tag{55.10}$$

Therefore, by measuring the pressures P_1, P_2, and P_3, we can determine the coefficient γ of the gas.

In this experiment, the pressures P_1 and P_3 can be determined by direct measurement without major difficulty. Their values are the average of the pressures measured before the syringe was compressed and after reaching equilibrium. However, determining P_2 is more complex. Since it is difficult to obtain a sufficiently fast, adiabatic process, compression always takes a few fractions of a second. As

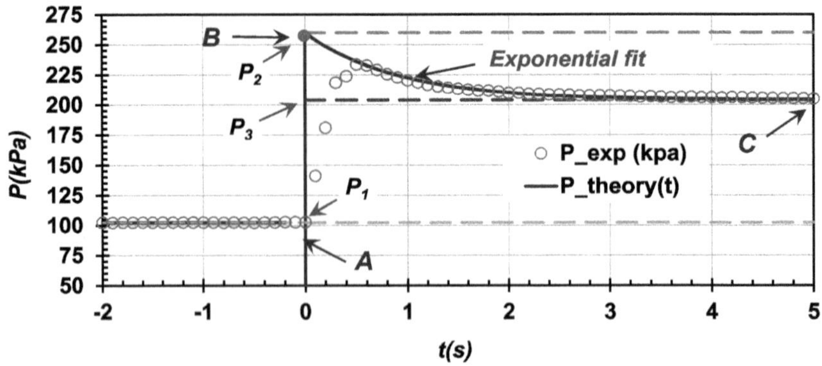

Figure 55.3 Schematic plot of the variation in pressure over time in a typical Clement–Desormes experiment. The circles correspond to the pressure data. The pressure P_2 can be obtained by extrapolating to $t = 0$ an exponential fit to the pressure data during the thermalization process from P_2 to P_3, i.e., from point B to C of Figure 55.2.

shown in Figure 55.3, experimental data do not directly provide the value of P_2. However, its value can be found by extrapolating the pressure trend as a function of time between $t = 0$ and the equilibrium situation, points B and C of Figure 55.2. If we assume that the container is cooled following Newton's law of cooling, we can assume that the shape of the curve of P as a function of time is also exponential. Therefore, if we fit an exponential function to the form

$$P(t) = (P_{20} - P_3)\exp(-\lambda t) + P_3, \tag{55.11}$$

where P_3 is the equilibrium pressure value and P_{20} and λ are two parameters whose values are determined by the interpolation that can be made of the experimental data of $P_{\mathrm{exp}}(t)$, as shown in Figure 55.3, then the value of P_{20} can be found by extrapolating Eq. (55.11) for $t = 0$.

55.2.1 *Project 188. Determining γ with the Clement–Desormes method*

> **Recommended equipment:** A pressure sensor connected to a data acquisition system. A glass or plastic container of approximately 250 ml to 1 l, equipped with a cap with connection for the pressure sensor and a plastic syringe to pressurize. Alternatively, the experiment can be performed with a syringe of approximately 25–50 ml connected to a pressure sensor associated with a PC data acquisition system (DAS).

There are many possible experimental arrangements for this experiment. This is one of the simplest and consists of a syringe connected to the pressure sensor, as shown in the right panel of Figure 55.1. Initially, the syringe plunger is expanded to its maximum volume and initial (atmospheric) pressure P_1. The data acquisition system is configured to perform approximately 100 measurements per second with a time interval of approximately 4 min ($\sim 240\,\mathrm{s}$). The initial pressure is recorded for approximately 10 s and the plunger is rapidly compressed to approximately half its initial volume. The plunger is held in this position until the equilibrium pressure P_3 is reached. This is when the gas temperature returns to its initial T_1 value (the ambient temperature). Figure 55.3 is a diagram of the expected results. A good estimation of the value of $P_2 \equiv P_{20}$ can be made by extrapolating the measured data.

Suggested tasks:

- ✓ Make at least five determinations γ for air, i.e., repeat the experiment at least five times. Find the best value and estimate your error.
- ✓ Compare your experimental results to the expected values of γ for air from the kinetic theory of gases and the values from tables.

55.3 Rüchardt's experiment

The aim of Rüchardt's experiment [2, 4–6] is to determine the adiabatic coefficient γ of a gas. In Rüchardt's original design, a metal sphere is inserted into a vertical glass tube connected to a container of larger volume, (Figure 55.4). If the coupling between the sphere and the tube is airtight and has little friction (such conditions are not always easy to obtain in practice), the sphere performs a series of oscillations before reaching its equilibrium position.

The difficulty with the original experimental arrangement, Figure 55.4, is that it requires a very precisely ground glass tube and a cylindrical or spherical mass made with similar precision to achieve perfect fit and low friction at the same time. These conditions can be achieved with an ordinary glass syringe, which usually meets both conditions (airtightness and little friction), in addition to being very inexpensive. This experiment therefore uses a glass syringe, in an arrangement like the one shown in the right panel of Figure 55.1.

A diagram of the experiment is illustrated in Figure 55.5, taking the horizontal axis as x, with its origin located to coincide with the equilibrium position of the internal tip of the syringe plunger. If the plunger travels a distance x, the internal pressure in the syringe will increase to a certain value P. The restoring force on the plunger will be

$$F = (P - P_0) \cdot A = \Delta P \cdot A, \tag{55.12}$$

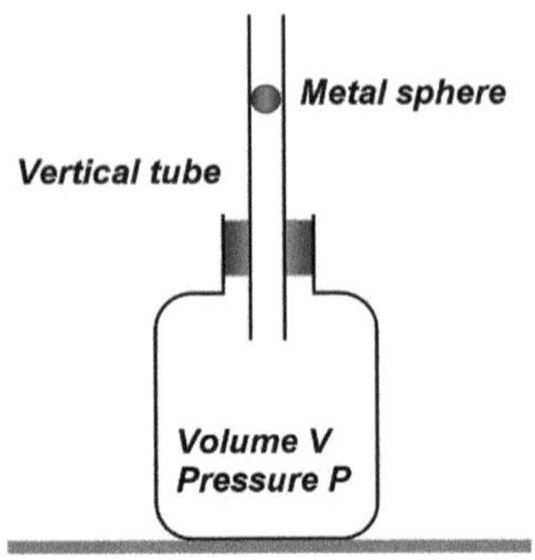

Figure 55.4 Rüchardt's original experiment.

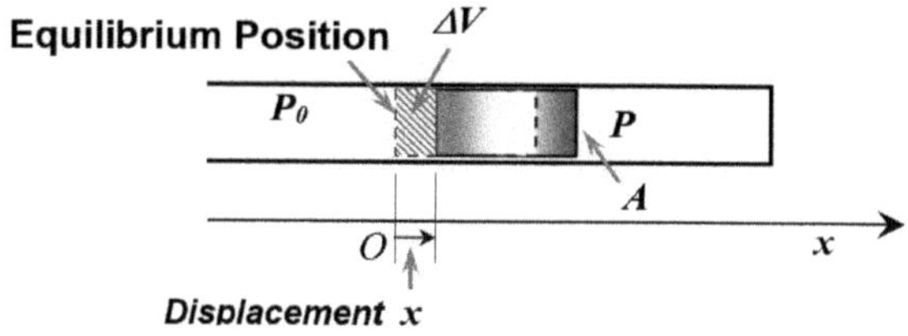

Figure 55.5 Diagram of the forces on the piston in the tube.

where P_0 is the external barometric pressure and A is the cross-sectional area of the plunger. This force tends to bring the piston (plunger) back to its equilibrium position, like a spring. The volume change produced by the displacement of the piston over a distance x is

$$\Delta V = xA. \tag{55.13}$$

If the compression is very quick, without heat flow between the gas in the system (inside the syringe) and the surrounding air, then according to the first principle of thermodynamics,

$$\Delta Q = \Delta U + \Delta W = 0. \tag{55.14}$$

In other words, it is an *adiabatic process*, and all the work done on the gas turns into internal energy and vice versa. For a reversible process, the relationship between volume and pressure is described by the Poisson equation, Eq. (55.7), which can be expressed in differential form as follows:

$$dP = -(\gamma P/V)dV. \tag{55.15}$$

The force F acting on the piston can be written as follows:

$$F = AdP = -A(\gamma P/V)dV = -A^2(\gamma P/V)x. \tag{55.16}$$

This elastic restoring force has the form $F = -kx$, where the "elastic constant" is $k = \gamma A^2 P/V$. The parameter k is not strictly a constant because it depends on the ratio P/V, but if the relative variations dP and dV are small in comparison to their values in equilibrium,

P_0 and V_0, then $P/V \approx P_0/V_0$, where V_0 is the volume of the system in equilibrium conditions. The equation of piston motion is in this approximation

$$m\frac{d^2x}{dt^2} = -\frac{\gamma PA^2}{V}x = -m\omega_0^2 x, \tag{55.17}$$

where $\omega_0 = \sqrt{k/m}$ is the natural frequency of oscillation, here m represents the mass of the piston or syringe plunger, therefore

$$\omega_0 = \sqrt{\frac{\gamma PA^2}{mV}} \approx \sqrt{\frac{\gamma P_0 A^2}{mV_0}}. \tag{55.18}$$

In general, the system has friction. If this friction can be approximated with $F_{\text{friction}} = bv$, where b is a constant and v is the speed, defining $\beta = b/2m$, the resulting motion equation is

$$\frac{d^2x}{dt^2} + 2\beta\frac{dx}{dt} + \omega_0^2 x = 0. \tag{55.19}$$

As we have seen in Chapters 15 and 38, the solution to this differential equation is in the form

$$x(t) = A_0 \exp(-\beta t)\sin(\omega t + \varphi), \tag{55.20}$$

with $\omega^2 = \omega_0^2 - \beta^2$. As before, A_0 and ϕ depend on the initial conditions of the system. The form of this solution is schematically illustrated in Figure 55.6. By measuring $x(t)$ or $dP(t)$ as a function of t, from the fit of the experimental data, with Eq. (55.20), the parameters A_0, ω, β, and ϕ can be obtained.

Once the parameters ω and β have been determined, using one of the techniques discussed in Chapters 15 or 38, we obtain

$$\omega_0^2 = \omega^2 + \beta^2 \approx \gamma(P_0 A^2)/(mV_0) \tag{55.21}$$

from which

$$\gamma \approx \frac{mV_0}{A^2 P_0}\cdot\omega_0^2 = \frac{mV_0}{A^2 P_0}\cdot\left(\omega^2 + \beta^2\right). \tag{55.22}$$

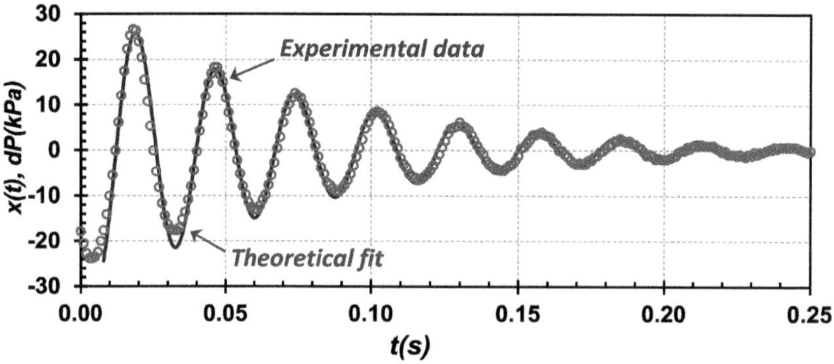

Figure 55.6 Change in position $x(t)$ of the plunger or pressure difference $dP(t)$ inside the syringe as a function of time. By fitting the experimental data (red circles) with theoretical expectation (blue line) of Eq. (55.20), the parameters: A_0, ω, β and ϕ can be obtained.

It is therefore possible to estimate the value of γ from the study of the mechanical oscillations of the plunger of a glass syringe. This experiment illustrates how the mechanical properties of a system are related to the thermodynamic properties of the vibrating elastic medium. In fact, it is left as an exercise for the reader to discuss what would happen if a soccer ball was inflated with a monatomic gas like He or Ar or if it was inflated with CO_2. How do you expect the ball to behave?

Furthermore, this experiment enables us to explore some interesting connections between thermodynamics and mechanics, such as the concept of *entropic damping* or *Entropic Force* [8, 9]. While the theoretical discussion presented above, in general, gives a good account of the frequency of the oscillation (ω), it might be tempting to ascribe the damping (β) to a linear frictional force, similar to viscous force, often found in mechanical problems such as the damped oscillator studied in Chapter 15. However, several authors argue that this damping is also partially associated with the irreversibility of the damped motion, which requires thermodynamical treatment involving dissipation (entropy production), hence the name of *Entropic Force*. Further discussion of this interesting topic can be found in Refs. [8, 9].

55.3.1 *Project 189. Determining γ for ideal gases — Rüchardt's experiment*

Equipment needed: A pressure sensor connected to a data acquisition system. A 20–50 ml glass syringe with little friction between its parts.

For this experiment, you can use a glass syringe [5, 7, 10], with little friction, with a volume of about 20–50 ml, with clearly visible graduation marks. Connect the syringe to the pressure sensor as indicated in the *right panel* of Figure 55.1. The pressure sensor does not need to have a very precise calibration but must have fast response and be sensitive to pressure variations of about 5% or less. There are several commercial devices available on the market that can be used for this type of experiment [11–13], or you can use a homemade sensor interfaced with Arduino [14, 15].

For this experiment, you can use air, which consists mainly of diatomic molecules, as it is a mixture of $\approx 80\%$ nitrogen and $\approx 20\%$ oxygen, both of which are diatomic gases. Other gases that can be used in this experiment are argon or helium (monoatomic gases), CO_2, or butane (the gas in a lighter), which are polyatomic, etc. [4].

With a typical device ($V \approx 30$ ml, $m \approx 10$ g, $A \approx 1$ cm^2), oscillations with a period of the order of $T \approx 0.2$ s are obtained, i.e., about five oscillations per second, which is fast enough to maintain the interior of the syringe in the conditions of "adiabaticity" that we assumed in the development of Eq. (55.22). For this type of experiment, it is advisable to acquire data at a sampling rate of about 500–2,000 Hz.

For the selected syringe, carefully measure the diameter and mass m of the plunger, so that A is known. Note that the diameter intervenes at the power 4, so it must be known with good precision. Also, determine the atmospheric pressure P_0 in your laboratory. If you don't have a barometer, you can check atmospheric pressure online from the nearest weather station.

Note that V_0 in Eq. (55.22) is the volume indicated by the syringe itself V_{Syr}, plus the volume of the tube, ΔV, which connects it to the pressure sensor. One way to find out ΔV is to use a second small plastic syringe, which can hold 1 – 5 ml (to have more sensitivity in the measurement of volumes). Connect this syringe to the pressure sensor with the same tube. Measure the volume when the system is at atmospheric pressure and again when the pressure has doubled. Let $V_m^{(i)}$ be the volume initially measured by the small syringe connected to the pressure sensor when the pressure is ambient pressure P_0. Similarly, $V_m^{(f)}$ is the final volume of the small syringe when the pressure is $2P_0$. From Boyle's law, $P_0(V_m^{(i)} + \Delta V) = 2P_0(V_m^{(f)} + \Delta V)$, from which $\Delta V = (V_m^{(i)} - 2 \cdot V_m^{(f)})$. The advantage of using a small syringe is that you can measure small volumes with greater accuracy [7].

Reconnect the glass syringe to the pressure sensor and practice priming the plunger to start its oscillation. Do not pull the plunger out too far. If the plunger hits the base of the syringe, it will break it. So, start by tapping the plunger gently, until you learn to make it oscillate a few times, preferably more than four complete cycles, see Figure 55.6. You can do this by either pushing or pulling the plunger. Try doing it while holding the syringe horizontally and vertically. The vertical position usually provides more oscillations.

Suggested tasks:

- ✓ Starting with the air in the small syringe, determine the volume V_0 that the gas occupies in equilibrium conditions. Note that V_0 includes the volume of the syringe V_{Syr} and ΔV, ($V_0 = V_{\text{Syr}} + \Delta V$).
- ✓ Connect the glass syringe and compress the gas by pushing or pulling the plunger, so that it begins to oscillate. Once you have mastered this technique, start data acquisition and set the plunger in oscillation. Measure the volume of air in the syringe, V_{Syr}, immediately before you disturb the plunger and again after it reaches its position of equilibrium after oscillating. V_{Syr} is the average of the values measured before and after the oscillation.
- ✓ Construct a plot of the measured values of $P(t)$ as a function of time.

✓ On the same graph, include the theoretical fit of the pressure data as a function of time with an equation that takes into account the damping, in the following form:

$$P(t) = P_0 + P_{20}e^{-\beta t}\cos(\omega t + \varphi). \tag{55.23}$$

✓ In this equation, P_0 is the equilibrium pressure, P_{20} is the amplitude of the oscillation, β is the damping constant, ω is the angular frequency of oscillation, and φ is the initial phase. Vary these four parameters to obtain the best possible fit of the experimental data, as shown in Figure 55.6.

✓ Using the results of the best fit to your experimental data and Eq. (55.23), determine ω and the damping constant β. From these results, and using Eq. (55.22), you can find the value of γ.

✓ Vary the initial value of V_{Syr} (and consequently V_0) and repeat the experiment. For each value of V_0, find a new value of γ. How do you expect this value to compare to the previously measured value? What do your experimental results indicate? Does the value of γ vary with V_0? Have your preliminary expectations been met?

✓ Compare your experimental results of γ for air to those quoted in the literature and to the predictions of the kinetic theory.

✓ ♣♣ From the fit to your data, estimate the uncertainty in parameters β and ω. From the estimated errors of the other measured parameters, which intervene in Eq. (55.22), estimate the uncertainty in the value of γ found in this experiment.

✓ ♣♣ Using another gas, such as the ones suggested above, repeat the experiment. For each gas, repeat the measurement, varying the initial volume V_0.

✓ ♣♣ For each gas used, find the best value of γ and its uncertainty. Compare your results to literature data and kinetic theory predictions.

55.4 Improved devices for studying the properties of gases and performing Rüchardt's experiment

It is interesting to note that there are some convenient commercial devices for performing a detailed study of the properties of gases as well as Rüchardt's experiment [5, 10]. These devices can often perform simultaneous measurements of *pressure, gas volume* and *gas temperature* as a function of time. For Rüchardt's experiment [12], the oscillating mass may be changed by adding extra loading masses, and the total gas volume involved in the process can also be varied by placing some ping-pong balls in the flask, in order to improve the accuracy in the measurement of γ. This setup is illustrated in Figure 55.7 [12].

With a typical setup ($V \approx 2{,}000 - 2{,}600\,\text{ml}$, $m \approx 50 - 100\,\text{g}$, $A \approx 6\,\text{cm}^2$), oscillations with a period of the order of $T \approx 0.5\,\text{s}$ are obtained, which are fast enough to maintain within the syringe the conditions of "adiabaticity" that we assumed in the development of Eq. (55.22). The manufacturer of this type of device usually provides more detailed information on its characteristics and operation. Here we will briefly describe some of its potentialities, and further details can be found in Refs. [12, 13]. The main advantage of this type of

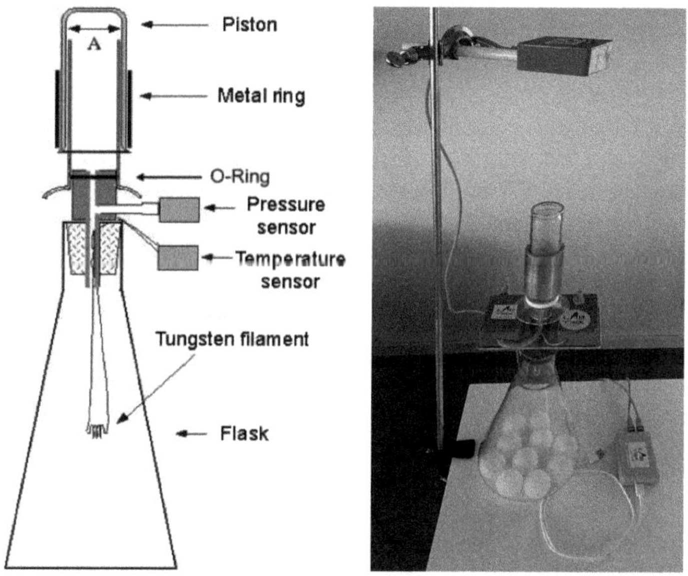

Figure 55.7 Left: Schematic setup of Rüchardt's experiment using a fast thermometer, a pressure sensor, and a sonar [12]. Right: a photograph of the whole setup, including the sonar and the data logger.

device is that since it is possible to measure all the thermodynamic variables of the system (V, P, and T) simultaneously as a function of the time, it is possible to explore in greater detail the variation of P and T as a function of V, and therefore verify the basic hypotheses regarding the adiabaticity of the processes. The device also provides greater flexibility regarding the variable (or combination of variables) to be analyzed as a function of time (e.g., both the oscillating mass and the buffer volume may be changed). Reference [5] discusses in detail several possibilities and potentialities of this type of device.

References

[1] F. Sears, M. Zemansky, H. Young and R. Freedman, *College Physics*, NY: Pearson Ed., 2009.

[2] M. W. Zemansky and R. H. Dittman, *Heat and Thermodynamics Subsequent Edition*, NY: McGraw-Hill, 1996.

[3] Y. Cengel and M. Boles, *Thermodynamics: An Engineering Approach*, 8th Ed., NY: McGraw-Hill Education, 2014.

[4] O. de Lange and J. Pierrus, "Measurement of bulk moduli and ratio of specific heats of gases using Rüchardt's experiment," *Am. J. Phys*, vol. 68, p. 265, 2000.

[5] G. Torzo, *et al.*, "A new microcomputer-based laboratory version of the Rüchardt experiment for measuring the ratio Cp/Cv in air," *Am. J. Phys.*, vol. 69, p. 1205, 2001.

[6] G. Severn and T. Steffenson, "A simple extension of Rüchardt's method for measuring the ratio of specific heats of air using microcomputer-based laboratory sensors," *Am. J. Phys.*, vol. 69, pp. 387–388, 2001.

[7] D. Jackson and P. W. Laws, "Syringe thermodynamics: The many uses of a glass syringe," *Am. J. Phys.*, vol. 74, no. 2, pp. 94–101, 2006.

[8] D.-Y. Kang, K.-H. Liou and W.-L. Chang, "Investigating friction as a main source of entropy generation in the expansion of confined gas in a piston-and-cylinder device," *J. Chem. Educ.*, vol. 92, no. 10, pp. 1667–1671, 2015.

[9] C. E. Mungan, "Entropic damping of the motion of a piston," *Phys. Teach.*, vol. 55, p. 180, 2017.

[10] J. L. Hunt, "Accurate experiment for measuring the ratio of specific heats of gases using an accelerometer," *Am. J. Phys.*, vol. 53, pp. 696–697, 1985.

[11] Vernier Software, "Gas pressure sensor," 2022. Available: https://www.vernier.com/product/gas-pressure-sensor/.
[12] LabTrek SRL, "Rüchardt experiment," 2022. Available: https://www.labtrek.it/experiment-of-ruchardt/.
[13] Pasco Scientific, "Thermal expansion apparatus — TD-8856," 2022. Available: https://www.pasco.com/.
[14] R. Salar, "An Arduino-based experiment designed to investigate gas pressure," *Phys. Teac.*, vol. 59, p. 202, 2021.
[15] The Arduino Team, "Arduino pressure sensor projects," 2022. Available: https://create.arduino.cc/projecthub/search?q=pressure+sensor+.

Chapter 56

THERMAL CONDUCTION AND CONVECTION — HEAT WAVES

This chapter studies the dynamics of heat propagation, more specifically, the dynamics of heat conduction and convection in a metal bar. When a heat source acts intermittently, patterns of temperature propagation are generated that can be described as heat waves. These propagation mechanisms can be used to determine the thermal conductivity of a material and its diffusivity constant. They also enable us to understand the thermal properties of the ground and how secular variations in surface temperature are recorded inside the Earth, providing a very interesting way to determine temperatures of the past and evaluate the occurrence and possible causes of global warming.

Goals

✓ Past temperatures
✓ Heat waves
✓ Determination of thermal conductivity
✓ Thermal waves in the ground
✓ Conduction and convection

56.1 Past temperatures — Global warming

One of the scientific challenges of greatest concern, importance, and interest at the beginning of the third millennium in which we live is understanding the evolution of the planet's climate, and elucidating, as precisely as possible, the origin and causes of global warming that has been occurring over the last two centuries [1, 2]. Our civilization developed during the millennia that followed the last ice age, which occurred about 10,000 years ago [3–5].

During the last ice age, the average temperature on Earth was about 5°C lower than it is today. Since the second half of the 19th century, the global surface air temperature has risen unequivocally [6–8], and the trend over the last 50 years indicates that the temperature is rising at a rate of about 2.2 ± 0.3°C/century, see Figure 56.1. If this trend persists, in two centuries there would be a temperature increase of a magnitude comparable to that which occurred in about 10,000 years, but in opposite direction (warming), posing a serious threat to humanity and civilization as we know it.

It is crucial for the future of our civilization to elucidate whether these temperature trends (see Figure 56.1) are a consequence of human activity or natural causes [4, 5, 12]. There is growing consensus in the scientific community that warming is related to the increase in CO_2 and other Greenhouse Gases (GHG) in the atmosphere, such as methane (CH_4), nitrous oxide (NO_2), chlorofluorocarbons, ozone, etc. [2, 6]. Human activities, mainly fossil fuel burning, change in land use, deforestation, methane emissions, etc., are increasing the concentrations of GHG in the atmosphere directly

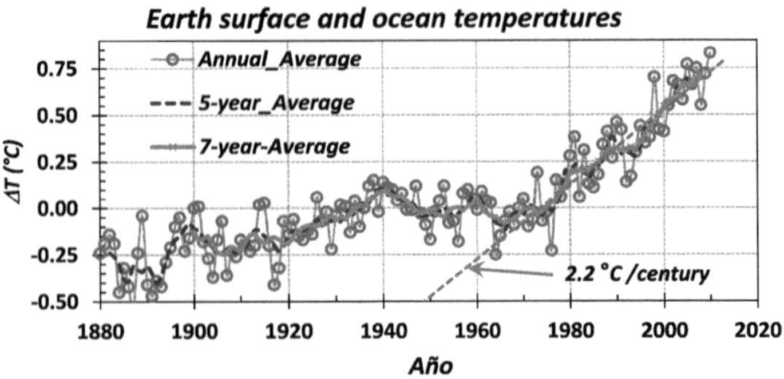

Figure 56.1 Variation of global temperatures. The average temperature increase in the world over the 50 years after 1965 was about 2.2°C/century [6, 9–11].

and indirectly, intensifying the greenhouse effect. The rise of CO_2 in the atmosphere as a consequence of the intensive use of fossil fuels increased dramatically during the industrial revolution and most notably since the second half of the 20th century. In other words, there is well-founded evidence supporting the hypothesis that the increase in average global temperatures is largely a consequence of human activity [6].

Of course, it is known that there are also natural causes that affect the Earth's climate, of which the occurrence of the last ice age provides evidence. Among the possible natural causes that can affect the Earth's climate are variations in the energy emission of the Sun; volcanic eruptions, which put a large amount of ashes and aerosols into the atmosphere [14], affecting the energy balance on Earth; and variations in the movement of the Earth around the Sun [4]. There is much evidence that the Earth has a history of quasiperiodic oscillations of climatic changes [4, 5], and periods of ice ages have been occurring with some regularity for the past 2 million years. Milutin Milankovitch (1879–1958) proposed an astronomical model to explain the causes [4]. This theory is based on the fact that in addition to its rotational and translational motion, the Earth moves in an elliptical orbit that has a variation in eccentricity of about 100 ka (1 ka = 1,000 years). Furthermore, the Earth's axis has an inclination of about 23° with respect to the normal to the plane of the ecliptic, which generates the seasons of the year. The Earth's axis also has a *precession* of about 26 ka, which means that every 13 ka, the seasons are reversed: in the southern hemisphere, January is summer during some parts of the cycle, and winter during others. On the other hand, the angle of inclination of the Earth (with respect to the normal of the ecliptic) oscillates between 21.5° and 24.5° over about 41 ka, a movement that is called *nutation*. As might be expected, there are couplings between these movements, generating the so-called Milankovitch cycles, with periods of approximately 22 ka, 41 ka and 98 ka, during which major alterations in the climate have been observed [5]. This theory is not free from criticism and is far from being universally accepted, but it does have important observational support.

To test climate models, it is necessary to be able to determine past temperatures. Standard temperature records based on terrestrial meteorological observatories are only reliable since the end of the 19th century and for few sites, mainly in Europe [9, 10]. Therefore, knowing temperatures from the past has become a challenge of great importance and scientific interest. Fortunately, there are several methods for studying past temperatures [15]. One of them is based on the isotopic abundance of oxygen (ratio $\delta\,^{18}\mathrm{O} = {}^{18}\mathrm{O}/{}^{16}\mathrm{O}$) and hydrogen (ratio $\delta\,^{2}\mathrm{H} = {}^{2}\mathrm{H}/{}^{1}\mathrm{H}$) in large blocks of ice in regions such as Greenland, Antarctica, etc.,[a] where the snow that falls does not melt but accumulates in layers. These layers vary with the seasons of the year and form stratifications, somewhat like the annual rings of a tree. By counting these layers, the time when the precipitation occurred can be determined. In addition, representative air of that time accumulates in these layers and can be used to determine the concentration of greenhouse gas (GHG) from the past. On the other hand, water formed with heavier isotopes ($\mathrm{H}_2^{18}\mathrm{O}$ or DHO) has a higher evaporation point than ordinary water. Thus, the concentration of these isotopes ($\delta\,^{2}\mathrm{H}$ or $\delta\,^{18}\mathrm{O}$) in clouds varies with ambient temperature, and the concentrations of the isotopes $^{2}\mathrm{H}$ and $^{18}\mathrm{O}$ in the different ice sheets of Antarctica and Greenland provide information on the planetary temperatures of the past. The proportions of natural isotopes on Earth are known: the natural abundance of $^{18}\mathrm{O}$ is 0.24% of all oxygen and the abundance of $^{2}\mathrm{H}$ is 0.015% of all hydrogen.

Another method for determining past temperatures is to study temperature profiles at great depths. Figure 56.2 illustrates how

[a]In geochemistry and paleoclimatology, the isotopic abundance of oxygen is designated by $\delta^{18}\mathrm{O}$. More specifically, $\delta^{18}\mathrm{O}$ measures the isotopic variation of the proportion of stable isotopes $^{18}\mathrm{O}$–$^{16}\mathrm{O}$, in parts per thousand, for a sample with respect to another sample taken as a reference: $\delta^{18}\mathrm{O} = [({}^{18}\mathrm{O}/{}^{16}\mathrm{O})_{\mathrm{sample}}]/[({}^{18}\mathrm{O}/{}^{16}\mathrm{O})_{\mathrm{reference}}) - 1] \times 1{,}000$. $\delta^{2}\mathrm{H}$ or $\delta\mathrm{D}$ is defined analogously; here $^{2}\mathrm{H} = \mathrm{D} =$ Deuterium.

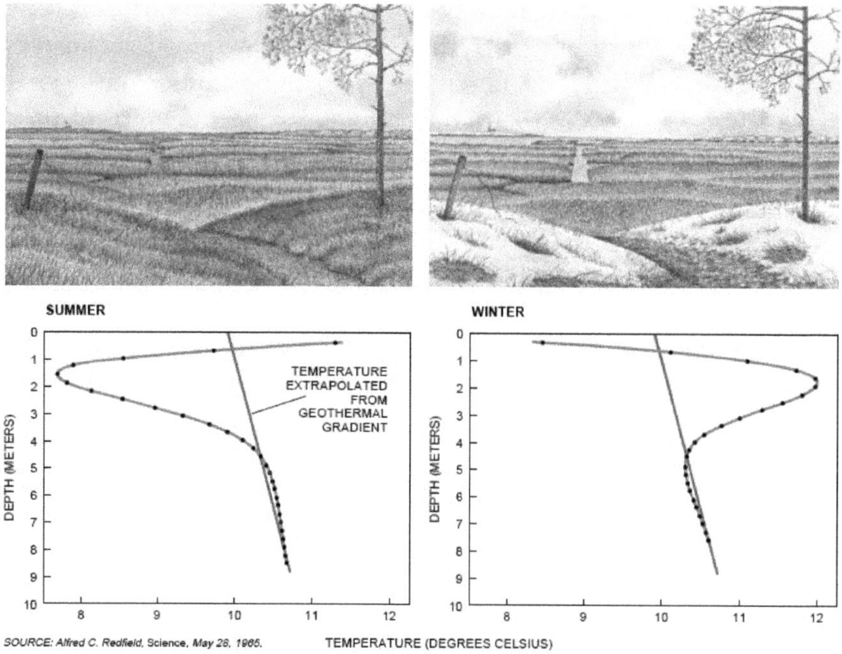

Figure 56.2 Temperature profile as a function of depth for two seasons: summer and winter. The soil surface follows the ambient temperature, but at depths between 1 m and 2 m, the temperature is representative of the previous season. Similarly, the thermal variations of a century of oscillation are recorded in the 100 m deep thermal profiles, and so on. The straight lines represent the Earth's thermal gradient of about $9.2 \times 10^{-2\circ}$C/m [12, 13].

past temperatures leave a clear signature with which they can be characterized [12, 13].

This chapter will study a one-dimensional analog that illustrates much of the physics involved in this problem.

56.2 Dynamic heat conduction-Fourier-Biot equation

Heat is transferred basically by three different processes: *conduction, convection, and radiation* [16–18]. In nature, all three transmission mechanisms intervene simultaneously with varying degrees of importance.

Liquids and gases are generally *not good* conductors of heat, but they can transmit heat efficiently by convection. A forced air heater, in which the air is heated and then distributed by a fan, is an example of *forced convection.* Convection also occurs naturally or spontaneously. For example, the hot air around a heat radiator rises and circulates around the room. Ocean currents such as the Gulf Stream are an example of large-scale natural convection. Wind is another example of convection. In the case of *natural convection*, a thin layer (boundary layer) of air forms around the hot object and propagates heat from the hot surface to the environment by conduction. The heat flux dissipated by natural convention can be written as follows [17, 18]:

$$H_{\text{conv}} = \frac{dQ_{\text{conv}}}{dt} = A_s \cdot h \cdot (T_{\text{Sup}} - T_{\text{ex}}), \tag{56.1}$$

which is *Newton's law of cooling*. Here, A_s is the surface of the body exposed to the medium, h is the *convection coefficient*, a characteristic constant that depends on the dissipation regime (laminar, turbulent, etc.), on the thermal conductivity of the fluid, and on the state of the surfaces [18] (see Table 56.1, Annex A). T_{sup} and T_{ex} are the surface temperatures of the body and of the external fluid or medium, respectively.

In solids, the most common transmission mechanism is heat conduction, which is described by *Fourier's law*. For one-dimensional problems, this law is [16, 17]

$$P = \frac{dQ}{dt} = -K \cdot A\frac{dT}{dx}, \tag{56.2}$$

where $P = dQ/dt$ is the power transferred through area A, transverse to the axis of the bar sample (see Figure 56.3), dT/dx is the thermal gradient, and K is the thermal conductivity constant of the material of which the bar is made (see Table 56.2, Annex A).

If we consider the infinitesimal element shown in Figure 56.3(b), we assume that there is an energy source inside it, generating heat at a power $g(x,t)$ per unit volume and that its temperature is $T(x,t)$. We further assume that the only heat transfer mechanism

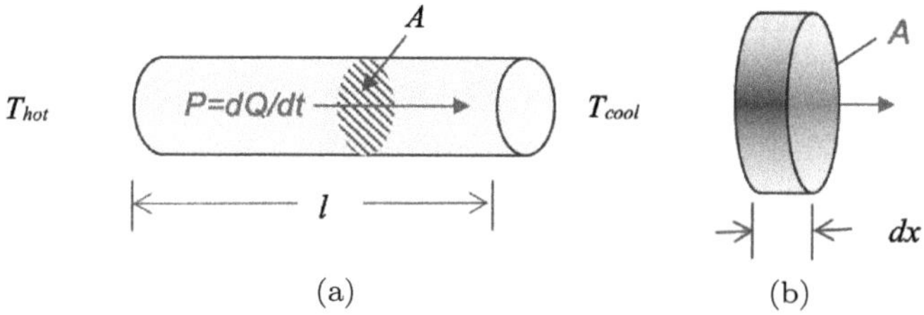

Figure 56.3 (a) Heat transmission along a bar with insulated sides of length l with a hot end ($T_{\rm hot}$) and a cold end ($T_{\rm cool}$). (b) Heat transmission through an element of matter of infinitesimal thickness dx.

is conduction. Let ρ and c be the density and specific heat of the sample, respectively. Performing an energy balance, we have

$$\frac{dQ}{dt} = (\rho A dx)c\frac{dT}{dt} = -K \cdot A\left[\left.\frac{dT}{dx}\right|_x - \left.\frac{dT}{dx}\right|_{x+dx}\right] + (\rho A dx)g(x,t), \tag{56.3}$$

which can be rewritten as follows:

$$\frac{\partial T}{\partial t} = \frac{K}{\rho \cdot c}\cdot\left[\frac{\partial^2 T}{\partial x^2}\right] + g(x,t), \quad \text{or} \quad \frac{\partial T}{\partial t} = k\frac{\partial^2 T}{\partial x^2} + g(x,t). \tag{56.4}$$

Here, $k = K/c\cdot\rho$ is the *diffusivity* of the material, which, as you can see, depends on the thermal conductivity K, the density ρ, and the specific heat c. In the three-dimensional case, this expression can be generalized to

$$\frac{\partial T(x,t)}{\partial t} = k\nabla^2 T(x,t) + g(x,t). \tag{56.5}$$

This is the fundamental equation of heat conduction with sources in three dimensions, known as the *Fourier–Biot equation* [18].

The fundamental equation, Eq. (56.5), without sources, $g = 0$, in one dimension, can be written as follows [18]:

$$\frac{\partial^2 T}{\partial x^2} = \frac{1}{k}\cdot\frac{\partial T}{\partial t}, \tag{56.6}$$

with $k = K/(c\,\rho)$.

56.3 Heat waves in one dimension

Model I

Consider the case of a bar in which the length L is much greater than the diameter $d = 2a$. In this section, we will assume that the only transmission mechanism is conduction along the bar (assume that there is no heat loss from its sides) and that the left end of the bar is in contact with a heat source that has a sinusoidally varying temperature:

$$T_a(t) = T_a^0 \cos(\omega_0 t + \varphi_0) + T_{00} = T_a^0 \cos\left(\frac{2\pi}{p_0} t + \varphi_0\right) + T_{00}, \quad (56.7)$$

where T_a^0 is the amplitude of the thermal oscillation around the mean value T_{00}, $\omega_0 = 2\pi/p_0$ is the angular frequency of the thermal variation, p_0 is its period, and ϕ_0 is the initial phase of the oscillation.

If $T_a^0 = 0$ (i.e., T_a = constant) so that the left end is at the constant temperature T_{00}, while T_e is the temperature on the right end, the bar will have a temperature distribution $T_b(x)$ given by

$$T_b(x) = T_{00} - \frac{(T_{00} - T_e)}{L} x = T_{00} - m \cdot x. \quad (56.8)$$

In this analysis, we have neglected heat losses by other mechanisms such as convection or radiation.

Going back to the case $T_a^0 \neq 0$, with the temperature at the left end oscillating according to Eq. (56.7), in steady state the mean temperature (time average) of the bar will be given by Eq. (56.8). Defining the new temperature as $\theta(x,t) = T(x,t) - T_b(x)$, Eq. (56.6) can be written as follows:

$$\frac{\partial^2 \theta}{\partial x^2} = \frac{1}{k} \cdot \frac{\partial \theta}{\partial t}, \quad (56.9)$$

since $\partial^2 T_b/\partial x^2 = 0$. This is a linear equation that can be solved by separating variables [19–21]. Its solution for periodic boundary conditions, Eq. (56.7), at $x = 0$ can be written as follows [22, 23]:

$$\theta(x,t) = A_0 \exp(-x/\mu_0) \cos(\omega_0 t \pm x/\mu_0 + \varphi_0). \quad (56.10)$$

This expression can be interpreted as an attenuated thermal wave that propagates with velocity v_0 to the right (−) or to the left (+), depending on the sign of the argument of the cosine function. Here, μ_0

represents the *penetration depth* of the wave. By replacing Eq. (56.10) in (56.9), we obtain:

$$\mu_0 = \sqrt{\frac{2k}{\omega_0}} = \sqrt{\frac{p_0 \cdot k}{\pi}} = \frac{\lambda}{2\pi} \quad \text{and} \quad \boldsymbol{\varepsilon_0} = \frac{\mathbf{2\pi}}{\boldsymbol{\lambda}} = \frac{\mathbf{1}}{\boldsymbol{\mu_0}}, \tag{56.11}$$

where $\lambda = \sqrt{4\pi \cdot kp_0} = 2\pi\varepsilon_0 = \text{wavelength}$ and

$$v_0 = \sqrt{4\pi \cdot k/p_0} = \sqrt{2k\omega_0} = \frac{\lambda}{p_0} = \text{velocity of propagation}. \tag{56.12}$$

Equation (56.10) can also be written as:

$$\begin{aligned}\theta(x,t) &= A_0 \exp(-2\pi \cdot x/\lambda)\cos(2\pi(x/\lambda \pm t/p_0) + \varphi_0) \\ &= A_0 \exp(-\varepsilon_0 x)\cos(\varepsilon_0 x \pm \omega_0 t + \varphi_0).\end{aligned} \tag{56.13}$$

The constants A_0 and ϕ_0 depend on the boundary conditions of the problem. The parameter $\mu_0 = \lambda/2\pi$ (*penetration depth*) has the units of length and is indicative of the length to which the thermal wave penetrates, λ is the wavelength and p_0 is the oscillation period. For $x = 0$, we have

$$\theta(x = 0, t) = A_0 \cos(2\pi t/p_0 + \varphi_0) = A_0 \cos(\omega_0 t + \varphi_0). \tag{56.14}$$

That is, at the point $x = 0$, we assume that the temperature oscillates periodically with frequency ω_0.

Equation (56.13) indicates that a thermal wave penetrates distances of the order of a few wavelengths with appreciable width. The amplitude attenuates to $1/e^2 \approx 0.135$ (13.5%) from its initial value for a distance $x = 2\mu_0 = 0.32\,\lambda$ [24, 25].

According to Eq. (56.11), as the period p_0 increases (greater wavelength), the greater the penetration of the thermal wave. Also, this wave penetrates further when the diffusivity of the material (k) is greater, Eq. (56.11). These results explain why daily temperature variations only penetrate the ground to about 20 cm (if $k_{\text{earth}} \approx 2.9 \times 10^{-7}\ \text{m}^2/\text{s}$ and $p_0 = 24$ h, then $\mu_0 \approx 9$ cm, $\lambda \approx 56$ cm), whereas annual variations ($p_0 = 365$ days) penetrate the Earth down to about 4 or 5 m ($\mu_0 \approx 2$ m and $\lambda \approx 1.7$ m) [24, 25]. The temperatures associated with the last ice age ($p_0 \approx 10{,}000$ years) would be at depths of between about 170 m and 200 m (see Figure 56.4) [12, 13].

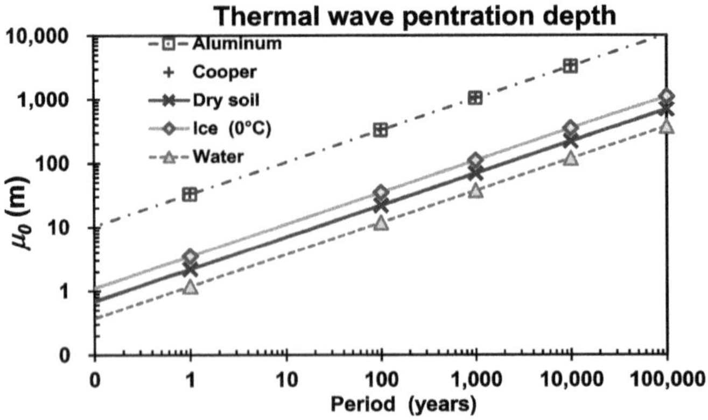

Figure 56.4 Variation of the penetration length of thermal waves, μ_0, in different materials. On land, daily variations are attenuated by about 50 cm, annual variations by a few meters, and secular variations by a few hundred meters.

This phenomenon of thermal waves — having a *penetration depth* μ_0 that depends on the wave period — is analogous to the skin effect of electromagnetism for waves that propagate in conductive media (see Chapter 40). Hence, secular temperature variations will penetrate deeper into the Earth. For this reason, the study of thermal profiles at great depths can provide information on temperature changes that occurred in the remote past [12, 25]. More specifically, if we assume that in Eq. (56.10) $\phi_0 = 0$, from this equation we deduce that for a later time $\tau = p_0/2$, at a depth $x = \lambda/2$, temperature will be

$$\theta(x = \lambda, \tau) = \theta(0, 0)/e^{\pi} \approx 0{,}043 \cdot \theta(0, 0). \quad (56.15)$$

Therefore, by measuring the temperature at depth $x = \lambda/2$, we can infer the surface temperature at a time in the past ($p_0/2$). More specifically, at this depth, the temperature will be 4.3% of its value at the surface at the past time $\tau = p_0/2$. This is an immensely valuable result for inferring the temperatures of the past, which can be reconstructed by studying the thermal profiles of perforations in the ground or in the ice [7, 26]. Figure 56.4 shows the variation of the penetration depth μ_0 of thermal waves in different materials as a function of the period p_0.

For the same time $\tau = p_0/2$, but for a distance $x = 0.38\ \lambda$, according to Eq. (56.13), we have that

$$\theta(\lambda, \tau) = -\theta(0,0)/e^{2\pi} \approx -0{,}067 \cdot \theta(0, \tau), \tag{56.16}$$

therefore, at this depth, the thermal wave is attenuated by 6.7% but has a phase opposite to the value of the wave at the surface ($x = 0$). Thus, in summer, at this depth ($x = 0.38\ \lambda$), the temperature is cooler than on the surface, while in winter it is warmer than on the surface. Therefore, by using a heat exchange system such as *buried tubes*, this property of the ground can be tapped to achieve natural temperature conditioning without consuming energy [27]. This technique is used in several regions of the world and makes caves particularly suitable regarding temperatures for living. Indeed, it was in cave dwellings that our ancestors endured the long periods of glaciation.

Using the reduced variables (dimensionless) $\tau = t/p_0$ and $\chi = x/\lambda$, Eq. (56.13) can be written as follows:

$$\theta(x, t) = A_0 \exp(-2\pi \cdot \chi) \cos(2\pi(\chi \pm \tau) + \varphi_0). \tag{56.17}$$

Its graphical representation for different values of τ and χ is shown in Figure 56.5.

The above treatment assumes a bar of infinite length. However, since the wave attenuates by a few wavelengths, a bar of length $l > 4\lambda$ can be considered infinite. This one-dimensional model can also be applied to a one-dimensional semi-infinite system, for example, the Earth, where the heat source is on the surface, and is heated by the Sun.

Notice, too, that the amplitude of the thermal wave decreases with depth. According to Eq. (56.17), we have for the amplitude:

$$\theta_{\max}(x) = A_0 \exp(-x/\mu_0) \quad \text{and} \quad \mu_0 = \sqrt{\frac{p_0 \cdot k}{\pi}}. \tag{56.18}$$

Therefore, we would expect that temperature variation with depth x along a semi-infinite bar, whose extreme is excited with a sinusoidal temperature variation of period p_0, or inside the Earth at a depth x, would have thermal amplitude $\Delta T(x) = (T_{\text{Max}}(x) - T_{\min}(x))/2 = A_0 \cdot \exp(-x/\mu_0)$, which would decrease exponentially

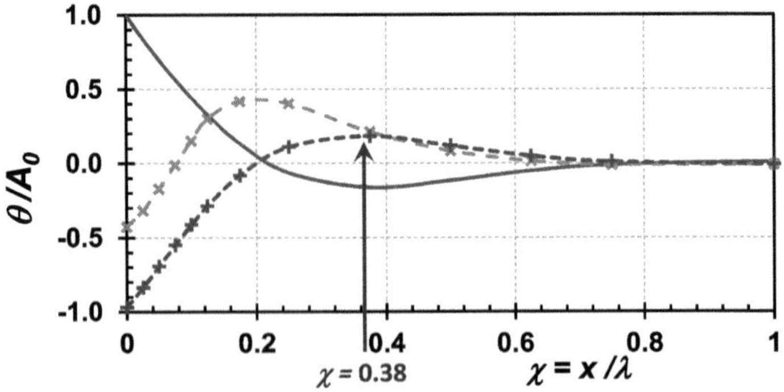

Figure 56.5 Temperature variation $\theta(\chi, \tau)/A_0$ of a bar at various reduced depths χ and at different reduced times τ. After a half-period ($\tau = 0.5$) and at a depth $\chi = 0.38$, the thermal wave has the opposite phase to its surface value, $\chi = 0$. Note the similarity of these curves to the soil temperature measurements in Figure 56.2.

with the penetration x. The parameter μ_o is representative of the *penetration depth* of the thermal wave, whose value, according to Eq. (56.18), increases with p_0 and k.

56.3.1 *Project 190. Thermal properties of the ground*

Recommended equipment: Four or more thermometers connected to a data acquisition system.

A problem of great practical importance in many fields is understanding the thermal properties of the ground. In agriculture, it is essential to know the soil temperature to ensure that seeds will germinate and grow properly. In engineering, it is essential to know the thermal behavior of the ground because gas or oil pipelines should not be subjected to large thermal variations, which could cause big changes in length due to thermal expansion, affecting their structure [24]. Moreover, the ground temperature alters the properties of the fuels carried by pipelines.

To study the thermal properties of the ground, you can measure the temperature at various depths as a function of time, over three or four days. For this experiment, it is advisable to choose clear days with wide temperature ranges and use several thermometers connected to a computer data acquisition system. It is advisable to have a sensor that registers the ambient temperature 1 cm or 2 cm above the ground. The depth of the deepest thermometer should not exceed 50 cm. The thermal sensors can be thermistors, thermocouples, Pt100, or sensors based on integrated circuits, such as the LM35 (see Chapter 50) [25].

Suggested tasks:

✓ Begin data acquisition by measuring temperatures at a rate of approximately once every 10–15 min.

✓ With the data obtained over two to four full days, plot temperature as a function of time for each thermometer.

✓ For each sensor, determine the thermal amplitude $\Delta T = (T_{\max} - T_{\min})/2$ for a full day and plot ΔT as a function of depth x, on linear and semi-logarithmic scales. If you observe that there is exponential dependence, such as that predicted by Eq. (56.18), determine μ_0 from your data [24]. Estimate the value of the ground diffusivity k and compare it to tabulated values. For a dry ground, $k \approx 4 \times 10^{-7}\,\mathrm{m^2/s}$ ($K = 1.2\,\mathrm{W/m \cdot K}$, $\rho = 2{,}500\,\mathrm{kg/m^3}$ and $c = 1{,}260$ J/kg · K), see Annex B.

✓ From the temporal position of the maxima of two thermometers, located at depths x_1 and x_2, determine the time interval $\Delta t_{1,2}$ at which each one reaches its maximum, see Figure 56.6. From these data, estimate the speed of propagation of heat waves in the ground $v_0 = (x_2 - x_1)/\Delta t_{1,2}$ for the different pairs of thermometers used and find the best values of v_0 and k, and their respective uncertainties. Compare your value to the one predicted by the theoretical model, Eqs. (56.11) and (56.12). A better method to obtain the average value of v_0 is to plot the position of each thermometer x_i as a function of phase shift Δt_{1i} of each of them with respect to the first one (the one on the ground closest to the surface is designated as 1). If the data points on the plot are aligned, the slope will be the best value of v_0 [25].

✓ Using all measured data from $T(x, t)$, try to fit your data within the theoretical model described by Eq. (56.13). Vary the value of the constant k until you get the best possible fit of the data.

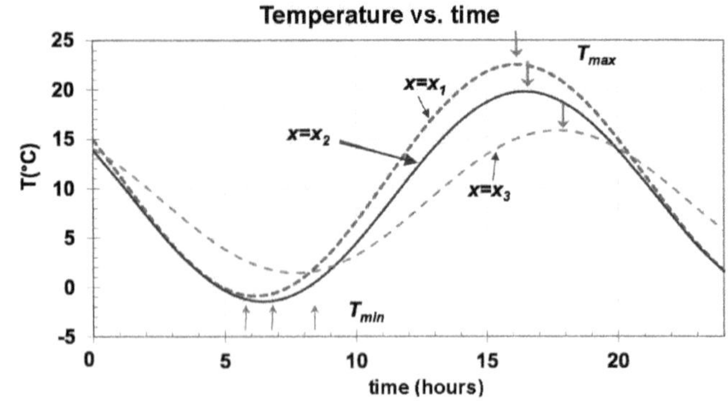

Figure 56.6 Variation of ground temperature at different depths (x), as a function of time. Note that the maximum and minimum values are reached at different times.

Discuss the goodness of the proposed model to explain your data. Annex B provides data on conductivities and diffusivities of different materials.

56.4 ♣♣ Forced heat waves

Model II (optional topic)

Let us now consider the case in which the variation of the heat source placed at the left end is periodic but not necessarily sinusoidal. By Fourier's theorem, the variation of the source can be broken down into a series of the form [19, 21, 28]

$$\theta(x=0,t) = \sum_{n=1} A_n \cos(\omega_n t - n\varphi_0), \tag{56.19}$$

with

$$A_n = \frac{2}{p_0} \int_0^{P_0} \theta(x=0,t) \cdot \cos(n\omega_0 t), \tag{56.20}$$

where $p_0 = 2\pi/\omega_0$ is the fundamental period of the wave. Using the variable separation and superposition method, the solution of the heat diffusion equation, Eq. (56.9), can be written in this case as follows:

$$\theta(x,t) = \sum_{n=1} A_n \exp(-\varepsilon_n x) \cos(\omega_n t - \varepsilon_n x + n\varphi_0) \tag{56.21}$$

or also

$$\theta(x,t) = \sum_{n=1} A_n \exp(-x/\mu_n)\cos(\varepsilon_n(v_n t - x) + n\varphi_0). \tag{56.22}$$

The constants ω_n and ε_n are obtained by replacing this expression in Eq. (56.22):

$$\omega_n = \frac{2\pi n}{p} = n \cdot \omega_0 \tag{56.23}$$

and

$$\varepsilon_n = \sqrt{\frac{\omega_n}{2k}} = \frac{\sqrt{n}}{\mu_0} = \frac{1}{\mu_n} \quad \text{and} \quad \mu_n = \frac{\mu_0}{\sqrt{n}}. \tag{56.24}$$

Therefore, the *penetration depth* μ_n decreases as n increases. The speed of propagation of the different harmonics is

$$v_n = \frac{\omega_n}{\varepsilon_n} = \sqrt{n} \cdot v_0, \quad \text{with} \quad v_0 = \sqrt{4\pi k/p_0}. \tag{56.25}$$

These parameters provide an idea of the penetration speed of the harmonics ($n > 1$) with respect to the fundamental ($n = 1$).

The constants A_n and ϕ_0 in Eq. (56.22) depend on the boundary conditions. Note that the penetration depth $\mu_n = \mu_0/\sqrt{n}$ of harmonic waves ($n > 1$) is less than the penetration depth of the fundamental μ_0. Thus, at a certain distance from the origin ($x \gg 0$), the thermal wave will be dominated by the fundamental wave ($n = 1$), while the harmonics will progressively fade as x increases. This means that at a certain distance from the origin or at a certain "depth", the thermal wave will be dominated by a sinusoidal wave whose frequency will be $\omega_0 = 2\pi/p_0$, regardless of the shape of the thermal signal at $x = 0$.

(a) If we assume that the boundary conditions for $x = 0$ are given by a square wave with period p_0 and amplitude θ_0, then

$$\begin{aligned}\theta(0,t) &= \sum_{n=\text{odd}} A_n \cos(\omega_n t + n\varphi_0) \\ &= \sum_{n=\text{odd}} \left(\frac{4\theta_0}{\pi n}\right)\cos(n(\omega_0 t + \varphi_0)),\end{aligned} \tag{56.26}$$

with

$$A_n = \frac{4\theta_0}{\pi n} \quad \text{for odd } n \quad \text{and} \quad A_n = 0 \quad \text{for even } n. \tag{56.27}$$

Here, ϕ_0 is the initial phase of the wave, which depends on the initial conditions, that is, the value of $\theta(0,0)$.

(b) Similarly, if we assume that the boundary conditions for $x = 0$ are given by a triangular wave of period p and amplitude θ_0, we have

$$\begin{aligned}\theta(0,t) &= \sum_{n=\text{odd}} A_n \cos(\omega_n t + n\varphi_0)\\ &= \sum_{n=\text{odd}} \left(\frac{4\theta_0}{\pi n^2}\right)\cos(n(\omega_0 t + \varphi_0)),\end{aligned} \tag{56.28}$$

with

$$A_n = \frac{4\theta_0}{\pi n^2} \quad \text{for odd}\, n \quad \text{and} \quad A_n = 0 \quad \text{for even}\, n. \tag{56.29}$$

By combining Eq. (56.28) with Eq. (56.13), we obtain

$$\theta(x,t) = \sum_{n=\text{odd}} \left(\frac{4\theta_0}{\pi n^2}\right)\exp(-\varepsilon_n x)\cos(\omega_n t - \varepsilon_n x + n\varphi_0). \tag{56.30}$$

(c) If the heat generating element is a resistance, when it is turned on, it generates a constant power input on the bar, which in turn generates a constant heat flow $H_0 = dQ/dt$, toward the interior of the bar. The heat conduction equation applied to this end of the bar leads to $H_0 = -KA\,dT_a/dt$. Here, T_a is the temperature of the end of the bar adjacent to the heat source. If we assume that $H = dQ/dt$ is a square wave, since: $dT_a/dt = -H_0/(KA)$, we can assume that $T_a(t)$ will be a triangular wave. Therefore, the condition that best approximates the case of a heat source at one end that turns on and off is the triangular thermal wave, Eq. (56.30), Figure 56.7.

In all cases, it is important to observe the behavior of the thermometer closest to the heater and verify which is the edge condition that best describes the system under study. One way to simulate the effect of the heater having a finite mass is to define the x coordinate (used in Eq. (56.30)) as follows:

$$x = x_{\text{real}} + \delta x_0, \tag{56.31}$$

where x_{real} is the measured position of the thermometer and δx_0 is a parameter that is set such that expression (56.30) reproduces the measured values of temperature for the thermometer closest to the heater.

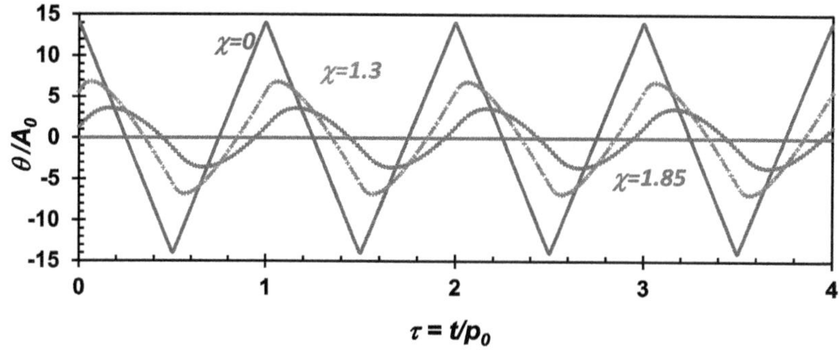

Figure 56.7 Variation of the temperature of a bar at different depths $\chi = x/\lambda$, as a function of normalized time $\tau = t/p_0$. If the thermal conduction decreases, the thermal wave attenuates more quickly. Also, the waves at greater depth (χ greater) are delayed and attenuated. Higher order components ($n > 1$) fade more quickly. These graphs were obtained using Eq. (56.30).

It is interesting to note that the maximum temperatures $\theta_{\max}$ for a given depth (x) will occur when the value of the cosine is equal to 1. That is,

$$\theta_{\max}(x) = \sum_{n=\text{odd}} \left(\frac{4\theta_0}{\pi n^2}\right) \exp(-\varepsilon_n x) \approx \left(\frac{4\theta_0}{\pi}\right) \exp(-x/\mu_0). \quad (56.32)$$

This is because the terms associated with $n > 1$ fade away at ever shorter distances as n increases. Therefore, for a metal bar, at a few centimeters away from the end in contact with the heater, the last part of Eq. (56.32) is a good approximation of the actual situation and indicates a possible way to determine k. The amplitude of the thermal wave $\theta_{\max}(x_i)$ is determined according to the depth x_i (thermometer position i). If the graph of $\ln[\theta_{\max}(x_i)/\theta_{\max}(x_1)]$ as a function of x_i is linear [see Eq. (56.32)], we can determine k. Here, x_1 is the position of the thermometer closest to the heater.

56.4.1 *Project 191. Heat waves in a bar*

> **Recommended equipment**: Two or more metal bars about 60–100 cm long, with diameters between 1 cm and 2 cm. Three or more thermometers connected to a data acquisition system. A soldering iron.

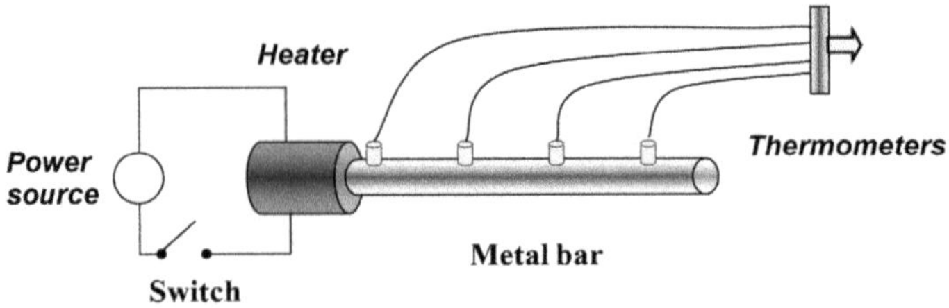

Figure 56.8 Experimental setup with a metal bar about 1 m long and about 1 cm in diameter, connected to a heater that can be turned on and off periodically [21, 22].

For this experiment, you can use a metal bar 60–100 cm long, with diameter 1–2 cm (see Figure 56.8). Use a lathe to thin down one end of each bar so that it can be connected to the heater of a standard tin soldering iron, thereby enabling close connection to the heat source. Every 5 cm along the bar, drill a hole about 5–8 mm deep, perpendicular to its axis, into which you can insert a temperature sensor [22, 25] connected to a data acquisition system. The heater will be turned on and off periodically, either manually, or more easily by connecting the soldering iron to an automatic on-and-off switch.

It is advisable to have two or more bars with similar geometric characteristics, made of different materials. One of them should be made of a metal with well-known thermal properties so that it can be used as a test bench for this experimental method. Thermal sensors can be thermistors, thermocouples, Pt100, or sensors based on integrated circuits (LM35, see Chapter 50) [22, 25].

The positions of the thermal sensors, x_i, indicate the coordinate or *depth* at which you will measure the temperature. It is advisable to choose an excitation period p_0 for the thermal source between 100 s and 500 s. To minimize heat loss by convection, the bar may be wrapped with 3–4 turns of 1/8″ insulation material (fiberglass or any other type, e.g., 1/8″ refrigerant tubing insulation), even though this is not crucial for Al or Cu bars in air. See the comment at the end of Section 56.5 of this chapter.

Suggested tasks:

✓ Begin cycling the heat source until a steady state is observed. This means that the average temperature of each thermometer remains constant. Operationally, it can be verified that the bar has reached steady state by plotting the maximum temperatures $T_{\text{max}}(x)$ and $T_{\text{min}}(x)$ as a function of x. If the relationship between these variables is linear, it is indicative that the bar is in stationary condition, see Eq. (56.8).

✓ Begin data acquisition by measuring temperatures at a rate of approximately 1 Hz, making sure to measure at least five temperatures $T(x)$ in a full cycle of time period p_0.

✓ With the data obtained, plot the temperature as a function of time, for each thermometer (i), $T(x_i, t)$, and verify that a stationary situation has been reached. Under these conditions, the signal oscillates maintaining the same maximums, minimums and average value.

✓ For each thermometer, plot the mean value $\langle T(x) \rangle [=(T_{\text{max}} + T_{\text{min}}/2]$ as a function of x. If the relationship is linear, the parameters T_{00} and m can be obtained from Eq. (56.8), $T_b(x) = T_{00} - mx$. Once this expression has been determined, calculate the normalized temperature $\theta(x,t) = T_{\text{measured}}(x,t) - T_b(x)$.

✓ Under stationary conditions, plot the maximum temperature θ_{max} (observed for each thermometer i) as a function of position x_i, on linear and semi-logarithmic scales.

✓ What can you conclude about the dependence of $\theta_{\text{max}}(x_i)$ as a function of x_i?

✓ If the dependence is exponential, use Eqs. (56.18) and (56.32) to determine μ_0 for each bar studied. From this information, use Eq. (56.11) to determine the diffusivity k and the conductivity K of each bar.

✓ From the temporal position of the maxima of two thermometers, 1 and j, determine the time interval $\Delta t_{1,j}$ in which each one reaches its maximum. From these data, estimate the speed of propagation of heat waves along the bar $v_0 = (x_j - x_1)/\Delta t_{1,j}$ for the different pairs of thermometers used and find the best values of v_0 and k, and their respective uncertainties. Compare them to the values predicted by the theoretical model, Eq. (56.11). A better method to obtain the average value of v_0 is graphing the position of each thermometer x_j as a phase shift function $\Delta t_{1,j}$ of each with respect to the first thermometer (the one closer to the heater). If the data are aligned, the slope will be the best value of v_0 [25].

✓ Use Eq. (56.30) to compare the predictions of those models with the measured data of $T(x,t)$. Fit the experimental data with these expressions and find the best value of the constant ε_0 or $\mu_0 = \sqrt{p_0 \cdot k/\pi}$ and an estimate of its errors.

✓ How do these values of μ_0 compare to those obtained previously and to the ones calculated using Eq. (56.11) and the value of k from the reported value for these parameters [29], for each material of which the bars are made?

✓ For each of the bars studied, use at least two or more driving frequencies, i.e., different periods p_0. Discuss how thermal waves are attenuated for the different frequencies using your experimental data. Does the model proposed, Eqs. (56.10) and (56.30), provide a good description of your experimental results?

✓ For a bar with known thermal properties (k), compare the value found for k from the experiments to the values provided in tables [29] for the same material. What can you conclude from this analysis?

✓ Using all measured data for $T(x,t)$ and the model described by Eq. (56.30) or Eq. (56.10), try to fit the measured data with this theoretical expression. Vary the value of the constant k until you get the best possible fit of the data. Discuss the goodness of the proposed model.

56.5 Bar with lateral cooling — Convection

The previous section studied the case of a bar in which the only transmission mechanism was the conduction along it. This section considers the case in which there are heat losses from the lateral face of the bar. Assume that the rod is in a fluid medium (air) at a temperature T_e and that it cools as a result of convection through the fluid and

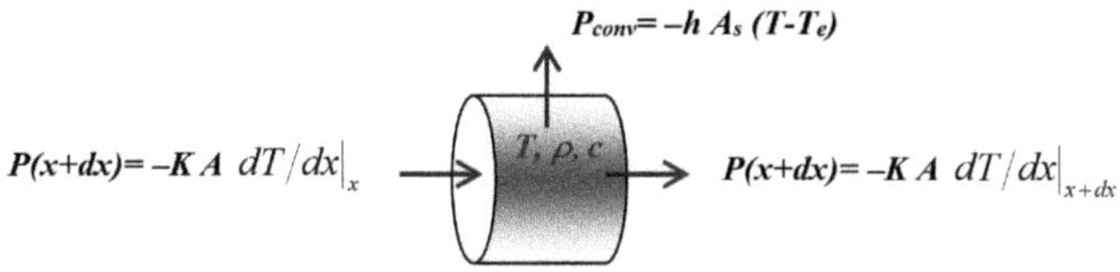

Figure 56.9 Energy balance for an infinitesimal volume element of a bar.

conduction through the bar. Convection is generally a very complex process. However, a useful approximation of this process is described by Newton's law of cooling, Eq. (56.1). Table 56.1 in Annex A provides approximate values for convection constants h [18].

Consider a bar of length l and radius a, like the one represented in Figure 56.3(a). If the convection effects are not negligible, by performing an energy balance on an infinitesimal element of length dx as illustrated in Figure 56.9, we have

$$c\rho A\delta x \frac{\partial T(x,t)}{\partial t} = -KA\left(\partial T/\partial x|_x - \partial T/\partial x|_{x+dx}\right) - hA_s(T(x,t) - T_e) \tag{56.33}$$

or

$$\frac{\partial T(x,t)}{\partial t} = +\frac{K}{c\rho}\frac{\partial^2 T}{\partial x^2} - \frac{h\cdot\phi}{c\rho A}(T(x,t) - T_e). \tag{56.34}$$

Here, $A_s\,(= 2\pi a dx \sim \phi\, dx)$ is the lateral surface and $\phi(= 2\,\pi a)$ is the perimeter of this surface. $A(= \pi a^2)$ is the cross-sectional area of the element. Defining $k = K/(c\rho)$ and $\beta = h\,\phi/c\rho A \approx 2h/c\rho\, a$, and $\theta(x,t) = T(x,t) - T_e$, we have

$$\frac{\partial\theta(x,t)}{\partial t} = +k\frac{\partial^2\theta}{\partial x^2} - \beta\cdot\theta(x,t). \tag{56.35}$$

In stationary conditions ($\partial\theta/\partial t = 0$), this equation is reduced to

$$\frac{\partial^2\theta}{\partial x^2} = \frac{\beta}{k}\cdot\theta(x,t) = \alpha^2\cdot\theta(x,t). \tag{56.36}$$

For a bar of circular section and radius a, we have

$$\alpha^2 = \beta/k = \frac{h}{K}\frac{\phi}{A} = \frac{h}{K}\frac{2}{a}. \tag{56.37}$$

Consider a bar in which the left end is at a fixed temperature T_a and the right end is at temperature T_b. By solving Eq. (56.36), we have

$$T(x) = T_e + (T_a - T_e)\frac{\sinh(\alpha(L-x))}{\sinh(\alpha L)} + (T_b - T_e)\frac{\sinh(\alpha \cdot x)}{\sinh(\alpha L)}. \tag{56.38}$$

It is useful to define the *Biot Number* (Bi) as the quotient between the convective to conductive effects in a body. More specifically, this number is the ratio of the heat transfer by conduction inside a bar and the heat loss by convection on its surface:

$$\begin{aligned} Bi &= \frac{\text{“Heat flow by convection”}}{\text{“Heat flow by conduction”}} \\ &= \frac{hA_s}{KA/dx} = \frac{h}{K}\frac{A_s dx}{A} = \frac{h}{K}\lambda \approx \frac{h}{K}2a. \end{aligned} \tag{56.39}$$

Here, λ is the quotient between a volume and an area, which for a cylinder with $dx \sim a$ is $\lambda \sim 2a$.

In practice, convective effects are considered negligible compared to conductive effects when $Bi < 0.01$ [18]. In Figure 56.10, the temperature variations are represented for a copper bar (good conductor $Bi \approx 10^{-4}$, upper panel) in air and a stainless-steel bar (poor conductor $Bi \approx 0.3$, lower panel) in water. In both cases, the diameter of the bar is $2a = 1.5$ cm. We see that in the second case it is not possible to neglect the effects of convection. In short, the *Biot number* is a dimensionless number that represents how important the energy exchange of heat is by *convection* from solid surface and by *conduction* in the solid interior. If $Bi \ll 1$, *heat conduction is dominant* (as in the top panel of Figure 56.10). On the other hand, if $Bi > 0.1$, *heat convection* on the surrounding surface cannot be neglected (as in the lower panel of Figure 56.10). This is the reason why, in Project 191, it is not crucial to use additional insulation

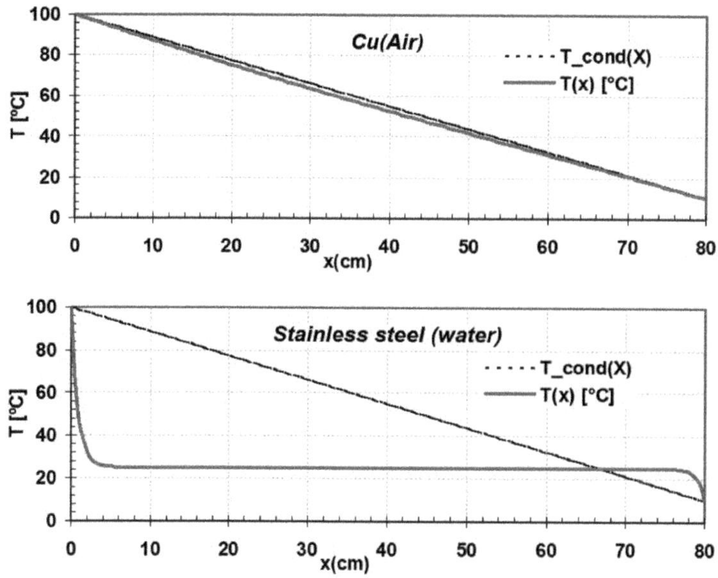

Figure 56.10 Effect of convection on a copper bar in air (top panel $Bi \approx 10^{-4}$) and a stainless steel bar in water (bottom panel $Bi \approx 0.3$), both in steady state and with a diameter of 1.5 cm. The thick solid red lines are the solution (56.38) considering convection and conduction. The thin black lines are the results obtained by neglecting convection. In general, when $Bi \leq 10^{-3}$, neglecting convection is a reasonable approximation; otherwise, the effect of convection is important. The external temperature was set at 25°C.

on the bar to minimize convection when you use a rod made of Al, Cu, or similar material in air, since the *Biot number* is generally very low.

56.5.1 *Project 192. Study of conduction and convection in a bar*

Recommended equipment: Two metal bars 60 to 100 cm long with diameter 1 cm to 2 cm. Two standard glass thermometers. A thermometer based on a thermistor or thermocouple. Two $^1/_2$ to 1 liter expanded polyurethane containers.

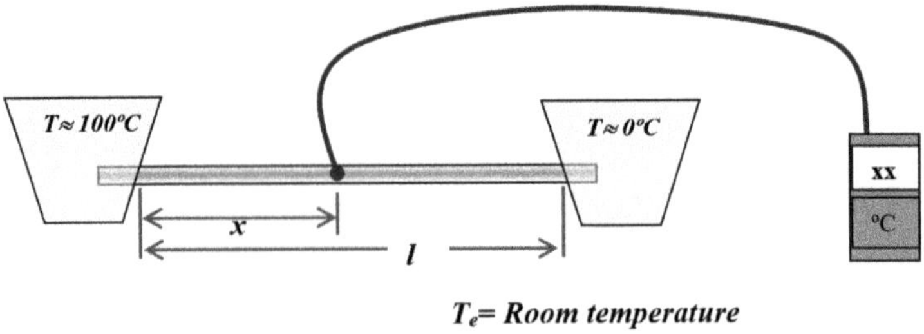

Figure 56.11 Experimental arrangement with a metal bar made of copper, aluminum, or stainless steel. The containers at each end of the bar act as heat reservoirs. A thermometer or thermocouple measures the temperature of the rod according to the position x.

For this experiment, it is proposed to use at least two metal bars. Take two approximately 1-liter expanded polyurethane containers (ice-cream containers work well). Drill holes in the sides of the containers so that you can insert the ends of the bar about 4 cm or 5 cm into the water contained in them, and seal with epoxy or an all-purpose silicone sealant, as shown in Figure 56.11.

Keep the temperature in each container as constant as possible, for example by placing ice in the cold reservoir and hot water in the hot reservoir (or making the water boil by means of an electric immersion heater). Place a thermometer in each container to monitor its temperature. It is also helpful (though not essential) to use thermistors or thermocouple thermometers to monitor the temperature at different points on the bar at about 1 cm away from the heat and cold sources.

Suggested tasks:

✓ When the temperature stops varying over time, a steady state will have been reached. Under these conditions, measure the temperature of the bar as a function of position x, that is, measure $T(x)$ for at least about 10 equidistant points between the two ends.

✓ For each bar, plot the temperature $T(x)$ as a function of x. In the same figure, include the curve corresponding to the theoretical model, Eq. (56.38). Vary the parameter α to find the one that provides the best fit for the experimental results.

✓ Using the conductivity values from Table 56.2 (Annex A), estimate the best value of the convection coefficient h for the bar in air. Compare this value to those in Table 56.1. This test would provide more sensitive results when using a low-conductivity bar, for instance, stainless steel.
✓ Repeat this analysis for all the bars you have.
✓ Discuss the goodness of the model to reproduce your data and, for each case, indicate the relevance of the convection phenomenon in each type of bar used.

56.6 Annex A: Thermal properties of materials

Table 56.1 Convection coefficient.

Type	**h[w/m^{2}°C]**
Free convection in gases	2–25
Free convection in liquid	10–1,000
Forced convection in gases	25–250
Forced convection in liquid	2,500–10^5

Table 56.2 Thermal conductivity and diffusivity coefficient [18]. More information is available in Ref. [29].

Material	**Conductivity *K* (W/m°C)**	**Diffusivity *k* (m^2/s) × 10^6**
Silver	429	177.6
Diamond	2300	149.0
Gold	317	127.0
Copper	401	113.0
Aluminum	237	97.5
Bronze (Cu63%–Zn37%)	109	33.6
Iron	80.2	22.8
Air (25°C) (g)	0.026	22.39
Bronze (Cu89%.Sn11%)	52	16.1
Mercury (liq.)	8.54	4.7
Stainless steel (25°C)	16.2	4.03

(*Continued*)

Table 56.2 (*Continued*)

Material	Conductivity K (W/m°C)	Diffusivity k (m^2/s) $\times 10^6$
Dry soil	1.5	2.778
Salt-Marsh	0.61	0.157
Soil (dry)	2.5	0.47
Ice	1.97	1.2
Expanded polystyrene	0.033	1.1
Soil (dry)	1.5	0.52
Brick	0.72	0.52
Glass	0.78	0.34
Fiberglass	0.043	0.23
Water (liq)	0.613	0.14
Oak wood	0.17	0.13

Table 56.3 Physical properties of heavy and light water.

Property	D_2O (heavy water)	H_2O (light water)
Melting point (°C)	3.82	0.0
Boiling point (°C)	101.4	100.0
Density (at 20°C, kg/l)	1.1056	0.9982
Temp. maximum density (°C)	11.6	4.0

References

[1] National Geographics, "What is global warming, explained," 2022. Available: https://www.nationalgeographic.com/environment/article/global-warming-overview.

[2] NASA-Global Climate Change, "Overview: Weather, global warming, and climate change," 2022. Available: https://climate.nasa.gov/resources/global-warming-vs-climate-change/.

[3] Wikipedia, "Ice age," 2022. Available: https://en.wikipedia.org/wiki/Ice_age.

[4] NASA Global Climate Change, "Milankovitch (Orbital) cycles and their role in Earth's cClimate," 27 October 2020. Available: https://climate.nasa.gov/news/2948/milankovitch-orbital-cycles-and-their-role-in-earths-climate/#:~:text=A%20century%20ago%2C%20Serbian%20scientist,glaciation%20periods%20(Ice%20Ages).

[5] R. A. Muller y G. J. MacDonald, *Ice Ages and Astronomical Causes: Data, Spectral Analysis, and Mechanisms*, NY: Springer-Verlag Telos, 2000.
[6] IPCC, "IPCC sixth assessment report, climate change 2022," 2022. Available: https://www.ipcc.ch/report/sixth-assessment-report-working-group-3/.
[7] M. Mastrandrea and S. Schneider, "Resource letter GW-2: Global warming," *Am. J. Phys.*, vol. 76, p. 608, 2008.
[8] S. E. Schwartz, "Resource letter GECC-1: The greenhouse effect and climate change: Earth's natural greenhouse effect," *Am. J. Phys.*, vol. 86, p. 565, 2018.
[9] NASA, "GISS surface temperature analysis (GISTEMP v4)," 2022. Available: https://data.giss.nasa.gov/gistemp/.
[10] Berkeley Earth, "Berkeley Earth — Data visualization," 2022. Available: http://berkeleyearth.org/data-visualization/.
[11] M. E. Mann, *et al.*, "Optimal surface temperature reconstructions using terrestrial borehole data," *J. Geophys. Res.*, vol. 108, no. D7, p. 4203, 2003.
[12] H. N. Pollack and D. Chapman, "Underground records of changing climate," *Sci. Am.*, vol. 265, no. 6, p. 44, June 1993.
[13] A. C. Redfield, "Terrestrial heat flow through Sal-Marsh peat," *Science*, vol. 148, no. 3674, pp. 1291–1292, 28 May.
[14] Berkeley Earth, "Temperature pattern and volcanic eruptions," 2022. Available: http://berkeleyearth.org/archive/volcanoes/.
[15] Le Treut, *et al.*, "Historical overview of climate change," In: Climate Change 2007: The Physical Science Basis. Contribution of Working Group I to the Fourth Assessment Report IPCC, Cambridge University Press, Cambridge, UK, 2007.
[16] J. D. Wilson, *College Physics*, 7th Ed., NY: Pearson, 2009.
[17] D. Giancoli, *Physics Principles with Applications*, 7th Ed., NY: Pearson, 2014.
[18] Y. Cengel, *Heat Transfer: A Practical Approach*, 2nd Ed., New York: McGraw-Hill, 2002.
[19] M. R. Spiegel, *Schaum's Outline of Fourier Analysis*, NY: Mc Graw Hill, 2011.
[20] M. Spiegel, *Advanced Mathematics for Engineers and Scientists*, NY: McGraw Hill, 1980.
[21] M. Anwar, *et al.*, "Fourier analysis of thermal diffusive waves," *Am. J. Phys.*, vol. 82, p. 928, 2014.
[22] A. Bodas, V. Gandía and E. López-Baezab, "An undergraduate experiment on the propagation of thermal waves," *Am. J. Phys.*, vol. 66, no. 6, p. 528, 1998.

[23] M. Sullivan and B. Thompson, "An experiment on the dynamics of thermal diffusion," *Am. J. Phys.*, vol. 76, no. 7, p. 637, 2008.

[24] G. McIntosh and B. S. Sharratt, "Thermal properties of soil," *Phys. Tech.*, vol. 39, p. 458, 2001.

[25] L. Iannelli and S. Gil, "Heat waves — Temperatures of the past," *Lat. Am. J. Phys. Educ.*, vol. 6, no. 1, pp. 82–88, 2012.

[26] NOAA Climate Gov., "Past climate," 2022. Available: https://www.climate.gov/maps-data/climate-data-primer/past-climate.

[27] Wikipedia, "Ground-coupled heat exchanger," 2022. Available: http://en.wikipedia.org/wiki/Ground-coupled_heat_exchanger.

[28] H. P. Hsu, *Applied Fourier Analysis*, NY: Int. Thompson, 1984.

[29] Engineering Tool Box, "Solids, liquids and gases — Thermal conductivities," 2022. Available: https://www.engineeringtoolbox.com/thermal-conductivity-d_429.html.

[30] P. Jones, D. Parker, T. Osborn and K. Briff, "Global and hemispheric temperature anomalies — land and marine instrumental records," *In Trends: A Compendium of Data on Global Change. Carbon Dioxide Information Analysis Center, Oak Ridge Nat. Lab. USA*, Oak Ridge, TN: Oak Ridge Nat. Lab. USA, 2005.

Chapter 57

DIFFUSION OF MATTER

This chapter studies the phenomenon of diffusion of matter, a very common, highly important effect in nature. Here we consider the diffusion of a drop of ink or a grain of potassium permanganate in water, which are two examples of diffusion in two dimensions. Based on these studies, we will try to determine the diffusion coefficients of the study systems and their variation with temperature.

Goals

✓ Diffusion of matter
✓ Diffusion in one and two dimensions
✓ Determination of the diffusivity constant

57.1 Fick's Laws of diffusion

When a drop of ink or tempera drips into a glass of water, we observe that it gradually spreads and enlarges. When a bottle of perfume is opened, after a certain time the smell spreads throughout the room. These examples are illustrative of a very interesting and prevalent physical phenomenon in nature, called diffusion of matter, which is often associated with the movement of atoms, ions, molecules, energy, etc., generally from a region of higher concentration to a region of lower concentration [1, 2].

In general, if the distribution of molecules in a medium is not homogeneous, flows will occur that tend to homogenize the system. Similarly, if the concentration of a solute in a solution is not homogeneous, the system will tend to evolve in order to homogenize it. Let n be the *concentration* of solute (which could be the number of moles per unit volume of solute) or the *number of molecules* per unit volume. Usually, n will be a function of the spatial and temporal coordinates, that is, $n(x, y, z, t)$. *Fick's first law* describes the relationship between particle fluxes per unit area, $\vec{J}$, and concentration gradients [1, 2]:

$$\vec{J} = -D \cdot \vec{\nabla} n. \tag{57.1}$$

Here D is the *diffusivity coefficient*, which will generally depend on the type of system under study and on the temperature. The minus sign indicates that the particle flows go from the regions of higher concentration to those of lower concentration. That is, these flows tend to homogenize the system, which is a consequence of the second principle of thermodynamics. This equation is analogous to *Fourier's law* for heat flow or diffusion of energy, Eq. (56.2).

On the other hand, if we consider an infinitesimal element of volume ΔV, the continuity or conservation of mass equation indicates that if there is a net flow toward this volume element, the density or concentration of this element will increase over time, that is,

$$\oiint \vec{J} \cdot d\vec{S} = -\Delta V \frac{\partial n}{\partial t} \quad \text{or} \quad \nabla \vec{J} = -\frac{\partial n}{\partial t}. \tag{57.2}$$

By combining this equation with Eq. (57.1), we obtain

$$\frac{\partial n}{\partial t} = D \cdot \nabla^2 n, \tag{57.3}$$

which is known as *Fick's second law* or *diffusion equation.*

It is interesting to note the analogy between Fick's second law and Fourier–Biot's law of heat diffusion, Eq. (56.6). In fact, both the diffusion of particles and the diffusion of heat are closely related to the second principle of thermodynamics. They are both examples of out-of-equilibrium thermodynamic effects, i.e., irreversible processes.

On the other hand, there are several microscopic theories that attempt to describe the value of the diffusion coefficient D in molecular or microscopic terms. In particular, Einstein's theory of Brownian motion establishes a simple connection for the non-turbulent diffusion of small spheres of radius a in a medium with viscosity η, from which [2]

$$D = \frac{k_B T}{6\pi\eta a}, \tag{57.4}$$

where k_B is the Boltzmann constant and T is the absolute temperature. Of course, this is a special relationship for a case of a small particle in a viscous medium, such as milk micelles in water.

57.2 Diffusion in one and two dimensions

For the one-dimensional case, Eq. (57.3) reduces to

$$\frac{\partial n}{\partial t} = D \cdot \frac{\partial^2 n}{\partial x^2}, \tag{57.5}$$

which can be solved by the method of separation of variables. It can be verified that a solution of this equation is given by [3, 4]

$$\begin{aligned} n(x,t) &= \frac{A_0}{2\sqrt{\pi D t}} e^{-\left(\frac{x^2}{4Dt}\right)} \\ &= \frac{A_0}{2\sqrt{\pi\tau}} e^{-\left(\frac{x^2}{4\tau}\right)} \quad \text{and} \quad \int_{-\infty}^{\infty} n(x,t)dx = A_0, \end{aligned} \tag{57.6}$$

where A_0 is a constant that depends on the boundary or initial conditions. The parameter $\tau = Dt$ describes the evolution over time of the diffusion process. The diffusivity coefficient D can thus be interpreted as *a parameter that determines the time scale in the process*. For $t = 0$, this solution leads to a singularity at the origin ($x = 0$). Physically, this solution would correspond to the case in which the ink drop that is initially dispersed is very small compared to the dimensions of the system, i.e., in which its original size can be considered negligible. Figure 57.1 illustrates the shape of the solution $n(x,\tau)$ for different values of τ.

Consider the expansion front of the drop, characterized by the coordinate point x_0, for which $n(x_0,t) = f \cdot n(0,t)$ is fulfilled at all times. Here, f is a factor $0 < f < 1$, i.e., the concentration of the solvent at this point is always a fraction f of the concentration at

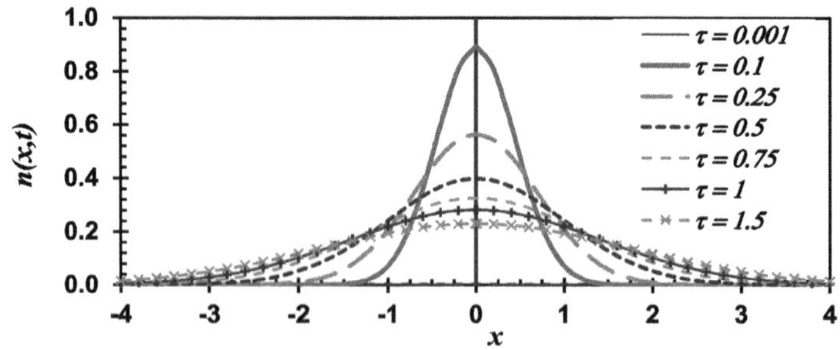

Figure 57.1 Solution of the one-dimensional diffusion equation for different values of the temporal parameter τ. The area of all these curves is the same, equal to A_0, Eq. (57.6).

the origin. From Eq. (57.6),

$$f = e^{-(x_0^2/4Dt)}, \text{ from where } x_0^2 = 4\tau \ln(1/f) = 4{\cdot}D{\cdot}t{\cdot}\ln(1/f). \quad (57.7)$$

Therefore, we expect the drop expansion front to advance in proportion to $t^{1/2}$.

Similarly, in the two-dimensional case, i.e., when a drop of ink diffuses within a flat, circular container, the cylindrical symmetry is preserved. Equation (57.3) can be written in cylindrical or polar coordinates as follows [3, 4]:

$$\frac{\partial n}{\partial (Dt)} = \frac{\partial n}{\partial \tau} = \frac{\partial^2 n}{\partial r^2} + \frac{1}{r}\frac{\partial n}{\partial r}, \quad (57.8)$$

where, as before, $\tau = Dt$ and the coordinate $r^2 = x^2 + y^2$ is the distance from any point on the plane to the coordinate origin. It can be verified that a solution of this equation is given by

$$n(x,t) = \frac{A_0}{2Dt} e^{-\left(\frac{r^2}{4Dt}\right)} = \frac{A_0}{2\tau} e^{-\left(\frac{r^2}{4\tau}\right)}. \quad (57.9)$$

This is the solution of the diffusion equation in two dimensions, with cylindrical symmetry and corresponding to the case of a drop of ink or a grain of solute with very small initial dimensions compared to the physical system under study. As in the one-dimensional case, the diffusion front propagates proportionally to $t^{1/2}$.

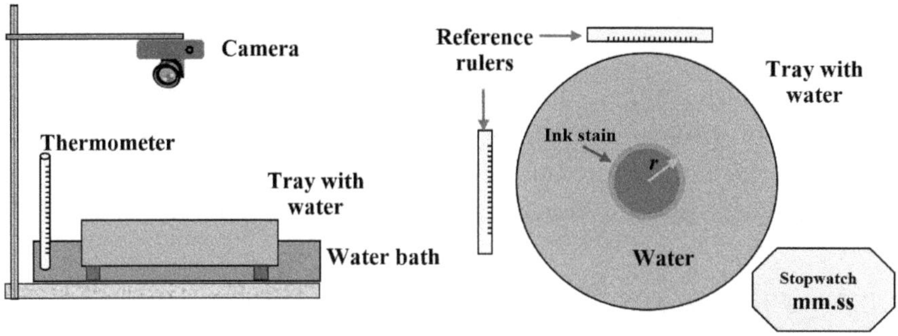

Figure 57.2 Experimental arrangement to study diffusion in two dimensions. The digital camera records the size of the expanding spot and the time displayed by the stopwatch. The thermometer measures the temperature of the water bath.

57.3 Diffusion in water

To test the conjectures discussed in the previous section, we can perform an experiment using an arrangement like the one in Figure 57.2. It consists of a glass or aluminum tray (or similar container), preferably round, containing water approximately 1 cm deep. This tray is placed inside another larger one that contains water that serves as a thermostat or thermal bath for the first tray. A thermometer is used to measure the temperature of the bath.

A scale (ruler), placed beside the tray, serves to calibrate the distances as discussed in Chapter 10. The stopwatch is used to record time. If the camera has an internal stopwatch that records the time in the frame, it can be used instead of the external stopwatch. The camera takes pictures as the ink stain, or a grain of potassium permanganate, dissolves and expands. Potassium permanganate is a non-toxic oxidant, disinfectant, and deodorizer, which is very soluble in water [5]. It has an intense fuchsia color, which is very easy to see and photograph. To prevent the outer edge of the tray from influencing the problem, it is advisable to take measurements only until the radius of the stain, r, is approximately half the radius of the tray. For sharp images, mount the camera on a tripod, and use a lamp to optimize the contrast. Once we have seeded a small grain of potassium permanganate or drop of ink, we can vary the position and orientation of the lamp until we get sharp, clear photos.

Potassium permanganate can be purchased from any chemical laboratory or pharmacy. Finding an appropriate ink is largely a matter of trial and error, but fountain pen ink (*Quink blue ink*) or stamp pad refill ink may be suitable for this experiment.

57.3.1 *Project 193. Diffusion of potassium permanganate or ink in water*

Recommended equipment: A preferably circular tray approximately 25 cm in diameter or larger. A stopwatch with clearly visible digits. An ordinary digital camera. A few tens of grams of potassium permanganate ($KMnO_4$), ink, or dark tempera. A water bath to stabilize the tray temperature and a thermometer.

Using an experimental arrangement like the one shown in Figure 57.2, optimize the lighting until you achieve photos in which the spot under study is clearly visible.

Suggested tasks:

- ✓ Measure the temperature of the thermal bath and make sure that it does not vary while the stain spreads.
- ✓ Place a small grain of permanganate or a small drop of ink near the center of the tray and start taking pictures of the system. At the beginning of the diffusion, it is advisable to take photos in more rapid succession than toward the end, because as the spot increases in size, it spreads more slowly. The idea is to have at least 15–20 frames from the start of the spread until it reaches the middle of the radius of the tray.
- ✓ For each frame, measure the radius of the spot, always using the same criteria to define its edge, and estimate its uncertainty. For each frame, record in a table the time t, the radius r, and the uncertainty Δr of each stain.
- ✓ Plot r and r^2 as functions of t. In each of these graphs, include the error bars in the determination of r and r^2.

✓ If the plot of r^2 as a function of t is linear, draw the line that best fits your data. From the slope, find the value of the diffusivity coefficient, D, estimate the error of its determination, see Eq. (57.7), and assume that $f = 0.5$. Does this assumption seem reasonable to you?

✓ Using different solutes — ink, tempera, or permanganate — repeat the experiment. For each case, determine the diffusivity coefficient D and its uncertainty.

57.3.2 *Project 194. Variation of the diffusivity coefficient with temperature*

Recommended equipment: A preferably circular tray approximately 25 cm in diameter or larger. A stopwatch with clearly visible digits. An ordinary digital camera. A few tens of grams of potassium permanganate, ink, or dark tempera. A water bath to stabilize the tray temperature, and a thermometer.

Use an experimental setup like the one shown in Figure 57.2, following the same instructions as in the previous project, but this time varying the temperature of the water bath by adding ice or hot water to it. Take a diffusivity coefficient measurement for at least five temperatures, as widely scattered as possible between 0°C and 90°C.

Suggested tasks:

✓ Measure the temperature of the thermal bath and make sure that it does not vary while the spot is spreading.

✓ Add a small grain of permanganate or a small drop of ink. Get at least 15 frames from the beginning of the spread until the stain reaches approximately half the radius of the tray.

✓ For each frame, measure the radius and estimate its uncertainty. For each photo, record in a table the time t and the radius r of each spot.

✓ Plot r^2 as a function of t.

✓ If this graph is linear, draw the line that best fits your data. From the slope, find the value of the diffusivity coefficient, D, and estimate the uncertainty of its determination.

✓ For each substance studied, graph the measured value of D and its error as a function of absolute temperature. Do your data follow a linear trend? What can you conclude from the model described by Eq. (57.4) to describe your experimental data?

57.3.3 *Project 195. Diffusion of potassium permanganate or ink in water*

For this task, you can use the frames you obtained in any of the previous activities. The objective here is to study the concentration profile of the ink or permanganate in the photos taken. As discussed in Chapter 47, there are different programs that can determine the intensity profile of a digital image. A digitized image can be analyzed using a program that can measure pixel intensity along a line (Intensity Profile) [7]. This type of analysis can be performed with specific programs, such as Able Image Analyser [6], MathWorks Matlab® [8], or ImageMeasurements [9], among others. An example of such a migration pattern is illustrated in Figure 57.3. The ruler in these images enables you to convert the pixel scale into real distances. With these tools, it is possible to obtain the intensity profile of a digital photo along a line, as shown in Figure 57.3.

Suggested tasks:

✓ For each series of photographs of the diffusion of the ink or permanganate taken in the previous activities, obtain the intensity profile of the figure of the stain, along a line that passes through the center.

✓ In each case, compare the intensity profile measured to the one predicted by the solution of the differential Eq. (57.9).

✓ What can you conclude about the model described to account for the profiles of the diffusion diagrams?

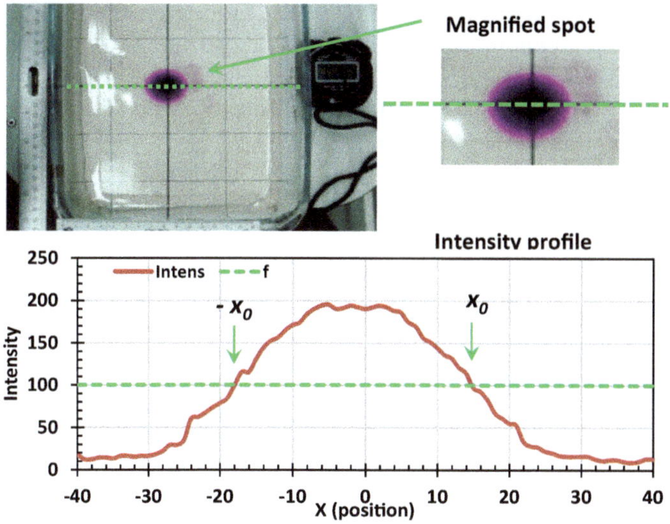

Figure 57.3 Top panel (left): Photo of the experimental setup, including the reference rulers and the chronometer to register the time. Top panel (right): Amplified image of the diffusion spot (permanganate). Lower panel: Intensity profile of diffusion spot (upper right) along the horizontal green line indicated in the upper panel [6, 7].

References

[1] Wikipedia, "Diffusion," 2022. Available: https://en.wikipedia.org/wiki/Diffusion.

[2] F. Reif, *Fundamentals of Statistical and Thermal Physics*, NY: McGraw-Hill, 1968.

[3] C. R. Wylie, *Higher Mathematics for Engineering*, 2nd Ed., NY: Mc Graw-Hill, 1996.

[4] L. A. Pipes, *Higher Mathematics for Engineers and Scientists*, NY: Mc Graw-Hill, 1970.

[5] Wikipedia, "Potassium permanganate," 2022. [Online]. Available: https://en.wikipedia.org/wiki/Potassium_permanganate.

[6] "Able image analyser," Internet Achive, 2021. [Online]. Available: https://archive.org/details/tucows_344088_Able_Image_Analyser.

[7] The MathWorks, Inc., "Intensity profile of images," 2022. [Online]. Available: https://www.mathworks.com/.

[8] A. Ramil, A. J. López and F. Vincitorio, "Improvements in the analysis of diffraction phenomena by means of digital images," *Am. J. Phys.*, vol. 75, p. 999, 2007.

[9] Vernier Software, "Logger pro and Vernier video analysis," 2021. [Online]. Available: www.vernier.com.

MODULE V

MODERN PHYSICS AND ASTROPHYSICS MODULE

Chapter 58

MICHELSON'S EXPERIMENT

This chapter studies the operation of a Michelson interferometer, both in its optical version and its acoustic analog known as Quincke's tube. These devices are then used to measure wavelength and small distances.

Goals

✓ Michelson interferometer
✓ Quincke's tube

58.1 The Michelson interferometer

The Michelson interferometer is one of the best-known interference instruments, both for its multiple applications and for its implication in the development of physics. Using this device, Michelson and Morley carried out their famous experiment in 1887, at today's Case Western Reserve University in Cleveland, Ohio, USA. This experiment originally attempted to detect the relative movement of the Earth through the stationary luminiferous *ether*, i.e., to detect the *ether wind*. The negative results they obtained showed the inconsistency of the *ether theory*. This negative result contributed to the development of the theory of relativity [1–3].

Figure 58.1 shows a schematic diagram of an interferometer, which can be hand-crafted from recycled components from other equipment [4, 5], or from components purchased specifically for the purpose [6]. Of course, there are also ready-to-use commercial models [7, 8].

This interferometer is based on the principle of dividing a light beam by a *beam splitter* or semitransparent mirror M_s. At M_s, the

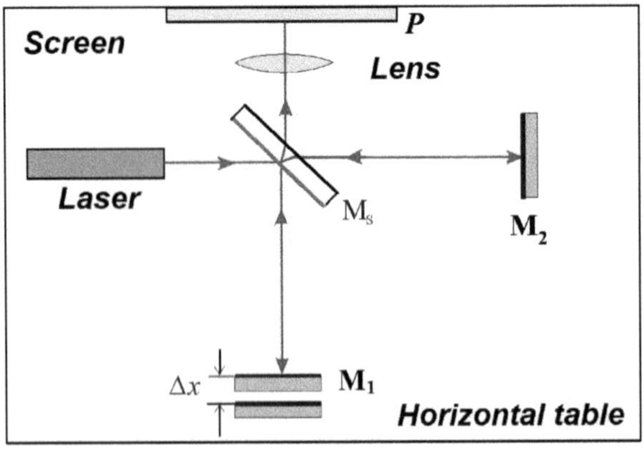

Figure 58.1 Schematic diagram of the Michelson interferometer. M_1 and M_2 are two first surface mirrors. M_S is a beam splitter (semitransparent mirror or simply a piece of glass). The beams are recombined on the screen, where the interference patterns can be seen. The distance Δx is the path difference of the two beams converging at P. The lens serves to increase the size of the image and make the interference image more visible.

incident beam is partly transmitted to the mirror M_2 and partly reflected to the mirror M_1, at which both beams are reflected. After being transmitted and reflected again at M_s, the beams are recombined on the screen P. Depending on the difference in paths these beams have traveled, the interference at P will be constructive or destructive. In fact, if the mirrors M_1 and M_2 are not exactly perpendicular — which is the usual case — each portion of the beam that is reflected in different parts of the mirrors will reach the screen after having traveled different lengths. This will result in a set of bright and dark lines appearing on the screen. Their orientation will depend on how the mirrors, M_1 and M_2, are fine-tuned. So, in order to achieve useful results with this instrument, it is important to have good mirror positioners that enable the positions and angular orientations of both mirrors to be adjusted precisely.

The mirrors M_1 and M_2 must be first surface mirrors, that is, the silver or aluminized surface must be on their front face. The mirrors and positioners can be handcrafted or purchased from commercial suppliers [6].

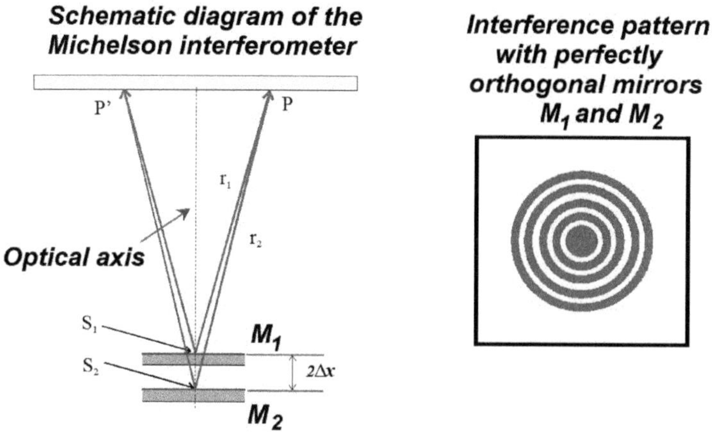

Figure 58.2 Left: Schematic diagram of the Michelson interferometer. S_1 and S_2 are the two images of the source formed by the mirrors M_1 and M_2. Right: Interference pattern when the mirrors are perfectly perpendicular [9].

When mirrors M_1 and M_2 are perpendicular, concentric bright and dark rings are seen on screen P. The reason why these rings form can be understood by observing that the mirrors M_1 and M_2 have the effect of producing two images of the source: S_1 and S_2, separated by a distance $2\Delta x$, as illustrated in Figure 58.2. Here, Δx is the difference in length of the interferometer arms delimited by M_s to M_2 and M_s to M_1. The origin of the rings is associated with the fact that the difference between the paths r_1 and r_2 varies with the position of point P on the screen. However, this difference in path is the same for any circumference centered on the optical axis of the interferometer, Figure 58.2 [9].

Considering only the point corresponding to the center of the interference pattern, that is, the point on the screen through which the optical axis passes, it is possible to predict the intensity of light at this point from the difference in the path of the light rays. More specifically, as discussed in Chapter 42, the phase difference of the rays that are reflected in M_1 and M_2 will be

$$\Delta\varphi = \frac{2\pi}{\lambda} 2\,\Delta x, \tag{58.1}$$

therefore, using the principle of superposition, the average intensity will be [3]

$$I = I_0 \cos^2\left(\frac{\Delta\varphi}{2}\right) = I_0 \cos^2\left(\frac{2\pi}{\lambda}\Delta x\right), \tag{58.2}$$

where Δx is the difference in length of the interferometer arms. If one of the mirrors M_1 or M_2 is mounted on a mobile linear positioner, it is possible to vary the distance Δx in a controlled and measurable way. If one of the beams passes through a medium with a variable refractive index, for example, if a glass plate is inserted into one of the arms of the interferometer, Eq. (58.2) is valid, provided that Δx is the difference between the optical paths ℓ_{op}, defined as follows:

$$\Delta x = \ell_{op}^{(1)} - \ell_{op}^{(2)} \quad \text{with} \quad \ell_{op}^{(i)} = \int_i n(l)dl, \tag{58.3}$$

where i $(= 1, 2)$ denotes each of the paths followed by the beams of each branch of the interferometer.

To visualize the interference images more easily, it is advisable to use a converging lens between the interferometer and the screen, to magnify the size of the interference pattern. Another precaution to take into account when preparing the interferometer is its correct alignment. A practical way to proceed is to first align the mirror M_2 (Figure 58.1). Do this by blocking reflection from M_1 and making the beam reflected by M_2 go back to the laser source itself, that is, the reflected image located at the exit point of the laser itself. Next, position M_s. The position of M_s must be such that the ray reflected at 90° reaches the central area of the mirror M_1. Once you have achieved this, adjust M_1 by ensuring that the beam reflected on M_1 falls on the screen and overlaps with the beam reflected by M_2. Only when this condition is achieved is it advisable to insert the lens that will increase the size of the image on the screen. This lens must be well centered with the beam, as shown by the fact that the center of the image with and without the lens does not vary in position. Once the images reflected by the two mirrors are clearly displayed overlapping with each other (the one reflected on M_1 and

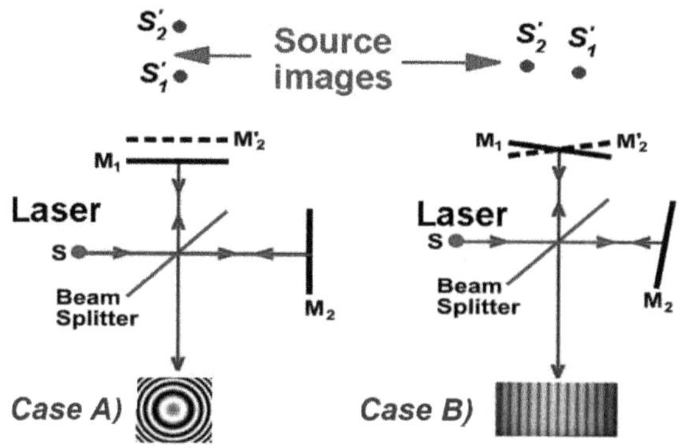

Figure 58.3 Schematic diagram of the Michelson interferometer. S_1' and S_2' are the two source images of S (laser) formed by M_1 and M_2. In Case (a), the mirrors M_1 and M_2 are perpendicular and the resulting interference pattern is the rings shown at the bottom. In Case (b), the mirrors M_1 and M_2 are not perpendicular, and the resulting pattern is a succession of light and dark lines, which can be interpreted as a portion of the rings seen far from their center. By varying the orientation of the mirrors and the beam splitter, you can switch from one case to the other.

the one reflected on M_2), adjust M_1 finely until the interference bands appear. Figure 58.3 illustrates the type of interference figures that can be obtained.

The final proof that you are indeed visualizing an interference pattern consists of alternately blocking one of the reflected beams. By blocking one of them, the interference bands disappear and only a uniformly illuminated region is observed, but by allowing both beams to reach the screen, the presence of the interference image characteristic of the interferometer will be noticed.

This simple operation provides evidence of one of the most fundamental principles of waves: the *superposition principle*, see Chapter 42. In order to obtain the final intensity of the superposition of two (or more) waves, we must add their amplitudes. The final intensity is proportional to the square of this amplitude. The superposition may thus result in *constructive* interference, in which case a given point or band will be brighter, or *destructive* interference, which produces a dark point or band. This explains why sometimes *light+light* results in more illumination (bright band), while other times *light+light* results in a dark band, as shown in this experiment.

58.1.1 *Project 196. Determining the wavelength of a laser*

Recommended equipment: A Michelson interferometer with one of its mirrors mounted on a graduated linear positioner and resolution of 10μm or less. A laser or laser pointer.

This experiment requires one of the mirrors (M_1 or M_2) to be mounted on a linear positioner so that it can be moved in a controlled, parallel manner with a micrometer or dial gauge. Thus, by means of a comparator or micrometric screw associated with the positioner, you can ascertain the changes made in Δx. For this experiment, it is advisable to have good visualization of the interference figures, for which it is useful to place a lens in front of the screen and adjust its position to optimize the interference pattern on the screen.

Suggested tasks:

✓ With the micrometric screw, which adjusts the position of the mirror M_1 (i.e., measures the distance Δx), make fine adjustments and observe how, by varying this distance, the bright areas occupy the positions of the dark ones and vice versa.

✓ Vary the position of the moving mirror M_1 (i.e., vary Δx) and count the number N of bright (or dark) bands that appear as you do so. Plot Δx as a function of N.

✓ According to Eqs. (58.1) or (58.2), two consecutive maximums (or consecutive minimums) are associated with $\Delta x = \lambda/2$. This means that the slope of the graph of Δx as a function of N will have a slope equal to $\lambda/2$. Using this property, determine the value of the wavelength of the laser used and its uncertainty. How does this λ value compare to the specifications of your laser?

✓ If the wavelength of the light is known, this procedure enables you to calibrate the micrometer that regulates the position of the mirror, taking the wavelength of the light used as a standard of length. In fact, a modern definition of meter is based on this type of technique [10]. This method can also be used to measure small dynamic displacements [11].

58.1.2 *Project 197. Effect of the state of polarization*

Recommended equipment: A Michelson interferometer. A laser or laser pointer. Two polarizing foils.

The aim of this experiment is to study the effect of the state of polarization of light on the interference patterns of a Michelson interferometer. In this experiment it is useful to have polarizing sheets mounted on rotating supports, to enable you to vary the state of polarization of a beam easily. To define the state of polarization, take as reference the direction normal to the plane of the interferometer. For this experiment, it is advisable that the incident light should not be polarized or that the state of polarization should form an angle of approximately 45° with respect to the direction taken as reference. Place the polarizers between the beam splitter and each of the mirrors M_1 and M_2. It is thus possible to choose a different polarization state for each of the incident beams on each mirror. Designate the state of polarization corresponding to the electric field that oscillates perpendicular to the plane of the interferometer (reference direction) as the state s. Designate the state of polarization in which the electric field oscillates parallel to the plane of the interferometer as the state p. The objective of this experiment is to study how the interference image is modified with the polarization state (s or p) of each beam. To ensure that any interference bands seen are genuine, slightly vary the position of one of the mirrors across the Δx and verify that the interference bands move.

Suggested tasks:

✓ First, align the interferometer without the polarizers until the interference bands are visible. Using a polarizer at the laser output, make sure that the laser is not polarized or, if it is, that it is neither in a pure s nor p, but has components in both states (preferably at 45° with respect to the reference direction).

✓ Study how the interferometer interference image is modified if both light beams are in the following polarization states:

- both in state s,
- both in state p,
- one beam in state s and the other in state p.

✓ You will notice that in the latter situation (overlap of modes s and p) no interference is observed. How can you explain your results?

58.2 Acoustic analog of Michelson's interferometer — Quincke's tube

Quincke's tube is an acoustic interferometer with an operating principle analogous to that of the Michelson interferometer [12, 13]. A schematic diagram of this acoustic interferometer is shown in Figure 58.4. It can be built with metal pipes (copper, bronze, etc.)

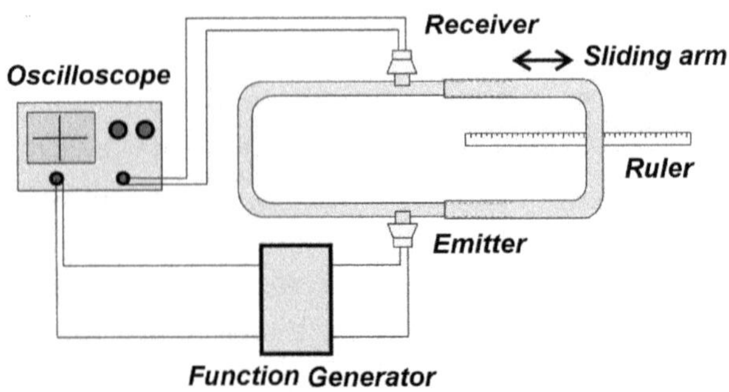

Figure 58.4 Schematic diagram of Quincke's tube acoustic interferometer. The right arm of the system can be moved on a graduated ruler, like a trombone slide [12].

or plastic pipes of the kind usually used in water installations. The internal diameters of the tubes should be about 1–3 cm. One of the arms of the system moves (glides) like a trombone slide.

The source of sound (emitter) must be monochromatic. It can be a loudspeaker powered by a function generator or a smartphone with some application (such as Phyphox [14]) that generates a monochromatic sound, as discussed in Chapter 42. The sound from the emitter follows two different paths, along the right arm and the left arm. The receiver picks up the combination of the two waves. The receiver could be either a microphone connected to an oscilloscope or a second smartphone acting as a sound detector [14]. The intensity detected by the receiver will depend on the difference in length between the arms, Δx. Therefore, we will have, as before, a phase shift given by

$$\Delta\varphi = \frac{2\pi}{\lambda} 2\,\Delta x. \tag{58.4}$$

and two consecutive maximums (or minimums) of intensity will be separated by a distance $\Delta x = \lambda/2$. By measuring the x positions for which maximums appear, you can determine the wavelength of the sound. Since the frequency f imposed by the function generator or smartphone is known or measurable, you can calculate the speed of sound, c, as follows:

$$c = \lambda\, f. \tag{58.5}$$

58.2.1 *Project 198. Acoustic analog — Quincke's tube*

Recommended equipment: A Quincke's tube. A function generator with a speaker. A microphone. A two-channel oscilloscope. Alternatively, this experiment can be performed using two smartphones and Quincke's tube.

Build an acoustic interferometer similar to that shown in Figure 58.4. It can be built with metal pipes or plastic pipes commonly used in water installations. One of the arms of the system moves like a trombone slide.

Suggested tasks:

✓ Vary the position of the mobile arm and verify the occurrence of maximum and minimum intensity. If you use an oscilloscope, observe both the input signal and the receiver response on it.

✓ Using the experimental setup shown in Figure 58.4 and audio frequencies in the emitter of about 2 kHz, determine the wavelength and speed of sound. Estimate the uncertainties associated with these measurements.

✓ Compare the value of c found to the values in the table at the corresponding temperature.

References

[1] F. Sears, M. Zemansky, H. Young and R. Freedman, *College Physics*, NY: Pearson Ed., 2009.

[2] M. Alonso and E. J. Finn, *Physics, Vol. II, Fields and Waves, and Vol. III, Quantum and Statistical Foundations*, NY, Addison-Wesley, 1967.

[3] E. Hecht, Optics, NY: Addison Wesley, 1990.

[4] Palisades Charter High School, "Michelson interferometer demonstration," 2021. Available: https://www.youtube.com/watch?v=VOipufe 6fGQ.

[5] C. R. D. Souza, "Homemade Michelson interferometer — USP," 2021. Available: https://www.youtube.com/watch?v=-SZSkIcEyCI.

[6] Edmund Scientifics Co., "Edmund scientifics," 2022.! Available: https://www.edsc.com.

[7] Pasco Scientfic Co., "Pasco scientific," 2022. Available: https://www. pasco.com/.

[8] Sargent Welch Co. Buffalo, NY., "Sargent Welch catalog," 2022. Available: https://www.sargentwelch.com/store/.

[9] TSG Physcis, "Michelson interferometer," 2022. Available: https:// www.youtube.com/watch?v=j-u3IEgcTiQ.

[10] Wikipedia, "Metre," 2022. Available: https://en.wikipedia.org/wiki/ Metre.

[11] J. Diamond, *et al.*, "Measuring small vibrations with interferometry," *Am. J. Phys.*, vol. 58, p. 919, 1990.

[12] Univ. of California-Berkeley, "Acoustic interference with Quincke (trombone) tube and sonalert," Pysics @ Berkeley, 2022. Available: http://berkeleyphysicsdemos.net/node/181.

[13] J. E. Fernández and E. Galloni, *Trabajos prácticos de Física*, Buenos Aires: Nigar, 1968.
[14] Physphox — RWTH Aachen University, "The smartphone as a mobile lab," RWTH Aachen University, 2022. Available: https://phyphox.org/.

Chapter 59

FERROMAGNETIC MATERIALS — PHASE TRANSITIONS

This chapter studies the properties of ferromagnetic and ferrimagnetic materials and, in particular, the phenomenon of hysteresis and the magnetic phase transition in these materials. It discusses ways to determine the transition temperature or Curie temperature of a ferromagnetic material.

Goals

- ✓ Ferromagnetic and ferrimagnetic materials
- ✓ Transition or Curie temperature

59.1 Ferromagnetic and ferrimagnetic materials

Some solid materials, such as iron, can form permanent magnets or be attracted to them. The magnetic interaction forces in these materials are remarkable and intense. The most notable properties of *ferromagnetism* (including ferrimagnetism) are observable with the naked eye. Some substances, such as *paramagnetic* and *diamagnetic* substances [1, 2], show a fainter response to magnetic fields. Paramagnetic (diamagnetic) materials are slightly attracted (repelled) by a magnetic field [3]. These effects can only be detected by sensitive instruments.

Ferromagnetic and *ferrimagnetic* [4] materials show some microscopic *ordering*, even in the absence of an applied magnetic field. Below a certain critical temperature, called the *Curie temperature* [5] T_C, ferromagnetic or ferrimagnetic samples have "spontaneous magnetization" and strongly interact with a magnet.

These transformations are examples of changes in microscopic atomic order, similar to the change of state of aggregation of matter. The difference is that in the latter there is a latent heat of transformation, and therefore the heat capacity ($C_p = (\partial U/\partial T)_P$) presents an infinite discontinuity at the transition temperature, characteristic of a *first-order phase transition.* In contrast, in ferromagnetic or ferrimagnetic phase transformations, there is no latent heat, but there is a finite discontinuity in Cp, characteristic of *second-order phase transitions* [2, 6, 7].

While ferromagnetic materials are good conductors of electricity, ferrimagnetic materials are ceramic materials that generally do not conduct electricity. They are commonly used as the core of transformers or coils. Magnetite (Fe_3O_4) is an example of material that has ferrimagnetic behavior.

Iron samples — which are ferromagnetic — sometimes show no perceptible interaction with each other, so they do not appear to be magnetized. However, they are strongly attracted by a permanent magnet. This is because in the absence of an applied field, energetic conditions favor these materials to subdivide into small, uniformly magnetized *domains* of macroscopic size, whose spontaneous magnetization vectors point in different directions [2, 6]. This produces a macroscopic cancellation of the magnetic moments of the individual domains. The presence of a magnetic field leads the domains to reorient and tend to align with the applied field, and to change in size (some grow at the expense of others decreasing in size). If the applied field is *weak*, when it is removed, the domains return to their original states and the sample behaves again as a "degaussed" sample. If the field is *strong*, the domains that have been favored by alignment in the direction of the field do not return to their primitive states [2, 6]. This gives rise to the phenomenon of *magnetic hysteresis* in this type of materials [2, 6, 7]. Thus, after a magnetic field is applied and removed, the sample can be left with remanent magnetization M_0 different from zero.

Moreover, the magnetic order decreases with increasing temperature, and spontaneous magnetization cancels out above a certain temperature, known as the *Curie temperature*, T_C. For $T < T_c$, the magnetization of the sample M_0 can be described by a potential law of the form [8]

$$M_0(T) \propto [(T - T_C)/T_C]^\beta, \qquad (59.1)$$

where β it is a parameter that ranges from 0.3 to 0.5. For ferrimagnetic materials [4], magnetization behaves as described by Eq. (59.1). References [6] and [7] provide a detailed discussion of the similarities and differences between ferromagnetic and ferrimagnetic materials.

59.2 Hysteresis curve

In the presence of material, it is helpful to introduce the *magnetic excitation* vector H, also called *magnetic field strength* or simply "H field", which generally describes the contribution to the magnetic field, associated with conventional or free currents, that is, those that circulate through cables and can be measured with ammeters, interrupted with switches, and so on. The *magnetic flux density or magnetic induction* field B depends on both the field H and the *magnetization* M of the material. The latter represents the contribution to the magnetic field produced by the atoms and molecules of the material. More specifically, M is the *magnetic dipole moment per unit volume* of the material. In a vacuum, or with paramagnetic or diamagnetic materials, the relationship between B and H is linear and can be written as follows [9, 10]:

$$\vec{B} = \mu_0\mu_r\vec{H} = \mu_0(\vec{H} + \vec{M}) = \mu_0(1 + \chi_m)\vec{H}, \qquad (59.2)$$

where μ_r $(= 1 + \chi_m)$ is the relative permeability of the material medium, and χ_m is the magnetic susceptibility. In the void, $\mu_r = 1$; for paramagnetic materials, $\mu_r \geq 1$ $(\chi_m > 0)$ and for diamagnetic materials, $\mu_r \leq 1$ $(\chi_m < 0)$. In ferromagnetic and ferrimagnetic materials, in addition to being non-linear, the relationship between B and H depends on the magnetic history of the sample, [5, 6] as schematically illustrated in Figure 59.1.

The value of the magnetic field when $H = 0$, B_r, is called *remanence* or *residual magnetization* and is represented by the point

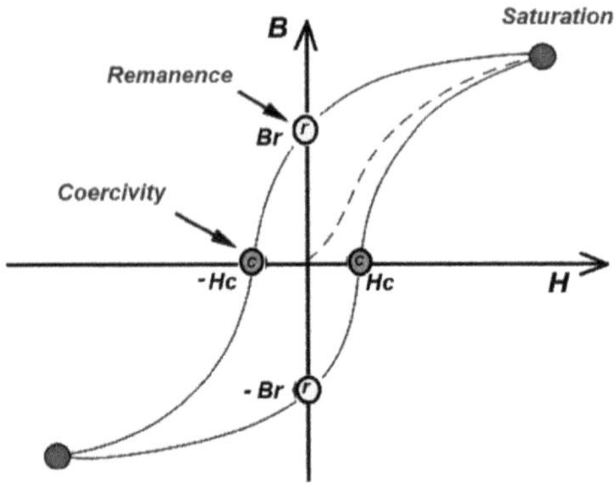

Figure 59.1 Hysteresis curve for a ferromagnetic material. Note that if the material starts from a point where it is originally demagnetized (dashed line) the magnetization path is different from the one it follows if it starts from non-zero magnetization.

"r" in Figure 59.1. Its value is a consequence of the spontaneous magnetization of the material. In other words, the value of the remanent field Br is closely linked to the magnetization of the material, M_0, in the absence of magnetic excitation.

Thus, if we wish to study how the spontaneous magnetization of a ferromagnetic or ferrimagnetic material varies with temperature, we can analyze how the hysteresis curve varies with temperature. When a material of this type is heated, at temperatures above the Curie temperature, T_c, the material behaves like a paramagnetic sample and the hysteresis curve disappears to give rise to a linear relationship between B and H.

This chapter discusses different ways to determine the hysteresis curve and then uses this technique to determine the Curie temperature of a ferrimagnetic sample [11, 12]. In principle, this technique could also be used for a ferromagnetic material, but the difficulty is that most common ferromagnetic materials have Curie temperatures higher than about 600°C.

59.3 Determining the hysteresis curve

A possible experimental arrangement to determine the hysteresis curve of a material is shown in Figure 59.2 [11, 12]. The magnetic

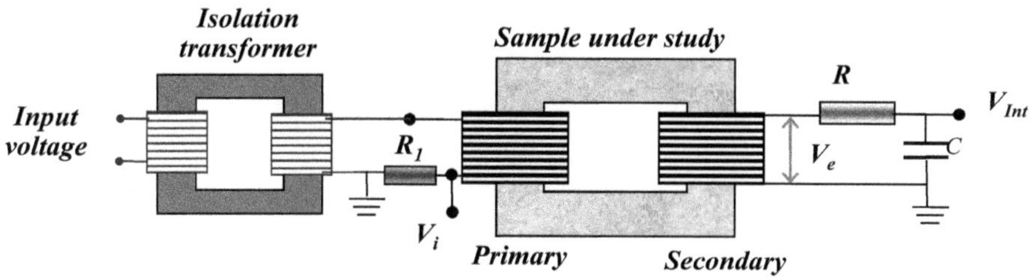

Figure 59.2 Diagram of an experimental setup to study the hysteresis curve of a sample. The isolation transformer is used to avoid electrical contact between the input voltage and the measuring instruments. It also serves to protect the experimenter from exposure to line voltage.

sample is a toroid made of ferrite or a ferromagnetic material such as iron or nickel. Ferrite cores in the shape of toroid are available from electronics stores [13].

The toroidal sample can be about 5–10 cm in diameter. In the case of iron, you could screw together four iron bars to build a yoke like the one shown schematically in Figure 59.2.

The isolation transformer is used to protect the experimenter and equipment from contact with input voltage. It is therefore important that this device should truly be a transformer with electrical isolation from the primary to the secondary. A 110 V or 220 V–12 V transformer of about 15 W may be suitable. The voltage that feeds the transformer can be line voltage or, better still, the output of an autotransformer [14] or Variac, so that the supply voltage can be regulated [15].

If you have a function generator with a power amplifier that can supply a current of the order of 1 A, this combination of instruments could replace the Variac and the isolation transformer.

The toroidal sample should have a primary of about 100–300 turns of enameled copper wire and a secondary of about 200–400 turns. An example of this device is illustrated in Figure 59.3. The number of turns of the secondary, N_2, should be greater than in the primary, N_1, so that the signal to be measured is amplified. The primary can be made with enameled copper wire of diameter $\phi = 0.2$–0.5 mm. The secondary can be made of thinner wire, of diameter $\phi = 0.1$–0.2 mm. Annex A discusses in detail the relationship between the

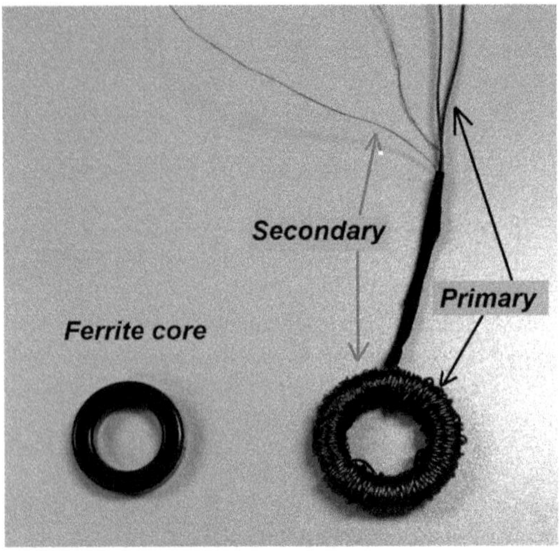

Figure 59.3 Ferrite toroidal core ("torus"), unwound and with superimposed primary and secondary windings. The secondary wire is thinner than the primary wire.

diameters of the wires, the number of turns of each winding and the dimensions of the torus.

The primary of the torus is connected to the secondary of the isolation transformer as illustrated in Figure 59.1. Do not connect the toroid winding directly to line voltage (110–220 V), as it could destroy the winding and endanger you or your colleagues.

The voltage drops V_{R1} on a resistor R_1, placed in series with the primary, provides a signal proportional to the current I_P in the primary ($V_i = I_P R_1$), and thus, a signal proportional to the value of H in the torus. The resistance R_1 can be a few ohms (2–10 Ω and Power > 5 W). Make sure that this resistor can dissipate the power that will be produced in it [16].

If you apply AC voltage to the primary, V_i, in the secondary there will be an induced *emf*, V_e, which according to Faraday's law, is

$$V_e = -N_2 \frac{d\Phi_B}{dt} = N_2 A_2 \frac{dB}{dt}, \tag{59.3}$$

where Φ_B is the magnetic flux in the sample (torus), N_2 is the number of turns of the secondary, A_2 is its area, and B is the mean field inside the torus. Equation (59.3) shows that the integral of emf V_e is proportional to the value of B in the torus. There are several ways to integrate the secondary signal and obtain a voltage proportional to B. The simplest is to use an RC circuit with a time constant $\tau \gg 2\pi/f$, where f is the frequency of the applied voltage. If $f = 50\text{–}60$ Hz is used, with $\tau \geq 0.1$ second, adequate integration is achieved (see Annex B). Of course, an active integrator based on op amps can also be used. The output of the integrator circuit is connected to the input of an oscilloscope or data collection system connected to a computer to measure the signal V_{Int} (integrated) $\propto B$. Thus, with the signal V_i connected to one channel of an oscilloscope (or data collection system) and V_{Int} to the other channel, using the oscilloscope in XY mode, the hysteresis curve of the sample can be observed. The value of V_{Int} when $V_i = 0$, (V_{Int}^0) is proportional to M_0, that is, to the spontaneous magnetization of the sample. By regulating the input voltage to the primary of the isolation transformer with a Variac, the conditions of study of this curve can be varied, approaching the saturation of the material.

59.3.1 *Project 199. Measuring the hysteresis curve for iron*

Recommended basic equipment: An iron torus approximately 5–10 cm in diameter. The torus can be substituted by a set of four iron bars forming a rectangular torus. Enameled copper wire with diameter 0.15–0.5 mm. A dual input oscilloscope or an acquisition system with two independent inputs. A transformer from 110 or 220 V–12 V. A Variac. Resistors 5 Ω @ 5 W.

The experimental device is shown schematically in Figure 59.2 [15]. The goal of this activity is to observe the hysteresis curve. Channels 1 and 2 of the oscilloscope are connected to measure the signals V_i and V_s in XY mode.

Suggested tasks:

- ✓ Study the variation of the hysteresis curve of the sample under study for different values of the input voltage.
- ✓ Increase the input signal so that it is as close as possible to the saturation conditions of the sample, consistent with the limitations of your circuit.
- ✓ Decrease the applied voltage and watch the hysteresis cycle decrease or change.
- ✓ Describe and qualitatively explain your observations.

59.3.2 *Project 200. Estimating the Curie temperature of a ferromagnetic sample*

Recommended basic equipment: A nickel sample weighing 3–15 g. A thin, flexible wire thermocouple. A powerful magnet. A gas Bunsen burner.

For this experiment, it is advisable to solder the end of the thermocouple to the nickel sample using a spot or silver solder [17]. Another alternative is to make a small hole in the sample and attach the tip of the thermocouple mechanically and thermally to the sample. One way to do this is to fasten the sample with a small brass bolt (i.e., a non-magnetic screw). The aim is to be able to hang the sample from the thermocouple. The wires should be flexible and as thin as possible, approximately 50 cm long, to allow a pendulum movement of the sample. Canadian 5¢ coins, which are ferromagnetic, can serve as a sample for this experiment [17]. A possible experimental arrangement is shown in Figure 59.4.

The reason for using nickel is that its Curie temperature, T_C, is about 360°C, so it is simpler to heat than iron, which has a Curie temperature close to 770°C [18]. When the sample is in ferromagnetic state, it is attracted by the magnet and its equilibrium position is position A in Figure 59.4. As it warms, the sample suddenly goes into paramagnetic state, and its position goes to B. The temperature recorded at that instant is the Curie temperature, T_C.

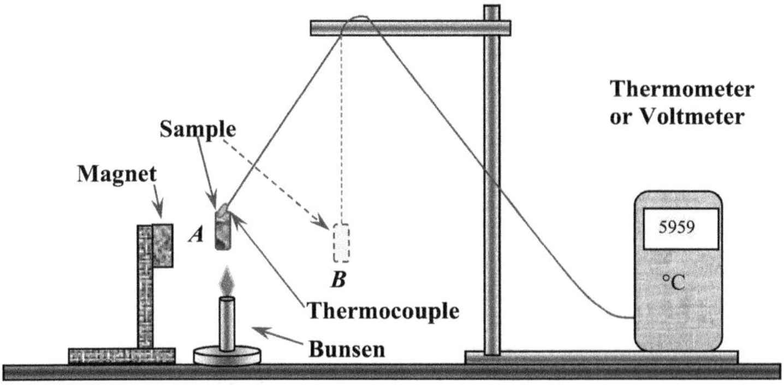

Figure 59.4 Diagram of an experiment to determine the Curie temperature of a ferromagnetic sample. When the sample undergoes the transformation from ferromagnetic to paramagnetic, it passes from position A to B. The temperature recorded at that instant is T_C.

Another way to record this temperature is to continue heating the sample up to about 20°C–30°C beyond the Curie temperature, leaving the sample in position B. While it is still hot, place a second small, powerful magnet on a platform a few millimeters away from the sample. As the sample cools, as soon as it reaches T_C, the small magnet will be strongly attracted to the sample. The temperature recorded at that instant is another way to estimate T_C. Magnets such as those in modern earphones or headphones, or inside a computer hard drive disk, are suitable for this test. To determine the appropriate platform height, find the position in which the magnet is spontaneously attracted to the cold sample.

Suggested tasks:

✓ Using the experimental device shown in Figure 59.4, repeat the test at least five times, recording T_C in each case. Find the best value of T_C and its uncertainty.

✓ Compare your result to the values reported in tables for this sample. Remember that commercial nickel is often alloyed with other metals that alter the value of T_C.

✓ Repeat the experiment using a ferrite sample and determine the best value of T_C. In this case, you can use copper wire to fix the thermocouple to the sample. Make sure there is good thermal contact with the thermocouple.

59.3.3 *Project 201. Measuring the hysteresis curve for ferrite core*

Recommended basic equipment: A ferrite torus approximately 3–7 cm in diameter. Enameled copper wire with diameter 0.15–0.5 mm. A dual input oscilloscope or a data acquisition system (DAS) with two independent inputs. A transformer from 110 or 220 V–12 V. A Variac. Resistors 5 Ω @ 5 W.

The experimental device is the same as in Project 199, shown in Figure 59.2.

Suggested tasks:

- ✓ Repeat the same analysis suggested in the previous project.
- ✓ Observe how the hysteresis curve of the sample changes for different values of the input voltage. Move the sample in and out of saturation conditions by increasing or decreasing the applied voltage with the input autotransformer or Variac, i.e., the current in the primary of the torus.
- ✓ Observe how the hysteresis curve of the sample changes for different values of input voltage. Move the sample in and out of saturation conditions.
- ✓ Describe and qualitatively explain your observations.

59.3.4 *Project 202. Determining the Curie temperature of a ferrite sample*

Recommended basic equipment: The same as in the previous project, plus a thermocouple or thermistor to measure temperature up to 250°C. A ferrite torus approximately 3–7 cm in diameter. A dual input oscilloscope or a data acquisition system (DAS) with two independent inputs. An oven or oil bath that can heat up to about 150°C. Optional: A Dewar with liquid nitrogen.

Before winding the torus to be used, apply the technique described in Project 200 to ensure that the Curie temperature of your sample is within the thermal range of the oil bath. Once you have verified this point, you can proceed to wind your sample.

Using the same device as in Project 199, but with a fixed input voltage, study how the hysteresis curve varies at different temperatures [11]. To do so, use a thermocouple to measure sample temperature. Connect the temperature sensor so that you get the best possible measurement of the temperature, T, of the torus. It is important that the sensor should have good thermal contact with the sample. To vary the temperature, you can use a controlled temperature oven, or even better, a thermostat with an oil bath [19–21], taking care that the temperature does not exceed 250°C, above which the wire insulation may begin to burn.

An oil bath heated by an electric heater can be used to heat the ferrite sample to a temperature of the order of about 200°C. There are commercial systems that perform this procedure safely [19–21], but you can also build your own system using a 1/2–1 liter container with an electric heater. For temperatures below 200°C, you can use mineral oil. For temperatures higher than 200°C (but lower than 300°C), use silicone oil. Always make sure that the oil does not contain toxic components.

Take care that the hot oil bath does not generate smoke, indicating that it is approaching the flash point, which can lead to ignition of the oil. An oil bath should always be monitored with a thermometer to ensure that its temperature does not exceed its flash point. Commercial devices generally have a temperature controller that does this automatically.

Suggested tasks:

✓ Study the behavior of the hysteresis curve at different temperatures. If you use a digital oscilloscope, record the curve at different operating temperatures. A DAS does this automatically.

✓ Perform this operation for temperatures above and below T_C.

✓ To characterize M_0, use the integrator voltage value V^0_{Int} (equivalent to Br) for $V_i = 0$, which corresponds to zero current in the primary and therefore to an applied field $H = 0$ (see Figure 59.1).

✓ **Optional:** To study the behavior of the ferromagnetic material at low temperatures, slowly introduce the device into liquid nitrogen in a Dewar, trying to cool the sample first with the rising cold vapor from the evaporation of the liquid. This will avoid sudden changes in temperature that could embrittle and break the ferrite. Study how the hysteresis curve changes at low temperatures. Analyze the behavior of M_0 at low temperatures.

✓ With all the data obtained, construct a graph of V^0_{Int} as a function of temperature, which is representative of the function $M_0(T)$. Try to determine the Curie temperature T_c and the parameter β using the graph and Eq. (59.1).

✓ If you are able to determine the Curie temperature T_c, construct a graph of V^0_{Int} as a function of $(T_c - T)$ in log-log scale and improve the estimation of the parameter β of Eq. (59.1).

59.4 Annex A: Number of turns of the toroid coils

For this experiment, a magnetic field must be applied to the torus such that it produces magnetization close to saturation in the material. For most ferromagnetic and ferrimagnetic materials, this is achieved at values of $B \approx 5{,}000$ G $= 0.5$ T. Suppose that the relative magnetic permeability [10] of the material is $\mu_r \approx 1{,}000$. If the mean radius of the torus is R_t and the number of turns in the primary is N_1, when a current i circulates through the primary,

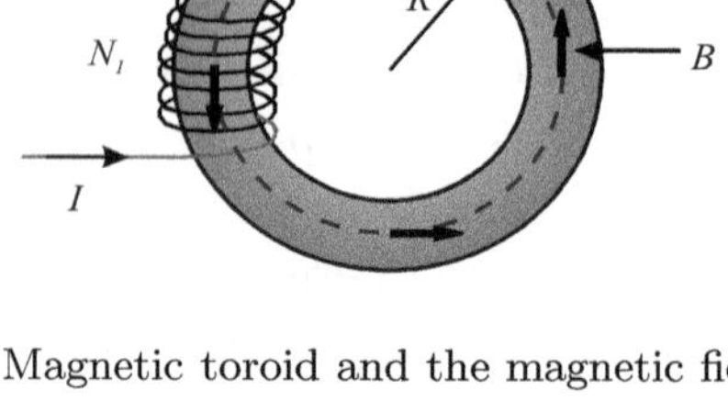

Figure 59.5 Magnetic toroid and the magnetic field B within it.

according to Ampere's law:

$$\oint \vec{B} \cdot dl = \mu_0 \mu_r N_1 i \tag{59.4}$$

Using the symmetry of the problem, we can solve the integral (59.4):

$$B = \frac{\mu_0 \mu_r}{2\pi} \frac{N_1 i}{R_t}. \tag{59.5}$$

To make $B \approx 0.5$ T, we have to satisfy

$$\frac{iN_1}{R_t} \approx 2500\text{A}/m. \tag{59.6}$$

If enameled copper wire is used for the primary winding, the diameter of the wire should be selected such that it can carry the current necessary to produce this value of B. A conservative criterion is to use approximately 1 mm^2 for every 3 A of current. If the current is of the order of 0.5 A, the enameled wires that form the winding should be of diameter $\phi \approx 0.5$ mm. In this case, if $i \approx 0.5$ A, the ratio of the number of turns N_1 to R_t of the torus should be

$$\frac{N_1}{R_t(\text{cm})} \approx 50. \tag{59.7}$$

A number of turns greater than N_1 (by a factor of 1–3) can be used for the secondary winding, so that the signal from the secondary is clearly visible on the oscilloscope. The diameter of the secondary wire can be smaller (e.g., 0.1–0.2 mm diameter). The greater the number of turns in the secondary (N_2), the greater the induced voltage. A reasonable criterion may be $N_2 \sim 2\ N_1$. Remember that no power is required from the secondary.

59.5 Annex B: Integrator circuit

The circuit in Figure 59.6 is fed with a sinusoidal signal. The relationship between the input signal amplitude $V_i(t)$ (V_i^0) and the

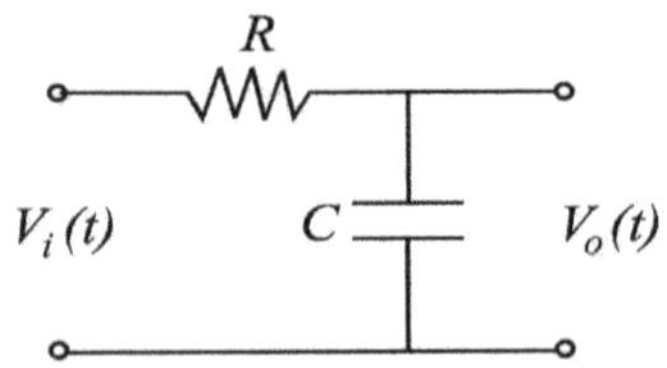

Figure 59.6 Diagram of an integrator circuit formed by a resistor and a capacitor.

exit $V_o(t)$ (V_o^0) is given by

$$V_o^0 = V_i^0 \frac{1}{1 + \mathrm{jRC}\omega} = V_i^0 \frac{1}{1 + j\tau\omega}, \tag{59.8}$$

where $\tau = RC$. In particular, if $\tau\omega \gg 1$, we can approximate

$$V_o^0 \approx V_i^0 \frac{1}{j\tau\omega}. \tag{59.9}$$

This means that the output signal is proportional to the integral of the input signal. In other words, the circuit in Figure 59.6 acts as an integrator of the input signal $V_i(t)$ if $\tau = RC \gg 1/(2\pi f)$, where f is the frequency of the input signal. For $f \cong 50$ Hz (frequency of linear voltage), a practical integrator can be built for this experiment using $R = 50k\Omega$ and $C = 10\mu F$ ($\tau = 0.5\,\mathrm{s}$).

References

[1] F. Sears, M. Zemansky, H. Young and R. Freedman, *College Physics*, NY: Pearson Ed., 2009.
[2] R. Feynman, *et al.*, *The Feynman Lectures on Physics, Boxed Set: The New Millennium Edition*, Vol. 1, Basic Books, 2011.
[3] uclaphysicsvideo, "Paramagnetism and diamagnetism," 2022. Available: https://www.youtube.com/watch?v=u36QpPvEh2c.
[4] wikipedia, "Ferrimagnetism," 2022. Available: https://en.wikipedia.org/wiki/Ferrimagnetism.
[5] Mopatin, "Curie temperature demonstration," 2022. Available: https://www.youtube.com/watch?v=haVX24hOwQI.
[6] C. Kittel, *Introduction to Solid State Physics*, John Wiley & Sons, 1996.

[7] N. N.W. Aschcroft and N. Mermin, *Solid State Physics*, Saunders College Publishing, 1976.
[8] Wikipedia, "Ferromagnetism," 2022. Available: https://en.wikipedia.org/wiki/Ferromagnetism.
[9] E. M. Purcell and D. J. Morin, *Electricity and Magnetism, Berkeley Physics Course*, 3rd Ed., NY: Cambridge University Press, 2014, p. 839.
[10] J. Reitz, F. Milford and R. Christy, *Foundations of Electromagnetic Theory*, 4th Ed., Addison Wesley, 2008.
[11] C. Lue, "A direct method for viewing ferromagnetic phase transition," *Phys. Teach.*, vol. 32, p. 304, 1995.
[12] Y. Kraftmakher, "Hysteresis loops of a ferromagnet," *Phys. Teach.*, vol. 43, p. 439, 2005.
[13] Amazon, "Ferrite toroid core," 2022. Available: https://www.amazon.com/Toroid-Core/s?k=Toroid+Core.
[14] Wikipedia, "Autotransformer," 2022. Available: https://en.wikipedia.org/wiki/Autotransformer.
[15] E. M. Kiess, "Circuit for displaying hysteresis loop of twoterminal inductors," *Am. J. Phys.*, vol. 58, p. 794, 1990.
[16] "Electronics forums, international electronics community," 2022. Available: http://www.forosdeelectronica.com/tutoriales/resistencia.htm.
[17] D. Wilson, "Curie point, again," *Phys. Teach.*, vol. 27, p. 374, 1989.
[18] S. Velasco and F. L. Román, "Determining the Curie temperature of Iron and Nickel," *Phys. Teach.*, vol. 45, p. 387, 2007.
[19] IKA, "Heating baths," IKA, 2022. Available: https://www.ika.com/en/Products-Lab-Eq/Heating-Baths-Heated-bath-Oil-bath-csp-206.
[20] "High-temperature-oil-bath," 2022. Available: https://www.thomassci.com/scientific-supplies/High-Temperature-Oil-Bath.
[21] IOWA, "Oil baths in laboratories," 2022. Available: https://ehs.research.uiowa.edu/oil-baths-laboratories.

Chapter 60

STATISTICAL NATURE OF RADIOACTIVE DECAY

This chapter will investigate the statistical nature of radioactive decay. In particular, it analyzes whether the Poisson distribution can explain some aspects of the statistical behavior of decay. It discusses the use of the *Multi-Channel Scaling* (MCS) instrument for data acquisition.

Goals

✓ Radioactive decay
✓ Statistics
✓ Poisson distribution

60.1 Radioactive decay

The individual decay of a nucleus or atom is a *stochastic* process [1, 2]. These processes take place randomly, over times that are not predictable microscopically. However, in a macroscopic assembly of decaying atoms (number of atoms $>10^{12}$), it is possible to predict the average number of decays and other properties of this process. For example, in a radioactive source with a very long half-life compared to the measurement time, successive determinations of the number of decays in a given time interval will not give exactly the same results. This *lack of definition or determinism* [3] is one of the intrinsic characteristics of all radioactive, nuclear, or atomic processes.

The issue of the lack of determinism in Quantum Mechanics appears quite clearly here. Assume that a radioactive source, during the time interval t_0 to $t_0 + dt$, has dn decays. Suppose you could magically roll back the universe to the exact same state at time t_0. If you again observed the same decay in the same atomic nucleus during the interval dt, you would not necessarily obtain the same number of decays dn. The decay process thus appears to be unpredictable and random [4]. Nonetheless, the values obtained will be distributed around a certain mean value $\langle n \rangle$ and present dispersion around this average. Furthermore, this number of decays may follow a statistical distribution. The objective of this chapter is precisely to study the nature of the statistical distribution associated with radioactive decay.

According to probability theory, if the distribution function is known for a random variable, all the moments associated with it can be known [1, 2]. Reciprocally, if all its moments of distribution are known, this is equivalent to knowing the distribution itself. Therefore, to determine a probability distribution associated with an experiment, there are two alternatives: to determine the distribution directly, or to determine all its moments or at least the most relevant ones, and from them, infer the distribution.

60.2 The Poisson distribution

This is a discrete probability distribution [5]. It occurs in randomized experiments where the possible outcomes are the set of natural numbers. The probability of the random variable taking the value n is given by the Poisson distribution function, defined as follows [6–8]:

$$P_\lambda(n) = \frac{\lambda^n}{n!} \cdot e^{-\lambda}, \quad n = 0, 1, 2, \ldots \tag{60.1}$$

where λ is a characteristic parameter of the distribution, with the following properties:

$$\langle n \rangle = \sum\nolimits_n n \cdot P_\lambda(n) = \lambda. \tag{60.2}$$

The variance of the distribution is given by

$$\mathrm{Var}(n) = \sigma^2 = \sum [(n - \langle n \rangle)^2 \cdot P_\lambda(n)] = \lambda, \tag{60.3}$$

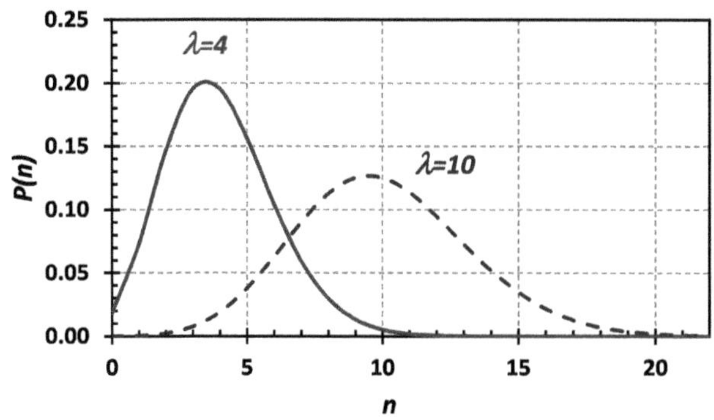

Figure 60.1 Poisson distribution for $\lambda = 4$ and $\lambda = 10$. One way to falsify the hypothesis that a given distribution follows the Poisson distribution is to determine its first two or four moments μ_r, and then check whether they obey the relations given by Eqs. (60.3) and (60.4).

with σ being the standard deviation of the distribution. The Poisson probability distribution has a maximum for $n \approx \lambda$. This probability distribution is not symmetric with respect to the mean value $\langle n \rangle$. In fact, the skewness and kurtosis of this distribution are [1, 2, 5]

$$\text{Skewness} = \frac{\mu_3}{\sigma^3} = \frac{1}{\sqrt{\lambda}} \quad \text{and} \quad \text{Kurtosis} = \frac{\mu_4}{\sigma^4} = 3 + \frac{1}{\lambda} \qquad (60.4)$$

Here, μ_r represents the moment of order r of the distribution, that is,

$$\mu_r = \sum\nolimits_n n^r P_\lambda(n) \quad \text{and} \quad \lambda \equiv \mu_1. \qquad (60.5)$$

Figure 60.1 illustrates the shape of this distribution. Note that λ is the mean value of the random variable n, that is, the first order moment, μ_1, of the distribution. A particularity of the Poisson distribution is that the mean value λ matches its variance and its mode. In terms of the results of this experiment, λ represents the average number of decays of a radioactive source in a given observation time interval.

60.2.1 *Project 203. Experimental study of the statistics of the radioactive process*

> **Recommended equipment:** A gamma radiation detector associated with a data acquisition system (DAS). Any radioactive source that emits gamma radiation. If you do not have this equipment, you can use the experimental spectra mentioned at the end of the activity.

This experiment uses a gamma radiation detector associated with a data acquisition system that enables you to measure the number of counts that the detector reaches in a given time interval: τ_d (*dwell time*). For this experiment, we need to be somewhat familiar with gamma radiation detection systems. There is extensive, excellent bibliography on this subject. Particularly, Refs. [6, 9, 10] provide very accessible and suitable presentations for γ, α, and β radiation detection systems.

In this experiment, a scintillation or solid state detector can be used [6, 9] in association with a multichannel analyzer with *Multi-Channel Scaling* (MCS). Many commercial multichannel models [9–11] include among their modes of operation the possibility of operating in MCS mode. Briefly, an MCS is a counter of pulses that occur in a predetermined time interval τ_d. This measurement sequence is automatically repeated N times and the pulses detected in each measurement interval k (=1,. . . ,N) are recorded as a sequence of numbers n_k. With this sequence of numbers, we can study the distribution of the random variable n = number of gamma rays detected within τ_d. For example, if the radioactive decay of a source were a deterministic process, it would be expected that $n_k = B\text{x}\tau_d$, with B being a constant of proportionality, independent of k. That is, all the n_k values in a measurement sequence would be the same. On the contrary, if the decay process were random, the values of n_k would reveal the statistics associated with the decay process.

A diagram of the proposed experimental device is shown in Figure 60.2. We have suggested a NaI-type gamma radiation detector [6], but any radiation detector (α, β or γ) can be used [12, 13].

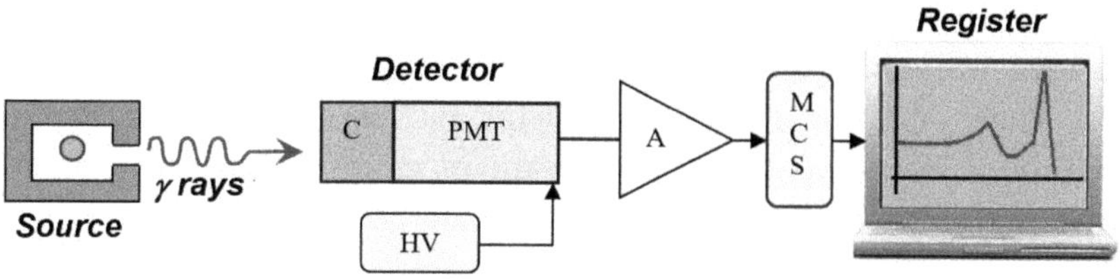

Figure 60.2 Diagram of the experimental device. The radioactive source produces the incident photons. The NaI (Tl) (C) scintillator crystal detects the γ rays. The high voltage source (HV) feeds the photomultiplier tube (PMT), which generates an electrical pulse with height proportional to the energy of the incident photon. The amplifier (A) shapes and amplifies this pulse, which enters a digitizer and Multi Channel Scaling (MCS) associated with the computer.

General recommendations: Perform measurements continuously to ensure that the physical and geometric characteristics of the experiment are as constant as possible. Keep the source, detector, electronics, and all experimental conditions unchanged throughout the experiment.

Suggested tasks:

✓ Place the radioactive source near the detector such that the count does not exceed about 100 counts per minute.

✓ Configure the MCS with $N = 512$ or more channels. That is, repeat the measurement 512 or more times.

✓ For each experiment, vary the measurement time τ_d (*dwell time*) of the MCS. In this way, each time you perform the experiment you obtain a sequence of values $n_k(\tau_d)$.

✓ For each value of τ_d, determine the moments of μ_r of the corresponding distributions, that is,

$$\mu_r = \frac{1}{N}\sum_{k=1}^{k=N}(n_k)^r, \quad r = 0, 1, 2, \ldots \tag{60.6}$$

✓ Knowing the first moments, determine λ and σ, and if possible, the skewness and kurtosis, for each value of τ_d. The idea is that the mean value λ should range from a few counts (less than ten) to a few thousand.

- ✓ For each repetition of the experiment (associated with each choice of measurement time τ_d), construct a histogram of n_k.
- ✓ For each of these histograms (that is, for each choice of measurement time τ_d), calculate the mean value of n, $\lambda(\tau_d)$, and the variance $\sigma^2(\tau_d)$.
- ✓ Using the data obtained, graphically represent σ^2 as a function of λ. What can you conclude from this graph about the relationship between σ^2 and λ?
- ✓ Is the Poisson distribution compatible with these results?
- ✓ Using the expression (60.1) for the Poisson distribution and the values of λ calculated for each choice of measurement time, construct a graph of the corresponding Poisson distribution, superimposed on the experimentally obtained histogram (properly normalized). What can you say about the experimentally obtained distributions?
- ✓ ♣♣ Using the data obtained, graph the skewness and, if possible, the kurtosis, as a function of λ on a log-log scale. Are your experimental results consistent with the theoretical expectations given by Eq. (60.4)? Is the Poisson distribution compatible with these results?

Note

In www.fisicarecreativa.com, in the section "Resources for Physics Experiments using Tic's", you will find radioactive decay data to analyze. The data was taken with solid-state detectors and NaI.

References

[1] S. Ross, *The Elements of Probability Theory and Some of Its Applications*, 10th Ed., 2018.
[2] H. Cramer, *The Elements of Probability Theory and Some of Its Applications*, 99a Ed., NY: Wiley, 1966.
[3] Wikipedia, "Determinism," 2022. Available: https://en.wikipedia.org/wiki/Determinism.
[4] F. Weinert, *The Scientist as Philosopher: Philosophical Consequences of Great Scientific Discoveries*, 2004th Ed., Springer, 2005.

[5] Wikipedia, "Poisson distribution," 2022. Available: http://en.wikipedia.org/wiki/Poisson_distribution.
[6] G. Knoll, *Radiation Detection and Measurement*, 3rd Ed., NY: John Wiley, 2010.
[7] R. Evans, *The Atomic Nucleus*, New York: McGraw Hill Book Co., 1955.
[8] P. Bevington and D. K. Robinson, *Data Reduction and Error Analysis for the Physical Sciences*, 3rd Ed., p. 230, 2003.
[9] ORTEC-AMETEX, "Advanced measurement tecnology," 2022. Available: http://www.ortec-online.com/.
[10] CANBERRA-PACKARD Industries Inc., "Nuclear measurements," 2022. Available: https://www.cpce.net/en/nuclear-environment/.
[11] J. Cardoso, *et al.*, "A very low-cost portable multichannel analyzer," in *Conference: Nuclear Science Symposium Conference Record, 2000 IEEE*, Coimbra, Portugal, 2000.
[12] P. Koczyk, P. Wiewiór and C. Radzewicz, "Photon counting statistics — Undergraduate experiment," *Am. J.Phys.*, vol. 64, p. 240, 1996.
[13] M. L. Martínez Ricci, *et al.*, "Photon counting statistics using a digital oscilloscope," *Am.J. Phys.*, vol. 75, p. 707, 2007.

Chapter 61

RELATIVISTIC DYNAMICS: COLLISIONS OF PHOTONS AND ELECTRONS — COMPTON EFFECT

This chapter proposes to study classical and relativistic models to describe the kinematics of the interaction of gamma rays with free electrons. Since the energies of the gamma rays are of the order of or greater than the rest mass of the electron, this study makes it possible to detect the relativistic effects of the collision. The proposed experiment also makes it possible to determine the rest mass of the electron. We will analyze the Compton effect and the gamma radiation spectra obtained with multichannel analyzers.

Goals

✓ Gamma radiation detectors
✓ Interaction of photons with matter
✓ Compton effect
✓ Relativistic dynamics

61.1 Relativistic kinematics

The special theory of relativity is based on the premise that the laws of Physics are the same for all inertial observers who move with constant speed relative to each other, and that the speed of light in vacuum, c, is the same for all of them [1, 2]. One consequence of this theory is that the momentum of a particle is given by

$$\vec{\boldsymbol{p}} = m \cdot \vec{\boldsymbol{v}} = \gamma \cdot m_0 \cdot \vec{\boldsymbol{v}}, \tag{61.1}$$

where $\vec{p}$ and $\vec{v}$ are the momentum and velocity of the particle, m is the mass, and m_0 is the rest mass [3]; $\beta = v/c$ and γ:

$$\gamma = 1/\sqrt{1-(v/c)^2} = 1/\sqrt{1-\beta^2}. \tag{61.2}$$

The total energy E of the particle is equal to the sum of its kinetic energy T and energy associated with its rest mass: m_0c^2, that is,

$$E \equiv c\sqrt{p^2+(m_0c)^2} = T + m_0c^2. \tag{61.3}$$

In relativistic collisions, the conservation of total energy and momentum always holds. For a zero-mass particle such as the photon,

$$E \equiv E_\gamma = cp_\gamma = \hbar\omega. \tag{61.4}$$

Since both the momentum p of a particle and its kinetic energy T are measurable parameters, the relationship between T and p can be used to test classical and relativistic predictions between these quantities. According to Eq. (61.3), the relativistic relation is

$$\frac{p^2}{m_0^2c^2} = \left(\frac{T}{m_0c^2}+1\right)^2 - 1, \tag{61.5}$$

while the corresponding classical or non-relativistic relation is

$$\frac{p^2}{m_{nr}^2c^2} = 2\left(\frac{T}{m_{nr}c^2}\right). \tag{61.6}$$

In Figure 61.1, the relativistic prediction, Eq. (61.5), is indicated with a thick red solid line, and the classical prediction, Eq. (61.6), is indicated with a thin blue dashed line. As can be seen, if it is possible to measure T and p for kinetic energies in the region, where $T \geq m_0c^2$, the classical and relativistic predictions can be tested.

There are several ways to measure the momentum and kinetic energy of a particle independently. For a charged particle in a uniform magnetic field, the radius of its orbit is proportional to p. On the other hand, the energy deposited in a particle detector is proportional to T [4]. In this chapter, you will try to explore the shape of the spectrum of a gamma radiation detector to determine these parameters for an electron [5, 6].

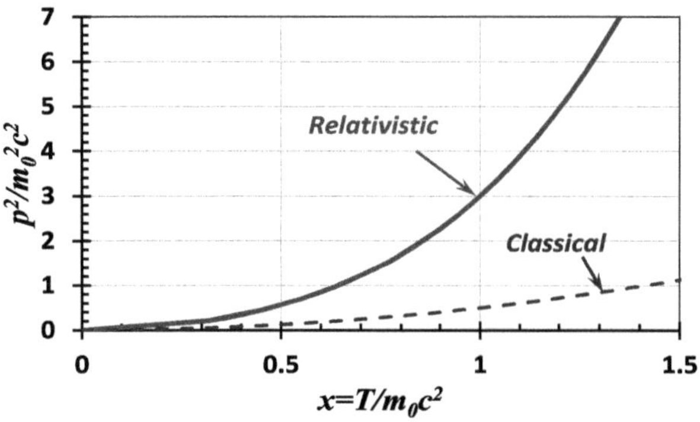

Figure 61.1 Classical and relativistic predictions of the variation of $(p/m_0c)^2$ as a function of $T/m_0^2c^2$. When T is of the order of or greater than m_0c^2, the theoretical expectations of the two models are quite different.

61.2 Interaction of radiation with matter — Compton effect

Electromagnetic radiation (photons) interacts with matter through several different processes. The relative importance of each process depends on the energy of the photons and the atomic number, Z, of the absorbing medium. At low energies, when the energy of the photons is less than the excitation energy of the atoms or molecules, or alternatively, if the wavelength is much greater than the dimensions of the atoms or molecules, the dominant mechanism is Rayleigh scattering. Chapter 66 analyzes this mechanism in greater detail.

At higher energies, when photons have energies comparable to those of the bonding of electrons in atoms, the dominant mechanism is the *photoelectric effect* [7]. Here, the energy and momentum of the photon are totally absorbed by the atom and the released electron. When the energy of the photons is much higher than the binding energy of the electrons, the dominant absorption mechanism is the *Compton effect.* In this process, the photon conserves part of its original energy and continues to move in a new direction. When the energy of the photons exceeds 1,022 keV, under the influence of the electromagnetic field of a nucleus, a new interaction mechanism becomes energetically possible: *pair production.* Here, the

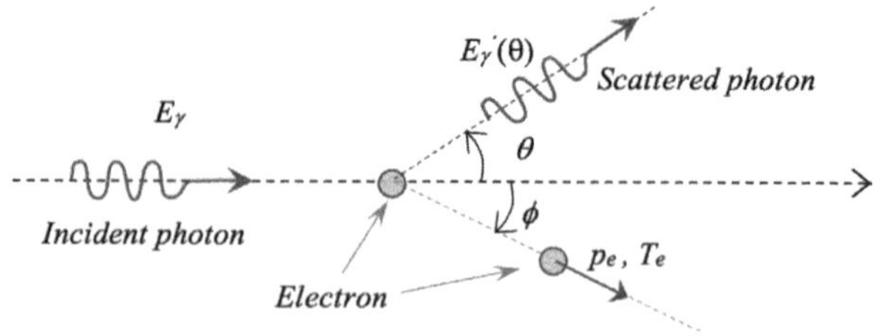

Figure 61.2 Diagram showing a Compton interaction.

photons can produce electron-positron pairs. The probability of pair production increases with the Z of the absorbing material and with the energy of the photon [7].

In the *photoelectric effect*, the absorption of the photon is complete (its energy and momentum are absorbed by an electron originally bound to the atom). This process cannot occur with free electrons because it is prevented by the simultaneous conservation of energy and momentum.

On the other hand, when a photon interacts with a free electron, in order to conserve energy and momentum in the interaction, the final state must have an electron and a new photon that share the energy and the momentum of the incident photon. This process of interaction of a photon with an electron is called the *Compton effect* and is shown in Figure 61.2.

Let p_e and T_e be the momentum and kinetic energy of the electron after the interaction. Let E_γ denote the energy of the incident photon and $E'_\gamma(\theta)$ the energy of the photon after the interaction, which we assume goes out in a direction that forms an angle θ with the direction of the incident photon. From conservation of energy and momentum in the collision [1, 5, 8],

$$\frac{1}{E'_\gamma(\theta)} - \frac{1}{E_\gamma} = \frac{1}{m_e c^2}(1 - \cos\theta) \qquad (61.7)$$

with m_e being the rest mass of the electron.

In particular, for a one-dimensional collision, that is, for the case in which $\theta = 180°$, from the conservation of momentum and energy [5, 8]:

$$E_\gamma/c = p_e^0 - E_\gamma^{'}/c \quad \text{and} \quad E_\gamma = T_e^0 + E_\gamma^{'}. \tag{61.8}$$

Since the scattered electron ($\theta = 180°$) exits in the opposite direction to that of the incident photon, by eliminating $E_\gamma^{'}$ from these equations,

$$p_e^0\, c = 2\, E_\gamma - T_e^0. \tag{61.9}$$

This relationship enables p_e^0 to be found if E_γ and T_e^0 can be measured, and it is valid both relatively and classically. Therefore, with this expression and the measurement of T_e^0 and E_γ, it is possible to construct an experimental graph like the one in Figure 61.1 for the electron, to test classical and relativistic predictions.

In everything that follows, we assume that the case of a one-dimensional collision is being considered unless explicitly stated otherwise.

The non-relativistic (or *classical*) connection between energy and momentum is

$$T_e = \frac{p_e^2}{2\, m_{nr}}. \tag{61.10}$$

Here, m_{nr} is the non-relativistic mass of the electron and T_e is its kinetic energy. Combining Eqs. (61.9) and (61.10),

$$m_{nr}\, c^2 = \frac{(2\, E_\gamma - T_e)^2}{2\, T_e}. \tag{61.11}$$

This expression makes it possible to obtain the non-relativistic mass of the electron according to the classical approximation, in terms of the energy of the incident photon E_γ, and kinetic energy, T_e, of electrons after a Compton interaction. The corresponding relativistic expression between momentum and kinetic energy is

$$T_e = \sqrt{p_e^2\, c^4 + m_e^2\, c^4} - m_e\, c^2, \tag{61.12}$$

where m_e is the rest mass of the electron. Combining Eqs. (61.9) and (61.12) gives the following expression for the rest mass of the electron:

$$m_e\, c^2 = \frac{2\, E_\gamma (E_\gamma - T_e)}{T_e} = \frac{2\, E_\gamma (E_\gamma - T_e^0)}{T_e^0}. \tag{61.13}$$

This equation is the relativistic expression analogous to the classical Eq. (61.11). It provides a way of testing the classical versus the relativistic models for the electron. If the second members of Eqs. (61.11) and (61.13) are plotted as a function of E_γ on the same graph, we can evaluate which of the models agrees better with the experimental results and which model should be rejected.

The relativistic parameters β and γ can also be written in terms of T_e and E_γ as follows:

$$\beta \equiv \frac{v}{c} = \frac{p_e c}{m_e.c^2} = \frac{2E_\gamma - T_e}{T_e + m_e c^2} = \frac{T_e (2\, E_\gamma - T_e)}{T_e^2 - 2\, E_\gamma\, T_e + 2\, E_\gamma^2} \tag{61.14}$$

and

$$\gamma = \frac{E_e}{m_e c^2} = 1 + \frac{T_e^2}{2E_\gamma (E_\gamma - T_e)}. \tag{61.15}$$

Thus, if we are able to measure E_γ and T_e^0 for a collision between a photon and an electron with $\theta = 180°$, from Eqs. (61.9) to (61.15), we can determine all the parameters required to falsify the classical and relativistic theoretical expectations.

61.3 Mechanisms of photon interaction in a gamma ray detector

There are various types of gamma radiation detectors, of which the most widespread are scintillation detectors and solid-state detectors. Among the first are the NaI (Tl) detectors [7, 9–11], consisting of a Tl-doped NaI scintillator crystal associated with a photomultiplier tube. When a photon interacts with the detector material, its energy is degraded (totally or partially) within the material, resulting in some atoms remaining in excited states. In turn, when these atoms decay, they emit visible photons, the number of which is proportional to the energy deposited on the crystal by the incident photon. The photomultiplier tube (PMT) detects these photons and generates a pulse whose height is proportional to the number of photons detected. Thus, the height of the voltage pulse is proportional to the energy of the incident photon on the crystal.

In semiconductor detectors such as high purity germanium detectors, Ge(Hp) [7, 10, 12], the energy of the incident photon is transformed into the creation of pairs of electrons and holes within the crystal of the detector. An electric field applied to the detector crystal collects these charges and generates a current pulse in the detector's external circuitry. As the number of electron-hole pairs is proportional to the energy of the incident photon, the pulse charge (integrated current) is also proportional to the energy of the photon. A preamplifier converts the signal from the collected charge into a voltage pulse.

Whatever the detector, the mechanisms of interaction of radiation with matter are the same. When X-radiation or gamma-radiation (γ) interacts with the atoms of a detector or any material, it does so through the mechanisms discussed above [7, 9, 10].

Associated with the detector (either semiconductor or scintillation), it is necessary to use a spectroscopic amplifier, which amplifies and shapes the detector pulses for analysis by the data acquisition system [7, 11, 12].

The most widely used data acquisition system is a *pulse height analyzer* associated with an analog-to-digital converter (ADC). With this electronic device, the pulses generated by the detector, which carry the information of the energy that the photons deposit in the detector, are converted into digital signals. These digitized signals are then taken by a multichannel that analyzes the pulse height and converts it into a number (channel) [10]. In this stage, a histogram is generated of the number of counts according to channel. Thus, the channel number is often proportional to the pulse heights, that is, to the energy of the γ-rays or X-rays. A more detailed description of the various components of the experimental device can be found in the literature [7, 10, 12].

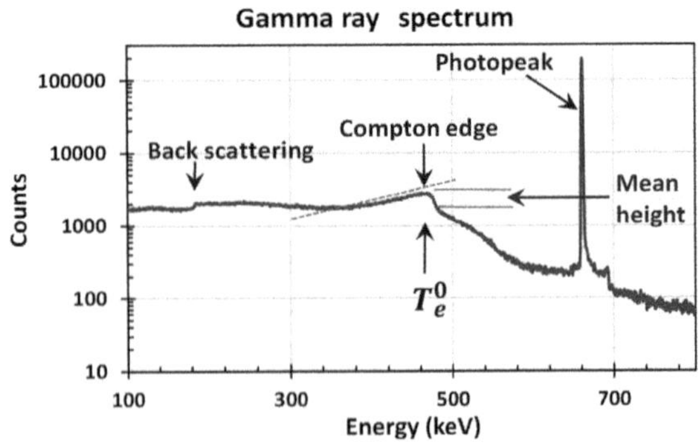

Figure 61.3 Typical gamma-ray spectrum obtained using a Ge (Hp) semiconductor detector. This spectrum corresponds to a monoenergetic source of ^{137}Cs. In addition to the main peak (*photopeak*, E_γ), two conspicuous characteristics are observed: the Compton edge, which corresponds to the maximum kinetic energy T_e^0 of electrons in a frontal collision with the incident photons, and the *back-scattering peak*, which corresponds to the energy of the photons that are backscattered in the surrounding medium and re-enter the detector. The value of T_e^0 can be obtained from the position of the *mean height* of the rising side of the Compton edge.

Figure 61.3 shows a typical gamma-ray spectrum obtained with a semiconductor detector. The spectra that result from a scintillation detector are similar, except that the resolution of the peaks is not as good as illustrated in this figure.

The main characteristics of a gamma spectrum are the *photopeak* of energy E_γ (corresponding to the case in which all the energy of the incident photon remains in the detector) and a *Compton plateau.* This continuous spectrum is due to the fact that when a Compton interaction occurs, more often than not, the electron leaves all its energy in the detector, while the scattered photon, which is produced in the interaction, escapes from it, as illustrated in Figure 61.4. In particular, if the secondary photon escapes at $\theta = 180°$ from the incident direction of the photon, the ejected electron will have the maximum possible energy (T_e^0) in this type of interaction. The fact that $T_e^0 < E_\gamma$ explains why the Compton plateau always appears at lower energies than the photopeak. To summarize, the energy of the recoiling electron can have kinetic energy T_e that

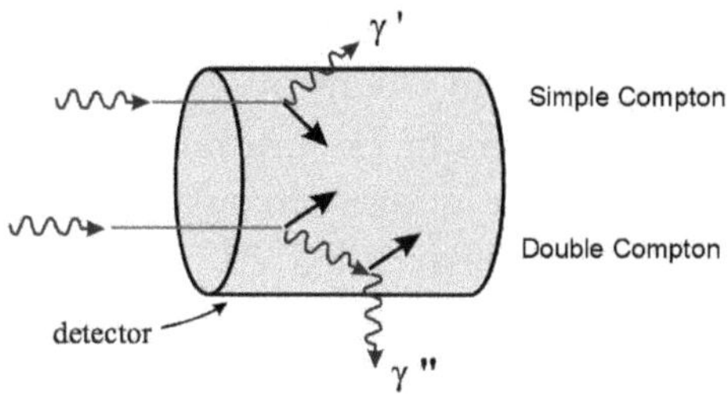

Figure 61.4 Schematic illustration of the origin of the Compton plateau (single Compton) and double Compton. In all cases, the energy of the electrons remains in the detector. In the first case, the gamma-ray (γ') escapes from the detector; in the second case, the secondary photon γ'' is not detected.

ranges from 0 ($\theta = 0°$) to T_e^0 ($\theta = 180°$); therefore, the *Compton plateau* ranges from 0 to the *Compton Edge*, which coincides with T_e^0.

The reason why the Compton edge is not steep is partly due to the fact that the detected electrons are not free, and the finite resolution of the detector. A practical way to determine the value of T_e^0 is to identify the mid-height point of the rising flank of the Compton edge, as illustrated schematically in Figure 61.3 [7].

The relative importance of the Compton plateau with respect to the photopeak depends, among other factors, on the *size of the detector*. The larger the detector, the lower the probability of escape of secondary photons and the smaller the magnitude of the Compton plateau with respect to the photopeak, and vice versa.

The presence of counts between the Compton edge and the photopeak is associated with the possibility that the photons produced in a Compton interaction make a second or third Compton collision in the detector (Figure 61.4). A more detailed discussion of the different types of interactions that occur in gamma-ray detectors can be found in Refs. [6, 7].

From the above discussion, it can be concluded that by studying the gamma-ray spectra obtained using semiconductor detectors or scintillators, it is possible to study the kinematics and dynamics of the interaction of photons with the electrons of the detector. More specifically, the photopeak energy of the gamma-ray spectrum is associated with the energy of the incident photons (E_γ), while the energy associated with the Compton edge (T_e^0) is the maximum energy of the ejected electrons in the interaction, that is, the energy of the electrons that make a one-dimensional collision with the incident photons, which we have called T_e^0 in Eq. (61.9). The project proposed here thus consists of performing an experimental study of the relationship between E_γ and T_e^0. Using this experimental information, we attempt to falsify the classical and relativistic theoretical expectations for describing the kinematic behavior of the electron.

61.3.1 *Project 204. Experimental study of the photon–electron collision — Compton effect*

Recommended equipment: A gamma radiation detector associated with a data acquisition system. Radioactive sources with few lines but covering a wide spectrum of energy (100 keV–1.3 MeV, approximately). If you do not have this equipment, you may use spectra that have already been obtained (see the note at the end of this project).

For this experiment, it is advisable to use an arrangement such as the one outlined in Figure 61.5 [11, 12] and a set of gamma radiation sources so that it covers as wide a range of energy of gamma rays as possible (from 100 keV to 1.3 MeV, approximately). Radioactive sources, such as ^{137}Cs, ^{60}Co, ^{22}Na, ^{52}Cr, ^{133}Ba, ^{57}Co, ^{241}Am, ^{198}Au, and ^{207}Bi, may be suitable for this experiment. With the detector you have available, collect well-defined gamma ray spectra for these sources.

Carry out an energy calibration of the detector by using a set of sources with known energy, and preferably, few spectral lines (discrete gamma rays), so that they are easy to identify in the spectrum. Construct a plot of the photopeak energy as a function of the channel where its centroid is located. Fit a linear relationship to this graph (you can use a polynomial if necessary). This gives us an energy channel calibration curve [7]. Using a source not used in the calibration, obtain your spectrum and check whether your calibration

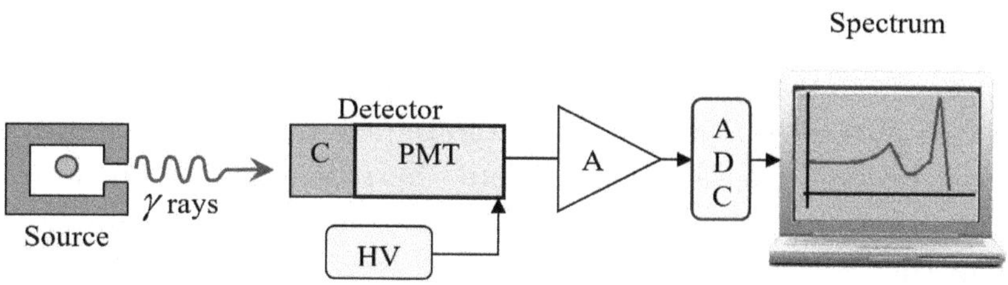

Figure 61.5 Diagram of a gamma ray detector. The radioactive source produces the incident photons. The NaI (Tl) scintillator crystal, C, detects γ rays. The high voltage power supply (HV) feeds the photomultiplier tube (PMT) and generates an electrical pulse of height proportional to the energy of the incident photon. The amplifier reshapes and amplifies this pulse, which enters the analog-digital converter (ADC). From there, these signals go to the multichannel analyzer (MCA) associated with the computer, which produces the energy spectra [7, 11, 12].

adequately predicts the position of the peaks corresponding to this new source.

Once the detector is calibrated, using the photo peaks (E_γ) from the sources used, proceed to identify the Compton edges (T_e^0) associated with each peak of the spectrum.

From the analysis of these spectra, construct a table of E_γ and T_e^0 for as many peaks as possible that you can identify. To facilitate this task, it is advisable that the sources should either be monoenergetic or have few peaks.

Make sure your chosen sources cover a photon energy range from about 100 keV to about 1.3 MeV. This range allows electrons that collide head-on to exit with lower and higher kinetic energies than the rest mass of the electron ($m_e \cdot c^2 \approx 511$ keV).

Safety note: Ask your instructor about the precautions to be taken when handling the experimental device (high voltage sources, radioactive sources, etc.). Do not apply voltage to the detector or connect anything to the multichannel until an instructor or supervisor has checked the electronic circuitry. Exercise caution in handling radioactive sources and, after use, place them in the appropriate lead containers. See Appendix B for Laboratory Safety Standards.

Suggested tasks:

If you have gamma-ray detectors, set up a device like the one shown in Figure 61.5 to detect gamma rays from radioactive sources, following the precautions described above and the recommendation of your instructors. If you do not have this type of instrument, you can use spectra that have already been obtained (see the note at the end of this activity).

✓ **Multichannel calibration:** Once the experimental setup is ready to collect spectra of the radioactive sources, using sources of well-known energy [7], carry out an energy calibration of your data acquisition system. That is, construct a plot of photon energy as a function of the channel in which the centroids of the photo peaks occur. This graph constitutes the energy calibration curve of your spectroscopic system.

✓ For each gamma ray, determine the position of the corresponding Compton edge T_e^0. One technique to locate this edge [7] consists of taking the intermediate position between the maximum and minimum of the edge. Briefly discuss the criteria used to characterize the value of T_e^0. References [6, 7, 13, 14] provide detailed analyses on how to characterize the position of the Compton edge. Using the technique described here or any other that you consider appropriate, determine the value of T_e^0 for each photon of energy E_γ.

✓ Make an energy plot of the Compton edge (T_e^0) as a function of the corresponding gamma-ray energy E_γ (photopeak).

✓ Using your data for E_γ and T_e^0, plot:

- $m_e c^2$ as a function of T_e^0, Eq. (61.11),
- $m_n c^2$ as a function of T_e^0, Eq. (61.13).

✓ Discuss which theoretical model, Eq. (61.11) or (61.13), best describes your observations. Clearly, one would expect rest mass not to depend on T_e^0. Use this expectation to evaluate your results. From your data, find the best value of the rest mass of the electron, m_e, and its uncertainty. How does your result compare to the best values reported for this parameter, m_e?

✓ Using Eq. (61.5), and employing your experimental results for T_e^0 and E_γ, estimate the value of p_e^0,

$$\left(\frac{p_e^0}{m_0 c}\right)^2 = \left(\frac{T_e^0}{m_0 c^2} + 1\right)^2 - 1. \tag{61.16}$$

Plot $\left(\frac{p_e^0}{m_0 c}\right)^2$ as a function of $\frac{T_e^0}{m_0 c^2}$. Compare your results to the relativistic relation, Eq. (61.5), and the classical expectation, Eq. (61.6). Discuss which theoretical model best describes your observations.

✓ Using your data of E_γ and T_e^0, and Eqs. (61.14) and (61.15),

$$\beta = \frac{2E_\gamma - T_e^0}{T_e + m_e c^2} = \frac{T_e^0(2\,E_\gamma - T_e^0)}{T_e^{0^2} - 2\,E_\gamma\,T_e^0 + 2\,E_\gamma^2} \tag{61.17}$$

and

$$\gamma = \frac{E_e}{m_e c^2} = 1 + \frac{T_e^{0^2}}{2E_\gamma(E_\gamma - T_e^0)}, \tag{61.18}$$

plot the following:

- γ as a function of β.
- In these graphs, include the predictions from the theoretical models, Eq. (61.2).
- How do your results compare to relativistic expectations? What can you conclude from this analysis?

✓ Discuss the validity of the classical and relativistic approaches to interpret the kinematics of the electron, according to your results.

Note

In www.fisicarecreativa.com, under "Physics Experiment Resources," you will find spectral data that are useful for this experiment. These spectra were obtained with an experimental device like the one shown in Figure 61.5 using semiconductor detectors.

References

[1] R. Feynman, *et al.*, *The Feynman Lectures on Physics, Boxed Set: The New Millennium Edition*, Vol. 1, Basic Books, 2011.

[2] W. Rindler, *Relativity — Special, General and Cosmological*, Oxford, 2001.

[3] E. Hecht, "Einstein on mass and energy," *Am. J. Phys.*, vol. 77, p. 799, 2009.

[4] J. Luetzelschwab, "Apparatus to measure relativistic mass increase," *Am. J. Phys.*, vol. 71, p. 878, 2003.

[5] J. Higbie, "Undergraduate relativity experiment," *Am. J. Phys.*, vol. 42, p. 642, 1974.

[6] L. Jolivette and N. Rouze, "Compton scattering, the electron mass, and relativity: A laboratory experiment," *Am. j. Phys.*, vol. 62, p. 266, 1994.

[7] G. Knoll, *Radiation Detection and Measurement*, 3rd Ed., NY: John Wiley, 2010.

[8] R. Evans, *The Atomic Nucleus*, New York: McGraw Hill Book Co., 1955.

[9] R. Maruyama, "Phenomenology of gamma ray and charged particles interactions," In: de *Encyclopedia of Nuclear Energy*, 1st Ed., NY: Elsevier, 2021, p. 3656.

[10] E. Henley and A. Garcia, *Subatomic Physics*, 3rd Ed., NJ: World Scientific publ. Co., 2007.

[11] CAMBERRA-PACKARD Industries Inc., "Nuclear measurements," 2022. Available: https://www.cpce.net/en/nuclear-environment/.

[12] ORTEC-AMETEX, "Advanced measurement technology," 2022. Available: http://www.ortec-online.com/.

[13] M. Lee, K. Veghese and R. Garner, *Nucl. Inst. Meth.*, vol. A262, p. 430, 1987.

[14] D. E. Di Gregorio, *et al.*, "Search for a neutrino mass component in the internal bremsstrahlung of 71Ge," *Phys. Rev. C*, vol. C47, p. 2916, 1993.

[15] C. Lederer, *et al.*, *Table of Isotopes*, 7a Ed., New York: John Wiley and Sons, 1978.

[16] S. Chu, L. Ekström and R. Firestone, "The Lund/LBNL nuclear data search," Version 2.0, February 1999. Available: http://nucleardata.nu clear.lu.se/toi/index.asp.

Chapter 62

INTERACTION OF ELECTROMAGNETIC RADIATION WITH MATTER

This chapter looks at some of the characteristics of the interaction of gamma and X radiation with matter, in particular, the ways in which electromagnetic radiation is attenuated as it passes through a material. The chapter analyzes the dependence of these phenomena on the energy and the atomic number Z of the absorbent material and discusses the different interaction mechanisms that predominate in each energy range.

Goals

✓ Interaction of radiation and matter

✓ Bouguer–Lambert law

✓ Absorption coefficient

62.1 Passage of electromagnetic radiation through matter

When a beam of electromagnetic radiation of intensity I_0 hits a material sample of thickness x, as illustrated in Figure 62.1, the beam is attenuated. The emergent intensity $I(x)$ is described by the *Bouguer–Lambert law* (discovered independently by Bouguer and Lambert in 1729 and 1760, respectively), which establishes that [1, 2]

$$I(x) = I_0 \exp(-\mu\, x), \tag{62.1}$$

where μ is the linear absorption coefficient and x is the thickness in units of length. This relation sometimes is simply called *Lambert's*

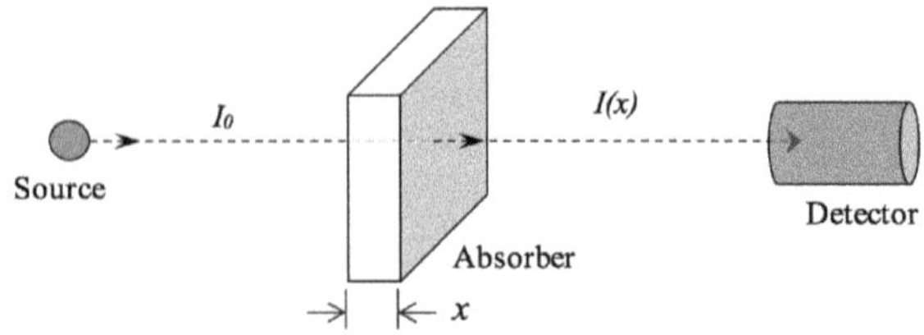

Figure 62.1 Attenuation of electromagnetic radiation when it passes through a material of thickness x.

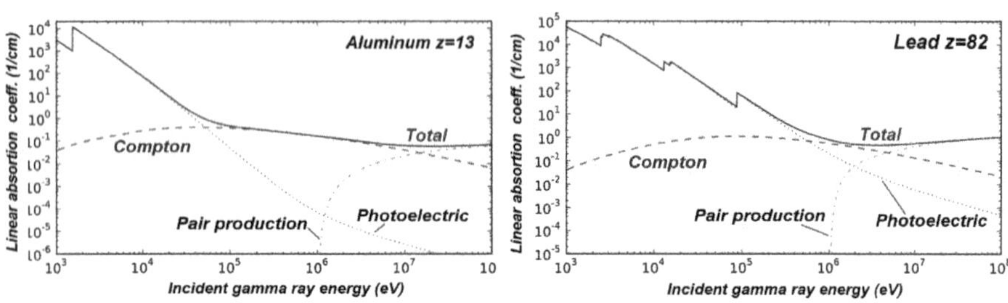

Figure 62.2 Linear absorption coefficient μ of electromagnetic radiation for aluminum (left) and lead (right). In both graphs, the dotted lines indicate the contribution of each attenuation mechanism, and the solid blue line indicates its total value [1].

law. It is often helpful to express thickness in units of mass per unit area.

Defining $t = \rho x$, where ρ is the density, *Lambert's law* can be expressed as follows:

$$I(x) = I_0 \exp(-\mu\, t/\rho) = I_0 \exp(-\lambda t), \tag{62.2}$$

where $\lambda = \mu/\rho$ is known as the *mass absorption coefficient.*

The total attenuation coefficient is the result of the three interaction mechanisms between radiation and matter discussed in Chapters 61 and 66, which are represented graphically in Figure 62.2 for aluminum and lead:

- **Photoelectric effect:** This is the dominant effect at low energies $E_\gamma \leq 200$ keV. It has a strong dependence on the photon energy and the Z of the material $\mu_{fe} \propto Z^n/E_\gamma^3$, with n ranging from 3 to 4.

- **Compton effect:** This is the dominant effect at intermediate energies, that is, between 100 KeV and 5 MeV. It depends on the number of electrons in the absorbent material, which in turn depends on the density of the absorbent material and the number of electrons per unit mass. Since in most elements, the number of electrons per unit mass is approximately the same, it is concluded that the probability of a Compton effect is almost independent of the atomic number, Z, of the absorber. This effect decreases as the photon energy increases.
- **Pair creation or electron–positron pair creation:** This effect becomes possible at $E_\gamma \geq 1022$ keV and prevails at high energies. It increases with the energy of the photon and with the atomic number Z of the absorber. This dependency is roughly proportional to the square of Z.

62.2 Absorption coefficients — Photopeak areas

Figure 62.3 shows an experimental design that can be used to study the attenuation of electromagnetic radiation, gamma rays, or *X-rays* when they pass through a sample of thickness x [3–5].

For this experiment, it is important to have a method for normalizing one measurement with respect to another. Data acquisition systems (the combination of detector, amplifier, ADC, multichannel and other elements, see Chapter 61– [6, 7]) have a finite processing time for each pulse [1]; therefore, not all photons registered by the detector are processed by the multichannel. The fraction of the total photons processed by the multichannel with respect to the photons identified by the detector is called FA (*Fraction Alife*). It is also useful to define the *dead time* [1, 8] TM = (1−FA) · 100.

Many modern multichannel devices can determine these parameters directly or indirectly. Some multichannel models report the real time of the measurement T_{real}, and the live-time T_{life}, that is, the fraction of the real time in which the detector was available for processing a signal. From these values, it is possible to estimate FA as $\text{FA} = T_{\text{life}}/T_{\text{real}}$.

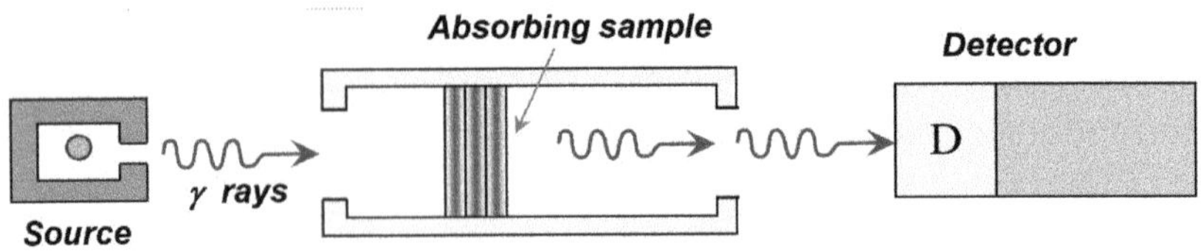

Figure 62.3 Experimental device for determining the attenuation coefficient of a sample.

Another method to determine FA consists of using a pulse generator (pulser) at a known frequency, f_{pulser}. The pulses from the pulser are injected into the detector so that a pulse is generated in some region of the spectrum free of real peaks. The number of pulses in the spectrum associated with this peak, N_{pulser}, and the number of pulses generated by the pulser, $N_{\text{real}} = f_{\text{pulser}} \times T_{\text{real}}$, are determined, where T_{real} is the real time that the experiment lasts. Thus, $\text{FA} = N_{\text{pulser}}/N_{\text{real}}$.

To determine the *absorption coefficient* of a sample, the general idea is to build a device like the one illustrated in Figure 62.3. Place the source at a fixed distance from the detector, and place between them sheets of material of known thickness and composition, whose absorption coefficient is to be studied. The effective thickness of the sample is denoted by x. The sample holder must ensure that all the rays leaving the source and reaching the detector pass through the same thickness of the absorbent sample.

For each gamma ray from the source, of energy E_γ, determine the photopeak area $A_\gamma(x)$. Furthermore, to minimize errors, it is advisable that the duration of the measurement, T_{real}, for the different values of x should not all be the same, since as the absorption increases, it is advisable to extend the measurement time to compensate for the lower rate of gamma rays reaching the detector. The *corrected effective area* of each measurement, for a given thickness x, is calculated as follows [1, 4]:

$$A_c(x, E\gamma) = \frac{A_\gamma(x, E_\gamma)}{T_{\text{life}}} = \text{FA} \cdot \frac{A_\gamma(x, E_\gamma)}{T_{\text{real}}}. \qquad (62.3)$$

Once the measurement has been made for at least four or five different thicknesses of the same material, plot $\ln[A_c(x, E_\gamma)/A_c(0, E_\gamma)]$ as

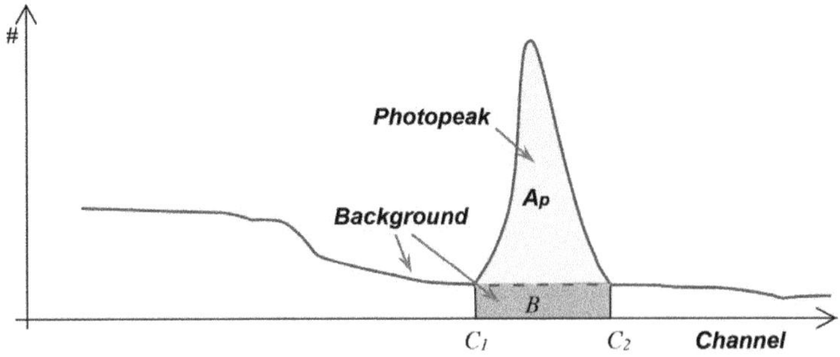

Figure 62.4 Photopeak and background area. The photopeak is delimited by the channels C_1 and C_2. The areas measured are the total area A_T and the background area between C_1 and C_2, denoted by B. The area of the photopeak is $A_p = A_T - B$.

a function of x. If in this graphical representation, the measured points are aligned, it means that Eq. (62.1) adequately describes the observed attenuation. The slope of this line is the linear attenuation coefficient $\mu(Z, E_\gamma)$.

Photopeak area: Another aspect that must be taken into account is the determination of the area of the peaks. Since these peaks are always mounted on a background (B), as shown in Figure 62.4, we need to be able to subtract this background.

The area of interest is the photopeak area, $A_p = A_T - B$. Since radioactive decay follows the Poisson statistic, the standard deviations of A_T and of B are given by $\sqrt{A_T}$ and $\sqrt{B}$, respectively. Thus, the uncertainty in the photopeak area will be [1, 4]

$$\Delta A_p \approx \sqrt{A_T + B}. \qquad (62.4)$$

62.2.1 *Project 205. Varying the dead time of the acquisition system*

> **Recommended equipment:** A gamma radiation detector associated with a data acquisition system. A radioactive source with activity less than 1 μCi.

The aim of this activity is to study the dependence of FA and TM on the counting rate of the detector. The simplest way to do this is to vary the distance between the source and the detector. According to the characteristics of your data acquisition system, choose the method that best suits the available equipment to determine the FA and TM of the available system.

Suggested tasks:

✓ Plot FA versus count frequency or count rate (number of photons detected per second). As the thickness of the absorbent increases, the number of photons reaching the detector decreases, so the FA value will also generally change as absorbents are introduced or varied. This effect could cause a systematic error in the determination of μ. The same happens if the source-detector distance is varied. Therefore, you need to account for these corrections in your measurements of the absorption coefficient.

62.2.2 *Project 206. Determining the linear absorption coefficient*

Recommended equipment: A gamma radiation detector associated with a data acquisition system. A set of radioactive sources of activities less than 1 μCi having gamma rays in the range of 100 keV–1.5 MeV. It is advisable that the gamma rays should be well separated in energy to avoid peak overlap.

Using an experimental setup like the one illustrated in Figure 62.3, investigate the dependence of the attenuation of electromagnetic radiation on the thickness of a material sample. Build a device that enables you to comfortably vary the thickness of the absorbers while keeping the source-detector distance constant. For this operation, it may be advisable to have a set of absorbers of the same shape. The effective thickness of the sample is easily changed by placing several absorbers together. The aim is to study the variation of the attenuation (cross sections) with the atomic number Z of the sample material and the energy E_γ of the incident photons, i.e., $\mu(Z, E_\gamma)$.

Some useful sources for this project are ^{241}Am, ^{57}Co, ^{22}Na, ^{60}Co, ^{137}Cs, and ^{138}Ba because they have few gamma rays, are easy to identify, and cover an energy range of about 59 KeV to about 1.3 MeV. Regarding absorbent sheets, the following are interesting to study: Al, Cu, Fe, Cd, and Pb. Explore the possibility of other absorbers, either pure substances or compounds of known composition. The dimensions of the absorber must be such that, when interposed between the source and the detector, they prevent radiation from reaching the detector directly from the source. Sheets of approximately 6×6 cm^2 are adequate. It is desirable that the sheets of low Z materials should be thicker than those of higher Z materials. As an example, 2 mm thick aluminum and 0.5 mm thick lead sheets are suitable.

Suggested tasks:

✓ Study the attenuation effect for at least three energies as widely spaced as possible. An interesting possibility is to use ^{60}Co, ^{137}Cs, ^{138}Ba, and ^{57}Co. Choose the radioactive sources you will use so that the gamma rays achieved enable you to isolate as much as possible the different interaction mechanisms that you anticipate and that may occur.

✓ To study attenuation variation with Z, include at least three materials whose atomic numbers are as widely spaced as possible, and at least five different thicknesses for each sample. You can vary thickness by superimposing several sheets of a given material.

✓ For each thickness of the absorber, measure the spectrum of gamma radiation reaching the detector. It is advisable to extend the measurement times as the thickness of the absorber increases so that the statistics of the different spectra are comparable.

✓ In each run of your experiment, observe, and answer the following questions:

- Does the energy and/or area of the photopeak vary as it passes through the material?
- Is the width of the photopeaks, i.e., the resolution in energy, modified as they pass through increasingly thick samples?

✓ For each gamma ray under consideration, determine the area of the photopeaks in the spectrum, subtracting the background. Most spectrum analysis programs perform this operation and estimate its errors. Refs. [3] and [4] discuss techniques for performing this operation. With this information, for each gamma ray and each absorbing element:

- ✓ Plot $\ln(A_c(x, E_\gamma)/A_c(0, E_\gamma))$ as a function of the thickness of the absorbent x. Here, $A_c(0, E_\gamma)$ is the *corrected effective area* of the photopeak, i.e., the peak area of each photopeak gamma ray divided by the live count time (T_{life}). $A_c(x)$ is the corresponding area divided by T_{life} when an absorber of thickness x is interposed.
- ✓ From this plot, as long as the dependence of $\ln(A_c(x, E_\gamma)/A_c(0, E_\gamma))$ as a function of x is linear, fit a line to your experimental data, and from the slope, find $\mu(Z, E_\gamma)$ as deduced from Eq. (62.1).
- ✓ Compare your results to the tabulated values for the absorption coefficients [9–12].
- ✓ For each substance tested, graph $\mu(E_\gamma, Z)$ (mass absorption coefficient) as a function of E_γ.
- ✓ For each photon energy E_γ, graph $\mu(E_\gamma, Z)$ as a function of Z.

✓ **Optional analysis**. If for the dependence of the mass absorption coefficient μ as a function of E_γ and of Z, the following relation is proposed:

$$\mu(E_\gamma, Z, \ldots.) = c\frac{Z^n}{E_\gamma^k}, \tag{62.5}$$

where c, n, and k are constants to be determined in the energy ranges $E\gamma < 200$ keV:

- ✓ Study the validity of this parameterization and estimate the values of the constants c, n, and k for the data obtained.
- ✓ If possible, compare the observed dependencies of μ to theoretical predictions [11].

Note

In www.fisicarecreativa.com, under "Physics Experiment Resources," you will find spectral data that are useful for this experiment. These spectra were obtained with an experimental device like the one shown in Figure 61.5 using solid-state detectors.

References

[1] G. Knoll, *Radiation Detection and Measurement*, 3rd Ed., NY: John Wiley, 2010.
[2] K. Siegbahn, *Alpha, Beta, and Gamma-Ray Spectroscopy*, Vol. I and II, Amsterdam: North Holland Publ. Co, 1966.
[3] Wikipedia, "Gamma rays," 2022. Available: http://en.wikipedia.org/wiki/Gamma_ray.
[4] B. Kerur, *et al.*, "Identification of elements by X–rays interaction — A laboratory experiment," *Am. J. Phys.*, vol. 57, p. 1148, 1989.
[5] EG & G ORTEC, "Experiments in nuclear physics, laboratory manual AN34 3rd ed.," 1984. Available: https://www.ortec-online.com/service-and-support/library/educational-experiments.
[6] ORTEC-AMETEX, "Advanced measurement technology," 2022. Available: http://www.ortec-online.com/.
[7] CAMBERRA-PACKARD Industries Inc., "Nuclear measurements," 2022. Available: https://www.cpce.net/en/nuclear-environment/.
[8] S. Usman and A. Patil, "Radiation detector deadtime and pile up: A review of the status of science," *Nucl. Eng. Tech.*, vol. 50, no. 7, p. 1005, 2018.
[9] C. Lederer, *et al.*, *Table of Isotopes*, 7a Ed., New York: John Wiley and Sons, 1978.
[10] S. Chu, L. Ekström and R. Firestone, "The Lund/LBNL nuclear data search," Version 2.0, February 1999. Available: http://nucleardata.nuclear.lu.se/toi/index.asp.
[11] W. Heitler, *The Quantum Theory of Radiation*, New York: Dover Publ. Inc., 1954.
[12] E. Segre, *Nuclei and Particles*, New York: The Benjamin Publ. Co., 1977.

Chapter 63

^{40}K HALF-LIFE — NUCLEOSYNTHESIS

This chapter discusses the use of gamma radiation detectors to determine the half-lives of naturally occurring radioactive elements. Of course, their half-lives are extremely long, of the order of the half-life of the universe, and based upon this, it is possible to estimate the time when the nucleosynthesis of the elements that make up the Earth occurred. The chapter also discusses some interesting aspects of natural radioactivity.

Goals

✓ Gamma radiation detectors
✓ Nucleosynthesis of elements
✓ Measurement of long half-lives

63.1 Nucleosynthesis

The origin of the universe and life itself are questions that have always captivated humanity. Thus, perhaps it is not an accident that the first book in the Bible is precisely Genesis. Over time, various cosmological theories have been developed, endeavoring to explain the origin and current state of the universe. One of the oldest cosmological theories is that the universe is static and infinite [1–4]. This is one of the simplest theories and is consistent with the Newtonian theory of gravitation.

If the universe were static (neither expanding nor collapsing), it should also be infinite. Thus, the gravitational force on each region of the universe would be zero on average and this would prevent gravitational collapse. It was also believed that the universe was eternal, i.e., that it had always existed in its current state, including the elements it consists of.

We now know that there are several serious objections to these theories. Perhaps the most elementary objection is the existence of the night (*Olbers' paradox*) [1, 2, 5]. If the universe were eternal (had always existed) and infinite, every single line of sight from the Earth to the sky would ultimately land on a star, so the night sky should be as bright as the surface of the Sun. Yet night exists and is dark!!

Another way of understanding this paradox is to calculate the intensity of radiation reaching the Earth's surface from the stars (remember that radiation intensity is the radiant energy that reaches us per unit area and unit time). Assume that the universe is infinite and its density of stars per unit volume is n. Assume that each star has an average emission power of $\langle P \rangle$. The intensity dI that comes from all the stars that are in the spherical layer of radius r and thickness dr, with the Earth at the center of this sphere, will be

$$dI = 4\pi r^2 n \frac{\langle P \rangle}{r^2} dr. \tag{63.1}$$

By integrating this intensity for all spherical layers from the minimum radius r_{min} (associated with the distance of the nearest stars) up to the maximum radius r_{max}:

$$I = 4\pi n \langle P \rangle (\boldsymbol{r}_{\text{max}} - r_{\text{min}}). \tag{63.2}$$

If it is assumed that the universe is infinite and has always existed, then $r_{\text{max}} = \infty$, with which $I \to \infty$, which is precisely Olbers' Paradox [5]. Note that this divergence persists for any value of n and of $\langle P \rangle$. The origin of this divergence is associated with $r_{\text{max}} \to \infty$. How can this problem be solved? If the universe were finite in time, i.e., if there were a "horizon" in time T_{horiz} (this time being the instant when the stars began to shine), then all the stars that we "see" would only be those that are within the sphere radius $R_{\text{horiz}} = c.\ T_{\text{horiz}}$ (c being the speed of light). So, in Eq. (63.2), we would have to associate $r_{\text{max}} = R_{\text{horiz}}$ and the divergence would be solved.

It is interesting to note that the possible absorption by particles that may exist between the Earth and the stars, which form the interstellar medium, does not solve the problem of divergence in intensity. In an eternal universe, this interstellar material would be in thermal equilibrium with the radiation it received and would emit as much as it absorbed. Thus, by *Kirchhoff's law*, this interstellar material would be as bright as the stars, making absorption irrelevant, and the problem of divergence would persist even with such interstellar material.

Another major objection to an eternal universe is the existence of *natural radioactivity*. As will be discussed in this chapter, the existence of natural radioactivity can easily be verified. Natural radioactivity is also the cause of the high temperature that exists in the interior of the Earth: the temperature at the center of the planet is about $5,000°$C [6, 7]. These high temperatures are related to volcanic activity and the movement of tectonic plates. The *natural radioactivity* inside the Earth is also manifested by the presence of helium inside the Earth, and it is believed that the energy released in these decays is an important contributor to these high internal temperatures. Helium would quickly escape from the atmosphere since its thermal velocity is close to the gravitational escape velocity from the Earth. The origin of helium inside the Earth (found in some natural gas and oil perforations) is a consequence of the alpha decay (He nuclei) of natural radioactive chains. If the universe and the Earth had always existed, the elements it consists of would not be radioactive. Radioactive elements would have already decayed and there would be *no natural radioactivity*, no helium on Earth, no volcanoes, no hot Earth, etc. The only way to observe radioactive phenomena would be through artificial sources. Therefore, the mere existence of natural radioactivity (non-man-made radioactivity), which this chapter proposes to study, is evidence that the elements have not always existed.

According to the current view of science, the universe is believed to have originated in an event known as the Big Bang (BB) that occurred about 13.8×10^9 years ($T_{\rm BB}$) ago [8]. This theory is based on many empirical observations that accumulated throughout most of the 20th century. These studies and observations are still ongoing today. One of the first indications of the Big Bang came from the discovery by E. Hubble in the 1920s that the universe is expanding and that galaxies are moving away from each other [8]. According to Hubble's Law, the speed v with which galaxies move away from an observer is directly proportional to the distance d at which they are located:

$$v = H \cdot d, \tag{63.3}$$

where H is the Hubble constant, whose value is directly related to the life of the universe. Its value is [9]:

$$H = 71 \pm 4(\text{km/s})/\text{Megaparsec} \approx 1/T_{\text{BB}},$$
$$\text{with} \quad T_{\text{BB}} \approx 14 \times 10^9 \text{years.}^{\text{a}} \tag{63.4}$$

Another consequence of the Big Bang Theory is that during the first three minutes of the universe [2, 3], almost all the light elements were formed (^{1}H, ^{2}D, ^{4} He, ^{7}Li, ^{9}Be). In fact, a necessary test of any theory that attempts to predict primordial nucleosynthesis is to account for the relative abundance of these elements. The stars were formed later, and during their final phase, they generated the heavier elements ($A > 4$). Thus, according to this model, the heavier elements that exist on Earth ($A > 4$) were created through nucleosynthesis processes during the life and death of those stars [1–3]. Thus, Sun, the Earth, the planets, and we ourselves are the "ashes" of those primordial stars.

From the nuclear perspective, an easily observable consequence is that the natural radioactive elements (i.e., those that were produced naturally, are radioactive, and are still present on Earth) must have half-lives of the order of or greater than 10^9 years. If they had shorter half-lives and had been generated together with the rest of the elements before the Earth was formed, they would have already been exhausted. By looking at a table of isotopes [10, 11], it is easy to verify that this statement about the half-life of natural elements is indeed true, i.e., all natural radioactive elements indeed have

[a]1 Megaparsec is equal to 3.0857×10^{22} m

half-lives of the order of or greater than 10^9 years. Another test of the Big Bang Theory was to predict the existence of cosmic background radiation (CBR), which was discovered by A. Penzias and R. Wilson in 1965 [12].

A gamma radiation detector placed almost anywhere to collect the background gamma radiation (without any artificial radioactive source nearby) will soon detect signs of conspicuous radiation [13] coming from the disintegration chains of $^{235}U(T_{1/2} \approx 7.04 \times 10^8$ years), $^{238}U(T_{1/2} \approx 4.47 \times 10^9$ years), $^{232}Th(T_{1/2} \approx 1.41 \times 10^{10}$ years), $^{40}K(T_{1/2} \approx 1.28 \times 10^9$ years), etc. These radioactive elements are very prevalent in almost any place.

This chapter will focus on determining the half-life of ^{40}K. It is an isotope that is present in a proportion of 1.2×10^{-4} in natural potassium (^{39}K (93.26%) and ^{41}K (6.73%)). Potassium ^{40}K emits a characteristic gamma ray, with energy $E_\gamma = 1461$ KeV, which is very easy to detect. Potassium is found almost everywhere on Earth; in walls, in the ground and in many other common places such as some fruits like bananas.

63.2 Gamma-ray spectroscopy

An experimental design for detecting gamma radiation is outlined [14] in Figure 63.1. This is a basic gamma spectroscopy system [15]. The objective of the spectroscopic technique is to determine the energy and intensity of the incident photons. The detector can be a sodium iodide (NaI) scintillator or a hyper-pure germanium (HPGe) detector [15–17].

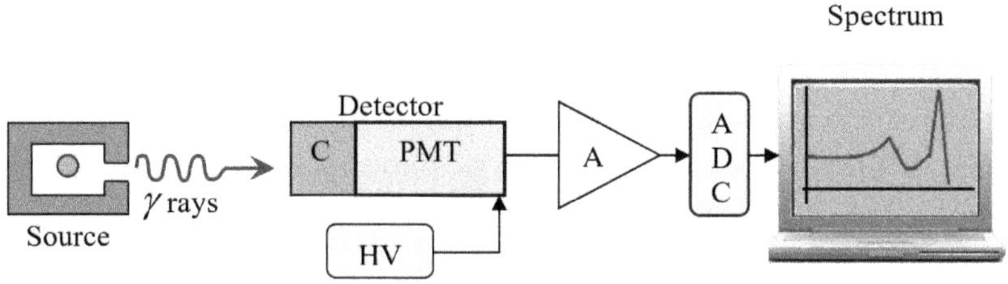

Figure 63.1 Diagram of the experimental device for obtaining gamma spectra from a source, using a scintillation detector.

As discussed in Chapters 60 and 61, a multichannel device analyzes pulse height, converting an analog signal (voltage) into an equivalent digital number (channel number). Finally, a computer program associated with the multichannel produces a histogram of the heights of the pulses, which represents the energy of the gamma rays. A more detailed description of the various components of the experimental device can be found in the literature [16–18]. In this type of detector, an energy calibration must be conducted first, as described in Chapters 60 and 61.

63.3 Efficiency of a gamma ray detector

Once the energy calibration has been performed, it is useful to determine the absolute photopeak efficiency. To do so, the geometry where the samples will be located must be defined. It is important to ensure that these positions are as reproducible as possible. Once the geometry to be used has been defined, we can perform the calibration of the detector's efficiency [15, 16]. To determine the detector's absolute photopeak efficiency, we need radioactive sources calibrated in intensity, which are available from many manufacturers.

The *absolute photopeak efficiency* is the fraction of the number of counts recorded by the detector in the photopeak during a given time interval, relative to the number of gamma rays of that energy emitted by the source during this same time interval. It is calculated as the net area of the peak in the spectrum divided by the number of photons emitted by the source in the same time interval.

If the absolute activity of the source when it was calibrated and its half-life are known, it is possible to calculate the current activity $A_f(t)$. If the gamma ray of interest is E_γ and the fraction of emissions of this ray for each decay of the source is p_γ (p_γ is the absolute intensity of the gamma ray, and the values can be found in the tables in Refs. [19–21]), then the number of photons of this type emitted by the source per unit of time will be

$$\begin{aligned} N_{\text{emit}}(E_\gamma) &= p_\gamma A_f(t_{\text{current}}) \\ &= p_\gamma A_f(t_{\text{calibration}}) \times 2^{(t_{\text{currenl}} - t_{\text{calibration}})/T_{1/2}}. \end{aligned} \quad (63.5)$$

Here, t_{current} is the time when the measurement is performed and $t_{\text{calibration}}$ is the time when the source was calibrated, $T_{1/2}$ is the half-life of the source. If the number of these gamma rays accumulated by the detector in the photopeak of the spectrum per time unit is $N_{\text{spectr}}(E_\gamma)$, the absolute efficiency of photopeak to energy E_γ will be

$$\varepsilon_{\text{abs}}(E_\gamma) = \frac{N_{\text{spectr}}(E_\gamma)}{N_{\text{emit}}(E_\gamma)}. \tag{63.6}$$

Clearly, the absolute efficiency depends on the distance from the source to the detector, therefore geometry of the measurement must be well defined. If it is not possible to determine the absolute efficiency experimentally, you can consult tables or the documentation provided by the detector manufacturer to find the values of the photopeak efficiencies at the energies of interest. For NaI (Tl) detectors, there are many references providing the absolute photopeak efficiencies for detectors of a certain size and at a certain distance from them [22].

The experimental device illustrated in Figure 63.1 can be used to find the gamma-ray spectra for a set of known sources covering a range of photon energies that include the gamma rays of interest ($E_\gamma = 1461$ KeV). Reference sources such as ^{22}Na, ^{60}Co, ^{133}Ba, ^{137}Cs, and ^{207}Bi may be adequate for carrying out the energy and efficiency calibration of the detector in the data acquisition system, for which the absolute activity of the reference sources must be known. This consists of making one graph of the positions of the centroids of the peaks (expressed in channels) as a function of the energies of the gamma rays emitted by each source, and another graph of the photopeak efficiency as a function of the energy of the gamma rays. As mentioned above, the efficiency should be calibrated [15, 16] by positioning the reference sources in the same place as will be used to position the samples to be investigated.

63.4 Determination of long half-lives

One of the most conspicuous elements, in terms of its natural radioactivity, is ^{40}K (see Figure 63.2), which emits a characteristic gamma ray of $E_\gamma = 1461$ keV.

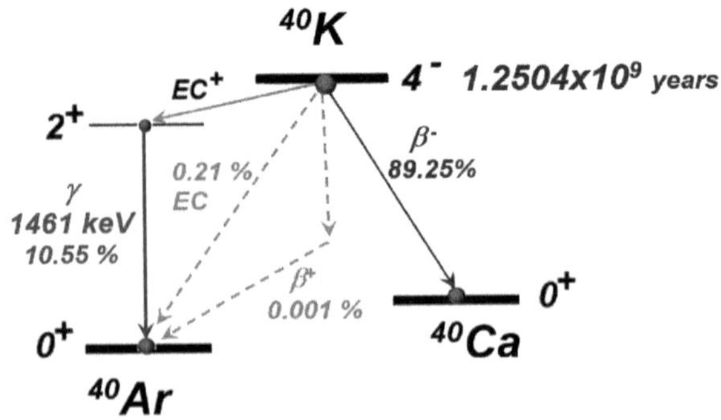

Figure 63.2 Decay scheme of ^{40}K, Refs. [20, 21]. The absolute intensity of the 1461 keV peak is 10.55(±0.11)%. The half-life reported for ^{40}K is 1.2504(±0.003) × 10^9 years. EC stands for electron capture [1].

The purpose of this experiment is to determine the half-life of this isotope (~10^9 years). The relative abundance (to the other isotopes of K) is designated as $q(^{40}\text{K})=$ 0.01167%. The experiment is based on determining the activity $A(t)$, of a sample of mass m of *KCl*, which is an inexpensive, non-toxic, easily purchased substance [23]. The number of decays per unit time, $A(t)$, of the sample in question is [1, 14]

$$\text{Activity} = A(t) = -\frac{\text{dN}(t)}{\text{dt}} = \lambda N(t) = \frac{0.693}{T_{1/2}} N(t), \tag{63.7}$$

where $N(t)$ is the total number of nuclei of ^{40}K that decay. $T_{1/2}$ is the half-life ($\lambda = \ln(2)/T_{1/2}$) of ^{40}K. The number of atoms of $^{40}K, N(t)$, can be determined from the mass of KCl, m_{KCl}:

$$N(t) = N(^{40}K) = \frac{m_{\text{KCl}}}{M_{\text{KCl}}} N_A\, q(^{40}K), \tag{63.8}$$

where $M_{\text{KCl}}(= 74.551\,\text{g/mol})$ is the molecular mass of KCl. Let $q(^{40}\text{K})(\approx 0.0001167)$ be the proportion of ^{40}K in natural potassium. N_A is Avogadro's number. In this way, Eq. (63.7) can be written as follows:

$$\text{Activity} = A_{\text{exp}}(m_{\text{KCl}}) = \frac{0.693}{T_{1/2}} \frac{N_A\, p(^{40}K)}{M_{\text{KCl}}} \cdot m_{\text{KCl}}. \tag{63.9}$$

Since the half-life of the substance in question is very long, activity A is almost independent of time, but it does depend on the mass of the *ad hoc* sample, that is, m_{KCl}. Therefore, if we can measure the A activity of ^{40}K as a function of the mass of the KCl sample m_{KCl}, we can determine the value of $T_{1/2}$. More specifically, if we measure A as a function of m_{KCl}, by plotting these variables we can test whether the hypothesis made in the deductions, Eq. (63.9), are indeed consistent with our observations. If these variables do show linear dependence, we can estimate $T_{1/2}$ from the slope of this line.

In addition to enabling verification of whether the hypotheses established in the derivation of Eq. (63.9) are correct, this procedure would enable us to recognize the possible contribution of background radiation to the detection of ^{40}K gamma rays. Radiation could come from the laboratory walls, the ground, yourself, and so on, in addition to coming from the *ad hoc* samples of KCl. If we were to measure a single sample, the result of the activity would be due to the sample plus the background K, and we would not be able to identify each contribution. On the other hand, if we measure A as a function of m_{KCl}, the value of the y-intercept ($A(0)$, for $m_{\text{KCl}} = 0$) provides the contribution of the background, i.e., the activity of ^{40}K (E_γ = 1461 KeV) in absence of the *ad hoc* samples of KCl. It is also important to measure the case of $\boldsymbol{A_{\text{exp}}(m_{\text{KCl}} = 0)}$ because it provides information on the background contribution to the activity.

63.4.1 *Project 207. Half-life of* ^{40}K

Recommended equipment: A gamma radiation detector associated with a data acquisition system. About 250 g of KCl, separated into fractions of known weight. If you do not have this equipment, you can use the spectra available on the Internet (see the note at the end of this chapter).

Note: Do not apply voltage to the detector or connect anything to the computer's multichannel until an instructor has reviewed your circuit. Exercise caution in handling radioactive sources and, after use, place them in the appropriate lead containers. Read the safety notes in Appendix B.

The objective of this activity is to measure the half-life of ^{40}K, using its characteristic gamma ray of $E_\gamma = 1461$ keV, which is easy to detect and identify with scintillation detectors such as NaI (Tl) or germanium detectors. The experiment is based on determining the activity $A_{\text{exp}}(m_{\text{KCl}})$, i.e., the number of decays per unit time of various KCl samples of different known mass m_{KCl} [14]. To do so, it is helpful to divide the sample into identical containers of approximately 10, 20, 50, 75, 100, 150, and 200 g of KCl, and also the case $m_{\text{KCl}} = 0$, background.

If you measure the photopeak area ($A_{\text{php}}(\gamma)$) of the gamma ray of $E_\gamma = 1461$ keV with a detector for which you know the absolute photopeak efficiency $\varepsilon_{\text{abs}}(E_\gamma)$ for this gamma ray and collect data for a time t_{run}, the experimental activity A_{exp} will be

$$A_{\text{exp}} = \frac{A_{\text{php}}(\gamma)}{t_{\text{run}}} \frac{1}{\varepsilon_{\text{abs}}(\gamma)}. \tag{63.10}$$

The goal is to measure A_{exp} as a function of m_{KCl}. If the hypothesis proposed are valid, a relation $A_{\text{exp}}(m_{\text{KCl}}) = b\, m_{\text{KCl}} + A_{\text{exp}}(0)$ would be expected. Here $A_{\text{exp}}(0)$ is the contribution to the activity due to the background, i.e., the contribution of the ^{40}K, present in the materials in walls, ground, etc.

It is useful to define the specific activity A_e:

$$A_e(m_{\text{KCl}}) = \frac{A_{\text{exp}}(m_{\text{KCl}}) - A_{\text{exp}}(0)}{m_{\text{KCl}}}. \tag{63.11}$$

According to Eq. (63.9), A_e would be expected to be independent of m_{KCl}. However, due to the self-absorption of gamma radiation in the sample itself, as its mass increases, so does the self-absorption. If upon plotting A_e as a function of m_{KCl}, the relationship is approximately linear, then extrapolating this dependence to $m_{\text{KCl}} = 0$ would provide the specific activity for zero mass $A_e(0)$, i.e., when there is no self-absorption, but making use of all the measured values of $A_{\text{exp}}(m_{\text{KCl}})$. We can thereby circumvent the effect of self-absorption and estimate the value of $A_e(0)$ with "zero" self-absorption. By using this extrapolated value in combination with Eq. (63.9), it is possible to find a better value of the half-life of ^{40}K, that is,

$$T_{1/2} = p_\gamma \frac{\ln 2}{A_e(0)} q(^{40}K) \frac{N_A}{M_{\mathrm{KCl}}}, \tag{63.12}$$

where p_γ is the fraction of 1461 keV gamma rays emitted for each decay of ^{40}K ($p_\gamma = 0.1055 \pm 0.0011$) and $q(^{40}\mathrm{K}) = 0.0001167$ is the fraction of ^{40}K in natural potassium. Consult the updated bibliography to verify that the values of the parameters indicated in Figure 63.2 provide a good description of the decay.

Suggested tasks:

- ✓ Perform an energy calibration of the gamma detector (photon energy as a function of the channel number).
- ✓ If possible, obtain detector efficiency calibration $\varepsilon_{\mathrm{abs}}(E_\gamma)$, that is, efficiency as a function of the energy of the photons.
- ✓ After the gamma spectrometer has been calibrated, obtain spectra without sample or source so that the background radiation is observed, i.e., $A_{\mathrm{exp}}(m_{\mathrm{KCl}} = 0)$. Acquire a background spectrum with enough statistics to clearly see the peak of $\mathrm{E}\gamma = 1461$ keV. Here, it may be useful to make a long measurement of a few hours to get a good statistic. What other peak of radiation do you see in the spectrum?
- ✓ Measure activity A_{exp} as a function of the different *ad hoc* samples of KCl, i.e., for all the samples available. Suggestion: Start by measuring the samples with highest masses, and leave the samples with lowest masses for later, or leave the lowest mass and the background for an overnight run.
- ✓ The main goal of this experiment is to determine the *activity* $A_{\mathrm{exp}}(m_{\mathrm{KCl}})$ and the *specific activity* A_e (activity per unit mass) as a function of m_{KCl}. To do so, it is suggested to construct the following plots:
  - $A_{\mathrm{exp}}(t)$ as a function of the mass of the m_{KCl} (*ad hoc* samples). If this relationship is linear, find the half-life value of ^{40}K from the slope (b).
  - A_e as a function of m_{KCl}. If $A_{\mathrm{exp}}(0) \approx 0$, estimate the half-life of ^{40}K from $A_e(0)$.
- ✓ In both cases, estimate the half-life of ^{40}K and its uncertainty.

Note: On the website www.fisicarecreativa.com (Resources for experiments) you can find background spectra taken in Buenos Aires City (Argentina) with a high-resolution detector. From these spectra, you can try to identify the peak of ^{40}K and the peaks of other elements. Gamma-ray spectra for various KCl samples can also be found using a high-resolution germanium detector.

References

[1] E. Henley and A. Garcia, *Subatomic Physics*, 3rd Ed., NJ: World Scientific publ. Co., 2007.
[2] J. Audouze and S. Vauclair, *An Introduction to Nuclear Astrophysics*, Holland: D. Reidel Pub. Co., 1980.
[3] S. Weinberg, *The Three First Minutes*, New York: Basics Books, 1977.
[4] Wikipedia, "Static Universe," 2022. Available: https://en.wikipedia.org/wiki/Static_universe.
[5] Wikipedia, "Olbers'Paradox," 2022. Available: https://en.wikipedia.org/wiki/Olbers%27s_paradox.
[6] "Why is the earth's core so hot? And how do scientists measure its temperature?" Scientific American, 6 10 1997. Available: https://www.scientificamerican.com/article/why-is-the-earths-core-so/.
[7] S. Earle and K. Panchuk, *Physical Geology*, 2nd Ed., Vancouver: UBC, 2019.
[8] Wikipadia, "Age of the universe," 2022. Available: https://en.wikipedia.org/wiki/Age_of_the_universe.
[9] Wikipedia, "Hubble's law," 2022. Available: https://en.wikipedia.org/wiki/Hubble%27s_law.
[10] National Nuclear Data Center (NNDC) at Brookhaven National Laboratory, "NuDat 3.0," 2022. Available: https://www.nndc.bnl.gov/nudat3/.
[11] Standford University, "Table of isotopes," 2022. Available: https://galprop.stanford.edu/download/plots/TableofIsotopes3.html.
[12] Wikipedia, "Microwave background radiation," 2022. Available: https://en.wikipedia.org/wiki/Cosmic_microwave_background.
[13] P. Siegel, "Gamma spectroscopy of environmental samples," *Am. J. Phys.*, vol. 81, p. 381, 2013.
[14] K. Gopal, *et al.*, "A laboratory experiment for determining the partial half-life of 40K for beta emission," *Am. J. Phys.*, vol. 40, p. 721, 1972.
[15] EG & G ORTEC, "Experiments in nuclear physics, laboratory manual AN34 3rd ed.," 1984. Available: https://www.ortec-online.com/service-and-support/library/educational-experiments.

[16] G. Knoll, *Radiation Detection and Measurement*, 3rd Ed., NY: John Wiley, 2010.
[17] CANBERRA-PACKARD Industries Inc., "Nuclear measurements," 2022. Available: https://www.cpce.net/en/nuclear-environment/.
[18] ORTEC-AMETEX, "Advanced measurement technology," 2022. Available: http://www.ortec-online.com/.
[19] S. Chu, L. Ekström and R. Firestone, "The Lund/LBNL nuclear data search," Version 2.0, February 1999. Available: http://nucleardata.nuclear.lu.se/toi/index.asp.
[20] C. Lederer, *et al.*, *Table of Isotopes*, 7a Ed., New York: John Wiley and Sons, 1978.
[21] R. Firestone and J. Tauren, "WWW table of nuclear structure EVTEC,V. 1.0b," December 2003. Available: https://www-nds.iaea.org/relnsd/vcharthtml/VChartHTML.html.
[22] Mirion Technologies, "GAMMA AND X-RAY DETECTION," 2022. Available: https://www.mirion.com/learning-center/nuclear-measurement-fundamental-principles/nuclear-measurement-fundamental-principle-gamma-and-x-ray-detection.
[23] Wikipedia, "Potassium chloride," 2022. Available: https://en.wikipedia.org/wiki/Potassium_chloride.
[24] Wikipedia, "Hubble's Law," 2022. Available: https://en.wikipedia.org/wiki/Hubble%27s_law.

Chapter 64

DETERMINATION OF THE FORBIDDEN ENERGY BAND OF SEMICONDUCTORS

This chapter looks at some characteristics of semiconductor materials. In particular, it discusses some experiments that enable the determination of the forbidden energy band or band gap of semiconductors using Si and Ge diodes.

Goals

✓ Forbidden band or band gap in semiconductors

✓ Measurement of band gap in Si and Ge

64.1 Forbidden energy band of semiconductors

A remarkable characteristic of some materials is that at low temperatures, they have high electrical resistivity and, in contrast to what happens in metals, this resistivity decreases as the temperature increases. Although the behavior of these materials, called *semiconductors*, had been known for a long time, their characteristics were not fully understood until the development of the *band theory* of a solid, around 1930. In the framework of this theory, in a solid, instead of the electrons having discrete energy levels as they do in free atoms, the allowed energy states form bands. The lower band corresponds to the states of the electrons that are most tightly bound to the atoms, which is called the *valence band*. The upper band, where the electrons can move through the solid, and therefore participate in electrical currents, is called the *conduction band*. In *insulators*, the electrons in the valence band are separated by a large band gap from the conduction band, so there are few or no electrons in the conduction band that can carry electricity. In contrast, in *conductors* (metals) the valence band overlaps the conduction band, and the conduction band always has electrons that can move and transport electricity. In semiconductors, there is a relatively small band gap between the valence and conduction bands. In a pure semiconductor, at zero absolute temperature ($T = 0$ K), all the orbitals in its valence band would be occupied, while the conduction band would be empty. In such a situation, no electric current can flow; therefore, at $T = 0$ K, the semiconductor behaves as an *insulator*. Electrical conductivity of non-metals is determined by the possibility of electrons being excited from the valence band to the conduction band. In a semiconductor, the two energy bands are separated by an energy gap (forbidden band or *band gap*) that an electron must overcome in order to be able to conduct electricity [1, 2].

At higher temperatures, $T > 0$ K, the conductivity of a semiconductor is not zero, due to the presence of electrons in the conduction band and holes in the valence band. This alteration in the occupancy of the band occurs by two mechanisms. One is the thermal excitation of electrons from the valence band to the conduction band. Another way of promoting the electrons to the conduction band can be achieved by an external excitation such as irradiation with photons that deliver enough energy to the electron to overcome the band gap. In any of these processes, pairs of charge carriers are created: electrons (negative charge, n) and holes ("vacancy of electron" with positive charge, p).

A third way to achieve non-zero conductivity is by adding impurities to the material, for example, by doping the semiconductor with traces of other atoms. If a trivalent atom is introduced (usually indium or aluminum) into the semiconductor material, a *p-type* semiconductor is obtained. If the impurity is from a pentavalent substance (such as arsenic or phosphorus), an *n-type* semiconductor is formed [1, 2].

A diode is created when a region of p-type semiconductor is placed next to a region of n-type semiconductor. This is why diodes are also known as *p–n junctions.* A diode has an abrupt variation in doping in one direction, with one part doped with p material and the other with n material [1]. Similarly, transistors can be made by assembling a *npn* or a *pnp* junction.

There are several experimental techniques to determine the band gap of a semiconductor, including using the volt-ampere characteristics of an n–p junction (of the diode or transistor) at several temperatures [3, 4] or measuring the optical transmittance of a thin sample of the material, using a spectrophotometer [5–7], to mention just two of them.

64.2 Semiconductor diodes

Semiconductor diodes are two-terminal electronic components that are very useful in multiple applications [8]. They allow the flow of current in only one direction. Chapters 24 and 50 discussed some of their characteristics, which are analyzed in greater detail in this section.

Chapter 50 showed that a diode can be used as a simple, inexpensive way to measure temperatures. The volt-ampere characteristic of a diode strongly depends on temperature. Indeed, the difference in potential, V, between the terminals of the diode is related to the current, I, that passes through it, and to the (absolute) temperature, T, which is described by the *Shockley diode equation* or *Shockley's equation* [9–11]:

$$I = I_0 \left[\exp\left(\frac{eV}{\eta kT}\right) - 1\right], \tag{64.1}$$

where e is the charge of the electron ($e = 1.6 \times 10^{-19}$C), V is the applied voltage in volts, η is a *non-ideality factor* whose value varies depending on the joint, k is Boltzmann's constant ($k = 8.617 \times 10^{-5}\text{eV} \cdot \text{K}^{-1}$), and T is the absolute temperature in Kelvin. The non-ideality factor takes into account other physical phenomena such as surface effects, recombination, tunnel effect, etc., and its value usually ranges from 1 to 2 [3]. At room temperature, $\eta kT \approx 0.025$ eV.

If $V \geq 0.1\text{V}$, $eV/\eta kT \geq 4$ and the 1 in the second member of Eq. (64.1) can be neglected at ambient temperatures or below. The factor I_0 is called the inverse saturation current of the p–n junction and depends on the temperature according to [9–11]

$$I_0 = \text{BT}^3 e^{-\frac{E_g(T)}{\eta kT}}, \tag{64.2}$$

where $E_g(T)$ is the value of the energy gap between the bands or *band gap* of the semiconductor, which generally depends on the temperature: $E_g(T) \approx E_g(0) + \alpha T + \cdots$ [3], and B is a constant that depends on the densities of the carriers n and p, and on their mean free path. If $\eta eV \gg kT$, then Eq. (64.1) can be approximated by

$$\begin{aligned} I = I_0 \cdot \exp\left(\frac{eV}{\eta kT}\right) &= Be^{(\alpha/\eta k)} T^3 \exp\left(\frac{eV - E_g(0)}{\eta kT}\right) \\ &\approx B_0 T^3 \exp\left(\frac{eV - E_g(0)}{\eta kT}\right). \end{aligned} \tag{64.3}$$

The goal of this chapter is to explore ways to estimate the value of the *energy-gap* or *band gap* $E_g(0)$ in semiconductors by means of simple experiments studying the variation in the volt-ampere characteristics (I–V curve) at different temperatures. The value of $E_g(T)$ should not be confused with $E_g(0)$, which is the extrapolation of the band gap at $T = 0$ K. The term *band gap* is reserved for $E_g(0)$ [3]. Equation (64.3) shows that if I–V curves are measured at *different temperatures*, the value of $E_g(0)$ (*band-gap*) can be determined. On the other hand, if the current I_c through the diode (forward-biased) remains constant, from Eq. (64.3),

$$I_c \approx B_0 T^3 \exp\left(\frac{eV - E_g(0)}{\eta\, k\, T}\right) \propto \exp(C_0) \rightarrow eV \approx C_0 \eta\, k\, T - E_g(0), \tag{64.4}$$

where C_0 is a constant [3]. Therefore, if the current through a diode is kept constant, the voltage drop across the diode, V, is proportional to the temperature, which is the property used in Chapter 50 to employ a diode as a thermometer. However, if for a diode (or other *pn* junction), the forward bias current is kept constant ($I_c =$ constant), then according to Eq. (64.4), a plot of eV as a function of the absolute temperature T should be a straight line in which y-intercept gives the value of $E_g(0)$.

P–N junction diodes are the most common semiconductor diodes. Silicon diodes (such as 1N4001, 1N4148, or equivalent) and germanium diodes (1N34A, AA112, or equivalent) are suitable for these types of experiments [3, 4, 11, 12].

Figure 64.1 shows two possible circuits to determine the *I–V* characteristic of a diode, although any current source could be used to keep the current fixed throughout the test.

The following projects present two possible proposals for estimating the $E_g(0)$ or *band gap* of diodes by studying the volt-ampere characteristics of a *pn* junction (of the diode or transistor) at several temperatures [3, 4].

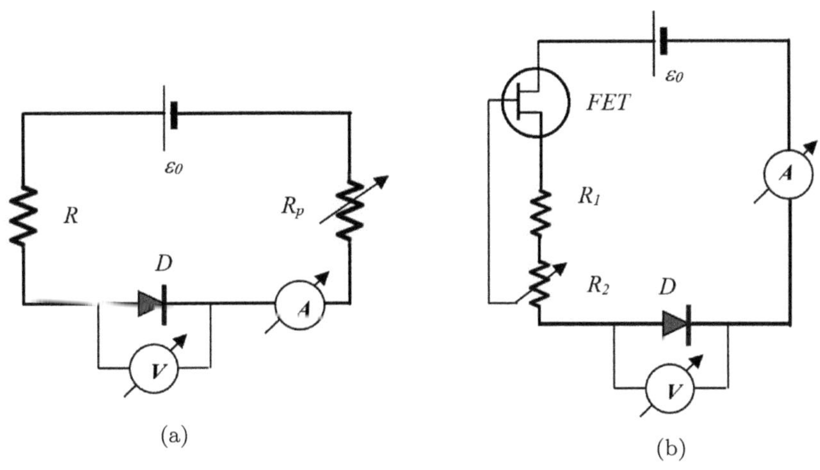

Figure 64.1 Possible circuits that can be used to characterize the *I–V* curves of a diode and estimate its *band gap*. In both cases, it is a matter of generating a constant current circuit and measuring the voltage drop across the diode. In circuit (a), the resistances R and R_p are high enough to limit the current. In circuit (b), an FET (for example, 2SK30A-GR or similar) and two resistors are used to control the current ($R_1 \approx 150\text{k}\Omega, R_2 \approx 100\text{k}\Omega$, and $\varepsilon_0 \approx 9\text{V}$). Of course, any current source could be used to keep the current fixed throughout the experiment.

64.2.1 *Project 208. Determining the band gap of Si and Ge by means of electrical measurements*

Recommended equipment: Si (1N4001 or 1N4148 or equivalent) and Ge (1N34A or AA112 or equivalent) diodes. Current source of approximately 1 mA and a voltmeter or a computer data acquisition system. A mini cryogenic Dewar flask or thermos with liquid nitrogen. A thermometer that measures temperatures ranging from the temperature of liquid nitrogen to room temperature, PT100, or thermocouple.

The goal of this activity is to study how the I–V curve of a diode changes with temperature [4, 10, 11]. For this experiment, it is advisable to fix the diode and the thermometer to a copper or aluminum bar with good thermal coupling between the bar, the diode, and the thermometer so that the temperature measured by the thermometer is representative of the temperature of the diode, but at the same time there is no electrical contact between the thermometer and the circuit of the diode.

When the bar with the diode and the thermometer approaches the thermos with liquid nitrogen, the temperature of the system changes. You can use a calibrated RDT (PT100 or other) or a thermocouple that measures temperatures ranging from 77 K (liquid nitrogen) to room temperature (see Chapter 50). Almost any commercial stainless steel insulated water bottle can be used as the thermos for these experiments.

Suggested tasks:

- ✓ Obtain I–V characteristics of diodes at different temperatures. To do so, for a given temperature, vary the applied current I (between 0 μA and 500 μA) and measure the voltage V, trying to cover the widest range of their possible values.
- ✓ Plot I as a function of V for each temperature. Also plot $\ln(I)$ as a function of V. Note that if the conditions leading to Eq. (64.1) hold, this representation is linearized according to

$$\ln(I) = \ln(I_0) + \frac{eV}{\eta kT}. \qquad (64.5)$$

For each temperature T, from the plot of $\ln(I)$ versus eV/kT, find the best value of I_0 and build a table of I_0 as a function of T.

✓ Based on these representations, $\ln(I)$ as a function of eV/kT, if Eq. (64.5) holds, the slope of the fitting line could provide information on the value of η. Determine the best value of the parameter η for each temperature and each type of diode. If possible, find the best values of η for Si and for Ge, and their uncertainty.

✓ From Eq. (64.3), it can be seen that

$$\ln(I_0) = \ln(B) + 3\ln(T) + \frac{\alpha}{\eta k} - \frac{E_g(0)}{\eta kT} \approx K_0 - \frac{E_g(0)}{\eta kT}. \quad (64.6)$$

Therefore, if $z \equiv \eta kT \cdot \ln(I_0)$ is plotted as a function of T, it would be expected that

$$z \equiv \eta kT \cdot \ln(I_0) \approx \eta kT \cdot K_0 - E_g(0) = B_0 \cdot T - E_g(0) \quad \text{or}$$
$$z \approx B_0 \cdot T - E_g(0). \quad (64.7)$$

Therefore, if the assumed hypotheses are valid, the graph of z as a function of T should be a line whose y-intercept is precisely $E_g(0)$. The tabulated values of this parameter for Si and Ge are $E_g(\text{Si}) \approx 1.17$ eV and $E_g(\text{Ge}) \approx 0.75$ eV, respectively.

Not that if $\ln(I)$ is plotted versus $V/\eta T$, according to Eq. (64.5), a straight line would be expected, whose slope could provide an estimation of the ratio of e/k, which are very important universal constants in Physics.

64.2.2 *Project 209. Determining the band gap of Si and Ge II*

Recommended equipment: Si (1N4001 or 1N4148 or equivalent) and Ge (1N34A or AA112 or equivalent) diodes. Current source of approximately 1 mA and a microammeter. A thermos with liquid nitrogen. A thermometer that measures temperatures ranging from the temperature of liquid nitrogen to ambient temperature, PT100, or thermocouple.

The aim of this experiment is to use a constant current through the diode to measure the voltage drop across the diode as the temperature varies [3], in other words, to calibrate the diodes as a thermometer, as discussed in Chapter 50. It is advisable to use a circuit like the one illustrated in Figure 64.1(b) or a current source. Adjust the current to the order of about 0.1 mA. Again, the thermometer and the diode should be thermally coupled to a metallic copper or aluminum bar without electrical contact between the thermometer and the diode circuits. Keeping the current I_c through the *pn* junction constant, Eq. (64.4), measure the voltage drop across the diode V_D for different temperatures T.

When the bar with the diode and the thermometer approaches the liquid nitrogen inside the thermos, the temperature of the system changes.

Suggested tasks:

✓ Keeping the current through the diode constant, perform a voltage drop calibration across the diode, V_D, as a function of the temperature, T. An appropriate temperature range for this experiment can be between about −50°C and room temperature, or −50°C to 70°C. For temperatures higher than room temperature, you can use a hot water bath to heat the metal bar.

✓ Plot V_D as a function of T, for T measured in Kelvin. Discuss whether these graphs are linear.

✓ If the plot of eV_D as a function of T is indeed linear, find the best value of $E_g(0)$ for each type of diode (Si and Ge) and estimate their corresponding errors or uncertainties.

✓ Compare the values you find to those for the same semiconductor reported in the literature [13, 14].

64.3 Optical absorption of a semiconductor

One way to study the behavior of a semiconductor consists of analyzing its optical transmission spectrum. Naively, one would expect the photon with less energy than the band gap, $E_\gamma < E_g$, to travel freely through the crystal without much interaction, and to have high transmissivity ($T \sim 1$), and those with higher energy than E_g to interact much more with the sample and maybe have lower transmission. Therefore, the optical transmission of a thin sample of semiconductor may be expected to present a jump in its optical transmission when the photons have energy comparable to E_g.

Nowadays, many modern laboratories have *spectrophotometers* [15, 16], whose spectrum range extends the infrared (IR), and which are very appropriate for studying the transmittance of sample [15]. There are also other ingenious ways to build a device to carry out this measurement using a monochromator [16]. This is an interesting and extensive field of knowledge, of which a brief account is provided below.

The principle of the experimental technique consists of irradiating a thin semiconductor sample with a beam of monochromatic radiation and measuring the transmissivity as the function of the wavelength of the beam. Modern spectrophotometers include a monochromator to filter light of a well-defined range of wavelengths. When the mechanism of interaction of the incident photons changes as the photons pass through the sample, a change in the response to the transmission of radiation is observed.

Semiconductor samples (for example, Si and Ge) must be thin to increase the sensitivity of the measurements. For this project, the samples can be cut from a block with a diamond saw and polished until they are less than about 0.5 mm thick.

When a monochromatic beam of light of wavelength λ and intensity I_0 goes through a sample of thickness x, and I is the transmitted intensity; the transmission coefficient T is defined as $T = I/I_0$. It is also related to the optical absorption coefficient α, as defined by [15]

$$T = \frac{(1-R)^2 \exp(-\alpha x)}{1 - R^2 \exp(-2\alpha x)}, \tag{64.8}$$

where R is the optical reflection coefficient at the air-semiconductor interface.

The predominant mechanism of absorption comes from electronic transitions from the valence band separated by an energy band gap E_g. If the energy of the photon $E\gamma > E_g$, then the incident photon creates one electron–hole pair, and the absorption coefficient α is expected to be [16]

$$\alpha = A(E_\gamma - E_g)^{1/2} \tag{64.9}$$

Therefore, if the transmission coefficient T is measured as a function of the energy of the incident photon, and the coefficient α is found from Eq. (64.8), from the dependence of this coefficient on energy or wavelength, the nature of the forbidden energy band, i.e., the band gap, Eg, can be estimated. There are several methods to carry out this type of study. If you are interested, see references [7, 16, 17]. This section only proposes to scratch the surface of this technique and to study only the variation of T with the wavelength or energy of the incident photons.

64.3.1 *Project 210. Optical absorption of a semiconductor*

Recommended equipment: Si, Ge, and GaAs samples. A spectrophotometer to measure the optical transmission coefficient of a material.

Suggested tasks:

✓ Prepare the semiconductor sample from the thin samples of Si and Ge. Preferably, they should be less than half a millimeter thick.

✓ Using the spectrophotometer, obtain the transmission curve T as a function of the wavelength of the incident radiation. The range of wavelength variation should be about 850 nm (1.46 eV)–1,300 nm (0.95 eV), in steps of about 20 nm (0.14 eV).

✓ To do this, choose the appropriate range of variation of the wavelength of the incident radiation and the variation step, in order to obtain a reasonable number of points. What can you say about the optical absorption of the sample used for the different wavelengths of the incident radiation? Make a preliminary determination of the value of Eg by noting where there is a significant change in optical transmission.

✓ Using the $T(E_\gamma)$ values and the sample thickness x, find the absorption coefficient.

✓ Plot the values of $T(E_\gamma)$ of the E_γ. Do you observe a change in the behavior of T? At what energy do these changes occur?

References

[1] N. N. W. Aschcroft and N. Mermin, *Solid State Physics*, Saunders College Publishing, 1976.

[2] C. Kittel, *Introduction to Solid State Physics*, John Wiley & Sons, 1996.

[3] A. Sconza, G. Torzo and G. Viola, "Experiment on a pn junction," *Am. J. Phys.*, vol. 62, p. 66, 1994.

[4] J. Precker and M. da Silva, "Experimental estimation of the band gap in silicon and germanium from the temperature–voltage curve of diode thermometers," *Am. J. Phys.*, vol. 70, p. 1150, 2002.

[5] I. Mártil and G. González Díaz, "Undergraduate laboratory experiment: Measurement of the complex refractive index and the band gap of a thin film semiconductor," *Am. J. Phys.*, vol. 60, p. 83, 1992.

[6] J. Dharma, *et al.*, "Simple method of measuring the band gap energy value of TiO2 in the powder form using a UV/Vis/NIR Spectrometer," 2012.

[7] A. Zanatta, "Revisiting the optical bandgap of semiconductors and the proposal of a unifed methodology to its determination," *Nat. Sci. Rep.*, vol. 9, p. 11225, 2019.

[8] Wikipedia, "Diode," 2022. Available: http://en.wikipedia.org/wiki/Diode.

[9] F. Inman and D. Woodruff, "The thermometric properties of a diode," *Phys. Teach.*, vol. 33, p. 120, 1995.

[10] B. Sukheeja, "Measurement of the band gap in silicon and germanium," *Am. J. Phys.*, vol. 51, p. 72, 1983.

[11] P. Collings, "Simple measurement of the band gap in silicon and germanium," *Am. J. Phys.*, vol. 48, p. 197, 1980.

[12] C. Fisher, "Elementary technique to measure the energy band gap and diffusion potential of pn junctions," *Am. J. Phys.*, vol. 50, p. 1103, 1982.

[13] HyperPhysics, "Semiconductor band gaps," 2022. Available: http://hyperphysics.phy-astr.gsu.edu/hbase/Tables/Semgap.html.

[14] Wikipedia, "Band gap," 2022. Available: https://en.wikipedia.org/wiki/Band_gap.

[15] Wikipedia, "Spectrophotometer," 2022. Available: https://en.wikipedia.org/wiki/Spectrophotometry.
[16] BioNetwork, "How does a spectrophotometer work?," 2022. Available: https://www.youtube.com/watch?v=pxC6F7bK8CU.
[17] Perkin-Elmer, "Spectrum 3 tri-range MIR/NIR/FIR spectrometer," 2022. Available: https://www.perkinelmer.com/.
[18] A. Sconza and G. Torzo, "Spectroscopic measurement of the semiconductor energy gap," *Am. J. Phys.*, vol. 62, p. 732, 1994.
[19] J. M. Essick and R. T. Mather, "Characterization of a bulk semiconductor's band gap via a near absorption edge optical transmission experiment," *Am. J. Phys.*, vol. 61, p. 646, 1993.
[20] L. Kirchup and F. Placido, "Undergraduate experiment: Determination of the band gap in germanium and silicon," *Am. J. Phys.*, vol. 54, p. 918, 1986.

Chapter 65

HEAT CAPACITY OF A SOLID AT LOW TEMPERATURES — EINSTEIN AND DEBYE MODELS

This chapter looks at the heat capacity of a solid at low temperatures. It discusses the quantum nature of vibrations in a solid and the concept of phonons in a crystal. To explain the behavior of the heat capacity of solids at low temperatures, a summary is provided of the Einstein and Debye models of heat capacity. This chapter proposes an experiment for measuring the heat capacities of solids based on a calorimetric approach, and a simple method for determining the Debye temperature. It also discusses the Leidenfrost effect.

Goals

- ✓ Quantum nature of vibrations in a solid
- ✓ Heat capacity of a solid at low temperatures
- ✓ Einstein and Debye models
- ✓ Leidenfrost effect

65.1 Heat capacity of a solid at low temperatures

In 1819, Dulong and Petit discovered that the molar heat capacity[a] C_v of many simple metals was about 6 *cal/K.mol.* The theoretical explanation of this result was one of the first achievements of the

[a]The molar heat capacity of a chemical substance is the amount of energy required to increase the temperature of one mole of this substance by one unit of temperature. It is also obtained as the product of the *specific heat* c_v times the atomic or molecular weight M of the substance, i.e., $C_v = c_v.M$.

classical kinetic theory, and in essence, it is a consequence of the *equipartition principle of energy* [1, 2]. Consider a *mole* of a solid, with its N_A (Avogadro's number) atoms. If we assume that each atom vibrates like a harmonic oscillator around its equilibrium position in all three possible directions (x, y, z), according to the classical equipartition principle [1–4], each atom will have an average energy $\langle\varepsilon\rangle$ given by

$$\langle\varepsilon\rangle = 3 \cdot 2 \cdot \left(\frac{1}{2}k_B T\right), \tag{65.1}$$

where the first factor (3) comes from the three degrees of freedom; the second (2) comes from the two contributions to the energy of the harmonic oscillator (potential and kinetic), the term $1/2 k_B T$ is the energy associated with each degree of freedom and each energy form, and k_B is Boltzmann's constant [1, 4]. Thus, the vibration energy of the crystal will be:

$$U(\text{mol}) = N_A 3 k_B T = 3RT. \tag{65.2}$$

Here, $R = k_B N_A$ is the universal gas constant ($R = 8.3145$ J/K $\cdot$ mol $= 1.987$cal/K $\cdot$ mol). Thus,

$$C_v = \frac{dU(\text{mol})}{dT} = 3R \approx 6\,\text{cal}/K \cdot mol, \tag{65.3}$$

as established by the Dulong–Petit law [5]. However, later studies showed that this law constitutes only an upper limit of C_v. In some metals, particularly those of low atomic weight ($A < 40$), deviations from this law occur at room temperature or lower. More specifically, it was found that at low temperatures, C_v is less than $3R$, and at very low temperatures [3, 4, 6],

$$C_v \propto T^3. \tag{65.4}$$

These limitations of classical theory inspired Einstein in 1907 to develop one of the first applications of *Plank's quantum theory* to explain this effect. The basic idea of this theory is that each atom in a solid (lattice) vibrates. Due to the quantum nature of atomic oscillators, not all modes of vibration are active, particularly at very low temperatures. Therefore, due to these quantum effects, it is expected that at low temperatures the heat capacity will decrease with respect to its classical value, Eq. (65.1). Einstein attributed the specific heat of solids to the vibrations of the atoms in the lattice, each of which is regarded as an identical quantum mechanical oscillator with the same frequency. This theory was partially successful because it was able to derive Dulong and Petit's law at high temperatures and showed that the specific heat capacity goes to zero as the absolute temperature also goes to zero. Later, in 1912, this theory of the heat capacity of solids was improved by Debye [3, 4, 6].

The Debye model assumes that the system of N_A atoms has $3N_A$ normal modes of vibration. In this model, the atoms do not vibrate individually as they do in Einstein's model but collectively. Each mode of vibration of the solid is characterized by its own frequency ν and the energy of each mode of vibration is quantized. Like any quantum oscillator or mode of oscillation, its energy is given by

$$E = \left(n_\nu + \frac{1}{2}\right) \cdot h.\nu, \tag{65.5}$$

where n_ν is an integer number, which constitutes the number of *phonons* associated with each mode of vibration. Moreover, the model assumes that the number of vibration modes in each frequency interval ν is given by a frequency distribution function $g(\nu)$, which is proportional to the available phase space, that is [3, 4, 6],

$$g(v) = a \cdot \nu^2. \tag{65.6}$$

Since the total number of vibration modes is $3N_A$, the distribution described by Eq. (65.6) will extend up to a maximum frequency ν_D such that

$$\int_{\nu=0}^{\nu_D} g(\nu) \cdot d\nu = 3N_A. \tag{65.7}$$

Thus,

$$\frac{a}{3} \cdot \nu_D^3 = 3N_A. \tag{65.8}$$

Therefore, there is a distribution of normal frequencies given by

$$g(\nu) = \begin{cases} \frac{9N_A}{\nu_D^3} \cdot \nu^2 \text{ si } \nu \leq \nu_D \\ 0 \qquad\quad\;\; \text{si } \nu > \nu_D \end{cases}. \tag{65.9}$$

The total energy of the system, U, will be

$$U(\text{mol}) = \int_{\nu=0}^{\nu_D} g(\nu) \cdot \bar{E}_\nu \cdot d\nu, \tag{65.10}$$

where $\bar{E}_\nu$ is the average energy of each vibration mode given by [3, 4, 6]

$$\bar{E}_\nu = \frac{3 \cdot h\nu}{2} + 3 \cdot \left[\frac{h\nu}{\text{Exp}(h\nu/k_B T) - 1}\right]. \tag{65.11}$$

The *Debye temperature* is defined as follows:

$$T_D = \frac{h\nu_D}{k_B} \approx 4.80 \times 10^{-11} \times \nu_D\, K. \tag{65.12}$$

This shows that the *Debye temperature* is associated with the maximum vibrational frequency of the ν_D crystal. It is expected that ν_D should be related to the frequency of oscillation of an atom in the lattice, whose frequency of vibration is associated with the forces that bind it to its neighbors (*force constant*) and its mass. The forces that bind atoms to each other are related to the elastic properties of the solid, in particular, its modulus of rigidity. Hence, for atoms of comparable masses, T_D would be expected to be higher for stiffer materials. Thus, for example, T_D is expected to be higher for silicon ($M = 28$) than for aluminum ($M = 27$), which is softer than silicon. In fact, $T_D(\text{Si}) = 645$ K and $T_D(\text{Al}) = 428$ K. Moreover, lead (heavy atoms and very soft material) has one of the lowest values of T_D, ($T_D(Pb) = 105$ K). Likewise, rigid solids with low atomic numbers will have rather high Debye temperatures. Interestingly, these temperatures indicate that the value of ν_D is of the order of 10^{12} Hz. As the propagation speed of acoustic waves, C, in solids is generally between 4,000 and 7,000 m/s, the wavelength λ_D associated with the frequency ν_D will be of the order of $\lambda_D = C/\nu_D \approx 1$ nm, a distance that is comparable to the interatomic spacing. Therefore, the Debye frequency can also be interpreted as the minimum wavelength λ_D of a vibration mode of a crystal lattice that must be greater than the interatomic distance. By combining Eqs. (65.10) and (65.11), the following expression can be obtained for the heat capacity of the solid [3, 4, 6]:

$$C_v(T/T_D) = 3R\left(\frac{T}{T_D}\right)^3 \int_0^{\frac{T}{T_D}} \frac{y^4 \cdot e^y}{(e^y - 1)^2}dy$$
$$= 3Rx^3 \int_0^x \frac{y^4 \cdot e^y}{(e^y - 1)^2}, \quad \text{with} \quad x = \frac{T}{T_D}. \quad (65.13)$$

This expression has the observed limiting properties, i.e., at high temperatures $(T \gg T_D)C_V \to 3R$, and at low temperatures $(T \ll T_D)$ is proportional to T^3. Figure 65.1 shows the difference between the classic Einstein and Debye models.

Equation (65.13) is an example of a *principle of corresponding states* (PCS), that is, the characteristic that describes given properties of different substances can be expressed in a similar manner if a reduced variable, such as $x = T/T_D$, is used. In other words, if the heat capacity of any simple solid is plotted as a function of x, the same curve is obtained for all solids; see Figure 65.1. This property has been verified experimentally [3, 4, 6]. Furthermore, T_D is the only parameter of the model. If C_v is measured as a function of T, the only parameter that can be obtained from these data is, precisely, T_D. In particular, the energy required to heat a sample of mass m from the temperature T_i up to the temperature $T_f, U(T_i, T_f, m)$ will

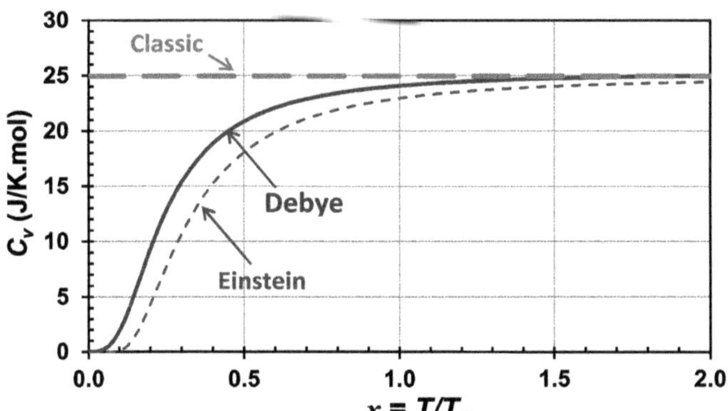

Figure 65.1 Expectations of the heat capacity of a solid according to the different theoretical models.

be given by

$$Q(T_i, T_f; m) = \frac{m}{M}(U(T_i, T_D) - U(T_f, T_D)), \tag{65.14}$$

where M is the mass of one mole of the solid in question and

$$U(T, T_D) = 3\text{RT}\left(\frac{T}{T_D}\right)^3 \int_0^{\frac{T}{T_D}} \frac{y^3}{e^y - 1}dy = 3\text{RT}x^3 \int_0^x \frac{y^3}{e^y - 1}dy. \tag{65.15}$$

It is also useful to define the average heat capacity $C_V(T_i, T_f)$ in the range from T_i to T_f as follows:

$$\begin{aligned}\bar{C}_V(T_i, T_f; T_D) &= \frac{Q(T_i, T_f; m, T_D)}{(m/M) \cdot (T_f - T_i)} \\ &= \frac{1}{(T_f - T_i)}(U(T_i, T_D) - U(T_f, T_D)). \end{aligned} \tag{65.16}$$

The objective of the following projects is to determine the value of T_D by using different experimental techniques and to observe the manifestation of the quantum nature of the crystal vibrations through a decrease in specific heat as the temperature decreases with respect to its classical value given by Eq. (65.1).

65.1.1 *Project 211. Determining Debye temperature I*

Recommended equipment: Metal samples of Cu, Al, Si, Pb, etc. Thermostat with a thermometer with a sensitivity of 0.1°C or better. One liter of liquid nitrogen.

This project requires samples of some pure metals ($\approx$ 99% purity) of approximately 100–300 g, and two expanded Styrofoam cups (or two ice cream containers) of approximately 300 ml — one with water and the other with liquid nitrogen (LN). The temperature of the cup with LN is at about $T_{LN} = 77$K. There must be enough water and liquid nitrogen (LN) to cover the sample completely, without overflowing. The mass of water (m_w) should be approximately 300 g($\pm$100 g). Let m_s be the mass of the metal sample. Assume that the water is originally at a temperature T_w, which should be about 10°C–30°C higher than room temperature [7, 8]. The body (sample) is originally under the liquid nitrogen at the temperature T_{LN}, tied to a fine cotton, nylon or steel thread. The thermometer is immersed in water and connected to a data collection system, so that temperatures can be acquired at approximately 10 Hz.

At the instant defined as $t = 0$, remove the body from the liquid nitrogen using the thread, and immerse it in the water. Wait until the temperature reaches equilibrium at final temperature T_f. The energy balance implies that

$$\frac{m_s}{M}\bar{C}_V(T_f - T_{\text{LN}}) = (m_w + M_E)c_w(T_a - T_f), \qquad (65.17)$$

where M_E is the water equivalent of the calorimeter (see Chapter 53), c_w is the heat capacity of water, and $\bar{C}_v$ is the average heat capacity of the sample. This expression enables us to determine C_V experimentally and compare it with Eq. (65.16), from which we can find T_D.

A numerical integration method is needed to make this comparison [9]. This can be done by using spreadsheets or math programs (Mathematica, MatLab, etc.), or by writing a dedicated program. Either way, it may be useful to have a plot of $C_V(T_i, T_f)$ as a function of T_D. Thus, once we have determined C_V experimentally, we can find T_D.

Suggested tasks:

✓ For each available sample, determine the value of T_D and estimate its uncertainty.
✓ Compare your results to the table values for the same substances.
✓ Discuss the degree of agreement found.

For this experiment, the value of M_E can be determined as discussed in Chapter 53.

65.1.2 *Project 212. Determining the heat of evaporation of liquid nitrogen (LN)*

Recommended equipment: One liter of liquid nitrogen (LN). A wire resistance of approximately 25 Ω and 10 W. A voltage source to supply the resistance, a voltmeter, and an ammeter. A scale with sensitivity 0.1 g or better.

The experimental method proposed here is similar to the one used in Project 184, Chapter 53. Place approximately 200 g of LN in an expanded polyurethane container (a coffee cup or a $1/4$ kg ice cream container) on a scale with sensitivity 0.1 g or better. You can use an electronic scale or a force sensor from which the container with LN hangs. Hang the resistance from a stand independent of the container. Immerse the resistance in the LN without it touching the walls of the container. Figure 65.2 shows a diagram of this experimental setup.

The first observation consists of measuring the variation in the mass of the container of LN with the resistance inside, but without passing current through it. This first measurement serves to characterize the mass variation due to the evaporation of LN in the system without disturbance. In general, but not necessarily, this variation in mass over time is linear. Next, fill the container with LN

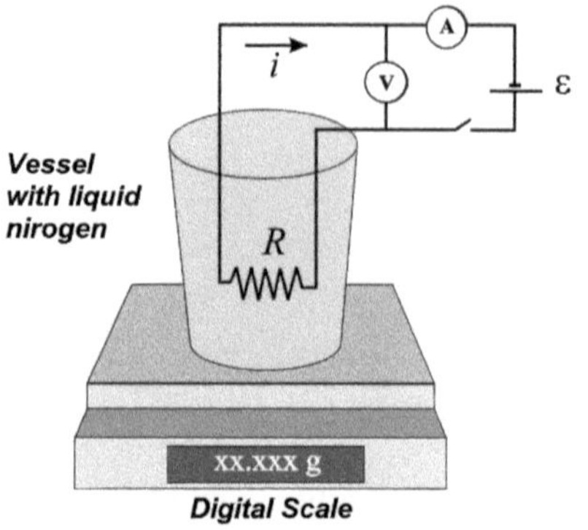

Figure 65.2 Schematic diagram of the proposed apparatus for measuring the latent heat of the LN.

again and repeat the experiment. After about 10 min of recording the mass, apply a current pulse for a time interval Δt which is small compared to the observation time. The idea is that, both before and after applying the pulse, the variation in the mass of LN over time is well defined. You will need instruments to measure the pulse current I, the voltage drop V in the resistance, and the duration (Δt) of the pulse, so that you can characterize the energy ΔU delivered by the resistance to the pulse.

$$\Delta U = I \cdot V \cdot \Delta t. \tag{65.18}$$

This energy is related to the amount Δm of LN evaporated in that interval and to the value of the latent heat of vaporization L_V of LN so that

$$L_V = \frac{I \cdot V \cdot \Delta t}{\Delta m}. \tag{65.19}$$

The determination of Δm is based on the graphical method (or its analytical correlate), illustrated in Figure 65.3.

Suggested tasks:

✓ Determine the change in the mass of LN without current, taking mass data every 5 s or faster if you have an automatic acquisition system. Plot the mass of LN as a function of time.

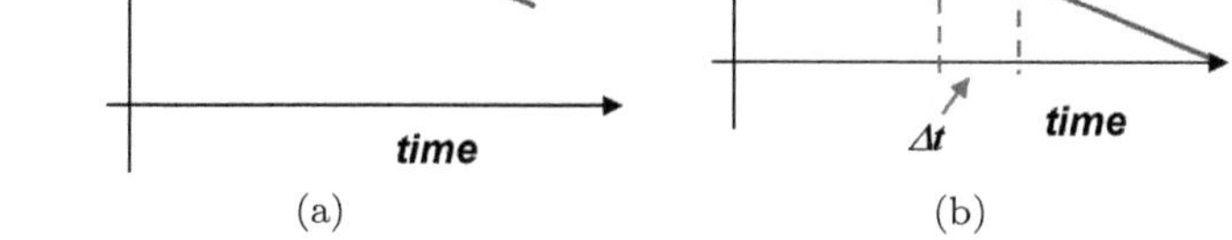

(a) (b)

Figure 65.3 Schematic diagram of the variation of the mass of LN with the resistance without current (a) and when a current pulse is passed through the resistance (b).

✓ Using a power dissipated in the resistor of approximately 30 W for a few minutes, determine the variation in the mass of LN as a function of time. Plot your results and determine Δm.

✓ Using Eq. (65.19), determine L_V and estimate its uncertainty. Compare your results to the table values.

65.1.3 *Project 213. Determining Debye temperature II*

Recommended equipment: Metal samples of Cu, Al, Si, Pb, etc. A scale with sensitivity 0.1 g or better. One liter of liquid nitrogen.

The technique proposed in this section is based on the same idea as in the previous project to determine L_V. In this case, an electronic balance or a calibrated force sensor can also be used. First, weigh the container of LN *together* with the sample that is outside the LN, so that the sample is originally at room temperature, T_a, and measure the mass variation over time. Once the trend of this variation has been well characterized, carefully place the sample inside the LN using a tweezer. Continue to measure the mass as a function of time. The recording of mass as a function of time will look something like the graph shown in Figure 65.3(b) [8, 10]. Wrapping the sample with a thin polystyrene film (kitchen wrapping film) should prevent the LN from bubbling (see the next project). It is a good idea to have a lid on the polyurethane container. It is important to note that all the same items must be weighed on the balance before and after immersing the sample inside the LN. The variation of the mass as a function of time will have a dependence as shown in Figure 65.2(b) [8]. This time, the mass variation Δm will be related to the heat delivered by the sample m_s going from room temperature T_a to the temperature of LN, T_{NL}. That is,

$$L_V \cdot \Delta m = \frac{m_s}{M} \cdot \int_{T_{\text{NL}}}^{T_a} C_V(T, T_D) \cdot dT. \qquad (65.20)$$

The values of T_a, m_s, and Δm can be calculated from the experiment. Then, using the formalism described earlier, Eqs. (65.15) and (65.20), it is possible to find T_D. Reference [11] describes a variation of the method proposed here to estimate heat capacity in narrower temperature ranges than those used in this project.

Suggested tasks:

✓ Determine the variation in the mass of LN with the sample at room temperature, taking mass data every 5 s or faster if you have an automatic acquisition system. Plot the mass of LN as a function of time, as illustrated in Figure 65.2(a).
✓ Refill the container with LN and repeat the study. After about 5 minutes, or when the change in mass over time is well defined, place the sample in the LN. Determine the change in the mass of LN as a function of time. Plot your results and determine Δm.
✓ Using Eqs. (65.15) and (65.20), determine T_D and estimate its uncertainty for the sample in question.
✓ Repeat this procedure for each sample you use.
✓ Compare the T_D values obtained experimentally to those tabulated for these same substances [3, 10]. Discuss the degree of agreement found.

Note: Excel files where the functions of Eqs. (65.13), (65.15), and (65.16) were calculated are available on www.fisicarecreativa.com (Experiment resources).

65.2 Leidenfrost effect

Around 1756, Leidenfrost published the observation of an effect which was later named after him [11, 12]. He noted that when a drop of water falls on a very hot surface, such as an iron skillet, it took longer to evaporate than when the same drop fell on the same surface when it was cooler. This is because in the first case, a vapor film is formed, which surrounds the drop of water, and prevents an optimum heat transfer between the hot surface and the drop [13]. This effect generally occurs when the temperature of the hot surface is much higher than the boiling temperature of the liquid. In the 19th century, Pierre Hippolyte Boutigny (1798–1884) studied this effect and proposed that the drops that were suspended on the surface of a hot plate constituted a new state of matter, which he called *spheroidal state.*

This effect occurs because gases and vapors are generally worse conductors of heat than liquids of the same substance. Therefore, the vapor layer surrounding the hot drop or object acts as a thermal insulator.

Some magicians and physics teachers rely on this effect to walk barefoot on a bed of embers (firewalking). Several demonstrations of these effects can be seen on YouTube [14]. The Leidenfrost effect is also observed when a drop of LN falls on the table or when a drop of water is thrown onto a very hot surface.

65.2.1 *Project 214. Observing the Leidenfrost effect*

Recommended equipment: Metal samples of Cu, steel, Al or Pb, etc. A low-temperature thermometer (thermocouple, RTD, diode, etc.) connected to a data acquisition system. Half a liter of liquid nitrogen.

The aim of this project is to conduct a semi-quantitative study of the Leidenfrost effect. It is useful to have a sample of copper or aluminum of approximately spherical shape (though the shape is not crucial) of about 20–50 g. Make a hole in the midde of the sample in which to place a thermometer. The range of the thermometer must include the boiling temperature of liquid nitrogen (77K). You can use a thermocouple, a diode, or an RTD (see Chapter 50) to measure the temperature of the sample. The thermometer must be connected to a data acquisition system with a measurement frequency of the order of 10 Hz. To achieve good thermal contact between the thermometer and the sample, you can use silicone grease of the kind used in electronic applications.

The experiment consists of placing a body suddenly into a container of liquid nitrogen and recording its temperature history — once without any kind of covering, and a second time, with the body covered in thin thermal insulation, such as Teflon tape, grease, or polyethylene wrap. The heat flux H (heat flowing per unit time) can be expressed as follows:

$$H = \frac{dQ}{dt} = m \cdot c_V \cdot \frac{dT}{dt} = -k.A.\Delta T, \tag{65.21}$$

where m is the mass of the body, c_V the specific heat, and T the temperature. Here, k is the effective thermal transfer constant of the medium-body interface, A is its area, and ΔT is the thermal difference between the body and the surroundings. If the body's temperature data are derived as a function of time, it is possible to estimate H. Remember that C_V generally varies with temperature.

Suggested tasks:

✓ Measure the thermal history, $T(t)$, of the body when a "naked" sample is placed in liquid nitrogen.
✓ Carry out the same observation when the same sample is placed in liquid nitrogen with thermal protection (insulator), i.e., the sample wrapped with thin polyethylene film.
✓ On a single graph, plot the evolution of temperature as a function of time for both experiments.
✓ Plot the derivative dT/dt as a function of time for both experiments on the same graph.
✓ Plot derivative dT/dt as a function of body temperature T, for both experiments on the same graph.
✓ Develop a semi-quantitative explanation for these results [11, 12].

References

[1] F. Sears, M. Zemansky, H. Young and R. Freedman, *College Physics*, NY: Pearson Ed., 2009.
[2] R. Feynman, *et al.*, *The Feynman Lectures on Physics, Boxed Set: The New Millennium Edition*, Vol. 1, Basic Books, 2011.
[3] C. Kittel, *Introduction to Solid State Physics*, John Wiley & Sons, 1996.
[4] D. Goodstein, *States of Matter*, NJ: Prentice Hall Inc., 1975.
[5] Wikipedia, "Dulong–Petit law," 2022. Available: https://en.wikipedia.org/wiki/Dulong%E2%80%93Petit_law.
[6] T. Hill, *Introduction to Statistical Thermodynamics*, MA: Addison Wesley Pub. Co., 1970.

[7] E. Langendijk, "A simple determination of Einstein temperature," *Am. J. Phys.*, vol. 68, p. 961, 2000.
[8] C. Deacon, J. de Bruyn and J. Whitehead, "A simple method of determining Debye temperatures," *Am. J. Phys.*, vol. 60, p. 422, 1992.
[9] S. Gil, "Fisicarecreativa (Resourses of the Book Experiment of Physcis)," Here you can find a program in Excel that allows you to calculate the heat capacity given by Eqs. (65.13), (65.14) and (65.16)., 2022. Available: https://www.fisicarecreativa.com.
[10] W. Mahmood, M. Anwar and W. Zia, "Experimental determination of heat capacities and their correlation with theoretical predictions," *Am. J. Phys.*, vol. 79, p. 1099, 2011.
[11] F. Curzon, "The Leidenfrost phenomenon," *Am. J. Phys.*, vol. 46, pp. 825–82, 1978.
[12] G. Guido Lavalle, *et al.*, "A boiling heat transfer paradox," *Am. J. Phys.*, vol. 60, p. 593, 1992.
[13] PBS Digital Studio, "The Leidenfrost effect," 2022. Available: https://www.youtube.com/watch?v=9tlIWlGvkRc.
[14] Myth Busters, "Leidenfrost effect molten led.," Youtube, 2012. Available: http://www.youtube.com/watch?v=UZio0f7fP04.
[15] C. Tompson and H. White, "Latent heat and low-temperature heat capacity for the general physics laboratory," *Am. J. Phys.*, vol. 51, no. 4, pp. 362–364, 1983.

Chapter 66

THE SOLAR CONSTANT — ATTENUATION OF LIGHT IN THE ATMOSPHERE

This chapter studies solar radiation on the Earth and the power radiated by the Sun. The latter is highly relevant to understand how the Sun and the stars work. Knowing the intensity of solar energy that the Earth receives, that is, the *solar constant*, is crucial for taking advantage of this important natural energy resource that makes life possible on our planet. In addition, studying the variation of solar radiation throughout a day enables the determination of the power which the Sun radiates, that is, its *luminosity*. The variation in solar irradiation with the solar *zenith angle*, which is the angle that Sun rays form with the vertical direction, allows us to know how this radiation attenuates in the Earth's atmosphere. These effects are of fundamental importance both in astronomy and in the variation of temperature with the seasons and the latitude of our planet.

Goals

- ✓ Determination of the solar constant
- ✓ Attenuation of radiation in the atmosphere
- ✓ Brightness of the Sun
- ✓ Maximum visibility distance on a clear day

66.1 The luminosity of the Sun — Solar constant

The issue of energy is often referred to as "the last frontier" of the 21st century. On the one hand, without energy, the production of goods and services is not possible. On the other hand, unlike other resources such as water and paper, energy itself cannot be recycled.

Energy is necessary to respond to many people's legitimate aspiration for a better quality of life. In 2022, almost 10% of the world's population still did not have access to electricity. However, there is increasingly clear evidence that global warming has anthropogenic causes, largely due to the use of fossil fuels [1], and that greenhouse gas emissions therefore need to be reduced. The efficient use of energy and the use of renewable energy, in particular solar energy, are some of the most promising responses to this energy challenge. To make use of solar energy, its characteristics need to be understood. Moreover, there is clear indication that small changes in the Sun's activity could produce large changes in the living conditions on Earth [2], so it is important to elucidate the possible contribution of variations in the luminosity of the Sun, if any, to climate change. Although the effect of these variations seems to have been important over the geologic time scale, its role in recent warming appears to be insignificant [1, 2].

The physics explaining how the Sun shines is crucial to understanding the evolution of the Sun and the stars. The nature of the Sun's brightness has been the subject of speculation since ancient times. Anaxagoras (500–428 BC), the teacher of Pericles, was imprisoned for suggesting that the Sun was an incandescent stone. During antiquity and the Middle Ages, the Sun was the paradigm of the perfect fire, until Galileo discovered sunspots. Among the many heresies of which Giordano Bruno was accused and led to his burning at the stake, was his proposition that the stars were suns. During the 19th century, several theories were proposed to explain how the Sun works, none of which were satisfactory [3–5]. The Sun's power and its age are two properties that scientists think they know very well, but which have been very difficult to reconcile.

In 1920, Arthur Eddington suggested the hypothesis that nuclear fusion is the process by which the Sun and stars generate energy.

However, efforts to confirm this hypothesis were not without difficulties. The nuclear reactions that occur inside the Sun simultaneously generate energy and a well-defined number of neutrinos, which travel unchanged to Earth [5]. Therefore, the amount of energy that reaches the Earth should be related to the amount of these particles received.

For more than thirty years, scientists have been trying to unravel the mystery of why the number of solar neutrinos detected on Earth was much lower than predicted by the model of the Sun based on Eddington's hypothesis [4–6]. A good model must make predictions that agree with the observations. Wouldn't this inconsistency of the nuclear fusion model have been an indication that perhaps our hypothesis of how the Sun works was wrong, as has occurred in the past with other models? [4, 6]. It took a little over thirty years to solve this enigma. The pioneers in solving it were awarded the Nobel Prize in Physics in 2002 and in 2015.

In general, the methods for determining the *luminosity of the Sun,* i.e., the total power output that the Sun radiates into space, consist of measuring the *solar constant,* I_{Sun}. This parameter is defined as the amount of energy received per unit area and per unit time when the Sun's rays strike perpendicular to the detection area and discount the attenuation effects caused by the Earth's atmosphere. Its value is of the order of 1.36 kW/m^2. Since the Sun-Earth distance (d_{ST}) is known, multiplying the area of a sphere of radius equal to d_{ST} by the solar constant gives the luminosity of the Sun, L_{Sun}:

$$L_{\text{Sun}} = 4\pi d_{\text{ST}}^2 I_{\text{Sun}}. \tag{66.1}$$

In this expression, it is assumed that the attenuation effects of the Earth's atmosphere in the determination of I_{Sun} have been correctly taken into account.

There are several techniques to determine I_{Sun}, some of which are highly sophisticated, while others are simple and inexpensive. This chapter presents some simple experiments to measure solar irradiance at the Earth's surface and the extinction of light in the Earth's atmosphere. With the combination of both results, the *solar constant* and the *Sun luminosity* can be found.

66.2 Attenuation of light in the atmosphere

Even though it may seem that to find the value of the solar constant, I_{Sun}, according to its definition, should be measured from a satellite located outside the Earth's atmosphere, it is not in fact necessary to do so. This chapter analyzes a simple radiation attenuation model that enables the value of solar irradiance measured on the Earth's surface to be related to the value of I_{Sun}. To do this, we need to understand how sunlight is dimmed as it travels through the Earth's atmosphere.

As it passes through the atmosphere, light is attenuated by scattering and absorption in the molecules that make up the atmosphere, and by the particles contained in it, which are generically called aerosols (*aerosol*: solid or liquid). Thus, in a clear sky, the greater the mass of atmosphere that the light passes through, the greater the attenuation. Consequently, the Sun and celestial bodies appear brighter at the zenith than at the horizon.

Most of the attenuation of sunlight occurs in the lower and denser layer of the atmosphere (the troposphere, which contains 75% of its mass), which is about 11 km thick. The radius of curvature of the troposphere is the radius of the Earth, that is, 6,371 km. Therefore, for the atmosphere, its radius of curvature is about 600 times greater than its thickness.

To analyze the problem of attenuation of the atmosphere, consider a simplified model of light attenuation in the atmosphere, assuming that the Earth and atmosphere are flat. Once this model has been developed, it is easy to generalize it to the case of a spherical Earth [7]. The approximation of the flat atmosphere is a good approximation, as long as the Sun is not close to the horizon, since in this case, the curvature effects are important. The *plane atmosphere model* is applicable to the case in which the angle formed by the Sun's rays with the vertical of the place, i.e., the *zenith angle*, is less than about 75°C.

Consider a beam of light or any other electromagnetic radiation of intensity I that passes through a sample that has N scattering centers (atoms or molecules) per unit volume. Each of these

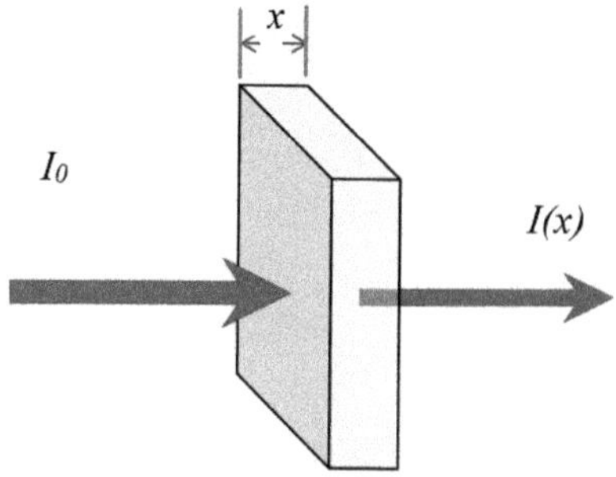

Figure 66.1 Attenuation of a beam of light when passing through a sample of thickness x, whose absorption coefficient is $\alpha(\lambda)$. I_0 is the incident intensity, and $I(x)$ is the intensity after passing through the sample.

centers has a probability σ(cross-section) of interacting with the light beam [8, 9]. As discussed in Chapter 62, when traversing an infinitesimal thickness dx of this material, the attenuation of the radiation will be proportional to the amount of interactions that take place along dx, so the intensity reduction dI in this region will be proportional to the product of the incident intensity $I(x)$ times $N\sigma dx$. In other words,

$$dI = -I(x) \cdot N\sigma dx = -\alpha \cdot I(x)dx. \tag{66.2}$$

Here, $I(x)$ is the incident intensity at position x in height in the atmosphere, and dI is its decrease after passing through a thickness dx. The parameter $\alpha = N\sigma = 1/\Lambda$ is known as the *linear attenuation coefficient* and Λ is the *characteristic extinction length*. For a distance $x = \Lambda$, intensity is attenuated by a factor of $1/e$. Equation (66.2) can be integrated to give

$$I(x) = I_0 \exp(-\alpha \cdot x) = I_0 \exp(-x/\Lambda). \tag{66.3}$$

This expression is known as the *Lambert–Beer law* (or Beer–Lambert–Bouguer law), similar to Eq. (62.2), which describes the variation in intensity of a radiation beam whose initial intensity is I_0.

In general, both α and σ depend on the frequency or wavelength of the incident radiation. It is often useful to define the mass thickness $t = \rho x$, where ρ is the density of the material, and the *mass absorption coefficient* $\mu = \alpha/\rho$.

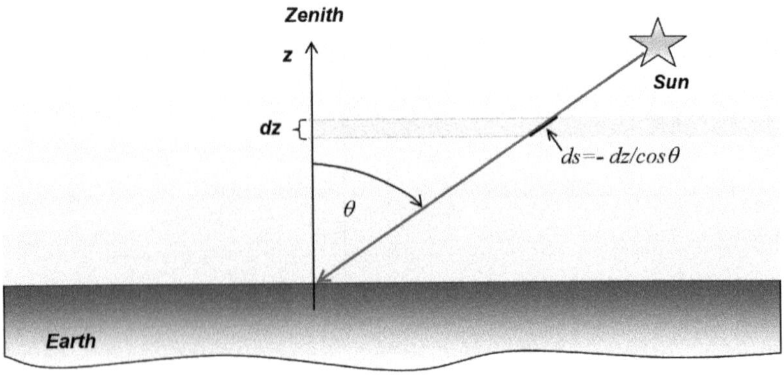

Figure 66.2 Flat atmosphere model. Here, ds is the infinitesimal distance traveled by sunlight between the heights $z, z + dz$.

Returning to the flat atmosphere model, assume that the density of air varies only with height, the z-coordinate in Figure 66.2. Air is a mixture of different gases, and σ_T and $n(z)$ represent the total absorption cross-section and the number of moles per unit volume of air at the height z above the Earth's surface. According to the Lambert–Beer law [8, 9],

$$dI = -\sigma_T n(z) N_A ds \cdot I(z) = +\sigma_T n(z) N_A I(z) \cdot dz/\cos\theta, \qquad (66.4)$$

where N_A is Avogadro's number and dI represents the variation in the intensity of sunlight passing through the longitude $ds = -dz/\cos\theta$ of air; thus, $\alpha = \sigma_{T.} nN_A$. The sign change in the third member of Eq. (66.4) is associated with the direction adopted for z in Figure 66.2.

As the penetration s of the light ray into the atmosphere increases, z decreases. Let θ be the *zenith angle* of the Sun, that is, the angle formed by the Sun's rays with the vertical direction at the site. The integral of Eq. (66.4) can be written as follows:

$$\begin{aligned}\int_0^\infty dI/I(z) &= \ln(I(\infty)/I_0) = +(N_A\sigma_T/\cos\theta)\cdot\int_0^\infty n(z)dz\\ &= +K_{\text{ef}}/\cos\theta, \qquad (66.5)\end{aligned}$$

with $K_{\mathrm{ef}} = N_A\sigma_T \int_0^\infty n(z)dz$. Since atmospheric pressure, P_0, is the weight of a column of air of unit area, then $P_0 = M_{\mathrm{Air}} g \int_0^\infty n(z)dz$, where M_{Air} is the (average) molecular weight of air ($M_{\mathrm{Air}} \approx 28.9$ g/mol) and g is the acceleration of gravity so that $K_{\mathrm{ef}} = \mathrm{N}_A\sigma_T P_0/M_{\mathrm{Air}}$. In Eq. (66.4), $I(\infty)$ represents solar radiation outside the atmosphere, that is, I_{Sun}, while I_0 is the solar irradiance on the surface of the Earth, after undergoing the attenuation of the atmosphere. Thus,

$$I_{\mathrm{Sun}} = I_0 \exp(+K_{\mathrm{ef}}/\cos\theta) \quad \text{or} \quad I_0 = I_{\mathrm{Sun}} \exp(-K_{\mathrm{ef}} \cdot \sec\theta). \quad (66.6)$$

This expression relates the Solar constant I_{Sun} with the solar irradiance I_0 measured on the surface of the Earth. One way to test the hypotheses made in this section about the attenuation of radiation in the atmosphere, which are implicit in Eq. (66.6), would be to measure the solar irradiance at the Earth's surface, I_0, as a function of the parameter $\mu \equiv \sec\theta$. If these hypotheses are correct, according to Eq. (66.6),

$$\ln(I_0) = \ln(I_{\mathrm{Sol}}) - K_{\mathrm{ef}} \cdot \mu. \quad (66.7)$$

Therefore, when plotting $\ln(I_0)$ as a function of $\mu = \sec\theta$, if the hypotheses were correct, we should obtain a line whose slope would be K_{ef} and its y-intercept or the ordinate to the origin, $\ln(I_{\mathrm{Sol}})$. Therefore, if the plot of $\ln(I_0)$ versus μ is linear, on the one hand, it would validate the hypotheses and, on the other hand, provide a way to find I_{Sun}.

Note that the value of K_{ef} depends on the value of σ_T and on the atmospheric pressure P_0. The effective section σ_T depends on the attenuation of light by air molecules (mainly N_2 and O_2) through Rayleigh scattering and particles in the atmosphere or aerosols (*Mie scattering*) [10]. If this experiment is performed on a clear, cloudless day, then air pollution by dust or aerosols would affect the slope of the line through the value of K_{ef}, but not the y-intercept. This provides a robust method for determining I_{Sol}.

The attenuation of radiation in the atmosphere, at a given pressure $P = n\mathrm{RT}$, where n is the number of moles per unit volume, is described according to Eq. (66.3) by

$$\alpha \equiv 1/\Lambda = N_A\sigma_T n = N_A\sigma_T P/\mathrm{RT} \quad \text{or}$$
$$\Lambda = (\mathrm{RT}/gM_{\mathrm{air}}/K_{\mathrm{ef}} = h_0/K_{\mathrm{ef}}, \tag{66.8}$$

with $h_0 = \mathrm{RT}/gM_{\mathrm{air}} \approx 7.79$ km for $T = -10°\mathrm{C}$. These expressions relate the value of K_{ef}, determined in the experiment, to the characteristic extinction length Λ of light in the atmosphere, which is indicative of the maximum visibility distance of an object in a clear atmosphere.

The value of the zenith angle depends on the *time* and *latitude* of the site and there are several expressions and programs for calculating it [11]. But perhaps the easiest way is to access the U.S. website Naval Observatory (USNO) [12] and enter the latitude and longitude of the measurement site, from which it is possible to obtain the value of the zenith angle θ as a function of local time. When using these tables, it is important to verify how local time is related to universal time (UTC).

66.3 Determination of solar irradiance

There are several techniques for measuring solar irradiance I_0. Its measurement has received considerable attention in the recent past, especially because of its implications for harnessing solar energy. The techniques vary in complexity from very simple and suitable for carrying out in a backyard to the most elaborate and precise [13, 14].

A simple, interesting approach is to use two identical black surfaces, one of which is protected from direct solar radiation. Both surfaces need to be connected to a temperature sensor and an electrical resistance that can be heated by passing current through it. In equilibrium, the surface exposed to the Sun will have a higher temperature than the surface that is protected or shielded from the Sun. Next, a current is passed through the resistance until the temperature of both plates is balanced. The measurement of the electrical power required for both plates to have the same temperature provides an estimate of solar radiation on the exposed plate [13].

This chapter discusses two alternatives for this technique, both equally simple and inexpensive, and capable of providing an estimate of solar irradiance with acceptable precision for a teaching laboratory.

66.3.1 *Project 215. Simple method for measuring solar irradiance*

Recommended equipment: A dozen expanded polystyrene disposable cups of about 200–300 ml (coffee dispenser cups). Two 5 cm diameter circular aluminum plates. A scale with sensitivity of 1 g or less and a range of about 400 g. Alternatively, a graduated cylinder or plastic syringe of about 20 or 30 ml can be used. A freezer in which to freeze water. An ordinary stopwatch (approx. 0.2 s accuracy).

This experiment calls for two equal circular aluminum plates, about 5 cm in diameter and about 1 mm thick. Polish one side to make the aluminum surface clean and smudge-free. Paint the other side with matte black paint. It also requires a dozen expanded polystyrene cups of about 200–300 ml (coffee vending machine or party cups) [14].

Fill a cup with water to a height such that when you put one of the disks into it, it sinks a few millimeters into the water. Mark this

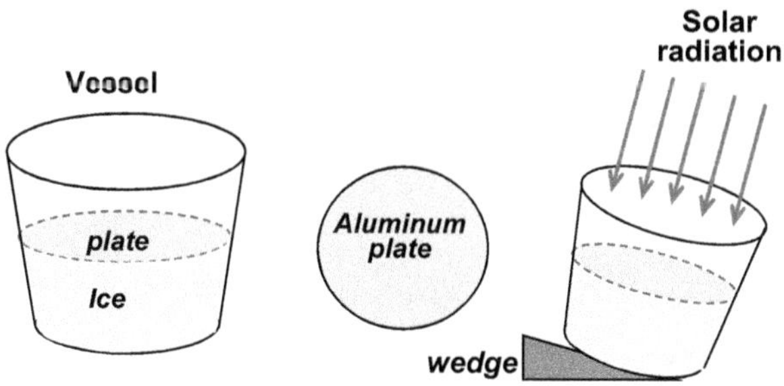

Figure 66.3 Setup of the experiment. First, mark the height on the cup where the metal disks fit into it. Fill two equal cups with water up to that height and freeze them. Then, place a plate in each cup, with the blackened surface facing upwards in one cup and the shiny surface facing upwards in the other. Point both of them toward the Sun. After about 10 min of exposure, measure the amount of water produced in each cup [14].

height and fill 10 or 12 glasses with water up to the same height. Freeze them for several hours, without the plates. You could do this the night before performing this activity. The experiment should be performed on a clear, cloudless day. The measurements should be made around noon. Remove two cups of frozen water from the freezer and allow them to rest at room temperature for about 15 min before beginning the experiment. The temperature of ice should be close to 0°C (they may be colder in the freezer).

Remove any melted water from the cups. Place a disk on the ice in each cup, one with the polished metal side up and the other with the blackened side up. It is important that the metal disks should make good contact with the ice throughout the measurement. Place both cups on a table or on a wedge so that the solar radiation reaches the metal plates perpendicularly. Expose both cups to the Sun for a time t_0, measured with a stopwatch.

The proposed method is based on the fact that a glossy aluminum plate has a reflectivity of $r_{\mathrm{Al}} \approx 95\%$, while a matte black surface has a reflectivity of $r_B \approx 5\%$. Let A_p be the area of the plates (approximately equal between them). The solar radiation absorbed by each plate will be

$$Q_i = r_i\, A_p I_0\, t_0, \qquad i = \text{ShinyAluminum, black}. \tag{66.9}$$

Let M_{Al} and M_B be the masses of water (melted) produced in the cups with the shiny and black plate faces, respectively, immediately after exposing them to the Sun. From the energy balance,

$$L_{\mathrm{water}} M_{\mathrm{Al}} = r_{\mathrm{AL}} A_p I_0 t_0 + \text{heat medium} \tag{66.10}$$

and

$$L_{\mathrm{water}} M_B = r_B A_p I_0 t_0 + \text{heat medium}, \tag{66.11}$$

where L_{water} is the latent heat of fusion of water (≈ 79.9 cal/g $=$ 333.9 kJ/kg). The term *heat medium* designates all sources of heat not associated with the metal plates, which are assumed to be the same for both cups. Subtracting these two equations,

$$I_0 = \frac{L_{\mathrm{water}}(M_{\mathrm{Al}} - M_B)}{(r_{\mathrm{Al}} - r_B) A_p t_0}. \tag{66.12}$$

Suggested tasks:

- ✓ On a clear, cloudless day, around noon, expose the two cups of ice with the aluminum plates in place to solar radiation, as described above, for about 10–15 min. If the mass of melted water is very small, you can extend the exposure time, but make sure that the sun's rays reach the plates perpendicularly.
- ✓ Measure the melted water masses during exposure (M_{Al} and M_n) with the balance or a test tube or syringe.
- ✓ Find I_0 from Eq. (66.12).
- ✓ Repeat the experiment two or three times, spaced no more than one hour apart, and calculate the average. If you take longer, the change in the position of the Sun will cause the radiation to differ.

Of course, this is a semi-quantitative method to find an order of magnitude of the local solar irradiance.

66.3.2 *Project 216. ♣ Attenuation of solar radiation in the atmosphere*

> **Recommended equipment**: A solar cell (or a piece of it) with its corresponding electrical connections. A milliammeter.

Pyranometers or solarimeters [15] are commercial instruments used to accurately measure incident solar radiation. They are generally sold calibrated to provide the solar radiation flux density in (W/m^2). Unfortunately, they are expensive and not usually available in teaching laboratories. It is possible, however, to build a pyranometer using a solar cell (or a piece of it) with its corresponding electrical connections. There are also very inexpensive commercial photometers or lux meters on the market, which can measure illumination in a wide range of up to 200,000 Lux, that can also be used for this type of study. Finally, applications can be downloaded from the Internet to transform a smartphone into a light meter [16].

If a solar cell is connected to a milliammeter, the value of the current is proportional to the intensity of the radiation it receives. To validate this conjecture, some of the proposals discussed in Annex A

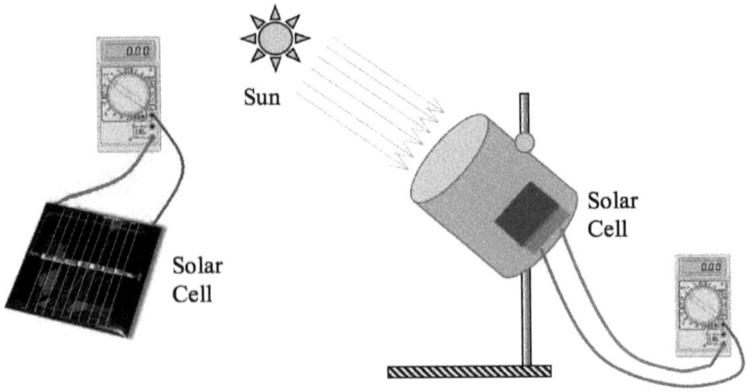

Figure 66.4 Diagram of a simple pyranometer based on a solar cell. The solar cell is connected to a milliammeter to measure current. In order to ensure that the light reaching the cell is the direct radiation from the Sun, the cell can be placed inside a tube or can painted with flat matte black that acts as a collimator, minimizing the diffuse radiation. Thus, by orienting the tube in the direction of the Sun, the light arrives perpendicularly to the cell and stray light is minimized.

can be used to calibrate the solar cell in intensity. Of course, the solar cell can be replaced by one of the photometers suggested in Annex A. It is useful to use an opaque cardboard or plastic tube, with the detector in its center, as shown in Figure 66.4, to filter the diffuse radiation, and by pointing the axis of the tube to the Sun, detect mainly the direct radiation from the Sun. Photoresistors are not recommended, given their nonlinear response in a wide range of intensity.

Suggested tasks:

✓ On a clear (cloudless) day, expose your pyranometer to solar radiation. Make sure the incident radiation is perpendicular to the detector. The shadow of the tube on the bottom of the tube can be used to ensure such perpendicularity and detect mainly the direct radiation.

✓ Measure the magnitude of the signal at different times of the day. Record the signal value I_{rel} and the time. It is not necessary for the pyranometer to have absolute calibration, that is, to give the intensity in lm/m^2 or Lux, since this experiment only requires a signal proportional to the intensity, that is, a relative intensity I_{rel}.

✓ Take measurements every 15 min for at least 5 h.

✓ Using one of the techniques described above, determine the value of the zenith angle θ for each time you perform measurements [12].

✓ Plot $\ln(I_{\text{rel}})$ as a function of $\mu = \sec\theta$.

✓ Are the theoretical expectations, Eq. (66.7), consistent with your experimental results?

✓ What can you conclude from the flat atmosphere model to describe your data?

66.4 ♣ Heating a plate exposed to the Sun

Consider a rectangular metal plate, made of aluminum or copper, for example, of area A and thickness t_p, exposed to the Sun and with its surface perpendicular to the incident radiation.

Let I_0 be the *solar irradiation*, that is, the total intensity of solar energy at ground level. The mean reflectivity of the face of the plate exposed to the Sun is denoted by r.[a] Assuming that the plate is prepared to minimize heat conduction with the medium, if its temperature does not differ significantly from that of the surrounding medium, the main heat dissipation mechanism will be natural convection [7]. As a first approximation, we can use Newton's law of cooling and characterize this energy dissipation:

$$\left[\frac{dQ}{dt}\right]_{\text{convection}} = m_p c_p \frac{dT}{dt} = -h\, m_p c_p (T - T_0). \tag{66.13}$$

Here, T represents the temperature of the plate, T_0 is room temperature, m_p and c_p are the mass and specific heat of the plate, respectively, and h is the heat transfer coefficient of the plate to the medium, t is the time and dQ/dt is the rate of energy loss from the plate. When the plate is exposed to the Sun, from the energy

[a]This is the effective reflectivity, taking into account all the wavelengths present in solar radiation.

balance,

$$\frac{dQ}{dt} = m_p c_p \frac{dT}{dt} = -h\, m_p c_p (T - T_0) + I_0 \varepsilon\, A, \tag{66.14}$$

where $\varepsilon = 1 - r$ is the effective emissivity of the face of the plate exposed to the Sun. This expression can be written as follows:

$$\frac{dT}{dt} + h\,T = h\,T_0 + \frac{A}{m_p c_p} \varepsilon\, I_0, \tag{66.15}$$

where

$$m_p = \rho_p A t_p \tag{66.16}$$

is the mass of the plate and ρ_p is the density of the plate material. By rearranging Eq. (66.14),

$$\frac{dT}{dt} + h\,T = h\,T_0 + \frac{1}{\rho_p c_p t_p} \varepsilon\, I_0. \tag{66.17}$$

This expression can be written as follows:

$$\frac{dT}{dt} + h\,T = h\,T_\infty, \tag{66.18}$$

where

$$T_\infty = \frac{1}{\rho_p c_p t_p} \frac{\varepsilon\, I_0}{h} + T_0 \tag{66.19}$$

represents the equilibrium temperature of the plate if it were exposed to the Sun long enough to reach thermal equilibrium. The solution of the differential equation, Eq. (66.18), with the initial condition $T(0) = T_i$, is

$$T(t) = [(T_i - T_\infty)] \cdot \exp(-kt) + T_\infty. \tag{66.20}$$

Given the relation $T_\infty > T_0$, if the plates are initially colder than room temperature, then $T_0 > T_i$. So for a certain time $t_0 > 0$, for which $T = T_0$ is fulfilled, then from Eqs. (66.15) and (66.17), it follows that

$$\left.\frac{dT(t)}{dt}\right|_{t=t_0} = \frac{\varepsilon}{\rho_p c_p} \frac{I_0}{t_p}. \tag{66.21}$$

Therefore, the slope of the curve $T(t)$, when the temperature of the plate is equal to the ambient temperature T_0, it is directly proportional to the solar irradiation I_0 and inversely proportional to the thickness of the plate. This relationship can be used to estimate the solar constant, regardless of the heat transfer coefficient.

A critical analysis of these arguments shows that Eq. (66.20) will hold even if radiation effects or other energy transfer mechanisms are important. The reason is that when the temperature of the plate is equal to the ambient temperature, the net energy flux between the plate and its surroundings is zero. At this temperature, the rate of change in temperature of the plate is due solely to the influx of energy from the Sun. In this sense, Eq. (66.21) is a more robust result than Eq. (66.20) because the asymptotic temperature T_∞ depends in detail on the mechanisms of energy dissipation, whereas in Eq. (66.21), it does not. For temperatures close to T_0,

$$\left.\frac{dT}{dt}\right|_{T\approx T_0} = \frac{\varepsilon}{\rho_p c_p}\frac{I_0}{t_p} - h\,(T - T_0). \tag{66.22}$$

The last term on the left side of this equation, at $T \approx T_0$, will affect the second derivative of the function $T(t)$.

66.4.1 *Project 217. ♣ Simple method to measure solar irradiance*

Recommended equipment: Two or three aluminum plates of approximately 15 cm × 15 cm and thicknesses t_p 1.5–3 mm, with one face painted flat matte black. An NTC thermistor connected to the back of the plates to measure their temperature. An expanded polystyrene foam (Styrofoam) box to house the plates and isolate them from drafts. An alcohol thermometer to measure the temperature in the box. A computer data acquisition system to register the temperature of the plates. 1/2 kg of dry ice.

Thermistors are inexpensive, highly sensitive devices for measuring temperatures. Their disadvantage is that they provide signals that do not vary linearly with temperature. To use them as thermometers, they must be previously calibrated (see Chapter 50 on thermal sensors). Attach the thermistors to the back of the plates, using

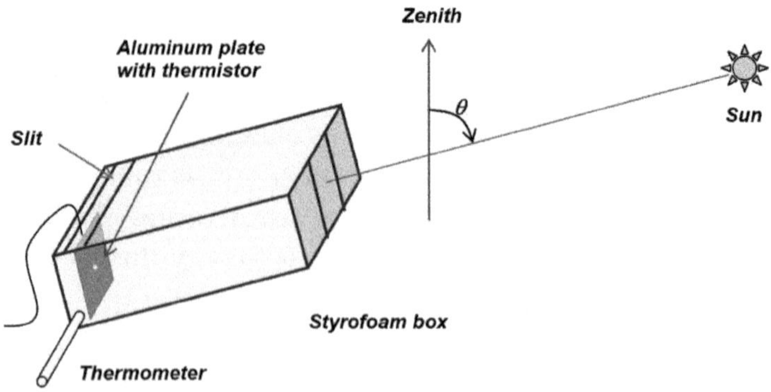

Figure 66.5 An experimental setup to measure the heating of metal plates exposed to the Sun. The thermometer measures the ambient temperature. The expanded polystyrene foam (Styrofoam) box isolates the plates from air currents.

silicone grease to ensure good thermal contact. The temperature of the plate is recorded as a function of time with a data acquisition system connected to a computer. Since temperature does not change rapidly, data can be acquired at a rate of approximately one data per second. The side of the plates exposed to the sun is painted matte black, which has a known reflectivity coefficient ($r \approx 5\%$).[b]

Build an expanded polyethylene box in which to house the plates and isolate them from drafts Figure 66.5, with an open side that allows the Sun's rays to enter. The box also serves as a collimator and positioner for the plates perpendicular to the incident solar radiation. Use an alcohol thermometer to measure the temperature inside the box, near the plates. It is advisable to make a slit a couple of centimeters wide in the box, through which to remove and insert the plates and to serve as an aerator, so that the temperature of the box is not too different from the ambient temperature. Use wedges to position the box in the direction of the Sun's rays. The dry ice is to cool the plates prior to each measurement. This makes the initial temperature of the plates lower than room temperature ($T_i < T_0$).

[b]Sherwin-Williams black floor spray lacquer paint has reflectivity $r = 5.4 \pm 0.5\%$. The reflectivity of other matte paints can be obtained from the manufacturer or by direct measurement with a reflection spectrophotometer.

To obtain more data on each measurement, more than one plate can be placed in the box, without interfering with each other.

Suggested tasks:

✓ On a clear day (cloudless), expose the plates to solar radiation. Make sure that the radiation in each measurement is perpendicular to the plates. The shadow of the tube on the bottom of the tube (or box) can be used to monitor this perpendicularity (see Fig. 66.4).

✓ Before each measurement, cool the plates with dry ice. To do so, it is advisable to remove the plates and place them horizontally with a little dry ice on their surface. The advantage of dry ice is that it cools the plates without leaving a film of water like ordinary ice would. When their temperature is about 10°C lower than room temperature, place the plates in the box, orient them perpendicular to the Sun's rays and start measuring their temperatures with the data acquisition system.

✓ Monitor the temperature of the box, and at each measurement, record the temperature and the time at which the plates reach the temperature of the box thermometer. Continue data acquisition until plate temperatures are about 10°C higher than ambient.

✓ Repeat the measurement approximately every half hour, for about 4–5 h. For each measurement, plot the temperature of each plate as a function of time and determine the slope dT/dt when the temperature of the plates is close to room temperature T_0, that is, $T \approx T_0$.

✓ Using Eq. (66.21), find the value of solar irradiation I_0 for each measurement. Note that this time, the values of I_0 are absolute determinations of irradiance. Using the local time corresponding to each experiment, obtain the zenith angle θ of the Sun for each measurement.

✓ Plot ln (I_0) as a function of $\mu = \sec\theta$.

✓ Is the theoretical expectation, Eq. (66.7), consistent with your data?

✓ Using Eq. (66.6), find the value of the solar constant I_{Sun} and estimate the uncertainty of the determination.

✓ Compare the known values for this constant and discuss the degree of agreement achieved.
✓ Using Eq. (66.1), estimate the value of the Sun's luminosity and its uncertainty. How does the value obtained from your measurements compare to the accepted values for this parameter? Discuss your results and the theoretical model proposed to determine the luminosity of the Sun.

66.5 Maximum visibility distance in air — Turbidity

In 1873, J.C. Maxwell was visiting Darjeeling, India, where he was enjoying a splendid view of Mount Everest, which was located about 150 km away from his hotel. Maxwell wrote to his colleague Lord Rayleigh, asking what was, according to the electromagnetic theory of light, the maximum distance of visibility at which Mount Everest could still be seen, due to the attenuation of light in the atmosphere. This section attempts to answer this question, following the spirit of V. Weisskopf's argument [17]. This reasoning deals with the interaction of electromagnetic radiation with atoms in the air and estimates the Rayleigh scattering cross-section in a simplified heuristic manner.

There are three basic hypotheses: (1) the molecules (or particles) have spherical symmetry and their size is much smaller than the wavelength of the incident light, (2) the molecules are polarizable, and (3) the radiation emitted or scattered by the molecules can be described by an oscillating dipole.

Imagine an atom of atomic number Z. For the moment, assume that the electron cloud around the nucleus is a sphere of radius a with a uniform charge density. If we apply an external electric field, the nucleus moves a distance x from the center of gravity of the electronic cloud, and the electric field generated by the electronic cloud will exert a restoring force given by

$$F(x) = -Ze.E_n(x) = -Ze \cdot \frac{1}{4\pi\varepsilon_0} \frac{Ze}{a^3} \frac{x^3}{x^2} = -k \cdot x, \qquad (66.23)$$

where $k = (1/4\pi\varepsilon_0)Z^2e^2/a^3$ and E_n is the field generated by the electron cloud at a distance x from its center [18, 19]. Therefore, when an external electric field E is applied, a dipole p is induced, given by

$$p = Ze \cdot x = \beta \cdot E. \tag{66.24}$$

Here, it was implicitly assumed that the applied field on the molecule is uniform. This is true if the wavelength $\lambda \gg a$. In turn, β is the polarizability given by $\beta = 4\pi\varepsilon_0 a^3$. This relationship between the induced bipolar moment and the applied field can be extended to the case of molecules. The polarization P of the sample, that is, the dipole moment per unit volume, can be written in terms of the dielectric constant of the medium K, as follows [18, 19]:

$$P = N \cdot \beta \cdot E = \varepsilon_0(K-1) \cdot E, \tag{66.25}$$

where N is the number of molecules per unit volume. The dielectric constant K is related to the refractive index n for $K - 1 = n^2 - 1 = N.\beta/\varepsilon_0$ [18, 19]. In Eq. (66.25), it was assumed that neighboring molecules have a negligible effect on the local electric field of a molecule, a condition that is generally met in gases at low pressures (less than 5 atm), such as air.

If a plane electromagnetic wave, of frequency $\omega = 2\pi f(\lambda \gg a)$ and amplitude E_0, is applied, the dipole induced in each atom in the sample will oscillate with this same frequency. One of the consequences of classical electromagnetic theory (Maxwell's equations) is that an oscillating dipole at a frequency ω radiates electromagnetic energy at an average power s, given by [18–20]

$$s = \frac{\omega^4}{12\pi\varepsilon_0 c^3} p_0^2, \tag{66.26}$$

where $p_0(= \beta \cdot E_0)$ is the peak value of the induced dipole moment and c is the speed of light in vacuum. Thus, the total mean power S radiated by the sample per unit volume will be:

$$S = N\frac{\omega^4}{12\pi\varepsilon_0 c^3} p_0^2 = N\frac{\omega^4}{12\pi\varepsilon_0 c^3}\beta^2 E_0^2. \tag{66.27}$$

In this expression, it was assumed that the molecules of the substance act independently of each other, that is, incoherently. As the average intensity I (power per unit area) of a plane wave of frequency ω is given by [18, 19], $I = c\varepsilon_0 E_0^2/2$, where E_0 is the amplitude of the electric field of the incident wave, then

$$\frac{dI}{dx} = -S = -N\frac{\omega^4}{12\pi\varepsilon_0 c^3}\beta^2\frac{2I}{c\varepsilon_0} = -\alpha \cdot I. \tag{66.28}$$

This expression is identical to the Lambert–Beer law, Eq. (66.3). Considering that $\lambda = 2\pi c/\omega$, then

$$\alpha = \frac{1}{\Lambda} = \frac{(2\pi)^4}{6\pi}\frac{(n^2-1)^2}{\lambda^4}\frac{1}{N} = \frac{8\pi^3}{3}\frac{(n^2-1)^2}{\lambda^4}\frac{1}{N}, \tag{66.29}$$

where $N(= N_A P/RT = P/k_B \cdot T)$ is the number of molecules per unit volume, N_A is Avogadro's number, $k_B = R/N_A$ is Boltzmann's constant, and P and T are absolute pressure and temperature. For air, under normal pressure and temperature conditions ($P_0 = 1$ atm and $T_0 = 273.15$K), the value of $n = n_0$ is weakly dependent on wavelength, with $n_0 - 1 \approx 28 \times 10^{-4}$. Thus,

$$n^2 - 1 = (n_0^2 - 1) \cdot \frac{T_0}{P_0}\frac{P}{T}, \tag{66.30}$$

and by combining this expression with Eq. (66.29),

$$\alpha_R = \frac{1}{\Lambda_R} = N\sigma_R = \frac{8\pi^3}{3}\frac{k_B(n_0^2-1)^2}{\lambda^4}\left(\frac{T_0}{P_0}\right)^2\left(\frac{P}{T}\right), \tag{66.31}$$

where N is the number of molecules or particles per unit volume, σ_R is the Rayleigh scattering cross-section and Λ_R is the *Rayleigh extinction length*, that is,

$$\Lambda_R = \frac{3}{8\pi^3}\frac{\lambda^4}{k_B(n_0^2-1)^2}\left(\frac{P_0}{T_0}\right)^2\left(\frac{T}{P}\right). \tag{66.32}$$

For air, under normal pressure and temperature conditions, $n_0 \approx 1$, thus, $(n_0^2 - 1)^2 = (n_0 + 1)^2(n_0 - 1)^2 \approx 4 \cdot (n_0 - 1)^2$. From Eq. (66.32),

$$\Lambda_R = \frac{1}{N\sigma} = \frac{3}{32\pi^3}\frac{\lambda^4}{k_B(n_0-1)^2}\left(\frac{P_0}{T_0}\right)^2\left(\frac{T}{P}\right). \tag{66.33}$$

Figure 66.6 shows the variation of the Rayleigh extinction characteristic length as a function of wavelength for two pressures: $P = 1$ atm and $P = 0.69$ atm (height 3,000 m), predicted by Eq. (66.32). At an

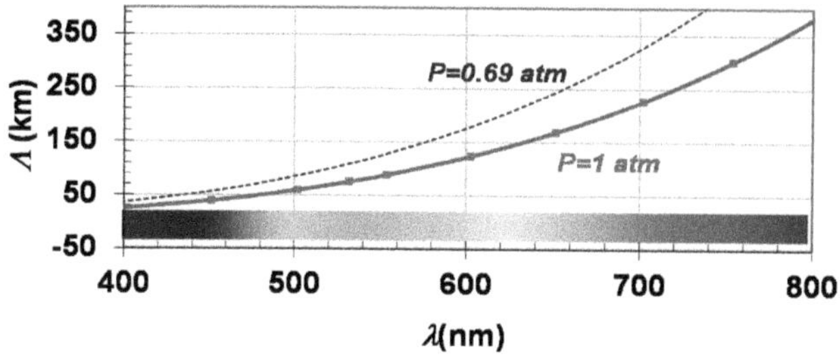

Figure 66.6 Variation of the characteristic Rayleigh extinction length for light in the atmosphere at two different pressures: (a) $P = 1$ atm and (b) $P = 0.69$ atm, corresponding to a height of 3,000 m above sea level, as function of wavelength, according to Eq. (66.33).

average height of about 3,000 m above sea level, this length, for the wavelength of about 600 nm, is about 150 km, which corresponds well to the distance from Mount Everest to Darjeeling and is consistent with Maxwell's observation of the year 1873.

Other effects, such as the presence of aerosols, water vapor, etc., increase the attenuation of light rays, causing the observed values of Λ_{exp} to be generally less than Λ_R. The values given by Eq. (66.33) are an upper limit for Λ, since in the previous calculation it was assumed that the only scattering effect in the atmosphere was due to Rayleigh scattering. Generally, the atmosphere contains water vapor, dust, particles, aerosols, and so on, in varying amounts. In fact, the relationship between Λ_{R} and Λ_{exp} determines the turbidity index of the atmosphere, which indicates the degree of air pollution.

Note that the Rayleigh scattering cross-section, Eq. (66.31), varies with λ^{-4}. Therefore, short wavelengths (blue) are more strongly absorbed and re-emitted by the atmosphere than long waves (red). This is the origin of the blue color of the sky, as well as the reason why, as we move away from objects such as mountains, they appear increasingly bluer until they disappear from sight.

It is interesting to note the strong influence of particles or aerosols in the atmosphere that strongly attenuate the solar radiation that reaches the surface, and therefore, the equilibrium temperature. Various natural events can put large amounts of dust into the atmosphere, including volcanic eruptions, large meteor impacts, and so on. There are several records of volcanic winters caused by the explosion of different volcanoes throughout history. During the three years following the 1883 explosion of Krakatoa (Java), the planet's average temperature decreased by more than 1°C. Likewise, there is evidence that mass extinctions have occurred on Earth due to the impact of large meteorites [21].

66.5.1 *Project 218. ♣ Maximum visibility distance in the air*

Recommended equipment: A solar cell (or a piece of it) with its corresponding electrical connections. A milliammeter.

For this activity, it is suggested to use a photometer or pyranometer, using the same setup as proposed in Project 216.

Suggested tasks:

- ✓ Follow the same recommendations as in Project 216.
- ✓ On a clear day, expose your pyranometer to solar radiation. Make sure the incident radiation is perpendicular to the detector. The shadow of the tube on the bottom of the tube can be used to ensure such perpendicularity.
- ✓ Measure the magnitude of the signal at different times of the day. Record the signal value I_{rel} and the time. It is not necessary for the pyranometer to have absolute calibration, since for this experiment you only need a signal proportional to the intensity, that is, a relative intensity I_{rel}.
- ✓ Take measurements every 15 min for at least 5 h.
- ✓ Using one of the techniques described above, determine the value of the zenith angle θ for each measurement performed [12].
- ✓ Plot ln (I_{rel}) as a function of $\mu = \sec\theta$.
- ✓ Are the theoretical expectations — Ec. (66.7) — consistent with your data?
- ✓ From the plot of $\ln(I_{\text{rel}})$ as a function of $\mu = \sec\theta$, extract the value of K_{ef} — Eq. (66.6) — and its uncertainty. Using this value

and Eq. (66.7), find the best value of Λ_{exp} and compare it to the expectations of Eq. (66.33).

✓ Is the theoretical expectation, Λ_{teor} exceeds $\Lambda_{\text{exp}}(<\Lambda_{\text{teor}})$, fulfilled? Discuss your results. If you define turbidity as $\Lambda_{\text{teor}}/\Lambda_{\text{exp}}$, what is its value? Do you think you can use this experimental method to determine the level of pollution in the atmosphere?

✓ If you have access to a high-rise building, on a clear day, similar to the day on which you took measurements, go up to the top and take pictures of distant sites that you can identify. Then, using Google Earth, determine the distance from the building to the landmarks. Do you find a relationship between the values of Λ_{exp} and the maximum distance at which you can identify places or their geographic coordinates at a distance?

66.6 Annex A: Photometers

A photometer measures light intensity and can be made from electronic components that are sensitive to light, including photodiodes, phototransistors, photoresistors, and solar or photovoltaic cells [22, 23]. This section only summarizes the characteristics of photodiodes, phototransistors and photoresistors, which are the simplest to use.

A very simple, inexpensive, linear photometer can also be constructed using a phototransistor, as shown in Figure 66.7. The linearity and range conditions are adjusted by varying the value of

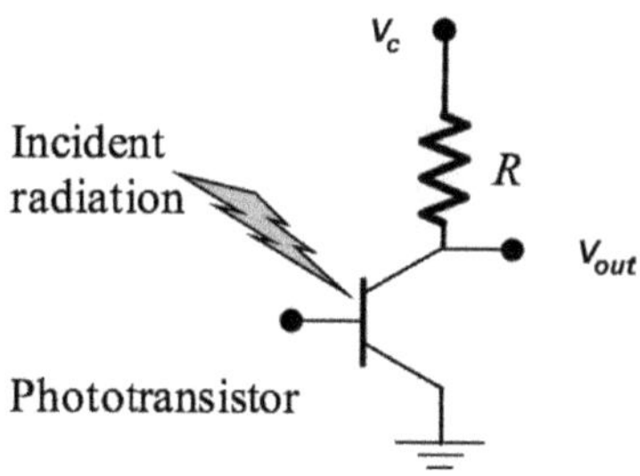

Figure 66.7 Photometer built using a phototransistor. A source of tension V_c powers the circuit. The signal V_{out} contains the illumination information received by the phototransistor.

the resistance R and the applied voltage V_C. The voltage V_{out} contains the illumination dependent signal. For a photometer, it is advisable to use a phototransistor without a lens, so that the signal does not vary with the angle of focus.

A photoresistor (also known as LDR, for *Light Detector Resistor*) is a semiconductor element in which the electrical resistance varies with the intensity of light due to the phenomenon of photoconductivity. Unlike the phototransistor, the photoresistor requires an electrical current i to be circulated through it. By measuring the voltage drop V across a resistor of known value R, the voltage drop, $V = R \cdot i$, provides information on the radiation that falls on the photoresistor. This variation in resistance can be large. CdS photoresistors, for example, which are very common and inexpensive, have a resistance in the dark between 0.1 M Ω and 10 M Ω, which falls below $\approx$100 Ω when illuminated with a 40 W incandescent lamp located at a distance of about 10 cm. A limitation of the photoresistors is their nonlinear response, often they are linear in a small range.

When using a photometer, it is important to keep in mind that its frequency response corresponds to or is compatible with the frequencies of the incident radiation. If you are interested in using the photometer with visible light, its sensitivity range should be between 400 nm and 800 nm. For example, there are photoresistors that have maximum sensitivity in the central zone of the visible spectrum ($\lambda = 550$ nm), and some that are more sensitive to infrared or ultraviolet radiation [24, 25].

There are several vendors of very inexpensive commercial light meters that can be used for the experiments proposed in this chapter [26]. The lux meter must have a sensitivity range of up to 200,000 Lux. Here are several applications that transform a smartphone into a light meter that can be downloaded from the Internet [16]. Strictly speaking, lux meters are not the same as photometers. The latter measure physical power, with all wavelengths measured equally. Lux meters take into account the response of the human eye. However, phototransistors and photodiodes also have a response curve that performs frequency filtering. In any case, for measurements that do not require high precision, the responses of photometers and lux meters to solar radiation are approximately proportional and they can be used for many experiments, such as those proposed here.

66.7 Calibration of a photometer

Whether you use a commercial [27, 28] or a home-made photometer [23, 29], it is always advisable to calibrate it reliably before performing quantitative experiments. Calibration consists of obtaining a response curve (voltage, resistance, or other electrical magnitude) of the photometer as a function of the intensity of light that falls on it. This section describes three methods of calibrating a photometer.

(A) Using the variation of light intensity with distance: In this case, a light source as similar as possible to a "point source" is used. Straight filament lamps, such as those used in some car lights, may be suitable. The calibration procedure consists of measuring the signal variation provided by the photometer with the source-photometer distance. Assuming that the light intensity of the source varies following the law of the inverse of the square of the distance, $I \propto r^{-2}$, a relative calibration of the photometer is possible. In experiments, you must judge the validity of the law $1/r^2$, that is, whether you can take this law as a good approximation taking into account your experimental design (size of the source and range of distances).

(B) Using known transmission light filters: This method consists of inserting known transmission light filters between the source and the photometer. You can use filters purchased from suppliers of laboratory items, which are essentially semi-silver glasses that allow a well-known percentage of the incident intensity to pass (5%, 10%, 20%, etc.).

(C) Using two polarizers: This method consists of inserting between the source and the photometer two polarizers supported in rotating mounts and equipped with a goniometer. As usual, the first polaroid is the polarizer and the second is the analyzer. The goniometer of the analyzer is set to zero and the polarizer is rotated until the transmitted intensity is maximum. This is called intensity I_0. The analyzer is then rotated through well-defined angles and the photometer readings are recorded as a function of the angle. The calibration in this case is based on *Malus's law*, that is, on the fact that the transmitted intensity varies as follows:

$$I(\theta) = I_0 \cdot \cos^2 \theta. \tag{66.34}$$

See Chapter 47 for further details.

References

[1] IPCC, "IPCC sixth assessment report, climate change 2022," 2022. Available: https://www.ipcc.ch/report/sixth-assessment-report-working-group-3/.

[2] Wikipedia, "Solar activity and climate," 2022. Available: https://en.wikipedia.org/wiki/Solar_activity_and_climate.

[3] J. Bahcall, "How the Sun shines," *J. R. Astr. Soc. Can.*, vol. 94, no. 6, pp. 219–227, 2000. Available: http://www.sns.ias.edu/~j.

[4] S. Gil, "Neutrinos solares ¿Por qué brillan las estrellas? — Solar neutrinos, why do stars shine?," *Ciencia Hoy*, vol. 14, no. 79, pp. 52–57, 2004.

[5] J. Bahcall, "How the sun shines," The Nobel Prize. Available: https://www.nobelprize.org/prizes/themes/how-the-sun-shines/.

[6] A. B. McDonald, *et al.*, "Solving the solar neutrino problem," *Sci. Am.*, vol. 288, no. 4, pp. 40–49, April 2003.

[7] S. Gil, M. Mayochi and L. J. Pellizza, "Experimental estimation of the luminosity of the Sun," *Am. J. Phys.*, vol. 74, no. 8, pp. 728–733, 2006.

[8] Wikipedia, "Beer-Lambert Law," 2022. Available: http://en.wikipedia.org/wiki/Beer-Lambert_law.

[9] D. E. Gray, *American Institute of Physics Handbook*, NY: McGraw-Hill Co., 1966.

[10] Wikipedia, "Mie theory," 2022. Available: http://en.wikipedia.org/wiki/Mie_theory.

[11] Wikipedia, "Airmass of the atmosphere," 2022. Available: http://en.wikipedia.org/wiki/Airmass.

[12] U.S. Naval Observatory (USNO), "Sun or Moon altitude/Azimuth tables," 2022. Available: https://aa.usno.navy.mil/data/AltAz.

[13] U. Ganiel and O. Kedem, "Solar energy, how much do we receive?" *Phys. Teach.*, vol. 21, no. 8, p. 573, 1983.
[14] B. Eaton, R. DeGeer and P. Freier, "The solar constant: A take home lab," *Phys. Teach.*, vol. 42, pp. 51–52, 2004.
[15] Wikipedia, "Pyranometer," 2022. Available: https://en.wikipedia.org /wiki/Pyranometer.
[16] Google Play, Light Meter — Lux & Kelvin, Light Meter Application by Borce Trajkovski, 2022. Available: https://play.google.com/store/apps/details?id=com.bti.lightMeter&hl=en.
[17] V. Weisskopf, "Sear for simplicity: Maxwell, Rayleigh, and Mt. Everest," *Am. J. Phys.*, vol. 54, no. 1, pp. 13–14, 1986.
[18] R. Halliday, D. Resnick and M. Krane, *Fundamentals of Physics*, NY: John Wiley & Sons, Inc., 2002.
[19] F. Sears, M. Zemansky, H. Young and R. Freedman, *College Physics*, NY: Pearson Ed., 2009.
[20] E. M. Purcell and D. J. Morin, *Electricity and Magnetism, Berkeley Physics Course*, 3rd Ed., NY: Cambridge University Press, 2014, p. 839.
[21] Berkeley Earth, "Berkeley Earth surface temperature," 2022. Available: http://www.berkeleyearth.org/index.php.
[22] P. Horowitz and W. Hill, *The Art of Electronics*, 3rd Ed., NY: Cambridge University Press, 2015, p. 1220.
[23] G. McIntosh, "A simple photometer to study skylight," *Phys. Teach.*, vol. 44, p. 540, 2006.
[24] Newark, "These data can be obtained from suppliers of electronic elements," 2022. Available: http://www.newark.com.
[25] National Semiconductors, "These data can be obtained from suppliers of electronic elements," 2022. Available: http://www.national.com/.
[26] Amazon, "Best sellers in science classroom optics kits," 2022. Avail able: https://www.amazon.com/Best-Sellers-Science-Classroom-Opti cs-Kits/zgbs/industrial/6651007011.
[27] Pasco Co., "Light & optics," 2022. Available: https://www.pasco.com/products/lab-apparatus/light-and-optics.
[28] Vernier Technology, "Vernier software," 2022. Available: https://www.vernier.com/.
[29] D. Levenson, "Sensitive small area photometer," *Am. J. Phys.*, vol. 38, p. 987, 1970.

Appendix A

GUIDELINES AND SUGGESTIONS FOR WRITING SCIENTIFIC AND TECHNICAL REPORTS

This appendix discusses some guidelines for preparing a laboratory or technical and scientific report, and provides some suggestions on how to format and organize it.

Goals

✓ Scientific and technical report
✓ Tips for writing a report
✓ Format of a report

A.1 Introduction

A laboratory report should endeavor to communicate, clearly and concisely, the objectives and achievements of a project, the hypotheses proposed, and, in light of the data presented, the reasons for which the author accepts or rejects those hypotheses. The report should be easy and pleasant to read, and persuasive on the main topics discussed.

The report is an intrinsic part of a scientific or technical project. A project is not concluded until a careful review has been written recording and/or analyzing the working hypotheses, the objectives, and the evidence collected from trials and observations; providing theoretical interpretations of the data obtained and drawing conclusions.

Report writing is an important professional skill in many fields of work. Reports are often used to inform supervisors and managers of a proposal or project. It is important to remember that projects are assessed largely based on the quality and clarity of the reports presented.

There is no mandatory structure for reports, but some formats have become international standards and are used by many journals and publications around the world. In particular, subdividing the report into sections, each with a well-defined objective, usually contributes to clarity. Some of these features are described in the following.

Writing a report: Sentences are best kept short, with each one conveying a well-defined idea. Avoid complicated or ambiguous sentences. Try to present the information and results synthetically, without repetitions or irrelevant data. Remember that you will not be there to provide explanations when someone reads your report. As far as possible, the text should be self-contained. If you introduce a concept or term that you consider important but — for the sake of brevity — will not make explicit, include a bibliographic reference where an interested or uninformed reader can find the necessary explanation.

A report should not be considered as a document prepared with the sole purpose of the teacher assessing your work. Try to aim at a wider audience than the lab setting. The text should communicate your ideas and results and persuade the reader of your arguments.

Before writing a report, it is useful to draft the main ideas that you want to communicate, highlighting the most important points to be included in the paper. Then, select the plots and figures that most clearly show how the data obtained support your conclusions. Highlight any novel, rare or surprising results, and/or interpretations. It is helpful to prepare a synthetic list of the points or main ideas that you want to highlight and convey. Once you have this preliminary outline, you will be ready to start drafting the report. The aims and conclusions of the work should be stated both in the Abstract and in the Final Conclusions.

There are many good general guides for writing scientific reports [1]. In addition, several scientific journals, such as *Physics*

Teacher [2], *American Journal of Physics* [3], *European Journal of Physics* [4], and *Latin American Journal of Physics Education* [5], provide good suggestions and useful guidelines, some of which are summarized in the following.

A.2 Report organization — Format

The usual sections in scientific and technical publications are the following:

Report header

✓ Title
✓ Authorship
✓ Abstract

Report body

✓ Introduction
✓ Materials and methods, description of the experiment
✓ Results and graphs
✓ Discussion
✓ Conclusions
✓ Bibliography
✓ Appendices

Report header

Title: The title should be clear and provide information on the content of the report. If possible, try to select a sharp, provocative title to attract readers' interest. Avoid using generic titles such as "Lab Report No. 5", which does not provide any information on the content. If a heading such as this is required, it could be used as a subtitle.

Authorship: Include the *names of all the authors*, including their affiliations, that is, the institution to which they belong or at which the work was carried out, as well contact information, such as *email address*, *telephone number*, or *postal address*. If the authors belong to a course, provide the name of the course, the institution, the teachers responsible, and the date of the activity.

Abstract: The abstract is the summary of the report. It should provide a preview of the information in the body of the report and motivate the reader to read the full text. An abstract should not exceed approximately 200 words. It should state concisely the subject of the work, the methodology used, and the most important or novel results obtained. It does not include any reference to figures, tables, or equations that are provided in the body of the text. It should give a preview of the conclusions and articulate logically and harmoniously with them.

Report body

Introduction: This section should introduce the goal of the report and guide the reader toward it. It should clearly state what you hope to learn from the experiment. It must also provide a justification of the method or procedure used. Why did you select this way of performing the experiment? What do you expect to learn? Why it is important? Is there any benefit to be gained from this knowledge?

- In this section, it is useful to present a brief synthesis of the "state of the art" on the subject under discussion. The background of the research should be succinctly discussed and the connection with previous works should be highlighted. The novelty of the current contribution should be made clear. An additional goal for the introduction is to provide a motivation and brief introduction to the subject of the project.
- State the goal of the work clearly. Describe the theoretical/experimental framework of the problem under study, including the relevant references (see References or Bibliography), that is, the description of the framework or conceptual model that will be used to interpret the results or solve the riddle. The construction of the model must include the justification of how the observation or experiment was carried out. It should state clearly which are the relevant parameters to be measured. Avoid repeating what is regularly stated in standard textbooks.
- **Materials and methods, and description of the experiment:** In this section, describe the procedures followed and the instruments used. It is useful to include an outline of the experimental setup, for which you can use schematic diagrams or

sketches of the experiment, which must be well labeled. You could also use photos of the experimental setup, if pertinent. Record the precautions taken during the measurements. It is important to indicate which variables were measured directly, which were obtained indirectly, and which data were taken from other sources. It is also important to analyze the sources of errors and identify the most critical parameters, that is, those that contribute the most to the final errors. If the analysis of errors is very complex, it should be transferred to an appendix.

Results: Results should be presented, if possible, in graphical form. Remember that a picture is worth a thousand words. Tables rarely have the same eloquence and impact, so limit them only to cases where they are absolutely necessary and cannot be replaced by graphics, or when the reader can benefit from easily viewing the exact numerical data. Even then, include them only if the amount of data they contain is small. If large tables are needed, it is preferable to place them in an appendix or annex. The data obtained in your work should be differentiated from data taken from other sources, which can be included for comparative purposes. For example, they can be differentiated by using different symbols in the graphics and indicating their meanings in the legends. The data collected in an experiment or measurement *must be expressed with their respective uncertainties.* Where possible, indicate how the errors or uncertainties were evaluated.

✓ A widespread convention in the literature is to use *symbols* to represent measured data, without connecting them with lines, whereas theoretical predictions or interpretations are represented by *solid or dashed lines*, using a different style (or color) for different models.

✓ Graphs and figures are a kind of universal language for data visualization. If they are well organized and presented, they can be easily interpreted and understood by everyone. The results of the experiment should be recorded with their corresponding units and uncertainties.

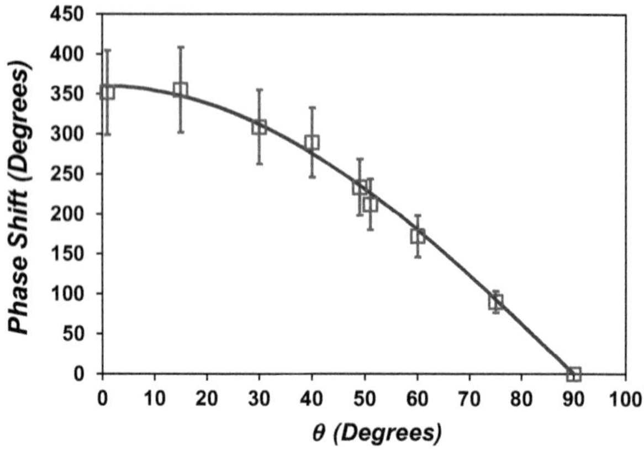

Figure A.1 Graphic representation of the variable "Phase shift", measured in degrees, as a function of the variable "θ", also measured in degrees. Square symbols are the results of measurements made. The vertical error bars show their corresponding uncertainties. The solid line represents the theoretical expectation of a relevant model which attempts to explain these data.

Discussion: This is one of the most important parts of the report, along with the conclusions. It presents the conceptual understanding of the results obtained. Alternative models or theories are compared and, if the evidence allows, the ones that are more appropriate or in accordance with the data obtained and/or established models are selected. Theories or hypotheses that do not pass the experimental test are discarded. It is useful to compare your results with those obtained by other authors in similar experiments. Similarities and discrepancies observed with respect to the results from other sources should be analyzed. If there are discrepancies, how can you explain them? How do the errors, precisions, costs, and experimental difficulties compare to other trials? If models from other works are used, their origin must be cited. Is there an interesting application or practical implication of the results obtained? Sometimes, for reasons of space, the discussion can be combined into a single section with the results or conclusions.

Conclusions: This section presents *what you have learned* from the work conducted and synthesizes the consequences and implications associated with the results. The conclusions should not be confused with the results. The conclusions are associated with the implications that arise from analyzing the data and the discussions. They are about answering the following question: What have we learned from the work done?

It is suggested to state first the conclusions linked to the goals stated in the introduction, followed by any that emerged unexpectedly from the analysis. In a way, it could be said that a good report is one that achieves the greatest number of (correct) conclusions or falsifies the greatest number of hypotheses based on the data obtained.

Bibliography or References: The bibliographic citations are arranged, preferably, in numeric sequence at the end of the report, although there are other ordering schemes.

The bibliographic citations must contain the following:

✓ the *names of the authors* of the publication (articles, magazines, or books),
✓ *the title of the work*,
✓ *the name of the journal or publisher*,
✓ *the year (and city or country)* of publication,
✓ in addition, *the data that help identify the sources* should be included: Internet address where they can be found, chapter, page, date of publication, etc.

An example is provided in the following:

1. F. Sears, M. Zemansky, H. Young and R. Freedman, *College Physics*, Vol. II, Addison Wesley Longman, 2008.
2. R. Halliday, D. Resnick and M. Krane, *Fundamentals of Physics*, New York: John Wiley & Sons, Inc., 2002.
3. Vernier Software & Technology, 2022. Available: http://www.vernier.com.
4. A. P. Schuetze, W. Lewis, C. Brown and W. J. Geerts, "A laboratory on the four-point probe technique," *Am. J. Phys.*, vol. 72, p. 149, 2004. Available: http://ajp.aapt.org/.

Appendices: This part may or may not be present. An appendix may be included to improve understanding of some part of the report or to include additional information. It is generally not advisable to distract the reader with too many calculations or error propagation in the body of the text, so an appendix may be a suitable section in which to include them. The main text should guide the reader to consult any appendices.

A.3 Additional comments

Like any art, writing scientific and technical reports requires practice. Reading reports of this kind carefully in specialized journals and magazines that are accessible to beginner students, such as *Phys. Teach.*, *Am. J. Phys.*, *Lat. Am. J. Phys. Ed.*, and *Eur. J. Phys.*, among others, can contribute to a better understanding of the structure described above.

The report should not be confused with the *laboratory logbook.* The latter is a notebook where all the data and details of an experiment are recorded. The logbook is mainly a notebook for use of the author and his/her collaborators, where all the details of the experiment are documented whenever possible. For example, the day and date of completion, the name of the files where the data of each test performed are located, and so on. The report, in contrast, is a final refined version intended for a reader who did not necessarily perform the experiment. Any superfluous details should be avoided in the report.

A good habit is to ask a colleague — if possible, someone who is unfamiliar with the subject of the report — to read it and make suggestions and comments. Ask him/her what he/she understood from it or what points were not clear so that you can improve it. Whatever the case may be, once the report has been written, it should be reread and edited carefully before presenting it.

Finally, I would like to draw your attention to the saying *Brevity is the soul of wit.* Here is a paradigmatic example:

"The Sniper" — *Every day, while I was waiting for the bus, a child would point his finger at me from a balcony and shoot his imaginary weapon, like a ritual, shouting, "Bang! Bang!" One day, just to follow the routine game, I pointed my finger back at him, yelling, "Bang! Bang!" The boy fell to the street as if struck down.*

I ran to him and saw that he half opened his eyes and looked at me dazedly. Desperately, I said, "But all I did was repeat what you did to me." Then he replied ruefully: "Yes sir, but I didn't shoot to kill."

For this story, Armando Macchina was awarded the 2013 International Short Story Award by the César Egido Serrano Foundation, based in Quero (Toledo), Spain.

Figures, graphs, tables, and equations: Figures and graphs must be numbered sequentially with the same denomination, for example, Figure A.1, Figure A.2, and so on. The number indicates its order

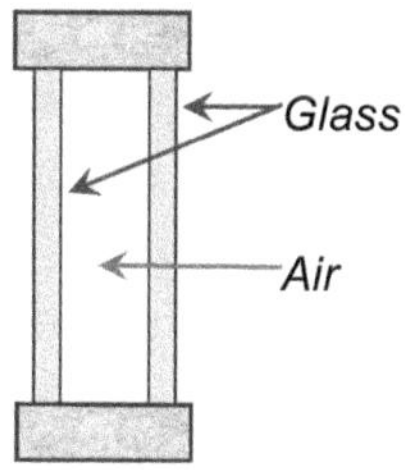

Figure A.2 Double pane window: two panes of glass with an air chamber in between. (*Note: In this case, the letter A refers to the fact that the figure is in Appendix A.*)

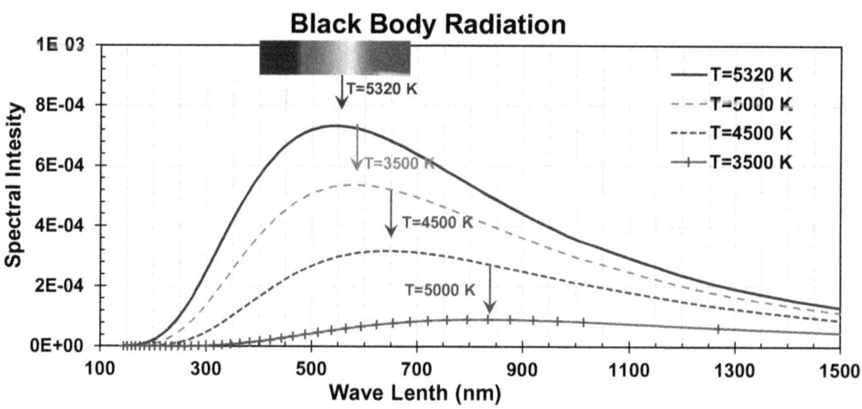

Figure A.3 Spectrum of electromagnetic radiation from a black body at different temperatures T. The arrows indicate the position of the maximum radiation at each temperature. The colors at the top illustrate the colors of light corresponding to the visible wavelengths λ, indicated on the horizontal axis.

of appearance in section or Chapter X (if the report is divided into chapters). Each figure should have a short explanatory legend to make it understandable, as illustrated in Figures A.1, A.2 and A.3.

The numbers in each figure help to refer to them elsewhere in the text. For the same reasons, the tables are also numbered sequentially. For equations, it is advisable to use the *equation editors* available in most word processors. The equations are centered on the line and sequential numbering is placed on the right, as illustrated in the following:

$$H = \frac{dQ}{dt} = -\frac{k \cdot A}{l} \cdot (T_C - T_F). \tag{A.1}$$

This equation can be referred to by writing: using Eq. (A.1).

Finally, the report pages should be numbered.

Some of the publications mentioned above [2, 3, 5, 6] provide many examples of scientific reports in the style described herein.

References

[1] The Writing Center-University of North Carolina at Chapel Hill, "Scientific reports," 2022. Available: https://writingcenter.unc.edu/tips-and-tools/scientific-reports/.

[2] American Physical Association (AAPT), "The physics teacher," Available: http://tpt.aapt.org/2022.

[3] American Association of Physics Teachers (AAPT), "American Journal of Physics," Available: http://ajp.aapt.org/.2022.

[4] IOP (Institute of Physics) of Great Britain, "European Journal of Physics," 2022. Available: http://iopscience.iop.org/0143-0807.

[5] Center for Research in Applied Science and Advanced Technology of the National Polytechnic Institute of Mexico, "Latin American Journal of Physics Education," 2022. Available: http://www.lajpe.org/.

[6] Website with reports from students from various Latin American universities, "Fisica reCreativa," 2022. Available: http://www.fisicarecreativa.com.

Appendix B

STANDARDS AND SAFETY RECOMMENDATIONS IN LABORATORIES

This appendix discusses some physical safety standards and guidelines that should be followed in laboratory work.

Goals

✓ Safety regulations
✓ Electrical safety
✓ Tips for using lasers
✓ Suggestions for the use of chemicals and radioactive and cryogenic substances

B.1 Safety first

In any laboratory activity, there are risks that can cause accidents, so it is advisable to carefully follow certain precautionary measures. These measures are often common sense rules, but some of them are part of regulated standards. As it would be too extensive to list all the risks that can arise in the laboratory and all the precautionary measures, this appendix will only mention some very general ones that must be taken into account when working with electricity, lasers, cryogenic liquids, chemicals, and radioactive elements. In general, each university establishes a set of safety regulations that it recommends to its staff. These regulations should be made known in a timely manner to students and teachers [1–3].

B.2 General recommendations

✓ Know the location of the safety equipment in the laboratory: fire extinguishers, alarms, emergency exits, etc.
✓ Place fire extinguishers in visible places.
✓ Keep a first aid kit in an easily accessible place.
✓ Learn about the risks of electricity to the human body.
✓ Make sure that your laboratory has a suitable electrical installation with correct ground connections. Also make sure there are circuit breakers to prevent electric shock.
✓ Keep the electrical installation in good condition and avoid poor connections.
✓ Use voltage transformers and high-voltage sources with special care.
✓ If gas outlets have been used, verify that they are closed.
✓ If you use strong magnets, read carefully some safety [4] measures provided in Refs. [5] and [4].
✓ Work with two or three people so that you can help each other in the event of an accident. Avoid working alone in the labs. If you do need to work alone, be sure to notify someone responsible.
✓ If you used computers, remember to record your data on removable media. Turn off the computer and any equipment you have used when you have finished your work.
✓ Maintain a clean, orderly workplace. Each person must be responsible for the area assigned to him/her and all the common areas. Before you leave the laboratory, it should be in perfect order and ready for others to work in without needing to tidy or clean up.

B.3 Personal care and recommendations

✓ Avoid wearing ties, scarves, chains, watches, bracelets, and rings.
✓ Work clothes should fit snugly, with no dangling straps.
✓ Hair should be tied back.
✓ Wear comfortable clothing and low-heeled, rubber-soled shoes. If you wear sandals or other open shoes, it is advisable to wear thick socks.

✓ Do not clean or oil machines while they are in operation.
✓ Do not leave a running machine alone.
✓ Wear the appropriate face shield that comes with the machine (especially if the machine projects chips or particles, e.g., abrasive wheels).
✓ If you use compressed air or gases, never point the nozzle at anyone because it may blow foreign particles into their eyes or cause damage.
✓ Leave no obstacles in the way.
✓ If possible, confine work areas and mark them with signs.

B.4 General recommendations for using a laser

✓ Never use a laser in open environments with people passing through.
✓ Keep work areas as bright as possible so that the observer's pupil is as closed as possible.
✓ The walls of the work area should not be shiny or allow significant reflection.
✓ If possible, position the laser so that the beam is not at eye level (for example, place it above 2 m and below about 80 cm).
✓ Targets at which the beam is directed should be made of light-absorbing material to prevent reflections. If the experiment requires a high-power laser, the target should be heat- and fire-resistant.
✓ Never align the beam using the eye. It can damage your retina.
✓ Do not look at the reflections produced by a laser beam.
✓ Predict the path of the beam and stay clear of it.
✓ Do not use chains, watches, or objects that might unwantedly reflect a beam.
✓ See an ophthalmologist if you notice that your eye sees a "ghost image" on the retina.
✓ Notify others when you are about to operate a laser. Do not allow onlookers in the work area.

B.5 General recommendations for using cryogenic liquids

- ✓ Observe the recommended standards for the use of different cryogenic liquids (helium, liquid air, and liquid nitrogen).
- ✓ When using cryogenic liquids, work in ventilated rooms. Remember that if the liquid evaporates suddenly, a large volume of gas will be produced that will dislodge the air from the environment.
- ✓ Store and transport cryogenic liquids in safe containers (stainless steel Dewar's with thermal insulation). If you use glass flasks, protect them from shocks.
- ✓ When transferring liquid nitrogen, avoid splashing on the skin. Protect your eyes from splashes. Wear suitable gloves and protective goggles.

B.6 General rules for using chemicals

- ✓ Keep all laboratory chemicals identified. The label should indicate the name of the product, date of purchase, expiration date, and other relevant information.
- ✓ Do not store chemicals on high shelves. Try to keep them at low levels (under a counter, for example).
- ✓ Handle acids and highly volatile liquids under a hood.
- ✓ Wear gloves when necessary.
- ✓ Do not pipette by mouth.
- ✓ Wash hands and face with soap and water after using chemicals.
- ✓ Never put chemical waste down the drain. Place it in authorized containers and consult about the procedures for its disposal.

B.7 General rules for using radioactive substances

- ✓ Every facility that uses radioactive sources must have an authorized manager who must instruct users on the safe use of radioactive material. Make sure you know who is responsible and authorized to use radioactive material in your laboratory.
- ✓ The radioactive sources used in the experiments proposed in this book are all low-activity gamma radiation sources ($A < 1\mu$Ci). It is recommended that these be *sealed* sources, that is, sources

where the radioactive material is not exposed but encapsulated in a sealed metal or plastic enclosure. It is important that the sources should come from a well-known licensed supplier. The source seal must present physical integrity, without cracks or breaks. With these precautions, these sources can be manipulated with the hands, though unnecessary contact should always be avoided. Remember, the exposure dose decreases with the square of the distance to the source and increases linearly with the exposure time and source activity, so keep sources at a distance and minimize handling time.

✓ Beta or alpha radiation sources (which are not used in the experiments proposed in this book) should be used with greater caution because the radioactive material is exposed. Under no circumstances should you touch these active areas because they are contaminated with radioactive material, and moreover, touching damages the source. Always wear rubber or latex gloves when working with these sources.

✓ If sources are used in liquid solutions, under no circumstances should pipetting by mouth be used. Wear rubber or latex gloves and an apron when working with these sources.

✓ Never flush radioactive material down the drain. Place the waste in special containers according to local regulations.

✓ When you have finished working, always wash your hands thoroughly with soap and water.

✓ If you have an accident, a source break, a radioactive material spill, etc., immediately notify your instructor or facility manager. The local regulatory authority can also help in these emergencies.

✓ In all cases, minimize the exposure time.

✓ Store supplies in proper storage areas. Remove only the material to be used in each experiment, and after completing the measurement, return the material to its storage location.

✓ Never remove radioactive material from the laboratory.

✓ Each country, city, province, or state may have its own legislation for handling radioactive substances. It is important to know what this legislation says, including the necessary requirements to acquire, transport, store, and handle this type of material. Some regulatory agencies for nuclear activity are as follows:

- International Atomic Energy Agency (http://www.iaea.org/worldatom/).
- Canadian Nuclear Safety Commission — Canada (http://www.nuclearsafety.gc.ca/).
- Lawrence Berkeley National Laboratory — California–USA (http://www.lbl.gov/).
- The Berkeley Laboratory Isotopes Project's Exploring the Table of Isotopes (http://ie.lbl.gov/education/isotopes.htm).
- US Department of Energy Laboratories (http://home.doe.gov/people/peopnl.htm).
- Australian Radiation Protection and Nuclear Safety Agency — Australia (http://www.arpansa.gov.au/).
- Direction de la Sureté des Installations nucléaires Health and Safety Executive — France (http://www.asn.gouv.fr/).
- Swiss Nuclear Safety Inspectorate — Switzerland (http://www.hsk.psi.ch/)
- Consejo de Seguridad Nuclear — Spain (http://www.csn.es/).
- Autoridad Regulatoria Nuclear Argentina (http://www.arn.gov.ar/).

References

[1] University of Stanford Environmental Health & Safety, "Laboratory safety," 2022. Available: https://ehs.stanford.edu/.

[2] University of Washington -Seatle WA, "UW laboratory safety manual (LSM)," 2021. Available: https://www.ehs.washington.edu/resource/laboratory-safety-manual-510.

[3] AAPT American Association of Physics Teachers, USA, "Safety in physics education," 2001. Available: https://www.aapt.org.

[4] Magfine Corporation Italy, "Handling precautions for large magnets," 2022. Available: https://www.magfine.it/en/user_data/handling.php.

[5] Hitachi Metals, Ltd., "Basic precautions before using magnets," 2022. Available: https://www.hitachi-metals.co.jp/e/products/auto/el/pdf/pre_mag_e.pdf.

Appendix C

LINEAR CORRELATION

This appendix extends the discussion of least squares, begun in Chapter 7. It analyzes the statistical significance of the parameters obtained from a fitting and their uncertainties, and presents a brief digression on confidence bands [1] in a linear regression, and the case of a linear regression in which the two variables contain errors.

Goals

✓ Least-squares data with errors
✓ Uncertainty of parameters of a setting
✓ Statistical significance
✓ Prediction bands
✓ Data with error in the two variables

C.1 Linear regression — Data with errors

Consider the case of a set of measurements (X_i, Y_i), where the uncertainty in the value of Y_i is given by σ_i (see Figure C.1). The objective of this section is to extend the discussion of least squares begun in Chapter 7 to the case of data with errors. It analyzes the procedure for finding the line that best fits the experimental data and the associated uncertainties in determining the fit parameters.

As in Chapter 7, assume that the data can be described by a linear equation of the form:

$$Y(x) = a \cdot x + b. \qquad \text{(C.1)}$$

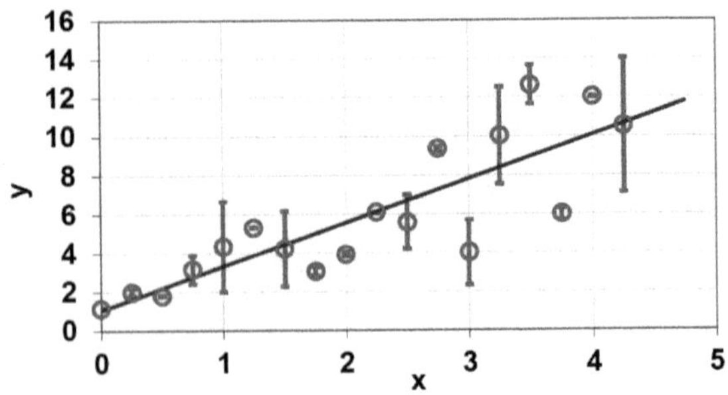

Figure C.1 Plot of an experimental data set (X_i, Y_i) with errors in the axis Y given by the values σ_I(red circles). The blue line is obtained by least squares regression.

The value of *chi square*, χ^2 is defined as follows:

$$\chi^2 = \sum_{i=1}^{N} w_i \cdot (Y_i - a \cdot X_i - b)^2. \tag{C.2}$$

Here, w_i is a weighting factor that can be defined in different ways, depending on the problem under study. A common way to weight the data is to use their respective errors as follows:

$$w_i = \frac{1}{\sigma_i^2} \tag{C.3}$$

with

$$W = \sum_i w_i. \tag{C.4}$$

If all the data are equally weighted, that is, if $w_i = 1$, then $W = N =$ *total number of data*. Of course, Eq. (C.3) is only one of the many ways in which data can be weighted.

The *method of least squares* consists of choosing as the best values of a and b those values that minimize the value of χ^2 — Eq. (C.2). This procedure results in [2–5]

$$\begin{aligned} a_0 &= \frac{W \cdot (\sum YX) - (\sum X) \cdot (\sum Y)}{W \cdot (\sum X^2) - (\sum X)^2} \\ &= \frac{\langle YX \rangle - \langle X \rangle \cdot \langle Y \rangle}{(\langle X^2 \rangle - \langle X \rangle^2)} = \frac{\mathrm{Cov}(YX)}{S_X^2} \end{aligned} \tag{C.5}$$

and

$$b_0 = \frac{(\sum Y) \cdot (\sum X^2) - (\sum X) \cdot (\sum Y \cdot X)}{W \cdot (\sum X^2) - (\sum X)^2}$$
$$= \frac{\langle X^2 \rangle \cdot \langle Y \rangle - \langle X \cdot Y \rangle \cdot \langle X \rangle}{(\langle X^2 \rangle - \langle X \rangle^2)} \tag{C.6}$$

or

$$b_0 = \langle Y \rangle - a_0 \cdot \langle X \rangle. \tag{C.7}$$

Where we have used the notation

$$\sum Y \equiv \sum_{i=1}^{N} w_i \cdot Y_i, \sum X^n \equiv \sum_{i=1}^{N} w_i \cdot X_i^n,$$
$$\sum Y \cdot X \equiv \sum_{i=1}^{N} w_i \cdot Y_i \cdot X_i, \tag{C.8}$$

and so on. The mean values of Y and X are defined as follows:

$$\bar{Y} \equiv \langle Y \rangle \equiv \frac{1}{W} \sum_{i=1}^{N} w_i \cdot Y_i \quad \text{and}$$
$$\bar{X} \equiv \langle X \rangle \equiv \frac{1}{W} \sum_{i=1}^{N} w_i \cdot X_i. \tag{C.9}$$

The standard deviations and variance are given by

$$S_x^2 \equiv \frac{1}{W} \sum_{i=1}^{N} w_i \cdot \left(X_i - \bar{X}\right)^2$$
$$= \left(\frac{N-1}{N}\right) \cdot \mathrm{Var}(X) = \langle X^2 \rangle - \langle X \rangle^2 \tag{C.10}$$

and

$$S_Y^2 \equiv \frac{1}{W} \sum_{i=1}^{N} w_i \cdot \left(Y_i - \bar{Y}\right)^2$$
$$= \left(\frac{N-1}{N}\right) \cdot \mathrm{Var}(Y) = \langle Y^2 \rangle - \langle Y \rangle^2. \tag{C.11}$$

Correlation coefficients are defined similarly:

$$S_{YX} \equiv \mathrm{Cov}(Y,X) = \frac{1}{W}\sum_{i=1}^{N} w_i \cdot \left(X_i - \bar{X}\right)\cdot\left(Y_i - \bar{Y}\right)$$
$$= \langle YX\rangle - \langle Y\rangle\langle X\rangle \tag{C.12}$$

and

$$R \equiv \frac{\langle YX\rangle - \langle Y\rangle\langle X\rangle}{S_X \cdot S_Y} = \frac{\mathrm{Cov}(Y,X)}{S_X \cdot S_Y}. \tag{C.13}$$

The standard error of the estimation of Y is related to the *normalized chi-square value*, χ^2_N, by

$$\chi^2_N = \frac{\sum w_i \cdot (Y_i - Y(X_i))^2}{W} = \text{typical error}\,(YX)^2. \tag{C.14}$$

It is also useful to define the chi-square value by degrees of freedom, χ^2_v:

$$\chi^2_v = \frac{N}{(N-2)} \cdot \frac{\sum w_i \cdot (Y_i - Y(X_i))^2}{W} = \frac{N}{N-2}\cdot \chi^2_N. \tag{C.15}$$

The total variance, $S_t \equiv S_Y$, provides a measure of how the points Y_i are distributed around the mean value of Y. S_t is defined as follows:

$$S_t^2 \equiv \frac{1}{W}\sum_{i=1}^{N} w_i \cdot \left(Y_i - \bar{Y}\right)^2 = S_Y^2. \tag{C.16}$$

The *unexplained* variance [6], S_{unex}, is a measure of the quality of the proposed linear model, $Y(X_i) = aX_i + b$, to explain the observed data (Xi, Y_i). This name arises from the fact that if the proposed linear model were adequate, the values $\varepsilon_i = (Y_i - Y(X_i))$ should have a random statistical distribution S_{unex}, defined as follows:

$$S^2_{\text{unex}} \equiv \frac{1}{W}\sum_{i=1}^{N} w_i \cdot (Y_i - Y(X_i))^2 = \chi^2_N. \tag{C.17}$$

The *explained* variance, S_{exp}, is defined by

$$S^2_{\text{exp}} = \frac{1}{W}\sum_{i=1}^{N} w_i \cdot \left(Y(X_i) - \bar{Y}\right)^2. \tag{C.18}$$

By adding Eqs. (C.16), (C.17), and (C.18) member to member, it can be shown that [7, 8]

$$S_t^2 = S_{\text{exp}}^2 + S_{\text{inex}}^2, \tag{C.19}$$

therefore,

$$R^2 = \frac{S_{\text{exp}\,l}^2}{S_t^2} = 1 - \frac{S_{\text{inex}}^2}{S_t^2} = 1 - \frac{\chi^2}{N \cdot S_Y^2} \leq 1. \tag{C.20}$$

An important property of estimators a and b is that if the residuals of the estimates ε_ι

$$\varepsilon_i^2 = (Y_i - Y(X_i))^2 \tag{C.21}$$

have a normal distribution, then the values a and b will in turn have a statistical distribution and their mean values will be given by $\langle a \rangle = a_0$ and $\langle b \rangle = b_0$, and their standard deviations, or *standard errors*, denoted by Δa_0 y Δb_0, respectively, are [5]:

$$\Delta a_0 = \frac{a_0}{\sqrt{N-2}} \cdot \sqrt{\left(\frac{1}{R^2} - 1\right)} \tag{C.22}$$

and

$$\Delta b_0 = \Delta a_0 \cdot \sqrt{\langle X^2 \rangle}. \tag{C.23}$$

If the errors ε_ι have a normal distribution, the variable t, defined by:

$$t = \frac{(a - \langle a \rangle)}{\Delta a_0}, \tag{C.24}$$

will present *Student's t-distribution* [7, 9] with N-2 degrees of freedom.

The uncertainty in the estimation of $a_0(\Delta a)$ from a sample of size N, with a confidence limit of $P\%$, is calculated from the value t_p, which is obtained from *Student's t-distribution* with (N-2) degrees of freedom [7, 9]:

$$\text{Probability\_Student-}t\ (t < t_p) = P\%. \tag{C.25}$$

In Microsoft Excel®, this value of t_p is calculated using the INV T.DIST ((1-P), N-2) function. Uncertainty Δa is calculated as follows:

$$\Delta a(P\%) \equiv \Delta a_p = t_p \cdot \Delta a_0 = t_p \cdot a_0 \cdot \sqrt{\frac{(1/R^2 - 1)}{(N-2)}}. \quad \text{(C.26)}$$

The error in b_0 is given by:

$$\Delta b(P\%) \equiv \Delta b_p = \Delta a(P\%) \cdot \sqrt{\langle X^2 \rangle}. \quad \text{(C.27)}$$

C.2 Statistical significance of parameters of a best fit

A usual test is to evaluate whether the value of a slope or other parameter obtained from a given experiment is significantly different from zero. Ultimately, what needs to be evaluated is the *null hypothesis*[a] H_0: $a_0 = 0$ versus the hypothesis H_1: $a_0 \neq 0$. In this case, it is analyzed whether the interval defined by $(a_0 - \Delta a_0, a_0 + \Delta a_0)$ contains the value zero. If this interval contains zero, it is clear that a value of $a_0 = 0$ is consistent with the result and therefore the result $a_0 \neq 0$ is not statistically significant. The same criteria can be applied for any other parameter, for example, b_0 — Eq. (C.23).

C.3 Case of data with error in the two variables

Error in both variables: The statistical techniques to consider these cases are the subject of discussion among different authors and experts on the subject. This section proposes an approximate scheme, based mainly on Refs. [7–9].

If measurements (x_i, y_i) contain errors Δx_i and Δy_i, respectively, and there are indications that the relationship between x and y is linear, of the form $y = ax + b$, the weight factors for each point are

[a] A hypothesis is often formulated with the sole purpose of rejecting it. For example, to decide whether a coin is rigged, it can be hypothesized that the coin is good. These hypotheses are called *null hypotheses*[7] and are designated with H_0. The maximum probability with which a hypothesis may be rejected when it should have been accepted (*Error type I*) is called the *significance level* and is designated by α. A frequent value of α is 0.005 (5%) or 0.01 (1%).

defined as follows:

$$W_i = 1/\sigma_i^2, \tag{C.28}$$

where

$$\sigma_i^2 = a_0^2 \cdot \Delta x_i^2 + \Delta y_i^2. \tag{C.29}$$

In general, if the weights of the variable x and y are $w_{x,i}$ and $w_{y,i}$, respectively, then

$$w_i = \frac{w_{x,i} \cdot w_{y,i}}{a_0^2 \cdot w_{y,i} + w_{x,i}}, \tag{C.30}$$

where a_0 is the slope of the regression line, Eq. (C.5). The difficulty is that to determine a_0, the regression problem must be solved, for which the weight factors w_i are needed, which in turn depend on a_0. This problem can be solved by proceeding iteratively. Use as initial weight only the values of $w_{yi} (= 1/\Delta y_i^2)$. With these coefficients, using Eq. (C.5), find the value of a_0, with which the weights w_i can be calculated. Then, using Eq. (C.30), again determine the coefficients w_i and from Eq. (C.5) the new coefficients a_0. By iterating until the successive values of a_0 do not change significantly, the parameters of the linear regression are found, that is, the linear regression for the case of data with errors in the two variables.

These ideas can be extended to the nonlinear case, where the function $f(x;a,b,c,\ldots)$, whose parameters a, b, c, ... are to be determined, depends on a nonlinear mode of x. In this case, the generalization of Eq. (C.30) leads to the concept of effective error:[8]

$$\sigma_i^2 = \left(\frac{df}{dx}\right)^2 \cdot \Delta x_i^2 + \Delta y_i^2. \tag{C.31}$$

References

[1] Wikipedia, "Confidence and prediction bands," 2022. https://en.wikipedia.org/wiki/Confidence_and_prediction_bands.

[2] P. Bevington and D. K. Robinson, *Data Reduction and Error Analysis for the Physical Sciences*, 3rd Ed., 2003, p. 230.

[3] D. C. Baird, *Experimentation: An Introduction to Measurement Theory and Experiment Design*, 3rd Ed., NY: Pearson, 1995.

[4] S. L. Meyer, *Data Analysis for Scientists and Engineers*, NY: John Willey and Sons., 1975.
[5] J. Higbie, "Uncertainty in the linear regression slope," *Am. J. Phys.*, vol. 59, p. 184, 1991.
[6] Wikipedia, "Fraction of variance unexplained," 2022. https://en.wikipedia.org/wiki/Fraction_of_variance_unexplained.
[7] R. Wrede and M. Spiegel, *Advanced Calculus (Schaum's Outlines)*, 3rd Ed., NY: McGraw Hill, 2010.
[8] G. Blom, *Probability and Statistics: Theory and Applications*, NY: Springer, 2011.
[9] J. Orear, "Least squares when both variables have uncertainties," *Am. J. Phys.*, vol. 50, p. 912, 1982.
[10] D. Barker and L. Diana, "Simple method for fitting data when both variables have uncertainties," *Am. J. Phys.*, vol. 42, p. 242, 1974.
[11] B. C. Reed, "Linear least-squares fits with errors in both coordinates II: Comments on parameter variances," *Am. J. Phys.*, vol. 189, p. 1, 1990.
[12] B. C. Reed, "A spreadsheet for linear least-squares fitting with errors in both coordinates," *Phys. Educ.*, vol. 45, no. 1, p. 93, 2010.

Appendix D

LEAST SQUARES METHOD — NONLINEAR CASE

This appendix extends the discussion of least squares, begun in Chapter 7, to the case of nonlinear relationships. It briefly discusses how to perform simulations using the Monte Carlo technique.

Goals

- ✓ Parameter setting in nonlinear cases
- ✓ Simulation using the Monte Carlo method

D.1 Least squares method in nonlinear relationships

Suppose we take a series of measurements of two quantities whose relationship we wish to determine. The result of your N measurements will give rise to a set of N triads of the form (x_i, y_i, σ_i), where σ_i is the uncertainty associated with the determination of y_i. Here, assume that the uncertainty of x_i is negligible. Suppose also that the model that fits the data is given by the function $f(x;a,b,c,\ldots)$, where a, b, c, etc. are the n_{par} model parameters. Designate the estimator of the value of y given by the model by $y(x_i) = f(x_i;a,b,c,\ldots)$. It is said that $y(x_i)$ represents the deterministic variation of y with x.

In this case, the *chi-square* function is defined as follows [1]:

$$\chi^2 = \sum_{i=1}^{N} \frac{(y_i - y(x_i))^2}{\sigma_i^2} = \sum_{i=1}^{N} w_i \cdot (y_i - y(x_i))^2, \qquad \text{(D.1)}$$

where the w_i values are the weight factors of each data triad (x_i, y_i, σ_i); in this case, $w_i = 1/\sigma_i^2$. The number of degrees of

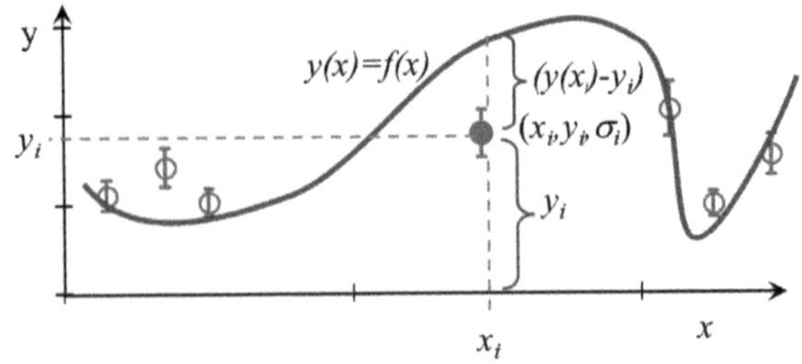

Figure D.1 Example of a nonlinear model represented by the function $f(x)$. The result of the measurements is the triad (x_i, y_i, σ_i). $\cdot\sigma_i$ represents the absolute error associated with each observation y_i. $y(x_i)$ is the model prediction for the value of x_i.

freedom, v, of the model is defined as follows:

$$v = N - n_{\text{par}}. \tag{D.2}$$

The mean value $\overline{y}$ of y_i is defined as follows:

$$\overline{y} = \frac{\sum_{i=1}^{N} w_i \cdot y_i}{\sum_{i=1}^{N} w_i} = \frac{\sum_{i=1}^{N} w_i \cdot y_i}{W}. \tag{D.3}$$

If all values y_i have the same weight (the errors σ_i are equal), this expression reduces to the standard expression discussed in Chapter 7. The *total variance* is also defined as follows:

$$S_t^2 = \left(\frac{N}{N-1}\right)\frac{1}{W} \cdot \sum_{i=1}^{N} w_i \cdot (y_i - \overline{y})^2. \tag{D.4}$$

S_t is a measure of the dispersion of the data around the mean value $\overline{y}$. This value does not depend on the model (function $f(x)$), that is, S_t ignores any deterministic variation of y with x.

The *variance of the fit* S_f^2 is also defined as follows:

$$\begin{aligned} S_f^2 &= \left(\frac{N}{N - n_{\text{par}}}\right) \cdot \frac{1}{W}\sum_{i=1}^{N} w_i \cdot (y_i - y(x_i))^2 \\ &= \left(\frac{N}{N - n_{\text{par}}}\right) \cdot \frac{1}{W}\chi^2. \end{aligned} \tag{D.5}$$

The variance of the fit, S_f^2, similar to χ^2 or χ_v^2 (see Appendix C), measures the deviation of the measured values from the values predicted by the model. Each term of the sum of Eq. (D.5) is the value of this deviation, weighted by the factor $w_i = 1/\sigma_i^2$, that is, it indicates how large this deviation is compared to the error associated with each point. If each term $w_i[y_i - y(x_i)]^2$ were of the order of a unit, it would be indicative of an adequate fit of the data by the proposed model. So, $\boldsymbol{S_f^2}$ and $\boldsymbol{\chi}^2$ are adequate measures of the *goodness of fit* of $y(x_i)$ to the measured values y_i. If the proposed model $f(x)$ were adequate, its value would be associated with the statistical fluctuations of y_i with respect to its value $y(x_i)$. A value of $\boldsymbol{S_f^2}$ of the order of a unit or less is indicative of a model that may be adequate to describe the data.

Sometimes it is useful to define the *regression coefficient* as was done in Appendix C (which also provides an idea of the quality of fit or goodness of the model) as follows:

$$R^2 = \left(\frac{S_t^2 - S_f^2}{S_t^2}\right). \tag{D.6}$$

If the model $y(x_i)$ is a good representation of the data, S_f will be much smaller than S_t, from where it follows that $R^2 \approx 1$. Otherwise, $S_t \approx S_f$; $R^2 \approx 0$ [2].

D.2 Estimation of the uncertainties of the model parameters

As in the case of the linear model discussed in Chapter 7 and in Appendix C, the best values of the model parameters are obtained from the minimization of the function $S_f^2(a, b, c, \ldots)$, that is, the best value of the parameter a of the model, a^*, will be given by

$$a^* \Leftrightarrow \left.\frac{\partial S_f^2(a, b, c, \ldots)}{\partial a}\right|_{a=a^*} = 0 \tag{D.7}$$

so that $\boldsymbol{S_f^2} = \boldsymbol{S_f^2}(\boldsymbol{a}^*, \boldsymbol{b}^*, \ldots)$ is the minimum of $\boldsymbol{S_f^2}$.

Determination of the uncertainties in the parameters (a^*, b^*, c^*, ...) is a sophisticated procedure on which there are various theories and opinions [1–3]. An approximate practical method to calculate these uncertainties graphically [2] is shown in Figure D.2.

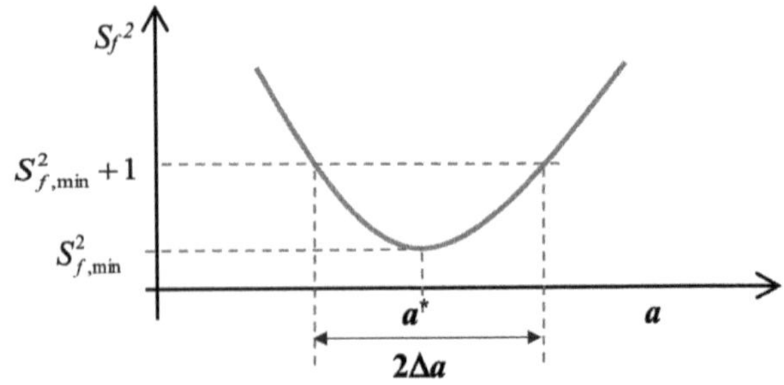

Figure D.2 Graph illustrating an approximate procedure for finding the uncertainties of the parameters of a nonlinear model.

For the case of a variable, a, the technique consists of graphing $\boldsymbol{S_f^2}$ as a function of a. $\boldsymbol{S_f^2}$ will pass through a minimum (a^*) which determines the best value of the parameter a. At this point, the value of $\boldsymbol{S_f^2}$ will be $\boldsymbol{S^2_{f,\mathrm{min}}}$. Then the width of the interval defined by the ordinates that make $S_f^2 = \boldsymbol{S^2_{f,\mathrm{min}}} + 1$ determines [2] the uncertainty interval Δa of the best value a^*.

D.3 Simulation of experimental results — Monte Carlo method

It is often useful to simulate the characteristics of an experiment before performing it. This enables you, for example, to decide the size of the errors allowed to observe a given effect. The *Monte Carlo technique* is a probabilistic formalism to generate numbers with a predetermined probability distribution that simulates the results of a physical variable. There is a very large family of commercial programs and spreadsheets that include random number generators with pre-established probability distributions, greatly facilitating the task of conducting Monte Carlo simulations.

To exemplify, imagine that you want to simulate data from an experiment in which each measurement will result in the pair (x_i, y_i). Suppose further that the expected relationship between x and y is linear of the form $y = ax+b$. Assume that only the values of y_i have a dispersion that is described by a normal distribution, whose standard deviation is characterized by a predetermined dispersion parameter (disp%).

To make the example clearer, assume that you are working with a spreadsheet. In the first column of the spreadsheet, define the range of x values in which you are interested. In the second column, enter the values of y found through the analytical expression, with the values of a and b which are assumed to be representative of the problem in question. These y values are designated as $y_{\text{teor}}(= ax+b)$. In the third column, calculate the values that will characterize the dispersion of the data (Δy_{teor}) given by

$$\Delta y_{\text{teor}}(x_i) = y_{\text{teor}}(x_i) \cdot \frac{\text{disp\%}}{100}. \tag{D.8}$$

Next, enter the random character of the experiment using the Monte Carlo method. To do this, in a new column, using the random number generation function of the spreadsheet, enter the random numbers, rnd, chosen so that they are normally distributed with mean 0 and standard deviation *1*, that is, $N(0,1)$. With these values, you can define the simulated values or synthetic data for the variable y_{Sim} as follows:

$$y_{\text{Sim}}(x_i) = y_{\text{teor}}(x_i) + \Delta y_{\text{teor}}(x_i) \cdot \text{rnd}. \tag{D.9}$$

The values of y and Δy thus obtained have the pre-established dispersion characteristics (characterized by disp%). This technique can be generalized to simulate real situations quickly and inexpensively. The website www.fisicarecreativa.com (Physics Experiment Resources) provides some simple examples of the implementation of this technique.

References

[1] P. Bevington and D. K. Robinson, *Data Reduction and Error Analysis for the Physical Sciences*, 3rd Ed., p. 230, 2003.

[2] S. L. Meyer, *Data Analysis for Scientists and Engineers*, NY: John Willey & Sons, Inc., 1975.

[3] W. Press, S. Teukolsky, W. Veetterling and B. Flanner, *Numerical Recipes: The Art of Scientific Computing*, 3rd Ed., Cambridge University Press, 2007.

Appendix E

LOCK-IN AMPLIFIER

In this appendix, we discuss the lock-in amplifier. This experimental technique makes it possible to accurately measure very small signals in the presence of large noises.

Goals

✓ *Lock-in amplifier*, operating principle
✓ Applications of a *lock-in*

E.1 The lock-in amplifier

The *lock-in amplifier* (LIA) [1–8] is a very useful instrument in laboratories and in industry. It is particularly suitable when very small signals are being measured in the presence of large noises. A basic requirement to be able to use a *lock-in* is that the input signal is *modulated* at a known frequency, or in other words, we know in advance the characteristic frequency of the small signal we are trying to measure [9].

To fix ideas, suppose that the *input signal* $V_{\rm in}(t)(= V_0\ \cos\omega_0 t))$, consists of a sinusoidal signal of known frequency ω_0 in the presence of a noise $V_{\rm noise}(t)$:

$$V_{\rm in}(t) = V_0\cos\omega_0 t + V_{\rm noise}(t) = V_0\cos\omega_0 t + \int_0^\infty V_n(\omega)\cos\omega t\cdot d\omega. \tag{E.1}$$

Here $V_n(\omega)$ indicates the Fourier components of the noise. We further assume that the RMS value of the noise is much larger than the signal of interest V_0, that is to say,

$$\mathrm{RMS}_{\rm noise} \equiv \langle |V_{\rm noise}(t)|^2\rangle = \int_0^\infty |V_n(\omega)|^2 d\omega \gg V_0^2. \tag{E.2}$$

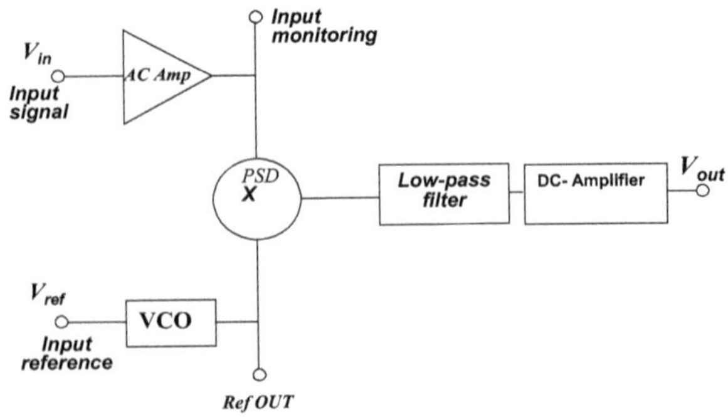

VCO = Voltage Control Oscillator	**Ref** OUT = Output to monitor the reference signal
PSD = Phase Sensitive Detector	**Input monitoring** = Output to monitor the amplified input signal
AC Amp = Linear AC amplifier	**Low-pass filter** = whose time constant can be varied from the control panel
DC amplifier = DC output amplifier	V_{OUT} = DC output proportional to V_0 signal

Figure E.1 Schematic diagram of a lock-in amplifier (LIA).

The symbol $\langle f(t)\rangle$ denotes time average. We will also assume that we can generate a *reference signal* V_{ref}, which can be sinusoidal, square, or triangular, but with the same frequency ω_0 as the signals of interest, i.e.,

$$V_{\text{ref}}(t) = V_{VCO}(t) = E_0 \cos(\omega_0 t + \phi). \tag{E.3}$$

A key component in the *lock-in* is the module named *Phase Sensitive Detector* (PSD), which multiplies the two input signals; in other words, the PSD generates a signal that is the product of the *input signal* and the *reference signal.* If we assume that the input amplifier (AC-Amp) of the *lock-in* has a gain G, then the signal at the output of the PSD is

$$V_{\text{PSD}}(t) = G\,V_{\text{in}}(t) \times V_{\text{VCO}}(t) = G\,V_0\,E_0 \cos(\omega_0 t)\,\cos(\omega_0 t + \phi) + GE_0 \int_0^\infty V_n(\omega)\cos(\omega t)\cos(\omega_0 t + \phi)\,d\omega. \tag{E.4}$$

From the trigonometric identity $\cos(\alpha)\cdot\cos(\beta) = (1/2)[\cos(\alpha-\beta)+\cos(\alpha+\beta)]$, we get

$$\cos(\omega_0 t)\cos(\omega_0 t+\phi) = \frac{1}{2}\left\{\cos\phi + \cos(2\omega_0 t+\phi)\right\}. \tag{E.5}$$

As the signal at the output of the PSD passes through a *low pass* filter (which in a first approximation gives as output the time average of the input), the high-frequency AC component will be cut, and we can write for the filter output:

$$\begin{aligned}\langle V_{\mathrm{PSD}}(t)\rangle = \langle V_{\mathrm{in}}(t)\cdot V_{\mathrm{VCO}}(t)\rangle &= \frac{G\,V_0}{2}E_0\cos(\phi)\\ &\quad + G\,E_0\int_0^{\infty} V_n(\omega)\,\delta(\omega-\omega_0)\,d\omega,\end{aligned} \tag{E.6}$$

or also

$$V_{\mathrm{out}} = \langle V_{\mathrm{PSD}}(t)\rangle = \frac{GV_0}{2}E_0\cos(\phi) + GE_0V_n(\omega_0). \tag{E.7}$$

Note that the only noise component present in the output signal V_{out} of the *lock-in* is the component of noise that has exactly the same frequency ω_0 as the input signal, that is, the entire infinite spectrum of frequencies present in the noise in Eq. (E.1) has been filtered, except the single component at frequency ω_0. As in general this component is only a signal within the noise continuum, it is to be expected that it will be less, or much smaller, than the signal of interest V_0, as $|V_n(\omega_0)|^2 \ll \int_0^\infty |V_n(\omega)|^2\cdot d\omega$.

In general, the phase difference φ between the input signal and the reference is variable. In practice, this phase is chosen so that the signal of interest can be maximized. A more detailed discussion of how a lock-in works can be found in Refs. [3, 5].

There are several commercial models of lock-in amplifiers that allow us to vary the phase φ manually until the signal of interest is maximized (V_0). In the market, there are excellent lock-in amplifiers, but more often than none, these commercial devices are expensive and not accessible for most Physics student labs around the world. Fortunately, there are several alternatives far more accessible. One alternative is to use an inexpensive Sound Board as a data acquisition and transform it into a basic LIA [10]. There is also the possibility of converting a conventional data acquisition system (DAS) into an LIA [11, 12]. More recently, has arose the possibility of building an LIA using a popular Arduino microcontroller and the Processing programming environment that allows one to have an LIA for less than U$S 100 [4, 13, 14]. It is interesting note that the algorithm described above can also be applied, by means of a software program, to signals measured with an oscilloscope or with a data acquisition system [2].

Suppose that the problem under study is to determine the voltage induced in a secondary "*search*" coil at a certain distance from another primary coil that generates a variable magnetic field, which we assume is excited with a known frequency ω_0:

$$V_{\mathrm{ref}}(t) = E_0 \cos(\omega_0 t + \phi), \tag{E.8}$$

this signal we refer to as the *reference signal*, $V_{\mathrm{ref}}(t)$, which we assume is stored in a channel, let's say channel A of a DAS. The induced signal in the secondary coil: $V_{\mathrm{in}}(t)$, in general, will be polluted with noise, especially if the two coils are far away from each other, similar to the signal indicated in Eq. (E.1), we assume that $V_{\mathrm{in}}(t)$ is stored in channel 2. By software we can make the product of these two signs ($V_{\mathrm{in}}(t)$ and $V_{\mathrm{ref}}(t)$) and then, also by software, we can perform the time average. Clearly, the result would be the one described by Eq. (E.7). Furthermore, if we multiply the input signal by a new artificial reference signal, of slightly different frequency, let's say $\omega' = \omega_0 + d\omega$, according to Eq. (E.7), we could obtain, by averaging, an estimate of the noise component at frequency ω':

$$V_{\mathrm{out}} = G\, E_0\, V_n(\omega'). \tag{E.9}$$

If the noise has a smooth spectrum without sudden changes, Eq. (E.9) could be subtracted from the output signal for the excitation frequency ω_0, or:

$$\begin{aligned} V_{\mathrm{out}} &= \frac{G\, V_0}{2} E_0 \cos(\phi) + G\, E_0\, V_n(\omega_0) \\ &\quad - G\, E_0\, V_n(\omega') \approx \frac{G\, V_0}{2} E_0 \cos(\phi). \end{aligned} \tag{E.10}$$

In this way, we can measure the value of the voltage induced in the secondary, even in the presence of a large source of noise. Note that the key point in the LIA is the fact that what we are trying to undig and small signal from a huge pile the noise, but we know the signal we are looking for has a well-known frequency. This is the essential point that makes the whole strategy work.

If in particular the input signal is related to the reference signal by a complex admittance, that is,

$$V_{\text{in}}(t) = \chi(\omega)E_0 \cdot e^{j\omega.t} = [\chi_1(\omega) + j \cdot \chi_2(\omega)]E_0 e^{j\omega t}. \tag{E.11}$$

If the reference signal is

$$V_{\text{Ref}}(t) = E_0 \cdot e^{j(\omega.t+\phi)}. \tag{E.12}$$

thus,

$$V_{\text{in}}(t) = \chi(\omega)V_{\text{Ref}}(t) \cdot e^{-j\phi} = [\chi_1(\omega) + j \cdot \chi_2(\omega)]E_0 e^{j\omega t}. \tag{E.13}$$

Therefore, the product of these signals will be

$$V_{\text{PSD}} = G.V_{\text{in}}^* x V_{\text{Ref}} = G.E_0^2(\chi_1 + j\,\chi_2).e^{i\phi}. \tag{E.14}$$

whose real part is

$$\text{Re}\{V_{\text{PSD}}\} = G.E_0^2(\chi_1 \cos\phi - \chi_2 \sin\phi). \tag{E.15}$$

Therefore, by adjusting the value of the phase ϕ, we can obtain the real (χ_1) or the complex component (χ_2) of the admittance. Most commercial LIAs already have two outputs that provide these two components simultaneously the in-phase:

$$V_X^{\text{Out}} = G.E_0^2(\chi_1(\omega) \cdot \cos\phi) \tag{E.16}$$

and out-of-phase components

$$V_Y^{\text{Out}} = -G.E_0^2(\chi_2(\omega) \cdot \sin\phi). \tag{E.17}$$

Some model also provides the module and the phase

$$V_{\text{Mod}}^{\text{Out}} = G.E_0^2\sqrt{\chi_1^2(\omega) + \chi_2^2(\omega)} \tag{E.18}$$

and

$$\phi = \text{Atan}\left(\frac{V_Y^{\text{Out}}}{V_X^{\text{Out}}}\right). \tag{E.19}$$

In this way, we can determine experimentally the complex admittance of a system $\chi(\omega) = \chi_1(\omega) + \iota\chi_2(\omega)$ using the lock-in amplifier.

In most two-channel lock-ins, the in-phase and out-of-phase components of the reference signal can be measured simultaneously. In other words, in these devices, we can measure the real part $\chi_1(\omega)$ and the imaginary part $\chi_2(\omega)$ of the admittance can be measured simultaneously.

References

[1] P. Temple, "An introduction to phase-sensitive amplifiers: An inexpensive student instrument" *Am. J. Phys.*, vol. 43, p. 801, 1975.

[2] R. Wolfson, "The lock-in amplifier: A student experiment" *Am. J. Phys.*, vol. 59, p. 569, 1991.

[3] J. Scofield, "Frequency-domain description of a lock-in amplifier" *Am. J. Phys.*, vol. 62, p. 129, 1994.

[4] S. Sengupta, J. Farnham and J. Whitten, "A simple low-cost lock-in amplifier for the student laboratory" *J. Chem. Educ.*, vol. 82, p. 1399, 2005.

[5] Stanford Research Instruments, "Instruction manuals of lock-in amplifiers from SRI" Stanford Research Systems, 2021. Available: http://www.thinksrs.com.

[6] G. Delfitto and G. Torzo, "The lock-in amplifier: What is it for? how to build one?" *Revista Brasileira de Ensino de Física*, vol. 44, p. e20220072, 2022.

[7] J. D. Pintle, "Fourier description of lock-in," *Revista Mexicana de Física*, vol. E 59, p. 1–7, 2013.

[8] Wikipedia, "Lock-in amplifier" 2022. Available: https://en.wikipedia.org/wiki/Lock-in_amplifier.

[9] P. Horowitz and W. Hill, *The Art of Electronics*, 3rd Ed., NY: Cambridge University Press, 2015, p. 1220.

[10] M. G. Gonzalez, G. Santiago, V. Slezak and A. L. Peuriot, "Simple synchronic detection at audio frequencies through a PC sound card," *Rev. Sci. Instrum.*, vol. 78, pp. 055108-1–4, 2007.

[11] I. Bhagyajyothi, P. Bhaskar and C. Parvathi, "Design and development of advanced lock-in amplifier and its application" *Sens. Transducers*, vol. 153, no. 6, pp. 22–28, 2013.

[12] J. Denker, "Computer-based lock-in amplifier" 2005. Available: http://www.av8n.com/physics/lockin.htm.

[13] Arduino, "Arduino project web site" 2022. Available: http://www.arduino.cc.

[14] K. D. Schultz, "Phase-sensitive detection in the undergraduate lab using a low-cost microcontroller" *J. Phys.*, vol. 84, pp. 557–561, 2016.

Index

Milton Keynes UK
Ingram Content Group UK Ltd.
UKHW050631160624
444182UK00001B/18

9 789811 277757